The Economics of Biodiversity

We are part of Nature, not separate from it. We rely on Nature to provide us with food, water and shelter; regulate our climate and disease; maintain nutrient cycles and oxygen production; and provide us with spiritual fulfilment and opportunities for recreation and recuperation, which can enhance our health and well-being. Nature's constituents such as ecosystems and the biodiversity that are embodied in them are therefore assets. Yet Nature is more than an economic good: many recognise its intrinsic worth and argue that it has moral worth too. This landmark report explains the current state of play in relation to biodiversity loss and outlines a sustainable path to deal with this problem, one that will require us to change how we think, act and measure success. The report was originally commissioned and published by HM Treasury.

Partha Dasgupta is the Frank Ramsey Professor Emeritus of Economics at the University of Cambridge and Fellow of St John's College, Cambridge. He is a Fellow of the Royal Society and of the British Academy, a Foreign Associate of the US National Academy of Sciences, a Foreign Member of the American Academy of Arts and Sciences and of the American Philosophical Society, and has received the Volvo Environment Prize, the Blue Planet Prize, the Tyler Prize for Environmental Achievement, the Kew International Medal of the Royal Botanical Gardens, Kew, and the BBVA Foundation Frontiers of Knowledge Award. In 2022 Dasgupta was named by the United Nations a Champion of the Earth, the first economist to be so honored, and in 2023 was made Knight Grand Cross of the British Empire by King Charles III for "services to economics and the natural environment".

The Economics of Biodiversity
The Dasgupta Review

Partha Dasgupta
University of Cambridge

CAMBRIDGE
UNIVERSITY PRESS

CAMBRIDGE UNIVERSITY PRESS

Shaftesbury Road, Cambridge CB2 8EA, United Kingdom

One Liberty Plaza, 20th Floor, New York, NY 10006, USA

477 Williamstown Road, Port Melbourne, VIC 3207, Australia

314–321, 3rd Floor, Plot 3, Splendor Forum, Jasola District Centre, New Delhi – 110025, India

103 Penang Road, #05–06/07, Visioncrest Commercial, Singapore 238467

Cambridge University Press is part of Cambridge University Press & Assessment, a department of the University of Cambridge.

We share the University's mission to contribute to society through the pursuit of education, learning and research at the highest international levels of excellence.

www.cambridge.org
Information on this title: www.cambridge.org/9781009494335

DOI: 10.1017/9781009494359

© Crown Copyright and Partha Dasgupta, 2024. This copyright work is an adaptation of Dasgupta, P. (2021), *The Economics of Biodiversity: The Dasgupta Review.* (London: HM Treasury) and published by Cambridge University Press & Assessment under the terms of the Open Government Licence v3.0.

This publication is in copyright. Subject to statutory exception and to the provisions of relevant collective licensing agreements, no reproduction of any part may take place without the written permission of Cambridge University Press & Assessment.

When citing this work, please include a reference to the DOI 10.1017/9781009494359

First published 2024

Printed in Great Britain by CPI Group (UK) Ltd, Croydon CR0 4YY

A catalogue record for this publication is available from the British Library

Library of Congress Cataloging-in-Publication data
Names: Dasgupta, Partha, author.
Title: The economics of biodiversity : the Dasgupta review / Partha Dasgupta, University of Cambridge.
Description: Cambridge, United Kingdom ; New York, NY : Cambridge University Press, 2024. | Includes bibliographical references and index.
Identifiers: LCCN 2024000169 | ISBN 9781009494335 (hardback) |
ISBN 9781009494304 (paperback) | ISBN 9781009494359 (ebook)
Subjects: LCSH: Biodiversity conservation – Economic aspects. | Biodiversity – Economic aspects. | Environmental economics. | Ecosystem health. | Environmental ethics.
Classification: LCC QH75 .D358 2024 | DDC 333.95/16–dc23/eng/20240206
LC record available at https://lccn.loc.gov/2024000169

ISBN 978-1-009-49433-5 Hardback
ISBN 978-1-009-49430-4 Paperback

Cambridge University Press & Assessment has no responsibility for the persistence or accuracy of URLs for external or third-party internet websites referred to in this publication and does not guarantee that any content on such websites is, or will remain, accurate or appropriate.

Contents

Foreword   xv

Preface to the CUP Edition   xvii

Preface   xix

Part I – Foundations   1

Chapter 0   How We Got to Where We Are   3
    0.1 Economic History Since Year 0   4
    0.2 Economic Growth and Sustainable Development   7
    0.3 Historical Success and Failures   10
    0.4 Understanding Humanity's Contemporary Overshoot   12

Chapter 1   Nature as an Asset   17
    1.0 Portfolio Management   17
    1.1 Classification of Capital Goods   20
    1.2 Rates of Return and Arbitrage Conditions   22
    1.3 Public Asset Management and the Wealth/Well-Being Equivalence Theorem   24
    1.4 Two Types of Comparison   26
    1.5 The Earth System and Economic Growth   27
    1.6 Total vs. Marginal Values   29
    1.7 Institutions and the Character of Natural Capital   30
    1.8 Anthropocentric Value of Biodiversity   31

Chapter 2   Biodiversity and Ecosystem Services   33
    2.1 Biodiversity in Ecosystems   33
    2.2 Primary Producers   36
    2.3 Ecosystems do not Maximise NPP   42
    2.4 Ecosystem Goods and Services: Classification   43
    2.5 Invisibility and Silence of Regulating and Maintenance Services   46
    2.6 Conservation vs. Pollution   48
    2.7 Ecosystem Productivity and Resilience   50
    2.8 Biodiversity and Ecosystem Productivity/Resilience: The Causal Connection   51
    2.9 Modularity as a Spatial Feature of Ecosystems   55
    2.10 Biodiversity and Ecosystem Productivity: Summary   56

    Annex 2.1 Community Structure   57
    Annex 2.2 Measuring Biodiversity   59
    Annex 2.3 Quantity and Quality of our Stock and How it has Changed   62

Contents

Chapter 3	Biospheric Disruptions	67
	3.1 Fragmentation as Disturbance	67
	3.2 Stability Regimes	70
	3.3 History Dependence	73
	3.4 Conservation Ecology and Tipping Points	75
	Annex 3.1 The Phosphorus Recycling Mechanism in Lake Systems	77
	Annex 3.2 Tipping Points	79
	Annex 3.3 Hysteresis and Irreversibilities in Lake Dynamics	81
	Annex 3.4 Instantaneous Elimination of Phosphorus Inflow	82
	Annex 3.5 Morals	83
Chapter 4	Human Impact on the Biosphere	85
	4.1 Depreciating the Biosphere	85
	4.2 Demand and Supply	98
	4.3 The Impact Inequality	101
	4.4 Two Notions of Inequality	103
	4.5 The Impact Equation	105
	4.6 Technology and Institutions	107
	4.7 Ecosystem Complementarities and the Bounded Global Economy	110
	4.8 Core of the Review	113
	Annex 4.1 Biodiversity Loss and Climate Change	115
	Annex 4.2 How Many People Can Earth Support in Comfort?	119
Chapter 4*	The Bounded Global Economy	123
	4*.1 Substitutes and Complements	124
	4*.2 Modelling the Global Economy	125
	4*.3 Contemporary Models of Economic Growth	129
Chapter 5	Risk and Uncertainty	133
	5.1 Portfolio Choice under Uncertainty	134
	5.2 Independent vs. Correlated Risks	136
	5.3 Reducing Risks and the Losses from Risks	143
	5.4 When to Stop Business-as-Usual	144
	5.5 The Value of Keeping Options Open	147
	Annex 5.1 Fat Tails and Unbounded Utilities: St. Petersburg Paradoxes	152
	Annex 5.2 Catastrophes and Ambiguities	155
Chapter 6	Laws and Norms as Social Institutions	157
	6.1 Societal Trust and Economic Progress	159
	6.2 The Idea of Trust	160
	6.3 The Basis of Trust, 1	161
	6.4 The Basis of Trust, 2	163
	6.5 Social Capital as the Basis of Societal Coherence	171

	6.6 Social Capital and Identity	173
	6.7 The Primacy of Integrity	176
	Annex 6.1 Corruption	177
Chapter 7	**Human Institutions and Ecological Systems, 1: Unidirectional Externalities and Regulatory Policies**	**179**
	7.1 Property Rights and Wealth Distributions	181
	7.2 Externalities and Rights	182
	7.3 Taxing and Subsidising Externalities	184
	7.4 Quantity Restrictions	187
	7.5 Markets for Externalities	189
	7.6 Payment for Ecosystem Services	193
Chapter 8	**Human Institutions and Ecological Systems, 2: Common Pool Resources**	**197**
	8.1 Open Access Resources	197
	8.2 Common Pool Resources (CPRs)	199
	8.3 CPRs and the Poor World	200
	8.4 Fragility of CPRs	203
	8.5 Property Rights to Land	206
	8.6 Property Rights and Management: A Schemata	208
	Annex 8.1 Estimating Subsidies	209
Chapter 8*	**Management of CPRs: A Formal Model**	**213**
	8*.1 A Timeless World	213
	8*.2 Mutual Enforcement of Optimum Herd Size	216
	8*.3 Privatising the CPR	217
	8*.4 Extensions	217
Chapter 9	**Human Institutions and Ecological Systems, 3: Consumption Practices and Reproductive Behaviour**	**221**
	9.1 Socially Embedded Consumption Preferences	222
	9.2 Consumption Practices	225
	9.3 Induced Behavioural Changes	227
	9.4 Factors that Slow Fertility Transition	229
	9.5 Socially Embedded Reproductive Behaviour	232
	9.6 Importance of Investment in Family Planning and Reproductive Health	235
	9.7 Unmet Need, Desired Family Size, and the UN's Sustainable Development Goals	237
	Annex 9.1 Socially Embedded Preferences for Consumption: Formulation	239
Chapter 10	**Well-Being Across the Generations**	**243**
	10.1 Classical Utilitarianism	244
	10.2 Utilitarian Reasoning behind the Veil of Ignorance	249
	10.3 Discounting Future Generations	250

10.4	Intuitionism and Pragmatism	253
10.5	Discounting in Arbitrary Futures	254
10.6	Directives on Discounting	258
10.7	Social Rates of Return on Investment	260
10.8	Accounting Prices	263
10.9	Should Environmental Projects be Evaluated Using Lower Discount Rates?	265
10.10	The Idea of Investment	266
10.11	Population Ethics	269
10.12	The Repugnance of Existential Risks	273
	Annex 10.1 A Simple Exercise in Optimum Saving	276
	Annex 10.2 Uncertainty and Declining Discount Rates	278
	Annex 10.3 Three Tiers of Ethical Reasoning	280

Chapter 11 The Content of Well-Being: Empirics — 289

11.1	Objective Measures of Well-Being	290
11.2	Measuring Well-Being by Asking People	292
11.3	Intranational and International Comparisons of Well-Being	294
11.4	Measurement and Interpersonal Comparisons of Subjective Well-Being	297
11.5	Determinants of Well-Being	299
11.6	Nature and Well-Being	301

Chapter 12 Valuing Biodiversity — 305

12.1	Estimating Accounting Prices: General Observations	306
12.2	Stated Preference for Public Goods	308
12.3	Revealed Preference for Amenities	309
12.4	Productivity as Accounting Price	310
12.5	Human Health	311
12.6	Nature's Existence Value and Intrinsic Worth: Sacredness	313
12.7	Nature's Intrinsic Worth: Moral Standing	314
	Annex 12.1 Valuing Ecosystem Services in China: Gross Ecosystem Product	317

Chapter 13 Sustainability Assessment and Policy Analysis — 327

13.1	Capital Goods	327
13.2	Enabling Assets	328
13.3	The Idea of Inclusive Wealth	330
13.4	Inclusive Wealth and Intergenerational Well-Being	331
13.5	Inclusive Wealth and the Substitutability of Capital Goods	334
13.6	Six Questions	335
13.7	SDGs and the Idea of Sustainable Development	335
13.8	Economic Progress as Growth in Inclusive Wealth	337
13.9	Total Factor Productivity Growth	339
13.10	Growth in GDP and Inclusive Wealth	342
13.11	Net Present Values	343
13.12	Inclusive Wealth and the Present Value Criterion	346

	13.13 Lower Discount Rates for Environmental Projects?	350
	13.14 Optimum Development	351
	13.15 Production and Consumption Targets	353
	13.16 Internal Rate of Return	354
	Annex 13.1 Economic Growth and Sustainable Development	354
	Annex 13.2 Saving the Blue Whale	357
	Annex 13.3 The Significance of GDP	360
Chapter 13*	Accounting Prices and Inclusive Wealth	363
	13*.1 The Model	364
	13*.2 The Optimisation Problem	364
	13*.3 Arbitrage Conditions and the Structure of Accounting Prices	365
	13*.4 Arbitrage Condition Between Produced Capital and Natural Capital	367
	13*.5 Inclusive Wealth and the Long Run	368
	13*.6 Counterfactual Futures	368

Part II – Extensions 371

Chapter 14	Distribution and Sustainability	373
	14.1 Global Variation in Demand and Supply	373
	14.2 Distribution of Humanity's Demands	374
	14.3 Distribution of the Biosphere's Supply	378
	14.4 Interactions Between the Biosphere and Societal Inequalities	382
Chapter 15	Trade and the Biosphere	385
	15.1 Trade and the Impact Equation	385
	15.2 Trade Expansion and Pressures on the Biosphere	388
	15.3 Enhancing Trade Practices and Policies to Support Sustainability	394
Chapter 16	Demand for Provisioning Goods and Its Consequences	403
	16.1 Current Harvest of Provisioning Goods and Future Prospects	403
	16.2 Trade-Offs Between Provisioning and Regulating Services	412
	16.3 Technology to Increase Efficiency in Our Use of the Biosphere	418
Chapter 17	Managing Nature-Related Financial Risk and Uncertainty	423
	17.1 Nature-Related Financial Risks	424
	17.2 Uncertainty and Short-Termism	431
	17.3 Assessing Nature-Related Financial Risks	433
Chapter 18	Conservation of Nature	441
	18.1 Ecosystem Assets	441
	18.2 How Much Ecosystem Stock Do We Need?	443
	18.3 What Kind of Stock Do We Need?	445
	18.4 How Can We Improve and Increase Our Stocks?	446

Contents

18.5 Conservation Planning and Evaluation	455
18.6 Multilateral Environmental Agreements	455

Chapter 19   Restoration of Nature   457

19.1 The Role of Ecosystem Restoration to Improve and Increase Our Stocks	457
19.2 Rewilding	460
19.3 Nature-Based Solutions	462
19.4 Sustainable Production Landscapes and Seas	466
19.5 Invasive Non-Native Species	467
19.6 Bringing Natural Capital into Spatial Planning	469

Chapter 20   Finance for Sustainable Engagement with Nature   473

20.1 Public Finance	474
20.2 Private Finance	480

Part III – The Road Ahead   491

Chapter 21   Options for Change   493

21.1 Address the Imbalance Between Our Demand and Nature's Supply, and Increase Nature's Supply	494
21.2 Changing Our Measures of Economic Progress	499
21.3 Transforming Our Institutions and Systems	500

Appendix   505

Acronyms	507
Glossary	509
References	517
Acknowledgements	609

Author Index   613

Subject Index   635

List of Boxes

Chapter 1

Box 1.1 Managing Assets as Daily Chores
Box 1.2 Valuing Nature's Stocks
Box 1.3 Arbitrage Conditions

Chapter 2

Box 2.1 Biomes
Box 2.2 Fishery Dynamics
Box 2.3 Counterfactual Reasoning in Asset Management
Box 2.4 Selection Pressures and Regeneration Rates
Box 2.5 The Soils
Box 2.6 Carbon Storage and Seed Dispersal by Large Fauna
Box 2.7 Watershed Services by Tropical Forests
Box 2.8 Non-Anthropocentric Perspectives
Box 2.9 Mix of Species in Grazing Lands
Box 2.10 Plantations as Ecosystems
Box A2.2.1 Global Biomass Census

Chapter 3

Box 3.1 Damming Rivers
Box 3.2 Fencing the Grasslands
Box 3.3 Multiple Stability Regimes – Fisheries
Box 3.4 Regime Shifts in Space
Box 3.5 How is Society to Know that an Ecosystem is Near a Tipping Point?
Box 3.6 Human Body as an Ecosystem

Chapter 4

Box 4.1 Deforestation and Species Extinction
Box 4.2 Deoxidation of the Oceans
Box 4.3 Soil Biodiversity Loss
Box 4.4 Impact of the Fast Fashion Industry
Box 4.5 The Idea of Indefinite Economic Growth
Box 4.6 Reaching the UN SDGs
Box 4.7 Reducing the Impact Inequality by Restoring the Peatlands
Box 4.8 Land-Use Change and the Spread of Viruses

Chapter 4*

Box 4*.1 Nature's Goods and Services in Mathematical Ecology

Chapter 5

Box 5.1 Expected Utility Theory
Box 5.2 Independent and Fully Correlated Risks: Comparison
Box 5.3 Short-Run vs. Long-Run Predictability

Box 5.4 The Value of Information
Box 5.5 Mexican Coastal Zone Management Trust
Box 5.6 When to Stop Business-As-Usual
Box 5.7 Option Values
Box 5.8 Translating Ecological Risks into Economic Risks
Box A5.1 The St. Petersburg Paradox

Chapter 6

Box 6.1 Civic Virtues
Box 6.2 The Grim Norm
Box 6.3 Reputation as an Asset
Box 6.4 The *Iddir* and ROSCAs
Box 6.5 Social Capital as Societal Coherence
Box 6.6 Dark Sides of Social Capital

Chapter 7

Box 7.1 International Trade, Wealth Transfers, and the Character of Technological Change
Box 7.2 Pigouvian Taxes and Subsidies
Box 7.3 Rationale for Environmental Regulations
Box 7.4 New Zealand's Tradable Permit Scheme for Fisheries
Box 7.5 Payments for Wildlife Services in Zimbabwe
Box 7.6 Auctions for Providing Temporary Habitat to Migrating Birds

Chapter 8

Box 8.1 Biosphere as the Common Heritage of Mankind
Box 8.2 Ownership Rights
Box 8.3 Plantations for Palm Oil

Chapter 9

Box 9.1 Prying Open Behavioural Anomalies
Box 9.2 The World Under Faster Demographic Transition
Box 9.3 The Matlab Experiment, 1977 to 1996

Chapter 10

Box 10.1 Weights and Measures
Box 10.2 The Value of a Statistical Life
Box 10.3 Zero Discounting and the Never-Ending Potlatch
Box 10.4 Consumption Discount Rates (CDRs): Basics
Box 10.5 Optimum Saving Principle
Box 10.6 Accounting Prices in a Dynamic Economy
Box 10.7 Discounting When Preferences are Endogenous
Box 10.8 Optimum Population

Chapter 11

Box 11.1 Day Reconstruction Method
Box 11.2 Relative Income Matters

Box 11.3 Are International Comparisons of Subjective Well-Being Meaningful?
Box 11.4 New Zealand's Living Standards Framework
Box 11.5 The Well-Gardened Mind

Chapter 12

Box 12.1 Contingent Valuation Methods: Problems
Box 12.2 Valuing Tourism
Box 12.3 Social Costs of Shrimp Farming
Box 12.4 Pharmaceuticals, Traditional Medicines, and Social Returns on R&D

Chapter 13

Box 13.1 The UK Government's Green Book
Box 13.2 Wealth and Well-Being: The Formal Connection
Box 13.3 Sustainable Development as Growth in Inclusive Wealth
Box 13.4 Composition of Inclusive Wealth
Box 13.5 Trade, Externalities, and Wealth Transfers
Box 13.6 Global Inclusive Wealth Change 1992 to 2014
Box 13.7 Restoring the Ganges
Box 13.8 The Equivalence of Wealth and Present Values
Box 13.9 Appraisal of the Restoration of the Exmoor Mires
Box 13.10 The Plight of Slow Growth Forests
Box 13.11 Project Complementarities and Lumpy Investments

Chapter 14

Box 14.1 Environmental Impact of Different Diets
Box 14.2 Urbanisation and Nature
Box 14.3 The Distributional Effects of Ecosystem Collapse

Chapter 15

Box 15.1 The Global Resources Outlook
Box 15.2 Exploring the UK's Global Ecological Footprint
Box 15.3 Trade Induced Biodiversity Loss Across Regions
Box 15.4 The US Shrimp/Turtle Dispute
Box 15.5 The Convention on International Trade in Endangered Species
Box 15.6 Trade Standards and Invasive Non-Native Species
Box 15.7 Integration of Environmental Issues into Trade Standards and Agreements
Box 15.8 Verified Sourcing Areas

Chapter 16

Box 16.1 Food Loss and Waste

Chapter 17

Box 17.1 Examples of Physical Risks
Box 17.2 Examples of Transition Risks
Box 17.3 Climate and Nature Sovereign Index
Box 17.4 White, Black and Green Swans

Box 17.5 CDC Biodiversité and the Global Biodiversity Score
Box 17.6 ASN Bank's Biodiversity Footprint
Box 17.7 Task Force on Climate-Related Financial Disclosures
Box 17.8 Biodiversity Loss and Risks for the Dutch Financial Sector

Chapter 18

Box 18.1 Costs and Benefits of Expanding Protected Areas
Box 18.2 Cases of Management by Indigenous Peoples and Local Communities
Box 18.3 The Coastal Cloud Forest of Loma Alta, Ecuador
Box 18.4 International Trade in Endangered Species
Box 18.5 Trade in Vicuña Fibre in South America's Andes Region
Box 18.6 Conservation of Biodiversity in the High Seas

Chapter 19

Box 19.1 The Urgent Case for Conservation and Restoration
Box 19.2 Restoration of Drylands and Rangelands in the Zarqa River Basin, Jordan
Box 19.3 Coral Reef Restoration
Box 19.4 Natural Regeneration or Active Restoration
Box 19.5 Restoration of Forests to Store Carbon – Good or Bad for Biodiversity?
Box 19.6 Examples of Nature-Based Solutions
Box 19.7 Job Opportunities From Nature Conservation and Restoration
Box 19.8 Environmental Land Management in England
Box 19.9 Working for Water in South Africa
Box 19.10 Building Natural Capital into Spatial Planning
Box 19.11 Biodiversity Offsets for a High-Speed Railway Line, Eiffage, France

Chapter 20

Box 20.1 Denmark's Pesticide Tax
Box 20.2 Agricultural Subsidies Reform in Switzerland
Box 20.3 PES in Colombia
Box 20.4 Seychelles Blue Bond
Box 20.5 Fiduciary Duty and the Biosphere
Box 20.6 EcoEnterprises Fund
Box 20.7 Intrinsic Value Exchange
Box 20.8 An EU Taxonomy for Sustainable Activities
Box 20.9 Spatial Finance

Foreword

We are facing a global crisis. We are totally dependent upon the natural world. It supplies us with every oxygen-laden breath we take and every mouthful of food we eat. But we are currently damaging it so profoundly that many of its natural systems are now on the verge of breakdown.

Every other animal living on this planet, of course, is similarly dependent. But in one crucial way, we are different. We can change not just the numbers, but the very anatomy of the animals and plants that live around us. We acquired that ability, doubtless almost unconsciously, some ten thousand years ago, when we had ceased wandering and built settlements for ourselves. It was then that we started to modify other animals and plants.

At first, doubtless, we did so unintentionally. We collected the kinds of seeds that we wanted to eat and took them back to our houses. Some doubtless fell to the ground and sprouted the following season. So over generations, we became farmers. We domesticated animals in a similar way. We brought back the young of those we had hunted, reared them in our settlements and ultimately bred them there. Over many generations, this changed both the bodies and ultimately the characters of the animals on which we depend.

We are now so mechanically ingenious that we are able to destroy a rainforest, the most species-rich ecosystem that has ever existed, and replace it with plantations of a single species in order to feed burgeoning human populations on the other side of the world. No single species in the whole history of life has ever been so successful or so dominant.

Now we are plundering every corner of the world, apparently neither knowing or caring what the consequences might be. Each nation is doing so within its own territories. Those with lands bordering the sea fish not only in their offshore waters but in parts of the ocean so far from land that no single nation can claim them. So now we are stripping every part of both the land and the sea in order to feed our ever-increasing numbers.

How has the natural world managed to survive this unrelenting ever-increasing onslaught by a single species? The answer of course, is that many animals have not been able to do so. When Europeans first arrived in southern Africa they found immense herds of antelope and zebra. These are now gone and vast cities stand in their stead. In North America, the passenger pigeon once flourished in such vast flocks that when they migrated, they darkened the skies from horizon to horizon and took days to pass. So they were hunted without restraint. Today, that species is extinct. Many others that lived in less dramatic and visible ways simply disappeared without the knowledge of most people worldwide and were mourned only by a few naturalists.

Nonetheless, in spite of these assaults, the biodiversity of the world is still immense. And therein lies the strength that has enabled much of its wildlife to survive until now. Economists understand the wisdom of spreading their investments across a wide range of activities. It enables them to withstand disasters that may strike any one particular asset. The same is true in the natural world. If conditions change, either climatically or as a consequence of a new development in the never-ending competition between species, the ecosystem as a whole is able to maintain its vigour.

But consider the following facts. Today, we ourselves, together with the livestock we rear for food, constitute 96% of the mass of all mammals on the planet. Only 4% is everything else – from elephants to badgers, from moose to monkeys. And 70% of all birds alive at this moment are poultry – mostly

Foreword

chickens for us to eat. We are destroying biodiversity, the very characteristic that until recently enabled the natural world to flourish so abundantly. If we continue this damage, whole ecosystems will collapse. That is now a real risk.

Putting things right will take collaborative action by every nation on earth. It will require international agreements to change our ways. Each ecosystem has its own vulnerabilities and requires its own solutions. There has to be a universally shared understanding of how these systems work, and how those that have been damaged can be brought back to health.

This comprehensive, detailed and immensely important report is grounded in that understanding. It explains how we have come to create these problems and the actions we must take to solve them. It then provides a map for navigating a path towards the restoration of our planet's biodiversity.

Economics is a discipline that shapes decisions of the utmost consequence, and so matters to us all. The Dasgupta Review at last puts biodiversity at its core and provides the compass that we urgently need. In doing so, it shows us how, by bringing economics and ecology together, we can help save the natural world at what may be the last minute – and in doing so, save ourselves.

David Attenborough

Preface to the CUP Edition

That economic policies should be evidence-based is, or should be, an incontrovertible requirement; but it is of no use if the evidence is obtained from a misleading conception of the human condition; for faulty models produce spurious evidence. Systems of thought that do not acknowledge humanity's embeddedness in Nature when used to project present and future possibilities open to us mislead. They mislead so hugely that policies based on them damage future generations, and in many instances, they damage some of the world's poorest communities.

The global standard of living has improved enormously since the end of World War II. Per capita income has increased nearly 5-fold to some 16,000 dollars PPP annually (at 2011 prices), life expectancy at birth has increased from 46 years to 72 years, and the proportion of people in extreme poverty has declined from approximately 60% to 10%. This enormous success has been achieved by an accumulation of *produced capital* (roads, buildings, ports, machines) and *human capital* (health, education, skills, character), but it has also been accompanied by the decumulation of *natural capital* (wetlands, grasslands, mangroves, coral reefs, woodlands, forests, lakes, and such biomes as the atmosphere, the oceans, the soils, and sub-soil resources). And these statistics reflect a rare contact between economics and the Earth sciences, for Earth scientists have dubbed 1950 as being the year we entered the Anthropocene, a human dominated Earth. In 1950 the global economy was small in comparison to the Earth system, today it is huge.

In the Anthropocene, expansion in our demands for Nature's 'provisioning goods' (food, water, timber, fibres, pharmaceuticals, non-living materials – i.e., the ingredients that, with human effort, go to shape the final products reflected in GDP) has eaten into Nature's ability to supply 'regulating and maintenance services,' such as climate regulation, decomposition of organic waste, nutrient recycling, nitrogen fixation, air and water purification, soil regeneration, pollination, and maintenance of the biosphere's gaseous composition.

Technological advancements have repeatedly shown ways to substitute provisioning goods among one another (fossil fuels replacing timber, solar panels and wind farms substituting for fossil fuels, and so on). Resource economists in their studies of production possibilities have thus emphasised substitution possibilities between natural resources and produced and human capital. In contrast, Nature's regulating and maintenance services have been found by Earth scientists and ecologists to be complementary to one another: disrupting one sufficiently disrupts the others. The mutual influence of climate change and the processes underlying the oceans is an example. Those complementarities tell us that we are embedded in Nature, we are not external to her. The biosphere is not exactly a house of cards, but we humans are now so ingenuous that we would be able to reduce it to one if we put our mind to it.

Even though the literature on environmental and resource economics has repeatedly exposed the harm done by the practice in contemporary economics of ignoring Nature, it hasn't done so from top to bottom. If contemporary economics is to be reconstructed, we would have to study our embeddedness in Nature at all levels: from the individual person, through households, communities, nations, regions, to the global economy. This book studies salient problems in each of those levels (Ch. 4 onward). The global economy is the scene where growth and development economics of the long run is fashioned, so the needed reconstruction would also refashion macroeconomic models of the long run. It would read contemporary economic growth as being countered by depreciation of the biosphere, which is a self-organizing regenerative entity. Chapters 4* and 13* contain a prototype of the kind of macroeconomic

Preface to the CUP Edition

model of the long run that is now needed. But it is only a prototype. Much work remains to be done, extending its coverage, and estimating the parameters of such a model.

The *Review of the Economics of Biodiversity* was launched at the Royal Society, London, on 2nd February 2021. The UK Treasury displayed exquisite courtesy in retaining part of my team for the remainder of the calendar year to help its dissemination. Since its launch I have engaged in more than 200, mostly virtual events, involving lectures, Q&As, panel discussions, and interviews. They have included not only professionals from environmental charities, government departments, international organisations, scientific associations, think tanks, academic journals, literary magazines, research institutes, NGOs, business schools, corporate bodies, and groups representing indigenous people, but numerically even more, financiers, bankers, insurers, farmers, ecologists, legal scholars, agronomists, journalists, statisticians, clerics, politicians, Earth scientists, and national and international civil servants.[1] There has no doubt been self-selection at work, but the level of interest in the economics of the biosphere among people at large feels unbelievable to me when I place it in comparison to the interest among editorial boards of leading economics journals.[2]

Which is why in preparing this edition of the Review, I have had graduate students in economics especially in mind. Although the version I prepared for the Treasury was both technical and detailed, there were several missing items of interest to students. I have prepared a few of the most important such items in new sections, boxes, and annexes. In such material, where necessary, I have used more contemporary estimates of such indices as global GDP and the human ecological footprint. The new material appears as Annex 4.2, Box 4*.1, Box 5.8, Sect. 10.11-10.12, Box 10.8, and Annex 10.3.

I am most grateful to Phil Good, economics editor of Cambridge University Press, for his encouragement, advice, and help in preparing this published edition.

To authors it is the final stages of publication that count most. So, I am particularly grateful to Claire Sissen, who has edited the typescript with enormous skill, encouragement, and patience. She has not balked once when I have asked to be allowed to make changes, even at the very last minute. This is editorial leadership at its finest.

Partha Dasgupta
St John's College, Cambridge

---

[1] Based on what I took away from those events, I prepared a paper – "The Economics of Biodiversity: Afterword" (Dasgupta, 2022) – listing those among the Review's recommendations that are very likely to be adopted by countries, those that are appreciated to be important but are felt to be unworkable under current circumstances, and those that are not for discussion because of alleged political sensitivities. The paper was published in *Environmental and Resource Economics*, in a symposium on the *Review*, edited by Ingmar Schumacher, for whose encouragement I am very grateful.

[2] An article-length presentation of the ecological economics underlying the *Review* (Dasgupta and Levin, 2023) has been published in a symposium on "detecting and attributing the causes of biodiversity change: needs, gaps, and solutions," in *Philosophical Transactions of the Royal Society B* (Vol. 378, Issue, 1881, 17 July 2023).

Preface

Economics, like I imagine other scientific disciplines, normally moves in incremental steps, and always without a central guide. Much like practitioners of other disciplines, we economists work with models of those features of the world we want to study in detail. That involves keeping all else in the far background. Models are thus parables, some say they are caricatures, which is of course their point.

Economics is also a quantitative subject. Finance ministers need estimates of tax revenues if they are to meet intended government expenditure; environment ministers today cannot but ask how much farmers should be paid to set aside land for 'greening' the landscape, and whether fossil-fuel subsidies should be eliminated; health ministers look to convince cabinet colleagues that investment in health is good for economic growth; and so on. Which is why economic models are almost invariably cast in mathematical terms.

That is also why the models that appear in economics journals can appear esoteric, unreal, and even self-indulgent. Many would argue as well that to model human behaviour formally, let alone mathematically, is to tarnish the human experience, with all its richness. And yet, economists in governments, international organisations, and private corporations find those models and their adaptations essential for collecting and analysing data, forecasting economic trajectories, evaluating options and designing policy. Perhaps, then, it should be no surprise that those same models go on to shape the conception we build of our economic possibilities. In turn, our acceptance of the economic possibilities those models say are open to us encourages academic economists to refine and develop them further along their tested contours. And that in turn further contributes to our beliefs about what is achievable in our economic future. The mutual influence is synergistic.[3]

That has had at least one unintended and costly consequence. Not so long ago, when the world was very different from what it is now, the economic questions that needed urgent response could be studied most productively by excluding Nature from economic models. At the end of the Second World War, absolute poverty was endemic in much of Africa, Asia, and Latin America; and Europe needed reconstruction. It was natural to focus on the accumulation of produced capital (roads, machines, buildings, factories, and ports) and what we today call human capital (health and education). To introduce Nature, or natural capital, into economic models would have been to add unnecessary luggage to the exercise.[4]

Nature entered macroeconomic models of growth and development in the 1970s, but in an inessential form.[5] The thought was that human ingenuity could overcome Nature's scarcity over time, and ultimately (formally, in the limit) allow humanity to be free of Nature's constraints (Chapter 4*). But the

---

[3] It will be asked who is represented in the collective 'we' and 'our' that I am using here. It is not everyone in the world, and certainly not restricted to those who agree with the claims I am making about the mutual influence of academic economic models and a general reading of economic possibilities. The group I have in mind is not fixed by designation but through invitation – for example, people who read this Review – to consider why and how we need to break the cycle and revise the conception we hold of humanity's place in the biosphere.

[4] The significance of the years immediately following the Second World War for the economics of biodiversity is shown repeatedly in the Review (see especially Chapter 4). I am referring to the evolution of economic thinking in the West. However, to the best of my knowledge the economic models that shaped state policy in the Soviet Union, and the ones developed by prominent academics in Latin America, also did not include Nature. In the Review, the terms Nature, natural capital, the natural environment, the biosphere, and the natural world are used interchangeably.

[5] See, for example, the special issue in the *Review of Economic Studies* (1974) on the economics of exhaustible resources.

Preface

practice of building economic models on the backs of those that had most recently been designed meant that the macroeconomics of growth and development continued to be built without Nature's appearance as an essential entity in our economic lives. Historians of science and technology call that feature of the process of selection 'path dependence'.[6] That may be why economic and finance ministries and international organisations today graft particular features of Nature, such as the global climate, onto their models as and when the need arises, but otherwise continue to assume the biosphere to be external to the human economy. In turn, the practice continues to influence our conception of economic possibilities for the future. We may have increasingly queried the absence of Nature from official conceptions of economic possibilities, but the worry has been left for Sundays. On week-days, our thinking has remained as usual.[7]

Biodiversity is the diversity of life. We will find that the economics of biodiversity is the economics of the entire biosphere. So, when developing the subject, we will keep in mind that we are embedded in Nature. The Review shows (Chapter 4*) that although the difference in conception is analytically slight, it has profound implications for what we can legitimately expect of the human enterprise. The former viewpoint encourages the thought that human ingenuity, when it is directed at advancing the common good, can raise global output indefinitely without affecting the biosphere so adversely that it is tipped into a state far-removed from where it has been since long before human societies began to form; the latter is an expression of the thought that because the biosphere is bounded, the global economy is bounded.

I imagine the person reading the Review is doing so because she wants to understand our place in Nature as a citizen. She is curious to know what sustainable development should mean; what criteria governments and private companies should use when choosing investment projects; what rules private investors such as herself should use to compare alternative asset portfolios; what she should insist be the practices of companies producing the goods and services she purchases and consumes; whether the social returns on investment in family planning and reproductive health to meet the unmet needs of millions of the world's poorest women are so low that the European Union assigns less than 1% of their international aid budget to them; and so on. Depending on the context, I call her the 'social evaluator', or the 'citizen investor'. The social evaluator recognises that her perspective as a citizen is different from the one she assumes as she goes about her daily life. And she wants to understand why that is so.

In the chapters that follow, the natural world is studied in relation to the many other assets we hold in our portfolios, such as the vehicles we use for transport, the homes in which we live, and the machines and equipment that furnish our offices and factories. But like education and health, Nature is more than a mere economic good. Nature nurtures and nourishes us, so we will think of assets as durable entities that not only have use value, but may also have intrinsic worth. Once we make that extension, the economics of biodiversity becomes a study in portfolio management.

That should be no surprise, for we are all asset managers pretty much all of the time. Whether as farmers or fishers, foresters or miners, households or companies, governments or communities, we manage the assets to which we have access, in line with our motivations as best as we can. But the best each of us is able to achieve with our portfolios may nevertheless result in a massive collective failure to manage the global portfolio of all our assets. The analogy of each of a crowd of people trying to keep balance on a hanging bridge and bringing it crashing down speaks to that possibility.

---

[6] A clear statement is in P. A. David (1985), "Clio and the Economics of QWERTY," *American Economic Review*, 75(2), 332–337.

[7] Over the years the absence of Nature's essentiality from macroeconomic models of growth and development has been remarked upon by scholars outside the mainstream of economic thinking and practice. But while it is all too easy to criticise existing practices, it is a lot harder to develop alternative models of comparable analytical depth and empirical reach to ones that have been honed by years of patient work. That may be why the criticisms have not been taken seriously by mainstream economists.

Preface

The Review has been prompted by a growing body of evidence that in recent decades humanity has been degrading our most precious asset, Nature, at rates far greater than ever before. Simultaneously, the material standard of living of the average person in the world has become far higher today than it has ever been; indeed, we have never had it so good. In the process of getting to where we are, though, we have degraded the biosphere to the point where the demands we make of its goods and services far exceed its ability to meet them on a sustainable basis. That is ominous for our descendants and suggests we have been living at both the best and worst of times.

The Review demonstrates that in order to judge whether the path of economic development we choose to follow is sustainable, nations need to adopt a system of economic accounts that records an inclusive measure of their wealth. The qualifier 'inclusive' says that wealth includes Nature as an asset. The contemporary practice of using Gross Domestic Product (GDP) to judge economic performance is based on a faulty application of economics. GDP is a flow (so many market dollars of output per year), in contrast to inclusive wealth, which is a stock (it is the social worth of the economy's entire portfolio of assets). Relatedly, GDP does not include the depreciation of assets, for example the degradation of the natural environment (we should remember that 'G' in GDP stands for gross output of final goods and services, not output net of depreciation of assets). As a measure of economic activity, GDP is indispensable in short-run macroeconomic analysis and management, but it is wholly unsuitable for appraising investment projects and identifying sustainable development. Nor was GDP intended by economists who fashioned it to be used for those two purposes. An economy could record a high rate of growth of GDP by depreciating its assets, but one would not know that from national statistics. The chapters that follow show that in recent decades eroding natural capital has been precisely the means the world economy has deployed for enjoying what is routinely celebrated as 'economic growth'. The founding father of economics asked after The Wealth of Nations, not the GDP of nations. The idea of wealth that is developed in the Review is, not surprisingly, a lot richer than the one Adam Smith was able to fashion, but his identification of assets as the objects of interest was exactly right.

Acknowledgement that by economic progress we should mean growth in inclusive wealth brings the Review back full circle to where it begins, which is that just as the private investor manages his portfolio with an eye on its market value, the citizen investor appraises the portfolio of global assets with an eye on their social worth. Wealth maximisation in its various guises unites microeconomic reasoning with its macroeconomic counterpart.

The Review makes use of this unification to develop the idea of sustainable development. It constructs a grammar for understanding our engagements with Nature – what we take from it, how we transform what we take from it and return to it, why and how in recent decades we have disrupted Nature's processes to the detriment of our own and our descendants' lives, and what we can do to change direction.

As this is a global Review, I often speak of the demands humanity makes on Nature. But much of the time the Review is obliged to look closely at smaller scales and local engagement with Nature. Differences in the way communities are able to live tell us that people do not experience increasing resource scarcity in the same way. Food, potable water, clothing, a roof over one's head, clean air, a sense of belonging, participating with others in one's community, and a reason for hope are no doubt universal needs. But the emphasis people place on the goods and services Nature supplies differs widely. To farmers in South Asia and Sub-Saharan Africa, it could be declining sources of water and increasing variability in rainfall in the foreground of global climate change; to indigenous populations in Amazonia, it may be eviction not just from their physical home, but from their spiritual home too; to inhabitants of shanty towns everywhere, the worry may be the infections they are exposed and subjected to from open sewers; to the suburban household in the UK, it may be the absence of bees and butterflies in the garden; to residents of mega-cities, it could be the poisonous air they breathe; to the multi-national company, it may be the worry about supply chains, as disruptions to the biosphere make

Preface

old sources of primary products unreliable and investments generally more risky; to governments in many places, it may be the call by citizens, even children, to stem global climate change; and to people everywhere today, it may be the ways in which those varied experiences combine and give rise to environmental problems that affect us all, not least the COVID-19 pandemic and other emerging infectious diseases, of which land-use change and species exploitation are major drivers. Degradation of Nature is not experienced in the same way by everyone.

Nature has features that differ subtly from produced capital goods. The financier may be moving assets around in his portfolio, but that is only a figure of speech. His portfolio represents factories and ports, plantations and agricultural land, and mines and oil fields. Reasonably, he takes them to be immobile. In contrast, Nature is in large measure mobile. Insects and birds fly, fish swim, the wind blows, rivers flow, and the oceans circulate, and even earthworms travel. Economists have long realised that Nature's mobility is one reason the citizen investor will not take the market prices of natural capital to represent their social worth even when markets for them exist. The Review studies the wedge between the prices we pay for Nature's goods and services and their social worth (the Review calls their social worth 'accounting prices') in terms of what economists call 'externalities'. Over the years a rich and extensive literature has identified the measures that can be deployed (the forces of the law and social norms) for closing that wedge. The presence of the wedge is why the citizen investor will insist that companies disclose activities along their entire supply chain. Disclosure serves to substitute for imperfect markets.

But in addition to mobility, Nature has two properties that make the economics of biodiversity markedly different from the economics that informs our intuitions about the character of produced capital. Many of the processes that shape our natural world are silent and invisible. The soils are a seat of a bewildering number of processes with all three attributes. Taken together the attributes are the reason it is not possible to trace very many of the harms inflicted on Nature (and by extension, on humanity too) to those who are responsible. Just who is responsible for a particular harm is often neither observable nor verifiable. No social mechanism can meet this problem in its entirety, meaning that no institution can be devised to enforce socially responsible conduct.

It would seem then that, ultimately, we each have to serve as judge and jury for our own actions. And that cannot happen unless we develop an affection for Nature and its processes. As that affection can flourish only if we each develop an appreciation of Nature's workings, the Review ends with a plea that our education systems should introduce Nature studies from the earliest stages of our lives, and revisit them in the years we spend in secondary and tertiary education. The conclusion we should draw from this is unmistakable: if we care about our common future and the common future of our descendants, we should all in part be naturalists.

The Review builds on six previous publications of mine, each directed at a particular class of problems that belong to the economics of biodiversity.[8] I may have been trying in those publications to develop a grammar for the subject, but I did not know it then, at least not consciously. For that reason, the exercises there now read like acts of reconnaissance. Each has informed the Review in essential ways, but taken together they did not sum to an economics of biodiversity. Which is why I am particularly grateful to Lord (Philip) Hammond, who as Chancellor of the Exchequer of the UK Government, invited me to lead the Review in Spring 2019.

---

[8] P.S. Dasgupta and G.M. Heal (1979), *Economic Theory and Exhaustible Resources* (Cambridge: Cambridge University Press), and P. Dasgupta: *The Control of Resources* (Cambridge, MA: Harvard University Press, 1982); *An Inquiry into Well-Being and Destitution* (Oxford: Clarendon Press, 1993); *Human Well-Being and the Natural Environment* (Oxford: Oxford University Press, 2004); *Economics: A Very Short Introduction* (Oxford: Oxford University Press, 2007); and *Time and the Generations: Population Ethics for a Diminishing Planet* (New York, NY: Columbia University Press, 2019).

My education in what is the substance of the Review began in the late 1970s in conversations with Karl-Göran Mäler. He encouraged me to develop my ideas on the links between rural poverty and the state of the local environmental resource-base in the world's poorest countries – a subject that was then notably absent from mainstream development economics, and which remained absent until well into the 1990s. I was further encouraged by Lal Jayawardena, Director of the World Institute of Development Economics Research (WIDER), Helsinki, who invited Mäler and me in 1988 to establish a programme at his institute on the environment and emerging development issues.[9]

But it wasn't until 1991 when, as the newly appointed Director of the Beijer Institute of Ecological Economics, Stockholm, Mäler asked me to serve as Chair of the Institute's Scientific Advisory Board, that we were able to pursue the programme jointly with ecologists. The Institute's mandate made it possible, which was unusual at that time, for ecologists and economists to conduct a regular series of workshops in ecological economics. In this, Mäler and I were aided greatly by the intellectual authority of Kenneth Arrow, Bert Bolin, Paul Ehrlich, and Simon Levin. The Institute's activities have continued with the same exacting standard under Carl Folke, who assumed the Directorship when Mäler retired.

As these developments were confined to Continental Europe, it was natural for us to imagine regional networks of ecological economists in developing countries. That was made possible by a grant from the MacArthur Foundation, Chicago. It enabled Mäler and me in 1999 to establish the South Asian Network for Development and Environmental Economics (SANDEE) and simultaneously the journal Environment and Development Economics (Cambridge University Press). Our idea was to offer not only encouragement, but also financial help and a journal based in the West where university teachers of economics in developing countries could publish their research findings. We were able soon after to help colleagues in Eastern and Southern Africa and in Latin America to establish their own networks.[10]

Mäler and I received further help. This time from Miguel Virasoro, Director of the Abdus Salam International Centre for Theoretical Physics (ICTP), Trieste, who invited us in 2001 to create a programme in ecological economics at ICTP. We used the opportunity to invite economists in our newly formed networks to the Centre, so that they could prepare their findings for publication with help from members of the journal's editorial board. Readers will find that the Review has been much influenced by the rich body of work by colleagues in those networks.

The economics of biodiversity requires attention to local socio-ecological details. I was introduced to the idea of social capital and its relevance for ecological economics at the biannual retreat that Ismail Serageldin convened for an advisory panel he had constituted in the mid-1990s at the Sustainable Development Vice Presidency of the World Bank.[11] My understanding of the subject has deepened at the annual teaching workshop that SANDEE has organised since its inception, from discussions with my fellow lecturers Rabindranath Bhattacharya, Randall Bluffstone, Enamul Haque, Karl-Göran Mäler, Pranab Mukhopadhyay, M.N. Murty, Mani Nepal, Subhrendu Pattanayak, Priya Shyamsundar, E. Somanathan, Jeff Vincent, and participants from Bangladesh, Bhutan, India, Nepal, Pakistan, and Sri Lanka, too numerous to mention individually. On the science of complexity, I have learnt enormously from discussions over a period of fifteen years with fellow members of the Scientific Advisory Panel of

---

[9] The programme's proceedings were published in P. Dasgupta and K.-G. Mäler, eds., The Environment and Emerging Development Issues, Vols. I and II (Oxford: Clarendon Press, 1997), and P. Dasgupta, K.-G. Mäler, and A. Vercelli, eds., *The Economics of Transnational Commons* (Oxford: Clarendon Press, 1997).

[10] Resource Accounting Network for Eastern and Southern Africa (RANESA) and the Centre for Environmental Economics and Policy in Africa (CEEPA), Pretoria; and Latin American and Caribbean Environmental Economics (LACEEP) and the Tropical Agricultural Research and Higher Education Center (CATIE), Costa Rica. SANDEE is based at the International Center for Integrated Mountain Development (ICIMOD), Kathmandu.

[11] The Panel's proceedings were published in P. Dasgupta and I. Serageldin, eds. (2000), *Social Capital: A Multifaceted Perspective* (Washington, DC: World Bank).

Preface

the Programme on Complex Systems at the James S. McDonnell Foundation, St. Louis, and from the Foundation's successive Presidents, John Bruer and Susan Fitzpatrick.

My understanding of the social embeddedness of individual preferences improved greatly by the many discussions I had with Dale Sutherland and Alistair Ulph during my tenure in 2008–2013 as Visiting Professor at the Sustainable Consumption Institute of the University of Manchester.

Before beginning work on the Review, I asked Simon Beard, John Bongaarts, Simon Levin, Tom Lovejoy, and Peter Raven to prepare essays for me on subjects I knew to be essential but on which I was inexpert. The ideas they developed are reflected in the present work.

The Review has been much influenced by Scott Barrett and Aisha Dasgupta, who assumed the lead in collaborative works that form the basis of some of the central ideas here.

During the Review's preparation, I have gained enormously from correspondence and discussions with Inger Andersen, Robert Aumann, Scott Barrett, Ian Bateman, Simon Beard, Simon Blackburn, Caroline Bledsoe, John Bongaarts, Stephen Carpenter, William Clark, Mary Colwell, Diane Coyle, Aisha Dasgupta, Shamik Dasgupta, Zubeida Dasgupta, Paul Ehrlich, Carl Folke, Patrick Gerland, Roger Gifford, Lawrence Goulder, Ben Groom, Andy Haines, Geoffrey Heal, Cameron Hepburn, Girol Karacaoglu, Phoebe Koundouri, Pushpam Kumar, Tim Lenton, Simon Levin, Justin Lin, Tim Littlewood, Georgina Mace, Robert Macfarlane, Shunsuke Managi, Eric Maskin, Henrietta Moore, Tid Morton, Ilan Noy, Gustav Paez, Charles Perrings, Stuart Pimm, Peter Raven, Martin Rees, Fiona Reynolds, Marten Scheffer, Ingmar Schumacher, V. Kerry Smith, Denise Spinney, Will Steffen, Nicholas Stern, Thomas Sterner, William Sutherland, Nicola Tagart, Alistair Ulph, Ruut Veenhoven, Jeff Vincent, Robert Watson, Gavin Wright, Anastasios Xepapadeas, Menahem Yaari, and Aart de Zeeuw.

I am especially grateful to HM Treasury for enabling Sandy Sheard to assemble an exceptionally gifted team who have helped me think through the economics of biodiversity. Drawn from across the public sector and based in HM Treasury, they have provided me with invaluable support over the course of the Review, including Mark Anderson, Heather Britton, Abbas Chaudri, Dana Cybuch, Rebecca Gray, Haroon Mohamoud, Robert Marks, Emily McKenzie, Diana Mortimer, Rebecca Nohl, Felix Nugee, Ant Parham, Victoria Robb, Sandy Sheard, Sehr Syed, Thomas Viegas, Ruth Waters, and Lucy Watkinson. They have gathered evidence from a wide range of experts from around the world, arranged for me to meet many of them, supported my Advisory Panel, prepared a wealth of case studies, edited the Review, and made vital contributions to drafting elements of the Review itself – particularly, the chapters in Part II. Even more, they queried every intellectual move I made; to a professor, there can be no greater reward.

Above all, I am grateful to Carol Dasgupta, on whom I have tested pretty much every idea in the Review. Her suggestions on what to emphasise and what is superficial have been invaluable.

The influence of Amiya Dasgupta, Kenneth Arrow, Paul Ehrlich, Peter Raven, John Rawls, and Robert Solow on the way I frame economics has become increasingly evident to me.

Partha Dasgupta
St John's College, Cambridge

Part I – Foundations

Chapter 0
How We Got to Where We Are

Introduction

On our 4.5-billion-year old planet, life is perhaps as much as 3.7 billion years old, with photosynthesis and multi-cellularity (appearing dozens of times independently) around 3 billion years old. Oxygen levels began to rise some 650 million years ago or even earlier (coinciding with the Metazoan stage); plants, animals, and fungi emerged on land perhaps 480 million years ago; forests appeared around 370 million years ago; and modern groups such as mammals, birds, reptiles, and land plants originated about 200 million years ago. The geological record shows that there have been five global mass extinction events, the first of them about 540 million years ago. The records also suggest that 99% of the species that have ever existed (perhaps 5 billion in number) have become extinct. The last major extinction event occurred about 66 million years ago, and the number of species on Earth and the complexity of their communities and ecosystems have increased steadily since that time.[12]

Over the past 66 million years, the number of species has grown to around 8 million to 20 million (possibly more) species of eukaryotic organisms – ones with cells that have a distinct nucleus – and an unknown and much larger number of prokaryotes (Archaea and bacteria). Our lack of knowledge is enormous. Only about 1% of the species that have existed during the history of life on Earth live in the ecosystems into which humans evolved and live now. From the time that human beings evolved, our dependency on *biodiversity*, that is, the diversity of life, has remained complete. Indeed, we ourselves are a part of biodiversity.

Within the global ecosystem, the first members of humanity's evolutionary line split from the other African apes about 6 to 8 million years ago. Our closest relatives, a group that we call hominids, appear in the fossil record about 2.7 million years ago, also in Africa. One of these, *Homo erectus*, was the first to migrate out of Africa to the north, starting around 2 million years ago, where it, along with the Neanderthals, the Denisovans, and a few more local species, represented humanity until the occurrence of another significant migration out of Africa. This event occurred at least 60,000 years ago, when the hominids present in Eurasia were joined by modern *Homo sapiens*, which had originated in Africa about 200,000 years ago. By about 30,000 years ago, *Homo sapiens* had conquered and killed the other hominids that had preceded them in the Northern Hemisphere, after interbreeding with Neanderthals and Denisovans when they came into contact with them.

For tens of thousands of years after *Homo sapiens* reached Eurasia, they lived as hunter-gatherers. Over the years, they began to create artistic works and make weapons and musical instruments; but because they were frequently on the move in search of food, necessarily carrying their babies with them, there was little opportunity for them to develop what we today call civilisation. Dogs were domesticated in Eurasia at least 20,000 years ago, and crops were being cultivated by about 12,000 years ago. Domestication, therefore, took place in a period of rising temperatures following the end of the preceding cold period.

---

[12] The Introduction to this chapter has been taken from notes prepared by Peter H. Raven, President Emeritus of the Missouri Botanical Gardens, for a joint meeting of the Pontifical Academy of Sciences and the Pontifical Academy of Social Sciences in Vatican City in 2017. The proceedings of the conference were subsequently published as *Biological Extinction: New Perspectives* (Cambridge: Cambridge University Press), 2019, edited by Partha Dasgupta, Peter H. Raven and Anna L. McIvor.

Chapter 0: How We Got to Where We Are

The intercontinental migration of *Homo sapiens* took place during a period of glacial expansion that lasted from 110,000 to about 10,000 years ago. Human dispersal from Eurasia to Australia (about 80,000 years ago) occurred long before there was any domestication of plants and animals, a practice that never developed in that continent. Dispersal to North America (via the then existing Bering Land Bridge connecting northern Siberia and Alaska) seems to have occurred some 18,000 years ago (possibly even earlier), after the domestication of dogs, which they brought with them. No crops were carried to the New World until modern times. Both in North and South America, crop agriculture was developed independently.

Along with domestic animals, cultivated crops (the first appearance being some 12,000 years ago in the Fertile Crescent, a crescent-shaped region in the Middle East) provided a major source of storable food, one that could see humans through droughts, winters, and other unfavourable times. At that time, the entire global population of humans is estimated to have been about 1 million people, with only about 100,000 in Europe. Agriculture allowed a single person to feed more than themselves and their family, and made possible a rapid increase in population. Farmers from the Fertile Crescent swept into Europe, displacing the sparse population that had existed there earlier. In these cultivated lands, the numbers of people who could live together in a village, town or city increased greatly. The first cities were built in Southern Mesopotamia between the Tigris and Euphrates Rivers some 7,000 years ago. The economic surplus enabled most aspects of what we call civilisation to develop in that region. Individuals could learn to become toolmakers, soldiers, tradesmen and priests, and the various elements of what we consider civilisation began to develop rapidly. A very important invention was writing. Sumerian writing and Egyptian hieroglyphs, understood to be the earliest writing systems, were invented around 5,500 years ago; the earliest texts about 4,000 years ago. The Sumerians are understood today to have also invented a number system, some 4,000 to 5,000 years ago.

As our human numbers grew, our impact on the planet increased with them. By about 3,000 years ago, pastoralists, agriculturists and hunter-gatherers had transformed large areas as they gathered and grew food for their increasing numbers. The roughly 300 million people who lived at the time of the Roman Empire had grown to 500 million around the year 1500 CE, near the beginning of the Renaissance, and today has reached nearly 7.8 billion (Table 0.2).

If human history is a mere blink in the history of the biosphere, economic history is only a point in time. Drawing on material objects uncovered from archaeological sites, sketches of quantitative history reach about 5,000 years into the past; while quantitative economic history looks back at best to the start of the Common Era.[13] In this chapter, we present data on changes (or lack of changes) over time in regional living standards, and global population numbers and health status since year 1 of our Common Era (Figures 0.1–0.2). We also report findings on various successes and failures of past societies to overcome the environmental stress they faced. In current understanding, those stresses arose from population pressure, climatic changes and defective land management (soil erosion being a prominent result). The global evidence, in its aggregative form, speaks to a long stretch, until about 1500 CE, of stagnant population numbers, living standards and health status, rising slowly until the start of the Industrial Revolution (round about 1750), growing somewhat more rapidly from then, and taking a sharp and accelerated increase from around the middle of the last century until now.

0.1 Economic History Since Year 0

The economist Angus Maddison spent much of his professional life uncovering past living standards across the world. To do that, he chose gross domestic product (GDP) per capita as a measure of the

---

[13] Finley (1982) offered a few estimates of economic indicators in Homeric Greece (approximately 1200 BCE), but they have been found by historians to be not without problems.

standard of living in a society. He chose that because GDP is the index in most common use today for assessing the performance of economies and for evaluating macroeconomic policy. GDP is the market value of the flow of all final goods and services produced within a country in a given year. It includes the monetary value of aggregate private consumption (consumer spending, as it is often called), gross investment (including the capital expenditures of businesses), the sum of government expenditures, and the difference between exports and imports. GDP is a measure of an economy's aggregate output. As the value of output has to reach *someone's* pocket, we will use the terms 'output' and 'income' interchangeably.[14]

Although it is routine today to study the performance of economies in terms of GDP, Maddison's work is especially interesting and important because it is on deep economic history. Peering into the past 2,000 years with a measuring rod, which is what Maddison did, takes courage, but Maddison used whatever record he could find that gave clues to wages, food consumption, clothing, housing, land rents, and so on. Table 0.1 reproduces figures constructed in ongoing work by others in what is now known as the 'Maddison Project' (Maddison, 2018; Bolt et al. 2018). The project is designed to improve upon the estimates Maddison (2001) offered in his now-classic work. The table presents output per capita in five regions of the world. The final row presents time series of global output per capita. The figures are expressed in 2011 international dollars.[15]

Table 0.1 Deep History, 1 – GDP Per Capita (2011 International Dollars)

	1 CE	1000	1500	1700	1820	1900	1950	2000	2016
Western Europe*	914	676	1,232	1,630	2,313	4,904	6,078	32,956	40,364
Western Offshoots*	636	636	636	755	2,070	8,027	14,867	44,331	51,342
Latin America*	636	636	660	843	999	1,822	3,048	8,728	13,470
Asia	725	747	904	909	939	1,099	1,201	5,286	11,102
Africa	747	676	660	668	774	1,444	1,596	2,889	4,680
World	747	723	898	978	1,132	2,446	3,277	9,456	14,574
World Bank (World)								10,346	15,080

Source: Maddison (2018), Bolt et al. (2018). Note: 'Western Offshoots' include what are today US, Canada, New Zealand and Australia.

Note: The Maddison Project Database (MPD) provides long-run data on GDP per capita for comparisons of relative income levels across countries. The measure of real GDP per capita is based on multiple benchmark comparisons of prices and incomes across countries and over time. The figures between the dates 1 CE to 1950 are updated from a combination of the 2010 and 2013 MPD releases, which were in 1990 prices. To account for the change in price level, a simple GDP deflator is used to adjust all regions for 2011 prices. The differences between the rebased regions' GDP and the newly calculated GDP do not significantly change the overall trends at the regional and global level, relative to the 1990 price level estimates. The figures between 1950 to 2016 are taken from the most recent release from the Maddison Project Database, in 2011 prices, apart from the regions denoted with a star where 1913 data was also available in their updated database. For comparison, the World Bank's estimate of GDP per capita PPP (2011 prices) for 2000 and 2016 are given, which are both within 10% of the latest Maddison Project data.

---

[14] GDP is to be distinguished from GNP (Gross National Product). The latter is GDP plus incomes earned by residents from overseas earnings, minus incomes earned within the economy by overseas residents. For our purposes in the Review, the difference between the two indices is inconsequential. See Chapter 13 for a fuller discussion of GDP.

[15] In constructing international dollars (i.e. dollars at purchasing price parity, PPP), the official exchange rates of various currencies with respect to the US dollar are converted so as to bring the purchasing power in the regions on par with one another. In what follows, it should be read that 'dollars' mean 'international dollars'.

Chapter 0: How We Got to Where We Are

The first thing to note about the figures in Table 0.1 is that the average person in the world was very poor in terms of income right up to the beginning of the modern period (approximately 1700 CE). In Late Antiquity and the Middle Ages, average income in most regions everywhere was not much above 1.90 dollars a day (a few even below it) – the figure that was taken by the World Bank in 2015 to be the line below which spells *extreme* poverty. Regional variations became prominent in the beginning of the Early Modern period (roughly, 1500), by which time Western Europe had begun to diverge from the rest of the world. But Maddison's estimates suggest that even in 1700 the average person in Asia languished in near-extreme poverty, at around 2.5 dollars a day. As tourists we are dazzled by the art, architecture and technology of past eras. We refer to them as great civilisations and imagine that those must have been prosperous times as well. Table 0.1 says we should imagine otherwise. So long as there is a ruling class to tax poor subjects, we have the beginning of the arts, humanities and the sciences. The Taj Mahal, for example, which is today the most renowned construction of the Early-Modern era, was built in the mid-17th century on the orders of a tyrant in memory of his favourite wife, on the backs of extremely poor subjects.

Figure 0.1 GDP Per Capita from Year 0 to 2016 CE

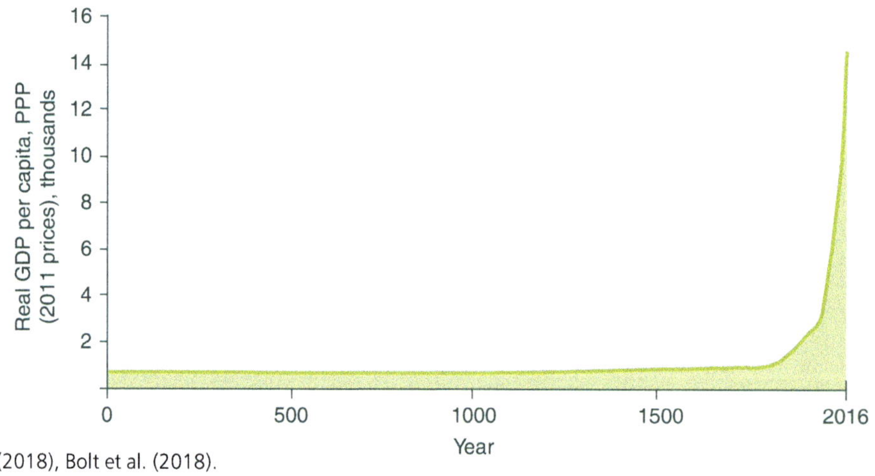

Source: Maddison (2018), Bolt et al. (2018).

Great art, great architecture, great literature, and even great scientific and technological discoveries can coexist with general squalor and widespread deprivation of the means available for a reasonable existence. And they have coexisted for nearly all of history. Average world income in 1820 CE, which in many economic historians' reckoning was about the time of the Industrial Revolution, was only about 50% higher than in 1 CE. That means the growth rate of world income per capita over the 1,820-year period, when averaged, was indistinguishable from zero. Table 0.1 confirms that significant increases in the standard of living took place only in the 20th century, mostly in the West and what Maddison called Western Offshoots (US, Canada, Australia and New Zealand). In a matter of a little under 70 years (1950 to 2016), GDP per capita increased nearly seven times in Western Europe. It is true that in 1945 those nations were in a devastated condition, meaning that the potentials for growth were large. But as Table 0.1 tells it, we should not imagine that the poorer a nation, the greater its potential for advancement. Western Europe had institutions in place, an educated population, and a social coherence that enabled them to take advantage of their potentials. In contrast Africa, which at the end of the Second World War was a lot poorer than the West, continues to languish in poverty. Average income in Africa in 2016 was barely over 11% of that in Western Europe.

To be sure, Table 0.1 hides social improvements taking place in various places from time to time during the 1,700 years following Year 0. An interval of 1,000 years (0–1000 CE) hides fluctuations of fortunes. Temin (2013) has suggested that GDP per capita in the Roman Empire in 2nd century CE was about the same as in India in 1990. But in time the Empire fell, and incomes dropped. We also know of the Black Death and Europe's revival after it, both of which are hidden from view in the thick, 500-year spell between 1000 and 1500 CE. Fouquet and Broadberry (2015) peered closer into Europe for the period 1200 to 1870 and found that there were periods when some regions enjoyed considerable growth in GDP per capita, while others declined. But none would appear to have enjoyed *sustained* growth in incomes. Recent work by historians of Mughal India suggests that per capita income there rose considerably by year 1600. But the empire fell into disarray by the first half of the 18th century and the economy skidded. Broadberry, Custodis and Gupta (2015) have estimated that in 1600 per capita income in India was more than 60% of that in England, but by 1870 had fallen to less than 15%. Notwithstanding the caveats, the pioneering Maddison estimates of GDP per capita in deep economic history are a stark reminder that for nearly all of history the average person in the world was extremely poor.

0.2 Economic Growth and Sustainable Development

Writing at the very end of the 18th century, the Rev. Thomas Malthus postulated that population size and the standard of living had kept each other in check throughout history in what we would today call a low-level equilibrium. The world he read was composed largely of "organic economies" (Wrigley, 2004), where not only food but also most raw materials needed for manufacturing artefacts were either animal or vegetable in origin. Production was subject to diminishing returns. Given the reproductive norms of societies (although this is not quite how Malthus put it), population grew whenever living standards rose above the equilibrium level, bringing living standards down. But whenever living standards fell below the equilibrium level, more people died (wars and pestilence) and the system equilibrated. As a matter of common observation, the equilibrium living standard was low.

Both population and living standards in Malthus' theory, like any good theory, were determined by factors operating at a deeper level. So, he identified various possible causes that had perturbed economies throughout history (wars and crop failure were two proximate drivers), from which they returned to equilibrium.[16]

Table 0.2, reproduced from Maddison (2001, 2018) and UNPD (2019a), provides estimates of population size and expectancy of life at birth over the past 2,000 years.

Table 0.2 Deep History, 2 – Global Health and Numbers

Year	0 CE	1000	1820	1900	1950	2000	2020
Life expectancy at birth (years)	24	24	29	31	46	66	73
Population size, rounded figures (million)	230	270	1,000	1,600	2,500	6,100	7,800

Source: Maddison (2001), Tables 1-5b and B-10, for columns 1-5; UNPD (2019a), for columns 6 and 7. See also the table of life expectancy at birth since 1800 in Riley (2005), which reports that global life expectancy at birth in 1800 was 29 years.

Note: Life expectancy at birth for 1 CE is Roman Egypt, 33–258, estimate. Data from 1950 to 2020 from UN Population Division.

---

[16] For a mathematical formulation of Malthus' theory, see Day (1983), who also drew attention to possible fluctuations away from equilibrium, depending on the parameters characterising organic economies.

Taken together, the two tables suggest that the global experience until 1820 CE was pretty much in line with the Malthusian theory. Global income per capita in 1820 was about 3 dollars a day, world population was about 1 billion, and a new-born was expected to live for at best 30 years.[17]

Figure 0.2 Global Population from Year 0 CE

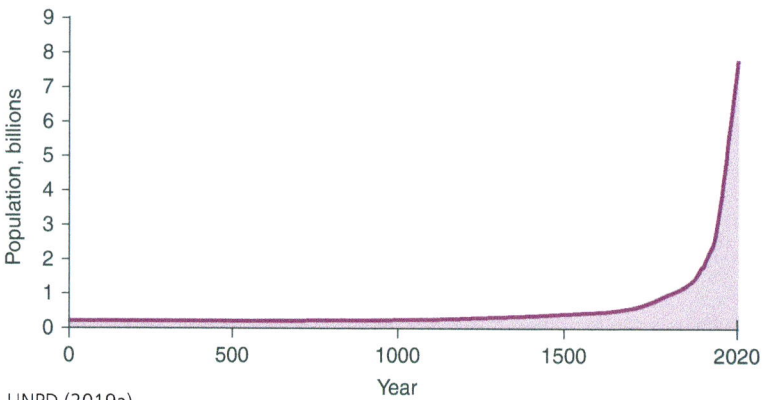

Source: Maddison (2001), UNPD (2019a).

While Malthus' theory would appear to have fitted the *global* economy at the time he published his work, it had begun to unravel in the West earlier; many experts believe as early as the 16th century, with the seeds having been laid even earlier. Identifying the factors that led to the great divergence between the West and the rest of the world has been a major subject of research among scholars studying deep history.[18]

Landes (1998) and Pomeranz (2000) are modern classics on the 'Great Divergence' and have generated enormous debate.[19] It is not in contention that a series of societal changes took place in the West in the Early Modern period (perhaps even earlier) which unleashed the innovatory forces that account for the emergence of the modern world; the debate is over the factors behind the divergence and its timing. Experience with reading evidence that bear on socio-economic processes tells us that, as with ecological processes (Chapters 2 and 3), monocausal explanations should be discounted. A multiplicity of factors can act on one another synergistically, but they can also act on one another in a discordant or inharmonious way. The former would be read in due course as societal success, the latter as societal failure. Landes (1998) laid stress on multi-faceted cultural factors[20] and noted as well that Europe in the Early Modern era was not a monolithic state. Rivalry, competition, and differences in power and beliefs among dukes, princes, and clerical eminences enabled ideas to flourish. If a scientist was in disfavour in one state, he (it was always a 'he') could find service in a rival court.[21]

The subsequent Industrial Revolution unleashed growth in labour productivity from technological advances and the production scale economies that came with those advances, and by the beginning of

---

[17] See also the table of life expectancy at birth since 1800 in Riley (2005), which reports that global life expectancy at birth in 1800 was 29 years.

[18] China was the exception, it was not only a contender with the West in 1500 CE, it is believed by historians of technology to have been ahead of it. Among other inventions, paper, printing, gunpowder, and the compass had been made in China long before they reached the West. But China fell behind in the centuries that followed because, or so it has been argued by historians, of a shift toward an inward-looking national philosophy. For example, maritime trade was stopped by Imperial decree by the mid-15th century.

[19] See O'Brien (2010) for a review of the debate.

[20] Max Weber, in his work on Protestant ethic and the spirit of capitalism, had invoked a far narrower explanatory variable.

[21] Johannes Kepler was a prominent case (Boorstin,1983).

the 20th century, Malthus' theory began to unravel elsewhere too, barring Asia (Japan excluded) (Maddison, 2001) and Africa. By then world population had risen to 1.5 billion, life expectancy at birth had advanced to Maddison's estimate of 31 years, and average income had grown to around 2,000 international dollars a year (at 2011 prices).

During the 20th century, several key dimensions of life improved greatly (and as we confirm in Chapter 4, in the second half of the century they improved spectacularly). Global income per head more than quadrupled, life expectancy of a new-born rose from 31 to 66 years (relatedly, people enjoyed ever greater protection against water- and airborne diseases, greater use of potable water and, since the end of the Second World War, antibiotics), even while world population grew by a multiple of four, to 6 billion. Even the brief period 2000–2016 was remarkable: global income per person grew by over 40%. By 2016, global income per head had reached nearly 15,000 international dollars (at 2011 prices), life expectancy at birth had risen to 73 years, and population had grown to nearly 7.8 billion. In Chapter 4, we draw attention to the 70-year span beginning in 1950. Tables 0.1 and 0.2 confirm its exceptional character. Prominent Earth scientists regard the middle of the 20th century as the period we entered the Anthropocene (Voosen, 2016).

In a famous 1930 essay 'Economic Possibilities for Our Grandchildren', John Maynard Keynes described a past that was consonant with the deep economic history of Tables 0.1 and 0.2. He concluded that humanity in his time had never remotely had it so good (Keynes, 1931). The world's living standard today is a lot higher than it was even when Keynes made his observation. The average person not only enjoys far higher income and lives longer, the proportion of the world's population in absolute income-poverty has fallen so dramatically (it is below 10% of the world's population, down from around 50% in 1970) that enthusiasts predict that within a generation the blight will have been eliminated for the foreseeable future (Jamison et al. 2013).[22]

These successes have inspired a number of intellectuals to draw the attention of the general public to the remarkable gains in the standard of living humanity has enjoyed over the past century.[23] The authors collated data on growth in scientific knowledge and the accumulation of our produced capital and human capital to argue that humanity has never had it so good. But with the exception of rising carbon concentration in the atmosphere, trends in the state of the biosphere accompanying those advances have gone unnoted by the authors. We note in Chapter 4 though, that global climate change is but one of a myriad of environmental problems we face today. And because it is amenable to technological solutions (innovating with cheap non-carbon sources of energy and, more speculatively, firing sulphur particulates into the stratosphere to reflect sunlight back into space (Pinker, 2018)), it is not representative. Global climate change attracts attention among intellectuals and the reading public not only because it is a grave problem, but perhaps also because it is possible to imagine meeting it by using the familiar economics of commodity taxation, regulation and resource pricing without having to forego growth in material living standards in rich countries. The literature on the economics of climate change (e.g. Stern, 2006; Nordhaus, 2007; Lomborg, 2013) has even encouraged the thought that, with only little investment in clean energy sources over the next few years (say 2% of world GDP), we can enjoy indefinite growth in the world's output of final goods and services (global GDP).

That is a thought that should be resisted. It will be shown that, when looking at the wider scope of the economics of the biosphere (based on an understanding of ecology and earth sciences (Chapters 1, 2, 3,

---

[22] Global poverty is likely to have risen sharply in 2020 due to the COVID-19 pandemic, partially reversing some of the improvement over recent decades. In October 2020, the World Bank suggested that the COVID-19 pandemic would push an additional 88 million to 115 million into extreme poverty in 2020 (World Bank, 2020c).

[23] Micklethwait and Wooldridge (2003); Ridley (2010); Lomborg (2013); Norberg (2016); and Pinker (2018). We will discover, though, that time series of subjective measures of well-being, such as happiness and life satisfaction, tell a different story (Chapter 11).

4 and 4*), our economic possibilities are circumscribed – even if several steps removed via technological progress – by the Earth-System's workings. We are embedded in Nature; we are not external to it. No amount of technological progress can make economic growth as conventionally measured an indefinite possibility. Ours is inevitably a finite economy, as is the biosphere of which we are part. Although there has been some recent recognition among a few economists and ecologists of these issues (we highlight leading work in Chapter 1), this understanding remains far from widespread.

Nevertheless, there is the temptation to say that because natural resources can be shifted around today with relative ease, dwindling supplies in one place can be met by imports from another (see Chapter 15 for more on trade). Intellectuals have been known to say that because of 'globalisation' location does not matter. The view emphasises the prospects offered by trade and investment and says if they are not enough, technological progress can be relied upon to solve the problems arising from resource depletion and environmental degradation. Today Malthus, the 'pessimistic parson', is seen as a 'false prophet', remaining as wrong as ever (*The Economist*, 15 May 2008).[24]

0.3 Historical Success and Failures

In the past, when communities faced exceptional resource stress (droughts, pests and soil erosion are only three causes), they introduced new practices and fashioned new arrangements. If migration to better locations was a possibility, communities would be expected to have tried that, if all else failed. We should not imagine people taking impending disasters lying down if they saw them coming. Boserup (1965) collated evidence from agrarian societies to argue that resource stress generates societal responses that not only fend off disaster but can even lead to prosperity. Exceptional scarcities may raise exceptional 'problems', but as the saying goes today, they offer exceptional 'opportunities' as well. Boserup's work countered a widespread fear in the early 1960s that our capacity to produce food was being overtaken by growth in human numbers. She saw population growth as a spur to innovations. The Green Revolution that came soon after her publication matched her narrative. Population was dropped from public discourse even as Boserup came to replace Malthus.[25]

Boserup's case studies were about organic economies. Inevitably, there was sample bias in her choice of examples. Societies that had not made the cut would have disappeared or moved to blend themselves among communities that survived; they would be absent from such records as those that Boserup studied. In a study of a modern-day society, Turner and Shajaat Ali (1996) put together the contrasting concerns of Malthus and Boserup by demonstrating that in the face of rising population and a deteriorating resource base, small farmers in Bangladesh expanded production by intensifying agriculture practices and, with government help, collectively strengthening drainage systems and flood and storm defences. The farmers have not been able to thrive, they still live in poverty, but they staved off collapse (they have not abandoned their villages en masse for cities), at least for now. The metaphor that comes to mind is of a crowd walking up an escalator that is coming down at the same speed. Studies with a similar flavour for agricultural prospects in Africa have been reported in Christiaensen (2017) and Juma (2019). Historically, migration has been a coping strategy against especial ecological stress. Petraglia et al. (2020) have collated archaeological findings in south-eastern Arabia to show that

---

[24] The Review argues, however, that the COVID-19 pandemic can in large measure be traced to weaknesses in commodity supply chains and to biodiversity loss.

[25] Economic historians refer to our need for energy to make the same point. Human societies have over millennia improved their living standard by finding new sources of energy in the face of rising costs of established resources. The succession of human sweat, animal power, rivers and streams, wind, timber, coal, oil and gas, and most recently the nucleus of radioactive matter, is a frequently cited example of the global success in finding ways to harness energy. Barbier (2011) contains a wide-ranging demonstration of ways in which societies have historically depended on natural resources for growth and well-being.

ancient droughts during the Holocene corresponded with population movements from inland regions to the coast that were rich in resources.

If Boserup is a counterpoint to Malthus, Jared Diamond is a natural counterpoint to Boserup. Techniques for reading archaeological records have improved since the 1960s. Reviewing a series of case studies drawn from the early-to-middle second millennium CE, Diamond (2005: 6) classified the processes through which those collapsed societies he had studied had undermined themselves by damaging their ecosystems. He identified eight categories: (i) deforestation and habitat destruction; (ii) soil erosion, salinisation and fertility losses; (iii) water management problems; (iv) overhunting; (v) overfishing; (vi) effects of introduced species on native species; (vii) human population growth; and (viii) increased per capita impact of people.

Societies that are unearthed by paleo-ecologists were universes unto themselves. Those for whom transportation was costly, trade was relatively small in comparison to domestic output, and communities had to rely entirely on their own ecosystems. Communities that were under resource stress demanded more from their ecosystem than it was able to supply on a sustainable basis. Diamond's category (viii) can be read as the 'ecological footprint' (see below) of the organic economies he had studied.

In Chapters 2, 3 and 4, we find that items (i)-(vii) on Diamond's list lead to (viii). For example, Diamond reported that a number of societies that had deforested their land had been able to develop successful forest management practices and population measures, but that in contrast there were others – most notoriously in the public imagination, Easter Island – that had failed to develop successful management practices, and had collapsed as a result. He also found a common pattern in past collapses: population growth that followed access to an abundant ecosystem made people intensify the means of food production (irrigation, terracing, double-cropping) and expand into marginal land. Growing populations led to a mining of their ecosystems. That left communities vulnerable to climatic variations, as there was little room left for either mistakes or bad luck.[26]

Proceedings of the National Academy of Sciences of the United States of America (PNAS, 2012) published a Special Feature on historical collapses. Contributors reported 12 studies of past societies that had faced environmental stress. Seven were found to have suffered severe transformation, while five had overcome them through changes in their practices. Butzer (2012) reported the ways in which a number of societies in 14th-18th centuries Western Europe displayed resilience by coping with environmental stresses through innovation and agricultural intensification. Like Diamond, he concluded that collapse is rarely abrupt.

That collapse is rarely abrupt suggests that even robust socio-ecological systems become less resilient in withstanding shocks and surprises when they experience continual stress (Chapters 3 and 5). In a study of the European Neolithic societies that began some 9,000 years ago, Downey, Haas and Shennan (2016) found that the introduction of agriculture spurred population growth, but societies in many cases experienced demographic instability and, ultimately, collapse. The authors also uncovered evidence of warning signs of eventual population collapse, reflected in decreasing resilience in socio-ecological systems. Scheffer (2016) has given further support to the thesis by reporting that there had been warning signs of reduced resilience prior to the great drought in the late 1270s that destroyed the communities that had built the iconic alcove sites of Mesa Verde.

Reviewing findings on past societal collapses, Beach, Luzzadder-Beach and Dunning (2019) suggest that pioneer communities often caused soil degradation in conjunction with terrestrial species collapse

---

[26] The present section is an elaboration of Section 5 of Dasgupta, Mitra and Sorger (2019), which contains a formal model of the socio-ecological processes identified by Diamond in his study of the success and failure of organic economies.

Chapter 0: How We Got to Where We Are

because of unfamiliarity with the ecosystem they had entered. They cite the case of the Norse who first entered Iceland in about 875 CE. Vegetation covered about 65% of the island then, but the cover dropped to 25%, of which forest cover shrank to 1% of its original extent. The authors report that the decline in biodiversity was caused by deforestation and overgrazing, mainly by sheep. More than half the island's soil had experienced considerable to severe erosion.

Greenland Norse communities disappeared in the 15th century, during a period known as the Little Ice Age. Recent findings show that the communities were fishermen, and also hunted seals and walrus tusks (Kintisch, 2016). Tusks were exported to the Continent. Over time, harvesting costs increased, as severe storms over the sea occurred more frequently with a climate that had become colder. The economic downturn in the Continent during the Little Ice Age also led to a decline in the demand for tusks; the export price of tusks fell. The Norse experienced a fall in the standard of living, below tolerable levels. The prevailing population size proved to be unviable under the new climatic conditions.

As illustration of past societal successes and failures, it is useful to contrast the fate of the Norse people with the Mayas, who emerged in Meso-America round 3,000 years ago. The Classic Period of the Mayas was 250–900 CE. It saw widespread growth of infrastructure including reservoirs and agricultural fields, which together can be called 'landscape capital'.[27] This produced intensive, polyculture farming systems, water management systems, terraces, wetland fields and extensive forest garden systems.[28] Their construction boosted output while conserving soil. Some terraced systems are thought to have been adaptations to eroded or depleted landscapes from earlier periods, for some terraces were perched on slopes above depressions containing sediments derived from previous earth-slope erosion, and some others were built on bare bedrock, with soils formed behind the walls only since the Mayas built them. Evidence suggests that the wetland fields were constructed on what had previously been dryland agricultural fields, as the water table rose on account of sea level rise. This was the kind of societal adaptation to changing circumstances that had excited Boserup. But a combination of drought and warfare in what is now known as the Terminal Classic period of the 9th century CE led to an inability of the Mayas to maintain their landscape capital even as it led to soil erosion and the eventual demise of their civilisation.

Loss at war can destroy a civilisation rapidly, in contrast to ecological decline which takes place gradually until a tipping point is reached (Chapter 3). The very fact that societies had allowed their socio-ecological system to come near a tipping point tells us that they probably could not read the signs of their impending collapse until perhaps it was too late. Paleo-ecologists have an advantage. They can read the archaeological record to uncover a society's doings over an extended period of time, long before they collapsed.

0.4 Understanding Humanity's Contemporary Overshoot

Inevitably, paleo-ecologists study societies that had tight geographical boundaries. A community that failed because of population overshoot or bad resource-management practices no doubt destroyed their natural resource base, but it was their *local* resource base they destroyed; societies until modern times were incapable of affecting the Earth System as a whole. Matters are different today. The human presence is so dominant that the Earth System is no longer as modular as it was until recently. Disturbance in one location today gets transmitted to other parts in short order. Movements of people and trade in goods have created a transmission mechanism with a long and quick reach. The mechanism's medium has, however, remained the same: Nature is *mobile*. We weaken the Antarctica

---

[27] This account of the Mayans is taken from Beach, Luzzadder-Beach and Dunning (2019).

[28] Beach et al. (2019) have unearthed further evidence from below contemporary forest canopies of the extensive construction of wetland fields for their food production.

Chapter 0: How We Got to Where We Are

ice sheet without ever going there; phosphorus discharge from farms in Minnesota contributes to a deadening of the Gulf of Mexico; emissions of soot from kitchens in the Indian sub-continent affect the circulation patterns of the monsoons; the Green Revolution's demand for water, fertilisers and pesticides pollute the rivers and ground waters of the Indo-Gangetic Plain; fish in the North Sea eat microplastic originating in markets in the Bahamas; and so on.[29]

Much of Nature and the processes governing it are also silent and invisible (Chapters 2 and 4). The three pervasive features – *mobility*, *silence* and *invisibility* – make it impossible for markets to record adequately the use we make of Nature's goods and services (Figure 0.3). That inadequacy extends naturally to the goods and services we produce. There is thus an inevitable wedge between the market prices of goods and services and their social scarcity values. This has far-reaching implications for our conception of our place in Nature. Low market prices for Nature's goods and services (we will discover that many come with a *negative* price tag) has encouraged us to regard ourselves as being external to Nature.

Figure 0.3 Nature's Properties

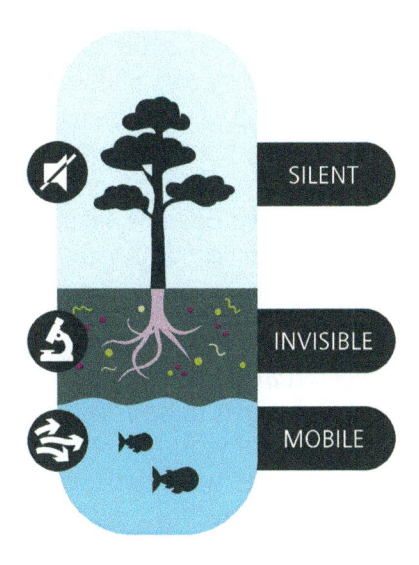

The *biosphere* (Nature and the biosphere are used interchangeably in the Review) is the part of Earth that is occupied by living organisms (Chapter 2). It pays to let our imagination roam and imagine the biosphere as a self-regenerative asset, a gigantic version of forests and fisheries. But there is a difference. We are a part of the biosphere; we are not external to it. What we take from our neighbourhood over a period of time and put back in is known as our *ecological footprint*. It is also known as our *impact* (Ehrlich and Holdren, 1971). We may even borrow from the economist's language and call our impact our 'demand'.

Humanity's impact does not have to equal the biosphere's regenerative rate. That is because the difference would automatically be accommodated by a change in the biosphere's stock (S). A world rich in a healthy biosphere could, on utilitarian grounds, choose to draw down the biosphere and use the

---

[29] A new species of crustacean, discovered deep in the Marina Trench in 2014, has been appropriately named Eurythenes plasticus for the contents of its stomach (reported in The New Yorker, 2020, May 18, p.15).

goods and services it supplies so as to accumulate produced capital (roads, buildings, machines, ports) and human capital (health, education, aptitude). That is what economic development has come to mean among many people (Chapters 1 and 4). But that view and the practices the view has encouraged have meant that in recent decades our global impact on the biosphere has exceeded the biosphere's regenerative rate. That is the Anthropocene. As a result, the biosphere has been, and is increasingly being, drawn down. In our figurative way of speaking, the biosphere has shrunk. But that has meant a decline in the biosphere's regenerative rate, which in its turn has meant an increase in the gap between demand and supply.

Contemporary models of long-run economic possibilities envisage that scientific and technological progress can be relied upon to sustain an ever-increasing growth in global output of final goods and services. That requires us to imagine that, in the long run, we can break free of the biosphere when investing in further scientific and technological progress (Chapters 4 and 4*). And that is the sense in which contemporary economic thinking on sustainable development assumes humanity to be external to Nature. The Review concludes instead that the global output of goods and services is inevitably bounded.

We should therefore ask whether the biosphere could support on a sustainable basis a global population of between 9.4 and 12.7 billion, which is the error bar round the UN Population Division's population median projection of 10.9 billion for year 2100 (UNPD, 2019b) at the material standard of living we are encouraged to seek. In effect we are asked in contemporary growth and development economics and the economics of climate change to imagine that the population numbers being projected today will be able to enjoy, at the very least, the current global living standard, even while making smaller demands on the biosphere than we do currently. No study in the economics of technological change has explored whether that is possible, let alone the question of what lifestyles that would involve. As of now we should be more than circumspect that the scenario is plausible, because at least as grave a danger facing humanity as global climate change is the unprecedented rate of loss in biological diversity now taking place (Chapter 4).

Largely as a result of human activities, species are becoming extinct much more rapidly than in the past. As currently calculated, extinction rates are judged to be 100 to 1,000 times higher than their background rate over the past tens of millions of years (0.1–1 per million species per year) and are continuing to rise (Chapter 4). Continued species extinctions will damage the biosphere irreparably, involving unknown numbers of tipping points, which should tell us that potential cascades cannot be staved off by mere technological fixes (Chapter 3). Politics has intervened to prevent even the relatively small global investment that economic experts only a few years ago suggested was required to contain climate change. So we should expect the problem of biological extinctions to remain off the table, at least until citizens take the matter seriously.[30]

In any event, talking in percentage terms when pointing to reductions in our demand for the biosphere's goods and services through technological progress, as people often do, can be misleading. The Earth's life support system does not calculate percentage changes; it responds at each instant to the *absolute* demands we make of it. That will be the message of Chapters 2, 3, 4 and 4*. If, as is nearly certain, our global demand continues to increase for several decades (Chapter 16), the biosphere is likely to be damaged sufficiently to make future economic prospects a lot dimmer than we like to imagine today. What intellectuals have interpreted as economic success over the past 70 years may thus have

---

[30] The really hard problem in the political economy of global climate change involves using the latter's special features to frame the way we should explore the prospects for international agreements. Barrett (2003), Barrett and Stavins (2003), Barrett and Dannenberg (2012), Barrett and Dannenberg (2014a) and Barrett and Dannenberg (2014b) are incisive analytical and empirical studies on this.

been a down payment for future failure. It would look as though we are living at the best of times and the worst of times.

The Review (Chapter 4) calls the excess of impact (I) over the biosphere's regenerative rate (G), the *Impact Inequality*.[31] I is in turn decomposed into three factors: human population numbers, global GDP per person, and the efficiency with which we convert the biosphere's goods and services into GDP.[32] The efficiency factor reflects not only technology but also institutions. Moreover, the factors are not independent of one another. The remainder of the Review probes into the drivers of the three factors. Our inquiry points to the types of international and national policies that would help to convert the Impact Inequality into an *Impact Equality*; that is, to bring about balance between I and G at a healthy stock (S) of the biosphere. That, the Review argues, is what *sustainable development* should be taken to mean.

There is a risk that the Impact Inequality and the decomposition of the impact we have chosen to work with will be read as a piece of Malthusian arithmetic. In fact, there is a risk that *any* study of the overshoot in the global demand for the biosphere's goods and services that includes population as a factor is read as a Malthusian tract. But that would be to misread the Review entirely. The Review applies the tools of modern economics to study the workings of the socio-ecological world as they are currently understood. In the process, the Review tries to explain how individual and group actions over the years have led globally to the Impact Inequality. It reads the source of the Impact Inequality in the Anthropocene as analogous to each of a crowd of people trying to keep balance on a hanging bridge, with a risk of bringing it crashing down. The Review then identifies the options humanity has for reversing the sign of the Impact Inequality.

The choices are hard, they involve a lot more than a tax here and a set of regulations there. Unlike the economics of climate change, at least as it is currently presented, the economics of biodiversity we construct in this Review requires not only national and intergovernmental engagement, but engagement by communities and civil societies throughout the world. The economics we construct here is neither entirely top-down nor entirely bottom up; it is also lateral. It advocates institutions that encourage information and directives to flow in every direction. Above all, it calls for changes in our sensibilities, because the silence and invisibility of Nature make it utterly vulnerable to our activities, which neither communities nor states can wholly address. Those changes can be realised only when our sensibilities toward Nature are acquired from the earliest stages of our lives. And that is all the more reason we citizens need to attend to them.

---

[31] Barrett et al. (2020) introduced the Impact Inequality so as to understand the socio-ecological processes that are shaping the Anthropocene. The Review builds on their analysis to construct the economics of biodiversity.

[32] The decomposition of I follows Ehrlich and Holdren (1971).

Chapter 1
Nature as an Asset

Introduction

We are all asset managers. Whether as farmers or fishermen, hunters or gatherers, foresters or miners, households or companies, governments or communities, we manage the assets we have access to in line with our motivations, as best as we can. This Review pays close attention to a class of assets we call Nature and studies it in relation to the many other assets in our portfolios.

It is commonly accepted that growing urbanisation accompanying economic growth has created a distance between people and the natural world. There is evidence of that (Chapter 11). Rural communities in low income countries are a lot closer to Nature than are urban households in high income countries. Daily engagements in rural Africa require goods and services extracted from the local landscape, in contrast to the daily lives in urban Scandinavia, where people depend equally on and extract more from Nature, but do so at several steps removed, often drawing on natural resources from distant parts of the world. Households in villages in Niger, in contrast to households in towns in Germany, may not have water on tap to drink, wash and cook with, nor access to electricity. One measure of resource degradation facing rural communities in poorer regions is the increased time needed for daily household production (Box 1.1). But exit is not an option, for neighbouring villages also face increasing resource scarcity and out of necessity are not welcoming.

In contrast, degradation of nature in distant lands has little to no bite on the lives of people in high income countries, for there are alternative sources of supply from other parts of the world, at least for now. Pendrill et al. (2019) for example have estimated that about one-sixth of the carbon footprint of the average diet in the European Union can be linked directly to deforestation in tropical countries.

As this is a global Review, we will often speak of the demands humanity makes on Nature. But much of the time we will peer closely at smaller economic units. Differences in the way communities are able to live tell us that people do not experience increasing resource scarcity in the same way. Food, potable water, clothing, a roof over one's head, clean air, a sense of belonging, participating with others in one's community and a reason for hope are no doubt universal needs, but the emphasis people place on the goods and services natural assets supply differs widely. To farmers in South Asia and sub-Saharan Africa it would be declining sources of water and increasing variability in rainfall in the foreground of global climate change; to indigenous populations in Amazonia it would be eviction not just from their physical home, but also from their spiritual home; to inhabitants of shanty towns everywhere the worry would be the infections they are subjected to from open sewers; to hunter-gatherers in the African grasslands it would be their shrinking resource base; to the suburban household in the UK it may be the absence of bees and butterflies in their garden; to residents of mega-cities it would be the poisonous air they breathe; to the multi-national company it could be the worry about supply chains as disruptions to the biosphere makes old sources of primary products unreliable and investments generally more risky; and to governments everywhere it would be the call from citizens, including children, to stem global climate change. Degradation of Nature is not experienced in the same way by everyone.

1.0 Portfolio Management

Nevertheless, their varied experiences have a feature in common: each of the actors – or *agents*, as we may call them – is responding to an asset management problem. Which is why it has become customary among economists to refer to natural resources generically as *natural capital*. It is convenient

Chapter 1: Nature as an Asset

(even natural) to create a two-way classification of natural capital in terms of whether the assets are *renewable* such as fisheries (we will use the term 'self-regenerative') or *non-renewable* (fossil fuels, minerals). As this is a review of the economics of biodiversity, we pay attention almost entirely to living systems, which are self-regenerative unless they have been degraded beyond repair.[33]

Biodiversity means the diversity of life. In Chapter 2, it is shown that biodiversity resembles diversity in the portfolios held by manufacturers and financial companies, and that it does so in two ways. First, biodiversity is akin to the complementarities among inputs in factory production, meaning that all inputs are significant in production. Second, biodiversity plays the same role in natural capital as diversity does in financial portfolios: it reduces variability (uncertainty) in yield. For many people the diversity of life has value independent of human wants and needs, but we will find that it pays to build the study of biodiversity's value from an anthropocentric perspective and then add non-anthropocentric perspectives to give further urgency to repairing our relationship with Nature (see Chapter 12). We will confirm, however, that by biodiversity we should not mean a headcount of species – the concept is a lot richer.

By the 'agents' facing asset management problems we do not simply mean individuals. The agent could be a person, household, village, company, the state, a nation or even an international body whose management strategies are reached on the basis of their respective decision-making processes (personal welfare, respect for tradition, bargaining, majority vote, rank order rule, the sway of their Chair and so on). The coin with which the agent chooses her portfolio or commends a portfolio to others depends on her role in the decision-making process. In the local garden centre, she values goods in terms of her personal needs, the quoted prices and her budget; as member of the local council, she would be expected to value maintenance expenditure on the village common in terms of the welfare of the local community; as civil servant, she would be charged, when comparing a wetland reclamation scheme to a highway construction project, to evaluate the options on the basis of a conception of the common good, or more broadly social well-being (Chapters 10–13). It could be that the agent chooses (or recommends) portfolios in a well-functioning economy; it could be that she faces an asset management problem in a society in turmoil; or it could be that she operates in one of the many societies that lie in between. But no matter where and no matter what the context happens to be, she would wish to allocate assets so as to achieve a portfolio that is best under the circumstances. The account we offer below covers all agents – from the individual household, to the corporate fund manager, to the government decision-maker.

Box 1.1
Managing Assets as Daily Chores

Some 65–75% of people in the world's poorest regions are rural (World Bank, 2020a and c). In semi-arid regions of South Asia rural households have been found to spend four to five hours a day collecting water from water holes, gathering firewood, and picking fruit, berries and medicinal herbs from the local vegetation (Cain, 1977). The cooking area in the family hut among the rural poor is organised as what economists call a 'vertically integrated industry'. Daily work there requires women to work from raw materials. There are no pre-cooked meals on offer, nor even processed ingredients. Food preparation can take up to five to six hours a day. Rural women in Bangladesh, for example, have been found to spend 50–55% of their day cooking (Chowdhury et al. 2011).[34]

---

[33] Throughout, we will use Nature, the natural world, the biosphere, natural resources, and natural capital interchangeably. The context will, we hope, make clear which aspect of Nature we are emphasising.

[34] In an account of daily life of village people in a micro-watershed in the central Himalayas in India, the Centre for Science and Environment (1990) recorded that of the total number of hours worked in a day, 30% was devoted to cultivation, 20% to fodder collection and about 25% was spread evenly between fuel collection, animal care and grazing. Some 20% of time was spent on household chores, of which cooking took up the greatest portion, and the remaining 5% was involved in other activities, such as marketing.

> Mention agricultural land, threshing grounds, grazing fields, village tanks and ponds, woodlands and forests, rivers and streams, coastal fisheries, mangroves, or coral reefs, and the importance of the *local* natural resource base becomes self-evident. Details differ across regions, but such tasks as collecting water, gathering fuel and minding domestic animals are often the responsibility of children. Children have been found to work from as young as six years old. Material needed for repairing homes is prepared using such resources as timber, straw, stone and mud, which are collected locally. Herbs from plants in the vicinity serve to contain illness. It is not that rural households in low income countries do not exchange goods in the market; it is more that, unlike households in high income countries, they produce much of their daily requirements in the home. In fact, the local ecosystem offers more to the household. Pattanayak and Sills (2001) have provided quantitative evidence of the importance of non-timber products as a buffer against agricultural shocks in the Brazilian rainforests. Preserving the local resource base has the features of purchasing insurance against hard times. Village life in the world's poorest countries continues to be experienced in what the historical demographer Tony Wrigley (2004) has called 'organic economies'. To exclude the local natural resource base when studying the lives of the world's rural poor is not to know how the poor live.[35] Chapters 8 and 14 provide more on this.

The term *asset* has an evaluative hint to it – we say, for example, that our children are our greatest assets. Assets are desired or are recognised to be desirable. The asset manager places a positive value to them. A recurring societal problem we point to in the chapters that follow is that no matter who she is and no matter where she resides, many types of natural capital and the services they provide come free to her. But a free good does not appear to a person as scarce any more than sunlight. She could of course extract or harvest those resources for immediate use, or even store them for later use, but they do not appear to her as limited in quantity. *The demands she makes are limited only by the extraction and harvesting costs that she has to bear; she is not required to pay for the resources.*[36] That, as we will see, has profound consequences. We will also confirm (Chapter 8 and Annex 8.1) that matters have been made even worse by governments subsidising the use of what were previously free goods. If we include government subsidies, the previously free goods have a negative price.

Such institutional imperfections create a gap between the prices we face for the goods and services we produce and consume, and the social worth of those same goods and services (Chapters 7–9). That creates a tension between our motivations in private life and our hopes and aspirations as citizens. We realise that market prices of Nature's goods and services often do not reflect their social worth, and we understand that the criteria we use to manage our personal assets differ from the criteria that we as citizens would want to use. The Review studies this tension and tries to uncover ways in which private incentives can be brought into alignment with public aspirations.

Assets are durable objects, producing streams of services. Durability does not mean everlasting. Durable goods depreciate (machines suffer from wear and tear, plants wither, skills are lost through neglect, indigenous knowledge disappears, people die), but unlike services they do not disappear instantly. Assets acquire their value from the services they provide over their remaining life. A refrigerator preserves food products by keeping them cold. It provides that service until it breaks down beyond

---

[35] For a book-length account of the place of the local natural resource base in rural life, see Dasgupta (1993). For pioneering studies on the embeddedness of life in the organic economy that is rural India, see Jodha (2001). Tallis et al. (2011) is a fine essay on rural poverty as seen through the reliance of the poor on the goods and services available to them from their immediate landscape.

[36] A formal demonstration of this feature of our use of natural resources is set out in Dasgupta, Mitra and Sorger (2019).

repair. The refrigerator's worth is a measure of the benefits it provides over its remaining life. An asset's future performance is built into it today.[37]

1.1 Classification of Capital Goods

It is tempting to call all assets *capital goods*. This term has proved to be so attractive that it now stretches to include public knowledge ('knowledge capital'); the law, the market system, and financial institutions ('institutional capital'); mutual trust and solidarity ('social capital'); culture and norms of behaviour ('cultural capital'); and even religion ('religious capital'). Economists have been a lot more reticent; they confine the use of the term to assets that are measurable. In the past, economists used to reserve the term 'capital goods' even more stringently than they do now, for they only included assets that are material (tangible) and alienable (i.e. whose ownership is transferable). Roads, buildings, machines and ports are ready examples. As patents held by a firm are part of the firm's asset base, they appear in its balance sheet. So intangible and alienable assets are included on the list of capital goods. Taken together, they are called *produced capital*.

The range of capital goods in the economist's lexicon has broadened over the years to include intangible and non-alienable assets such as health, education and skills, which, taken together form *human capital*. Economists include human capital as a category of capital goods because they have discovered ways to measure its value, not only to the individuals who acquire it, but also to society at large.

In the past decades, economists have developed methods for measuring the value individuals place on natural resources, so we now have a third category of capital goods: *natural capital*. The methods can be involved, for natural capital ranges over plants (tangible and alienable), pollinators (tangible and often non-alienable), the view from one's sea-front home (intangible and alienable) and the global climate (intangible and non-alienable). Interactions among produced, human and natural capital are depicted in Figure 1.1.

As this Review studies reasons for the growing disparity between private incentives and public aspirations, we pay particular attention to the wedge between market prices of capital goods, especially natural capital, and what we may call their social scarcity values, known as *accounting prices* or 'shadow prices' (Figure 1.2). By a capital good's *accounting price*, we mean the contribution an additional unit of it would make to the flow of social benefits (Chapter 10). Accounting prices reflect an accommodation between the socially desirable and the socio- ecologically possible. There are cases where market prices approximate accounting prices, but for reasons we explore in the Review (Chapter 7), many kinds of natural capital simply do not have markets. They are free to the user. So special methods have to be devised for estimating accounting prices (Box 1.2). Moreover, measurement problems are also rife in estimating the *stock* of many kinds of natural capital (even the number of species today is thought to lie in a wide range of between 8 and 20 million, possibly more), but it is far better to work with rough and ready figures than to ignore whole swathes of capital goods by pretending they do not exist. Unfortunately, the macroeconomic growth and development theories that have shaped our beliefs about economic possibilities and our understanding of the progress and regress of nations do not recognise humanity's dependence on Nature. One purpose of this Review is to correct that mistake.[38]

---

[37] Pollution can last as well, but as we confirm in Chapter 2, pollutants can be viewed as assets with a negative value.

[38] Prominent representations of modern growth and development economics are Aghion and Howitt (1998), Barro and Sala-i-Martin (2003), Helpman (2004), Acemoglu (2008), and Galor (2011). The absence of Nature is also prominent in the models that inform government finance ministries and central banks. Chapter 4* contains a contrasting model of global economic possibilities.

Chapter 1: Nature as an Asset

Figure 1.1 Interaction Between the Capitals

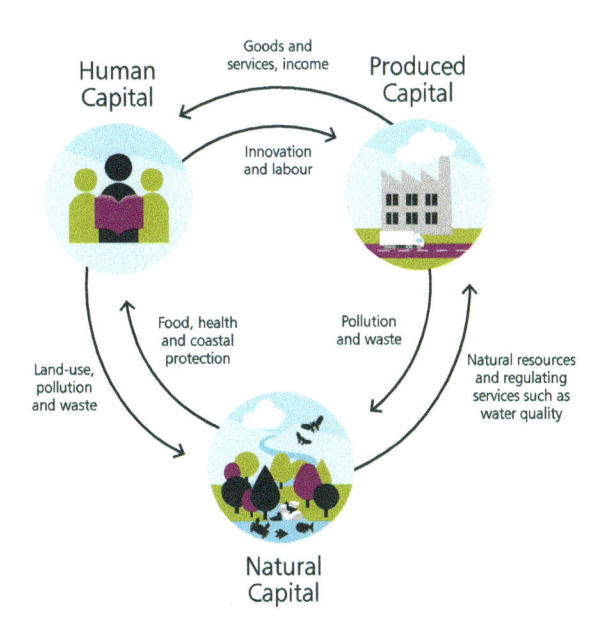

Figure 1.2 Market Prices and Accounting (or Shadow) Prices

An accounting (or shadow) price is the price that reflects the true value to society of any good, service or asset.

A market price is the price at which a good, service or asset is exchanged for in a market.

The Economics of Biodiversity: The Dasgupta Review

Chapter 1: Nature as an Asset

> **Box 1.2**
> **Valuing Nature's Stocks**
>
> Estimating natural resource stocks is difficult (Annexes 2.2 and 2.3 and Chapters 12–13). Aerial surveys provide information about forest cover and soil quality, and sonar technologies further enable fishing companies to estimate fish stocks in the oceans. But as they generate aggregate figures, important details are missed.[39] Economists have worked less on ways to measure stocks of natural capital (it has been the object of interest among environmental scientists) and more on ways to measure the benefits we derive from them (Chapter 12). The latter also poses problems, because many forms of natural capital are free, meaning there are no market prices that could be used as proxy. The benefits we derive from natural resources can be direct (e.g. the pleasure of a walk in the park) or they can be indirect, often several steps removed from experience (e.g. filtration of water by wetlands, or the natural regeneration of soil). Valuation exercises on natural resources that are of direct benefit now form a rich, informative literature.[40] The exercises frequently involve asking people in subtly devised questionnaires to disclose the value they place on those benefits (Chapter 12). Valuation exercises on natural resources that are of indirect benefit (i.e. they are factors in the production of goods and services that are of direct benefit) require an understanding not only of the processes that regenerate them but also their role as factors of production (Chapters 12–13*).

1.2 Rates of Return and Arbitrage Conditions

Of central importance to asset management is the concept of *yield* on investment (also known as an *own rate of return*). Formally, the yield on investment in a capital good is the increase in its size that can be expected tomorrow if a unit more of it were added to a portfolio today. The additional unit today is the investment in question. An example would be the additional biomass of a fishery less a unit that would be expected tomorrow if the biomass in the fishery were increased by a unit today. A further example would be the increase in a tree's biomass per unit of biomass if we waited a while. Waiting suggests that an asset's yield is the *growth* one obtains from investing in it.

The yield on investment in produced capital is its marginal product. But these contrasting examples suggest that investment has a wider meaning than electric drills and workers in hard hats applying tarmac to a road. Investment can be passive. If restoration of a wetland is investment, then so is conservation: *investment can mean simply waiting*.[41]

Yield is a pure number of per unit of time. Its dimension is therefore the inverse of time (i.e. $time^{-1}$). An example is the return the UK government offers, which has historically averaged approximately 4% (or 0.04) a year, for its long-term bonds (Thomas and Dimsdale, 2017).[42] So 4% a year is the yield.

When comparing assets in a portfolio, however, own rates of return are not enough. Unless the economy is in a stationary state, assets' relative prices can be expected to change over time. To illustrate, suppose a household values assets in pounds sterling. The *rate of return* on an asset (as opposed to the asset's *own* rate of return) is its yield plus the capital gains it enjoys over a unit of time. Portfolio

---

[39] See the lively interchange between Zhang et al. (2020) and Feng et al. (2020) on estimates of the size of water bodies in China.

[40] Freeman (2003) and Haque, Murty, and Shyamsundar (2011) are prominent publications.

[41] Solow (1963) reinstated the place of own rates of return in the economics of growth and distribution. His analysis, which is what we follow here, covered investment in both its active and passive senses.

[42] More recently, this has been closer to 1% (Bank of England, 2020).

management requires that the household chooses a portfolio with the maximum value among all the portfolios that are available to it. Of course, yields would typically be uncertain, as would future prices. Value maximisation would reflect the uncertainty and the household's attitude toward risk and uncertainty (Chapter 5).

The portfolio decisions individual households make do not influence rates of return in the economy: they are negligible in size. So, the prices and risks the individual household faces are exogenous to it. At the other extreme are agents serving government, whose choices over macroeconomic policies influence yields and the future accounting prices of goods and services. As elsewhere in macroeconomics, there is circularity here: government policies influence yields and accounting prices, even while, as portfolio managers, governments are required to choose policy on the basis of yields and (accounting) prices. These mutual influences come together when the choices made are optimal. Good governance exploits the circularity (Chapters 7, 13).

An asset that has a lower rate of return than another will not be chosen. A portfolio is the best for the agent *only* if the assets in it have the same rate of return. Rules governing portfolio selection are summarised in *arbitrage conditions* (Box 1.3). But people differ in the way they read the world; they differ in their attitudes to risk and uncertainty; and they differ by way of the opportunities open to them. That is why not everyone chooses to hold the same portfolio. The value of an agent's portfolio to her is her *wealth*.

However, even the arbitrage conditions are not sufficient for the task facing an asset manager. The conditions tell her how to choose the right mix of assets to hold in her portfolio, but they do not say what she should do with the returns. They do not say what proportion of returns she could justifiably put aside for consumption and how the remainder should be allocated so as to add to her portfolio. The study of optimal mixes of consumption and investments leads us to discuss ideas underlying well-being across time and the generations (Chapters 10–13). We confirm that optimum programmes of consumption and investment necessarily satisfy the arbitrage conditions, and that they also necessarily satisfy a further set of arbitrage conditions involving the agent's valuation of the present in comparison to the future. Those considerations are summarised in ethical objects such as 'social discount rates' (Chapter 10). The Review studies reasons behind humanity's failure to manage our portfolios well and explores ways in which we could shift direction. The concept of inclusive wealth is crucial for the exercise (Chapter 13).

Box 1.3
Arbitrage Conditions

An individual in a deterministic world is considering whether to place £500,000 in an investment bank that offers an annual 5% yield or whether to purchase an apartment at that price and rent it at the going market rate of £15,000 a year. Under the first option the person's wealth in a year's time would be £525,000, which would seemingly trump the second option but for the capital gains she may enjoy in a year's time from owning the apartment. Imagine that the capital gains were £10,000. Then the return she would enjoy from the apartment would be £25,000 (£15,000 rental income + £10,000 capital gains). Because 25,000/500,000 = 5%, the person would be indifferent between the two options. That is a requirement of an efficient capital market. If capital gains on her apartment were to be either less or more, the two markets would be in imbalance. If it were less, she and others in her situation would place their funds in the bank; if it were more, they would purchase apartments and avoid the bank. For both markets in the example to exist, rates of return on the two assets must be equal. That equality is the *arbitrage condition* in the example. The condition identifies a process in which investors arbitrage their portfolios in such a way that at the margin they are indifferent as to the mix of the assets they hold in their portfolio. If asset markets were functioning well, their prices would adjust so as to equalise the rates of return.

Chapter 1: Nature as an Asset

> To add to the example, suppose the same £500,000 could buy an agent a tract of timberland and that sales of timber would generate £20,000 in net profit to him over a year. The agent would be indifferent between purchasing the tract and investing in the government bond if the market value of the tract were to increase by £5,000 over the year, because the rate of return from the forest would then be 5%. That is another example of arbitrage conditions.
>
> To study arbitrage conditions in a formal way, we continue for simplicity to consider a market economy in a deterministic world. Let us choose an asset that is to serve as the unit of account. That is the *numeraire*. The price of a unit of our numeraire is therefore 1.
>
> We imagine that time is continuous. It is denoted by t. Let the yield on the numeraire be $r(t)$. Let $p_i(t)$ be the price of asset i at t. If $r_i(t)$ is the yield on i at t, then for the two assets to be equally attractive to the agent, the arbitrage condition reads
>
> $$r(t) = r_i(t) + [dp_i(t)/dt]/p_i(t) \qquad (B1.1)$$
>
> The second term on the right-hand side of equation (B1.1) is the percentage rate of change in $p_i(t)$. That is capital gains in asset i (it could of course be losses, in which case the sign of the term is negative). Repeated use of equation (B1.1) tells us that for any pair of assets i and j in the agent's portfolio, it must be that
>
> $$r_j(t) + [dp_i(t)/dt]/p_i(t) = r_i(t) + [dp_j(t)/dt]/p_j(t) \qquad (B1.2)$$
>
> Define the price of asset j relative to the price of asset i as p_{ij}. Then equation (B1.2) can be expressed as
>
> $$r_j(t) + [dp_{ij}(t)/dt]/p_{ij}(t) = r_i(t) \qquad (B1.3)$$
>
> Equation (B1.3) is a formalisation of the numerical examples with which we started.
>
> Because economies suffer from distortions, asset holders do not all face the same prices. That means even if each agent were to allocate the assets at her command efficiently, the economy in the aggregate would be inefficient. Environmental externalities are a prime cause behind economy-wide inefficiencies (Chapters 7–8). Cases of particular interest are assets that are open to all to use as each sees fit, free of charge. They are known as 'open access' resources. The atmosphere as a sink for our carbon emissions is the most well-known example. Marine fisheries are another. The classic 'tragedy of the commons' speaks to them (Hardin, 1968). In those cases, equation (B1.1) does not hold, because being free ($p_i = 0$), open access assets appear to each individual as unlimited in size. The only thing that prevents people from drawing on them at an infinite rate are harvesting and extraction costs.
>
> Public bodies, whose remit would not be the same as that of private investors, would also want to choose their portfolios efficiently. They too would seek to choose their portfolios so as to satisfy the arbitrage conditions. But unlike the agents in the examples we have just studied, the prices they would use would be *accounting prices*. The enormous literature on valuation of environmental resources alluded to in the text is about ways to estimate accounting prices (Chapter 12). Equations (B1.1)–(B1.2) have been derived for a deterministic world. Investors typically add discounts on assets to correct for risk and uncertainty in their returns. These issues are discussed in Chapter 5.

1.3 Public Asset Management and the Wealth/Well-Being Equivalence Theorem

We have identified three categories of assets that can be called *capital goods*: produced capital, human capital and natural capital. The sum of the accounting values of a society's capital goods is known as *inclusive wealth*, the qualifier signalling that by wealth we mean not only the accounting price of produced capital and human capital, but also of natural capital. The Review explains why inclusive wealth should be the coin with which *citizens* would wish to evaluate economic change – but it will take us all of 12 chapters to get there. In Chapter 13 we show by way of what may be called the *wealth/ well-being equivalence theorem*, that social well-being is maximised *if and only if* inclusive wealth is maximised. Accounting prices provide the link between wealth and well-being, which is why the theorem is valid no matter which conception of well-being is adopted by the portfolio manager. The theorem will bring us back full circle to where we began, that the task of asset managers is to maximise the value of their portfolios, and that inclusive wealth is the social value of an economy's portfolio of capital goods. In ideal circumstances the market value of a portfolio would equal its accounting value, but for reasons the Review unravels, that ideal cannot be reached. Private incentives and social imperatives inevitably differ, so a government's task is to put into practice policies that bring the two into alignment as close as possible.

The equivalence theorem is fundamental for economic evaluation. To see why, consider the demand citizens could make of their government that it should only select policies that advance the quality of their lives. The problem is that the demand does not offer guidance on what should be selected. Even if restoring a degraded woodland advances the quality of life, there would be contending projects, each with its own characteristics. Moreover, the same resource may be an input in alternative projects. There would also be projects that may not appear to be life-affirming but would contribute to their quality of life indirectly. As always, there are alternative ways to allocate goods and services, each with its own set of likely consequences. And goods and services do not come marked with 'quality of life' stamps. Accounting prices are the necessary stamps. The equivalence theorem says using inclusive wealth to evaluate society's options is in line with the requirement that its portfolio reflects societal ends, no matter what the ends happen to be. And because it is an equivalence theorem, we know there can be no measure other than inclusive wealth that can serve the purpose.

What about all those assets that are not on our list of capital goods? Quantifying such assets as public knowledge, institutions and mutual trust raises insuperable difficulties. Try, for example, to estimate the accounting price of differential calculus, or good governance, or the extent of trust among citizens, and the stumbling block becomes apparent. So we create a separate category named *enabling assets*, for they help societies to allocate capital goods. We will find (Chapter 12) that the value of enabling assets is reflected in the accounting prices of capital goods. A classroom in a society at peace can function in ways it cannot in a country at civil war. That alone means its accounting price is not the same in the two contexts. A society could raise its inclusive wealth and thereby social well-being simply by improving its institutions and practices.[43]

Biodiversity is a characteristic of natural capital, as diversity of aspirations, talents and drives are features of human capital. In Chapter 2, we review a literature that has found biodiversity to be a factor influencing the productivity of natural capital, or more concretely, ecosystems. Biodiversity is an enabling asset. Which is why environmental and resource economists estimate the accounting prices of items of natural capital – for example ecosystems – not biodiversity. The value of biodiversity is embedded in the accounting prices of natural capital.

---

[43] The partitioning of a society's durable entities into capital goods and enabling assets was proposed in Dasgupta and Mäler (2000), who also stated and proved the equivalence between societal well-being and inclusive wealth in a general setting. The equivalence result was extended by Arrow, Dasgupta and Mäler (2003a,b). Chapter 13 contains a detailed account. The term 'inclusive wealth' was introduced in UNU-IHDP and UNEP (2012, 2014). Arrow et al. (2012) used the term 'comprehensive wealth'.

1.4 Two Types of Comparison

Portfolio management involves making two types of comparisons. We illustrate them by considering public decision-makers.

One type of evaluation involves evaluating the change to a portfolio brought about by a decision at a point in time, the hallmark of *policy analysis*. An example would be to evaluate a proposal to change the government's tax schedule (Meade (Report), 1978; Mirrlees (Review), 2011). Evaluation is necessary because the government would not otherwise know whether the proposed change is desirable. A particular type of policy evaluation is *cost-benefit analysis*, or *project evaluation*, which offers a methodology for evaluating investment projects. The exercise involves evaluating alternative uses to which capital goods and their yields can be put – for example, judging how much of the yield should be consumed and how much should be reinvested, and in what form. The word 'social' is added to the term 'cost-benefit analysis', as in *social cost-benefit analysis*, when the agent chooses on behalf of a public body.[44]

Another type of comparison involves valuing the change a portfolio displays over time. This is the hallmark of *sustainability assessment*, which responds to such questions as, is our country more prosperous today than it was a year ago? There is no presumption that by prosperity the person asking the question is looking for GDP figures (there are in fact very good reasons why he should not do so); it is more likely that prosperity is taken by him to mean the quality of life, possibly even *well-being* (Chapters 10–11). However, during the year in question there could have been an accumulation of produced capital and human capital and a decumulation of natural capital. That is the experience of most countries in recent years (Managi and Kumar, 2018). The problem then is to weigh the changes in the asset structure so as to judge whether well-being today and projected well-being in the future is greater now than it was previously. The exercise involves inferring the extent to which one's 'ends' have been met from changes in the 'means' to those ends.

There is a third type of comparison, which is related to the above two but differs from them in important ways. It involves comparisons of the state of affairs in different economies. For 'economy' we could, for example, read 'country'. We may then ask whether country A is more prosperous than country B. At the formal level, the question falls under the domain of sustainability assessment. The difference is that it involves cross-country comparisons at a moment in time rather than comparisons of the state of affairs in one country across time. Neither involves policy choice. But problems arise in making such comparisons because political cultures differ across countries. The use of a common metric is questionable. The current practice among international agencies such as the World Bank is to make comparisons in terms of features that are commonly thought to speak to human dignity, independent of differences in political cultures. Life expectancy at birth, the maternal mortality rate, literacy and the standard of living as measured by market prices are commonly used measures. There have been attempts to aggregate them into a single index – for example, the United Nations' Human Development Index (HDI) – but those moves can be questioned because they apply weights to the various components of the index that are independent of political cultures. Cross-country experiences can nevertheless be of enormous use to individual countries as they search for policies that bring about or maintain prosperity. So, although the economics of biodiversity is concerned with all three types of comparisons, the Review for the main part develops the first two: policy analysis and sustainability assessment.

---

[44] Little and Mirrlees (1968, 1974), Arrow and Kurz (1970), and Dasgupta, Marglin, and Sen (1972) developed the theory of social cost-benefit analysis in imperfect economies.

But these are *two* lines of enquiry. The wealth/well-being equivalence theorem shows that, fortunately, policy analysis and sustainability assessment involve the same considerations: *both require estimating the value of changes to the stocks in our portfolio of capital goods*. Because the value of a portfolio represents *inclusive wealth*, both policy analysis and sustainability assessment involve wealth *comparisons* (Chapter 13).

To illustrate the use of inclusive wealth, imagine someone asks whether the United Nations' Sustainable Development Goals (SDGs) would be sustainable if attained. Suppose also that a national government produces a plan for attaining the SDGs, which requires accumulating produced capital and human capital in sufficient amounts while distributing them appropriately. Meanwhile, however, biodiversity loss and rising mean global temperature are found to be depreciating natural capital. The government recognises that growth in produced capital and human capital needs to be balanced against the depreciation of natural capital. The study of inclusive wealth enables the government to do that. Assessing the progress or regress of national economies requires one to study movements over time of the inclusive wealth of nations, not the GDP of nations, nor the HDI of nations. The wealth/well-being equivalence theorem assures that.[45]

Inclusive wealth accounts for a country correspond to the balance sheets of firms. But we should not expect countries to move from their current systems of national accounts to a comprehensive system of wealth accounts. Measuring the value of natural capital stocks, not to mention the quantity and quality of stocks, is notoriously difficult (Annexes 2.2 and 2.3, and Chapters 12–13). The moves currently being made in individual countries such as the UK and New Zealand involve the preparation of satellite accounts of sectors such as forests, fisheries and ground water. They involve estimating accounting prices of their stock. There are countries that are at a more advanced stage and have satellite accounts in which the value of natural capital is presented in an aggregate form.[46]

1.5 The Earth System and Economic Growth

Natural capital is essential for our existence (the air we breathe, the water we drink are immediate examples); of direct use as consumption goods (fisheries); of indirect use as inputs in production (timber, fibres); and essential for our emotional well-being (green landscape, sacred sites). Many have multiple uses (forests, rivers, the oceans). We are embedded in Nature; we are not external to it. But until relatively recently, influential writers on economic development saw natural capital only as luxuries. An unnecessary debate took place between those who expressed environmental concerns in low income countries and those who saw the need for economic growth there above all else. Well-meaning writers tried to reconcile the two viewpoints. An editorial in the UK's *Independent* (4 December 1999), for example, observed that "[economic] growth is good for the environment because countries need to put poverty behind them in order to care," and a column in *The Economist* (4 December, 1999: 17) insisted "trade improves the environment, because it raises incomes, and the richer people are, the more willing they are to devote resources to cleaning up their living space."[47]

---

[45] Arrow et al. (2004, 2012), UNU-IHDP and UNEP (2012, 2014) and Managi and Kumar (2018) contain quantitative studies of movements of the inclusive wealth of nations. The publications covered periods between 1995 and 2010. See Chapter 13 and Annex 13.1.

[46] See Bright, Connors, and Grice (2019) and a literature, published by scholars involved in several 'natural capital projects', that presents estimates of the monetary value of various forms of natural capital, mostly at the national level. They include Kareiva et al. (2011), Fenichel et al. (2016), Kumar (2010), Natural Capital Committee (2019) and the ongoing work of the UN Statistical Division as represented by their most recent, 2017, publication. We report on these works in Chapters 12–13.

[47] Visions of a prosperous world in which Nature plays no part continues to thrive. Criticising the young climate activist Greta Thunberg for her speech at the United Nations in September 2019, the economics editor of Sky News wrote in *The Times* (27 September, 2019: p. 30): "Eternal economic growth is not a phrase one spits out in derision; it is precisely what we should be aiming for."

Chapter 1: Nature as an Asset

The origins of this limited view of the place of nature in economic life can be traced to the World Bank (1992), which reported that in cross-country studies the emission of sulphur oxides had been found to be related to GDP per head capita in the form of an inverse-U. Emissions were found to increase with GDP per capita when countries are poor, but to decline with GDP per capita when countries are rich. Inevitably, the relationship was named the 'environmental Kuznets curve' in honour of the economist Simon Kuznets, who had observed an inverse-U relationship between GDP per capita and income inequality (Kuznets, Epstein and Jenks, 1941).[48]

Emissions of sulphur oxides are unrepresentative of environmental harm. The oxides are emitted by industry and automobiles and blow away to become someone else's problem when emissions cease. But if a company destroys a mangrove forest to make way for shrimp farms, the protection it had afforded neighbouring villages against storms is lost irretrievably (Chapter 2). At an extreme is loss of species, which is irreversible.[49]

In fact, a piece of natural capital can be a luxury for some even while it is a necessity for others. Many goods and services that are provided by watersheds are necessities for local inhabitants (forest dwellers, downstream farmers, fishermen), some are sources of revenue for commercial firms (timber companies), while others are luxuries for outsiders (eco-tourists). Some benefits accrue to nationals (agricultural crops), while others spill over across national boundaries (carbon sequestration). Watersheds offer joint products (protection of biodiversity, flood control, household goods; Chapters 2, 9–10), but they also offer services that compete against one another (commercial timber, agricultural land, biodiversity).

Competition among rival services has been a prime force behind the way the biosphere has been transformed. Moreover, commercial demand frequently trumps local needs, especially under non-democratic regimes (Chapters 8–9). International public opinion and pressure from the country's elite are often tepid. These complex interrelationships have generally been ignored by growth and development economists (the newspaper quotes above are after all, only two decades old, the last only some months old). In the event, the economics of biodiversity was left to be studied by a few groups of economists and ecologists working together.[50]

---

[48] See also Cropper and Griffiths (1994) and Grossman and Krueger (1995).

[49] See the comments in Arrow et al. (1995) on the environmental Kuznets curve and the responses it elicited in symposia built round the article, in Ecological Economics, 1995, 15(1); Ecological Applications, 1996, 6(1); and Environment and Development Economics, 1996, 1(1). See also the special issue on the subject in Environment and Development Economics, 1997, 2(4).

[50] Prominent among institutions that have laid the groundwork for the economics of biodiversity (although that was not the term in use) are Resources for the Future (RFF), Washington, DC; the World Resources Institute (WRI), Washington, DC; the Beijer Institute of Ecological Economics, Stockholm; the South Asian Network of Development and Environmental Economists (SANDEE), Kathmandu; Resource Accounting Network for Eastern and Southern Africa (RANESA) and the Centre for Environmental Economics and Policy in Africa (CEEPA), Pretoria; and Latin American and Caribbean Environmental Economics (LACEEP) and the Tropical Agricultural Research and Higher Education Center (CATIE), Costa Rica. RFF in the 1960s-70s studied the economics of irreversible investment, persistent pollutants, material balances (Chapters 4* and 13*) and valuation methods for environmental amenities; while WRI in the 1970s produced among the first economic estimates of the degradation of tropical rainforests. At an institutional level, however, the economics of biodiversity has found its greatest expression since the early 1990s in research and teaching networks elsewhere. The Beijer Institute, reinstated in 1991, has brought ecologists and economists together in a manner unthinkable previously. As will become apparent, Chapters 2–4 have been greatly influenced by their work. SANDEE, RANESA/CEEPA, and LACEEP/CATIE, established later in the 1990s, have organised regular teaching and research workshops and funded work by young economists in South Asia, Eastern and Southern Africa, and Central and South America, respectively, so as to help develop the interface of poverty and the local natural capital base there. Members of all four groups shaped the pioneering Millennium Ecosystem Assessment (MA 2005 a-d).
The journal Environment and Development Economics, which was established in 2000 with support from the Beijer Institute, has actively sought and published papers from South Asia, sub-Saharan Africa and Latin America. As will become apparent, Chapters 7–8 and 12–13 have leant greatly on their work. On the closely related subject of reproductive behaviour in regions that are still far from experiencing a demographic transition, the Population Council, New York, the London School of Hygiene and Tropical Medicine, the Guttmacher Institute and the United Nations Population Division, New York, have consistently produced work of great relevance for the population-environment nexus, Population and Development Review being a prominent quarterly publication. Chapter 9 has leant greatly on their work.

There is a deeper, more general point that lies behind the thought that because there is no obvious bound on human ingenuity, technological progress and institutional improvements can enable the global output of final goods and services to grow indefinitely. This is to imagine the human enterprise as being *external* to Nature; it is to see humanity dipping into the biosphere for its goods and services, transforming what is taken for production and consumption, and depositing the residue back into it as waste. We show that this view allows us to claim that in due course human ingenuity can enable us to increase global output indefinitely without making anything but vanishingly small further demands on the biosphere. Never mind that indefinite growth will require continuing investment in research and development and equipment, the hope is that those further investments will require vanishingly small inputs from Nature.

To entertain that hope today is more than ironic. Over the past 70 years, global GDP has increased in real terms by a factor of nearly 15, even while our global demand for the biosphere's goods and services – our ecological footprint – now far exceeds the biosphere's ability to supply its goods and services on a sustainable basis (Chapter 4). Which is why a group of Earth scientists have identified mid-20th century as the point at which we entered the Anthropocene.[51]

The economics of biodiversity takes its cue from the environmental sciences to build on the fact that we are *embedded* in Nature (Chapter 4*). We will not be able to extricate ourselves from the Earth System even if we try to invest continually for indefinite economic growth. This somewhat metaphysical distinction between being 'external' to the biosphere and being 'embedded' within it has potent force. The viewpoint adopted in the Review says that the finiteness of Nature places bounds on the extent to which GDP can be imagined to grow. It also places bounds on the extent to which inclusive wealth can grow.

1.6 Total vs. Marginal Values

Asset management involves *comparisons* of portfolios – of portfolios across time or of portfolios at a point in time (Section 1.2). In contrast, the absolute value of a portfolio carries no information. The value of a marginal change to the biosphere is meaningful because it is presumed that humanity will survive the change to experience it, but the matter is different when it comes to valuing the biosphere as a whole. It may be that because growth and development economists ignored our place in Nature that environmentalists some years ago were tempted to value the entire biosphere, presumably to show that it is of great economic worth. In a widely cited publication in *Science*, the authors estimated that the global flow of the biosphere's services was, toward the end of the 20th century, worth US$16-54 trillion annually, with an average figure of US$33 trillion (Costanza et al. 1997). As that figure was larger than global Gross National Product (GNP) in the mid-1990s (estimated by the authors at the time to be approximately US$18 trillion annually) we were meant to appreciate the economic significance of the biosphere.

The estimate is a case of misplaced quantification. If the biosphere was to be destroyed, life would cease to exist. Who would then be here to receive US$33 trillion of annual benefits if humanity were to exchange its very existence for them? Economics, when used with care, is meant to serve our ethical values. The language it provides helps us to choose in accordance with those values. But the authors of the paper sought to persuade us that the biosphere is valuable *because* it can be imputed a large monetary value. That is to get things backward.[52]

---

[51] The term 'Anthropocene' was popularised by Crutzen and Stoermer (2000) to mark a new epoch in which humans dominate the biosphere. Ehrlich and Ehrlich (2008) study the evolutionary consequences of our growing dominance. Dasgupta and Ehrlich (2012) read the human overreach of the biosphere in terms of adverse externalities associated with consumption and reproductive externalities.

[52] Formally, we have a case where the value of an entire something has no meaning, and is therefore of no use, even though the value of marginal changes to that same thing – expressed as differences – not only has meaning, but also has use. Examples abound in economics (cardinal utility) and physics (potential field).

Chapter 1: Nature as an Asset

At smaller levels of aggregation, total values of Nature's services can be meaningful and yet not be useful for policy. It is tempting, for example, to cite the estimate that pollination contributes an annual £510–690 million to the UK's agricultural production as providing a reason for restoring the population of pollinators (Breeze, Roberts, and Potts, 2012). But should we regard it to be a large or small figure? Based on 2019 data from the UK Office of National Statistics, as a proportion of the UK's annual agricultural output, it is approximately 5%. As a proportion of the UK's GDP, it is 0.03%, a negligible figure. So why care whether any pollinators are left?

The reason we should not be dismissive of pollinators is that proportional figures do not signal worth. National asset management requires that pollinators enter projects with their accounting prices. Chapters 2 to 4 demonstrate that pollinators may be of great value even if their measurable services to GDP are of negligible worth.

1.7 Institutions and the Character of Natural Capital

Processes driving a wedge between our demand for the biosphere's goods and services and its ability to supply them without undergoing decline harbour *externalities*. These are the unaccounted-for consequences for others, including future people, of actions taken by one or more persons. The qualifier 'unaccounted for' means that the consequences in question follow without prior engagement with, or adequate consideration towards, those who are affected.

Human activities give rise to externalities because *property rights* to large segments of the biosphere are either weakly defined or inadequately enforced. And a common reason for the latter is that Nature is *mobile*: the wind blows, rivers flow, fish swim, the oceans circulate, and insects and birds fly. One consequence of this is that no one can contain the atmosphere they befoul, the soil they contaminate, the rivers they pollute. Moreover, the harm they cause is *non-excludable*, meaning that it is not possible for the person or agency to whose action the harm is traceable to pick and choose those who are affected.

That Nature is mobile is a familiar fact and easy to appreciate. What is a lot harder to appreciate is that both Nature and its processes are in large part *silent* and *invisible*, and so they are not easily detectable. One way to detect them is to infer their presence from their detectable effects. That is no easy matter, and requires hard science, and it provides the reason the rudiments of that science are studied in the following three chapters.

These features of Nature make it hard for anyone to trace the adverse effects of many of our actions back to us. Unlike point sources of pollution, such as the factory chimney, pollution such as nitrogen and phosphorus that discharge into an estuary are an aggregate of leakages from innumerable agricultural fields, factories and households long distances away. Polluters do not pay for using the estuary as a sink for their pollution.

By property rights we do not simply mean private property rights, we include group rights, for example, community rights and national rights. There are no property rights to the oceans beyond national jurisdiction – they are open to all, free of charge – which is why no one has an incentive to protect them from contamination or overfishing.

Externalities can be beneficial of course. An extreme form of beneficial externalities is provided by *public goods*, which are goods that are *non-rivalrous* (use by one person does not diminish the amount available to others) and non-excludable (use cannot be confined to any particular person or group). Public goods are thus a mirror image of open access resources. If the formula for manufacturing a vaccine were to be made freely available to all, the discovery would be a public good (the formula is by its very nature non-rivalrous, but if it were to be made free to all to use it as they wish, it would be non-excludable as well, making it an enabling asset). A reasoning identical to the one that was deployed above for explaining why open access resources are overused can be used to explain why public goods are underproduced.

Patents are a social contrivance for providing incentives to people to make discoveries. A patent for a new vaccine would make the formula for manufacturing it excludable: no one other than the patent owner would have the right to use the formula while the patent was in force. But the issue of a patent on the discovery, however, would award the owner a monopoly over vaccine production. Monopoly pricing would create a distortion in the market. A combination of the need to create incentives for scientific and technological establishments to make discoveries and to avoid creating distortions in the product market poses a social dilemma. The compromise that is practised everywhere is to limit the duration of patents (currently 20 years in the UK as defined in Section 25 of the Patents Act (1977)).[53]

Not all common property resources are 'global commons'. Geography plays a large role. The open oceans are extensive, in contrast to the village commons in England. Which is why the institutions that have evolved for managing what we may call *local commons* differ widely across the globe (Chapters 7–8). The village commons in England are under the jurisdiction of local authorities – they are subject to the laws of the land. The law's authority resides with the State. Village tanks (artificial ponds) in India and grazing fields in sub-Saharan Africa are also local commons, but their use is usually subject to social norms, whose force comes from mutual enforcement by the villagers themselves. The locus of mutual trust differs in the two cases (Chapter 6). However, there are many reasons community practices have been known to break down where they were once thriving. In some places, government rules replaced community norms; in other places, outsiders encroached on the inhabitants' land (worse, evicted the inhabitants); in still others, local knowledge was displaced by modern technology, and so on. For global commons like the atmosphere and the oceans as sinks for our waste, institutions never got a foothold to limit their use. The economics of biodiversity enquires why societies fail to manage their assets well, and it seeks to identify institutional changes that would improve management practice.

1.8 Anthropocentric Value of Biodiversity

In the chapters that follow, we mostly adopt an anthropocentric viewpoint – the value of biodiversity is studied in terms of its contributions to humanity, that is human well-being. This is an altogether limited point of view, and to many even a repugnant point of view (see Chapter 12 for other perspectives). Surely, it will be insisted, biodiversity has an *intrinsic value*, beyond what we humans impute to it. They would ask, for example, whether the biosphere had value before modern humans appeared on the scene some 200,000 years ago. The anthropocentric perspective could affirm its value even in that distant past on grounds that it furnished the environment in which we humans were able to emerge, but that alone shows the perspective's limitations.

There are nevertheless good reasons for concentrating on what one may call the *instrumental value* of biodiversity. One reason is that there are innumerable systems of thought that go beyond an anthropocentric perspective. Many people argue that life itself has intrinsic value, never mind that only a few among the 8 to 20 million species (of eukaryotes) on Earth are known to feel, never mind to have self-awareness.[54] There are also many systems of belief – alas, all too readily overridden by cosmopolitan society – in which objects that to the cosmopolitan are inanimate, are sacred. They may house life, but they are not life; nevertheless, they are sacred. Uluru in Australia is a famous example. It is sacred to the

---

[53] For a formal account of the tension that societies face between the need to create incentives for people to make discoveries and inventions and to create an environment in which discoveries and inventions are widely used, see Dasgupta and Stiglitz (1980).

[54] In a widely noted work, Singer (1975) presented a Utilitarian argument for awarding rights to animals. That animals can feel pain plays a justifiably crucial role in the viewpoint he developed in his book.

Chapter 1: Nature as an Asset

Pitjantjatjara, the Aboriginal people of the area surrounding it. And there is the river Ganges, sacred to Hindus (Box 2.8). But the narratives underlying their sacredness differ.

There is a second reason. If we are able to show, as we intend to in the Review, that biodiversity is of the utmost value to humanity, and that because we are embedded in Nature, gradual biological extinction will hasten our own extinction, then for purely anthropocentric reasons we would wish to preserve and promote it. But if biodiversity is worth preserving and promoting for purely anthropocentric reasons, it would be even more deserving of protection and promotion if it had sacred status. Therein lies the advantage of a limited point of view.

Chapter 2
Biodiversity and Ecosystem Services

Introduction

The *biosphere*, which is the part of Earth that is occupied by living organisms, is a self-organising, regenerative entity. Its rhythms, for example the seasons, shape the regeneration patterns of the living world. But living systems in turn make use of the non-living, or abiotic, constituents of the biosphere and transform them. Water, carbon and nitrogen cycles are expressions of that. Because the ability to regenerate is a characteristic of living systems, the biosphere's regeneration is key to the sustainability of the human enterprise. Regenerations of the living world at various scales and periodicity are synchronised via natural processes that are still not understood well.[55]

Biological diversity, or *biodiversity* for short, means the diversity of life. Its decline disrupts biospheric processes, for example, the processes governing the climate system. The sustainability of our engagement with Nature is thus ultimately about the functioning of the biosphere, not just the living part of it. Which is why the economics of biodiversity is the economics of the biosphere.[56]

Biodiversity can be read at various levels (Figure 2.1), for genes combine to form species, species combine to form assemblages of populations, and assemblages combine to form communities that interact with the physical world to create ecosystems and, at a larger scale, biomes (see Box 2.1 on the latter). So biodiversity does not only mean the diversity of genes and species, it also means the diversity of ecosystems.

Plants, algae and many bacteria capture energy from the sun, which is why they are called *primary producers*. The energy they capture, along with other abiotic materials, flows through ecosystems and enables a wide range of natural processes to function, including biomass production, nutrient cycling and water dynamics. These processes support biodiversity, but the influence is mutual, for biodiversity strengthens the processes, enabling Nature to renew constantly.

2.1 Biodiversity in Ecosystems

Traditionally, environmental and resource economists studied Nature's resources, such as fisheries, forests, lakes, airsheds, coastal waters, minerals and fossil fuels. For the economics of biodiversity, it proves useful to think of the biosphere instead in terms of constituents we call *ecosystems*. Ecosystems combine the abiotic environment with biological communities (plants, animals, fungi, microorganisms) to form self-organising, regenerative *functional* units, by which we mean combinations of life forms that control such fluxes as that of energy (e.g. photosynthesis), nutrients (e.g. nitrogen fixation) and organic matter (e.g. decomposition of organic waste).

Ecosystems are capital goods, like produced capital (roads, buildings, ports, machines). As in the case of produced capital, ecosystems depreciate if they are misused or are overused. But they differ from produced capital in three ways (Chapter 3): (i) depreciation is in many cases irreversible (or at best the systems take a long while to recover); (ii) it is not possible to replicate a depleted or degraded ecosystem; and (iii) ecosystems can collapse abruptly, without much prior warning.

---

[55] Strogatz (1994, 2004) are excellent technical and non-technical expositions, respectively, of the principles underlying synchronisation among simple dynamical systems.

[56] In this Review, we use the terms biosphere and Nature interchangeably.

Chapter 2: Biodiversity and Ecosystem Services

Figure 2.1 From the Micro to the Macro

Ecosystems regenerate as part of Nature's rhythmic cycles. New forests emerge from the ashes of fires, rising from self-sown seeds and shoots from the roots of plants. We confirm below that biodiversity enables that regeneration to occur. It affects both living and non-living parts of ecosystems, which are connected through nutrient cycles and energy flows. Ecosystems also differ enormously depending on a range of factors, such as the underlying geology, climate, nutrient and chemical status of the soils, hydrology, prevailing winds and season. About 85% of plant species inhabit entirely within just over a third of Earth's land surface. Some ecosystems are highly diverse, such as tropical rainforests, while others have low diversity such as polar ecosystems. But the latter hold ice sheets, whereas the former do not. Protecting a million square miles of tropical rainforest has different consequences from protecting the same area of a polar landscape.[57]

Biodiversity is the variety of life in all its forms, which is why it is not uncommon to regard biodiversity to be the number of species of organisms that inhabit Earth. Today there are around 8 to 20 million (possibly more) kinds of organisms with cells in which the genetic material in the form of chromosomes that are contained within distinct cells (they are called eukaryotes). Of them, only about 2 million have been recognised and named (Raven, 2020). There are, in addition, an unknown and much larger number of prokaryotes, consisting of archaea and bacteria (Larsen et al. 2017) – our lack of knowledge is enormous.[58] But biodiversity has several dimensions, including the diversity and abundance of living organisms, the genes they contain and the ecosystems in which they live. The chemical reactions of Earth's plants, animals, bacteria and fungi sustain life by converting sunlight and nutrients into food, energy and the building blocks of life, as well as recycling waste. The activities of these organisms are often both silent and hidden from view, but they enable ecosystems to function and provide a multitude of services on which we rely. Most species occur only in one ecosystem, but many occur in more than one.

---

[57] Pimm, Jenkins, and Li (2018) elaborate on the differences.

[58] Hubbell (2015), for example, notes that a recent estimate that there are between 40,000 and 53,000 tree species in the tropics is based on a very small sample of forest patches. Peter Raven has emphasised to us in correspondence of the enormous uncertainty in our knowledge of the number of species, let alone tree species in the tropics. Moreover, the distribution of species numbers has a thick tail: a very large proportion of species are of small population size. That makes it even more difficult to estimate species numbers. Pimm and Raven (2019) contains a summary of what we know.

Chapter 2: Biodiversity and Ecosystem Services

Biodiversity is key to the processes governing ecosystems. However, an ecosystem's productivity and resilience to disturbances depend less on the taxonomy of plants, animals and microorganisms than on them performing particular functions. The biodiversity in a wetland that filters water effectively differs from the biodiversity needed in a woodland that supplies timber, which in turn differs from the biodiversity in a grassland that supports wildlife. In each case, ecosystem productivity is determined by root structure, biomass above ground, leaf display, soil quality, crown architecture and wood production and composition. The role of biodiversity in preserving what we may broadly call an ecosystem's *integrity*, is studied in this chapter.[59]

Biodiversity is thus a multi-faceted feature of ecosystems, including variations among genes, species and functional traits among and across species. Generally speaking, wetter and warmer regions harbour greater biodiversity than drier and colder regions.

Three features of the diversity of life forms are significant: the number of unique life forms (richness); the flatness of the distribution of life forms (evenness); and dissimilarities in the life forms (heterogeneity). Common diversity measures are based on richness and evenness. Because biodiversity includes the diversity of the functional characteristics of an ecosystem's species populations, a mere headcount of species can mislead. Basing biodiversity exclusively on measures of genetic diversity would mislead even more. Moreover, traits among the constituents of an ecosystem are so closely related to its functioning, that judgment is required for identifying the aspect of biodiversity (e.g. species or functional trait) that is closely related to the ecosystem's productivity. That may explain why ecologists have avoided relying on a single measure of diversity, a move that has proved useful in other contexts such as information (the Shannon Index) and the distribution of socio-economic variables (e.g. the Gini Coefficient). In Annex 2.2 we discuss the various ways in which biodiversity is measured and collate the work of ecologists that provides a map of the global distribution of biodiversity.[60]

Individual actors in ecosystems include organisms that pollinate, decompose, filter, transport, redistribute, scavenge, fix gases and so on. Nearly all organisms that help to produce those services are hidden from view (a gram of soil may contain as many as 10 billion bacterial cells), which is why they are almost always missing from popular discourses on the environment (Dasgupta and Raven, 2019). But their activities enable ecosystems to maintain a genetic library, preserve and regenerate soil, fix nitrogen and carbon, recycle nutrients, control floods, mitigate droughts, filter pollutants, assimilate waste, pollinate crops, operate the hydrological cycle and maintain the gaseous composition of the atmosphere.

Ecosystems are not defined in a sharp manner from rigid principles. Watersheds, wetlands, coral reefs and mangrove forests are ecosystems, as are agricultural land, fisheries, freshwater lakes, coastal fisheries, estuaries and the ocean. As a general rule, ecosystems are not tightly knit entities, they blend into one another. But there are ecosystems that have strong interactions among their own constituents and weak interactions across their boundaries. The boundaries may harbour discontinuities, such as in the distribution of organisms, soil types, depth of a body of water and so on.

Ecosystems differ in their spatial reach (a hedgehog's gut is an ecosystem, as is a tropical rain forest) and rhythmic time (minutes for bacterial colonies, decades for boreal forests). Some ecosystems are of near-continental size (the Amazon rainforest), there are those that cover regions (the Ganga-

---

[59] In their excellent review of the value of biodiversity, Hanley and Perrings (2019) refer to ecosystem integrity as 'ecosystem functioning'. Perrings (2014) is a deep meditation (there is no other word for it) on the role of biodiversity in our lives.

[60] Hanley and Perrings (2019) contains an excellent account of measures of species diversity that have been deployed by ecologists. Groom (2020) has constructed a taxonomy of possible biodiversity measures.

Chapter 2: Biodiversity and Ecosystem Services

Brahmaputra river basin), many are volcanic islands (the islands comprising Micronesia), others involve clusters of towns (micro-watersheds in the Ethiopian highlands), while yet others are confined to a village (village ponds in Norfolk, UK).

Box 2.1

Biomes

Biomes refer to the large, distinct biological communities that have formed on the planet in response to similar physical environments. They are characterised by the most common growth forms of plants distributed across large geographic areas. A biome can thus occupy disjoint regions, indicating responses to similar temperatures and precipitation. In their most admirable text, Bowman, Hacker and Cain (2018) have compiled a list consisting of nine terrestrial biomes: tropical rainforest, tropical seasonal forest and savannah, desert, temperate grassland, temperate shrubland and woodland, temperate deciduous forest, temperate evergreen forest, boreal forest, and tundra.

Tropical rainforests are of particular interest to the economics of biodiversity, as they contain an estimated 50% of Earth's species and some 40% of the terrestrial pool of carbon in just over 10% of Earth's terrestrial vegetation cover. Logging and the conversion of rainforests to pasture and croplands have altered approximately 50% of the biome (Cuff and Goudie, 2009).

In contrast to the poor soil that is characteristic of tropical rainforests, the soils of temperate grasslands are fertile because of an accumulation of rich organic matter that the extensive root system of grasses leaves behind. As a consequence, most of the fertile grasslands of North America and Eurasia have been converted to agricultural and pastoral development.

If they are peered at closer, biomes themselves are a patchwork of distinct landscapes, each when peered at closer still, are found to be a patchwork of even smaller landscapes, a feature known as 'fractal'. We return to this essential feature of biomes in Chapter 3, for they point to the fact that the processes governing the biosphere are non-linear. Biomes are self-organising regenerative assets. Levin (1999) is an excellent guide to the science underlying the processes that have led to the spatial characteristics of ecosystems we see around us.

Over the past 150 to 200 years, land-use changes have transformed Earth's land surface. Approximately 60% has been altered, primarily by agriculture, forestry, livestock grazing, mining and quarrying, and a smaller amount (2–3%) by urban development and transportation corridors (Bowman, Hacker, and Cain, 2018). The actual distribution of biomes thus differs markedly from their potential distribution. See Chapter 14 for more on the distribution of our stocks of natural assets and Chapter 16 for more on current and future land and ocean use change.

2.2 Primary Producers

Biomass in any particular location is the total mass of living material in it, measured, say, in kilograms. The regenerative rate of a stock of biomass is the net addition to it in a period of time.[61] Box 2.2 presents the simplest mathematical formulation of an ecosystem, that of a single species in an

---

[61] The corresponding term for a piece of produced capital is its yield, expressed in its own unit (Chapter 1).

environment that supplies a constant flow of nutrients. The model presented – that of a fishery – forms the basis on which theoretical models of communities of organisms that populate ecosystems have been constructed by ecologists (Annex 2.1).[62]

Peering inside ecosystems enables us to uncover units that are a lot smaller, for example, individual organisms. Ultimately, it is practical convenience (what the data can shoulder) that determines how deeply we should peer into ecosystems in order to understand their workings and the role that biodiversity plays in them. Among ecologists there is thus a continual to-ing and fro-ing between studies of macro manifestations of ecosystems and of the micro content and behaviour of various constituents of ecosystems.

Box 2.2
Fishery Dynamics

We consider a lake harbouring a species of fish of interest to the fishing community. Let G be the regenerative rate of the fish population. G is the birth rate minus the death rate. For ease of exposition we do not distinguish between the fish stock's biomass and its population. Let the stock of fish be denoted by S. G is a function of S, so we write it as $G(S)$. It follows that the slope of the G-function, $dG(S)/dS$, is the marginal product of S. It is also the yield (i.e., own rate of return) on the fish stock (Chapter 1). Organisms other than fish are taken to be the background against which the fish population exists. Time is denoted by t (≥ 0) and is taken to be a continuous variable. The stock of fish at t is $S(t)$.

The lake provides a constant flow of nutrients to the fish population. If the fish stock were to be very small, the lake would in effect be of unlimited size, so the stock would grow at an exponential rate. We write that rate as $r > 0$. r is called the *intrinsic growth rate* of the fish stock. But if the stock were to be non-negligible, the finiteness of the supply of nutrients would bite – the larger the stock, the smaller would be the nutrient supply per unit of fish population – and that would impose increasing limits on population growth. Beyond a size, the fish stock's regenerative rate is thus a declining function of the stock. If the stock were to be sufficiently large, the birth rate would fall short of the death rate and the regenerative rate would be negative. The simplest way to represent the dynamics involving these features is to suppose that $G(S)$ is the quadratic function, $rS(1-S/K)$, in which both r and K are positive numbers (Figure 2.2). We thus assume

$$G(S) = rS(1 - S/K) \qquad r, K > 0 \qquad (B2.2.1)$$

The dynamics of the fishery is governed by the differential equation

$$dS(t)/dt = rS(t)[1 - S(t)/K], \qquad (B2.2.2)$$

Integrating equation (B2.2.2) yields the famous logistic growth curve (Figure 2.3),

$$S(t) = KS(0)e^{rt}/\{K + S(0)[e^{rt} - 1]\} \qquad (B2.2.3)$$

$S(0)$ is the stock at the initial date, $t = 0$.

Equation (B2.2.3) says that $S(t)$ tends asymptotically to K as t tends to infinity (Figure 2.3).

---

[62] Subsequently (Chapters 4-4*), we will use the fishery's model as a heuristic device for representing the biosphere as a whole. We could equally use forests and the growth and death of plants and trees as our heuristic device.

Chapter 2: Biodiversity and Ecosystem Services

NB: The logistic growth curve has also been used to study the growth of trees from seedling to maturity. The curve explains why forest biomass grows at its fastest rate at an intermediate stage of succession. Growth slows as the forest matures.

Notice that equation (B2.2.2) possesses two stationary points: $S = 0$ and $S = K$. The former is unstable, the latter is, as we have just confirmed from equation (B2.2.3), stable. In the absence of human predation, the system would therefore settle at $S = K$. K is called the *carrying capacity* of the fishery.

Let $R(t)$ denote the rate at which fish is harvested by the human population. Then the dynamics of the fishery would be governed by the differential equation

$$dS(t)/dt = rS(t)[1 - S(t)/K] - R(t) \tag{B2.2.4}$$

Consider any S in the interval $(0,K)$. If the harvest rate is so chosen that $R = rS[1 - S/K]$, then S remains constant over time. We then say the fishery is at a *stationary state*. Notice that $G(S)$ attains its maximum at $S = K/2$. The *maximum sustainable yield* is $G(K/2) = rK/4$. It will be noticed that the yield from the fishery (i.e. the fishery's own rate of return) is positive as long as $S < K/2$, but is negative if $S > K/2$. This fact will prove to be significant when we discuss the optimal management of fisheries. Yield is zero at $S = K$.

Figure 2.2 A Single Fishery

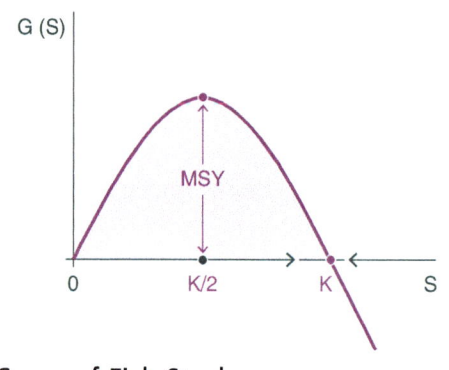

Figure 2.3 Logistic Growth Curve of Fish Stock

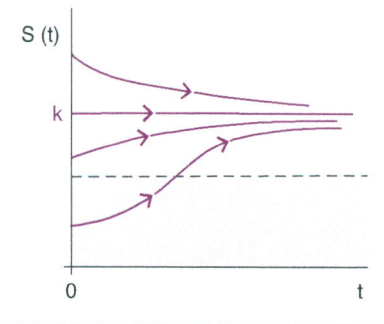

A fruitful way to peer inside ecosystems is to look for organisms that obtain energy directly from the sun to produce their own food. They are known as *primary producers*. They consist of plants, algae and many bacteria. Nearly all other organisms depend on them for their energy.[63] The regenerative rate of a stock of primary producers is called its *net primary productivity* (NPP).

NPP in an ecosystem is the spatial distribution of organic compounds that are fixed by primary producers there, minus respiration per unit of time. During respiration, organic compounds are broken down to fuel the processes that govern a primary producer's activities. It is the fixed carbon as measured by NPP that becomes primary producer biomass, that is, the producer's carbon pool. The fixed carbon then either remains as primary producer biomass, or is consumed by herbivores and thence up the food chain, or enters the detrital pathway and becomes part of the detrital carbon pool. The detrital carbon in turn is either recycled by decomposers and detritivores or stored as refractory carbon. Figure 2.4 depicts the carbon flow in and out of primary producers and detrital pools. Annex 2.1 presents a simple model to illustrate the distribution of biomass among trophic levels and the influence of that distribution on NPP.

Figure 2.4 Carbon Flow in and out of Primary Producers and Detrital Pools

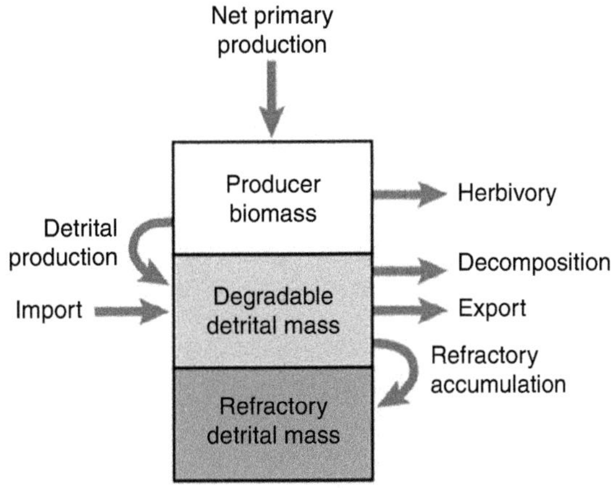

Source: Lartigue and Cebrian (2009). *Permission to reproduce from Princeton University Press.*

NPP is a *flow* (kilograms of biomass per unit of time). In contrast, the biomass of primary producers is a *stock* (kilograms of biomass) and is the locus of the goods and services it generates. Although the diversity of life covers far more than the diversity of primary producers, we study primary producers here because all life is dependent on it.

In both aquatic and terrestrial ecosystems, the transfer of fixed carbon from primary producers to herbivores and decomposer/detritivores provides pathways for the flow of energy and nutrients. These flows have consequences not only for carbon storage but also for nutrient cycling and herbivore and decomposer/detritivore populations.[64] We humans have co-evolved with other species over evolutionary time, as they have with one another. That is as trite an observation as there is, but we will return to it periodically because it will remind us not only that the ability of communities of organisms to adapt to rapidly changing environmental conditions is limited, but also that some of the most draconian environmental changes have been unleashed by we humans in what can only be regarded as a blink in evolutionary time (Chapters 4–5).

---

[63] Nearly all, because there are organisms deep down in the ocean floor, at openings of hydrothermal vents. Their source of energy is not primary producers.

[64] Lartigue and Cebrian (2009) offer a fine account of carbon flows.

Chapter 2: Biodiversity and Ecosystem Services

By focusing on primary producers here we will be able to align the economics of biodiversity with the ecologist's tactic of tracing the goods and services the biosphere produces to processes governing material and energy flows at a micro level. To be sure, there are also organisms higher up in the food chain, as well as non-living material that must be included. But to start with primary producers is rather like the educationist focusing first on primary education, while fully aware that its value is not independent of the quantity of produced capital and natural capital it works with, nor independent of secondary and tertiary education.

The simplest way to create a link between micro- and macro-ecology is to construct a grid of the biosphere. Each box in the grid is to be thought of as containing a bounded, self-regenerative resource, as in the case of a fishery (Box 2.2) or forest. The regeneration rates of the resources across the grid are of course related: they influence one another. Our interest here, however, is in primary producers. To model spatial heterogeneity we suppose not only that biomass of primary producers is non-uniformly distributed across the grid, but also that the parameters of the self-regenerative function of primary producers are not necessarily the same.[65] The biomass in some boxes are near zero either because the boxes are in infertile geographical terrains or because they have been overharvested; in others the biomass is large because they are in a fertile terrain and have not been harvested, and so on.

We begin by adopting the perspective of humanity as a whole. The accounting price of a unit biomass of a primary producer is the contribution it makes to the 'common good' (Chapters 10–12). The price may be thought of as the unit's 'social worth.' Imagine now that an accounting price has been estimated for the biomass of primary producers in each box in the grid (Chapters 1 and 10–13). Our construction therefore respects the fact that the accounting price of a unit of biomass is a function of location as well as time: a unit of biomass in a neem tree (Azadirachta indica) at a particular location (e.g. Uttar Pradesh, India) and time (e.g. May) differs in value from a unit of biomass in a baobab tree (genus Adansonia) in the Okavango Delta, Botswana, in December.[66]

Not all biomass of even primary producers has a positive accounting price. The acres of water hyacinths on the surface of Lake Cariba in northern Zimbabwe prevent sunlight from penetrating the lake surface. Which is one reason fishermen spend much time and effort removing them.[67]

The accounting value of primary producers in an ecosystem is the weighted sum of the units of biomass of primary producers comprising them, the weights being accounting prices. Thus, the spatial distributions of primary producer biomass and NPP at a point in time represent a feature of an ecosystem in the same way as the spatial distributions of human capital (education, health, aptitude) and income (wages and salaries) represent a feature of an economy. In Box 2.3 we provide an informal argument what suggests that humanity has been mismanaging its portfolio of assets badly, accumulating produced capital (buildings, machines, roads, ports, dams) and running down natural capital (ecosystems). What we are interested in addition is the regenerative character of the biosphere. Therefore, in the remainder of the chapter, we return to the larger conception with which we began, that of ecosystems and a feature of theirs of central importance: *biodiversity*.

The analysis in Box 2.3 is a simple exercise. It is designed to display the regenerative character of the biosphere by concentrating on one aspect of it, namely, NPP. In the remainder of the chapter, we return to the larger conception with which we began, that of ecosystems and a feature of theirs of central importance: *biodiversity*.

---

[65] A box could be likened to a pixel in an image.

[66] Of course, a neem tree could be thought of being an overly aggregate object. An obsessive ecological economist would distinguish a unit of biomass in a leaf from a unit inside the trunk, and so forth. In Chapter 5, where we develop choice under uncertainty, it will be found that in addition to location and time, the state of Nature (amount of rainfall) must be included in identifying a unit of biomass.

[67] That the accounting price of organisms higher up the food chain can be negative is self-evident.

Chapter 2: Biodiversity and Ecosystem Services

Box 2.3
Counterfactual Reasoning in Asset Management

In valuing (or pricing) assets, it is common practice to choose an aggregate of produced capital as *numeraire* and value all other assets in terms of it. We tend to think of income as *numeraire*, but that is because we usually have in mind that produced capital is the source of that income, and that the income is the yield generated by produced capital.

An asset's yield is also called its own rate of return. Moreover, an asset's rate of return is its yield plus capital gains (or losses) relative to the *numeraire*. Efficient asset management requires that all assets in one's portfolio earn the same (risk-adjusted) rate of return (Ch. 1, Box 1.3). Jordà et al. (2019) have estimated that the long run rate of return (rent or dividend) on housing and equities in the US has been round 5% a year. If we choose that to be the *numeraire*, then 5% should be regarded as its yield.

Consider a public agency that has shares in US housing and equities and depends also on an ecosystem in the public domain. Imagine that the ecosystem's yield (or own rate of return) to society is estimated to be ρ per year. Efficient management of the public portfolio would have it that if $\rho < 0.05$, it must be that the ecosystem enjoys a (risk adjusted) capital gains of $(0.05-\rho)$ a year; on the other hand, if $\rho > 0.05$, it must be that the ecosystem suffers a (risk adjusted) capital loss of $(\rho-0.05)$. In what follows, we conduct a counterfactual form of reasoning. We begin by assuming that the public agency is managing its portfolio efficiently. We then check whether the two assets offer the same rate of return. If we find they don't, we conclude that the agency does not manage it portfolio efficiently.

Markandya and Murty (2000) conducted a social cost-benefit analysis of an Action Plan of the Government of India to undertake measures to raise the quality of the Ganges River water to bathing standard. The Ganges is sacred to Hindus but is also one of the most polluted rivers in the world (see Ch. 13, Box 13.7 for details). The authors elicited responses to questionnaires for estimating the willingness to pay by people living in the Ganges basin for a cleaner river, and so estimated the annual social benefits to be realised by the Action Plan. The investment outlay and recurring costs were taken from the Plan documents. Using that data, the authors estimated that the own rate of return on the project to be round 15%. That tells us that the implicit value of Ganges River water should be declining relative to public income by some 10% a year.

Patently, that would be absurd. Rapid urban development on the Ganges Basin was worsening the quality of the river, even while income per capita in the Basin was increasing. Taken together, they implied that the Ganges was becoming *scarcer* relative to produced capital, not more abundant (in fact, the Ganges Action Plan was soon abandoned). That is a rough and ready way of establishing the imbalance in economic development: the asset portfolio was inefficient.

The example is representative of many unmanaged ecosystems. It points especially to the enormous imbalance we have created between produced and human capital on the one hand, and natural capital on the other. In Chapter 4 we visit more familiar evidence of the imbalance in humanity's portfolio of assets.

Subsequent chapters enquire into why this has happened. Briefly put, we will find that much of Nature's worth to society are not reflected in market prices. That's because the many forms of natural capital are open to all at no charge (Ch. 1, 8–9). The *private* rate of return on investment in many forms of natural capital remains low, even zero. Worse, widespread government subsidies in agriculture, energy, water, and fisheries mean that critical biospheric services come with a *negative* price (Annex to Ch. 8). Because of these price distortions, we invest relatively more in other assets, such as produced capital, that yield lower rates of return.

Chapter 2: Biodiversity and Ecosystem Services

2.3 Ecosystems do not Maximise NPP

Any mention of a financial portfolio's yield and the thought turns to finding ways to maximise it. Giving intention to Nature's processes (Nature 'abhors' a vacuum; light 'seeks to' minimise the time it takes to travel between any two points, and so on) is not an uncommon practice.[68] So it could be thought that ecosystems maximise their regenerative rate.

They do not. Nature does not even maximise NPP per unit area. The abiotic environment (precipitation, temperature, altitude, nutrients and so on) is not the same across the globe, so there are wide differences in NPP across biomes. Average NPP per square metre in a temperate grassland (0.75 Kg/m^2/year) is lower than the corresponding figure in a tropical rainforest (2.5 Kg/m^2/year), that in the ocean lower still (0.14 Kg/m^2/year).[69] But because biomass per unit area in the ocean is far smaller than that in temperate grasslands and tropical rainforests, the ratio of NPP to biomass in the oceans is far larger than in temperate grasslands and tropical rainforests. That means the unweighted average yield on oceanic biomass is higher than in temperate grasslands and tropical rainforests.[70]

Nature does not maximise planetary NPP even when controlling for the fact that the abiotic environment is not uniform across space. NPP in old growth forests is nearly zero. Prior to the dominance of humans, evolutionary stable distributions of life forms defined community architectures across the planet. But those distributions emerged under selection pressures at the genetic level, they did not give rise to what could in any sense be thought of as an efficient distribution of NPP. Box 2.4 confirms the claim and explains why Nature does not maximise planetary NPP.

Box 2.4
Selection Pressures and Regeneration Rates

To illustrate that Nature does not maximise ecosystem regeneration rates, even when controlling for the fact that the biosphere is spatially heterogeneous, consider the long-run state of the fishery in Box 2.2. The biomass (population) of fish there equals K. At K, however, the fishery's yield is zero, whereas it is positive at populations smaller than K. The example also shows that Nature does not maximise the population's yield – the marginal product of the stock of fish is *negative* at K. The biological mechanism that dictates this is as follows:

The lake provides a constant supply of nutrients to the fish population. Supply per fish declines with the size of the fish population. That can be read as 'crowding', which is an example of *inter-organismic externalities*, a phenomenon of prime importance also in human societies (Chapters 8–10). Because of crowding, the externalities in the fishery are *negative*. The negative externalities are so powerful that when the stock exceeds $K/2$ (Figure 2.2), an additional fish causes the fishery's regenerative rate to decline.[71]

---

[68] The famous Gaia hypothesis of James Lovelock (1995) has been read by critics as anthropomorphising Nature; in fact the author did no such thing. Lovelock was only concerned to argue that the Earth system is a self-regulating complex system with considerable stability, but capable of being disrupted to the point where it enters a different stability regime from the one it is in now. See Chapter 3 for a fuller discussion.

[69] The figures are taken from Bowman, Hacker and Cain (2018).

[70] Total oceanic NPP is about 48 trillion Kg/year, which is about 46% of global NPP. What the ocean loses out to terrestrial biomes in NPP per unit area are made up by its vastly greater surface area (70% of total surface area of the planet).

[71] A simple proof of the theorem on the generic inefficiency of evolutionary stable equilibria is in Binmore and Dasgupta (1986).

> Financial portfolios also harbour externalities if the (uncertain) returns on individual financial stocks are correlated. The externalities are positive if the correlations are positive, they are negative if the correlations are negative. Positive correlations among people facing global environmental uncertainties are discussed in Chapter 5.

2.4 Ecosystem Goods and Services: Classification

The Common International Classification of Ecosystem Services (CICES), which also identifies the contributions ecosystems make to human well-being, is built on the pioneering work of the Millennium Ecosystem Assessment (MA, 2005a-d). It consists of three categories of ecosystem services, contributing directly or indirectly to human well-being. They are:

(1) *Provisioning Goods*. This category includes the provision of materials and energy needs for the range of products we obtain from ecosystems. It includes food, fresh water, fuel (dung, wood, twigs and leaves), fibre (grasses, timber, cotton, wool, silk), biochemicals and pharmaceuticals (medicines, food additives), genetic resources (genes and genetic information used for plant breeding and biotechnology) and ornamental resources (skins, shells, flowers).

(2) *Regulating and Maintenance Services*. This category regulates and maintains ecosystem processes, including maintaining the gaseous composition of the atmosphere, regulating both local and global climate (temperature, precipitation, winds and currents), controlling erosion (retaining soil and preventing landslides), regulating the flow of water (the timing and magnitude of runoff, flooding and aquifer recharge), purifying water and decomposing waste, regulating diseases (controlling the abundance of pathogens, such as cholera, and disease vectors such as mosquitoes), controlling crop/livestock pests and diseases, pollinating plants, offering protection against storms (forests and woodlands on land, mangroves and coral reefs on coasts), recycling nutrients, and maintaining primary production and oxygen production through photosynthesis.

(3) *Cultural Services*. This category offers non-material benefits, including spiritual experiences and an identification with religious values. It is perhaps more appropriate to trace these experiences and values to Nature, rather than ecosystems, since the latter is a term of recent origin. The diversity of life has, in part, shaped the diversity of cultures (Chapters 6 and 12). Many religions attach spiritual and religious significance to particular flora and fauna. Nature's processes have influenced the systems of knowledge that epitomise cultures; they have provided the basis for education and instruction in societies, and have been the inspiration for art and architecture, folklore and epic poetry. People find aesthetic value in Nature, which gives expression in private gardens, public parks and protected areas (forests and coastlines). Ecosystems influence social relationships (social capital in coastal fishing villages takes a different form from social capital in nomadic herding and agricultural societies (Chapters 8–9). The local ecosystem offers people a sense of place, their cultural landscape (Chapter 12). And particular ecosystems attract tourism and recreation.

Cultural services and a variety of regulating services such as disease regulation contribute directly to human well-being, whereas other services (e.g. pollination) contribute indirectly. Ecosystem functions and processes are considered as the structures and processes that underpin the production of ecosystem services. Changes in the health of ecosystems would be expected to have an impact on the flow of ecosystems services (Figure 2.5).

Chapter 2: Biodiversity and Ecosystem Services

Figure 2.5 Ecosystem Service Cascade

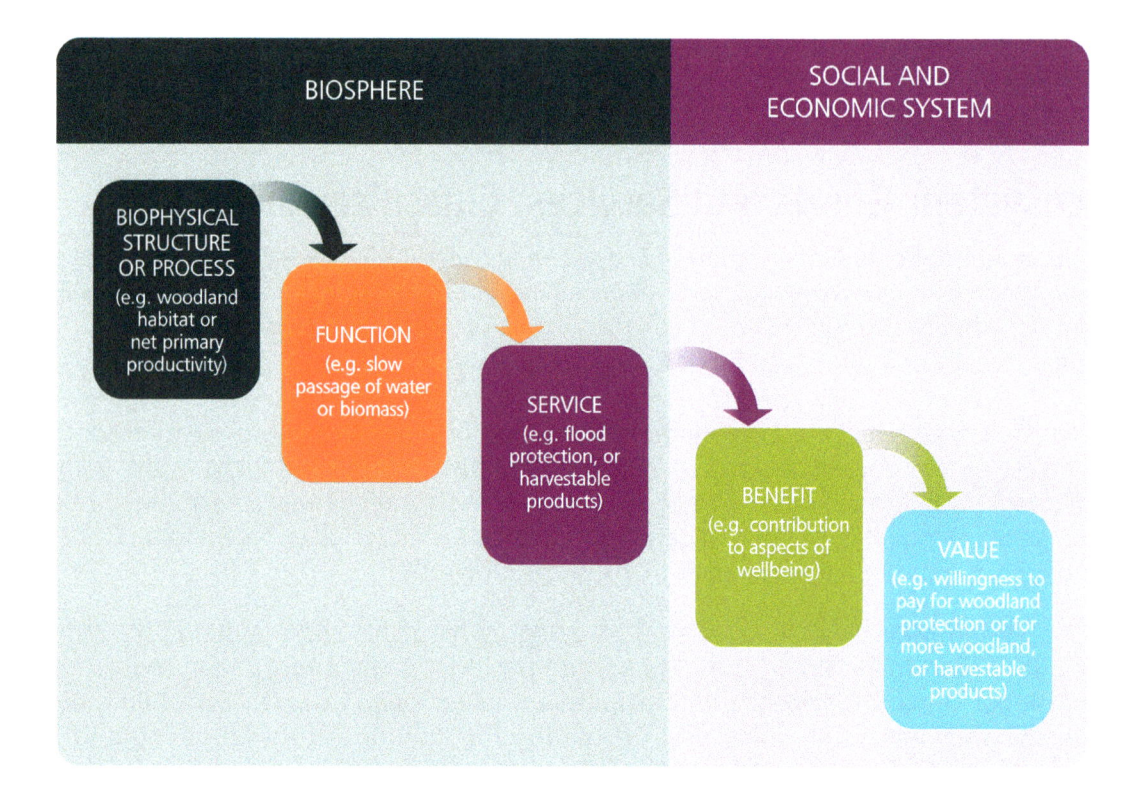

[Diagram: BIOSPHERE | SOCIAL AND ECONOMIC SYSTEM. Flow from BIOPHYSICAL STRUCTURE OR PROCESS (e.g. woodland habitat or net primary productivity) → FUNCTION (e.g. slow passage of water or biomass) → SERVICE (e.g. flood protection, or harvestable products) → BENEFIT (e.g. contribution to aspects of wellbeing) → VALUE (e.g. willingness to pay for woodland protection or for more woodland, or harvestable products).]

Source: Adapted from Potschin and Haines-Young (2016).

As always in ecology, there are no sharp contours separating ecosystem services. Thus, erosion control can be classified both under regulating and maintenance services, depending on the timescale involved. Similarly, climate regulation can be interpreted as a regulating service because ecosystem changes can have an impact on local and global climates over timescales that are relevant for human decision-making (decades or centuries), whereas production of oxygen gas (through photosynthesis) may be considered a regulating service because the changes in oxygen concentration take place over a very long time.[72]

Box 2.5

The Soils

The soils, called the pedosphere by ecologists, are similar to terrestrial and marine ecosystems in the forms and functions of their food webs and trophic levels, but appear far less in public discourses on biodiversity. Beach, Luzzadder-Beach and Dunning (2019) call the soils "dark matter biodiversity". The soils are living ecosystems, approximately half air and water, 45% minerals, and 5% organic matter. Of that 5%, only 10% is life, but that 10% contains some of the greatest biodiversity in the

---

[72] In an illuminating paper, Ferraro et al. (2012) have collated quantitative estimates of the magnitude of the services that tropical forests provide. The authors presented the estimates in various publications, in turn, for carbon storage (more than 10% of global carbon emissions are from ongoing deforestation there), ecotourism, hydrological services, pollination, human health and non-timber products.

biosphere. By one estimate, the soils contain nearly 80% of terrestrial carbon, with an inevitable accompanying estimate that up to 25% of our uncertainty in the flux of global carbon is traceable to soil erosion.

Soil organisms include earthworms, nematodes, arthropods, protozoa, fungi and bacteria, but by far the most abundant are fungi and bacteria. These organisms form food webs which drive soil ecosystem processes, including nutrient cycling, carbon sequestration, nitrogen storage and water purification, and are important components of global cycling of matter, energy and nutrients. Advances in molecular genetics have revealed the enormous diversity of fungi and bacteria associated with plant roots. They play diverse roles, for example promoting plant growth through enhancing plant nutrition and protecting plants from herbivores and pathogens (Orgiazzi et al. 2018). By one set of estimates, more than 25% of the Earth's species live only in the soil or soil litter (Beach, Luzzadder-Beach, and Dunning, 2019).

Soils themselves are diverse. 300,000 soil types appear in one classification, one class being the much-discussed Amazonian Dark Earth soils that create islands of fertility in a sea of low-fertility soils. The 2016 Global Soil Biodiversity Atlas was the first attempt to map life in soil at a global scale (Figure 2.6). Biodiversity in soil is highest in tropical rainforests, they are lowest in deserts. Soil has historically been the world's largest carbon sink. However, recent studies have suggested that soil could shift to being a net emitter of carbon, due to climate change-driven increases in heterotrophic respiration (Lugato et al. 2018).

The soils also supply most of the water needed by plants and for terrestrial biodiversity. Soil water makes up 65% of the world's fresh water, is the source of 90% of global farm output and provides over 99% of our food calories (Pimentel and Burgess, 2013). Like most ecosystems, the dynamics in the soils start from primary producers, who shed leaf litter on soil surface and the narrow band of soil that is influenced by root secretion and soil microorganisms. The enormous network of fungi that connects tiny roots (rootlets) to a wider array of soil nutrients and water helps to fix nitrogen (a regulating service). Nitrogen is a macronutrient, vital in the food chain.

The soils are also a major reservoir for medicines (a provisioning good). Pepper et al. (2009) have reported that over 75% of antibacterial agents and 60% of new cancer drugs approved between 1983 and 1994 had their origins in the soils, as did 60% of all newly approved drugs between 1989 and 1995.

When threats to soil biodiversity are mapped, areas at high-risk often correspond with areas of highest soil biodiversity (Figure 2.6). The risk index was generated by combining eight potential stressors of soil biodiversity: loss of above-ground diversity; pollution and nutrient overloading; overgrazing; intensive agriculture; fire; soil erosion; desertification; and climate change (Orgiazzi et al. 2018).

What happens when the diversity of life within soil is lost? If soil biodiversity performing its huge range of roles were completely lost, the land-based food system would cease to function. Wagg et al. (2014) found a strong linear relationship between all measured ecosystem functions and indicators of soil biodiversity. Reductions in soil biodiversity contribute to several environmental problems, such as the eutrophication of surface water, reduced above-ground biodiversity and global warming (Bender, Wagg and Van Der Heijden, 2016). Declines in soil biodiversity cause declines in performance of essential processes and land managers compensate by applying fertiliser at a significant economic and ecological cost. Using US farm-gate practices as a guide, global soil erosion annually costs US$33 to 60 billion in nitrogen application and US$77 to 140 billion for phosphorus (FAO and ITPS, 2015).

Chapter 2: Biodiversity and Ecosystem Services

Figure 2.6 Soil Biodiversity Index

Source: Orgiazzi et al. (2016).

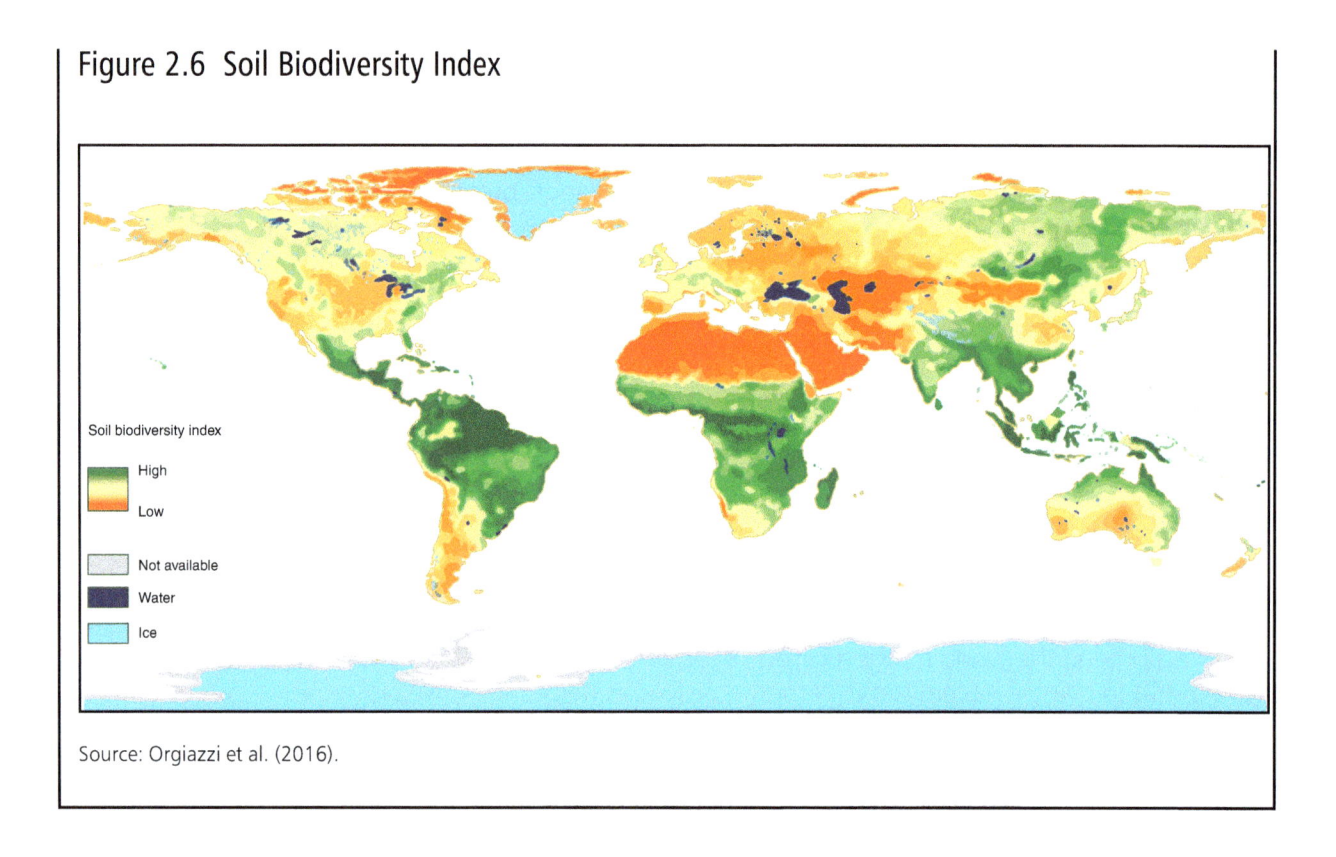

2.5 Invisibility and Silence of Regulating and Maintenance Services

It is significant that the processes governing regulating and maintenance services are either several steps removed from our direct experience or are felt by us only over the long run. Regulating and maintenance services are mostly hidden from view. They are also mostly silent. In contrast, provisioning and cultural services have readily detectable outcomes or are directly observable and can be felt even in the short run (e.g. in agricultural production, forestry and fishing).

We rarely appreciate the presence of regulating and maintenance services, it is only when they are *absent* that we feel their worth. Mangrove forests, for example, are a rich stock of timber and offer a habitat for a wide variety of fish species. But they also protect villages against storms. While timber and fish are visible, protection against even extreme storms such as cyclones is appreciated only when the mangroves have been degraded. Estimating the value of mangrove forests for the timber they represent and the fish they harbour is hard work, but it helps that the objects of interest are visible. Evaluating the forests for the protection they offer against cyclones is a far harder task. Entrepreneurs also have an eye on the potential for converting mangrove forests into shrimp farms, and that makes mangroves all the more vulnerable to human predation. It also means that when estimating the social cost of shrimp, one should include the carbon that had been sequestered as mangrove forests and the protection they offered communities in the neighbourhood. Strikingly, shrimp consumption is now estimated to be about as costly in terms of the carbon it embodies in its production as beef consumption.[73]

---

[73] There is now a rich literature on the protection mangrove forests offer against the devastation cyclones can cause. Das and Vincent (2009) is a classic dissection of a natural experiment: the Indian Super-Cyclone of 1999 that grew in the Bay of Bengal and played havoc in the State of

46   The Economics of Biodiversity: The Dasgupta Review

The competition we humans have created between the availability of provisioning goods on the one hand and regulating and maintenance services on the other is vividly illustrated by tropical rainforests. These ecosystems produce goods such as fuelwood, fodder, timber, leaf manure, food and medicines, and supply services such as sequestering carbon, offering a habitat for wildlife and more generally housing biodiversity. Upstream forests also regulate waterflow (e.g. groundwater recharge, flood control) and conserve soil. In a study of the conflicting nature of the demands inhabitants make of the watersheds of the Western Ghats in India, Lele et al. (2011) found that farmers in the catchment area cultivate irrigated paddy in both the wet and dry seasons. The fields are irrigated in the dry season from water stored in village tanks (artificial ponds), which get filled in the wet season. The Western Ghats have been degraded over the decades. But the authors find that regeneration of the forests would reduce water runoff to an extent that the tanks would not get recharged sufficiently to supply water in the dry season. Farmers are projected to lose income from their paddy output.

The competing demands are a reason it is easy to overlook the significance of regulating and maintenance services (Boxes 2.5 and 2.7); they are easy prey to economic development in the way it has been pursued. We study the significance of this for modelling growth and development possibilities in Chapters 4–4*. We will find that over the years, economic development has come to mean growth in the products we enjoy from provisioning goods and cultural services, but that the pursuit of economic growth (i.e. growth in GDP) has led to a decline in the ability of the biosphere to supply regulating and maintenance services. Paradoxically, NPP is highest where ecosystems have been converted either to agriculture or plantation forestry; or have been disturbed through other interventions. In fact, most investment in renewable resource systems, such as agriculture, is about increasing NPP of species valued for particular traits and reducing it in others.

That decline endangers the biosphere's ability to supply provisioning, maintenance and cultural services to our descendants. The literature on economic growth and development has shied away from acknowledging this conflict of interest between the current and future generations.[74] The present Review can be read as an attempt at framing the conflict and at looking for ways to remove the biases contemporary growth and development policies have created against our descendants. Chapters 4–13 develop the required analysis. Chapter 16 looks at current extraction of major categories of provisioning goods and their future prospects.

Box 2.6
Carbon Storage and Seed Dispersal by Large Fauna

Many large tropical trees making significant contributions to carbon stock rely on large vertebrates for their seed dispersal. These frugivores are threatened by hunting, illegal trade and habitat loss due to land invasion. Studies examining both small and large spatial scales have found that loss of large frugivores could cause carbon storage by these trees to decline. Bello et al. (2015) modelled local extinctions of trees depending on large frugivores in 31 Atlantic forest countries and found that this could significantly erode carbon storage potential, even when only a small proportion of large

---

Orissa (the cyclone is estimated to have inflicted losses of some US$4.5 billion). The authors found that villages that were protected by a greater stock of mangroves had suffered less property damage. Das (2011) contains a more detailed analysis of the data. (Sun and Carson, 2020) also find substantial protection against property damage that wetlands offer from tropical cyclones along the Eastern Seaboard of the US. A major, global study of the protection mangroves offer against cyclones is Hochard, Hamilton and Barbier (2019). Barbier and Cox (2003) explore the question whether the idea of economic development as it is currently conceived is exploitative of mangrove forests. Their conclusion fits with the review we conduct in Chapter 4* of contemporary models of growth and development.

[74] See, for example, Aghion and Howitt (1998), Barro and Sala-i-Martin (2003), Helpman (2004), Acemoglu (2008) and Galor (2011).

Chapter 2: Biodiversity and Ecosystem Services

> seeded trees are lost. These extinctions would occur if the frugivores trees depend on no longer inhabited these areas.
>
> Another study (Chanthorn et al. 2019) focused on a 30 hectare forest plot in Thailand. In this plot, an intact fauna of primates, ungulates, bears and birds of all sizes still exists. Tree species dependent on seed dispersal by large frugivores account for nearly one-third of the total carbon biomass on the plot. If all these animals were lost, carbon stored by the plot would be reduced by 2.4 to 3.0%.
>
> These studies show that if we want forests to continue to be the enormous carbon sinks they are – the Amazon is estimated to contain an amount of carbon equivalent to a decade's worth of human emissions (Lovejoy and Hannah, 2019) – we need to maintain the complex range of diversity within them.

2.6 Conservation vs. Pollution[75]

Not all of Nature's objects have a positive value for us. Pathogens and pests are obvious examples. They damage not only our health and that of our crops and animals, they also destroy pristine forests and fisheries, so we try to suppress them when they arise. That their emergence is an immanent feature of Nature's processes does not alter our attitude to them, for we too are an emergent feature of those processes.

The view that the biosphere is a mosaic of self-regenerative assets also covers its role as a sink for pollution. Conservation of mass means that the waste we create has to be discharged somewhere (Kneese, 1970; Kneese, Ayers and d'Arge, 1970; Mäler, 1974). The discharge contributes to the depreciation of the biosphere. More generally, we may view pollution as the depreciation of capital assets. Acid rains damage forests; carbon emissions in the atmosphere trap heat; industrial seepage and discharge reduce water quality in streams and underground reservoirs; sulphur emissions corrode structures and harm human health; and so on. The damage inflicted on each type of asset (buildings, forests, the atmosphere, fisheries, human health) should be interpreted as depreciation. For the broad class of assets we are calling natural capital (Chapter 1), depreciation amounts to the difference between the rate at which they are harvested and their regenerative rate, whereas the depreciation that pollutants cause on natural capital is the difference between the rate at which pollutants are discharged and the rate at which natural capital is able to neutralise the pollutants (the biosphere's regulating services). The task in either case is to estimate depreciation. Economists have tried to estimate the damage an additional ton of carbon in the atmosphere is likely to inflict over time on agricultural production, submerged coastal habitats, human health and so on. It is the accounting price of carbon emissions. Being negative, it is called the *social cost of carbon* (Chapters 10, 12–13). Estimates today range from US$15 to 100 for a ton of CO_2, and there are outliers well above US$100 (Tol, 2008). Differences in estimates of the social cost of carbon can be traced in large measure to differences among social scientists in the way they weight the well-being of future people relative to the well-being of present people (Chapter 10), and to differences in estimates of the damage likely to be suffered by future people owing to increases in carbon concentration in the atmosphere (Chapter 4).[76]

---

[75] This section has been taken from Dasgupta (1982), which developed ecological economics on the basis of the equivalence, modulo sign, between pollution and conservation.

[76] Modelling our lack of knowledge of the processes by which carbon emissions map into the damage we will suffer in the future is another reason the range of estimates of the social cost of carbon is large (Chapter 5).

In short, resources are 'goods', while pollutants (the degrader of the biosphere) are evils, or 'bads'. Pollutants are the reverse of natural capital and polluting is the reverse of conserving. The equivalence is used in Chapters 4–4* to construct a viewpoint that accommodates both the economics of biodiversity and the economics of climate change. The construction shows that each is a subdiscipline of the economics of the biosphere.

In Chapter 4, we report that enormous expansions in waste production accompanying economic growth since the middle of the last century have contributed to stretching the biosphere's capacity to supply regulating and maintenance services, including decomposing biodegradable material (e.g. food products) and maintaining the gaseous composition of the atmosphere. It is useful to remember that even the carbon we emit is biodegradable, for it is taken up by plants. Which is why it is a mistake to think that if somehow all waste could be produced in a biodegradable form (e.g. moving away from plastics to biodegradable containers), our demands on the biosphere as a sink for pollution could be made to shrink to zero. To be sure, the waste would be packaged in a different form, but that would merely shift the burden onto a different class of biospheric processes. And that in turn would give rise to a vastly different composition of life forms.[77]

Box 2.7
Watershed Services by Tropical Forests

Covering less than 2% of the planet's surface, tropical rainforests are the habitat for more than 50% of Earth's plants and animals (the largest diversity of forest animals are housed in the canopies). In contrast to temperate forests, which are often dominated by a half dozen species of trees, making up 90% of trees, a tropical rainforest may be home to 480 species of trees in a single hectare. It has been reckoned that a single bush in the Amazon rainforest may house more species of ants than the British Isles.[78]

Even though tropical rainforests are an enormous stock of carbon, continued deforestation and land degradation has transformed them into net annual emitters of carbon (Baccini et al. 2017). But ecologists and hydrologists have long stressed that in addition to being carbon reserves, forests upstream generate watershed services – they purify water and mitigate floods and drought downstream. The magnitude of these services depends on the type and quality of soils, vegetation cover, slope, rainfall and so on. A series of studies from south-east Asian tropical forests illustrate how watershed functioning can be mapped to human well-being.

In an early economic application, Pattanayak and Kramer (2001) combined watershed hydrology and farm level survey data in an econometric analysis to show that the dry season baseflow (a proxy, for droughts mitigated by upstream forests) reduces the time needed to farm coffee and rice in Flores, Indonesia. A companion study showed that this baseflow availability reduces the time needed by women and children to collect drinking water (Pattanayak, 2004).

Tan-Soo et al. (2016) examined data on land-use change over the period 1984 to 2000 for 31 river basins in Peninsular Malaysia. The conversion of forests into oil palm and rubber plantation significantly increased the number of flooded days during the wettest months of the year. Specifically, an additional square kilometre of conversion to oil palm or rubber was found to lead to 0.001 additional flood-related deaths and two additional evacuees per year.

---

[77] We are grateful to Paul Ehrlich and Peter Raven for correspondence on this.

[78] Wilson's (1992) classic opens with an unforgettable ode to the Amazon rainforest.

> In a parallel contribution from Malaysia, Vincent et al. (2016) have found robust evidence that protecting both virgin and logged forests against conversion to non-forest land uses reduces water treatment costs. Avoiding the conversion of 1% of a water treatment plant's catchment area from virgin forest to non-forest land uses was estimated to reduce treatment cost by about 1%. Subsequent research in neighbouring Thailand, which spans a larger range of climatic zones and forest types than Malaysia, detected similarly large and positive benefits of forest protection on water treatment costs (Vincent, Nabangchang and Shi, 2020).These benefits are real, but people place additional value to tropical forests. Vincent et al. (2014) used choice experiments to quantify the value Malaysian households place on reduced extinctions of species in the country's rainforest. After accounting for this value, they determined that protecting virgin forests generated benefits twice as great as the costs.

2.7 Ecosystem Productivity and Resilience

Ecosystems are subject to shocks and disturbances. That points to the need to include in an ecosystem's *productivity* – its yield in terms of the goods and services it produces – its stability against disturbances. We therefore enlarge the meaning of an ecosystem's productivity by its ability to withstand shocks and disturbances. A sharper notion is the speed with which the ecosystem recovers from a disturbance. A broader notion involves the magnitude of the disturbances it can absorb and recover from. The latter property is often called *resilience*, a notion we revisit in Chapter 3. An ecosystem's characteristic rhythm plays a role in the notion of resilience. Overexploited rocky intertidal shellfish may recover within a decade upon receiving protection, whereas old-growth temperate rainforests require at least the lifetime of multi-centennial trees to recover fully. So, by an ecosystem's *productivity* we shall now mean its productivity *and* resilience, considered as a joint attribute. In Section 2.9, where we study the spatial aspects of ecosystems, we will add a further attribute, that of *modularity*, which contributes to resilience.

Ecosystems differ as regards the services they supply. Even in this respect, a tropical wetland differs from a coastal fishery. The CICES classification has an anthropocentric basis. The term productivity is applied not only to regulating, maintenance and cultural services, but also to provisioning goods. The latter group consists of those services that go to supply food, fibre, timber, medicines and fresh water to the human population (services in category (1)). The global demand for provisioning goods, aided by the application of pesticides, insecticides and industrial fertilisers, has increased massively over the decades. Almost as an ecological corollary (we confirm this below and more extensively in Chapter 4) there has been a corresponding loss in the biosphere's overall productivity in terms of regulating, maintenance and cultural services (categories (2) and (3)). Our reliance on fossil fuels and the global climate change it has given rise to is a vivid illustration.

Then there are tensions among the provisioning goods themselves. Nitrogen is an essential element in living matter. Plants obtain their nitrogen not from the abundant store in the atmosphere, but from the soil, where bacteria convert nitrogen from the atmosphere into a form usable by plants.[79] Modern agriculture applies industrial fertilisers so as to enhance nitrogen supply in the soils. However, the nitrogen in fertilisers that are not required by agricultural crops sink into soils, often creating conditions that favour weeds rather than native plants. The excess nitrogen then washes away into waterways, ultimately into coastal zones, creating biological 'dead zones' of algae blooms (Chapter 4). By one estimate, in 2019 the seasonal dead zone in the Gulf of Mexico was about the size of the state of Massachusetts, and cost the fishing industry there at least half a billion US dollars a year (Union of Concerned Scientists, 2020), perhaps even as high as an annual US$2.4 billion.

---

[79] In natural conditions nitrogen is often the limiting factor in plant abundance and growth.

> **Box 2.8**
> **Non-Anthropocentric Perspectives**
>
> Both the MA classification and the CICES reworking of it are built on an anthropocentric perspective. The value ascribed to Nature is traced to the goods and services the biosphere supplies to we humans. There are many people who would reject a purely anthropocentric reasoning outright. They would say that Nature has value, period. Some would go further and say there are entities that are sacred. Among Hindus, the river Ganges is a goddess (she is called Ma Ganga), having risen from the head of Lord Shiva, his locks tempering her effusive energy at birth. The Ganges provides water to some 600 million people as she flows for more than 2,500 kilometres from Gomukh, an ice cave at the foot of the Gangotri glacier in the western Himalayas of Uttarakhand, India, to the Bay of Bengal. And yet, the Ganges is one of the most polluted rivers in the world, being used as the sink into which industrial, agricultural and household waste, urban sewage and cremation remains are discharged along her entire route. Material needs would seem to trump sacred duties in the Gangetic plain.[80]
>
> Many people say, however, that the pollution is illusory, that being celestial, Ma Ganga is incapable of being polluted. Accommodating conflicting imperatives is universal practice, and we frequently accomplish it by rationalising our beliefs. Which is why it will pay to build the economics of biodiversity on the back of an anthropocentric viewpoint. Doing so will, moreover, enable us to stay close to the way environmental and resource economics in general and the economics of climate change in particular has been fashioned by economists. In the chapters that follow, we will find it necessary to break away from the conceptual apparatus that has been deployed in environmental and resource economics. Adopting the anthropocentric perspective will serve as a good discipline when we do that. Later in the Review (Chapters 12, 13*), we will ask in which ways the framing of public policies would require additional considerations when non-anthropocentric perspectives are adopted.

2.8 Biodiversity and Ecosystem Productivity/Resilience: The Causal Connection

A large body of work involving field experiments (e.g. Tilman and Downing, 1994; Tilman, 1997; Tilman, Reich and Knops 2006), site studies (Box 2.8) and aerial surveys (Bowman, Hacker and Cain, 2018), complemented by mathematical modelling (e.g. Kinzig, Pacala and Tilman, 2002), has found that the number of functional groups in an ecosystem – that is, *functional diversity* – is strongly related to ecosystem productivity.[81] Functional diversity points to *complementarities* among traits. It resembles complementarities among inputs in industrial production (iron, coal and labour in steel production) and among consumption bundles (left and right shoes). If any one complementary factor is absent, the value of the enterprise is gone.[82]

---

[80] Markandya and Murty (2004) is a major study of the economic costs and benefits of repairing the Ganges waters. We report on their work in Chapter 13.

[81] Tilman et al. (1997) and Hooper and Vitousek (1997) are pioneering, back-to-back publications that reported the significance of functional diversity for an ecosystem's productivity.

[82] Hooper et al. (2005), Cardinale et al. (2012) and Tilman, Isbell and Cowles (2014) are major reviews of what is known about the relationship between biodiversity and ecosystem functioning. Our treatment of the link between biodiversity and ecosystem functioning is taken from them.

Chapter 2: Biodiversity and Ecosystem Services

Soil biodiversity is an example. Different groups of organisms act to maintain soil health in different ways. Archaea, bacteria and fungi act as chemical engineers, decomposing plant residues and soil organic matter, contributing to nutrient transitions and recovery of polluted soils. Other organisms act as biological regulators, controlling plant pathogens and contributing to food security. Larger organisms, such as earthworms, termites and small mammals act as ecosystem engineers, controlling the structure of the soil matrix. Without these diverse species playing different roles, the soils would fail to support the global food system.

Mutual dependence among the species is a reason diversity enhances ecosystem productivity. Many trees in tropical forests produce large, lipid rich fruits that are adapted for animal dispersal, which means the demise of fruit-eating birds can have adverse consequences for forest regeneration (Box 2.6). Looking elsewhere, about one-third of the human diet in tropical countries is derived from insect-pollinated plants. But that means a decline in forest-dwelling insects has an adverse effect on human nutrition. In their study in Costa Rica on pollination services, (Ricketts et al. 2004) discovered that forest-based pollinators increase the annual yield in coffee plantations within 1 km from the forest edge by as much as 20%. Ricketts et al. (2008) have analysed the results of some two dozen studies, involving 16 crops in five continents, and discovered that the density of pollinators and the rate at which a site is visited by them declines at rapid exponential rates with the site's distance from the pollinators' habitat. At 0.6 km from the pollinators' habitat, for example, the visitation rate drops to 50% of its maximum.[83]

In food webs, the relationships between populations affecting the state of an ecosystem are unidirectional. Primary producers in the oceans (phytoplankton, sea weeds) are at the bottom of the food chain, while species at higher trophic levels consume those that are below. Species whose impact on a community structure is large relative to their size and abundance are called *keystone* (Power et al. 1996). They are usually at the top end of the food chain. Examples include sea otters in the north-east Pacific, which, by preying on sea urchins, enable kelp forests to thrive. As kelp provides important habitat for many other species (kelp forests are among the most productive ecosystems in the world), an explosion of sea urchins can create kelp barrens. If the population of sea otters declines sizably, the coastal ecosystem flips to a state in which biomass production is greatly reduced (Chapter 3). Such a crash occurred in the 1900s when sea otters had been hunted to near extinction.[84] The Annex to this chapter contains a formal model that displays the importance of top predators in maintaining an ecosystem's architecture.

The reverse side of this are species below the top in food webs. If their population drops, populations dependent on them collapse, and the ecosystem cascades into a different state. Dramatic falls in populations of aquatic insects will have such an effect on their amphibian predators. The model presented in the Annex to this chapter speaks to both top-down and bottom-up influences on ecosystem architecture.

The introduction of non-native species changes the composition of ecosystems in a different way. The 'Nile' perch, a voracious predatory species of fish, is a classic example. Its introduction into Lake Victoria in the 1950s caused a local economic boom, but it decimated native species; by most estimates more than 200 species of fish went extinct in the lake. Selective fishing also serves as a driver of the future of a fishery. In their study of the optimum management of non- linear systems, Ludwig, Carpenter and Brock (2003) noted that adult trout in the Great Lakes of North America are able to survive attack by sea

---

[83] Pollination by bees and other animals increases the size, quality or stability of harvest for 70% of the world's leading crops (Klein et al. 2007).

[84] Gregr et al. (2020) estimate that reintroducing sea otters in the north-east Pacific will have amounted to an increase by 40% of annual biomass production and a sizeable increase in profits from the fisheries. An early study of the long-term dangers of habitat destruction to keystone species and the consequent extinction of species lower in the food chain was (Tilman et al. 1994).

lamprey, but that smaller trout find it hard to survive. Fishing, by selectively removing large trout left the trout populations vulnerable to invasion of the Great Lakes by sea lamprey and contributed to the extirpation of lake trout from all lakes but Lake Superior. Had fishing targeted smaller individuals and left the large ones alone, the populations may well have been resilient to sea lamprey invasion.

Nested measures such as the number of species within each functional group have proved to be useful. Combinations of native species that have co-evolutionary history of interactions have been found to display greater complementarities than combinations of exotic species with briefer histories of interactions. Box 2.9 points to the contribution diversity *within* functional groups makes to the resilience of ecosystems. Elmqvist et al. (2003) speak of the diversity of responses to environmental change among species contributing to the same ecological function, *response diversity*. Response diversity is an important feature of the resilience of ecosystems and resembles diversity of companies in the same economic sector represented in a financial portfolio. Returns from them are unlikely to be perfectly correlated. Just as diversifying a portfolio is a way to reduce *risk* in the portfolio's yield, response diversity increases ecosystem resilience.[85]

Diversity of functional traits that enables Nature to thrive appears also in behavioural ecology. Ducks, geese and swans are able to cohabit, not by sharing resources but by exploiting their adaptive physical advantages – for example in the shape and size of their bills – for grazing. Some skim the water surface and dive in the shallows, while others dive deep into the grasses, while still others graze in the banks. Each species in the community occupies a distinct niche. That has enabled Nature to find a balance among species. One way to read the cause of humanity's overreach of the biosphere (Chapter 4) is to recognise that we have invented ways to occupy nearly all niches and carry material from one to another. In an early study, Daily and Ehrlich (1996) identified the adverse consequences those inventions to human health with immune suppression, loss of biodiversity and indigenous knowledge, and the evolution of antibiotic resistance. In some cases, the adverse consequences have been unintentional in that we did not know the damage that would follow (carbon emissions in 19th century Europe and the US). Today, the adverse consequences are better known, but all too frequently unappreciated.

Box 2.9
Mix of Species in Grazing Lands

To appreciate the role of what we will call 'functional diversity' of species in maintaining an ecosystem's productivity, consider that sites on a rangeland that have been lightly grazed by cattle would be expected to have greater biodiversity than sites that have been heavily grazed, and so would function differently. In a study in the Australian rangeland, Walker, Kinzig and Langridge (1999) found that just a few species were dominant, making the bulk of the biomass, but there was a long tail of different species of low abundance. The dominant species were functionally dissimilar (they complemented one another in their traits – long versus short root systems, differences in mean height, leaf size and shape, and so on), while species that were functionally similar to them were in low abundance (they would be competitors and therefore potential substitutes). The authors reported that the heavily grazed site in contrast was less resilient to changes in conditions.

The minor species at the lightly grazed site could be regarded as 'waiting in the wings' to take over if required to do so. When conditions changed, some of these less abundant species were better

---

[85] Relatedly, biodiversity has been found to be a determinant of the efficiency with which an ecosystem uses limiting resources (e.g. nitrogen and phosphorus). Experiments on grasslands have found that biodiversity improves the ability of ecosystems to resist invasive species, the improvement being due to lower soil nitrate, greater abundance of neighbouring plants and lower abundance of light (Tilman, Reich and Knops, 2006; Tilman, Isbell, and Cowles, 2014).

able to respond to the change because of their variable traits. Without this variability across a wide range of species, ecosystems have fewer options, and would respond less well to unpredictable changes in the environment. The presence of the minor species offers the ecosystem what Elmqvist et al. (2003) have dubbed 'response diversity', which is defined as the range of reactions to environmental change among species contributing to the same ecosystem function. Rather like a team with substitutes on the bench, you are better equipped to make a good substitution if you have a range of substitutes with a range of skills to choose from. The study offers a vivid picture of the mix of traits to be found in ecosystems and of how we could interpret their presence. Biodiversity in general, and response diversity in particular, offer an ecosystem spare capacity, to be called upon when required.

These varied examples warn us that it is easy to misunderstand the claim that biodiversity enhances an ecosystem's productivity. Modern agriculture vigorously eliminates species diversity both above ground and in the soil beneath. Farm yield is kept from falling by application of the 'plough' and industrial fertilisers and pesticides. Monocultures among forest stands such as palm oil plantations are also directed at reducing species diversity. When, however, ecologists speak of productivity, they have in mind the ability of ecosystems to provide regulating, maintenance and cultural services (CICES categories (2) and (3)). They see regulating and maintenance services as being the often-hidden services that enable the items in provisioning goods (CICES category (1)) to be delivered on a sustainable basis. Over the past 70 years (since we entered the Anthropocene; Chapters 3–4), the huge increase in the demand for provisioning goods has meant that the ability of the biosphere to supply regulating, maintenance and cultural services has been increasingly compromised. On that reading, productivity and resilience increase with biodiversity.[86] And that brings us back full circle to a point that we made previously, that the feedback between the growth humanity has enjoyed in provisioning goods and the declines we are witnessing in regulating, maintenance and cultural services, lies at the heart of the difference in perception we read in magazines and newspapers, between those who see the world as a habitat that is getting better and better for humans, and those who fear for the planet's future – and therefore humanity's future.

Box 2.10
Plantations as Ecosystems

Cocoa cultivation provides a useful example of a cultivated ecosystem. Cocoa crop is sensitive to variations in temperature and moisture. Studies in Ghana have shown that longer dry seasons and lower rainfall that have come in tandem with other forms of global climate change have contributed to a drop in output in recent years (dropping, for example, from 970,000 tonnes in 2016–2017 to only a bit over 800,000 tonnes in 2018–2019). Conversion of forests into agricultural land has meant less shade for cocoa trees, and that has contributed to greater soil erosion and lower moisture retention. Cocoa plantations are now increasingly vulnerable to pests and diseases. As soil productivity has declined, fertilisers are now applied, further damaging soil biodiversity. Deforestation has altered microclimates. Farmers are in the main small shareholders (less than 2 hectares) and tenure is dependent on maintaining crop production. Replacing old trees is risky for the shareholder because it takes up to 10 years for trees to come to full maturity. Cocoa farming in Ghana illustrates the tight couplings between factors that drive socio-ecological processes. It also

---

[86] Tilman et al. (2002) is an illuminating study the positive feedback between modern agricultural practices and biodiversity loss, leading in turn to weakening the basis of crop production.

illustrates the source of concerns multinational companies today express over their supply chains. The factors that are contributing to the unreliability in cocoa supply is beyond the individual cocoa farmer's control. Chapter 17 discusses ways in which the uncertainty in supplies can be reduced. It would involve engagement not only with cocoa farmers as a group, but also with the wider community and government.

For the chocolate manufacturer in the UK, problems are compounded by the fact that in addition to the cocoa, it has to import soybeans (for manufacturing another ingredient, lecithin), very likely from the Amazon, and vanilla, very likely from Indonesia. Institutions governing each of these two primary products include farmers, government and local community (Chapter 5). Moreover, global climate change has made supply from those sources riskier. Taken together, production risks for the manufacturer have risen over the years.

2.9 Modularity as a Spatial Feature of Ecosystems

Biological communities can influence their abiotic environment. Trees of tropical forests take up large volumes of water from the ground and release it through transpiration. This governs the amount of water in the local atmosphere, maintaining the high levels of rainfall there. Rainfall returns water to the ground for trees to take up and transpire. Mathematicians call this a positive feedback.

Following large-scale deforestation, water's link between the soil and the atmosphere is broken and a tropical rainforest can be expected to transform itself in time into savannah. The Amazon, for example, generates about half of its own rainfall by recycling moisture five to six times as air-masses move from the Atlantic across the basin to the west. Unfortunately, nearly 20% of the Brazilian Amazon has been cleared for plantations, cattle grazing and mining (Lovejoy and Nobre, 2019). Deforestation of the Amazon would be expected to reduce rainfall and lead to a lengthier dry season in the region. One estimate has it that 20 to 25% further deforestation can be expected to flip the forests into savannah vegetation (Lovejoy and Nobre, 2018).

Tower, ground-based and satellite observations have found that deforestation of tropical rainforests result in warmer, drier weather at the local level. However, large-scale deforestation can be expected to shift weather patterns and thus agriculture, both locally and far from the areas where deforestation occurs. The spatial linkages are striking. Extreme deforestation in the Amazon would be expected to reduce rainfall at key growing seasons in the Midwest, north-east and south of the USA. As another example, the West African Rainforest (WARF) region accounts for more than 30% of global tropical rainforest cover.[87] Air masses coming across the WARF region account for nearly 50% of the moisture that is released in the highlands of Ethiopia. About 85% of the surface water reaching Aswan in Egypt originates in the Ethiopian Highlands. In the period 1990 to 2010, the WARF was reduced by 15%, owing to agricultural expansion, urbanisation, mining and logging (Gebrehiwot et al. 2019). That has reduced rainfall in the Highlands. Complete deforestation in Central Africa would be expected to trigger shifts in rainfall in the American wheat-belt and southern Europe. Interestingly though, it could boost rainfall in China, western Asia and the Arab peninsula (Lawrence and Vandecar, 2015).

Communities and ecosystems are thus not spatially uniform. Instead, they exhibit heterogeneity and patchiness on multiple scales.[88] In forests, local disturbances such as treefalls, and even burns, create

---

[87] The WARF region comprises about 13% of the area of Africa. It stretches from the Democratic Republic of Congo, Gabon, Congo, Cameroon, Equatorial Guinea and the Central African Republic.

[88] This section has been taken from Levin (2020).

Chapter 2: Biodiversity and Ecosystem Services

opportunities for appropriately named *opportunistic* species to colonise and thrive for limited times, only to be replaced by better competitors in a broadly predictable *successional* sequence, often called an *r-K* spectrum (the analogy is with *r-K* model in Box 2.2), where species trade off colonisation ability (*r*) with competitive advantage (*K*) until the so-called *climax* species take over. Under conditions that favour high rates of disturbance, one finds only the early successional species. On the other hand, if the disturbance is suppressed, only the climax species persist. This has led to the conjecture that diversity is maximised at intermediate levels of disturbance, a generalisation that has broad support across a range of systems. Humans, unfortunately, are increasing the level of disturbance in a wide variety of systems, driving those systems back towards the low-diversity early-successional phases (Chapter 4).

Healthy, diverse ecological communities are thus patchworks; that is, mosaics of assemblages in various stages of succession. The modular nature of these systems, and of ecosystems more generally, provides a further feature of ecosystems: *modularity*.[89] Modularity is the degree to which densely connected components of an ecosystem can be decoupled into separate communities, or clusters of communities, that interact more among themselves than with other communities in the ecosystem. In the absence of modularity, an ecosystem would be a house of cards – it would collapse in time if even a small component was compromised. Modularity thus reduces the likelihood that the ecosystem will break down irretrievably because of disturbances. Systems scientists call this feature *robustness*. Robustness is an aspect of the resilience of ecosystems; it is the spatial equivalent of the temporal features reported in experiments on the grasslands by David Tilman and his colleagues (Section 2.8).

In Box 2.9, we observed that resilient ecosystems harbour redundancy, where species 'sit' in the wings, like spare capacity in economic production systems, and attain prominence when dominant species have weakened due to disturbances. Modularity traces that feature of ecosystems to their spatial architecture. One often hears that a 'balance of Nature' stabilises ecosystems, but that is to some extent an illusion. Stasis in any system implies reduced variation, and reduced capacity for adaptation, and a consequent loss of resilience. For healthy ecological communities, the relative resilience at broad scales is due to its absence at local scales.

2.10 Biodiversity and Ecosystem Productivity: Summary

There are now literally hundreds of studies based on field experiments in which species were added to plots, or sites studies and aerial surveys that were undertaken to record species diversity and the functional traits in ecosystems. A consensus has emerged on a number of influences that biodiversity has on ecosystem functioning. Here we follow the summary in Cardinale et al. (2012: pp 60–61):

(1) Functional traits of organisms have large impacts on ecosystem processes. That gives rise to a wide range of consequences for an ecosystem's ability to function when it suffers from biodiversity loss.

(2) Biodiversity loss reduces the efficiency with which ecological communities capture biologically essential resources (nutrients, water, sunlight, prey) and produce biomass. As an example, diversity of plant litter enhances decomposition and recycling of biologically essential nutrients. These influences appear to be consistent across different groups of organisms, among trophic levels and across ecosystems. The consistency points to an influence of the organisation of organism communities on ecosystem functioning.

(3) Biodiversity increases the stability of ecosystem functioning over time against stresses the system is subjected to by fluctuations in the physical environment. Box 2.4 illustrated that.

---

[89] Levin and Paine (1974), Levin (1992, 1999) and Levin and Lubchenco (2008) are key publications on this.

(4) The influence of biodiversity on an ecosystem's processes is non-linear and saturating, such that the ecosystem's productivity declines at an accelerated rate as biodiversity loss increases. That says the own rate of return (Chapter 1) on an ecosystem's biomass increases as biodiversity declines.

The above finding should be interpreted with care. The accelerated rate of productivity decline in response to greater biodiversity loss may amount to regime shifts (Chapter 3). During such shifts, estimates of own rates of return are inevitably error strewn – the productivity decline is overly rapid.

(5) Diverse communities are more productive because they contain key species that have a large influence on productivity. They are more productive also because differences in functional traits among organisms (economists would call that 'complementarities' among traits) increase total resource capture.

(6) Loss of diversity across trophic levels has the potential to influence ecosystem functions even more strongly than diversity loss within trophic levels. Interactions in food webs mediate ecosystem functioning. Studies have found the loss of a few top predators to reduce plant biomass by at least as much as that due to transformation of diverse plant assemblages into species monocultures.

Cardinale et al. (2012: pp 61–62) list four further effects of biodiversity that are becoming increasingly evident from empirical studies:

(1*) The impacts of biodiversity loss on ecological processes may well be sufficiently large to rival the impacts of the other major global drivers of environmental change, such as fires, nitrogen overload and rising carbon concentration in the atmosphere.

(2*) The influence of biodiversity, and hence biodiversity loss, grow stronger with time and may be larger at larger spatial scales. Small scale studies may therefore underestimate the significance of biodiversity on ecosystem functioning.

(3*) Maintaining multiple ecosystem processes at multiple places and times requires greater biodiversity than does a single process at a single place and time.

(4*) Ecological consequences of biodiversity loss can be predicted from evolutionary history. This discovery offers an understanding of the biosphere's modularity and the tight coupling among its various components.

To these we may add the interesting finding reported in Renard and Tilman (2019) that even crop diversity can yield higher food production.

We may thus say that biodiversity is an enabling characteristic of ecosystems. It is vital for the health of ecosystems. Biodiversity corresponds to what we call *enabling assets* in human societies (Chapter 13). A possible analogy is mutual trust (Chapter 9), which lubricates human societies and enables them to thrive. Similarly, biodiversity lubricates ecosystems and enables them to be productive. Neither trust nor biodiversity is usefully measurable – but that doesn't matter – for they enable human societies and ecosystems, respectively, to function healthily. And the latter, as the CICES classification of ecosystem services tells us, can be measured.

Annex 2.1 Community Structure

We study ecosystem architecture by considering a three-level trophic web: plants, herbivores and carnivores. The model brings out the influence of populations at higher trophic levels on a system's NPP. In the community studied here, the presence of predators (carnivores) increases plant biomass relative to the same community lacking predators. The model, taken from Borer and Gruner (2009), simplifies by ignoring decomposers and detritivores.

Chapter 2: Biodiversity and Ecosystem Services

Time, assumed continuous, is denoted by t. Let $S(t)$ be plant biomass, $X(t)$ the biomass of a population of herbivores, and $Z(t)$ the biomass of carnivores. There are 5 constants ($r, K, \alpha, \beta, \gamma > 0$) in the system of differential equations governing the trophic web:

$$dS(t)/dt = rS(t)[1 - S(t)/K] - \alpha S(t)X(t) \tag{A2.1}$$

$$dX(t)/dt = \alpha\gamma S(t)X(t) - \alpha X(t)Z(t) - \beta X(t) \tag{A2.2}$$

$$dZ(t)/dt = \alpha\gamma X(t)Z(t) - \beta Z(t) \tag{A2.3}$$

The model is deliberately simple, in that it contains only five parameters. Equation (A2.1) says that plant biomass increases by logistic growth (at an intrinsic rate, r) with carrying capacity K (Box 2.2) but declines with the biomass of herbivores. Equation (A2.2) says herbivore biomass increases with plant biomass but declines with carnivore biomass, while equation (A2.3) says carnivore biomass increases with herbivore biomass. Both herbivores and carnivores experience the same intrinsic death rate β, the same attack rate α and the same conversion efficiency γ.

The ecosystem's total biomass at t is $S(t) + X(t) + Z(t)$. NPP is $rS(t)[1 - S(t)/K] - \alpha S(t)X(t)$. The own rate of return on the ecosystem (i.e. the ecosystem's yield) is thus $\{rS(t)[1 - S(t)/K] - \alpha S(t)X(t)\} / [S(t) + X(t) + Z(t)]$.

We are interested in stationary states, in which equations (A2.1)–(A2.3) reduce to

$$rS(1 - S/K) - \alpha SX = 0 \tag{A2.4}$$

$$\alpha\gamma SX - \alpha XZ - \beta X = 0 \tag{A2.5}$$

$$\alpha\gamma XZ - \beta Z = 0 \tag{A2.6}$$

Figure A2.1.1, taken from Borer and Gruner (2009: p.299), tracks the stationary biomass of plants, herbivores and carnivores by varying the ecosystem's carrying capacity K. The figure is drawn under the assumption that $r = 10$, $\alpha = 0.5$, $\gamma = 0.5$, and $\beta = 1$. Notice that for K in the range (0, 4), the ecosystem does not support herbivores, which means it does not support carnivores. Thus $X = Z = 0$ and $S = K$.

For K in the range (4, 5) the ecosystem supports herbivores, but is not expansive enough to support carnivores. Plant biomass is constant in the range, because higher values of K enable herbivores to thrive at an increasing rate. Beyond $K = 5$, carnivores (Z) make their appearance and increase with K (birth and death rates increase but remain in balance); while the biomass of herbivores remains constant.

A common illustration of such a model as this is a lake system where phytoplankton are the dominant primary producers. Phytoplankton are typically grazed by zooplankton, which are in turn consumed by small planktivorous fish (e.g. minnows). Whole-lake manipulations of piscivorous fish (e.g. bass) have shown that three-level chains lacking piscivorous fish result in green lakes, because planktivorous fish (third trophic level) limit zooplankton (second trophic level) and release phytoplankton (primary producers). Phytoplankton blooms give a lake its green colour. In contrast, lakes are blue where piscivorous fish are added as a fourth trophic level. A large number of empirical studies have uncovered an entire class of restoration techniques in which the removal or addition of trophic levels ('biomanipulation') achieves desired states of primary productivity and water clarity (Levin, 2009).

In Chapter 2, it was noted that coastal keystone species, such as sea otters in the north Pacific, prey on sea urchins and enable kelp forests to thrive. As kelp provides a habitat for many other species, an explosion of sea urchins can create urchin barrens. If the population of sea otters declines sizably, the coastal ecosystem flips to a state in which biomass production is greatly reduced. Such a crash occurred in the 1900s when sea otters had been hunted to near extinction. Non-linearities in ecosystem processes are discussed in Chapter 3.

Figure A2.1.1 Stationary Biomass of Plants, Herbivores and Carnivores by Varying Ecosystem Carrying Capacity

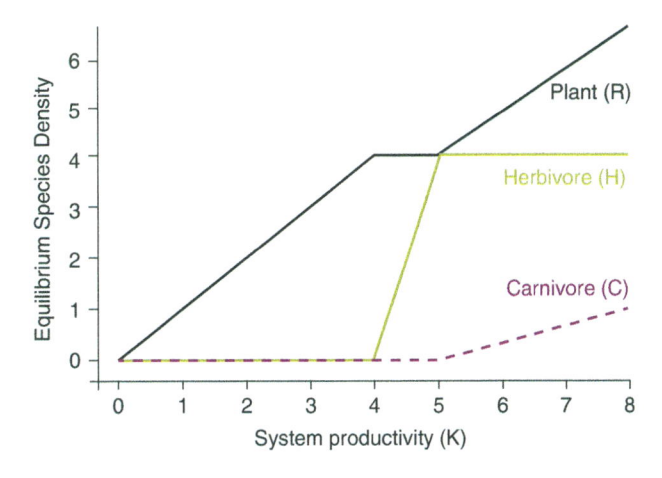

Source: Borer and Gruner (2009: 299).

Annex 2.2 Measuring Biodiversity

In Chapter 2 we noted that biodiversity is not the diversity of a single attribute. In the same way that we are unable to use one measure to describe our personal health, we can't use one measure to describe the state of global biodiversity. So, what are the most important attributes of biodiversity and what should we be measuring if we wish to understand how biodiversity is changing around the world?

A2.2.1 Biodiversity Indicators

There are many proposed indicators and variables in the literature for measuring biodiversity. Nearly 100 different indicators were suggested for the Aichi meeting of the Convention on Biological Diversity (CBD) (Pereira et al. 2013). Recognising this complexity, Pereira et al. (2013), under the auspices of the Group on Earth Observations Biodiversity Observation Network (GEO BON), developed a framework of Essential Biodiversity Variables (EBVs) that could form the basis of monitoring programmes worldwide. EBVs are designed to help prioritise indicators by seeking to define a minimum set of essential measurements that capture major dimensions of biodiversity change. The potential indicators fit under six broad classes that describe the breadth of biodiversity:

(1) Ecosystem structure is typically measured using habitat or biome extent and includes consideration of structural complexity and fragmentation.

(2) Ecosystem function relates to the functions and processes within the ecosystem and the interactions between biotic and abiotic factors that enable a flow of energy through the system. Within the formal EBV framework net primary productivity, secondary productivity and nutrient retention are identified as key variables under this class. Two variables are of particular interest when viewing biodiversity as a characteristic of natural assets: (i) NPP and (ii) carbon sequestration, i.e. the difference between CO_2 uptake by plants through photosynthesis and release by respiration, decomposition, river export and anthropogenic actions.

(3) Community composition describes the diversity of organisms making up the ecosystem. These vary depending on a wide range of variables, such as temperature, soil type, water availability and

Chapter 2: Biodiversity and Ecosystem Services

evolutionary time. Across the world, community composition is highly variable with changes across altitudinal and latitudinal gradients. For example, the highest number of terrestrial species per unit area occurs at the equator and the lowest at the Poles.[90] Species richness measures are relevant in this category, which are essentially the number of different species represented in an ecological community, landscape or region.

(4) Species populations seek to measure the abundance and distribution of species.

(5) Species traits relate to the structural, chemical and physiological characteristics of organisms, for example, body size, plant height and clutch size. The variation of traits within and among species will determine how the ecosystem is able to respond to perturbation and environmental change.

(6) Genetic composition relates to the genetic diversity which underpins variation within and between species. A diverse gene pool is essential to provide resilience to environmental change and to pressures such as disease or climate change.

While some of these attributes may seem removed from ecosystem service provision, they underpin the essential functions and processes that enable the delivery of benefits for people. For example, species traits relate to the provision of ecosystem services as they are associated with ecosystem functions (Lavorel et al. 2013). Ecosystem structure and function closely relate to biomass and productivity, which are core attributes of many ecosystem services.

Even though EBVs are designed to be the minimum set of essential variables, we can see that biodiversity measurement remains complicated and requires multiple measures. In a bid to come up with simpler measures, Mace et al. (2018) suggested three key indicators to understand change in biodiversity, which are useful when considering trajectories to international biodiversity targets and would adequately capture key dimensions of biodiversity (essentially extinction risk, abundance and composition):

(1) Conservation status: Estimating near-future global losses of species (extinctions) which can be measured using the IUCN Red List Index (RLI).

(2) Population trends: Trends in the abundance of wild species which can be measured using the Living Planet Index (LPI).

(3) Biotic integrity (community composition): This can be measured using the Biodiversity Intactness Index (BII), which measures the fraction of naturally present terrestrial biodiversity that still remains.

The IUCN Red List Index

The RLI is developed from expert assessments of species using information on life-history traits, population and distribution size and structure, and their change over time (IUCN, 2020). Red List assessors classify species into one of eight categories (Extinct, Extinct in the Wild, Critically Endangered, Endangered, Vulnerable, Near Threatened, Least Concern or Data Deficient). Over 100,000 species have now been assessed. The Red List Index uses this data to shows trends in survival probability over time (Butchart et al. 2007). It is only available for five taxonomic groups that have had repeat assessments: birds, mammals, amphibians, corals and cycads although species assessments are being added all the time (Annex 2.3).

---

[90] There are several species groups that deviate from this trend such as birds (Willig and Presley, 2018).

Chapter 2: Biodiversity and Ecosystem Services

The Living Planet Index

The LPI is a measure of the state of the world's biological diversity based on population trends of vertebrate species from terrestrial, freshwater and marine habitats. The LPI is based on trends of over 4,000 species and thousands of population time series collected from monitored sites around the world. Species groups are chosen based on data availability. Many groups of species do not have data collected systematically and regularly enough to measure change. The LPI has been adopted by the CBD as an indicator of progress towards its 2011 to 2020 target to "take effective and urgent action to halt the loss of biodiversity". The species population data that is collected goes into a global index, as well as indices for more specific biogeographic areas, referred to as realms, based upon distinct groupings of species. The extent to which this index can be disaggregated depends on the quantity and resolution of the data. For example, there is no LPI for the UK.

The Biodiversity Intactness Index

The BII represents the fraction of naturally present biodiversity that still remains in terrestrial ecological communities in a region (Scholes and Biggs, 2005). The BII aims to use a taxonomically and geographically representative data and consequently includes considerable amounts of insect and plant data, unlike the LPI and the RLI. This breadth of data makes it a useful indicator of ecosystem resilience and it has been used as one of the indicators in the Planetary Boundaries framework (see Chapter 4 for more on this; also Steffen et al. (2015)). It is also a 'core' indicator for IPBES and adopted by the CBD. The BII considers the impacts of land-use, land-use intensity and other pressures, such as human population density and proximity to roads in a modelling framework called PREDICTS (Purvis et al. 2018). It does not include impacts of climate change or delayed effects of land-use change and does not distinguish between plantations and natural forest. The Index ranges from 100–0% with 100 representing an undisturbed or pristine natural environment with little to no human footprint and 90% suggested as the lower safe limit.

A further metric that describes biotic integrity is mean species abundance (MSA), which calculates the mean abundance of species in disturbed habitat relative to their abundance in undisturbed habitat. The GLOBIO model uses the MSA to assess terrestrial biodiversity integrity in relation to anthropogenic impacts including climate change, land use, nitrogen deposition, infrastructure, human encroachment and habitat fragmentation (Schipper et al. 2016). The inclusion of some different anthropogenic drivers in the GLOBIO model makes it a useful addition to the BII results.

A2.2.2 Does Measuring Biomass and Net Primary Productivity Help?

Biomass can be one of the variables to help to describe ecosystem structure and, as discussed in Chapter 2, is very relevant to ecosystem service delivery. Variations in biomass will have implications for the types and quantities of ecosystem services available both in terms of people locally but also in relation to wider services such as climate regulation and carbon sequestration. Different ecosystems will be associated with different levels of biomass and associated NPP, depending on whether they are grasslands or woodlands, early succession or mature, freshwater or oceans, degraded or intact.

Closely related to biomass and NPP is the measurement of carbon sequestration in a system. The amount of carbon sequestered is the difference between the CO_2 uptake by photosynthesis minus the release from respiration, decomposition, river exports and human activities such as burning or harvesting. This measure of ecosystem function is particularly important in relation to climate change.

Farmed systems have been designed to try to optimise biomass, as crops for example, and have high levels of NPP. Measures of biomass and NPP have to be put in context if we wish to infer any measures of quantity or quality of biodiversity. High levels of biomass will tell you that you have high levels of one measure of stock but will not describe its variability, quality, composition or how well it is functioning.

Chapter 2: Biodiversity and Ecosystem Services

Similarly, high levels of NPP will say something about how the stock is functioning, but not about variability within the system, how resilient the system is or how much standing stock there is.

The miscellaneous attributes of biodiversity describing complex systems at varying scales means that one indicator will not do. But research suggests that a handful of well-chosen purposeful indicators can tell us quite a lot. The trick is to pin down what it is about the systems that we want to understand, at what resolution and at what intervals, and then choose indicators and variables appropriately. Continuing development of remote sensing, big data and modelling are all contributing to a rapidly developing field of biological data collection that we need to measure change meaningfully.

Box A2.2.1

Global Biomass Census

Bar-On, Phillips and Milo (2018) undertook a census of global biomass and found that there are ≈550 gigatons of carbon (Gt C) of biomass distributed among all of the kingdoms of life in the biosphere. Plants are the dominant kingdom (≈450 Gt C) and are primarily terrestrial, whereas animals (≈2 Gt C) are mainly marine, and bacteria (≈70 Gt C) and archaea (≈7 Gt C) are predominantly located in deep subsurface environments. Terrestrial biomass is about two orders of magnitude higher than marine biomass at approximately 6 Gt C of marine biota, although their NPP is roughly the same.

Human biomass (≈0.06 Gt C) and the biomass of livestock (≈0.1 Gt C, dominated by cattle and pigs) is far greater that the biomass of wild mammals today (≈0.007 Gt C). The same is true for wild and domesticated birds. The biomass of domesticated poultry (≈0.005 Gt C, dominated by chickens) is about threefold higher than that of wild birds (≈0.002 Gt C) (Bar-On, Phillips and Milo, 2018). The impact that humans have had on the biosphere has been estimated based on levels of biomass prior to our arrival. Human activity contributed to the Quaternary Megafauna Extinction between approximately 50,000 and 3,000 years ago, which claimed around half of the large (>40 kg) land mammal species (Barnosky, 2008). The biomass of wild land mammals is estimated to be ≈0.02 Gt C before this period whereas the present day biomass of wild land mammals is approximately sevenfold lower, at ≈0.003 Gt C. Exploitation of marine mammals and whaling have resulted in an approximately fivefold decrease in marine mammal global biomass (from ≈0.02 Gt C to ≈0.004 Gt C). While the total biomass of wild mammals (on land and sea) decreased by a factor of ≈6, the total mass of mammals increased approximately fourfold from ≈0.04 Gt C to ≈0.17 Gt C due to the vast increase of the biomass of humanity and its associated livestock. Plants, however, have declined approximately twofold relative to their value before the start of human civilisation (Erb et al. 2018).

Annex 2.3 Quantity and Quality of our Stock and How it has Changed

Our understanding of the quantity (extent), quality (condition) and distribution of ecosystems and associated biodiversity is still incomplete, and many gaps in the data exist (for more on distribution of the stock of Nature, see Chapter 14). This is compounded in the ocean, where the quantity and quality of marine ecosystems is particularly poorly understood. This is despite open ocean and the deep sea covering a huge area of 357.79 million square kilometres (mSqKm), dwarfing the extent of terrestrial and freshwater ecosystems combined (152.93 mSqkm) (Ichii et al. 2019).

The Intergovernmental Science-Policy Platform on Biodiversity and Ecosystem Services (IPBES) summarised our knowledge about the extent of the world's ecosystems using a unit of analysis which broadly represents biomes (Ichii et al. 2019). Table A2.3.1 presents a summary of the extent of these units of analysis on land (alongside information on net primary productivity (NPP), species richness and percentages of urban and cultivated lands within these units) (Ichii et al. 2019). It demonstrates the variability of NPP and species richness across the different assets, and shows how high NPP (productivity) is not correlated with high species richness (diversity), but is particular to the ecosystems and geography associated with those units. Large proportions are cultivated and influenced by humans, a story that is repeated in the oceans where only 13% are considered free from human influence (Jones, K. et al. 2018) compared to 23% on land (Watson et al. 2016). In fact, 85% of remaining wilderness areas – a vast proportion – is located in cold and dry biomes, rather than in more species-rich areas where conservation is particularly important (Ellis et al. 2010). More broadly, most global indicators show a net decline in the extent and condition of natural ecosystems since 1970 (Díaz et al. 2019). For example, in the marine environment, seagrass, mangroves and live coral cover are declining rapidly, (Waycott et al. 2009; Hamilton and Casey, 2016; Ortiz et al. 2018).

Table A2.3.1 Overview of Some of the Features of the IPBES Units of Analysis

Unit of analysis	Area (mSqkm)	NPP (gC/m2/year x 10^6)	Average relative species richness	Urban area %	Cultivated area %
Tropical, subtropical, and dry and humid forests	23.49	64	0.51	0.6	29.1
Temperate and boreal forests and woodlands	32.04	69	0.17	1.5	24.2
Mediterranean forests, woodlands and scrub	3.22	5	0.2	1.8	48.9
Arctic and mountain tundra	13.55	12	0.09	0.1	5.1
Tropical and subtropical grasslands	20.18	26	0.35	0.2	21.9
Temperate grasslands	11.19	14	0.2	0.9	56.0
Deserts and xeric shrublands	27.89	8	0.14	0.2	7.8
Cryosphere	17.71			0	0

Source: Ichii et al. (2019).

Biodiversity is an enabling characteristic of ecosystems, and species data is a useful proxy for ecosystem quality. Unfortunately, like our knowledge of ecosystems, our understanding of species is also far from complete. For example, we do not have a definitive list of species that exist on Earth because efforts to quantify and record species have been limited. We rely on estimates that are derived from patterns discerned from known taxa. Mora et al. (2011) noted that higher taxonomic classifications of species follow a predictable and consistent pattern. They used this pattern to predict species numbers in taxonomic groups. They estimated that there are approximately 8.7 million eukaryotic species (animals, plants and fungi, excluding bacteria and similar organisms) of which 2.2 million live in marine environments and 8.1 million are animals and plants (Mora et al. 2011). It has been estimated that 75% of eukaryotic species are insects (Chapman, 2009; Ichii et al. 2019). There are probably 8–20 million species of eukaryotes (Chapter 2), maybe more, and an additional unknown and much larger number of prokaryotes (Larsen et al. 2017) – our lack of knowledge is enormous. These predictions contrast with the numbers of known, recorded species, of which only about 2 million have been recognised and named so far (Chapter 2). For example, the total number of known vascular plants is estimated to be between 340,000 to 390,000, with around 2,000 plants new to science described each year (Lughadha

Chapter 2: Biodiversity and Ecosystem Services

et al. 2016; IPNI, 2020; WCVP, 2020; WWF, 2020a). One million insect species have been named (Stork, 2018) compared to just under 6,500 mammal species recorded (Burgin et al. 2018).

Broadly speaking, measures of species abundance, species richness and community composition are all currently declining, but the rates of these declines vary geographically and across different species, with some species increasing. A general pattern is that rarer species and habitat specialist species are declining, whereas some generalist species are stable or increasing (Marvier, Kareiva and Neubert, 2004). For example, invasive non-native species are on the increase globally (Seebens et al. 2017). Larger species seem to be particularly vulnerable to extinction with direct harvesting for consumption as the principal driver of declines (Ripple et al. 2019). A global review of 166 long-term surveys of insect assemblages found that on average there have been declines of terrestrial species abundance by around 9% per decade compared to increases in freshwater insect abundance by approximately 11% per decade since 1925 (van Klink et al. 2020). These patterns were dominated by trends in North America and some parts of Europe, and it is suggested that improvements in water quality in these regions explain the increasing freshwater insect numbers. The 2020 global LPI shows that the abundance of almost 21,000 populations of vertebrates has declined on average by 68% (in terms of animal population sizes) between 1970 and 2016 (WWF/ZSL, 2020) (see Figure A2.3.1). For freshwater vertebrates, the picture is worse, with average declines of 84% (WWF/ZSL, 2020).

Figure A2.3.1 Trends in Global Vertebrate Abundance as Measured by the LPI

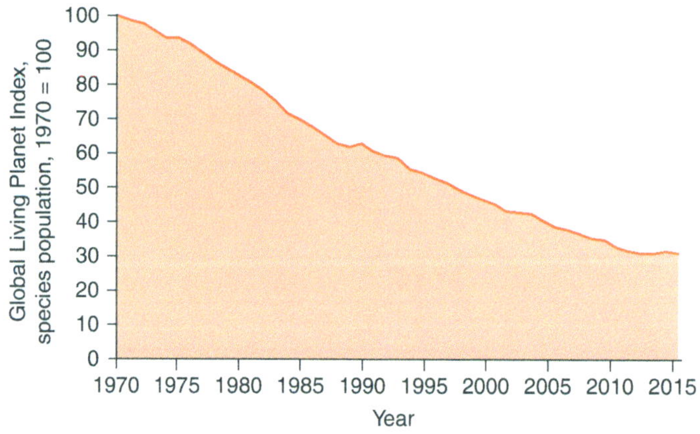

Source: WWF/ZSL (2020). Note: based on 20,811 populations of 4,392 vertebrate species.

Species extinction threat is increasing. Of the 120,000 species on the IUCN Red List, more than 32,000 species are threatened with extinction. This includes 26% of mammals, 41% of amphibians, 34% of conifers, 33% of reef building corals and 14% of birds (IUCN, 2020). The RLI shows that birds, mammals, amphibians, corals and cycads are moving towards extinction more rapidly (IUCN, 2020). Ichii et al. (2019) investigated whether species had sufficient habitat to support them. The analysis suggested that approximately half a million animals and plants may become extinct because of the loss and degradation of suitable habitat that has already taken place (Ichii et al. 2019). This so-called 'extinction debt' has important implications for conservation. It implies that species may become extinct due to past habitat destruction, even if those changes cease. Interventions such as habitat restoration may reverse extinction debt, but these interventions are required quickly.

As noted in Annex 2.1, the Biodiversity Intactness Index represents the fraction of naturally present biodiversity that still remains in terrestrial ecosystems and is a useful indicator for community composition and ecosystem resilience. The most recent estimates suggest that the global BII is 79% in 2015, with most biomes below 90%, far less than the proposed safe limit of 90% (Steffen et al. 2015; Hill et al. 2018). A model of tropical and subtropical forest biomes found their BII is both lower and declining more rapidly (De Palma et al. 2018). Biotic integrity is also seen to be falling when considering the MSA (Annex 2.1, Schipper et al. 2016). The biotic integrity of hotspots of rare and endemic species have a lower status and are declining more quickly than the global average trend as measured by the BII and MSA (Ichii et al. 2019). Indigenous lands have a better status and a slower rate of decline, although they are still below the proposed planetary boundary (Ichii et al. 2019).

The IPBES Global Assessment reported that indicators of ecosystem extent and condition, composition of ecosystems and species populations show net declines over recent decades (Ichii et al. 2019), all of which signals the declining quantity and quality of ecosystems assets and our stocks of natural capital.

Chapter 3
Biospheric Disruptions

Introduction

It is not uncommon to hear people referring to tipping points, positive feedback, path dependence, irreversibility, and even catastrophic risks when talking or writing about climate change and biodiversity loss. As the expressions point to features of the processes governing the biosphere, it is important to understand what they amount to. Otherwise the terms will remain in wide use even when their significance for economic policy remains unclear.

The source of those features is the *non-linearity* of the processes that govern the biosphere. Non-linearity is a ubiquitous feature of Earth System processes, so ubiquitous that it would not be an exaggeration to say that the economics of biodiversity is the study of non-linear processes.[91] The meaning of non-linearity is simple enough: if, to take an example, you were to replicate an ecosystem M times, the functions of the enlarged ecosystem would not be an M-fold replica of the original ecosystem. The findings that we reported in Chapter 2 on the links between biodiversity and ecosystem productivity pointed to non-linearities, but as we see below (Section 3.1), more direct evidence comes from fragmenting ecosystems.

Even though the meaning of 'non-linearity' is simple, its implications for the economics of biodiversity are by no means immediate. Details matter, and they are often technical. In this chapter we try to draw out the salience of non-linearities. We will discover that they shape ecosystem functions and are vital to the design of conservation and restoration projects. In later chapters, we will find that the pervasiveness of non-linearities is a reason *markets* are a woefully inadequate system of institutions for protecting the biosphere from overuse, let alone for ensuring that our consumption and investment decisions are based on a reflective balance between our own well-being and the well-being of our descendants. There are other prominent reasons, and they are discussed in Chapters 4–10.

3.1 Fragmentation as Disturbance

As providers of regulating and maintenance services (i.e. category (2) in the classification of ecosystem services that was offered in Chapter 2), ecosystems resemble near-indivisible entities; the functional traits of healthy ecosystems complement one another. If you divide an ecosystem into parts by creating barriers, the sum of the productivities of the parts will typically be found to be lower than the productivity of the whole, other things equal.[92] This could seem odd to someone versed in island biogeography, the study of which has found that the number of species in a habitat is, as a good approximation, a power function of the size of the habitat, the power being in the range 0.2–0.8 for species of birds, ants and plants in clusters of islands of varying size (Rosenzweig, 1995).[93] But the species in island studies had evolved over millennia, whereas the fragmentation of ecosystems into disjoint units that we are referring to here are

---

[91] Steffen et al. (2005) conclude from their most comprehensive study of Earth System processes that they were unable to identify a process that is not non-linear.

[92] See for example Loreau et al. (2001), Worm et al. (2006), and Sodhi, Brook and Bradshaw (2009).

[93] Formally, if A is the area of the island and S the number of species, then $S = aA^\beta$, where a is a constant specific to the island's characteristics (tropical, temperate and so forth) and β is a constant in the interval (0,1). The number of species per unit area is then $aA^{-(1-\beta)}$, which is a declining function of A. In the examples cited, β has been found to lie between 0.2 and 0.8.

a recent occurrence. Moreover, ecosystems harbouring a greater number of species are not necessarily more productive. The two sets of findings are not at odds.

When habitats are fragmented, abrupt boundaries appear between patches. The length of a fragmented boundary, called an *edge* by ecologists, increases with fragmentation. In a long-running study of the Amazon rainforest, Thomas Lovejoy and his colleagues have found that even large fragments of forest area (100 ha) can lose up to 50% of their species in a dozen years. Clearings as narrow as 80 metres have been found to hinder the recolonisation of fragments by birds, insects and tree-dwelling animals (Laurance et al. 2002).[94] In a review of this and other studies, Haddad et al. (2015) found that fragmentation reduces biodiversity by up to 75%.

Fragmentation hinders dispersal. Animals are wary of entering clearings, if for no other reason than that they evolved within large, continuous and climatically stable habitats. Fragmentation has also been found to expose species to harsh environmental conditions, including fires, diseases and invasive species. That amounts to reduction in resilience owing to a loss in biodiversity. Paleo-biologists have found fragmentation of natural habitats to be a good early-warning sign of biodiversity loss. Unfortunately, 70% of Earth's remaining forests today are within 1 km of the forest's edge (Haddad et al. 2015). That is fragmentation at a massive scale. Future losses to natural habitats owing to further extensions of land devoted to agriculture (estimated to be a near-20% increase by 2050; Tilman et al. 2001), plantations and mines will create further fragmentation. And that does not include the fragmentation that will be caused by increases in the number and sizes of towns and cities and transportation networks (see Boxes 3.1 and 3.2 on dams and fences).

When comparing the character of ecosystems, other things are never the same, which is why conclusions derived from their study are rarely unambiguous. The seeming oddities can only be understood by prising open ecosystems so as to look closer. There are disturbed ecosystems that contain as many (or even more) species than those that are pristine. One reason for this is that disturbed environments can contain exotic or non-native species that have invaded them. In their study of avian populations at the heart of the Amazon rainforest, Rutt et al. (2019) report that new residents comprise 13% of the current community of permanent residents. They also report that the invasive species are common generalist birds, favouring secondary forests as habitats. The composition of species changes with disturbance.

Box 3.1
Damming Rivers

Fragmentation can prevent migratory populations from conforming to their behaviour over the life cycle. That translates into species extinction. Dams are a major contributor to freshwater biodiversity loss. The construction of high dams is favoured by national economic planners because they expand irrigation, they supply energy, and they offer protection against floods. The problem is, they also disrupt the hydrology of freshwater ecosystems by fragmenting them. Fragmentation obstructs migration routes, which are essential for spawning and feeding; it limits dispersal.

Freshwater habitats cover only 0.8% of Earth's surface, nevertheless one-third of described vertebrates, including approximately 40% of fish species, are found in them (Balian et al. 2008; Mora et al. 2011; Dawson 2012). Barbarossa et al. (2020) have constructed a connectivity index of river water by combining species occurrence ranges with the hydrology and location of dams.

---

[94] Their continuing field study (it is now over 40 years old) has been dubbed the largest ecological experiment on Earth by Bowman, Hacker and Cain (2018), from which we have taken the material reported here.

Fragmentation can then be read as reduction of the index. The authors report that the approximately 40,000 high dams that currently exist worldwide have altered 50% of the volume of river water, by either regulation of water flow or fragmentation, and that the pending construction of some 3,700 more high dams will raise the figure to over 90%.[95] Current measures of fragmentation are highest in the US, Europe, South Africa, India and China, but increases in fragmentation due to future dams is estimated by the authors to be especially high in the tropics, with declines in the connectivity index of some 20–40% in the Amazon, Niger, Congo and the Mekong Basin.

If high dams are environmentally damaging, they would appear to be uneconomic even if environmental concerns were set aside. Ansar et al. (2014) reported findings from a study of large hydropower dams built for which they could find reliable data. The authors' sample of 245 projects initiated in the period 1934 to 2003, covered 65 countries. Construction costs on average were found to have been *double* the figures projected in their original feasibility studies. Smaller-scale dams, and fish ladders that allow fish to pass around dams, have been found to reduce risks to fisheries through fragmentation.

Box 3.2
Fencing the Grasslands

In terrestrial ecosystems, a major cause of fragmentation is the building of fences. They are prominent in the grasslands, because among other things they serve to delineate private property in cattle and sheep. But they prevent migration by animals in the wild. Proliferating fences, along with habitat loss and wildlife poaching, has sent ecosystems such as the Greater Mara in Kenya into ecological turmoil. A 2009 audit of Earth's greatest terrestrial-mammal movements found that of 24 large species that once migrated in their hundreds to thousands, six migrations had ceased (Harris et al. 2009). Africa is home to some of the world's most spectacular wildlife migrations, but of the 14 seasonal migrations of large-mammal species known to migrate en masse, five migrations are now extinct. The good news from the few cases where the migration barriers have been removed is that animals resume their movements if they are allowed to do so (Harris et al. 2009).

Laurance et al. (2015) have constructed a map of the ecological fragmentations that will be caused by Africa's 33 'development corridors'. The gigantic infrastructure project, designed to generate GDP growth, involves the constructions of roads, railways, ports and pipelines. If completed, they would in total exceed 50,000 km in length and criss-cross the continent, chopping ecosystems into bits. The corridors will also have a significant impact on existing protected areas: the authors report that the roads and railways at the heart of the corridors would bisect 408 protected areas while cutting through 5,740 km of protected habitats.

Vilela et al. (2020) have studied 75 road projects, totalling 12,000 km of planned roads, in the Amazon region. The authors estimate the resulting fragmentation of forests, but their purpose is to make a different point. They calculate that none of the proposed projects makes conventional economic sense: the net present value of profits (Chapter 13) is negative in 45% of proposed projects, meaning that the rates of return on them are far too low. The authors' analysis suggests that investing in the projects would amount to what some would call a 'lose-lose' move: the projects are both environmentally destructive and of negative economic value.

---

[95] High dams are defined as dams that are taller than 15 metres. The figure of 40,000 for the number of existing high dams worldwide is probably an underestimate, but it pays to work with conservative figures when even they correspond to massive disruptions.

Chapter 3: Biospheric Disruptions

3.2 Stability Regimes

Non-linear systems have further features that add complications to the economics of biodiversity. Lakes, rivers, estuaries and coastal waters suffer from eutrophication when subjected to an overload of nutrients. Phosphorus runoff from agriculture, forestry and urban development is usually the primary cause of nutrient overload in lakes.[96] Because the sources of phosphorus are spatially dispersed, the pollution is hard to regulate. What is measurable is the eutrophication itself, for it is accompanied by an increase in plant growth in the lake-bottom, toxic algae blooms, reductions in water transparency and oxygen, and fish kills. And to these negative effects we should include foul odour from the waters. A eutrophic lake is biologically active, but its water is unsuitable for drinking or bathing, or production of fishes.

Lakes are said to be in an oligotrophic state when phosphorus seepage from the external landscape has been low and continues to be low. Oligotrophic lakes display low primary productivity but supply high levels of provisioning goods, such as drinking water, and water for irrigation and industrial use. When a lake that was previously in an oligotrophic state becomes eutrophic, losses from the disappearance of those services can be huge. Wilson and Carpenter (1999) estimated that the benefits of policies that would achieve 'swimmable' water quality in all the US fresh waters at the end of last century was about US$58 billion per year.

In an illuminating body of work on the dynamics of lake systems, Stephen Carpenter and his colleagues have found that while nutrient addition to lakes may lead to immediate increases in eutrophication, decreases in nutrient input does not lead to immediate or complete reversal: lakes display hysteresis, even irreversibility.[97] The underlying reason is that lake ecosystems have memory. This is a feature of the non-linearities that are characteristic of not only freshwater ecosystems, *but ecosystems generally*, and is important in conservation ecology; for it says that other things equal *it is cheaper to maintain an ecosystem than it is to degrade it and to then restore it*. For lakes the authors trace the delayed response to reductions in nutrient input (in the extreme it can be lack of response) in lakes to phosphorus recycling. The process is roughly the following:

The phosphorus that seeps into a lake from outside partly disappears as outflow and plant sequestration and partly accumulates in the lake's sediments. Phosphorus in sediments is recycled to lake water. Whole-lake experiments have shown that recycling rates can build up to significant levels in a matter of years. Interestingly, recycling rates in eutrophic lakes commonly exceed the rate at which phosphorus seeps in from outside. Thus, phosphorus recycling interferes with attempts at mitigation and can sustain eutrophy long after phosphorus seepage from outside has been lowered. In some cases, eutrophication cannot be reversed by decreasing phosphorus inflow alone, but requires interventions that decrease recycling or accelerate sedimentation.

There is more. If you were to label lakes in terms of the rate of phosphorus inflow they experience, you would find that, other things equal, higher levels of eutrophication correspond to higher rates of phosphorus inflow; but you would also find that the relationship is not continuous. At discontinuities small differences in rates of phosphorus in-flow correspond to large differences in eutrophication. However, disturbances that amount to an increase in phosphorus seepage into a lake before the lake becomes eutrophic need not take place all at once, they could be accumulations of small increases in the rate of phosphorus inflow, until a further small increase kicks a lake into an eutrophic state. Annex 3.1 presents a formal account of phosphorus recycling in lakes.

---

[96] Garden ponds are notoriously hard to maintain in a state where the plants at the bottom are visible. That is because algae growth owing to phosphorus input from the surrounding area can make them eutrophic.

[97] See for example Carpenter and Cottingham (1997) and Carpenter et al. (1998).

One way to read this feature is to recognise that a lake system harbours more than one *stability regime*, by which is meant that the possible states of the lake can be partitioned into regimes, or 'domains', with the property that once the lake enters a regime, it is confined to it unless it experiences a large disturbance. The move from one stability regime into another is called a *regime shift*.[98] It is today a commonplace to say that regime shifts occur at *tipping points*. Recall the notion of 'resilience' that was introduced in Chapter 2. It can be rephrased thus: the *resilience* of an ecosystem in a stability regime is the magnitude of disturbance it can tolerate before it tips into a different stability regime.[99] If the domain in which the ecosystem happens to be becomes smaller, its resilience is reduced; a chance event that would previously have been absorbed by the ecosystem can trigger a sudden, dramatic change and loss of structural integrity.

The time involved in regime shifts differs enormously, depending on the character (e.g. size) of the ecosystem. It could take decades for a forest biome to tip over into Savannah, whereas grasslands have been known to tip into shrubland in years; and garden ponds have been known to tip into a eutrophic state in a matter of hours.[100] Figure 3.1, taken from Folke et al. (2004), presents a pictorial description of multiple stability regimes in various ecosystems.

Figure 3.1 Multiple Stability Regimes

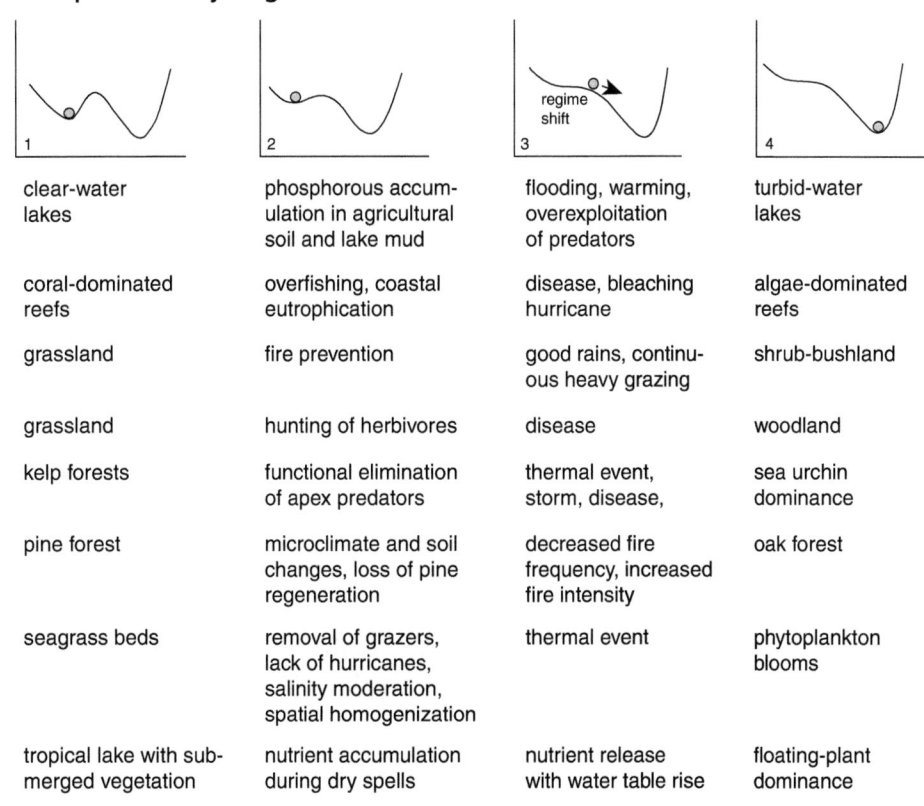

1	2	3	4
clear-water lakes	phosphorous accumulation in agricultural soil and lake mud	flooding, warming, overexploitation of predators	turbid-water lakes
coral-dominated reefs	overfishing, coastal eutrophication	disease, bleaching hurricane	algae-dominated reefs
grassland	fire prevention	good rains, continuous heavy grazing	shrub-bushland
grassland	hunting of herbivores	disease	woodland
kelp forests	functional elimination of apex predators	thermal event, storm, disease,	sea urchin dominance
pine forest	microclimate and soil changes, loss of pine regeneration	decreased fire frequency, increased fire intensity	oak forest
seagrass beds	removal of grazers, lack of hurricanes, salinity moderation, spatial homogenization	thermal event	phytoplankton blooms
tropical lake with submerged vegetation	nutrient accumulation during dry spells	nutrient release with water table rise	floating-plant dominance

Source: Folke et al. (2004). *Permission to reproduce from Annual Reviews, Inc.*

---

[98] Scheffer and Carpenter (2003), Scheffer (2009), and Scheffer et al. (2012) are key references. Stability regimes are also called 'basins of attraction'.

[99] Holling (1973) is the key reference for this notion of resilience.

[100] Dakos and Hastings (2013) is an introduction to a Special Issue of the journal *Theoretical Ecology* on regime shifts and tipping points in ecology.

Chapter 3: Biospheric Disruptions

Box 3.3 presents an example of an ecosystem containing more than one stability regime. It extends the model of the single fishery in Box 2.2 to include a minimal size of the fish stock, a *threshold*, such that if the stock falls below it the fishery is doomed (Figure 3.2). We may also use the example to introduce uncertainty in the location of tipping points (we do that in Chapter 5). In a famous metaphor, Ehrlich and Ehrlich (1981) likened the pathways by which an ecosystem can be tipped out of a stable productive regime into a stable unproductive regime to a flying aircraft from which rivets are removed, one at a time. The probability that it will crash increases very slowly at first, but then at some unspecifiable number rises sharply to 1. Their metaphor is converted into a formal model in Chapter 5 (Box 5.3).[101]

Figure 3.2 Regeneration Function of Fishery with Threshold

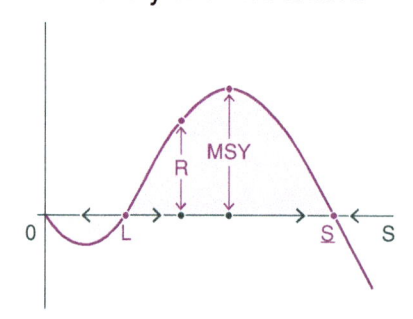

Box 3.3
Multiple Stability Regimes – Fisheries

Consider once again a lake harbouring a species of fish of interest to the fishing community. Let G be the regenerative rate of the fish population. As previously, G is the birth rate minus the death rate. The stock of fish is denoted as S. As G is a function of S, we write it as $G(S)$. But unlike the logistic function in equation (B2.2.1) in Box 2.2, there is a population threshold such that if the stock were to fall below it the fishery would be doomed. Here we use the example in Scheffer (2009: 332) to illustrate this. It is supposed that

$$G(S) = rS[1 - S/\underline{S}][(S-L)/\underline{S}], \qquad L, r, (\underline{S}-L) > 0 \qquad (B3.3.1)$$

The only difference between the G-function in equation (B3.3.1) and the G-function in equation (B2.2.1) in Box 2.2 is that in equation (B3.4.1) the logistic growth function has been multiplied by the term $(S-L)/\underline{S}$. Equation (B3.3.1) says that for the fish population to be viable, it must be of a minimum size L (Figure 3.2). The point is, there are other species 'lying in wait' in the lake; so, if the stock falls below L, it will be replaced by a competitor.

The dynamics of the fishery is then governed by the differential equation

$$dS(t)/dt = rS(t)[1 - S(t)/\underline{S}][(S(t) - L)/\underline{S}] \qquad (B3.3.2)$$

Figure 3.2 depicts the G-function in equation (B3.3.1). Notice that equation (B3.3.2) possesses three stationary points: $S = 0$, $S = L$, $S = \underline{S}$. As is shown in Figure 3.2, $S = 0$ and $S = \underline{S}$ are stable,

---

[101] Ecological processes, as is the case with all other processes, are subject to uncertain shocks. In Chapter 5, we study the various ways in which policies affecting the biosphere (or in other words any policy) can be reached when decision-makers admit uncertainty into their analysis. Econometricians have been known to study time series of global income and mean global temperature by modelling the human-climate system in terms of linear, stochastic processes. The moral is banal: limitations of empirical techniques should not determine how Nature is to be read.

while $S = L$ is unstable. In the absence of human predation, the system would settle at $S = \underline{S}$, which is the *carrying capacity* of the fishery.

If $R(t)$ is the rate at which fish is harvested, the dynamics of the system becomes,

$$dS(t)/dt = rS(t)[1 - S(t)/\underline{S}][(S(t) - L)/\underline{S} - R(t) \tag{B3.3.3}$$

L is the threshold size of the fish population. It is known as a *separatrix*. If $R(t)$ is maintained at a high level the stock declines. If the stock falls below L, the fish population is doomed.

The idea of stability regime is well-defined, what is not known usually is, for example, the location of the states that separate the regimes (i.e. the separatrix). In equation (B3.3.3) the uncertainty would be about the location of L. Good management practice of the fishery would demand that the stock be prevented from dropping below a safe distance from, say, the expected value of the threshold (that's the task of conservation ecology). That precaution amounts to adjusting for risk and uncertainty. The problem is formulated in Chapter 5 (Section 5.3).

There is always uncertainty also in the values of the parameters that define ecological processes. Suppose for example there is uncertainty in the value of $\underline{S}$ as well as in the values of r and L. Uncertainty aversion would encourage the harvester to base her harvesting policy on the basis of figures for r and $\underline{S}$ that are smaller than their expected values, and a figure for L that is larger than its expected value.

3.3 History Dependence

Non-linear processes have 'momentum' built into them. Stop the engines of an ocean liner and the ship will continue to plough the waters for several miles. That is not non-linearity at work, of course, but it does convey the idea of momentum. What non-linearities bring with them is something that is best thought of as 'memory'. The state of a non-linear system in the next instant is not simply a function of where it is at now, but also of where it has been in the past. In Annex 3.2, we use the lake models of Stephen Carpenter and his colleagues to show that ecosystems display hysteresis, even irreversibility. Our analysis can be extended to climate systems, which explains why climate scientists say that even if global carbon emissions were to come to a halt immediately, the global climate will continue to 'warm' for a long while.

Current technology and infrastructure are likely to be relatively ineffectual under such regime shifts. Investment in socio-economic infrastructure (technological constructions, institutions; Chapter 13) are mostly non-malleable. They contain specifications that are unlikely to prove adaptable in the hugely different circumstances that would prevail if critical thresholds were to be crossed. That is why it is important to develop an understanding of such seemingly technical features of the biosphere's processes as its non-linearities.

Studying the contained and rapid can be a help in gaining an understanding of the extensive and slow. Similarly, people's experiences today can inform us of what may lie ahead. Studying the experiences of communities in poor countries tells us that the human impact of biodiversity loss is not a problem solely of the future. The loss of local biodiversity is felt by communities in poor countries when their resource base degrades. They experience it when the shrubs and grasses recede and when residents in shanty towns suffer from waterborne diseases and lose their homes from landslides that follow deforestation of the slopes (Chapters 7–8). When the transition to a breakdown of the local natural resource base is swift, tipping points cease being simply forecasts of global happenings under runaway climate change and biodiversity loss, they become real experiences at the local level.[102] Large-scale migration, as we are

---

[102] See Scheffer et al. (2012), and Steffen et al. (2018) on climate driven tipping points for the Earth System.

Chapter 3: Biospheric Disruptions

witnessing today in some of the world's poorest countries, is often traced to political conflict, but the origins of political conflicts in some cases have been traced by scholars to degradation of the local resource base. When the local landscape tips into a state that cannot support a community, migration is the only option. But neighbouring communities also suffer from resource stress, so they are not welcoming.[103]

The examples of ecological transformations that have been collated here suggest that we are drawing down the biosphere, in that at the global level there is a mismatch between the demand we make of the biosphere's goods and services and the ability of the biosphere to supply them on a sustainable basis. In Chapter 4, we offer evidence of the mismatch and begin to unpick the demands we make of the biosphere. In Chapters 6–8, we study the socio-economic processes that have led to the mismatch.

Box 3.4
Regime Shifts in Space

The biosphere is a self-organising entity. Transitions from one stability regime to another occur not only temporally, they display themselves also spatially. As was noted in Chapter 2, tropical rainforests can shift to grassland (Savannah). This has significant consequences for the water cycle in the surrounding region.

In a systematic research programme, Simon Levin and his colleagues have explored the spatial characteristics of dynamical systems and have applied the findings to explain salient features of large-scale landscapes (Levin 1999; 2000). Forest and Savannah are alternative stability regimes for tropical land, separated by biotic and abiotic drivers, and maintained by feedback loops of fire (in the case of Savannah) and fire suppression (in the case of rainforest). When rainfall levels are low, vegetation is dominated by grasses with very limited wood cover to provide shade. The intensity of the sun can thus start fires easily, creating a 'trap' for any developing woody cover, and making the return shift to forest extremely unlikely. In contrast, high levels of rainfall enable wood cover to grow and develop into canopies, which then maintain the necessary water cycle (as described above) and provide shade which limits the possibility of fire breaking out. Such dynamics give rise to a tropical world that admits two broad stability regimes, one dominated by grass, the other dominated by trees, with the potential for sudden flips as conditions change (e.g. precipitation levels and hence the frequency of fire (Staver, Archibald, and Levin (2011); see also Hirota et al. (2011)). If deforestation goes beyond a certain point, rainforests are no longer able to maintain the level of moisture they need in the atmosphere and will no longer be able to shade their forest floor from fire. Widespread deforestation in the southern Amazon has already caused rainfall to decline there significantly (Lovejoy and Hannah, 2019).

In semi-arid regions, grasslands and deserts coexist, and the spatial transition from grassland to desert is often abrupt at the boundary. Zeng et al. (2004) have demonstrated the presence of such spatial transitions in a model in which the ecosystem was described in terms of living biomass, wilted biomass and soil wetness. The latter, which was taken to be the ratio of annual precipitation to potential evaporation, was the only external driver, the key mechanism in their model was vegetation-soil interaction. The ecosystem was found to divide into two spatially separated stability regimes: grassland and desert. Maintenance of grassland was found to require a minimum rate of precipitation, which was no surprise; however, the model also showed that grassland and desert can

---

[103] Homer-Dixon (1994) and Suhrke and Hazarika (1993) are empirical studies that trace violent conflicts to local resource scarcity, the latter focusing on experiences in the Indian sub-continent. This is an unusually hard subject, in as much as social conflicts, even violent social conflicts, have multiple causes. Bernauer, Böhmelt and Koubi (2012) stress that.

coexist when precipitation is within a range below the threshold. Wilted biomass was found to play an important role in shaping the transition from grassland to desert. The model identified a grazing strategy that maintains grassland and was tested against data on the Sahara and the Sahel region of Africa. By stretching our imagination, we may thus view the biosphere as a spatial configuration of multiple stability regimes of the Earth System (e.g. biomes).[104]

3.4 Conservation Ecology and Tipping Points

Conservation ecology uncovers ways to keep ecosystems from falling prey to large disturbances. The closer is an ecosystem judged to be to a tipping point, the more urgent is the case for conservation measures. Restoration ecology identifies ways to bring ecosystems back to health if it has suffered damage, in the extreme if they have undergone a regime shift. Restoration is costlier than maintenance, other things equal, and pre-emptive moves to conserve ecosystems are less costly than to wait and invest in restoration only when they have deteriorated badly. What makes both conservation ecology and restoration ecology particularly complex is that the processes that drive ecosystems have memory, meaning that they are *path dependent*, which is to say ecosystem dynamics display hysteresis. Just as a demagnetised iron displays hysteresis when it is re-magnetised (the path followed by the ferromagnet is not a retrace of the path it followed while being de-magnetised), restoration practices cannot retrace an ecosystem's path back to health. The lake model we study in Annex 3.2 displays the possibilities.

The dynamics of Earth's climate system is also understood to be path dependent. When climate scientists express concern about deploying geo-engineering in the future to bring carbon concentration in the atmosphere down to current levels, they have in mind the fact that even if concentration were brought down to today's level, the mean global temperature would not necessarily follow suit; more importantly they worry that the biomes would look very different from what they are today. There could even be tipping points lying in wait in those unchartered terrains. Identifying ways to protect ecosystems from damage is not easy science.

Box 3.5
How is Society to Know that an Ecosystem is Near a Tipping Point?

We should read large-scale fragmentation of tropical rainforests as a sign that they are locally near breakdown. Scheffer (2009: 282–296) lists a number of clues that should be read by society as warnings of impending regime shifts in ecosystems. One clue is a slowing down of recovery time from small disturbances (brief period of drought, warmer-than-usual winter, small outbreak of predatory insects, and so on). This is a clue because near tipping points the dynamics of ecosystems are slower (see also Scheffer et al. 2012; and Scheffer et al. 2015). Such slowing down can be measured experimentally, but also indirectly inferred from changes in natural fluctuations and spatial patterns.

It is not easy to test whether a system has become slower. A related clue, one that is less hard to test, is whether under the small disturbances an ecosystem is naturally subjected to, its state at any one moment is more like its previous state (i.e. there is greater step-1 autocorrelation among its states).

---

[104] Levin (1999) and Scheffer (2009) are excellent, book-length accounts of non-linear processes in socio-ecological systems and the multiplicity of regimes they can harbour. The latter studies systems that can tip over from one regime into another when it moves through *time*, while the former studies the spatial characteristics of systems where discontinuities in the spatial map of systems are manifestations of regime shifts across *space*.

Chapter 3: Biospheric Disruptions

The intuitive reason behind this is that if the recovery rate of the ecosystem from small disturbances has slowed down, its state in one period would be more similar to its state in the previous period. Long time series or satellite images of the state of the ecosystem would be the basis of inferring this.

For the climate system, a rise in the frequency of extreme events is a related sign. Climate scientists have long argued that rising concentration of CO_2 in the atmosphere should be expected to bring in its wake more frequent tornados, hurricanes, prolonged heat waves and other extreme forms of weather. Suppose the probability, say π, of the occurrence of an extreme event in any year is independent of whether there was an extreme event in the previous year. Then $1/\pi$ would be the expected date of the next extreme event. An increase in the frequency of extreme events would mean that ϖ has increased. What was a once-in-a-century event (e.g. a mega fire) becomes once-in-a-decade event, which reads as an increase in π from 0.01 to 0.10. The increase is a tell-tale sign that the system is getting closer to a tipping point.

A further test has been suggested by studies of vegetation clusters in semi-arid lands, where water is a limiting resource. Plants are known to create a local environment that, among other things, minimises water runoff. That facilitates the survival of other plants and seeds. That (positive) feedback has been found to give rise to size distributions of vegetation patches in which most patches are small but more than a few are very large. Field studies in the Kalahari and southern Europe reported by Scanlon et al. (2007) and Kéfi et al. (2007), respectively, found the distributions to satisfy power laws, that is, if S is the size of a patch, the proportion of patches of size S is proportional to S^{-a} in which the parameter a is in the range 1 to 2. The authors showed that animal grazing on the patches lead to a departure from the power law distribution. Proportionately the distribution in a degraded landscape contains many more patches of small size.

That the distribution of patches in ecosystems has the form of power functions is not a happenstance in the grasslands of semi-arid regions. Levin (1999) contains an excellent account of why we should expect power functions to reflect the distribution of clusters in complex adaptive systems. Power laws in the size distributions of patches imply that the geometry of the landscape containing the patches is fractal, that is, they are self-similar at all scales.[105]

Box 3.6
Human Body as an Ecosystem

The human body is an ecosystem. Someone experiencing a breakdown in health would at a formal level be exhibiting a regime shift. Non-linearities in ecosystem processes also lead to segmentations among ecosystem populations, as in the spatial dynamics reported in Box 3.4. Likewise, non-linearities in human metabolic processes lead, most prominently in poor countries, to a partitioning of human populations into groups that are healthy (for example, people who enjoy recommended body mass indices) and groups that are malnourished. A simple example illustrates the phenomenon:

An individual's capacity to do physical work, as reflected in such attributes as endurance is, other things equal, known to be a function of his nutritional status, involving physical attributes such as body mass index (BMI). The standard definition of BMI is weight (in kilograms) divided the square of height (in metres). The World Health Organization (WHO) regards someone to be in good health if

---

[105] For a simple proof that power laws give rise to fractal geometry, see Dasgupta and Mäler (2003).

his BMI is in the range 18–25. Extreme undernourishment would be reflected in a BMI less than 16, while obesity is a BMI above 30.

The relationship between nutritional status and work capacity has been found to be sigmoid: its gradient, although positive beyond a BMI equal to, say, 17 is very small but then increases with BMI rapidly until tapering off from values of 24 or so, and then is negative for people whose BMI exceeds 30 (Dasgupta, 1993).

Figure 3.3 Nutritional Status and Work Capacity

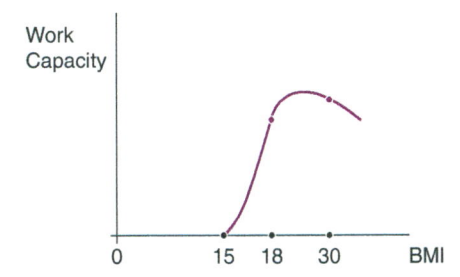

The sigmoid form of the relationship below the obesity level is a non-linearity that has far reaching consequences.

Imagine a country that is so poor that no economic system could ensure that everyone has a high nutritional status. At the same time, no efficient economic system in the country would countenance on grounds of equity that everyone should have a low nutritional status, for that would reduce aggregate labour productivity to very low levels and keep the country in poverty. The choice is cruel, but an efficient economic system would bifurcate the population into one whose nutritional status is low and another whose nutritional status is high. The former group is trapped in poverty, which is formally a stability regime (Dasgupta and Ray, 1986). Societal aim would be to tax the income of the rich to raise the nutritional status of the poor. In the world as we have come to know it, economic systems are not only inefficient, but all too often unjust too. Inequality in the distribution of wealth and power ensures that the proportion of people who are trapped in poverty is greater than it would be if socio-economic support from government or the community were to reduce the number of people from getting caught in a poverty trap (Dasgupta and Ray, 1987). Technically that would amount to a policy that keeps many people, though not all, from entering a stability regime that spells 'undernourished'. In practice, such cruel choices probably do not have to be made; undernourishment is an outcome of bad politics, but its prevalence is accentuated by non-linearities in our metabolic processes.

Annex 3.1 The Phosphorus Recycling Mechanism in Lake Systems

Regime shifts can occur in two ways. We call the set of possible states of an ecosystem a *separatrix* if it separates distinct stability regimes. Depending on the context, an ecosystem's state could be its biomass; it could be the quantity of phosphorus in a lake's water table, and so on. A separatrix is thus a set of ecosystem states that mark the border between stability regimes. An ecosystem can tip from one stable regime into another if an external disturbance kicks it across a separatrix. An example would be the case we cited in Box 3.4, the possibility that a further 20–25% deforestation allied to more fragmentation of the Amazon basin would tip the biome into a savannah-like landscape even without any further deforestation. Another example, which we study in detail below, would be the quantity of phosphorus in a lake's water table above which the lake would be in classified as eutrophic. Figure 3.1, taken from Folke et al. (2004), provides illustrations of tipping points in various biomes.

Regime shifts can also be caused by changes in the external environment influencing the parameters defining an ecosystem's dynamics. *Bifurcations* mark critical borders with respect to parameter values. If changes in the external environment cause the ecosystem's parameters to cross a bifurcation, the system moves into a different stability regime. The ecosystem's *map* (technically, its 'topology') undergoes a structural change (non-technically, a 'flip'). An example is a 'dead zone'. When subjected to excessive stress (e.g. run-off of slow but large-scale deposition of nitrogen and phosphorus from agricultural fields), once flourishing ecosystems (biologically rich estuaries) flip into dead zones. The stress could be occasioned by an invasion of foreign species or substance (as in the above example), it could be due to a loss of population diversity owing to changes in physical conditions affecting the ecosystem (see below), it could be triggered by the demise of a dominant species (Annex 2.1) or of species in the lowest rung of food chains, owing perhaps to technological advances in fishing equipment. And it could be, as it so widely happens with biodiversity in soils (Chapter 4), that the stress is caused by intensive farming.

In practice it could prove difficult to distinguish the two types of tipping phenomena. What is taken to be a parameter is often a slow-changing variable, sufficiently slow that it is to all intents and purposes a constant. The climate could be thought of as being exogenous to an ecosystem's processes, but a micro-climate could be affected by the workings of an ecosystem itself (Box 3.5). The difference between the two cases depends on the relative speeds at which the various sub-systems of the larger ecosystem operate.

A3.1.1 The Model

In Section 3.2 we reported that the reason lakes harbour both oligotrophic and eutrophic stability regimes is phosphorus recycling in the water. It pays to illustrate the phenomenon formally. We will encounter bifurcation points in the lake system.[106]

Oligotrophy is not a single state, it is an entire range of states, some less oligotrophic than others. The same of course is the case with the term eutrophic. A binary division of the possible states is a mere simplification. It could also be that there is more than one tipping point separating the water quality regimes in a lake. For simplicity of exposition we ignore the latter and study a lake that harbours bifurcation points separating regimes of varying water quality.

A key variable that determines which state the lake finds itself in is the quantity of phosphorus in the water table. Denote that by S. Steven Carpenter and his colleagues found from both field and natural experiments that the recycling rate, when expressed as a percentage of S is approximately proportional to S when S is small (that is a positive feedback in the recycling mechanism), but declines to zero for large values of S (that is a negative feedback arising from the finiteness of the lake). Let the recycling rate be denoted by G. As it is a function of S, we write it as $G(S)$.[107]

A simple form $G(S)$ that conforms to the asymptotic properties we have just recounted is (Figure A3.2.1)

---

[106] The lake system harbours separatrices as well, but it will prove more natural to discuss that in Chapter 13, where we study the optimal management of resources. Here we will be assuming that the inflow of phosphorus into the lake under study is exogenously given, it is not taken to be a policy variable.

[107] We are deliberately resorting to the same notation, G, to represent phosphorus recycling in the water table as the one that was used for representing the regeneration rate of fisheries in Box 2.2. By doing that we are able to signal formally that conservation is the reverse of pollution (Chapter 2).

$$G(S) = aS^2/(1+S^2), \qquad a > 0 \qquad (A3.1)$$

Notice that $G(S)$ is bounded above for large values of S. This too was found to be supported by evidence. In equation (A3.1) $G(S)$ asymptotes to a as S becomes bigger (Figure A3.2.1).[108]

Let t denote time, which we take to be continuous. We are to imagine that phosphorus seeps into the lake at the constant positive rate R from surrounding agricultural fields. Imagine too that phosphorus gets buried in the sediments or seeps out of the system altogether at a constant percentage rate $\lambda > 0$. It follows that the net rate of change in the stock of phosphorus in the lake's water column is given by the dynamical equation

$$dS(t)/dt = R + G(S(t)) - \lambda S(t) \qquad (A3.2)$$

Using equation (A3.1) in equation (A3.2) yields,

$$dS(t)/dt = R + aS^2(t)/(1+S^2(t)) - \lambda S(t) \qquad (A3.3)$$

$S(0)$ is the initial level of phosphorus in the lake's water table. We could imagine that $S(0) = 0$, that is, the lake was in a pristine condition before agriculture made its appearance in the surrounding landscape. The model says that starting from a pristine state at $t = 0$, equation (A3.3) can be used to track the history of the stock of phosphorus in the lake's water table.

Equation (A3.3) contains three parameters: R, a, λ. We would like to study their influence on the lake ecosystem. The problem is, it is not possible to derive an explicit solution of equation (A3.3). We know however that the equation possesses stationary values, that is, values of S at which $dS(t)/dt = 0$. So, we take a different route for understanding the influence of the three parameters on the lake's dynamics. What we do is to compare the lake in its various stationary states. Setting $dS(t)/dt = 0$, equation (A3.3) reduces to

$$R + aS^2/(1+S^2) = \lambda S \qquad (A3.4)$$

Solutions of equation (A3.4) are stationary states. Being a cubic equation, there are at most three real solutions.

Annex 3.2 Tipping Points

Notice that the parameters R and the parameter a play similar roles in equation (A3.4). Instead of considering a large R we could consider a large a without altering the map of the left-hand side of the equation. Similarly, R (equally, a) and λ counter each other in equation (A3.4), in that our analysis would not be affected if instead of considering a large R we were to consider a small λ. But R is the external driver of the system and can be altered by a change in economic policy. In contrast, a and λ are ecological parameters of the lake system. So we keep a and λ fixed and experiment with alternative values of R.

In Figure A3.2.1, the two sides of equation (A3.4) are plotted separately as functions of S. The right-hand side of the equation is the straight line λS. The (non-linear) function $aS^2/(1+S^2)$ has been lifted up by a distance R in the figure. The uplifted curve is the left-hand side of equation (A3.4). The idea is to determine the number of intersections of the left-hand side of equation (A3.4) with its right-hand side.

---

[108] As we confirm below, the simplicity of the non-linear function in equation (A3.1) ensures that the lake system harbours only two stability regimes. Note also that $dG(S)/dS = 0$ at $S = 0$.

Chapter 3: Biospheric Disruptions

Case 1: Large λ

We assume λ to be large, meaning that phosphorus flushes out of the lake at a fast rate.[109] Figure A3.2.1 considers five values of R, labelled R_1, R_2, R_3, R_4, R_5. R_1 should be understood as being small. The figure shows that if $R = R_1$ the pair of curves intersect at only one point, S_1, which represents the unique stationary state of the lake system. It is simple to confirm that S_1 is stable.[110]

Figure A3.2.1 Phosphorus Inflow and Lake Dynamics

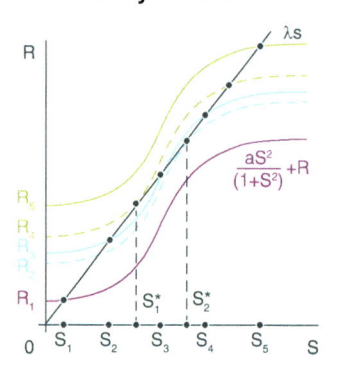

We now conduct a thought experiment. Let us raise the value of R from R_1 slowly, so slowly that the lake system equilibrates at each value of R. A little experimentation in Figure A3.2.1 confirms that there is value of R, which we label R_2, such that the upper turning point of the left-hand side of equation (A3.4), which we have drawn as a dashed curve, brushes the straight line λS at S_2^*. Notice that a small further increase in R, say to R_3, leads to the emergence of three stationary values of S. R_2 is therefore a critical value of the parameter R. It is a simple exercise to confirm that of the three stationary values of S at $R = R_3$, the first and third intersections, S_2 and S_4 respectively, are stable while the second intersection, S_3, is unstable. R_2 bifurcates the lake system: it separates two stability regimes, one of which contains a single stationary point, while the other contains three.[111] The former regime could be read as oligotrophic, the latter as eutrophic. As we confirm presently, past history of phosphorus inflow will have determined in which of the two stable states the lake happens to find itself.

Let us continue to conduct our thought experiment by raising R further. A little experimentation shows that the smaller of the two stable stationary values comes closer and closer to the unstable stationary value until, at R_4 they collapse into one. The stock at that point is labelled S_1^*. If R is raised any further, say, to R_5, the two curves intersect only once, at S_5. R_4 is thus another bifurcation point.

---

[109] By large λ we mean that $\lambda > aS/(1 + S^2)$ for all $S > 0$, which as can easily be checked, is equivalent to the condition

[110] To confirm, suppose S were to be slightly less than S_1. Equation (A3.3) says that $dS(t)/dt > 0$. S would then move to the right. Similarly, suppose S were slightly larger than S_1. In that case equation (A3.3) says $dS(t)/dt < 0$. So S would move to the left.

[111] To confirm that S_1 and S_3 are stable stationary points of the dynamical system (Equation (A3.3)), suppose S were to be slightly smaller than S_1 (resp. S_3). Equation (A3.3) says that $dS/dt > 0$ at that S. So, S would then move closer to S_1 (resp. S_3). Similarly, suppose S were to be slightly larger than S_1 (resp. S_3). Equation (A3.3) says in that case $dS/dt < 0$. So, S would move closer S_1 (resp. S_3). To show that S_2 is an unstable stationary point of the dynamical system suppose S were to be slightly smaller than S_2. Equation (A3.3) says that $dS/dt < 0$ at that S. So, S would move away from S_2, toward S_1. Similarly, suppose S were to be slightly larger than S_2. Equation (A3.3) says in that case $dS/dt > 0$. So, S would move away from S_2, toward S_3. We conclude that S_2 is a separatrix, and that the lake system harbours two stability regimes $0 \leq S < S_2$ and $S > S_2$.

Annex 3.3 Hysteresis and Irreversibilities in Lake Dynamics

Imagine now that we were to reverse the process in our previous thought experiment. Start with $R = R_5$ in Figure A3.2.1, at which $S = S_5$, and reduce R slowly. Figure A3.3.1 plots the stationary values of S we have just identified against the external driver R as the latter is reduced. The figure shows that on the return journey, S declines continuously along the upper arm so long as $R > R_2$. This means that for R in the interval $[R_2, R_4]$, S remains higher than it had been in that interval on the onward journey we have just studied. The ecosystem displays *hysteresis* here. However, at $R = R_2$ the ecosystem 'flips' to the lower arm of the curve in Figure A3.3.1. Further declines in S would occur continuously if R were reduced further, reversing the track that was followed on the onward journey.

Figure A3.3.1 Phosphorus Inflow and Hysteresis in Lake Dynamics

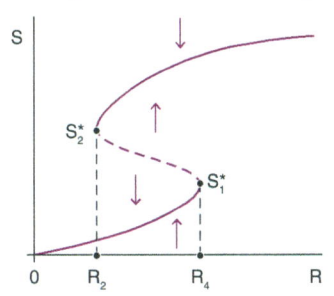

Even though the lake ecosystem displays hysteresis, the water quality of the water is *reversible*. Given enough time, S can be made to be as small as we like provided R were reduced sufficiently. We now show that this is not a possibility if the rate at which phosphorus gets flushed out of the lake is slow.

Case 2: Small λ

Figure A3.3.2 Lake Dynamics under Small λ

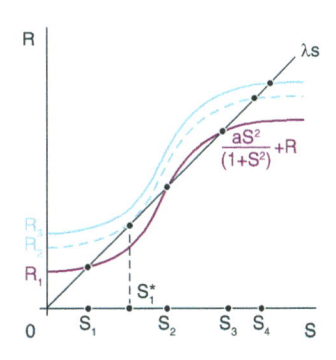

Figure A3.3.2 is constructed in the same way as Figure A3.2.1, but for values of λ that are small.[112] We conduct the same thought experiment as in the previous case. Start with a small value of R, say R_1. The left-hand side of equation (A3.4) cuts the right-hand side at three points, the first and third of which are stable, the second unstable. For concreteness suppose the lake is at $S = S_1$, that is, it is oligotrophic.

As previously, let us now raise the value of R from R_1 slowly, so slowly that the lake system equilibrates at each value of R. A little experimentation with Figure A3.3.2 confirms that there is value of R, which we

---

[112] By 'small', we mean there exists $S > 0$ such that $\lambda < aS/(1 + S^2)$. See footnote 111.

Chapter 3: Biospheric Disruptions

label R_2, such that the lower turning point of the left-hand side of equation (A3.4), which we have drawn as a dashed curve, brushes the straight line λS at S_1^*. Notice that a small further increase in R, say to R_3, leads to the emergence of a single stationary value of S. R_2 is therefore a bifurcation point of the lake system. It is obvious from Figure A3.3.2 that the lake harbours a unique stationary point at all $R > R_2$.

Suppose as we did in the previous case, we were reverse the process we have just described. Start with $R = R_3$ in Figure A3.3.2, at which $S = S_3$, and reduce R slowly. Figure A3.3.3. plots the unique stationary value of S against the external driver R as the latter is reduced. The figure shows that on the return journey S declines continuously *all* along the upper arm, even after R has been reduced to below R_1. The lake cannot be brought back to its oligotrophic state S_1, let alone to any state more oligotrophic. The dynamics of the lake system are *irreversible*, which is to say that *restoration of the lake system is not possible*.

Figure A3.3.3 Phosphorus Inflow and Irreversibility in Lake Dynamics

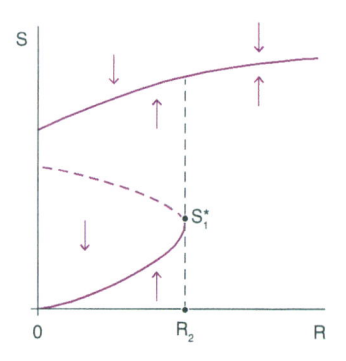

Annex 3.4 Instantaneous Elimination of Phosphorus Inflow

What though if draconian changes in government policy led to the surrounding agricultural land being transformed into a nature reserve? We are therefore to imagine that in place of a gradual reduction in phosphorus inflow (the case just considered), the inflow ceases quickly and entirely. Would the lake system return to its once-pristine state, $S=0$? The answer, alas, is 'no'. To see why, we first construct Figure A3.4.1, which displays the difference between the functions $aS^2/(1+S^2)$ and λS as a function of S. There are three intersections with the S-axis, the first of which is the origin. It is simple to check that the second, labelled S_2, is unstable, while the third, labelled S_3, is stable. And it is the presence of S_3 that is the problem. The lake was previously in a badly eutrophic state, in excess of S_3. From the moment $R = 0$, the lake improves in quality, but only up to S_3, which is however eutrophic. *Despite draconian measures the lake system is not restorable.*

Figure A3.4.1 Irreversible Lake Dynamics under Instantaneous Elimination of Phosphorus Inflow

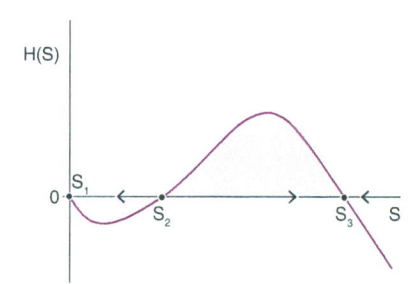

Annex 3.5 Morals

Ecosystem flips have been observed many times and at many scales (Turner et al. 2020). One should not think that the larger is a dynamical system, the slower is the process of abrupt change. It is true that shallow lakes have been known to flip from clear to turbid water in a matter of months, village tanks in a matter of weeks, garden ponds in a matter of hours; but ocean currents with important climate effects, such as the Gulf Stream that warms Europe, could shut down quickly and have done so in the past (Lenton et al. 2008). Insect populations have been known to crash or explode in a matter of days and undetectable viruses spread as pandemics in a matter of weeks. Grasslands in sub-Saharan Africa can take more than a decade to change into shrublands. The fossil records suggest that the interglacials and glacials of ice ages have appeared only occasionally but have arrived and departed 'precipitously' – the flips occurring over several thousand years. And so on.

In the lake system we have studied here in detail, there was a single external driver, R. In studying whether the system could be brought back to something like its original state, we did not study how the dynamics would unfold if the value of one of the other parameters of the system, for example λ, was altered through technological measures. If the rate at which phosphorus is flushed out was made to increase sufficiently, the lake system would still display hysteresis, but restoration to an oligotrophic state would be possible. The lake model is illuminating because it offers a simple way to illustrate not only the character of non-linear ecosystems, but also the levers they may provide for influencing its dynamics. The former is common to a wide category of ecosystems, but the latter is context-specific.

The economics of biodiversity has to be as aware of the commonality of the character of dynamical systems as it has to be sensitive to the particularities of the ecosystem under study (Chapter 2). It is a truism that the composition of an ecosystem, thus its character, is influenced by its external drivers. Details of what those external drivers are and what influences their features depends on the context.[113]

---

[113] Carpenter, Ludwig and Brock (1999), Brock and Starrett (2003), and Ludwig, Carpenter, and Brock (2003) have studied the *optimum* management of phosphorus inflow. They modelled the idea that reduction in phosphorus inflow has a social cost because the inflow is an inevitable by-product of crop production. As against that, the reduction would have a social benefit because the lake would be less polluted. The authors studied the optimum trade-off between the two features when the objective is to maximise the discounted sum of net benefits over time. They showed that the optimum long-run inflow depends on the initial state of the lake: if the lake was already polluted, the long run target would be a less oligotrophic lake than if the lake was initially oligotrophic. Such history dependence, that is, where we aim to reach in the long-run depends on where we are initially, is a general feature of non-linear systems. Kurz (1968) and Arrow and Kurz (2013) are the classics on the subject of history dependence of optimum economic policies. We return to these important matters in Chapters 7 and 13.

Chapter 4
Human Impact on the Biosphere

Introduction

World population in 1950 was around 2.5 billion and global output of final goods and services, at 2011 prices, a little over 9.2 trillion international dollars (dollars at purchasing price parity, PPP) (Figures 4.1 and 4.2). As noted in Chapter 0, the average person's annual income was about 3,300 dollars PPP, a high figure by historical standards (Maddison, 2018) (Figure 4.3). Since then the world has prospered beyond recognition. Life expectancy at birth in 1950 was 46; today it is above 72. The proportion of the world's population living in absolute poverty (currently 1.90 dollars PPP a day) has fallen from nearly 60% in 1950 to less than 10% today (World Bank, 2020a). In 2019, the global population had grown to over 7.7 billion even while global income per capita had risen to 15,000 dollars PPP (at 2011 prices). The world's output of final goods and services was a little above 120 trillion dollars PPP (at 2011 prices), meaning that globally measured economic activity had increased 13-fold in only 70 years, none of which had been remotely experienced before (Chapter 0).

This remarkable achievement has, however, come in tandem with a massive deterioration of the biosphere. This chapter collates scientific evidence that points to this deterioration (Section 4.1) and then constructs a heuristic device for expressing the global demand for the biosphere's goods and services per unit of time and the rate at which the biosphere supplies them (Sections 4.2, 4.3 and 4.4). The latter could be thought of as the biosphere's regeneration rate.[114] We call the difference between demand and supply the *Impact Inequality* (Expression 4.1). That difference has been widening in recent decades. Because demand is decomposed in Expression (4.1) into its several factors, the Impact Inequality points to policy levers that can help steer the global economy toward equality between supply and demand on a sustainable basis, which would convert the Impact Inequality into an Impact Equality. Attaining Impact Equality should be a minimum requirement of the United Nations' Sustainable Development Goals (SDGs). The chapter offers a way to construct quantitative estimates of what has to happen as a minimum if the SDGs are to be sustainable.

4.1 Depreciating the Biosphere

In Chapter 2 (Box 2.3) we offered a qualitative argument, based on crude estimates of own rates of return on primary producers were presented and compared with own rates of return on composite baskets of financial assets to signal that global investments in recent decades have been enormously skewed against Nature. Three related types of evidence have been offered by environmental scientists and one (also related) by economists that vastly enrich that finding. They show that there has been, for some decades, an enormous overshoot in the demands we make of the biosphere.

---

[114] We are taking literary licence here by speaking of the biosphere's regeneration rate, as non-living material does not regenerate, but rather is subject to processes that give rise to their rhythm over time. Steffen et al. (2011) present a synthesis of quantitative evidence since the middle of the last century of growth in humanity's demand for the biosphere's goods and services. The authors combine that with data on declines in the biosphere's health since then. Steffen et al. (2015b) describe that growth as the Great Acceleration.

Chapter 4: Human Impact on the Biosphere

Figure 4.1 Global Real GDP since 1750

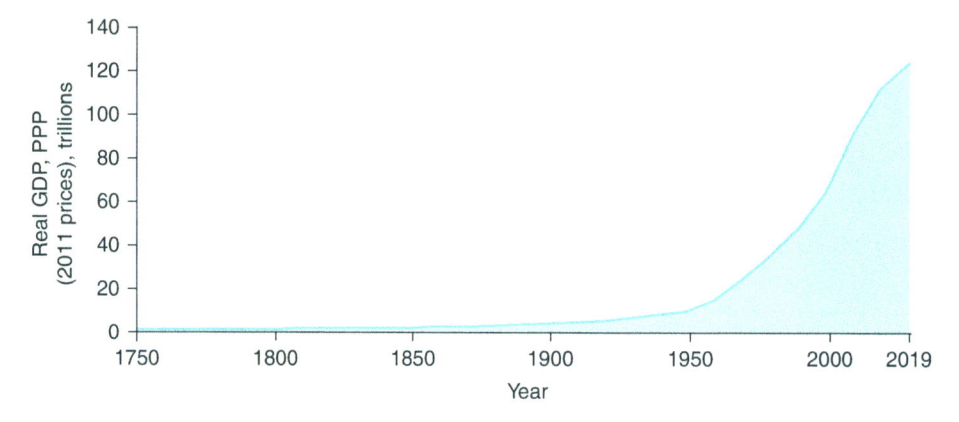

Source: Our World in Data based on World Bank (2020a), Maddison (2018), Bolt et al. (2018) and Review calculations.

Figure 4.2 Global Population since 1750

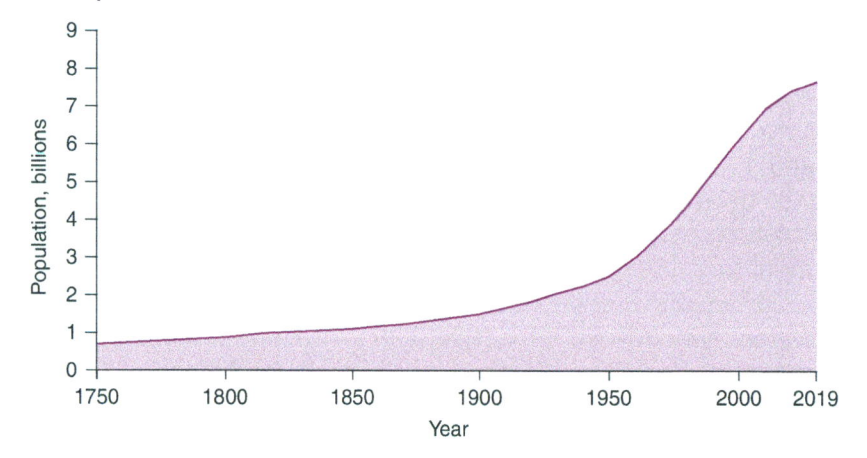

Source: Maddison (2010), United Nations Population Division (2019) and Review calculations.

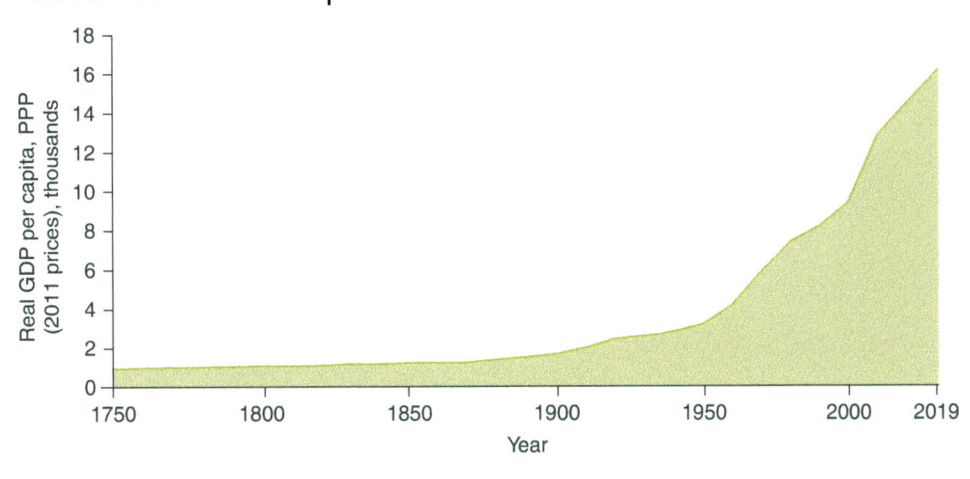

Figure 4.3 Global Real GDP Per Capita since 1750

Source: Our World in Data based on World Bank (2020a), Maddison (2018), Bolt et al. (2018) and Review calculations.

4.1.1 The Anthropocene and Species Extinction

One route to examining our demand overshoot involves the study of the Earth's biogeochemical signatures. In a wide-ranging survey, Williams et al. (2015) divided the evolution of the biosphere into three stages: (1) a microbial stage from about 3.5 billion years ago to about 650 million years ago; (2) a metazoan stage, evident by 650 million years ago when the oxygen level in the atmosphere had begun to rise; (3) the modern stage, starting with the use of stone stools by our ancestral Hominids some 2.6 million years ago and accelerating since the beginnings of agriculture some 14,000 years ago. The authors characterise the current face of the modern stage as (i) global homogenisation of flora and fauna; (ii) a single species, *Homo sapiens*, commanding 25–40% of net primary production (NPP) and also mining fossil fuels to overcome photosynthetic energy constraints; (iii) human directed evolution of other species; and (iv) a rising and less modular interaction of the biosphere with the global human enterprise. The authors suggest that these features of today's biosphere point to a new era in the planet's history that could persist over geological timescales.

In a review of evidence from the past 11,000 years (the Holocene), Waters et al. (2016) have taken a closer look, by tracking the human-induced evolution of soil nitrogen and phosphorus inventories, and carbon dioxide and methane in sediments and ice cores. The authors reported that the now-famous figure of the 'hockey stick' that characterises time series of carbon concentration in the atmosphere is also displayed by time series of a broad class of global biogeochemical signatures (IPCC, 2018). They display a flat trend over millennia until some 250 years ago, when they begin a slow increase that continues until the middle of the 20th century, when they show a sharp and continuing rise. The trends in global economic activity over the past 70 years that we have summarised above and displayed in Figures 4.1 and 4.2 are entirely consistent with these findings.[115] Waters et al. (2016) suggested that the mid-20th century should be regarded as the time we entered the Anthropocene. Figure 4.4 summarises the time profile of key anthropogenic markers that are indicative of the Anthropocene.[116]

---

[115] Steffen et al. (2011) provide a more detailed account of deterioration in the biosphere's health since the beginning of the Industrial Revolution.

[116] The Anthropocene Working Group has proposed that the immediate post-war years should be regarded as the start of the Anthropocene (Voosen, 2016). Kolbert (2013a) contains an account of how stratigraphers uncover geological signatures of abundance and disappearance of species in the distant past.

Chapter 4: Human Impact on the Biosphere

Figure 4.4 Magnitude of Key Markers of Anthropogenic Change Indicative of the Anthropocene

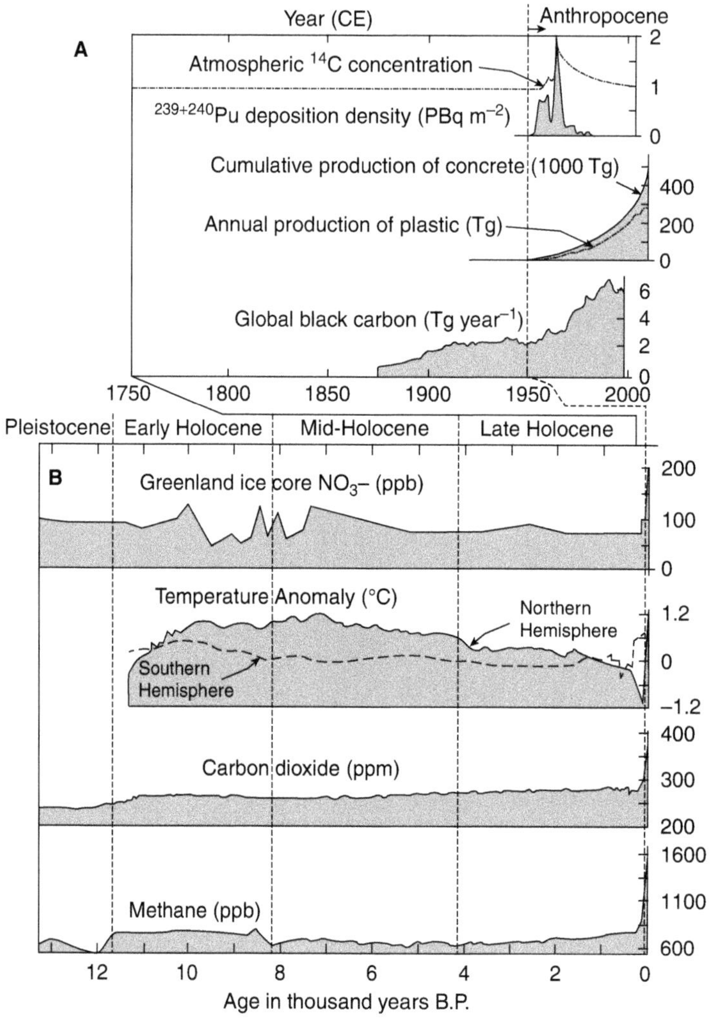

Source: Waters et al. (2016). *Permission to reproduce from The American Association for the Advancement of Science (AAAS).*

One indelible signature of the Anthropocene is species extinction. As noted previously, there are 8 to possibly 20 million or more species of eukaryotes, but only about 2 million have been recognised and named (Raven, 2020). Current extinction rates of species in various orders are estimated to have risen to 100–1,000 times the average extinction rate over the past tens of millions of years (the 'background rate') of 0.1–1 per million species per year (expressed as E/ MSY), and are continuing to rise.[117] In absolute terms, 1,000 species are becoming extinct every year if 10 million is taken to be the number of species and 100 E/MSY the current extinction rate.

Extinction rates are inferred from comparisons with fossil records in groups that have hard body parts, like vertebrates and molluscs, and from empirically drawn relationships between the number of species in an area and the size of the area (Box 4.1). But the latter relationships are known to vary

---

[117] See Wilson (2002, 2016), Sodhi, Brook and Bradshaw (2009), De Vos et al. (2014), Pimm et al. (2014), and Ceballos, Ehrlich and Ehrlich (2015).

substantially among communities and habitats, which is why, as the range shows, there are great uncertainties in the estimates. Despite the uncertainties, the figures put the scale of humanity's presence in the biosphere in perspective.[118] The figures also tell us why Earth scientists and ecologists say we are witnessing the *sixth* great biological extinction since life began.[119]

Judged by what is known about relatively well-studied groups (terrestrial vertebrates, plants), some 20% of the species could become extinct within the next several decades, perhaps twice as many by the end of the century. It is estimated that 84 mammal species have become extinct since 1500 and 32 species of mammal have gone extinct since 1900 (IUCN, 2020; Pimm and Raven, 2019).[120] In their recent survey of population data on nearly 30,000 species of terrestrial vertebrates, Ceballos, Ehrlich and Raven (2020) have estimated how many are on the brink of extinction. Their criterion was populations with fewer than 1,000 individuals. By this measure, 515 species are on the brink, representing 1.7% of the vertebrates on the authors' survey list. If extinction follows at the same rate, the population of terrestrial vertebrates will halve in about 40 years. But the rate is likely to increase at an accelerated rate, for several reasons. First, human pressure on the biosphere is increasing (see below); second, the distribution of those species on the brink coincides with hundreds of other endangered species, surviving precariously in regions with high human impact; and third, close ecological interactions among species tend to move other species toward annihilation – extinction breeds extinction. Assuming all species on the brink have experienced similar trends, the authors estimate that more than 237,000 populations of those species have vanished since 1900.

Box 4.1
Deforestation and Species Extinction

Human induced habitat destruction is today the leading cause of species extinction. A quarter of all tropical forests have been cut since the Convention on Biodiversity (CBD) was ratified 27 years ago. Pimm and Raven (2000) observed that generally speaking, many of the species found across large areas of a given habitat reside in small areas within it. That means habitat loss initially causes few extinctions, but the numbers rise as the last remnants of habitat are destroyed. At current rates of habitat destruction, the peak of extinctions may not occur for a long while, even decades.

The above reasoning follows also from species-area graphs familiar from island biogeography, which have the broad features of power functions. Writing the number of species by S and area by A, their relationship can be approximated as a two-parameter power function

$$S = \alpha A^\beta, \alpha > 0, 0 < \beta < 1 \tag{B4.1.1}$$

Rosenzweig (1995) reported that for birds, ants and plants β has been found to be in the region 0.2–0.8.[121] To see the salience of species-area relationships for estimating extinction rates, here

---

[118] Pimm and Raven (2019), in addition to providing the most recent estimates of extinction rates in various orders, includes an excellent overview of the methods that are deployed for estimating them. Wilson (1992), Wilson (2002, 2016) and Ehrlich and Ehrlich (2008) are expert dissections of the scale of the human presence in the biosphere. Dasgupta, Raven and McIvor (2019) is a collection of essays on biological extinctions, ranging from birds and mammals to microorganisms in the soils.

[119] The Sixth Extinction has also been extended in time and been called the Holocene Extinction, which began at the end of the last glacial period some 11,000 years ago, the signature being provided by the extinction of large land mammals. Kolbert (2014) is a narrative on the mass extinctions under way.

[120] We are grateful to Peter Raven for correspondence on this.

[121] The classic on this is MacArthur and Wilson (1967), which also provided an account of a process of immigration and emigration of which equation (B4.1.1) is an equilibrium.

Chapter 4: Human Impact on the Biosphere

is a rough estimate of extinctions that can be expected from the continuing destruction of tropical rainforests.[122]

Of the approximately 10,000 bird species today, some 5,000 inhabit tropical rainforests. As a reasonable approximation we set β = 0.25 in equation (B4.1.1). Suppose a further 50% of tropical forests were destroyed in the next 100 years. It would mean a loss of about 13% of bird species there, which would amount to 650 species. Other things equal, extinction of 650 species of birds in 100 years out of a total of 10,000 species of birds yields a figure of 650 E/MSY. That is either 65 times or 650 times the background extinction rate, depending on whether that rate is taken to be 0.1 E/MSY or 1 E/MSY.

Suppose, however, that humanity is able to restrain itself in the future and limits the destruction of tropical forests to only a further 25%. That would mean an eventual extinction of 6% of bird species, that is, 300 species. That is either 30 times or 300 times the background extinction rate, depending on whether that rate is taken to be 0.1 E/MSY or 1 E/MSY.

Suppose for a moment too, that humanity is able to come to grips with species extinction and limits tropical deforestation to only a further 0.8% over the next 100 years. That would mean an eventual extinction of 0.1% of bird species, that is, 10 species. Even that is 10 or 100 times the background extinction rate, depending on whether that rate is taken to be 0.1 E/MSY or 1 E/MSY. It is clear that destruction of tropical rainforests has to come to a complete halt if the extinction rates of birds are to be brought down to anything like background rates of species extinction. And we have not accounted for the millions of other, uncounted species that are being extinguished in those forests and elsewhere.

Species-area relationships allow one to estimate, albeit very crudely, extinction rates that follow habitat destruction. But one can flip the reasoning and ask what limits should be set on habitat destruction if bounds are set on further species extinction. There is a temptation to do that because one can then set the bounds by relating them to the background rate, as we have just done (Rounsevell et al. 2020).

But it is doubtful that the line of reasoning is fruitful. Even expert knowledge is so incomplete about species numbers and their distribution and mix, that setting extinction bounds would not provide a guide to policy. For example, the recorded number of species of mites is around 45,000 and there may perhaps be 1 million more; of nematodes around 25,000 and 500,000 more; and of fungi round 100,000 and 2.2 to 3.8 million more (Mueller et al. 2007; Kiontke and Fitch, 2013; Walter and Proctor, 2013; Hawksworth and Lücking, 2017).[123] There is vast uncertainty in these numbers. Moreover, unlike habitats, species numbers cannot be observed directly. So, it is not possible to place bounds on species extinction rates as policy targets when the number of species lies within a large range (perhaps 8 to 20 million). In contrast, habitat destruction can be observed and verified. The approach taken by the CBD in the Aichi Biodiversity Targets of 1992, which was to set limits on habitat destruction and specify Protected Areas is in line with this reasoning. That the targets are far from being met is not a fault in reasoning, it is, as in the case of international targets on carbon emissions, an inability of countries to design an enforcement mechanism.

---

[122] We are indebted to Stuart Pimm for it in correspondence.

[123] We are grateful to Peter Raven for correspondence on this.

4.1.2 Safe Operating Distances from Planetary Boundaries

Further evidence of the biosphere's degradation is adduced from a study of Earth System processes. The idea has been to identify processes of the biosphere that are critical for maintaining the stable state we experienced in the Holocene. Rockström et al. (2009) identified *nine* biophysical processes that are critical for Earth System functioning. The authors' proposal was to set quantitative boundaries for each, beyond which the Earth's Holocene state would be put at further risk, making the move to the Anthropocene firmer. The authors named the markers that may be used to check whether the processes are undergoing rapid change *planetary boundaries*. A planetary boundary is not equivalent to a global threshold or tipping point. In any case, not all nine key processes are known to possess single definable thresholds, and for those where a threshold is known to exist, there are uncertainties about where they might lie. Boundaries are placed upstream of these thresholds at the safe end of the zone of uncertainty.[124]

Although not all the nine processes have single identifiable markers, crossing the boundaries increases the risk of large-scale, potentially irreversible, environmental changes. Four of the nine processes have taken the planet into regions the authors regard as outside safe operating space, meaning that there is now increasing risk of a significant change from the biosphere's conditions in the Holocene. Biosphere integrity (for which one may read 'biodiversity') and nitrogen and phosphorus cycles have exceeded their boundaries farthest. But land-use change and climate change are also outside their safe operating space (Figure 4.5).

Figure 4.5 Critical Earth System Processes and their Boundaries

Source: Lokrantz/Azote based on Steffen et al. (2015). Note: P = phosphorus; N = nitrogen; BII = Biodiversity Intactness Index and E/MSY = extinctions per million species per year.

---

[124] Chapter 5 constructs the necessary language for quantifying these types of uncertainty.

Chapter 4: Human Impact on the Biosphere

Unravelling the notion of biosphere integrity has proved problematic. As Figure 4.5 shows, Rockström et al. (2009) had identified it with the extinction rate of species per million per year (E/MSY). One problem with the use of this metric is that extinction rates are estimated most often for vertebrate species (only amounting to <2% of described species). Mace et al. (2014) have argued moreover that extinction rates do not reflect the genetic library of life, nor the functional diversity of ecosystems, nor the conditions and coverage of Earth's biomes. The authors' observations speak to the markers of biodiversity we explored in Chapters 2 and 3. We see below that those markers reflect ominous features of the Anthropocene. A new boundary based on the Biodiversity Intactness Index (BII) is under development, but has yet to be quantified (Steffen et al. 2015a).

A further respect in which the idea of planetary boundaries has been extended is to study sub-global boundaries. This is important because, as was noted in Chapter 2, crossing a boundary at a regional level (e.g. destruction of the Amazon rainforest) can have implications for the whole Earth System. Regional level boundaries have now been developed for biosphere integrity, biogeochemical flows, land-use systems and freshwater use.

The idea of planetary boundaries has powerful heuristic appeal and has excited the public's imagination of the processes that govern the Earth System. It may have proved to be a problematic concept, but it is a useful classification of the Earth System's biogeochemical processes.[125]

Box 4.2
Deoxidation of the Oceans

To many people today the oceans are a source of cultural services. In fact, they are an essential part of the biosphere. They help to stabilise climate, produce oxygen, nurture biodiversity, store carbon and directly support us by providing food and nutrients. The OECD estimates that by 2030 the oceans will generate US$3 trillion of goods and services annually (OECD, 2016b). By the looks of the state of the oceans today, that is an ominous forecast, for that is very likely to increase further the burdens we have inflicted on them.

Oxygen is essential for life in the oceans, but alarmingly, the levels of oxygen in our oceans have been declining dramatically over the past 50 years. While it is natural to have some low oxygen areas in our seas, the size of these areas has expanded by 4.5 million km^2 – roughly the size of the European Union – and the volume of water with zero oxygen has quadrupled. In coastal waters, the number of sites with low oxygen has risen from 50 to 500, which is probably an underestimate due to a lack of comprehensive monitoring data around the world (Figure 4.6; Breitburg et al. 2018)).

---

[125] Steffen et al. (2018) have further advanced the concept.

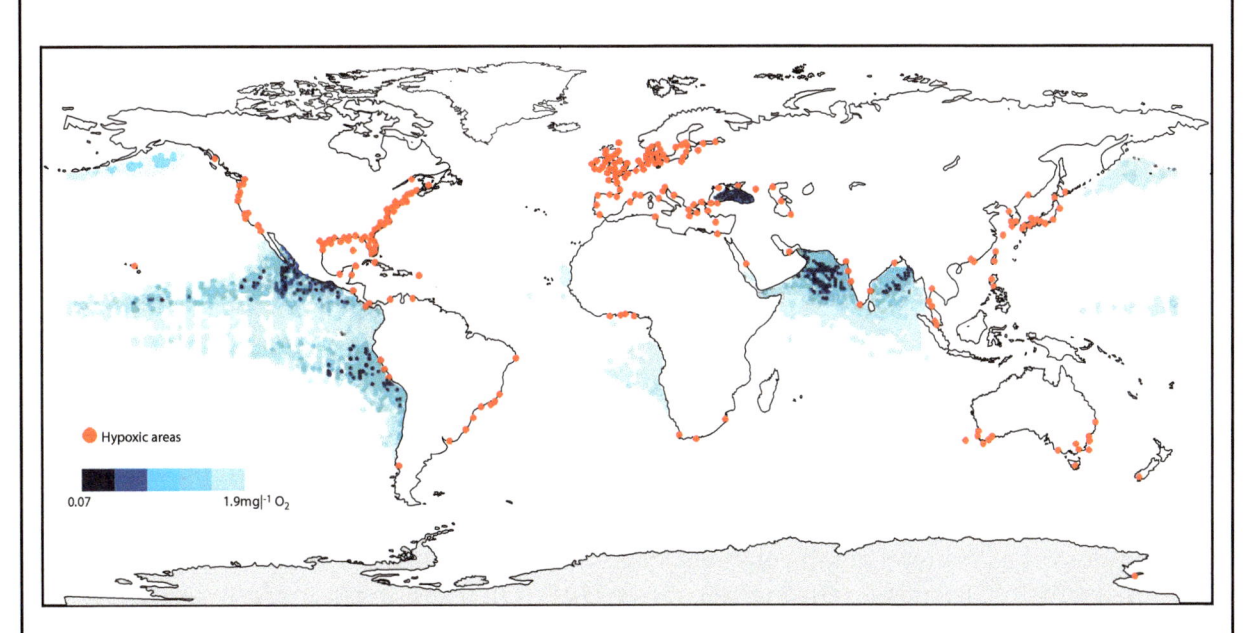

Figure 4.6 Low and Declining Oxygen Levels in Ocean and Coastal Waters

Source: Breitburg et al. (2018). *Permission to reproduce from The American Association for the Advancement of Science (AAAS).*
Note: Map created from data provided by R. Diaz, updated by members of the GO2NE network, and downloaded from the World Ocean Atlas 2009.

The reduction in oxygen in the oceans is largely due to human activities, including the global warming we are causing. Ocean warming reduces the solubility of oxygen in the water, but it also increases metabolic rates of organisms, thereby increasing oxygen consumption. This causes the water column to become more stratified, which in turn is likely to reduce the ventilation of oxygen into the ocean interior and reduce the availability of nutrients.

In coastal waters, oxygen declines are caused by increased levels of nitrogen, phosphorus and organic matter from agriculture and sewage, causing eutrophication. This increases the volume of organic matter reaching the sediments where microbial decomposition consumes oxygen. Once the oxygen levels are low the systems often do not return back to their original state.

What would happen to the biosphere if life in the oceans was to be extinguished? Here is a possible scenario:

Given that prokaryotes (bacteria, fungi) are masters of difficult environments, we could imagine they would be able to live, even thrive, in the oceans after everything else died. The biogeochemical cycles, planetary gas (for example CO_2) cycles and nutrient flows stemming from deaths in the oceans would rapidly cascade toward major, abrupt changes in ecosystems on land and in river systems; to an extent that one could reasonably envisage a steady major long-term decline in terrestrial biodiversity through extinctions. It could even be that the changes would be so enormous that life on land itself ceases for the vast majority of organismal lineages. Perhaps not total annihilation over time, but a mass extinction that has not occurred since life began.[126]

---

[126] We are grateful to Tim Littlewood of the Natural History Museum, London, for the scenario that is sketched.

4.1.3 Biodiversity Indicators

Erosion of natural capital usually goes unrecorded in official economic statistics because Gross Domestic Product (GDP) does not record depreciation of capital assets. Destroy biodiversity so as to build a shopping mall, and the national accounts will record the increase in produced capital (the shopping mall is an investment), but not the disinvestment in natural capital unless it commanded a market price. While industrial output increased by a multiple of 40 during the 20th century, the use of energy increased by a multiple of 16, methane-producing cattle population grew in pace with human population, fish catch increased by a multiple of 35, and carbon and sulphur dioxide emissions rose by more than 10. Human appropriation of terrestrial NPP has been variously estimated to be around 20–40%, and it is thought that over 50% are being appropriated in many of the most intensively farmed regions (Krausmann et al. 2013; Haberl, Erb and Krausmann, 2014; Williams et al. 2015). Human activity today deposits more nitrogen compounds into terrestrial and marine ecosystems than is generated in the natural nitrogen cycle (Vitousek et al. 1997). In Chapter 3, it was noted that discharges of phosphorus from agriculture into water bodies is a major ecological concern. Soil acidification, eutrophication of freshwater lakes and marine dead zones are among the consequences of nitrogen and phosphorus overload. These figures tell us much about the extent to which humanity is interfering with biogeochemical processes.

Biodiversity loss at a local level may be observable but tracking the state of the biosphere at the global level is no easy matter. Regulating and maintenance services (Chapter 2) are in any case hard to monitor. Which is why it is prudent to track biodiversity loss along as many routes as experience and evidence point to. Although the work of the Intergovernmental Panel on Climate Change (IPCC) on global climate change justifiably receives continuous global attention, the Millennium Ecosystem Assessment (MA, 2005a-d) recorded large-scale biodiversity losses in a wide range of ecosystems but has rarely been acknowledged by the public. The MA reported that 15 of the 24 ecosystems the authors had reviewed world-wide were either degraded or being exploited at unsustainable rates. The publication also reported that extraction of provisioning goods has increased, while regulating and maintenance services have declined. It also noted a decline in cultural services.

The MA noted, for example, that coastal zones account for some 20% of the Earth's surface but are inhabited by more than 45% of the world's population. An overwhelming majority of megacities are located there as well. The ecosystems in the zones include coral reefs, mangrove forests, salt marshes and other wetlands, seagrasses and seaweed beds, beaches and sand dunes, estuaries and lagoons, forests and grasslands (Turner, 2011). These ecosystems provide a range of services, including carbon, nutrient and sediment storage; water flow regulation; and quality control. They also serve as a buffer against storms and soil erosion. MA reported that over the previous three decades 50% of marshes, 35% of mangroves and 40% of reefs had been either lost or degraded.[127] In a decade-long study (2003 to 2014) that peers deeper into the state of the world's rainforests than satellite images are able to provide, Baccini et al. (2017) estimated that annually the forests captured around 435 million tonnes of carbon but lost over around 860 million tonnes, 70% of which was due to deforestation and, more generally, land degradation.

The publication by the Intergovernmental Science-Policy Platform on Biodiversity and Ecosystem Services (IPBES, 2019a) presents an extensive, spatially sensitive coverage of the biodiversity loss that is taking place today. The evidence collated there is even more disturbing than in the previous assessment by the MA. For example, since the early 1970s, there has been a decline in 14 of 18 categories of Nature's services, including purification of water, air quality, and disease regulation.[128] Reporting the

---

[127] See also Turner and Schaafsma (2015) for a fine collection of essays on the state of coastal ecosystems.

[128] MA (2005a-d) and IPBES (2019a) are reports on the state of the global environment. This Review should be seen as a complement to them. We allude to their findings, but do not rehearse them here in any detail. We use their findings as guides for framing the economics of biodiversity.

Chapter 4: Human Impact on the Biosphere

ongoing work on the Amazon by environmental scientists (see Lovejoy and Hannah, 2019), IPBES notes that the Amazon rainforest has shrunk by a sixth since the UN Convention on Biological Diversity was established in Rio de Janeiro in 1992. The publication also reports that the extent and condition of our ecosystems have declined by nearly 50% from their natural state and that only 23% of the land and 13% of the sea remain classified as 'wilderness'. Figure 4.7 presents the authors' findings in greater detail.

Figure 4.7 Global Trends in the Capacity of Nature to Sustain Contributions to Good Quality of Life from 1970 to the Present

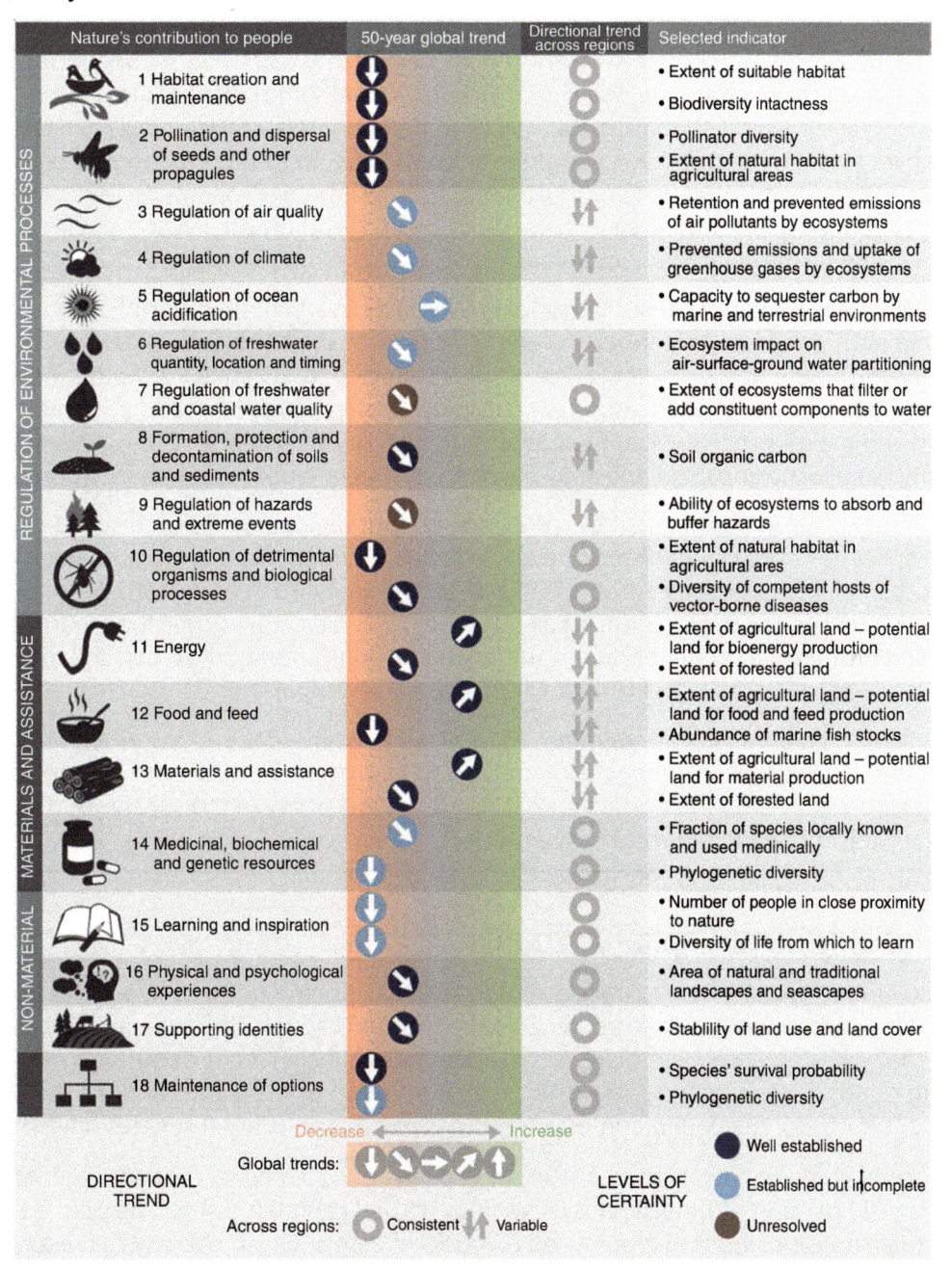

Nature's contribution to people	50-year global trend	Directional trend across regions	Selected indicator
1 Habitat creation and maintenance	↓ ↓	○ ○	• Extent of suitable habitat • Biodiversity intactness
2 Pollination and dispersal of seeds and other propagules	↓ ↓	○ ○	• Pollinator diversity • Extent of natural habitat in agricultural areas
3 Regulation of air quality	↘	↕	• Retention and prevented emissions of air pollutants by ecosystems
4 Regulation of climate	↘	↕	• Prevented emissions and uptake of greenhouse gases by ecosystems
5 Regulation of ocean acidification	→	↕	• Capacity to sequester carbon by marine and terrestrial environments
6 Regulation of freshwater quantity, location and timing	↘	↕	• Ecosystem impact on air-surface-ground water partitioning
7 Regulation of freshwater and coastal water quality	↘	○	• Extent of ecosystems that filter or add constituent components to water
8 Formation, protection and decontamination of soils and sediments	↘	↕	• Soil organic carbon
9 Regulation of hazards and extreme events	↘	↕	• Ability of ecosystems to absorb and buffer hazards
10 Regulation of detrimental organisms and biological processes	↓ ↘	○ ○	• Extent of natural habitat in agricultural ares • Diversity of competent hosts of vector-borne diseases
11 Energy	↗ ↘	↕ ↕	• Extent of agricultural land – potential land for bioenergy production • Extent of forested land
12 Food and feed	↗ ↓	↕ ↕	• Extent of agricultural land – potential land for food and feed production • Abundance of marine fish stocks
13 Materials and assistance	↗ ↘	↕ ↕	• Extent of agricultural land – potential land for material production • Extent of forested land
14 Medicinal, biochemical and genetic resources	↘ ↓	○ ○	• Fraction of species locally known and used medinically • Phylogenetic diversity
15 Learning and inspiration	↓ ↓	○ ○	• Number of people in close proximity to nature • Diversity of life from which to learn
16 Physical and psychological experiences	↘	○ ○	• Area of natural and traditional landscapes and seascapes
17 Supporting identities	↘	○	• Stablility of land use and land cover
18 Maintenance of options	↓ ↓	○ ○	• Species' survival probability • Phylogenetic diversity

DIRECTIONAL TREND — Global trends: Decrease ← ↓ ↘ → ↗ ↑ → Increase

Across regions: ○ Consistent ↕ Variable

LEVELS OF CERTAINTY: ● Well established ● Established but incomplete ● Unresolved

Source: IPBES (2019a).

The Economics of Biodiversity: The Dasgupta Review

Chapter 4: Human Impact on the Biosphere

Box 4.3
Soil Biodiversity Loss

Soil erosion is usually slow in stable ecosystems but accelerates with the removal of vegetation; for example, deforestation.[129] According to a 1998 estimate, we obtain more than 99% of our food calories from land-based products, even while loss of soil organic carbon through conversion to agriculture is significant (Pimentel, 2006; Sanderman, Hengl, and Fiske, 2017).[130] Studies suggest that some 80% of the globe's farmland has moderate to severe erosion, first (surprisingly, to the uninitiated) from water and second from wind. Wetlands hold specific types of soil, rich in carbon and nutrients (as in peatlands; Box 4.7). Nearly 90% of wetlands have been lost over the past 300 years; about 35% since 1970 (IPBES, 2018). Collating data on soil erosion, WWF (2017) reported that some half of all top soils have eroded in the past 150 years. A typical estimate is that 75 billion tonnes of soil erode annually at a rate 13 to 40 times the background rates of erosion that prevailed before the acceleration caused by human dominance of the biosphere (Pimentel and Kounang, 1998). The rate of soil erosion accompanying land-use change is judged to be the highest in the past 500 million years (Wilkinson and McElroy, 2007), and some regard it to be the greatest geomorphic agent on the planet today (Hooke, 2000).

What happens when the diversity of life within soil is lost? Wagg et al. (2014) found a strong relationship between ecosystem functions and indicators of soil biodiversity. Reductions in soil biodiversity contribute to eutrophication of surface water, reduced above-ground biodiversity and global warming. Declines in soil biodiversity cause declines in performance of a number of regulating and maintenance services (Bender, Wagg and Van Der Heijden, 2016). Alarmingly, if soil biodiversity were lost completely, the land-based food system would cease to function.

Soil biodiversity loss can be identified by combining quantitative estimates of the circumstances and substances that destroy soil organisms. They include habitat fragmentation, invasive species, climate change, urban sprawl over soils, soil erosion, and soil pollution such as industrial fertilisers and pesticides. Moreover, soil degradation accelerates runoff, and erosion moves the organic sediments, rich in macronutrients, to water bodies, resulting in eutrophication and oxygen collapse in aquatic ecosystems. Dead zones, as in the Gulf of Mexico, are an example.

Once lost, can soil biodiversity be restored? Reduced soil disturbance and increased organic matter as well as the use of deeper rooting crop varieties can help improve soil health, as can cover crops, changes to crop rotations, and no-till approaches. Such practices are the substance of 'organic farming', a subject that we return to in Chapter 16.

4.1.4 Global Natural Capital Accounts

National accounting systems do not track our use of the biosphere's goods and services. GDP does not record the depreciation (nor possible appreciation) of natural capital. The evidence we have cited here so far says that while modern technology has enabled humanity in recent decades to obtain provisioning goods at an increasing rate (the Green Revolution of the 1960s and the 1970s was the defining event for that), regulating and maintenance services and cultural services have shrunk.

If we were to think of the biosphere as a stock of capital – natural capital – its net regeneration rate would be its yield per unit of time. By this reckoning, natural capital is a stock, its yield is a flow. Of course, that stock is composed of a myriad of stocks of assets, which we may call *natural resources*,

---

[129] The material here is based on Beach, Luzzadder-Beach and Dunning (2019).

[130] Pimentel (2006) estimated that the costs of soil erosion in the US is over US$38 billion.

some being non-renewable (fossil fuels), while others are self-renewable with regeneration rates that can differ by orders of magnitudes (bacteria, minnows, whales, redwoods). Thus, the biosphere's yield in a period of time is not a single number, but the yields of a myriad of goods and services. Accounting for them provides the beginnings of a way to record the value of the biosphere's (net) regeneration rate. By 'value' we do not mean market value, for many forms of natural capital do not have markets at all – they are free to all who use them – and those that do reflect distorted values owing to institutional imperfections. So, by value we mean 'accounting value'.

The *accounting value* of the stock of an asset is its contribution to societal well-being; that is, it reflects the asset's social worth. The asset's *accounting price* is the accounting value of a (marginal) unit of the asset. The way natural capital accountants estimate the value of ecosystems is to estimate the accounting value of the flows of goods and services provided by ecosystems, and then estimate the corresponding accounting value of the ecosystems themselves by computing the present (discounted) value (PDV) of the flows. Chapter 13 shows how that conversion is made.

A country's *natural capital accounts* constitute a system that records the state of the economy's natural capital. The idea is to impute an accounting value to each type of natural capital and then add the values to reach the accounting value of the entire stock of natural capital to which the economy has 'claim'. The notion of claim, or the related notion of 'ownership', is the subject of Chapters 7–9.

That is all well and good in theory. In practice, estimating stocks and their accounting prices is so fraught with difficulty that the natural capital accounts that have so far been developed for sectors (e.g. Kareiva et al. 2011), national economies (e.g. Arrow et al. 2012) and the global economy (e.g. UNU-IHDP and UNEP, 2012, 2014; Managi and Kumar, 2018) are yet nowhere near as polished as the national accounts of the breakdown of the gross domestic product of national economies. Practical methods for estimating accounting prices cut through many of the problems, if only to provide simple but informative pictures of the time trajectory paths of the accounting value of natural capital, including sub-soil resources.

A decline in the accounting value of natural capital in an economy over a period of time would mean *depreciation* of natural capital. The estimates provide quantitative information on changes in the state of a country's natural capital, and are meant to be placed in parallel to estimates of changes in the value of produced capital and human capital. The aggregate value of a nation's produced capital, human capital and natural capital is called *inclusive wealth*, a concept that was introduced in Chapter 1.

As we noted in Chapter 1, inclusive wealth taken on its own is meaningless. It is *changes* in inclusive wealth that are not only meaningful, but of enormous use to anyone wanting to understand the meaning of sustainable development. The difference between inclusive wealth in one year and in the next measures *net inclusive investment* in the country's capital goods that year. The idea of inclusive wealth and inclusive investment extends naturally to the global economy. We develop these ideas more fully in Chapter 13.

Managi and Kumar (2018) tracked the accounting values of produced capital, human capital and natural capital over the period 1992 to 2014 in 140 countries.[131] The authors built their estimates using the UN data base. In their work, renewable resources include forest resources (stocks of timber and a selected group of non-timber resources), fisheries (stocks were estimated from past records of catch), agricultural land (cropland and pastureland); while non-renewable resources cover fossil fuels and a selected set of minerals. Figures for the social cost of carbon (a negative accounting price) of greenhouse gas emissions were used to address future losses from global carbon change (see

---

[131] The value of produced capital (PC) was obtained from official national accounts. Data limitations meant that natural capital (NC) was limited to minerals and fossil fuels, agricultural land, forests as sources of timber, and fisheries. Market prices were used to value them. The accounting value of human capital (HC) was estimated by using the approximations in Arrow et al. (2012) for both education and health.

Chapter 4: Human Impact on the Biosphere

Chapter 10). Figure 4.8 displays the authors' estimates of global per capita accounting values of the three classes of capital goods over the period 1992 to 2014. It shows that globally produced capital per head doubled and human capital per head increased by about 13%, but the value of the stock of natural capital per head declined by nearly 40%.

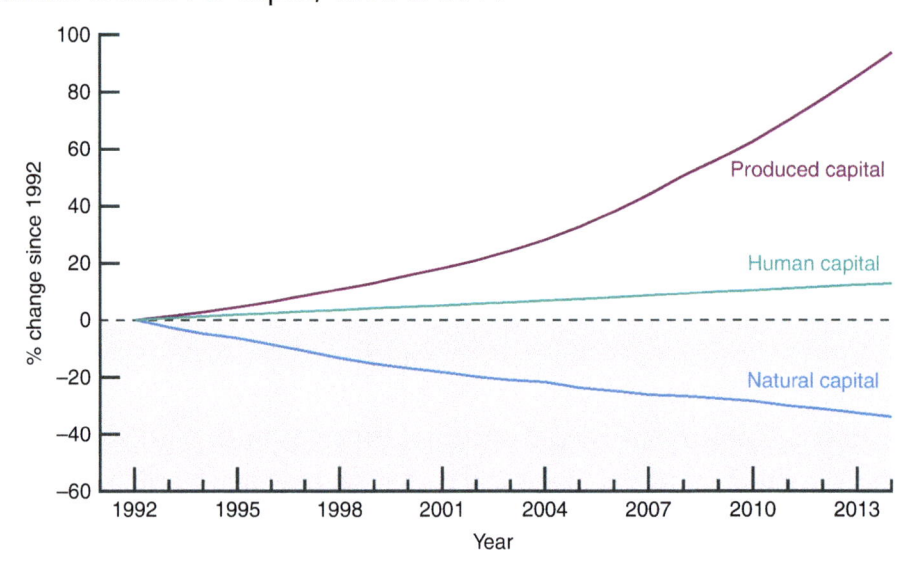

Figure 4.8 Global Wealth Per Capita, 1992 to 2014

Source: Managi and Kumar (2018).

4.2 Demand and Supply

In a classic decomposition of humanity's impact on the biosphere, Ehrlich and Holdren (1971) identified global population size, individual demands on the biosphere – as reflected in, say, living standards – and the technologies and institutions in play as shaping humanity's demand for the biosphere's goods and services. We build on their analysis by decomposing the demand in quantitative terms. We first decompose the demands that are made, but by smaller economic units – from national economies to village economies in poor countries. We then sum the demands to arrive at the aggregate demand of the global economy. The decomposition of demand allows us to identify the overarching factors that determine it. It is then possible to ask what steps need to be taken in order to alter the demand. We then compare that demand with the biosphere's regeneration rate. When humanity's demand exceeds the regeneration rate, the biosphere depreciates; if it were to be less, the biosphere would appreciate. Uncovering the relationship between demand and supply is the core of the Review.[132]

4.2.1 Aggregate Demand

Humanity's demands on the biosphere per unit of time – Ehrlich and Holdren called it *impact* – is known today as the *global ecological footprint*. We begin by defining the footprint of smaller economic units and then sum them to define the global footprint. For convenience we shall use the terms 'ecological footprint' and 'human impact' interchangeably.

---

[132] This section is based on Barrett et al. (2020).

Let us divide the global economy into distinct economic units, labelled by *i*, numbered as 1, 2, ... and so on. Depending on the context, the units are individuals (that is the relevant partition of population when sociologists study age-related consumption patterns), households (the relevant partition for national environmental policy), nations (the relevant partition in climate negotiations) or the world as a whole (the scope of this Review). Let N_i be the population size of *i* and y_i an index of *human activity* per person in *i* per unit of time. Then $N_i y_i$ is aggregate activity by members of *i*.

All human activity requires the biosphere's goods and services as inputs. So, we need to link y_i to the demands the average person in economic unit *i* makes of the biosphere. Estimating y_i poses huge measurement problems, so for tractability we suppose it corresponds to the standard of living as measured by income per capita in *i*. For example, if *i* is a household, y_i is income per head in the household; if *i* is a nation, y_i is GDP per capita in that country; and so on. Using income as a measure of human activity almost surely yields an underestimate of what we are after, for there are many human activities that are not captured in income as measured by economic statisticians. On occasion, national income statisticians offer estimates of the magnitude of economic transactions that are missing in gross domestic income (equivalently gross domestic product, GDP), for example, the size of the black economy, but they are too scanty to be of use here. And there are human activities that would not be covered even by those corrections. So even though we know income per capita in *i* is an underestimate of the activity of the average person in *i*, we shall use it as a proxy.

We now place our analysis on the global economy. Let *N* denote the global population, *y* per capita global GDP, and let *i* cover the world's population. Then

$$Ny = {}_i\Sigma N_i y_i \qquad (4.1)$$

(In equation (4.1) the incomes y_i are summed over the whole world's population.) We now trace *y* to the biosphere's goods and services.

The demands we make of the biosphere take two forms: (1) We harvest Nature's goods and use Nature's services for consumption and production. Fish, timber and fresh water constitute goods; whereas pollination, water purification, flood protection, and carbon sequestration and storage constitute services. (2) We use the biosphere as a sink for our waste products. Landfills, rivers carrying pollutants into estuaries and carbon concentration in the atmosphere are examples of our use of the biosphere as a sink for our waste.

Let *X* denote what we extract or harvest from the biosphere and let *Z* denote the demand we make of the biosphere as a pollution sink. As both are functions of *Ny*, we write $X = X(Ny)$ and $Z = Z(Ny)$. The *X*-function records that both production and consumption require the biosphere's goods and services as inputs, while the *Z*-function reflects the fact that waste products are inevitably associated with production and consumption and they impose a strain on the biosphere. Partitioning our ecological footprint into *X* and *Z* reconfirms that pollution is the reverse of conservation.

Let α_X be a numerical measure of the efficiency with which the biosphere's goods and services are converted into global GDP; and let α_Z be a numerical measure of the extent to which the biosphere is transformed by global waste products (the latter in part depends on the extent to which we treat our waste before discharging them). So, we have $X = Ny/\alpha_X$ and $Z = Ny/\alpha_Z$.

Define $\alpha = (\alpha_X + \alpha_Z)/(\alpha_X \alpha_Z)$. Then $(Ny/\alpha_X + Ny/\alpha_Z) \equiv Ny/\alpha$ is our proxy measure of the global ecological footprint.[133] Writing the latter as *I* (Ehrlich-Holdren's 'Impact') we have

---

[133] As noted previously, Ehrlich and Holdren (1971) had called what is today called 'ecological footprint', 'human impact'.

Chapter 4: Human Impact on the Biosphere

$$I = (Ny/\alpha_X + Ny/\alpha_Z) \equiv Ny/\alpha \qquad (4.2)$$

The distribution of global GDP affects the efficiency coefficients α_X and α_Z, but here we are concerned with global aggregates.[134] In Section 4.5, we study the distribution of ecological footprints across households, villages, regions, or other disaggregated groups of institutions. There we will replace the word 'biosphere' by the term 'ecosystem'.[135] The distribution of components of the Impact Inequality is considered further in Chapter 14.

Decoupling the global ecological footprint, Ny/α, also serves to remind us that measures to reduce environmental pollution, Z, can raise our demand for the biosphere's products (X). Solar panels require minerals such as aluminium, cadmium, and zinc. But to obtain those minerals usually requires fragmenting forests (Section 4.7 and Chapter 3).

Box 4.4
Impact of the Fast Fashion Industry

The global fast fashion industry relies on cheap manufacturing to encourage more frequent purchase. Garments are discarded well before their physical life span. This consumes large amounts of textiles (rising from 5.9 kg to 13 kg per capita between 1975 and 2018) and has significant environmental impact (Peters, Sandin and Spak, 2019; Niinimäki et al. 2020). The industry has periodically been subjected to adverse publicity, but it continues to grow. In 2019, the industry was worth around US$36 billion globally and is expected to be worth around US$38 billion in 2023 (Research and Markets, 2020). Its environmental impacts include water use, chemical pollution, carbon dioxide emissions and textile waste. It is estimated that the industry produces 8–10% of global emissions of CO_2 annually and uses over 79 trillion litres of water per year (Niinimäki et ki et al. 2020). The industry is responsible also for pollution from textile treatment and dyeing, amounting to around 20% of industrial water pollution (Kant, 2012) and is responsible for approximately 35% (190,000 tonnes per year) of oceanic microplastic pollution (United Nations Climate Change, 2018). Over 92 million tonnes of textile waste ends up in landfill or is burnt (Niinimäki et ki et al. 2020).

4.2.2 Aggregate Supply

Let G denote the accounting value of the biosphere's regenerative rate (Chapter 2). G is a function of the real accounting value of the stock of the biosphere, which we write as S. Thus $G = G(S)$. This requires a heroic (read impossible!) feat of aggregation, because the biosphere has a modular structure. Depending on the fineness of the grid with which we choose to define our spatial unit, we would need weights on biospheric material in every square on the grid, to measure the material in it and estimate the weighted sum of the material across the grid.[136]

---

[134] Equation (4.2) could be thought to be saying that for any given value of α, N and y are in inverse relation to each other – for instance, that if N was to double then y would have to halve if global impact Ny/α was to be held constant. Arithmetically that would be correct, but the thought could be misconstrued. The reason is that N and y are not independent of one another: we humans have not only mouths, but hands and brains too. Thus, y is a function of N. Chapter 4* presents a complete capital model from which 'trade-offs' between N and y can be estimated at various stages of economic development. Dasgupta and Dasgupta (2021) have constructed simple methods that can be deployed on the model to calculate trade-offs between N and y that are consistent with sustainable development.

[135] The decomposition of I in equation (4.2) when the impact I in question is global carbon emissions is known as the Kaya Identity. See Kaya and Yokobori (1997).

[136] For simplicity, it may help to interpret biospheric material as biomass. The relevant set of weights would then be different, of course, because biospheric processes involve interactions between non-biomass material and biomass.

That would be S. The weights to use are the accounting prices we encountered in Section 4.1.4. Invoking the function G(S) here serves only as a heuristic device for explaining humanity's overshoot in its demands on the biosphere. The function points to where policy can be directed; it is not meant for determining policy.

In Figure 3.2 and Box 3.3, we depicted the widely used, cubic form of G with a threshold L, such that if S were to cross it, the ecosystem would be a 'dead zone'. Here we go beyond the quadratic form and reconfirm its general features, such that G is a declining function of S at large values of S. The analogy is with the fishery, modelled in Box 3.3, which is bounded in extent and so has a finite carrying capacity. In the range of stocks we are concerned with here (stocks below the level capable of sustaining maximum sustainable yield), $dG/dS > 0$; that is, locally G increases if S increases. For simplicity of exposition we are assuming here that G is a deterministic function. In fact, the biosphere is governed by stochastic processes, meaning that G is a stochastic function. In Chapter 5, we show how policy can be designed in a stochastic world.

The G-function can be affected by policy. Investment in biotechnology is one general class of policies. A recent experiment in American Samoa found that transplanted heat-tolerant corals were more likely to survive a bleaching event than less tolerant local corals, enabling quicker recovery of the ecosystem after such an event – a technological intervention known as 'ecosystem engineering' (Morikawa and Palumbi, 2019). But importing foreign species into ecosystems has been known to have unintended, detrimental consequences. More familiar, still strangely controversial, interventions involve genetically modified crops, which can raise food production. In Chapter 16, we discuss sustainable food prospects, where genetically modified crops can play a vital role.

4.3 The Impact Inequality

Section 4.1 produced evidence that over several decades aggregate demand per unit of time, $Ny/\alpha_X + Ny/\alpha_Z$, has exceeded aggregate supply $G(S)$ per unit of time. That reads as

$$Ny/\alpha > G(S) \tag{4.3}$$

We call expression (4.3) the *Impact Inequality* (see also Figure 4.9).[137]

The Impact Inequality as presented in expression (4.3) applies to the biosphere as a whole. Although the notion of ecological footprint (the left-hand side of the Inequality) can be applied to any group of individuals – from the individual and the household, to nations and the global population – trade in commodities and services breaks the link between demand (Ny/α) and supply ($G(S)$) for economic units smaller than the world as a whole. The ecological footprint of a nation will not balance the regenerative rate of its ecosystems if its trade in the biosphere's goods and services does not balance, in units of biospheric material. Of course, it could be that a country pays for its imports, perhaps even at their appropriate prices, but that is a different matter. Here we are only formulating a way to break down the global imbalance of demand and supply of those goods and services into imbalances among groups in the global population; we are not discussing 'fair trade'. Trade is discussed further in Chapter 15.

---

[137] The Impact Inequality was formulated in this form by Barrett et al. (2020).

Chapter 4: Human Impact on the Biosphere

Figure 4.9 The Impact Inequality

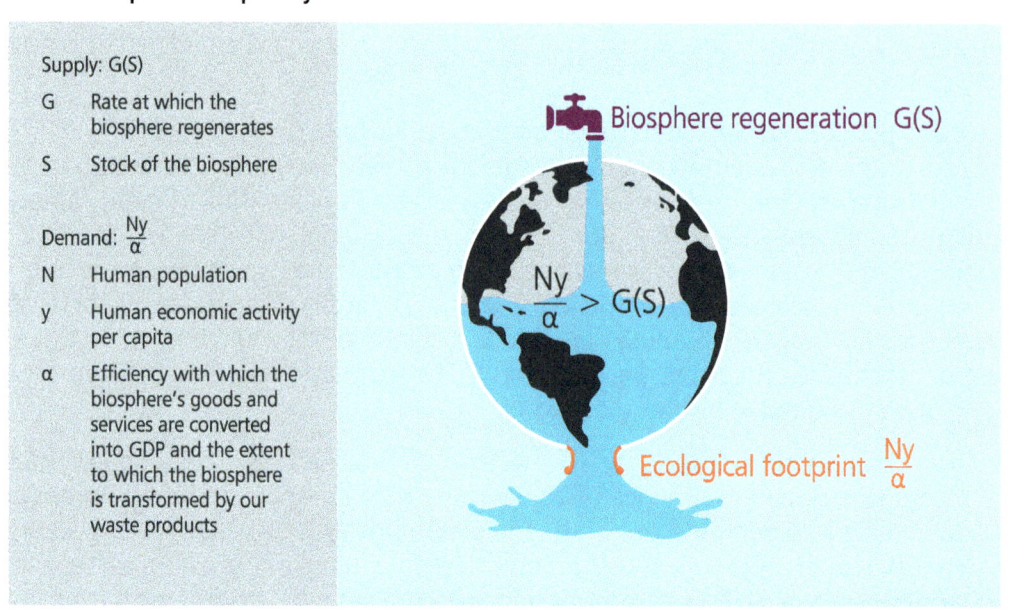

Supply: G(S)
G   Rate at which the biosphere regenerates
S   Stock of the biosphere

Demand: $\frac{Ny}{\alpha}$
N   Human population
y   Human economic activity per capita
α   Efficiency with which the biosphere's goods and services are converted into GDP and the extent to which the biosphere is transformed by our waste products

Biosphere regeneration G(S)

$\frac{Ny}{\alpha} > G(S)$

Ecological footprint $\frac{Ny}{\alpha}$

Source: Described in Barrett et al. (2020).

If the global ecological footprint, *I*, exceeds the biosphere's regenerative rate, *G*, the biosphere as a stock diminishes, and the gap between *I* and *G* increases. Similarly, if the footprint is less than the biosphere's regenerative rate, the stock increases, and the gap between *I* and *G* shrinks.[138] However, either global population (*N*) or global output per capita (*y*), or both, could increase without making additional demands on the biosphere *provided* either α_X or α_Z and thus α was to increase correspondingly. Improvements in technology (e.g. substituting degradable waste for persistent pollutants; decarbonising the energy sector) and institutions and practices (e.g. establishing Protected Areas; reducing food waste), and appropriate redistributions of wealth are among the means by which α can be raised.

The factors affecting our demand for the biosphere's goods and services, namely *N*, *y*, α_X and α_Z, affect one another. When a region of the Amazon rainforest is converted into cattle ranches, the transformation would be expected to raise food production by raising the efficiency with which land is used to grow crops (a rise in the corresponding α_X), but it lowers α_Z (industrial fertilisers and pesticides degrade the soils and water bodies). The transformation could be read as reducing *S*, or alternatively, because the composition of the biome changes, it could be read as a less productive *G*-function. The overall effect would be to widen the gap between $G(S)$ and $(Ny/\alpha_X + Ny/\alpha_Z)$.

In an ingenious set of exercises, Wackernagel and Beyers (2019) have estimated *G* by calculating the area of land and sea surface covering different categories of ecosystems (agricultural land, plantations, wetlands, fisheries, marshes, oceans, forests, and so on) that is needed, given existing technologies, to meet humanity's current demands for various provisioning goods, while leaving space to allow for other life forms to provide pollination, seed dispersal, fertilization, and decomposition of waste. Thus, the land-sea area required to meet our demand for natural fibres on a sustainable basis is taken to be Earth's bio-capacity for that demand. And so on for our other demands. However, ecosystems differ in their ability to provide the same service. For example, marshes sequester 10 times the carbon temperate forests do. A sq. m of marshland is therefore awarded a weight 10 as against 1 for a sq. m of a temperate forest, and

---

[138] Bifurcations leading to regime shifts and irreversibilities do not affect the formulation. Crossing a bifurcation amounts to a discontinuous change in the biosphere's regenerative rate.

so on. The authors obtain G by estimating the weighted sum of the areas required. Using that apparatus, they calculate that the ratio of the left-hand side to the right-hand side in expression (4.3) is currently about 1.7, whence the metaphor that we need 1.7 Earths to meet humanity's current demands from it. Whatever the term 'sustainable development' could mean, it must as a minimum mean transforming the Impact Inequality into an equality; that is, reducing our overreach of the biosphere to zero. (Net-zero emissions is the corresponding idea when restricted to the Earth system as a sink for our carbon emissions.)

Box 4.5
The Idea of Indefinite Economic Growth

The literature on the economics of climate change has mostly taken future projections of Ny as given and has focused instead on raising the efficiency parameters α_X and α_Z by decarbonising the economy and removing CO_2 from the atmosphere, and by raising the G-function by geo-engineering. Why then have α_X, α_Z, and the G-function not risen more to close the gap between carbon emissions and the biosphere's capacity for assimilating carbon? The reasons are low rates of innovation and investment in non-fossil fuel energy sources, carbon capture and carbon storage technologies. Those low rates, in turn, have been due to persistent and pervasive institutional failure to achieve global collective action in limiting climate change. Despite nearly 30 years of diplomatic effort, the world has been unable to overturn the tragedy of the climate commons.[139]

But the reason carbon concentration in the atmosphere has increased is not only that de-carbonisation and direct carbon removal have been slow, it is also that growth in both global GDP per capita (y) and world population (N) have been strong. Ironically, publications on the economics of climate change and international negotiations over carbon emissions have not only *not* questioned the desirability of continual global economic growth, they would appear to have taken as given that it is the only viable route for (i) reducing carbon emissions, (ii) eliminating global poverty, and more generally (iii) ensuring that development is sustainable. That has been the implicit assumption underlying the UN's SDGs.

That stance has given rise to a paradox: *growth in global output (Ny) is seen as necessary for providing the funds that will be needed for reducing our ecological footprint (Ny/α), even though growth in global output is known to increase the footprint*.

In the chapters that follow, we move away from the viewpoint that has given rise to the paradox. The viewpoint is built on the thought that humanity is *external* to the natural world. It sees us as dipping into the biosphere for its goods and services, transforming them for our production and consumption, and then discharging the residue into the biosphere as waste. The Review is in contrast built on a recognition that humanity is *embedded* in the natural world. It will be shown that this somewhat metaphysical distinction – being 'external to' and being 'embedded in' – has enormous implications for our conception of future economic possibilities. The conception we adopt here says that Ny cannot be increased indefinitely: it is instead *bounded* (Chapter 4*).

4.4 Two Notions of Inequality

The decomposition of Impact into N, y, and α has been expressed in aggregate terms. Quite obviously, it is decomposable into income groups. Let i, j, variously denote households. Households differ according to their incomes, y_i, y_j, and so on, but they differ as well with regard to the efficiency with which they convert the biosphere's goods and services into income. It is conventional to view inequality in terms of

---

[139] In a sustained research programme, Scott Barrett has explored reasons for the failure of climate negotiations and has proposed treaty architectures that would encourage compliance. See Barrett (2003, 2012, 2015) and Aldy, Barrett, and Stavins (2003).

Chapter 4: Human Impact on the Biosphere

the distribution of household incomes, but in the Review we are interested also in inequality among households in terms of the impact they have on the biosphere. The latter is reflected in the distribution of incomes when corrected for the efficiency with which the biosphere's goods and services are converted into income (i.e. y_i/α_i). And income and income in efficiency units are not the same. We may read y_i/α_i as household i's ecological footprint.[140]

Imagine that you label households by income and rank them in terms of increasing income. In an economy with N households, we then have $y_i < y_{i+1}$ for $i = 1, ..., N$. The IPCC (2014) reported from cross-national statistics that carbon emissions are an increasing function of income. There is a corresponding finding that says ecological footprint is an increasing function of income (IPBES, 2019a; Wackernagel et al. 2019). So we assume that $y_i/\alpha_i < y_{i+1}/\alpha_{i+1}$, for all i. In this reading, households enjoying higher income demand more from the biosphere.

A question arises whether the curve y_i/α_i is convex or concave. Consider an income interval where the function is convex. An egalitarian redistribution of incomes among households in that interval would lead to a smaller global ecological footprint, implying there is no conflict between income equality and the biosphere's integrity. But in a concave interval, the reverse holds: *egalitarian redistributions of incomes would lead to larger global ecological footprints and society would face a cruel choice between income equality and the biosphere's integrity* (A. Dasgupta and Dasgupta, 2017).[141] Figure 4.10, which displays a regression between the ecological footprint of nations and GDP per capita, shows that our demand for the biosphere's goods and services increases with affluence and development but that the efficiency with which we transform them so increases with affluence that ecological footprint is a concave function of income at all levels of incomes. In short, ecological footprint rises less than proportionately with income. That suggests, ominously, that egalitarian redistributions of incomes lead to larger global ecological footprints, other things the same.[142] Distribution is discussed further in Chapter 14.

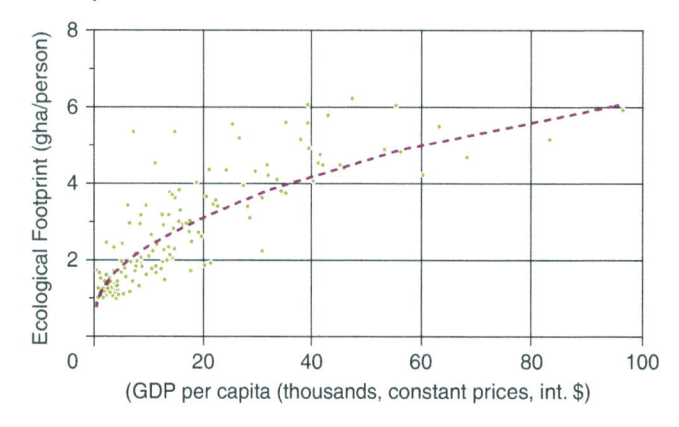

Figure 4.10 Ecological Footprint and Income

Source: World Bank (2020a), Global Footprint Network (2019), and Review calculations.

---

[140] For notational simplicity we integrate extraction and pollution and speak of a single efficiency index, α.

[141] The modern classics on the meaning of egalitarian income redistributions are Kolm (1969) and Atkinson (1970). We avoid details here, but readers should note that Figure 4.10 (below in the text) says the global ecological footprint would be higher than it is today if everyone in the world was to receive an equal share of today's global income.

[142] Figure 4.10 combines data from the World Bank on per capita GDP of nations and estimates from the Global Footprint Network on ecological footprint per capita (in global hectares). A total of 136 countries are in the sample, including both developed and developing countries. Reinterpreting the Impact Inequality, if I denotes ecological footprint and y is income per capita, the estimated functional form is $I = 0.97y^{0.41}$, with $R^2 = 0.71$.

4.5 The Impact Equation

Over the long run, global demand (per unit of time) must equal the biosphere's ability to meet that supply (per unit of time) on a *sustainable* basis. The widely discussed UN SDGs were formulated on the assumption that they can be attained, but the background documents did not probe the question of whether they are sustainable in a global economy that simultaneously enjoys growth in global GDP.

The economics of biodiversity involves a dynamic resource allocation problem (Chapter 4* and Chapter 13*). The demand we make on the biosphere per unit of time does not have to equal the biosphere's ability to supply goods and services per unit of time, because the difference is naturally accommodated by a change in the biosphere's stock S. A world rich in healthy ecosystems could, on utilitarian grounds, choose to draw down the biosphere and use the goods and services it supplies so as to accumulate produced capital and human capital. That is what economic development has come to mean among many thinkers, but the scenario comes in tandem with an overshoot in our demands from the biosphere. The overshoot cannot be maintained indefinitely because our life support system would be threatened.

We therefore work backwards, by first identifying a condition the global economy's treatment of the biosphere must satisfy if the SDGs are themselves to be sustainable. That condition tells us to find ways in which S can be stabilised. To *sustain* that stabilised value of S requires that the global ecological footprint equals the biosphere's regenerative rate, that is,

$$Ny/\alpha_X + Ny/\alpha_Z \equiv Ny/\alpha = G(S) \tag{4.4}$$

Equation (4.4), which we call the Impact Equation, applies to the biosphere as a whole.

The Impact Equation is a condition of *global sustainability*. The equation does not say what S should be. There is an entire range of values of S and corresponding sets of values of the remaining factors, N, y, α_X, α_Z, and the parameters of the G-function for which equation (4.4) can be expected to hold. The requirement that a state of affairs is *sustainable* is that it can persist indefinitely. That is different from a requirement we may insist on, that the state of affairs we should aim for should in addition be *desirable*. This distinction will be studied in Chapters 11–13.[143]

Suppose then that we have identified the most desirable value of S, say S^*, which is the state of the biosphere the global economy should aim for. The problem remains that there are potentially an infinite number of paths that would lead the biosphere from today's S to the target level S^*. And that is a dynamic portfolio management problem (Chapter 1). Therefore the most desirable way for getting from where we are today to where we ought to be must at each moment satisfy arbitrage conditions among all assets (produced, human and natural), and they must also include a set of arbitrage conditions that mediate between the present generation's well-being and the well-being of future generations (Chapter 10). In Chapter 13, we show that these arbitrage conditions would be satisfied if the allocation of assets toward different activities and engagements were to be governed by the rule that an inclusive measure of *wealth* should be maximised. Optimal portfolio management for an economy, be it a national economy or the global economy, involves wealth maximisation at each moment. The implication is striking: *economic progress should be read as growth in inclusive wealth, not growth in GDP nor growth in any of the other ad hoc measures that have been proposed in recent years such as the UN's Human Development Index*.

---

[143] Equation (B3.2.1) in Box 3.3 showed that in a fishery that exhibits a threshold A, any biomass S in the interval (A,K) is sustainable, so long as annual rate of harvest demand R equals the fishery's regenerative rate $G(S)$.

Chapter 4: Human Impact on the Biosphere

Box 4.6
Reaching the UN SDGs

The Impact Inequality offers a way to discover the policies and behavioural changes that will be required if the global economy is to achieve the UN SDGs. To illustrate, we consider the Goals related to reaching a sustainable use of the environment by year 2030.

We have defined the global ecological footprint by Ny/α. The Global Footprint Network (GFN) in contrast defines it as the ratio of the global demand for the biosphere's goods and services and the biosphere's current capacity to supply them on a sustainable basis (which we interpret here as G). The GFN's global ecological footprint is then $[Ny/\alpha]/G$. Wackernagel and Beyers (2019) report that the ratio increased from 1 in 1970 to 1.7 in 2019. That means the ratio increased at an average annual rate of 1.1%.[144] Moreover, global GDP at constant prices has increased since 1970 at an average annual rate of 3.4%.

We turn to the right-hand side of the Impact Inequality. As noted previously, Managi and Kumar (2018) estimated that the value of per capita global natural capital declined by 40% between 1992 and 2014. That converts to an annual percentage rate of decline of 2.3%. But world population grew approximately at 1.1% in that period. Taken together it follows that the value of global natural capital declined at an annual rate of 1.2%. Because there are no estimates of the form of the G-function, we assume for simplicity that local variation is a good approximation, meaning that G is proportional to S. So, G can also be taken to have declined at an annual rate of 1.2%.[145]

The estimates for the annual percentage rates of change of Ny, G, and $[Ny/\alpha]/G$ enable us to calculate that α had been increasing at an annual percentage rate of 3.5% in the period 1992 to 2014. Suppose we want to reach Impact Equality in year 2030. That would require $[Ny/\alpha]/G$ to shrink from its current value of 1.7 to 1 in 10 years' time, implying that it must decline at an average annual rate of 5.4%. Assuming global GDP continues to grow at 3.4% annually and G continues to decline at 1.2% (i.e. business is assumed to continue as usual), how fast must α rise?

To calculate that, let us write as $g(X)$ the percentage rate of change of any variable X. We then have

$$g([Ny/\alpha]/G) = g(Ny) - g(\alpha) - g(G) \tag{B4.6.1}$$

Equation (B4.6.1) can be re-arranged as

$$g(\alpha) = g(Ny) - g([Ny/\alpha]/G) - g(G) \tag{B4.6.2}$$

We now place the estimates of the terms on the right-hand side of equation (B4.6.2) to obtain

$$g(\alpha) = 0.034 + 0.054 + 0.012 = 0.1$$

In short, α must increase at an annual rate of 10.0%. As that is a huge hike from the historical rate of 3.5%, we consider a different scenario.

---

[144] GFN's estimates are based on data furnished by the United Nations Statistical Office. For an account of the methods that are deployed for estimating G, see Wackernagel and Beyers (2019). It should be noted that as an approximation they take G to be linear in S.

[145] While Managi and Kumar (2018) base their work on the UN data base, the questions they ask differ from the questions asked by the GFN (Wackernagel and Beyers, 2019). Moreover, because the Managi-Kumar study includes fossil fuels and minerals, we must assume for our purposes of illustration that the percentage rate of global decline in the accounting value of sub-soil resources equalled the corresponding figure for ecological resources. Using data from different systems of measurement in the numerical calculation we conduct here is a price we have to pay for continual neglect of the economics of biodiversity in international organisations. GDP estimates have been refined continually over the decades by thousands of experts, whereas the human footprint on the biosphere remains of interest only to a handful of people.

> Suppose global output was to remain constant from now to year 2030 and draconian steps were taken by us over our demands to limit the rate of deterioration of the biosphere to an annual 0.1%. What would be required rate of increase in α need to be? Using equation (B4.6.2) we have
>
> $g(\alpha) = 0.054 + 0.001 = 5.5\%$
>
> Even that is considerably larger than the 3.5% rate at which α has been increasing in recent decades.

4.6 Technology and Institutions

The expression $(Ny/\alpha_X + Ny/\alpha_Z)$ is the global ecological footprint in absolute terms. As noted earlier, the Global Footprint Network (Wackernagel and Beyers, 2019) defines the footprint instead as the ratio of the demand to supply; that is, $(Ny/\alpha_X + Ny/\alpha_Z)/G(S)$. The network's latest estimate (2020) is 1.6, which they read as a demand that can only be satisfied on a sustainable basis by, as a minimum, 1.6 Earths (Lin et al. 2020).[146] The estimate is very rough, but the point should not be to focus on the exact estimate. What should be uncontroversial is that $(Ny/\alpha_X + Ny/\alpha_Z)$ has exceeded $G(S)$ since the 1970s.

In subsequent chapters, we study ways in which the trajectories of y and N can be altered (see in particular Chapters 9 and 16). Here we consider a few ways by which α_X and α_Z could be raised so as to close the gap between $(Ny/\alpha_X + Ny/\alpha_Z)$ and $(G(S))$. The twin pillars of technology and institutions would be involved, for together they determine α_X and α_Z.

That institutions and technology influence one another is a commonplace assertion. Institutions are the seat of incentives, and incentives shape the production, dissemination and use of knowledge. A state that invests vigorously in life-saving technology and then applies it is able to transform society for the better. Likewise, technological possibilities shape institutions. Advances in mapping the geographical spread of natural capital and in methods to monitor its use can help enforce property rights. History is rife with examples where institutions and technology have influenced one another beneficially.

That each can be made at least partially to mitigate the other's failure is also widely recognised. While it is widely recognised that degradation of the biosphere is a manifestation of institutional failure, hope is often expressed that progress in science and technology can put things right. The economics of climate change has encouraged that thought. Development of cheap renewable energy sources would help to reduce carbon emissions to the point where carbon concentrations are kept within acceptable levels.

Nature saving technology, for example, substituting degradable waste for persistent pollutants and decarbonising the energy sector, is one class of ways in which technology can raise the aggregate efficiency index α. Institutional changes, such as improving the character and enforcement of property rights to natural resources points to another class of ways in which α can be raised. Directives that establish Protected Areas to conserve natural habitats; imposition of pollution taxes; and removal of subsidies on resource extraction and agricultural production make up another class of institutional changes that can be brought about by public policy.

Changes in behavioural norms, such as those that lead to a reduction in food waste, is yet another avenue. The incentives entrepreneurs have for developing technology are shaped by the systems of property rights in place. Remarkable post-war developments in sonar technology and advances in the technology for harvesting fish came about in large measure because ocean fisheries beyond national

---

[146] This is down from 1.7 in 2019 due to the impacts of the COVID-19 pandemic.

Chapter 4: Human Impact on the Biosphere

jurisdiction are free. Unbridled application of modern technology in clearing tropical rainforests has been made possible because governments have permitted it at a small price. Both forms of environmental destruction would have been avoided had institutions not failed. There is nothing good or bad about technology per se, it is the use to which it is put that affects α. Indeed, a wealth of examples of technologies that can be a force for sustainability in the food sector are given in Chapter 16. But no matter how effectively institutions are established to synergise with technological advances, unending growth in global output is an ecological impossibility. The biosphere is bounded and there are theoretical bounds on global output. Or so we argue in Section 4.7 and Chapter 4*.

Decoupling the demand humanity makes on the biosphere into X and Z in the left-hand side of equation (4.2) also serves to remind us that measures to reduce environmental pollution (Z) can raise our demand for the biosphere's goods and services (X). As noted already, solar panels offer a technology for reducing carbon emissions, but solar panels are built on aluminium, zinc, cadmium, and other minerals. And mining and quarrying usually require that forests be destroyed. Equation (4.2) also reminds us that the two sides are not independent of one another. A move away from intensive farming to methods that rely on mulch, among other practices, gives rise changes in α_X and α_Z (the latter parameter would increase) in food production, but it also leads to a change in the G-function.

In Chapter 8, we argue that effective institutional structures are *polycentric*.[147] The most well-known among such structures is a system of markets for private goods and services, in which a central authority supplies public goods and services, including measures that brings about a fair distribution of assets among people. The price system is the hallmark of markets. It serves not only to coordinate the choices people make, it simultaneously aggregates diffused information across the economy.[148] That system however cannot serve adequately in humanity's engagement with the biosphere, because Nature's processes do not satisfy the technical conditions on production possibilities that are required for markets to function well. Three conditions are especially pertinent:

(1) Because Nature is mobile, much of the biosphere consists of 'fugitive resources', meaning that it is impossible to establish property rights to them (Chapters 7 and 8). By property rights we mean not only private rights, but also group rights including those of communities and nations.

(2) Production and consumption possibilities involving the biosphere (in other words, *all* production and consumption possibilities!) are characterised by non-linearities (Chapters 3), a condition that is at odds with a requirement of any well-functioning market system (Chapter 7).[149]

(3) The risks to life and property that are associated with ecological degradation are *positively correlated* across people (Chapter 5). That means insurance premia cannot be set at fair odds by private firms. Insurance markets are inevitably imperfect, and national and supra-national institution are needed to fill the gap. Nature-related financial risks are covered further in Chapter 17.

In later chapters of this Review we study the character of institutions that can in principle implement the Impact Equation. We will discover that the polycentric structure requires *layered* institutions: global, regional, national and communitarian. Each layer however requires an authority at the apex to achieve coordination below.

---

[147] Ostrom (2010a,b) popularised the term.

[148] Stiglitz (1989) is an outstanding exposition of the place of the price system in a well-functioning economy.

[149] Debreu (1959) is the classic exposition of the theory of well-functioning market economies.

Box 4.7
Reducing the Impact Inequality by Restoring the Peatlands

Peatlands comprise peat soil and the wetlands that grow on its surface. Year-round waterlogged conditions slow the process of plant decomposition to such an extent that dead plants accumulate to form peat soil, which can be several metres thick. Peatland exists in almost every country in the world and covers around 3% of global land surface (>3 million km^2) (Joosten, 2010). Appearing in such diverse forms as open, treeless vegetation in Scotland and swamp forests in South Asia and the Congo Basin, peatlands together comprise the largest natural terrestrial store of carbon, harbouring more than 450 gigatonnes of carbon, which is more than 40% of all soil carbon (Joosten, 2010). Peatlands sequester nearly 0.4 billion tons of CO_2 annually, while regulating water flow and quality, lowering the risks of flooding and the effects of droughts, preventing sea-water intrusion, and offering habitat for numerous forms of wildlife. Local inhabitants harvest their peatland so as to grow and obtain food, fibre and other local products.

Unfortunately, some 15% of the world's peatlands (i.e. about 0.4% of global land surface) has been drained for intensive cultivation, animal husbandry and human habitation (Joosten, 2010). CO_2 emissions from drained peatland now contribute more than 5% of global anthropogenic greenhouse emissions, never mind the loss of biodiversity and the corresponding loss of a multitude of ecosystem services.

The UK's peatlands, covering 12% of the nation's land area, have been in the making for over 10,000 years (ONS, 2019). Today they are in a damaged state. Acid rains and more general pollution, overgrazing and burning, draining, and drying of the peatlands for our other demands have so affected the state of the peatlands that some are emitters of carbon rather than stores.

Peat restoration, especially in the UK hills, is now a tried and tested technique. It involves blocking drains and holding water in the hills. They represent ecological solutions to restoration problems. Peat restoration projects in the Peak District's 'Moors for the Future', and two lowland restoration projects (the National Trust's 'Fen Vision' and the Wildlife Trust's 'Great Fen Project') are now yielding promising results. A case study of the use of natural capital accounting is provided in Chapter 13.

Conservationists have speculated on an even wider set of policies for peat restoration. Imagine that intensive farming in the nation's peatlands was to be abandoned and farmers were deployed to act as stewards of the wetlands. The move would in principle make no dent on Ny (FAO estimates that approximately a third of food is wasted globally[150]), but it would raise α_X and α_Z and S, thus reducing the gap between $G(S)$ and $(Ny/\alpha_X + Ny/\alpha_Z)$.

Natural England (2010) has found that under mid-range assumptions on the social cost of carbon in the atmosphere (around US$40 per ton of CO_2), restoring the nation's peatlands would be a financially effective method of reducing greenhouse gas emissions. And that estimate is based on valuing peatlands solely as carbon stores. If we were to add the other services provided by peatlands, the case for restoration increases substantially. These considerations have prompted environmentalists to suggest that the nation's peatlands be restored in their entirety by retiring agriculture from them.

---

[150] This is a rough estimate due to lack of information at some stages of production, discussed in FAO (2019).

4.7 Ecosystem Complementarities and the Bounded Global Economy

Within bounds the biosphere is a self-regenerative asset, supplying us with a bewildering variety of services. In Ch. 2 (Sec. 2.4) a distinction was drawn between Nature's 'provisioning goods' and her 'regulating and maintenance services.' The former category includes food, water, timber, fibres, pharmaceuticals, and non-living material, which we transform into consumption and investment and record the transformation at the national level as GDP. The latter category, it will be remembered, includes among many other services, climate regulation, decomposition of waste, disease regulation, nutrient recycling, nitrogen fixation, air and water purification, soil regeneration, and pollination.

The evidence we have brought together in this chapter has shown that humanity has increasingly drawn on Nature's regulating and maintenance services to provide ourselves with provisioning goods. We have done that by mining ecosystems and transforming the landscape (land-use change, as it is called generically). The worldwide conversion of grasslands and forests (ecosystems with rich biodiversity) into farms, ranches, and plantations (assets with poor biodiversity) is an example. There is thus a tension between our desire for provisioning goods on the one hand and our need for regulating and maintenance services on the other. But regulating and maintenance services are fundamental, for without them there would be no provisioning goods, nor for that matter cultural services. Which is why the tension in question expresses itself today in our overshoot in demand for Nature's provisioning goods relative to her ability meet that demand on a sustainable basis (i.e., the Impact Inequality). Put another way, when we speak of a shrinking biosphere, we mean a decline in Nature's ability to supply regulating and maintenance services, caused by our ever-increasing demand for provisioning goods.

And here is another sobering finding: the processes governing the biosphere's ability to provide regulating and maintenance services are *complementary* to one another, meaning that if you draw down one such service (e.g., climate regulation) sufficiently, you will in due course draw down Nature's capacity to supply the others (e.g., as reflected in biodiversity). Nature is, to be sure, resilient – it has, after all, evolved over 4.5 billion years – but we humans today are so powerful, that we could if we put our mind to it, bring it down like a house of cards. Recent concerns over global climate change and biodiversity loss are an acknowledgement of that possibility.

The force of complementarities becomes evident when we study the components of objects that are indivisible. A steering wheel, for example, is of little use on its own, brake pads are of no use on their own, a gear appliance taken alone serves no purpose, and so on; but together they can be assembled to manufacture an automobile. Repair shops carry inventories of automobile parts, but that simply re-enforces the point that automobiles are indivisible capital goods. A car with worn out brake pads is not roadworthy. They have to be replaced.

The components of indivisible object are *perfect* complements. Complementarities among ecosystem services are not perfect, but they are far from being substitutes. In that less rigid sense, complementarities are an essential feature of the Earth System also at levels of aggregation higher than ecosystems. Within bounds the biosphere is a self-regulating entity. The bounds are defined by its stability regimes. Regulating and maintenance services, for example, are provided by the biosphere as *joint products*. Weaken any one sufficiently by overuse, and the biosphere would flip into a different stability regime.

Our ecological footprint is not only of the material we take from the biosphere, but also the transformed material we deposit into it; what is requisitioned for human use has to be returned. The macroeconomic models of growth and development in use in finance ministries and planning commissions, however, do not acknowledge that material must balance – from source to sink. Persistent pollutants such as plastics, nylon (fishing nets and synthetic textiles), toxic chemicals and metals provide examples of waste that have adverse consequences for the soils and water bodies especially. But even (perhaps, especially) biodegradable waste has to be accounted for. It does not do to imagine that if waste is biodegradable it leaves no footprint. If we overload Nature with such waste, the process of decomposition compromises other biospheric services. Pharmaceuticals such as antibiotics and fashion products such as cosmetics contaminate the soils and water bodies. They have an adverse effect on the food we eat, the water we drink, and the air we breathe. Chemical fertilisers and waste from livestock emerge at the other end of farms as waste, causing nutrient overload in streams and water bodies, disrupting the nitrogen cycle. Even the carbon dioxide emitted by our economies is a biodegradable waste: it is absorbed by primary producers for photosynthesis. But an overload compromises the ability of the biosphere to regulate climate. Global climate change will increasingly be a major cause of biodiversity loss (Lovejoy and Hannah, 2019). That will compromise the functional integrity of ecosystems. Rising concentration of carbon in the atmosphere is thus expected to bring about a chain of events that will radically alter the biosphere's workings. Regulating and maintenance services will move out of the bounds within which our economies have evolved. That is an expression of biospheric complementarities.

Those complementarities find expression in the growth of all forms of waste. It may not be possible yet to predict how that will in time affect biospheric services (Box 4.2) but what one can anticipate with a level of certainty is that the transformations will be adverse to the human economy because we did not evolve under them. If the mass of waste material continues to increase, the composition of the biota can be expected to undergo sufficient change to bring about biospheric regime shifts. No such shift could be expected to bring good news to us, for human activities evolved only under gradual changes in the biosphere's operations.

Biospheric complementarities point to a further truth: *The efficiency with which its goods and services can be converted into produced goods and services is bounded*. Formally, α is bounded (Equation (4.2)). As the biosphere's regenerative rate G is also bounded, global output is also bounded (the Impact Equality). That is the sense in which humanity is embedded in Nature.

That G is bounded finds vivid expression in the idea of *planetary boundaries* (Rockström et al. 2009; Section 4.1.1). Contemporary models of macroeconomic growth and development do not necessarily overlook the fact that Earth is bounded, what they explore instead are production possibilities in which, by exercising sufficient ingenuity (read technological progress), humanity will be able to free itself from the biosphere's constraints.

The increasing share of non-material goods in the GDP of high income countries is often cited as a move in that direction, miniaturisation in the production and use of information technology being a concrete example. One problem with the example is that the miniatures themselves are built with material goods; another is that the income drawn from a rising GDP can be, and is, spent on material goods. Currently global raw material consumption is estimated to be 90 gigatonnes a year (OECD, 2019c).[151] Mining and quarrying operations degrade ecosystems. Applying methods similar to ones deployed for estimating our global ecological footprint (Section 4.2), 50 billion tonnes a year is reckoned to be a sustainable rate. The OECD (2019c) has estimated that if the global population N in 2060 was to rise to 10 billion and per capita global income y was to rise to today's per capita income in OECD countries, raw material consumption would be about 180 billion tonnes a year. That's nearly four times the sustainable rate. Even if the idea of a weightless economy was to be believed, it would provide no solace if the biosphere was to flip to an uninhabitable state before it could be realised. We return to this issue in Chapter 9 in connection with population growth.

Contemporary models of macroeconomic growth may be interpreted as saying that the boundedness of the biosphere does not imply that the human economy has to be bounded. The existence of planetary boundaries would not necessarily preclude the possibility of perpetual economic growth, for we may feel we are entitled to imagine that with sufficient ingenuity humanity would be able to convert the

---

[151] The bulk of raw material consumption under the OECD's classification consists of sand, gravel and crushed rock, and metals. Timber is relatively small portion in weightage.

Chapter 4: Human Impact on the Biosphere

biosphere's goods and services into final products at an unbounded rate, that is, that there are no theoretical bounds on α. So, we need a further argument.

It is significant that a mechanical engine that converts heat into work at 100% efficiency is a theoretical impossibility. The biological counterpart is that it would not be possible even theoretically to convert our further waste into a state that makes no further demands on the assimilative services of the biosphere; for if we were able to do that, we would be able to break free of Nature. Chapter 4* presents a model of production possibilities for the global economy in which that dependence on the biosphere is represented by the idea that no matter how ingenious we are able to be, we cannot increase α to infinity (Figure 4.11).

Figure 4.11 The Economy is Embedded in the Biosphere

Box 4.8
Land-Use Change and the Spread of Viruses

It is customary to regard trade in goods as ways that smooth local disruptions across space and time. Globalisation is also applauded because it expands output and has been shown to have helped to reduce global poverty. But because globalisation has taken place when much of the biosphere is not merely free for all to use as we like but is also subsidised for our use (Annex 8.1, Chapter 8), it has increased the likelihood of societal crashes. It has done that by connecting economic units closely to one another via firms' supply chains and the movement of people. Close connections among its parts make the global economy less modular (Chapter 2): a crash in one part spreads to other parts.

There are further drivers of societal crashes. Our remarkable ability to enter every ecological niche has raised the chances of pandemics (Daily and Ehrlich, 1996; Jones et al. 2008). Humans now enter niches occupied by organisms with which we have not evolved. Intimate associations between humans and wildlife disease reservoirs have raised the risks of exposure to zoonotic viruses. Being unfamiliar pathogens, they are able to spread rapidly across the globe (Gottdenker et al. 2014).

Moreover, biodiversity loss creates niches for pathogens that are lying in wait in small numbers to explode in their populations, and for new pathogens to evolve.[152]

Enormous changes in land-use have taken place in recent history. Increases in logging and forest clearance for mining and extracting oil, cultivating oil palm and farming cattle and crops have come allied to increases in the volume of trade in bushmeat and exotic pets. These activities have disrupted vegetation and wildlife that are host to countless species of viruses and bacteria, mostly unknown so far, and also increase the number of available host species of diseases (Gibb et al. 2020). Those microbes, once released, can infect new hosts, such as humans and cattle (Jones et al. 2013). The spillovers are then transmitted via globalisation. An example is the human immunodeficiency virus, which would appear to have spread from chimpanzees and gorillas, who were being slaughtered for bushmeat in West Africa. By conservative estimates, some 33 million deaths have occurred due to the virus (UNAIDS, 2020).

Quantitative studies of the transmission of infectious diseases (e.g. Anderson and May, 1991) point to the analogous fact that wide-scale movements of people and goods make the socio-ecological world brittle in many ways. The questions epidemiologists therefore ask about the spread of an infectious disease include: Can the infection be stably maintained? Is it endemic or epidemic? What is the time course of the proportions of a population that are (i) susceptible, (ii) infected and (iii) recovered?

Mathematical models of the dynamics of infectious diseases in a host population (and the models are necessarily mathematical) in effect are the dynamics of the three categories of subjects in the host population.[153] Today the elaborate models that routinely incorporate new data to revise the values of parameters and measures of human behaviour are becoming familiar, at least several steps removed, to us all as we listen to daily reports on the spread of COVID-19 by some of the most distinguished epidemiologists of our time. But the underlying logic in the models is the three-way partitioning of a host population.[154]

Dobson et al. (2020) have made a concrete proposal, accompanied by estimates of how much it would cost globally to (i) halve the rate of tropical deforestation, (ii) monitor wildlife and embark on programmes to detect and control the spread of potentially deadly viruses and bacteria among domesticated animals, and (iii) stop illegal trade in wildlife. The authors estimate the net prevention costs of these actions to be in the range US$20–30 billion per year, a pittance when compared to the devastation pandemics are known to have brought. The world may lose at least US$5 trillion in GDP in 2020, not accounting for the willingness to pay for lives lost and deaths caused by disrupted medical systems. This makes the estimate of the present value of prevention costs for 10 years around 2% of the costs of the COVID-19 pandemic.

4.8 Core of the Review

Studying our aggregate demand ($Ny/\alpha_X + Ny/\alpha_Z$) and the biosphere's aggregate supply $G(S)$ allows us to unravel the proximate factors affecting our relationship with Nature. They consist of humanity's numbers (N), our wants and desires (summarised in y), the efficiency (α_X, α_Z) with which we make use of

---

[152] As elsewhere in the Review, we use expressions that could suggest organisms have intention. But as noted previously, such expressions are used routinely among scientists with no such intention, as for example when they say Nature abhors a vacuum.

[153] The host population may not be human of course. Laboratory experiments in animals, and models of the spread of transmittable diseases among farm animals are routine.

[154] Using a modified version of the model in Anderson and May (1991), Barrett and Hoel (2007) studied the theoretical underpinnings of the optimum management of infectious diseases. Their pioneering paper appeared in what today should be judge as a prescient symposium on the socio-ecology of infectious diseases in the journal *Environment and Development Economics*, 2007, Vol. 12, Issue 5. The ecologist and evolutionary biologist Simon Levin served as Guest Editor.

Chapter 4: Human Impact on the Biosphere

the biosphere's goods and services to provide us with our wants and desires, and the biosphere's supply of its goods and services ($G(S)$). These are, however, proximate factors. The Review peers into them so as to unravel the forces that shape those factors and the way they influence one another. Depending on the context, that will require us to study the socio-ecological systems that define in turn households, communities, national governments, and even the world as a whole.

In subsequent chapters, we discuss ways to influence the future trajectories of y and N. Our analysis shows that, fortunately, it may be possible to reduce both projected values of N and y without unacceptable human cost. We also study ways in which the G-function can be raised (e.g. by introducing GM crops).

To find a way to convert the Impact Inequality into an Impact Equality, it pays to imagine the reasoning to be iterative:

We could start by (i) further decoding the regeneration function $G(S)$, (ii) identifying states of the biosphere (S) within which the human enterprise ought to confine itself, and (iii) finding ways to influence our wants and desires (as expressed in y), our numbers (N), and the efficiency with which the goods and services produced by the biosphere (G) is converted into the realisation of our wants and desires (α_X, α_Z). The latter could be, for example, by reducing the enormous waste in the global food system by eliminating agricultural subsidies and deploying the released funds to restore and maintain ecosystems. The aim would be to bring our aggregate demand ($Ny/\alpha_X + Ny/\alpha_Z$) in line with aggregate supply ($G(S)$), or in other words, to find ways to satisfy the Impact Equation. We could then search for ways to raise α_X and α_Z while simultaneously study the trade-offs that are involved between the standard of living (y) and numbers (N). Iterating the procedure would require selecting a different value of S and conducting another round. The aim would be to continue the iterative process until we are able to reach what the philosopher John Rawls famously called a *reflective equilibrium* (Rawls, 1972), always bearing in mind that the search involves peering into possible states of affair far from where we may happen to be at (the tipping points). The programme of work involves thought experiments, model building, and empirical investigations.

To contemporary sensibilities, this mode of reasoning could appear strange, perhaps even repulsive. Some would invoke the language of rights. Should S not be determined by market forces? Whose business is it to choose y_i if not household i? Should N not be left to the personal choices of individual couples? And who other than entrepreneurs know how best to devise α_X and α_Z? And should the G-function not be left to be enhanced by agronomists, energy specialists and technologists?

There are several reasons these questions misread the socio-ecological world entirely. The stresses humanity has inflicted on the biosphere to the point where our mode of conduct is not sustainable are due to *institutional failure* writ large. That failure is not only due to malfunctioning markets, but also to households, communities and states. Ultimately, the finger should point to we citizens. Chapters 7 and 8 (environmental externalities) and Chapters 9 and 10 (reproductive externalities) provide an outline of the source of that overarching failure and relates it to fundamental properties of the biosphere we have studied in the previous chapter. When they are taken together, it is apparent that we are far removed from the model of the world that has shaped the contemporary reading not only of economic growth but also of economic development. Economics provides a remarkably effective language in which to read the socio-ecological world. The problem is not with economics, it is rather the fundamentally flawed reading of the structure of economic reasoning. The Review will use examples and illustrations to provide a language for identifying institutional arrangements that align the incentives facing various actors in an economy, so as to protect and sustain our place in the biosphere. It is a fundamental misconception of economists that we can continue to rely on models of growth and development in which our impact on the biosphere is of second-order importance (Chapter 4*). This

Review is an attempt at constructing a formulation of economic reasoning that has the biosphere always in sight. Much remains to be done in advancing the subject; this is only a start.

Annex 4.1 Biodiversity Loss and Climate Change

Climate change and biodiversity loss are intimately related. It is predicted that climate change could overtake land-use change as the leading cause of biodiversity loss by 2070 (Newbold, 2018). Biodiversity loss will in turn have huge implications for climate change: enormous amounts of carbon are locked within animal life and vegetation. The Amazon contains an amount of carbon equivalent to a decade of global human emissions (Lovejoy and Hannah, 2019). Therefore, mitigating against the worst effects of climate change will have significant benefits for biodiversity, and avoiding biodiversity loss will have a positive effect on climate change. As the climate changes, people will need to adapt to new conditions, and employ strategies that deliver for humanity, biodiversity, and the climate simultaneously.

A4.1.1 The Relationship between Biodiversity, Climate Change, and People

Human-induced climate change is leading to changes in precipitation, seasonality, storm intensity, and more.[155] All ecosystems, marine, terrestrial and freshwater, are affected; the ways in which people relate to Nature will be significantly altered. An integrated response to climate change and biodiversity loss is needed, which is made difficult by the fact that different institutions are responsible for each (for example, the UN Framework Convention on Climate Change (UNFCCC) and the Convention on Biological Diversity (CBD)). Ecosystems, and the biodiversity they contain, are being altered by climate change, but can also help us adapt to and mitigate its effects (see Chapter 19 for discussion on Nature-based solutions).

Climate change will alter the way we relate to Nature. People living in rural areas will find that the food they can grow – and where they can grow it – changes, people in cities will find that there are higher costs of importing fuel, and supply chains for consumer goods will have vastly different environmental impacts. We need to understand the relationships between people and Nature, and how they are affected by climate change, in order to help the best elements of that relationship to be maintained – or even improved, where new relationships with Nature will be forged because of climate change

How can positive outcomes be achieved for people and ecosystems in different contexts as the climate changes? For example, communities living with coral bleaching face lowered fishery output and reduced income from tourism. Life with climate change looks very different for this community: a planned response may maintain some fishery output through creating climate-adapted habitats for reef fish. Without a response that considers climate change, the community could face loss of all benefits from the reef.

Different aspects of an ecosystem are affected differently by climate change and have roles in its recovery. For example, after reef bleaching, recovery is a race between recolonising corals and the algae that grow on the dead reef. Parrotfish are among the most effective consumers of algae, and they are also negatively affected by climate change. Helping protect parrotfish will help the reef recover, and maintain reef fishery output and communities' livelihoods. So, a short-term reduction in fishing is needed in order to maintain the reef fishery's long-term productivity.

A4.1.2 Climate-Related Changes to Biodiversity are Already Being Observed

Climate change is already harming biodiversity in many ways. Here we discuss a selection: globally coherent patterns of species distribution shifts; impacts on marine ecosystems, particularly coral reefs; and genetic signatures of climate change.

---

[155] This section summarises from Lovejoy and Hannah's excellent 2019 work on biodiversity and climate change (Lovejoy and Hannah, 2019).

Chapter 4: Human Impact on the Biosphere

Species ranges are shifting towards the poles. This has been identified as a climate change 'signal'. Meta-analyses have shown that about half of species on which there were data had changed their distributions significantly over the past 20 to 140 years (Lovejoy and Hannah, 2019). Boreal species diversity in the Yukon, Canada, increased over a 42-year period, during which temperatures increased by 2°C (Research Northwest and Morrison Hershfield, 2017; Lovejoy and Hannah, 2019). The poleward shift trend is consistent across taxa. Marine species are moving towards poles more quickly than those on land. Meta-analyses show that the speed of the leading edge of marine species distribution change is many times greater than the terrestrial, at ~72 km/decade. Poloczanska (2013) compared dynamics in marine species range and found that the leading edges of poleward range shift were expanding nearly five times faster than the trailing edges were contracting, which matches the trend seen in European butterflies. 63% non-migratory European butterfly species expanded their northern range boundary, but only 3% contracted their southern range boundary (Parmesan et al. 1999).

Climate change is already contributing to rapid, broad-scale ecosystem changes, with significant consequences for biodiversity. Several changes have already occurred at spatial scales sufficiently broad to represent biome changes. For example, inland water systems have already been significantly altered, and the spatial scale of changes in fire and precipitation frequency cover large proportions of tropical and boreal biomes respectively (Gonzalez et al. 2010; IPCC, 2014). Rapid broad-scale changes differ from other patterns in vegetation dynamics in that they represent a 'crash' in one or more populations over large areas. Climate triggers these through megafires (as seen in Australia in late 2019 to early 2020), drought-triggered die-off, floods and hurricanes. Pest and pathogen outbreaks are frequently associated with these events, though the relationship between pests and pathogens and climate change is complex (Rosenzweig et al. 2001; Jactel, Koricheva, and Castagneyrol, 2019).

Studies show impacts on individual species and communities, as well as changes at the biome level. Parmesan (2006) reviewed studies that had focused on single species. For example, populations of the well-studied butterfly *Euphydryas editha* are observed to be declining due to warming causing their host plants to senesce before the insects diapause (a period of suspended development in their lifecycle), leading larvae to starve (Hellmann, 2002; Singer and McBride, 2012).

The changing climate is also altering marine biomes, including unique megadiverse systems such as coral reefs. The ocean plays a crucial role in stabilising the Earth System, mainly due to its huge capacity to absorb CO_2 and heat while experiencing minimal change in temperature. The ocean has absorbed 93% of the extra energy existing due to the greenhouse effect, and approximately 30% of human-generated CO_2. These absorptions have had an impact: sea levels have risen, sea ice extent has decreased and ocean pH has dropped rapidly, which is associated with concentrations of key ions such as carbonate and bicarbonate. Oxygen levels are being driven down in deeper areas of the ocean; oxygen-dependent organisms are beginning to disappear from these so-called 'dead zones'. Elsewhere, levels of productivity are rapidly increasing or decreasing due to retreating ice, changing winds, and altering nutrient compositions (Pörtner et al. 2014).

Coral reefs only occupy 0.1% of Earth's surface, but they provide habitat for 25% of known marine organisms (Hoegh-Guldberg, 2019). Coral reefs have experienced small shifts in temperature and ocean chemistry over the past 420,000 years at least. This is due to the fact that even large environmental changes, such as shifts between ice-ages and interglacial periods, are experienced as small changes over relatively long periods of time compared to the current pace of change. The great barrier reef has waxed and waned in the past as sea levels have risen and fallen, but today's pace of change makes current fluctuations more significant. As much as 75% coral reefs are threatened, and as much as 95% is in danger of being lost by 2050 (Hoegh- Guldberg, 1999; Hoegh-Guldberg et al. 2007). Reef-building corals have contracted over the past 30–50 years, which is associated with loss of reef 3D structure (Bruno and Selig, 2007). Reef-building corals rely on the symbiosis between coral and small photosynthetic organisms (*Symbiodinium*). Rapid temperature and pH changes cause this symbiosis to break down, leading to coral bleaching (Glynn, 1993; Hoegh-Guldberg, 1999). Around 1980, large-scale bleaching began in tropical regions, with no precedent in scientific literature; these bleaching events were associated with short periods where maximum sea temperatures rose by 1–2°C (Hoegh-Guldberg, 1999). Reefs occasionally recover from bleaching, but most do not. Protecting the remaining 10% of coral reefs will deliver huge benefits for biodiversity, ecosystem services and human well-being (IPCC, 2014). Climate change mitigation will form an important part of this response, but should take place alongside conservation and restoration efforts (such as the expansion of the marine protected area networks) (IPCC, 2014).

In addition to its impact on coral reefs, ocean acidification has been responsible for changes in animal behaviour: reductions in gastropod shell thickness (due to reduced calcification as a result of reduced pH) lead to enhanced escape activity when a predator is present, demonstrated in a study on *Littorina littorea* (Bibby et al. 2007). Another example is foraging in deep-sea urchins, which increased under lowered pH conditions, possibly to compensate for reduced ability to detect food (Barry et al. 2014).

As well as changing species distribution and altering ecosystems, climate change has already left signatures in organisms' genomes. Selection for genes enabling organisms to survive in warmer temperatures has been identified: an example is changes to the mitochondrial DNA (mtDNA) *NADH* gene in American pika (*Ochotona princeps*, a small relative of rabbits and hares), suggesting local adaptation to different thermal and respiratory conditions (Lemay et al. 2013). Spatially divergent selection of climate-associated genes in oak trees (*Quercus lobata*) has also been observed, including genes involved in bud burst and flowering, growth, and osmotic and temperature stress (Sork et al. 2016). Evolution is a stochastic process, dependent on the genetic options available and their underlying architecture, so we can never be entirely certain about how change will happen. As well as genetic evolution, environmentally determined plasticity is an important aspect of adaptation to changing temperatures: examples include increases in body size in marmots, and changes to clutch size in birds (Hoffmann and Sgró, 2011).

A4.1.3 What Does the Future Hold?

Biodiversity changes have been projected under different scenarios, combining climate change and mean species abundance, showing the negative correlation between temperature rise and species abundance for plants and warm-blooded mammals (Figure A4.1, Schipper et al. 2019).

Figure A4.1 Mean Species Abundance (MSA) Plotted Against Global Mean Temperature Increase (GMTI)

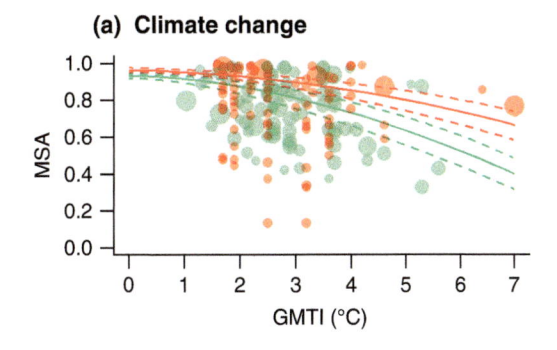

Source: Schipper et al. (2019). Note: Red: warm-blooded mammals; green: plants.

Chapter 4: Human Impact on the Biosphere

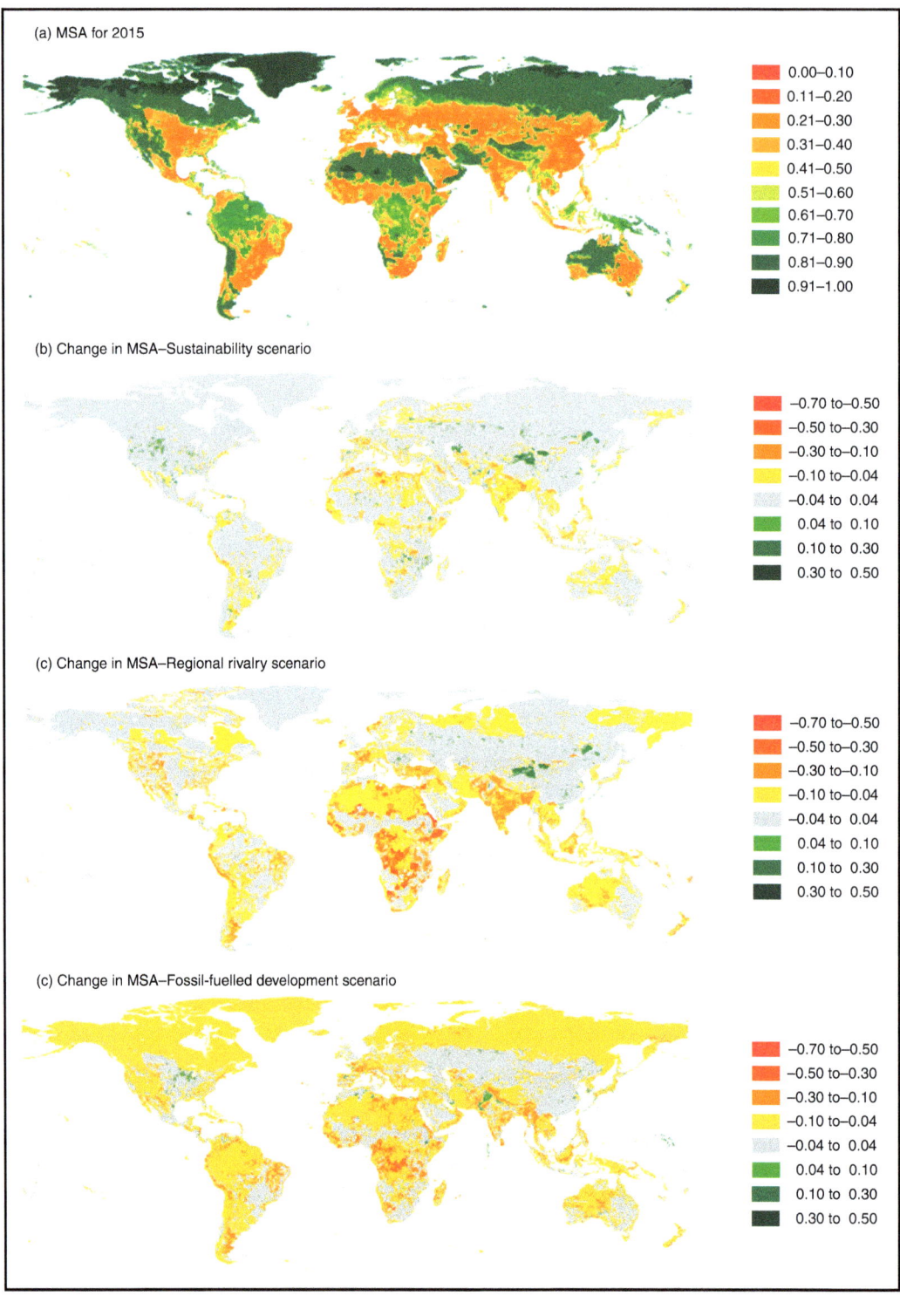

Figure A4.2 Change in Mean Species Abundance (MSA) under Different Scenarios

Source: Schipper et al. (2019). Note: Scenarios represent combinations of shared socio-economic pathways (SSPs) and representative concentration pathways (RCPs).

Schipper et al. (2019) used three shared socio-economic pathways (SSPs) to project changes in mean species abundance (MSA) to 2100, each associated with different pressures placed on the environment by human activity. Even in the most sustainable scenario[156], global changes in MSA are projected to be negative. The regional rivalry scenario projects high population growth and resource-intensive consumption, while the fossil-fuelled development scenario is also characterised by a consumption-oriented society, which is even more energy-intensive. Both lead to high levels of climate change, and widespread negative changes to MSA. Future impacts of climate change on marine biodiversity are discussed in Chapter 16.

Climate change is affecting, and will continue to affect, tropical forests. These forests play a huge part in regulating global climate, accounting for a third of land surface productivity and evapotranspiration (Malhi, 2012). Forest biodiversity will be affected, as tropical forests fall outside their historic climate variability ranges up to a decade faster than any other major terrestrial ecosystem (Mora et al. 2013). Mean land surface temperatures for tropical forest regions are expected to increase by at least 3°C this century (Zelazowski et al. 2011) which is similar to the warming extent of the Paleo-Eocene Thermal Maximum (around 55.5 million years ago), but occurring at an unprecedented speed. This will cause more intense dry seasons, stronger and more frequent droughts, and stronger, longer-lasting heat waves (Malhi et al. 2014). Eastern Amazonia in particular will experience significant declines in total rainfall (IPCC, 2014). There is evidence that tropical trees have narrower thermal niches than temperate trees (Araújo et al. 2013). The effect of increased productivity from increased CO_2 concentration may have a compensatory effect, however, the narrowness of the thermal niche will probably cause a marked decline in fitness for most tropical forest systems.

The decline in fitness and rainfall in the Amazon in particular will have serious consequences for the entire Earth System. At the moment, the Amazon acts as a planetary cooling system, influencing global circulation of air and vapour by evaporating vast amounts of water into the Earth's atmosphere. Even in long dry seasons, the deep-rooted Amazon trees absorb water many metres below the soil surface. Deforestation has reduced the amount of vapour in the atmosphere, leading to reduced rainfall in neighbouring areas, and climate-induced decline in forest fitness would have similar consequences. Where light penetrates to the forest floor, fires become more likely, releasing large amounts of carbon into the atmosphere, and contributing to the greenhouse effect. An estimated amount of carbon equivalent to a decade's worth of human emissions is stored within the wood of Amazon trees. Releasing it through burning would have disastrous consequences for global warming (Lovejoy and Hannah, 2019).

Annex 4.2 How Many People Can Earth Support in Comfort?

The Impact Inequality (Ch. 4) represents the imbalance between our demand for Nature's provisioning goods (i.e., our ecological footprint) and Nature's ability to meet that demand on a sustainable basis.[157] It represents the difference between demand and sustainable supply at a point in time. We expressed the inequality as $Ny/\alpha > G(S)$, where N is global population, y is global per capita final output, or global GDP, α is the efficiency with which Nature's provisioning goods are converted into final output, S is a scalar measure of the stock of biosphere, and $G(S)$ is the biosphere's net regenerative rate. Currently

---

[156] Figure A4.2 b, which is characterised by relatively low population and consumption growth, less resource-intensive lifestyles such as lower red meat consumption, increased use of resource efficient technologies, more protected areas, and improvements in agricultural efficiency which lead to reforestation.

[157] Because land use changes directed at increasing the supply of provisioning goods diminish Nature's ability to supply maintenance and regulating services, we could equally have interpreted the Impact Inequality as the gap between demand and supply of maintenance and regulating services.

Chapter 4: Human Impact on the Biosphere

the ratio of global demand to sustainable supply is approximately 1.7, whence the saying that we would need 1.7 Earths if our current demand was to be met on a sustainable basis.

We have presented the Impact Inequality for the global economy. The corresponding inequality for a national economy (or for that matter for a village community) would require a further term representing the impact on the biosphere arising from exchange with others, involving trade and environmental externalities. It is a simple matter to conceptualise that. So, for brevity we continue to adopt a global perspective.

To convert the inequality into an equality requires that either Ny/α be reduced or $G(S)$ be increased, or both. In Part II (Ch. 18–19) we study ways in which investment in Nature can be used to raise $G(S)$. In the chapters that follow in Part I, we uncover institutional changes and practices that would help to reduce Ny/α.

The Impact Inequality tells us that, if we are to hold our ecological footprint fixed, then other things equal, any increase in N would have to be compensated by a reduction in y, or, conversely, any increase in y would require that N be lower. In Ch. 4* we construct a dynamic model of global economic possibilities. Among other things, the model displays the relationships over time between N, y, α, and $G(S)$. It uncovers the choices over consumption and various forms of investment the global citizen (we previously referred to her as the 'social evaluator') faces over time. But because it is prudent to proceed step by step, the model in Ch. 4* does not specify human institutions, nor does it make assumptions regarding human behaviour; instead, it is couched exclusively in terms of *stocks* (of capital assets) and *flows* (of the stocks' yields and our consumption and investment rates). Ch. 13* rectifies this by uncovering the conditions our choices over consumption and various forms of investment must satisfy if they are to lead, in the social evaluator's judgment, to the ethically best future. Economists label such choices as *socially optimal choices*.

In Box 4.6 it was shown that if our collective aim is to close the gap between Ny/α and $G(S)$ by 2030, then on the assumption that the global economy will continue to enjoy the average annual growth rate in GDP of recent decades, we would require the efficiency parameter α to increase at an annual percentage rate some *four* times the rate that has been experienced in recent years. That is so unlikely a scenario that we now study the interplay of N and y in bringing about a sustainable state of affair today. To do that it is simplest to avoid studying social optimum policies, but instead ask a more restricted question: *How many people can Earth support in comfort*? The idea is to hold all else constant and determine the value of N at which the human ecological footprint equals the biosphere's net regenerative rate at a comfortable standard of living y.[158]

There is a commonplace intuition that the current imbalance in our demand for Nature's provisioning goods is due to high consumption among the world's rich people; with an accompanying corollary that large additions to human numbers over the past 70 years – N having risen from approximately 2.5 billion to 8 billion – has had little to do with it. The intuition is at variance with evidence. Suppose, for example, the 1.5 billion inhabitants of OECD countries (the *Economist* newspaper calls the OECD "a club of mostly rich countries") were to accept a *halving* of their annual incomes from the current figure of 40,000 international dollars per person to 20,000 international dollars. That's a huge drop in incomes, but as the Impact Inequality shows, the move would reduce the imbalance from the current ratio of 1.6 to 1.2. And 1.2 is a substantial figure, implying further erosion of the biosphere at a fast pace.

So, we now choose a value of y at which life is deemed to be comfortable. Per capita global GDP in 2019 was approximately 16,000 dollars at 2011 prices. As an exercise let us take the chosen y to be 20,000 international dollars at 2011 prices. As the figure falls in the range of per capita

---

[158] The analysis that follows is taken from Dasgupta, Dasgupta, and Barrett (2023).

incomes in the World Bank's list of high middle-income countries, we use it to represent a comfortable standard of living.

We assume that people apply their labour on produced capital and natural capital to produce an all-purpose commodity that can be consumed. As of now we have little quantitative knowledge of the biosphere's dynamics when viewed in the aggregate (i.e., we have no estimates of the G-function in the Impact Inequality). But as produced capital is complementary to natural capital in production (any expansion of the former makes further demands on natural capital), an expansion of the stock of the former depresses the stock of the latter, other things equal. We noted previously (Sec. 4.1.2) that Rockström et al. (2009) have found evidence in the Earth system's signatures that several planetary boundaries are so close to being breached, that further deteriorations in the state of the biosphere would take it into terrains that are unchartered and therefore should be avoided. Let us then regard K to be an aggregate measure of produced capital and natural capital and hold it fixed to ensure that there is no further deterioration of the biosphere. The idea is to stop K on its tracks by a global quota on what we are permitted to take from the biosphere.[159]

Let Q be aggregate output. In Ch. 4* we construct a complete model of economic possibilities involving produced, human, and natural capital. Here we present a truncated version of it. If global population is N and φ the proportion of N in production, we assume that output Q is a power function of K and N, that is,

$$Q = K^{(1-\rho)}[\varphi N]^\rho, \qquad 0 \leq \rho < 1, 0 < \varphi < 1 \qquad (A4.2.1)$$

We now estimate $K^{(1-\rho)}$ from the current size of the world economy.

For want of data to the contrary, we assume that the value of the world's production of final good and services draws proportionately on all ecosystem services. In 2019 world output was about 120 trillion dollars at 2011 prices. Using the model of production in equation (A4.2.1), we therefore have

$$K^{1-\rho}[\varphi N]^\rho = 120 \text{ trillion dollars} \qquad (A4.2.2)$$

World population was 7.8 billion in late 2019. The global dependency ratio, that is, the ratio of the sum of the number of people below age 15 and above age 65 to the number of people between 15 and 65, is today about 1.6 to 1. Thus $\varphi = 1/2.6$, which means $\varphi N = 3$ billion. A huge empirical literature in economics suggests that as a rounded figure, $\rho = 0.5$ is not unreasonable. Equation (A4.2.3) then says,

$$K^{0.5} = 120 \times 10^{12}/(3 \times 10^9)^{0.5} \text{ dollars per producer}^{0.5}$$

$$\approx 2.2 \text{ billion dollars per producer}^{0.5} \qquad (A4.2.3)$$

Having calibrated our model of global production, we compute the sustainable population size if $y = 20,000$ dollars. Let N^* denote the size of the sustainable global population. To err on the conservative side of the size of the Impact Inequality today, we assume that the global ecological footprint is currently 1.5. That means if the biosphere and the stock of produced capital were stopped on their tracks, their sustainable value would be $K/1.5$, which we denote by K^*. Using equation (A4.2.3),

$$(K^*)^{0.5} \approx 1.8 \text{ billion dollars per producer}^{0.5} \qquad (A4.2.4)$$

Using eq. (A4.2.2) – (A4.2.4), we have

---

[159] Quotas are applied routinely to fisheries and forestry, and for access to potable water in dry regions. The recent international agreement to limit the rise in mean global temperature to 1.5°C above what it was in pre-industrial times is tantamount to the use of quotas in emissions. Wilson (2016) has made an impassioned plea to leave half of Earth free of human encroachment. We follow that route to identify a sustainable socio-ecological state.

Chapter 4: Human Impact on the Biosphere

$$(K^*)^{0.5}(\varphi N^*)^{0.5} = [1.8 \times 10^9](\varphi N^*)^{0.5} = (20 \times 10^3)N^* \qquad (A4.2.5)$$

But $\varphi = 1/2.6$. From eq. (A4.2.5)) it follows that,

$$N^* \approx 3.3 \text{ billion} \qquad (A4.2.6)$$

Global population was about 3 billion in 1960 (Fig. 4.2); which means, in 3.3 billion we have arrived at a figure that prevailed only about 60 years ago. As the finding in Fig. 4.10 showed, if inequality in the distribution of incomes was judged to be inevitable, the size of global population that would support an average income of 20,000 international dollars would be smaller.

But even a global population of 3.3 billion seems so foreign to us today that the above exercise should probably be interpreted less as prescription than as a sign of how quickly we have overstrained Nature. The idea of sustainable development is meaningless unless it ensures that it does not carry with it the Impact Inequality. Subject to all the caveats we have stressed, our finding says that if humanity were to find ways to reside in the biosphere in a sustainable manner and to bring about economic equality, the human population Earth could support at a living standard of 20,000 dollars is approximately 3.3 billion. It is a simple matter to conduct the exercise with alternative figures for the living standard. We resist doing that.

It is informative to flip the question underlying the calculation by asking what living standard we could aspire to if world population was to attain the UNPD's near lower-end projection for 2100 of 9 billion (UNPD, 2019b). Equations (A4.2.4) – (A4.2.5) provide us with the tools needed to provide an answer. Sustainability requires that,

$$(1.8 \times 10^9)(\varphi N)^{0.5} = Ny \qquad (A4.2.7)$$

Set $\varphi = 1/2.6$ and $N = 9$ billion. That means equation (A4.2.7) reduces to

$$[(1.8 \times 10^9)(9 \times 10^9/2.6)^{0.5}]/9 \times 10^9 = y \qquad (A4.2.8)$$

Let y^* denote the solution of equation (A4.2.8). Then we have $y^* \approx 11,840$ dollars at 2011 prices. The figure falls within the range of middle-income countries. But 11,800 dollars at 2011 prices was the global living standard in about year 2000 (Fig. 4.3). At that time, however, world population was only a little over 6 billion. That 3 billion fewer people did not enjoy a higher living standard should not surprise, because the global stocks of produced capital and human capital were a lot less 20 years ago than it was in 2019 and our model was calibrated with the stocks in year 2019.

How should we read these exercises? It would be easy enough to dismiss them for their naivety, for example that they don't allow for the technological advances we should expect to be made to enable even 9 billion people to enjoy a standard of living a lot higher than 11,800 dollars. But that would be to overlook that unless the global economy finds ways to charge for our use of Nature's provisioning goods, technical advances will continue to be directed at economising on human capital and produced capital and will continue to be rapacious in the use of natural capital. If by some miracle it was possible to make us pay for Nature's services at something like their accounting prices (or social worth), our consumption patterns would be very different. Not only would our household budgets look different, but entrepreneurs would have the incentive to invest in the technologies that economise on the use of natural capital, not be rapacious in its use. The human economy would move in such a different direction that it could even be that our descendants would have a better life than the average person does currently. Today, much of Nature is free, and we add to that insult by subsidising its exploitation to the tune of some 4–6 trillion dollars annually. That makes Nature come to us with a *negative* price! Our efforts should be directed at so improving our institutions that these distortions are eliminated.

Chapter 4*
The Bounded Global Economy

Introduction

Mathematical models of economic growth and development could appear esoteric, unreal or even self-indulgent; but they shape, and are in turn shaped by, our conception of humanity's place in the biosphere. Economists, in planning commissions, ministries of economics and finance, international organisations and private corporations, use the models to analyse data, forecast economic trajectories, evaluate options, and design policy. The economic models that government decision-makers routinely use can be traced directly to academic journals. In turn, economic models in academic journals reflect evolving societal beliefs about what is achievable. The influence is mutual.

Perhaps without conscious design, the macroeconomics of growth and development has been built on the view that human society is external to Nature. In contrast, when constructing the economics of biodiversity, we will keep in mind that we are embedded in it. The difference has profound implications for what we can legitimately expect of the human enterprise. The former viewpoint encourages the thought that human ingenuity can raise global output indefinitely without affecting the biosphere so adversely that it is tipped out of the stability regime it has been in since long before human societies began to form. The latter is an expression of the thought that because the biosphere is bounded, the global *economy* is bounded (Chapter 4, Section 4.7).

The economics of climate change (Cline, 1992; Nordhaus, 1994; Stern, 2006) was developed in response to a growing realisation that rising carbon concentrations in the atmosphere will increasingly damage the global economy. But it has side-stepped the question of whether indefinite growth in global Gross Domestic Product (GDP) is possible. Moreover, it has not asked whether GDP is an appropriate measure of sustainable development; nor what should replace it with if it is found to be inappropriate. There is even an implicit suggestion in the literature that GDP growth is a necessary means to financing the investment needed to reduce net carbon emissions to zero. The social cost of carbon is routinely estimated in models that assume global GDP will increase over the foreseeable future without affecting other features of the biosphere.

Because the potentials for substitution among scarce resources lie at the heart of these questions, we begin (Section 4*.1) with a brief discussion of them. Macroeconomic models of economic growth focus on substitution possibilities among factors of production. The model we construct below (Section 4*.2) acknowledges those very substitution possibilities (it is in that respect wholly orthodox), but keeps a tab on the factors underlying the human impact on the biosphere (Equation (4.2)). So it tracks the waste products that follow *all* production and consumption (in that respect the model is unorthodox).

Our model provides a complete capital theoretic account of human activities, from source to sink. It builds on the fact that the processes governing the supply of regulating and maintenance services are *complementary* to one another, meaning that if one of them is disrupted sufficiently, the others will be disrupted as well. In Chapter 4 (Section 4.7), it was noted that those complementarities set bounds on the efficiency with which the biosphere's goods and services can be converted into output (GDP). That in turn implies that the global economy is bounded. In Chapters 10–13 and 13*, where the rationale is presented for using an inclusive measure of wealth in economic evaluation, the

Chapter 4*: The Bounded Global Economy

global model will be used to show how the accounting prices of capital goods should ideally be constructed.[160]

It proves simplest to present our model in terms of stock/flow identities only. So we do not mention institutions nor human behavioural rules. The idea is to study the *bounds* of human activities, nothing more. Breaking the bounds would amount to tipping the biosphere into a stability regime of which the human enterprise has had no experience. So we study global economic possibilities under the assumption that the biosphere remains within the stability regime in which it currently resides. That requirement circumscribes economic possibilities.

As our model has not yet been estimated, it cannot say when the global economy's bounds would bite so badly that further GDP growth would irreversibly tip the biosphere into a different, and for us, worse, stability regime. Some might argue that boundedness of the global economy could still leave room today for further GDP growth for years into the future even while we invest to repair the biosphere. In *theory* there could be room, but the evidence collated in the previous chapters runs counter to that possibility. Moreover, as is shown in Chapter 13, by economic growth we should mean growth in an *inclusive* measure of wealth, not GDP. Measuring economic prosperity by the size of GDP misleads hugely. The model constructed here shapes the Review.

In Chapter 13 we show that inclusive wealth increases *if and only if* aggregate consumption is less than net domestic product (NDP), that is, GDP less the depreciation of all capital assets. We will also show that to be the criterion that should be used to check that development has been, or is expected to be, sustainable. So, one is naturally led to ask: Is GDP growth compatible with sustainable development?

The question can be answered only within the context of complete macroeconomic models of the long run, in which natural capital plays an essential role – from source to sink. The model we construct here contains those features and so can serve as a prototype of the kind governments and international organisations should now construct. As the model economy is bounded, unbounded growth in output, consumption and inclusive wealth is not possible. Nevertheless, one may ask whether, while keeping consumption at politically acceptable levels it is possible for both GDP and inclusive wealth to grow indefinitely even as they tend to finite limits. The answer is "yes", *provided* the stock of natural capital is large.

Models with features of the prototype constructed below need to be estimated and fed with data concerning current stocks of produced, human and natural capital to answer the question we have raised. Economists will thus have to check using such a model whether GDP growth can be maintained for an extended period even while *politically acceptable* consumption levels remain below GDP net of depreciation of all capital goods. The evidence presented in Chapter 4 casts doubt that such a possibility exists for the global economy today, for humanity's ecological footprint is far in excess of the biosphere's ability to meet it on a sustainable basis.

4*.1 Substitutes and Complements

Classical economists, by whom we mean economists of the late-18th and 19th centuries, argued that the human economy would converge in the long run to a stationary state, in which output would be finite. Their models recognised substitution possibilities among the factors of production (David Ricardo's account of agricultural production saw farmers tilling land whose rents reflected their quality, the

---

[160] A previous literature (Forrester, 1971; Meadows et al. 1972) presented computer simulations of world models to show that there are limits to economic growth. As the models did not have prices even for allocating marketed resources, they were much criticised by economists (e.g. Nordhaus, 1973).

marginal land in use being just productive enough to be economically profitable), but they were not central to the concerns of classical economists.

In any event, the models did not entertain the possibility that output could grow indefinitely through technological advances.[161]

Economic possibilities for the future depend on the extent to which goods and services are substitutable among one another. Four kinds of substitution may be mentioned. First, there can be substitution of one thing for another in consumption (e.g. nylon and rayon substituting for cotton and wool; pulses substituting for meat). Second, produced capital can substitute for labour and natural resources (e.g. the wheel in place of raw labour; refrigeration as way to extend the durability of food). Third, novel production techniques can substitute for old ones (e.g. the discovery of effective ways to replace the piston by the steam turbine). Fourth, and for us here most importantly, natural resources themselves can substitute for one another (e.g. wind and solar power in place of fossil fuels.) All this involves the more general idea that as a resource is depleted, there are substitutes lying in wait, either at the same site or elsewhere, meaning that even as constraints increasingly bite on any one resource base, humanity should be able to move to other resource bases, either at the same site or elsewhere. The enormous additions to the sources of industrial energy that have been realised (successively, human and animal power, wind, timber, water, coal, oil and natural gas and, most recently, nuclear power) are a prime historical illustration of that possibility.[162]

Humanity has been substituting one thing for another since time immemorial. Even the conversion of forests into agricultural land is a form of substitution: large ecosystems have been routinely transformed to produce more food. But as noted in Chapter 4, the pace and scale of substitution in recent centuries have been unprecedented. The question today is whether it is possible for the scale of human activity to increase substantially beyond what it is today without placing undue stress on the major ecosystems that remain. The model we construct below takes substitution possibilities among factors of production seriously, but unlike contemporary models of macroeconomic growth and development it also takes biospheric complementarities seriously. That makes an enormous difference to what we can legitimately imagine economic possibilities to be over the long run.

4*.2 Modelling the Global Economy

The model we construct is highly stylised and can be disaggregated in obvious ways. We assume that there is a produced good, whose output (Y) can be consumed (C) or invested (I).[163] Four categories of capital goods are required for producing Y: produced capital (K), human capital (H), natural capital (S) – each of which is taken to be measurable as a scalar – and publicly available knowledge (A, also a scalar). Natural capital, however, makes its appearance in production in two forms: (i) as a flow of extracted provisioning service (R); and (ii) as a stock supplying regulating and maintenance services (S) in the form of a global public good. When referring to S we shall use the terms biosphere and natural capital interchangeably.

We are interested in studying economic possibilities in the biosphere's *safe zone*. Rockström et al. (2009) call the edges of that zone *planetary boundaries* (Section 4.1.2). If the biosphere was allowed to

---

[161] The pantheon of classical economists is usually taken to be Adam Smith, David Ricardo, Thomas Malthus and John Stuart Mill. Amiya Dasgupta (1985) included Karl Marx and explained why.

[162] Of course, these shifts have not been without unanticipated collective costs. Global warming, associated with the burning of fossil fuels, did not feature in economic calculations in earlier decades.

[163] It would be easy enough to extend the model by disaggregating output Y into various types of products, some less environmentally damaging than others.

Chapter 4*: The Bounded Global Economy

cross that zone, it would fall into an unproductive state (Equation (4*.5)). Global production Y is a power function of the four factors of production:

$$Y = AS^\beta K^a H^b R^{(1-a-b)}, \qquad \beta > 0;\ a, b, (1-a-b) > 0 \qquad (4^*.1)$$

Remark: But for the factor S^β, which we explain presently, the production function in equation (4*.1) is entirely orthodox. The combined factor $K^a H^b R^{(1-a-b)}$ is routine in environmental and resource economics. Its functional form, with the exponents summing to 1 (i.e. production is subject to constant returns to scale in K, H, and R), is known as the Cobb-Douglas form, in honour of the economists who first gave prominence to it in their work on agricultural production.[164] The carrier of technological progress and institutional reform is the factor A, which appears as a global public good. Growth economists refer to A as *total factor productivity* because it can be regarded as the productivity of the comprehensive factor of production $K^a H^b R^{(1-a-b)}$.[165]

Our first move away from orthodoxy is to include a further global public good, denoted in equation (4*.1) as S. The multiplicative factor $S^\beta(t)$, absent from nearly all contemporary models, captures the fact that the human economy is embedded in the biosphere, like a family sheltered in their home. As an example, if life in the oceans were to disappear, human life would disappear (Chapters 1, 4).

R(t) is what we take from the biosphere directly (the *provisioning goods* that yield fish, useable water, timber, fibres, food), whereas $S^\beta(t)$ reflects *regulating and maintenance* services (Chapter 2) – carbon and nitrogen cycles, disease control, climate regulation, soil regeneration and so on – without which production, even life, would not be possible. We do not 'harvest' regulating and maintenance services – their mere presence enables us to exist. In contrast, the economy 'draws on' provisioning (and cultural) goods. The distinction between *drawing upon Nature* and *dwelling in Nature* is all important.

Time $t \geq 0$ is continuous. We express output at t as

$$Y(t) = A(t)S^\beta(t)K^a(t)H^b(t)R^{(1-a-b)}(t), \qquad \beta > 0;\ a, b, (1-a-b) > 0 \qquad (4^*.2)$$

Remark: The economy's Total Factor Productivity (TFP) is $A(t)S^\beta(t)$. Even if A(t) recorded an increase over a period of time, TFP would have declined if $S^\beta(t)$ had declined at a faster rate. But the national income statistician would not know he had overestimated TFP if the model he uses assumes $\beta = 0$. In Annex 3 to Chapter 13, we will show that in standard national accounting practices TFP growth is overestimated: it is recorded as having been large even if it were to have been negative. There is more than irony here: *Under standard accounting practices, the faster the global economy degrades the biosphere, the higher is recorded growth in TFP.*

The Impact Equation (Equation (4.4)) was constructed on the argument that to conserve is the same, modulo sign, as to pollute. Human impact on the biosphere is $R(t) + Y(t)/\alpha_Z$. We imagine for simplicity that A plays a dual role, contributing positively to both Y and α_Z. The model is designed to provide the outlines of global economic possibilities on condition that the biosphere remains within a safe zone for we humans. There are limits to the amount of waste we can produce without tipping the biosphere

---

[164] Among production functions in which the elasticity of substitution (ES) is constant, the Cobb Douglas case is the most useful for our purposes here, because ES there equals 1. Production functions in which ES is greater than 1 are uninteresting because they would enable output to be positive even if one of the factors was reduced to 0; and those whose ES is less than 1 are uninteresting because the quantity available of each factor would limit how much could be produced, no matter how much more of the other factors was available. On the asymptotic properties of production functions, see Dasgupta and Heal (1979).

[165] It is an easy matter to introduce land in the model and distinguish agriculture from industry. But that would be an unnecessary luggage.

into an uninhabitable stability regime. That means there are limits to the extent to which α_Z can be increased by changing our practices and increasing A. We formalise this now.

The efficiency with which output is converted into waste that does not unduly tax the biosphere depends on both the state of knowledge (science and technology) and institutions. Both are reflected in the parameter A. Thus $\alpha_Z = \alpha_Z(A)$, which should be seen to be an increasing function. But even if we imagine that continual investment in research and development can raise A indefinitely, we have to acknowledge that $\alpha_Z(A)$ is bounded above, for otherwise, with sufficient human ingenuity Y/α_Z could be made to go to zero for any finite value of Y; and that would be to imagine we could eventually free ourselves of the biosphere's services entirely. Nor could $\alpha_Z(A)$ go to infinity with Y at the same rate, for that would be to imagine that the demands we make on the biosphere's ability to absorb our waste without experiencing breakdown is unbounded. If $\alpha_Z(A)$ tends to infinity with A at a slower rate than Y, then Y/α_Z would tend to infinity with Y, and that would not be theoretically possible on a bounded Earth. We may as well then suppose that $\alpha_Z(A)$ is bounded above, by say α^*.

As in Chapter 4, we assume heroically that the biosphere's goods and services can be so aggregated as to be measurable as a scalar quantity. So we suppose that the biosphere's net regeneration rate is a bounded function of S, $G(S)$. Thus

$$dS(t)/dt = G(S(t)) - R(t) - Y(t)/\alpha_Z, \qquad \alpha^* \geq \alpha_Z(A) \geq \alpha_Z > 0 \qquad (4^*.3)$$

$(R(t) + Y(t)/\alpha_Z)$ is humanity's impact on the biosphere. Using equation (4*.1), we may define α_X by the equation $R(t) = Y(t)/\alpha_X$.[166] Any efficiency gain in extraction processes would be subsumed in definition of A in equation (4*.1). Equation (4*.3) can then be expressed as

$$dS(t)/dt = G(S(t)) - Y(t)/\alpha_X - Y(t)/\alpha_Z, \quad \alpha_X > 0; \quad \alpha^* \geq \alpha_Z > 0 \qquad (4^*.4)$$

Remark: Equation (4*.4) says that materials must balance: what is taken out of the biosphere has to be put back in, albeit in a different (e.g. biodegradable) form (Section 4.7).[167] The global economy displays the Impact Inequality (Equation (4.3)) when $dS(t)/dt < 0$.

For concreteness, we return to the example in Box 3.3 and suppose

$$G(S) = rS[1 - S/\underline{S}][(S-L)/\underline{S}], \qquad L, r, (\underline{S}-L) > 0 \qquad (4^*.5)$$

Equation (4*.4) then reads as

$$dS(t)/dt = rS(t)[1 - S(t)/\underline{S}][(S(t)-L)/\underline{S}]$$
$$- Y(t)/\alpha_X - Y(t)/\alpha_Z, \alpha_X > 1; \quad \alpha^* \geq \alpha_Z > 0 \qquad (4^*.6)$$

Equations (4*.2) and (4*.6) say that if S was to fall below its safety zone (L), the human economy would collapse eventually. We now investigate economic possibilities inside the safety zone.

The global population size is denoted as $N(t)$. If human capital per capita is $h(t)$, $H(t) = h(t) N(t)$. We take it that population is not controllable directly, but that future population size can be influenced by investment in human capital. This influence is reflected in the number of children desired by adults in a household. Call that number J. As evidence that adults do have reproductive targets, Bongaarts and Cain (1981) noted that the rural population in Bangladesh lost some 1.5 million children in the famine of 1974, and the authors estimated that in the following year there were 1.5 million births in excess of what would have been expected.

---

[166] α_X is obtained by using equation (4*.1), which says $R = [Y/(AS^\beta K^a H^b)]^{1/(1-a-b)}$.

[167] As noted in Chapter 4, the requirement that material must balance in economic accounting is due to Kneese (1970), Kneese, Ayres, and D'Arge (1970) and Möler (1974).

Population size over time has a logistic shape (as in the model of the biosphere in equation (4*.5)), so that

$$dN(t)/dt = N(t)[J(h) - N(t)], \qquad J > 0, dJ(h)/dh < 0 \qquad (4^*.7)$$

Remark: *J* is the long run population size (the logistic specification has been taken from Arrow, Dasgupta, and Mäler (2003b)). That *J* is a declining function of *h* reflects the finding that women's education and knowledge of, and access to, modern family planning services reduce desired family size. We could, should we wish, make *J* a function also of *K* and assume that

(i) $\partial J/\partial K > 0$ for small *K* (better diet, hygiene etc. encourages the birth rate to exceed the death rate, perhaps even when adjusted for public investment in modern family-planning services) and

(ii) $\partial J/\partial K < 0$ for large *K* (the cost of time increases with accumulation of produced capital).

Remark: We could elaborate on equation (4*.7) by multiplying *J(h)* by a constant, say *a*, whose value would reflect the extent to which the economy suffers from externalities, that is, how far the desired number of children in a distorted economy differs from the desired number of children in a non-distorted world. Chapter 9 discusses this issue in detail.

Investment, I, can take three forms: (i) accumulating produced capital *K*, which we write as I_K; (ii) accumulating human capital, which we write as I_H; (iii) expenditure on research and development so as to increase *A*, which we write as I_A. National accounts then say

$$Y(t) = C(t) + I(t) = C(t) + I_K(t) + I_H(t) + I_A(t) \qquad (4^*.8)$$

We assume produced capital depreciates at a constant proportional rate λ. That means net investment in produced capital, d*K(t)*/d*t*, satisfies the condition

$$dK(t)/dt = I_K(t) - K(t) \qquad (4^*.9)$$

Combining equations (4*.8)-(4*.9) yields

$$dK(t)/dt = Y(t) - C(t) - I_H(t) - I_A(t) - K(t) \qquad (4^*.10)$$

Equation (4*.10) records the net accumulation of produced capital. Moreover,

$$dA(t)/dt = I_A(t) \qquad (4^*.11)$$

and

$$dH(t)/dt = N(t)dh(t)/dt + h(t)dN(t)/dt \qquad (4^*.12)$$

Equations (4*.10)-(4*.12) represent net investment in produced capital, knowledge and human capital. But in addition to the three forms of what one may call 'active investment' (Chapter 1), the economy admits a relatively passive form of investment, which can in extreme cases involve simply 'waiting' (given a chance, forests regenerate, fisheries re-stock and wetlands recover). This form of investment finds expression in equation (4*.6).

Equations (4*.2)-(4*.12) represent our model of the global economy when expressed in stock/flow identities. In Chapters 11 and 13, we insert human institutions into it. But even the stock/flow identities tell us that, as *G(S)* is a bounded function, *Y* must be bounded too (Equation (4*.6)). We can buy time by reducing population growth and by raising the efficiency parameters α_X and α_Z through technological change, institutional reforms, substitution among factors of production, and wider behavioural changes; but no amount of technological progress (*A*) and produced and human capital formation (*K*, *H*) can enable global output *Y* to grow indefinitely, for sooner or later the biosphere will cross its safety zone at *L*, and the human economy would in time cease to function. The idea of planetary boundaries (Rockström et al. 2009; Steffen et al. 2015, 2018) puts scientific flesh into the parameter *L*.

4*.3 Contemporary Models of Economic Growth

Contemporary models of macroeconomic growth and development are extreme special cases of our model, the extremities being that they imagine that boundedness of the biosphere does not mean the human economy is bounded. The models assume implicitly that α_Z can be made to rise to infinity, at a minimum at the rate at which Y increases. We confirm that here.

Solow (1956): (i) $A(t) = A$ (constant); (ii) $\beta = 0$; (iii) $h = 1$; (iv) 1–a–b = 0; (v) $\alpha_Z = \infty$; (vi) $J = J(t) = n + N(t), n > 0$.

In words, population grows exogenously at the rate n, and there is no technological progress, no human capital accumulation, and no biosphere to constrain the economy. Y can grow to infinity in the long run, but per capita output ($Y(t)/N(t)$) is bounded above.

Dasgupta and Heal (1974) and Solow (1974a): (i) $A(t) = A$ (constant); (ii) $\beta = 0$; (iii) $h = 1$; (iv) b = 0; (v) a > (1-a) > 0; (vi) $\lambda = 0$; (vii) $\alpha_z = \infty$; (viii) $r = 0$; (ix) $J = J(t) = N(t) =$ constant.

In words, population is constant and the biosphere is an exhaustible resource, which means it is even more bounded than is envisaged in our model. The share of income attributable to produced capital, K, exceeds the share attributable to the flow of natural resources in production, R, that is, a > (1–a) > 0. There are sufficient substitution possibilities between produced capital and exhaustible resources in production, so that even though $R(t)$ has to tend to 0 in the long run, Y is able to grow indefinitely to infinity, albeit at a rate that declines to zero in the long run (Dasgupta and Heal, 1979).

Mirrlees (1967): This model is the same as Solow (1956), but for one difference, which is that $[dA(t)/dt]/A(t) = g > 0$.

In words, the world economy enjoys a constant, exogenous rate of technological progress. That allows not only Y to grow indefinitely, but allows per capita Y also to increase to infinity.

Brock and Taylor (2010): The model assumes pollution is a by-product of output, but that there is an abatement technology whose efficiency grows at a constant exogenous rate. The model can be interpreted as assuming: (i) $\beta = 0$; (ii) 1–a–b = 0; (iii) $[dA(t)/dt]/A(t) = g > 0$; (iv) $J = J(t) = n + N(t)$, $n > 0$; (v) $h = 1$; (vi) $[d\alpha_Z/dt]/\alpha_Z = m > 0$, implying that $\alpha^* = \infty$.

It is shown by the authors that not only Y, but Y/N can grow to infinity.[168]

Stiglitz (1974): The model makes two amendments to Dasgupta-Heal-Solow. (i) $[dA(t)/dt]/A(t) = g > 0$; (ii) $J - N = n > 0$.

The author shows that technological progress can overcome the limitations imposed on an economy by the exhaustibility of resources (oil and natural gas, coal) and can allow the world to enjoy indefinite growth in Y. It is shown that if $g > (1-a-b)n$, per capita output can grow indefinitely.

Remark: Romer (1986) and the many papers that followed it assume that technological advances require investment. Which is why they are called 'endogenous growth models'. Despite the investment costs, indefinite growth in output is assumed to be possible. It follows that, other things equal, growth possibilities in endogenous-growth models are more circumscribed than growth possibilities in Mirrlees (1967), Stiglitz (1974) and Brock and Taylor (2010), because in the latter three publications technical progress is assumed to occur without cost at a positive exogenous rate.

---

[168] Xepapadeas (2005) is an excellent survey of contemporary models of growth and the environment.

Chapter 4*: The Bounded Global Economy

> **Box 4*.1**
>
> **Nature's Goods and Services in Mathematical Ecology**
>
> The biosphere is the part of Earth that is occupied by living organisms. It is an amalgam of interconnected, overlapping, often nested ecosystems, which combine the abiotic environment with biological communities to form self-organizing, regenerative entities that differ in their spatial reach and rhythmic pulse.
>
> In this Review we are building the economics of biodiversity on a classification of Nature's goods and services that was introduced in Ch. 2 and elaborated upon in Ch. 4 (Sec. 4.7).[169] Here we provide a reading that aligns the classification with mathematical models of ecosystem dynamics. Our aim is to show that concepts that are now routine in environmental policy are embedded in the equations of mathematical ecology.
>
> We are to distinguish *provisioning goods*, which include food, water, timber, fibres, pharmaceutical products, and non-living material, from *regulating and maintenance services*, which include climate regulation, decomposition of organic waste, nutrient recycling, nitrogen fixation, air and water purification, soil regeneration, pollination, and maintenance of the biosphere's gaseous composition.[170]
>
> Provisioning goods are the 'produce' of ecosystems. Fibres collected from wetlands are an example, as are fish caught from lakes, food grown on agricultural fields, water drawn from rivers and ponds, timber and pharmaceuticals harvested from forests, and sand dug from the seashore. They are what economists call 'primary products.' With human ingenuity provisioning goods are transformed into final products which, when aggregated using market prices, are read as GDP. Constituent parts of an ecosystem are *stocks* (ecosystems are themselves stocks), and the flow of regulating and maintenance services is the *mapping* that charts the stocks' trajectories over time. The mapping reflects not only Nature's workings, but also human agency in so far as the latter affects Nature's workings.
>
> Eq. (B3.3.3) in Box 3.3, reproduced below, is a stripped down, continuous-time illustration of ecosystem dynamics. As in eq. (B3.3.3) we interpret the ecosystem to be a lake. The lake nurtures provisioning goods which, in the case being considered, are fish. The dynamics are represented by,
>
> $$dS(t)/dt = rS(t)[1 - S(t))/\underline{S}][(S(t) - L)/\underline{S}] - R(t) \qquad (B4^*.1)$$
>
> In eq. (B4*.1) S is the stock of fish in the lake and R is the rate at which fish are harvested. r, L, and $\underline{S}$ are positive constants. It is also assumed that $\underline{S} > L$. Notice that L is a separatrix (fig. 3.2).
>
> The left-hand side of the equation is the rate at which the fish stock (in biomass or numbers, it does not matter here) changes at t. In contrast, the right-hand side of the equation denotes the flow of regulating and maintenance services supplied by the lake system at t. It is the mapping in question. If at any time t you estimate the right-hand side, you can integrate it over time to calculate the future movement of S.

---

[169] As elsewhere, we are using 'Nature' and 'biosphere' interchangeably.

[170] Notice that climate regulation is only one among the many regulating and maintenance services that characterize Nature's workings. The established economics of climate change has encouraged us to think otherwise. Understandably, it focuses on the place of climate regulation in the global economy (Nordhaus, 1994; Nordhaus and Boyer, 2000; Stern, 2006). But that focus has encouraged people to think that if we could attain net-zero emissions, say by substituting clean energy sources for fossil-fuels, global economic possibilities would be boundless. As has been shown in Ch. 4, that is a thought to be avoided.

Harvested fish has a value, say in the marketplace. In Ch. 10–12 we study methods that can be used by the social evaluator to estimate the value of the stock $S(t)$. They would involve relying on harvesting costs and projections of future values of R (the flow of the harvested provisioning good). However, the value of the lake ecosystem would be different from the value of the fish stock. In addition to harbouring fish, the lake ecosystem may have an amenity value, say, as a place where people enjoy leisure activities, which would be an example of a *cultural service* (Ch.2). The value of the lake ecosystem would be the sum of its amenity value and the market value of the fish stock.

If we were to read the global Impact Inequality (Ch. 4, Sec. 4.3) in terms of the lake ecosystem of eq. (B4*.1), it would appear as,

$$dS(t)/dt = rS(t)[1 - S(t))/\underline{S}][(S(t) - L)/\underline{S}] - R(t) < 0 \qquad (B4^*.2)$$

The sign of the two sides of eq. (B4*.2), being negative, says that the harvest rate at t exceeds the fish population's net reproduction rate. Extending this reading to the global level, we conclude that the Impact Inequality can be described either as a drawing down of Nature's regulating and maintenance services or as an excess demand for her provisioning goods. The two descriptions would be equivalent. A more common name for ecological overshoot is *ecological footprint*. A positive footprint in the case of the lake ecosystem is a negative sign for the two sides of eq. (B4*.2).

Humanity has drawn on regulating and maintenance services to provide us with increasing quantities of provisioning goods. We have done that by mining, quarrying, and otherwise transforming the landscape. Such land-use changes as forests razed to make way for agricultural fields and plantations, grasslands transformed into pastureland, wetlands drained for housing, roads, and shopping malls are examples. Biodiversity has been lost at each of those conversions, leading thereby to a decline in regulating and maintenance services. There is thus a tension between our demand for provisioning goods and our need for regulating and maintenance services. That tension is expressed in their mutual feedback. The stocks of natural capital, from which provisioning goods are drawn, influence the supply of regulating and maintenance services, while regulating and maintenance services drive the dynamics of the system, thereby determining future values of the stocks (eq. (B4*.1). The feedback can be explosive, as with climate regulation and biodiversity loss, leading to tipping points being crossed (e.g., crossing the separatrix L in eq. (B4*.1)).

When economists talk of substitution possibilities between different types of natural capital (e.g., wind farms, solar panels, and nuclear reactors substituting for fossil fuels), they have provisioning goods in mind (Dasgupta and Heal, 1979). In contrast, regulating and maintenance services are *complementary*, and that sets bounds on the extent to which human ingenuity can be exercised to transform natural capital into produced capital (roads, ports, buildings, machines) and human capital (health, education, skills, character). As noted previously (Sec. 4.7), the complementarities arise from the fact that ecosystems supply multiple regulating and maintenance services; that is, they supply 'joint products.' Tropical rain forests, rich in biodiversity, influence wind patterns, transport nutrients, decompose organic waste, recycle carbon, capture nitrogen, recycle and filter water, nurture pollinators, and so on. A desiccated forest is unable to offer any of these with vigour. Under extreme stress it flips into a waste land.[171]

---

[171] A direct way to model the tension between our demand for provisioning goods and our need for regulating and maintenance services is to view the biosphere spatially. Conversion of forests into plantations and cattle ranches converts land from one use to another, reducing biodiversity and so to a corresponding loss in critical ecosystem services.

Chapter 5
Risk and Uncertainty

Introduction

The framework we have constructed so far for building the economics of biodiversity has had but the briefest of looks into risk and uncertainty. Our account could almost be read as though the socio-ecological world is deterministic. But we know it is not of course. Even in everyday conversation it is customary to take as understood that when someone says, for example, that the next train to London is scheduled to arrive at the station in 50 minutes time, uncertainty is acknowledged, that the speaker says the train is *expected* to arrive in 50 minutes time.

The term 'expectation' has something like 'average' built into it. If you are given a probability distribution over the arrival date of, say, the monsoon rains in Varanasi, India, you would be using the expression 'expected arrival date' in the sense in which statisticians would use it; that is, the weighted sum of all possible arrival dates in the year, the weights being the probabilities that the monsoon rains will arrive precisely at those dates.

Chapter 3 discussed the extent to which we lack knowledge of the biosphere's processes. This ranges from probability distributions drawn from long-term records, as in the monsoon example above, to a reading of possibilities such as the consequences of an increase in mean global temperature in excess of 4°C since pre-industrial times. In the latter, the evidence is so confounding that experts are unable to read it with sufficient clarity to form even subjective judgements about the relative likelihood of possible events. Absence of knowledge in the former example is known as *risk*; in the latter it is called *ambiguity*. There is, however, a zone lying between the two, where the decision-maker feels able to ascribe weights to the various possible events, but the weights are drawn far more subjectively than from studies of long-time weather data. The absence of knowledge in that intermediate zone is known as *uncertainty*. In a classic work, Savage (1954) showed that if a decision maker's ranking of alternatives comply with a set of conditions of choice that many would regard as reasonable, the criterion of choice she would be found to use would be her *expected utility* of the alternatives she is faced with. The probability weights are *subjective*, as is the corresponding utility function. It is worth emphasising that the utility function derived from Savage's axioms is bounded.

In this chapter we offer a sketch of expected utility theory and, in view of Savage's work, use the terms 'risk' and 'uncertainty' interchangeably. However, a number of economic theorists (e.g. Weitzman, 2009, 2014; Pindyck, 2011) have made a move in their modelling of the economics of climate change that falls somewhere between choice based on Savage's principles and choice when the future is 'ambiguous'. Their work uses unbounded utility functions and probability distributions with fat tails to create instances of what are known as St. Petersburg Paradoxes. Annex 5.1 gives a brief account (Weitzman (2009) called the 'paradoxes' the Dismal Theorem). As the theory of choice under ambiguity remains unsettled, we discuss the subject only in Annex 5.2, and that too briefly. For clarity, we assume that the decision-maker (DM) is a concerned citizen: her viewpoint is *societal*.[172]

---

[172] Raiffa (1968) remains the outstanding treatise on the theory of decision under uncertainty. The literature on ambiguity is still being developed. Heal and Millner (2014) offers an excellent account of the subject as it relates to the economics of climate change.

Chapter 5: Risk and Uncertainty

5.1 Portfolio Choice under Uncertainty

Begin by assuming that DM follows statistical decision theory and ranks choices on the basis of the expected values of societal, or *social*, well-being (Box 5.1 summarises the theory).[173] We imagine that DM is able to associate with each possible portfolio of her society's assets (Chapter 1) a (subjective) probability distribution of yields. In DM's judgment social well-being depends on the realised yield (Chapters 10–11). But because yields are uncertain, social well-being consequent upon DM's choice of a portfolio is uncertain. DM therefore chooses from the available portfolios so as to maximise the *expected value* of social well-being (Box 5.1). So as to keep the exposition simple, we assume unless we say otherwise, that people are identical. As DM chooses on their behalf, she treats them identically. To her, social well-being corresponds to an individual's well-being.[174]

Although some individuals are known to have a flutter at the gambling table, discussions on environmental risks nearly always assume that people are risk averse. By that we mean that among risky portfolios with the same expected yield, people choose the one with the smallest risk. The literature on what it means to say that one portfolio is riskier than another is technical and involves statistical decision theory.[175] The examples with which we have chosen here to study choice under uncertainty allow us to bypass the technicalities in that literature. Many of the examples involve choice between a risky portfolio and a riskless one with the same expected yield. If the expected yields differ, DM will be assumed to trade off expected yield against riskiness in yield in the way the representative individual will be assumed to do.

Box 5.1
Expected Utility Theory

B5.1.1 The Theory in Outline

To formulate DM's decision problem under uncertainty, we assume that she can choose one from among several alternative courses of action, ranging from actions she can take to policies she can adopt or recommend. Let X be the set of available courses of actions. Each course of action, written as x, y, z, and so on, leads to one of a number of possible consequences. We have assumed that DM acts on behalf of society, which means she ranks consequences in line with her conception of social well-being. Often we will read 'consequence' as 'wealth' (or income, the two are conflated in a timeless setting), but here we present a general account of the theory of choice under uncertainty.

Which consequence will result from a course of action depends on the state of Nature. In statistical decision theory, a *state of Nature* is a description of the world so complete that, if true and known by DM, the consequences of every action would be known by her. By definition, which state of Nature will prevail is outside DM's control. DM is required to choose a course of action before the state of Nature has revealed itself (e.g. the farmer plants his crops before the rains arrive). An *event* (e.g. 'no rainfall on the first of May') is a union of states of Nature (e.g. 'no rainfall on the first of May, followed by rain on the second of May 2'; 'no rainfall on the first of May, followed by no rain on the second of May'; and so on).

Let s_i be the state of Nature i where $i = 1, ..., N$. N is the number of possible states of Nature. We write DM's subjective probability that s_i will occur by $\pi(s_i)$ and social well-being if she chooses course

---

[173] The idea of social well-being is developed in Chapters 10–12.

[174] None of the concepts that are developed in this chapter depends on the assumption that people are identical.

[175] In economics the classic on the subject is Rothschild and Stiglitz (1970).

of action x and s_i were to occur by $u(x,s_i)$. Then expected (social) well-being is $_i\sum[\pi(s_i)u(x,s_i)]$. In the way we have worded DM's decision rule, expected well-being defines the *preference* relationship among the elements of X on the basis of which DM's chooses from X or from any subset of X should she be restricted to choose from the subset. Formally, a *preference ordering*, or *ordering* for short, on X is a binary relationship ≽ between the elements of X that is (i) 'reflexive', meaning that for all x in X, x≽x; (ii) 'transitive', meaning that for all x, y, z in X, x≽y and y≽z implies x≽z; and (iii) 'complete', meaning that for all x and y in X, either x≽y or y≽x. We write $x > y$ if x≽y and not y≽x. (> represents strict preference).

Savage (1954), who constructed subjective probability theory, took the opposite route and began by assuming that DM is able to order the elements of X. The ordering ranks the consequences of her possible choices in line with her conception of social well-being as well as her subjective assessment of the likelihood of the occurrence of the various states of Nature. The axioms of rational choice Savage proposed say that if ≽ is DM's ordering over the elements of X, then there exists a probability distribution π over the state of Nature and a scalar function u of the course of action and state of Nature, (x,s_i), such that for all x, y in X, x≽y implies and is implied by $_i\sum[\pi(s_i)u(x,s_i)] \geq _i\sum[\pi(s_i)u(y,s_i)]$. Savage referred to the u-function as a *utility function* and $\pi(s_i)$ as DM's subjective probability that s_i will occur.

There is no absolute scale for measuring u. More particularly, u does not have a given sign, positive or negative, what matters in Savage's theory is the sign of the difference in the expected value of u when DM compares two courses of action. Savage's axioms of rational choice yield a measure of u that is unique up to 'positive affine transformations'. That says if u corresponds to DM's preference ordering on X and if β is a positive number and γ is a number of either sign, then $\beta u_i + \gamma$ corresponds to same preference ordering. To confirm, we note that

$$_i\sum[\pi(s_i)u(x,s_i)] \geq _i\sum[\pi(s_i)u(y,s_i)] \text{ if and only if}$$
$$_i\sum\{\pi(s_i)[\beta u(x,s_i) + \gamma]\} \geq _i\sum\{\pi(s_i)[\beta u(y,s_i) + \gamma]\}$$

In the latter scale β represents the unit in which utility is measured and γ is the level. We say u in that case is a *cardinal* measure.[176] In practical applications it is often necessary, as we will discover in Section 5.3, to modify Savage's language by considering courses of action that alter the probabilities of *events*, which are unions of states of Nature. When we do that the sign of u matters, which means γ is not arbitrary; u's level matters. When the level of a cardinal u-function has to be specified, we say u is *strongly cardinal*. We return to this in Chapter 10.

B5.1.2 The Measure of Relative Risk Aversion

Choice of an action in Savage's theory is frequently interpreted as choice of a *lottery* over wealth (or income, the difference between the two doesn't matter here). Suppose DM cares about wealth (we write that as w) and suppose that choice of an action can be read as choice of a lottery. The u-function is then defined over wealth, that is, $u(w)$ is the u-function. In Savage's theory u is an increasing function of w. To say that DM is *risk averse* over all portfolio choices is to say that $u(w)$ is (strictly) concave, that is, marginal utility is a declining function of w. Formally, that means $u'(w)$ – the first derivative of u – is positive and $u''(w)$ – the second derivate of u – is negative (Figure 5.1).

Let w_i be DM's wealth if the state of Nature is i. As there are N states of Nature, a lottery is an N-vector $(w_1, ..., w_i, ..., w_N)$. Intuitively it is apparent that the degree of risk aversion displayed by DM is

---

[176] Temperature scales are cardinal. The Fahrenheit and Centigrade scales are related in the way described in the text. Conversion from the Centigrade scale to the Fahrenheit scale requires multiplying by 9/5 (that is the β) and adding 32 (that is the γ).

Chapter 5: Risk and Uncertainty

reflected in the curvature of the u-function, a convenient measure of which is the elasticity of marginal u with respect to w. Denoting the elasticity of marginal u by η, we have $\eta = \eta(w) = -wu''(w)/u'(w) > 0$. For obvious reasons $\eta(w)$ is also known as the measure of DM's *relative risk aversion* (Pratt, 1964; Arrow, 1965).

In the economics of climate change it is near-universal practice to assume that relative risk aversion, η, is independent of wealth (Chapter 10). The family of u-functions for which η is a positive constant has the form

$$u(w) = w^{(1-\eta)}/(1-\eta), \quad \eta \neq 1,$$
$$u(w) = \log w, \quad \eta = 1 \tag{B5.1.1}$$

The larger is η, the greater is the curvature of u-function. Notice also that $u(w)$ is bounded above but unbounded below if $\eta > 1$, whereas $u(w)$ is bounded below but unbounded above if $\eta < 1$. If $\eta = 1$, $u(w)$ is unbounded at both 'ends.' The asymptotic properties of the u-function in equation (B5.1.1) are its weakness, for Savage's theory yields *bounded* u-functions. Annex 5.1 reveals that the use of unbounded u-functions gives rise to paradoxes, known as St. Petersburg Paradoxes.

So as to confine attention to bounded u-functions while retaining the empirical finding that, generally speaking, η is less than 1 for small w but greater than 1 for large w, Arrow (1965) proposed the use of u-functions that are bounded, and for which η is (i) an increasing function of w and (ii) less than 1 for small w and greater than 1 for large w. Condition (ii) implies that $\eta = 1$ at some intermediate value of w. Here is an example of a u-function with these properties:

$$u(w) = 1 - e^{-\lambda w} \tag{B5.1.2}$$

The measure of relative risk aversion for the u-function in equation (B5.1.2) is $\eta(w) = \lambda w$.

As an example of bounded u-function equation (B5.1.2) is attractive, because it is built on a single parameter, λ. But the functional form is not without a problem. To reflect the fact that to be destitute is the ultimate calamity for a person, it makes sense to postulate that $u'(w) \to \infty$ as $w \to 0$. In the example here, $u'(w)$ tends instead to λ.

Arrow and Priebsch (2014) have shown that the requirements that (i) $u(w)$ is bounded and (ii) $u'(w) \to \infty$ as $w \to 0$ cannot be integrated to yield a closed functional form for u. As the authors show, that does not prevent DM to study the economics of risk and uncertainty, but it is harder work.

In certain applications, it makes sense to extend Savage's theory and work with u-functions that are explicit functions of the state of Nature. In studying optimum expenditures on health, for example, one may take the state of a person's health, labelled as h, as a state of Nature and express the u-function as $u(w,h)$. The idea here is that a person's needs depend on his health.

5.2 Independent vs. Correlated Risks

It is useful to discuss extreme cases, and so we distinguish between risks that are independent across individuals from those that are perfectly positively correlated across individuals.[177] We begin with the case of independent risks. We then study perfectly (positively) correlated risks.

---

[177] Negative correlation in the risks people face is of course a possibility. Firm 1 could hold a patent on the genetic material of plant A, while firm 2 holds the genetic material of plant B and it is commonly known that one and only one of them is a cure for a disease, but not which.

5.2.1 Ideal Insurance Markets

The risks people face from being involved in road accidents are independent of one another, which is why insurance markets for road accidents are able to flourish. People *pool* their risks in the insurance market; in turn insurance firms *diversify* their risks by selling accident insurance to large populations. One way to interpret insurance markets is to imagine that all who purchase insurance mutually insure one another and use insurance companies as intermediaries. Under ideal market conditions competition among insurance firms drives the premium for a risk category down to the expected loss from an accident in that category. It would be as though people deposited their expected loss from a risk in that category into a common fund from which they were entitled to take what they had lost in case they had an accident.

Consider for simplicity the case where everyone belongs to the same risk category. Let w denote a person's wealth and let w^* be his wealth if he was not to suffer from an accident (Figure 5.1). If π is the probability that he will suffer from an accident, and m is the monetary loss from the accident, then πm would be the premium. m may be large, the accident could even be a catastrophe for the person, but if π is sufficiently small, the premium would be small relative to w^*.[178] More importantly, because the risks are independent of one another, people would be able to insure themselves fully against personal accidents at that price (at least in theory!). The reason they would be able to do that is that if the population were large, the Law of Large Numbers in probability theory says everyone could be 'guaranteed' the wealth (w^*-πm) if they mutually insured one another. As πm is small relative to w^*, people would be protected (almost) fully against loss. The insurance market in principle does that; and one says the market pools risks.[179]

This, in very broad terms, is the theory underlying ideal insurance markets. In practice any number of features of the world muddy the account. For example, in many places people do not have access to insurance markets even for personal risks such as illness and death in the family. In the absence of markets, rural households in poor countries form networks for pooling personal risks (Chapter 6). Village numbers being small, households cannot take advantage of the Law of Large Numbers, so they are able to insure one another only partially. But networks have two advantages insurance markets do not enjoy. Proximity among households and the fact they know one another well mean that rural insurance schemes suffer far less than insurance markets from 'moral hazard' and 'adverse selection'.

Moral hazard is the inability of insurance companies to ensure that the individual purchasing insurance takes due diligence. The point here is that companies are unable to observe insurees' actions. Without due diligence in our example the risk the insuree will have an accident would exceed π. Moral hazard is a ubiquitous problem in insurance markets. By introducing suitable incentives insurance companies reduce the moral hazard they would otherwise face. In automobile insurance, for example, companies reduce moral hazard by requiring applicants to produce an acceptable certificate each year that says their car is roadworthy. Moral hazard also lies behind the practice of lowering the premium if no claim had been made in the previous year. For property insurance against burglary, moral hazard is reduced by a requirement that the insuree installs safety features in his home (burglar alarm, sturdy locks, and so forth).

The other ubiquitous feature of insurance markets is *adverse selection*, which is the inability of an insurance company to fully determine an insuree's relevant characteristics. In the case of automobile insurance, insurance companies remedy this partially by requiring applicants to produce evidence that

---

[178] In Annex 5.2 we comment on the case where the risks are perfectly positively corrected, π is small, but the collective loss would be a catastrophe, meaning that πm is huge, maybe even unboundedly large. The expected utility theory will not hold in this case.

[179] Malinvaud (1972) is a standard source for an account of the economics of insurance.

Chapter 5: Risk and Uncertainty

they are safe drivers (e.g. they enjoy good eyesight). Thus, while moral hazard points to the problems insurance companies face in confirming that insurees *do* what is required of them under the terms of the contract, adverse selection points to the problems companies face in confirming that insurees *are* what they claim to be (e.g. regarding their risk category).

Moral hazard and adverse selection give rise to *externalities*, which are the unaccounted-for consequences for others of actions taken by someone. Externalities, which are discussed in Chapters 6–9, are a pervasive feature in the economics of biodiversity. They reflect a class of imperfections in the allocation of resources, which in the present context are imperfections in the production and distribution of risk. In what follows, we do not formally model either moral hazard or adverse selection, but we do point out on appropriate occasions the steps insurers take to reduce them. We therefore interpret π to be the average risk across the population.

5.2.2 Fully Correlated Risks

The examples that were studied in Chapters 2–4 tell us that the risks associated with biodiversity loss and global climate change are positively, often perfectly, correlated.[180] Losses from floods and tropical cyclones can significantly affect a community's wealth, particularly in low income countries. In extreme cases, communities can be made destitute by one such calamitous event (Chapters 14 and 17). Earthquakes are another class of natural occurrences. Because the risks are perfectly correlated, they are called *common risks*. The Law of Large Numbers does not apply to them. If the community is an entire country and insurance companies are unable to diversify their risks beyond the country's borders, they would not be able to offer cover against a common risk.

Because there is no scope for insurance, each person in the country would have to bear his entire risk. Suppose w^* would be a person's wealth if the catastrophe were not to occur (or at least not for many years from now) but would be (w^*-m) if it *did* occur. If $u(y)$ is the person's utility function, his expected utility would be $[(1 - \pi)u(w^*) + \pi u(w^* - m)]$, which is less than $u(w^* - \pi m)$, because the person is risk-averse. In Figure 5.1, income w is drawn on the horizontal axis and utility u is drawn on the vertical axis. u is an increasing function of w and, because DM respects people's aversion to risk, its shape is concave. If the person was not to suffer the loss, his utility would be $u(w^*)$, which in Figure 5.1 is BB'. If on the other hand he was to suffer the loss, his utility would be $u(w^* - m)$, which is AA'. In Box 5.2 we hold constant the potential loss m and the probability π that the loss will occur so as to study two extreme cases: (i) the risks are independent and people are fully insured against them; and (ii) the risks are common and people are fully exposed to them.

Box 5.2
Independent and Fully Correlated Risks: Comparison

Consider first the case where the risks are independent across people. The Law of Large Numbers applies. Under full insurance (fair odds), the premium is πm, and each person has a sure wealth amounting to ($w^* - \pi m$). This is drawn as CC' in Figure 5.1. If on the other hand, the individual risks are fully correlated, so that there is no insurance coverage, an individual's wealth is w^* with probability (1–π) and (w^*-m) with probability π. Expected utility of wealth in this case is CC", which is of course smaller than CC'. People are worse off under correlated risks than under independent risks, other things equal.

---

[180] We are talking of the commonality of the *risks*, not the losses suffered by households should a catastrophe occur.

Figure 5.1 Risk Averse Preferences

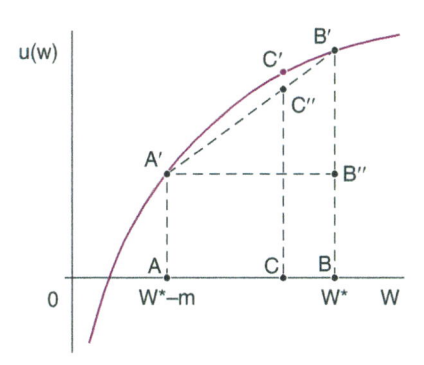

[Figure: Concave utility curve u(w) plotted against w, showing points A, A' at (W*−m, u(W*−m)), B, B' at (W*, u(W*)), with C, C', C'', B'' illustrating risk averse preferences via chord from A' to B'.]

Intermediate cases can be decomposed into the two extreme types of risk we are considering here. An increase in automobile emissions in a city raises the risk that inhabitants will suffer from bronchial disorders, but that additional risk is composed of a common increase all individuals experience, and an idiosyncratic increase for each person, depending on such personal factors as age, state of health, and so on.[181] The balance of common and idiosyncratic risks differ from risk to risk. It could be thought that the component of risk to agricultural production that is idiosyncratic is small, but as micro-climates matter and soils are not the same across farms, risks to crop output have a non-negligible farm-specific element.

Insurance companies are able to offer partial coverage for even common risks such as to food production. Crops Shortfall Insurance (CSI) in the UK is an example (Farmers and Mercantile, 2020). The CSI was instituted in 2019 to offer insurance to farmers against droughts, floods, windstorms, excessive rainfall, hail, heat waves, and plant diseases and pests. If the scheme purchased by a farmer covers total output (of cereals and oilseed), the policy he purchases triggers automatic pay-out if a natural disaster causes farm yields in the region to fall below the previous eight-year average.[182] A policy, after chosen deductibles, protects the farmer by up to 25% of the shortfall of his projected crop output from the following harvest. That crop production faces a common risk means that full protection against losses is unavailable.

Financial support for households from government following natural disasters to entire communities reduces the losses suffered, but the obligation to cover losses, if only partially, falls on the state, not the market, and usually falls only *after* the event, not *before* the event. The state does not usually offer citizens insurance coverage against natural disasters; rather, citizens expect the state to offer help should a disaster strike them. We may imagine that citizens do not know for sure how much help they will receive from the state.[183] That additional uncertainty creates a further wedge between 'insurance' and 'emergency relief'.

To the extent common risks are confined to a region of the country, national insurance policies suggest themselves. One possibility would be for the state to cooperate with private insurance companies so as to offer coverage through local councils, in addition to offering emergency aid following calamitous events to those in especial need.

---

[181] The literature on the estimation of health losses owing to urban pollution is extensive, but mainly confined to rich economies (Freeman III, 1982; Freeman III, Herriges and Kling, 1996). For an application to residents in a South Asian city, see Haque, Murty and Shyamsundar (2011).

[182] The average is determined by the UK's Department for Environment, Food & Rural Affairs (Defra).

[183] There are exceptions. In response to the COVID-19 pandemic, many governments have established wage support schemes.

Chapter 5: Risk and Uncertainty

The existence of common global risks from natural capital depletion means there are limits to the extent standard insurance practices can be effective. By definition, global negative externalities require coordinated global action. If entire countries are vulnerable to physical risks from natural capital depletion and costs from these risks crystallising are significantly large, national insurance schemes are likely to be unable to deal with future extreme events on their own. For low income countries, in particular, such events could in extreme cases wipe out significant parts of their produced capital (such as infrastructure) and productive capacity, and thereby the fiscal capacity of the country.[184] Put simply, when entire countries fall prey to the risks arising from global climate change and biodiversity loss, the insurance burden needs to be borne by the international community. In principle, a global risk pool – with contributions from all countries – could help protect vulnerable countries against such shocks following extreme Nature-related events (Chapter 17).

Box 5.3
Short-Run vs. Long-Run Predictability

Individuals and communities spend resources to reduce variance in the supply of food, water and other ecosystem services. That is a reflection of our risk aversion; we want greater predictability in supply rather than less. However, investment is often directed at decreasing variance (an index of variability) over short time intervals (2–4 years). If, for example, phosphorus discharge into a freshwater lake is from a point source (a single farm), a plausible public policy would be to ask the farm to limit its discharge to an acceptable, constant rate. In an important and interesting paper, Carpenter et al. (2015) show that when applied continuously, the constancy of short-term phosphorus intake can lead to a shrinkage in the lake's stability regime, and thereby increase the risk that it will cross a critical threshold. The shrinkage can occur because short-run smoothing removes the pressures on the ecosystem to build tolerance to unexpected stress. Essentially, smoothing reduces the system's modularity. The authors have found a similar characteristic in harvesting wild populations and in the yield of domestic herbivores in rangelands. Formally, short-run smoothing creates 'red noise', which is a stressor that is novel to the ecosystem. No existing component of the ecosystem can cope with the redder noise created by the smoothing. The ecosystem loses response diversity (Chapter 2). In short, if an ecosystem is made to be more resilient in the short run, it may become less resilient to novel disturbances. Reducing one form of risk can lead to an increase in another form of risk.

While currently there is no *global* insurance scheme for natural capital risks, there are examples of *regional/multi-country* insurance schemes for environmental disasters. These include the African Risk Capacity (ARC) and the Caribbean Catastrophe Risk Insurance Facility. Both schemes pool risk across participating countries and provide pay-outs to reduce the cost of disasters occasioned by climate change, earthquakes and hurricanes. Schoenmaker and Zachmann (2015) have proposed a global risk pool system to mitigate climate risk following the principles of reinsurance with a mix of contributions from donor countries and insured countries. Reinsurance is a common mechanism for sharing insurance risks.[185] It reduces an insurer's risk of loss by sharing the risk with one or more reinsurer. The practice generally works by either transferring a portion of a particularly large risk that has been taken on by an insurance company, or by transferring a portion of all the pool of risks to a reinsurer in return for a share of the original premium. In the event of a claim, the reinsurer compensates the insurer for its share of the risk.

---

[184] As the events will result in lower tax revenues and high rebuilding costs.

[185] The largest private firms conducting reinsurance operations include Lloyd's, Swiss Re and Munich Re (Schoenmaker and Schramade, 2018).

In principle, the idea of a layered global risk pool system could be applied for risks emanating from natural capital more broadly, such as biodiversity loss. The insured country would bear the first layer of loss itself so that it has an incentive to work on Nature conservation and restoration projects. The insurance premium could be partly related to a country's ecological footprint to provide an incentive to reduce their demands on the biosphere. As in the case of a climate risk-pool, the premium structure could be two-tiered, with the premiums for vulnerable countries based on risk of ecosystem hazards and their ecological footprint, and premiums for donor countries based on level of GDP and their ecological footprint.

Schoenmaker and Zachmann (2015) use a trend-line of global sea level rise to form the basis in their proposed system for assessing triggers for pay-outs and to determine the *extremeness* of events. But they note that there still appears to be a significant gap in knowledge about the regional impacts of climate change. For biodiversity, a similar approach would be inherently harder, given its complexity.

When entire countries fall prey to the risks arising from global climate change and biodiversity loss, the insurance burden needs to be borne by the international community. The ARC, comprising 34 member countries in sub-Saharan Africa, is one such agency. It was created in 2012 to insure against risks to agriculture from extreme weather events. The capital for the pool comes from participating countries' recurring premiums as well as one-time contributions. Countries select the level at which they wish to participate, by choosing the amount of risk they wish to bear themselves and the amount they would like to insure against droughts of varying severity. The maximum amount that can be claimed remains small – about US$30 million per country per season for droughts. This is a paltry amount, considering the stakes.

Box 5.4
The Value of Information

Information has value to the person acquiring it, which is why people pay for it. In what follows, we apply Savage's theory to quantify the value of information.

As in Box 5.1, there are N states of Nature $(s_1,..,s_i, ..., s_N)$. The (subjective) belief DM holds that s_i is the true state is given by probability $\pi(s_i)$. DM is unable to observe the state of Nature, but at a cost k is able to observe a signal from a set of M signals, indexed by j, that are correlated with states of Nature. Let the jth signal be denoted by ϑ_j. For example, s_i could be the weather and ϑ_j the weather report. We assume DM has a theory of her sampling process, meaning that she imputes a probability $\gamma(\vartheta_j | s_i)$ that she will observe ϑ_j conditional on s_i being the true state of Nature. DM now uses Bayes' theorem to construct her posterior probabilities of that the true state of Nature is s_i as

$$\mu(s_i|\vartheta_j) = \pi(s_i)\gamma(\vartheta_j | s_i)/ {}_i\Sigma\pi(s_i)\gamma(\vartheta_j | s_i) \tag{B5.4.1}$$

DM makes her decision on the basis of her posterior probabilities μ. Her expected utility would be higher if she were to make a decision after observing a signal (because signals are correlated with states of Nature). If the gain in expected utility exceeds k, she will want to purchase the right to observe a signal. Otherwise she will decide on the basis of her prior probabilities π.

In an interesting controlled experiment, a randomly selected group of households in a middle-class district in New Delhi were informed of whether their drinking water tested positive for fecal contamination, using a simple, inexpensive test kit (Jalan and Somanathan, 2008). Households that had so far not been purifying their water and were told that their water was possibly contaminated were found to be more likely by 11 percentage points to begin some form of home purification in

Chapter 5: Risk and Uncertainty

> the next eight weeks than households that were given no information. The former group spent $7.24 (Purchasing Power Parity, PPP) more on purification than control households. As $7.24 (PPP) was in the circumstances a non-negligible sum, we should conclude that the information was valuable. The finding suggests that estimates of the demand for environmental quality that assume full information may significantly underestimate it.

5.2.3 Insuring Nature

An alternative to insuring individuals against personal loss would be to insure against the environmental collapse itself. The idea would be to commit the insurance premium to restoring the ecosystem should it collapse. Such a scheme is in principle much like insuring buildings.[186] A common problem any such scheme would face is one that is familiar in the theory of public goods. As the benefits from restoration would be enjoyed by all, individuals will not have the incentive to purchase insurance: people could be expected to free-ride. We should then imagine such schemes to be installed by a collective, such as local government in the case of a spatially confined ecosystem. The premium would be a tax, so to speak, to be put aside in a restoration fund. In a wide-ranging essay, Kousky and Light (2019) cite a number of schemes that provide communities insurance against the common risks they face of ecosystem disruptions.

Several global funds have been established to implement mitigation and adaptation measures against climate change; more generally to reduce environmental degradation. They provide grants and mobilise a mix of public and private funds to support conservation and restoration projects across countries (Chapter 20).[187] The funds have been created by groups of countries, in partnership with international institutions (such as multilateral development banks), national entities, international Non-Governmental Organisations (NGOs) and the private sector. Contributions by members are pooled and channelled towards mitigation and adaptation projects, that is, they involve reducing π and m. The funds set criteria that projects must meet in order to receive a grant. Although the funds monitor and evaluate project outcomes, they do not (yet) have a systematic mechanism for *ex post* project evaluation.

> ### Box 5.5
> ### Mexican Coastal Zone Management Trust
>
> In 2005, Mexico's Caribbean coast was struck by two hurricanes, causing US$8 billion in damages and closing hotels and other businesses in Cancún (The Nature Conservancy, 2019). But some hotels and beaches suffered less damage than other areas in the state of Quintana Roo. Subsequent evidence suggested this was in part due to the protection provided by a stretch of the Mesoamerican Reef. In response, a Coastal Zone Management Trust was established in 2018 to build ecosystem resilience along 160 km of the Yucatan peninsula.[188]

---

[186] Ideally the payment would include the costs borne by the person suffering damage while the ecosystem/building is out of service.

[187] The schemes include the Global Environmental Facility, the Climate Investment Fund, the Green Climate Fund and the Global Facility for Disaster Reduction and Recovery.

[188] This project involved The Nature Conservancy, Quintana Roo State Government and the Cancún and Puerto Morelos Hotel Owners Association.

> The Coastal Zone Management Trust receives funds from an existing fee paid by beachfront property owners, among other private and public sources. These funds are used to finance the maintenance of reefs and beaches, which help protect produced capital, such as roads and buildings, from storm damage. The Trust purchases insurance to pay for the restoration of coastal ecosystem services after extreme weather events. The insurance policy is parametric, involving payments to the insured when a certain predefined triggering event occurs. Parametric insurance does not require assessor verification or economic valuation of the damage and eliminates any ensuing disputes that might occur over the extent of the damage. That enables swift provision of funds for restoration (Kousky and Light, 2019). In the case of the policy offered by the Trust, the pay-out depends on the severity and location of the event. Specifically, should wind speeds in excess of 100 knots hit a predefined area, an insurance pay-out will be made quickly to the trust fund, allowing swift damage assessments, debris removal and initial repairs to be carried out. The pay-out varies according to the registered wind speed, as the higher the speed the more damage is expected. The policy has a coverage limit of US$3.8 million (The Nature Conservancy, 2019).
>
> Although parametric insurance is not new, its application to protection and restoration of environmental public goods is novel and could be expanded to other regions and ecosystems.

5.3 Reducing Risks and the Losses from Risks

Generally speaking, policies aimed at protecting communities against a common risk, such as a collapse of their local ecosystems, take one of two forms: (i) reduce the loss m that would be incurred by each person should the ecosystem collapse; (ii) reduce the probability π of a collapse. But there are cases where (iii) both m and π are reduced. Conservation and restoration activities illustrate each of these avenues.

Examples abound of policies targeting m. Restoring a mangrove forest provides protection to communities against cyclones. The project would not reduce the probability that a cyclone will hit the coast, what it would do is to reduce the damage communities would suffer should a cyclone hit the coast. Restoring coastlines and wetlands are other examples. Suitably targeted, they reduce the risk of ecosystem collapse.

Investing in physical infrastructure against floods also targets losses should a flood occur, but this have been known to come with a sting. Embankments that are constructed to prevent overflow from rivers reduce loss of property and lives when the rivers rise, but they come at the cost of destroying the wetlands that once protected people and property against the less frequent storm surges that occur downstream.[189]

Deforestation upstream of watersheds not only increases risks of floods downstream (that is π) but also the frequency of floods (that is in effect m). In a study using data over 1990 to 2000 from 56 developing countries, Bradshaw et al. (2007) found that after controlling for rainfall, slope and degraded landscape, an arbitrary decrease in natural forest area of 10% led to an increase in predicted frequency of floods by 4–28% and a 4–8% increase in the total duration of floods. These findings come alive when note is made of the fact that each year extreme floods kill or displace hundreds of thousands of people in developing countries and result in billions of dollars in losses to property and infrastructure.

Broad categories of policies that target π also involve conservation and restoration measures. Conservation of wetlands reduces the risk of homes being flooded should the rainfall be unusually

---

[189] An example is Hurricane Katrina, which in 2005 made landfall at the estuary of the Mississippi River and devastated the city of New Orleans and its surroundings. The resulting damage was all the greater than it would have been because the wetlands along the river had given way to levees built in the past as a protection against the regular flooding of the Mississippi.

heavy. Restoring hedgerows on farmlands raises the probability that pollinators will return. The motivation behind the many conservation and restoration programmes under way may be similar, but they can differ in the way they are financed. This is illustrated in Chapter 6 under the heading, 'payment for ecosystem services'.

5.4 When to Stop Business-as-Usual

When the rate at which a community consumes an ecosystem's goods and services exceeds its regenerative rate over a period of time, the ecosystem shows signs of stress. Ecologists then warn us of the possibility of sharp declines in its productivity, perhaps even a catastrophe, not only for the local community but also for other communities that rely on it (Chapters 3–4). Even though ecosystems display signs when they are under stress – perhaps even display signs that they are near a tipping point (Chapter 3) – no one can tell at what point the system will reach a state from which restoration will prove to be prohibitively costly. As in all matters concerning complex systems, uncertainty prevails. But sceptics have been known to exploit that uncertainty and say they have been there before; they complain that environmental scientists cry wolf too often.[190]

In fact, environmental scientists have been cautious in their professional publications, not impulsive. Accumulating evidence over the years, the Intergovernmental Panel for Climate Change (IPCC), for example, have systematically revised their estimates of both the speed of climate change and its consequences for the biosphere's processes under business-as-usual; but the revisions have invariably been upward, not downward. And their readings have always been qualified by uncertainty (IPCC, 2014).

The influence of environmental sceptics on public opinion has diminished in recent years. More frequent and more extreme weather events (intense storms and prolonged droughts), demonstrable harm to marine life from an accumulation of plastics in the oceans, photographs of dead zones in what were once thriving estuaries, televised evidence of melting glaciers, recorded rise in species extinction, and most recently the global pandemic that is COVID-19 have brought home the fact that various parts of the biosphere are now overstretched. But an ecosystem could show signs of stress without having tipped over (Chapter 3).

When should a society change course and reduce its demands on an overstretched ecosystem? Changing course would have costs. At the very least there would be costs of reducing dependence on a cheap source of goods and services, possibly also from having to abandon complementary assets (e.g. a network of roads leading to the ecosystem). However, not to change course would expose society to the possibility of a regime shift (Chapter 3) that would inflict even greater social losses. These are contending considerations.

So then, when should DM advocate a change in course? It is no good saying "before it is too late", for until the ecosystem tips over we would not know that it was too late. Nor would it be of any use to say "it is never too early", for if the stress that the ecosystem displays is mild, the likelihood of a regime shift will be read by experts to be small.

---

[190] Lomborg (2001) is the most widely cited among authors in recent years to have expressed scepticism of the environmentalist's concerns. Vigorous endorsement of the book by *The Economist* newspaper helped to foster public scepticism. It took time, perhaps starting with a 2005 cover story in *Time* magazine, and Vice President Al Gore's 2006 documentary, *An Inconvenient Truth*, both on global climate change, for the public to appreciate the economic consequences of carbon overshoot. That concern has not yet extended to biodiversity loss. But as our citations in Chapter 2 show, contemporary threats to global biodiversity have been identified in professional journals since at least the 1970s and collated periodically for the general public by ecologists (e.g. Ehrlich and Ehrlich, 1981; Wilson, 1992, 2016; Levin, 1999) and writers (e.g. Kolbert, 2013, 2014). The pioneering work was Rachel Carson's 1962 book, *Silent Spring*, which drew attention to the way industrial chemicals and insecticides exterminate birds and insects (Carson, 1962). And that work faced not just scepticism in much of the media, but fierce resistance from chemical companies.

Kriegler et al. (2009) reported responses from 43 scientists to a question about their (subjective) assessments of the likelihood that under global climate change there will be major restructuring of the Atlantic circulation, the Greenland/West Atlantic Ice Sheets, the Amazon rainforest, and El Niño-Southern Oscillation (ENSO).[191] The authors arrived at a conservative lower bound for the probability, 16%, that at least one of the events would occur for a rise in mean global temperature of 2–4°C above the mean temperature in the year 2100, and a probability of 56% for a rise of over 4°C. In an interesting and important paper, Cai et al. (2016) have constructed a stochastic integrated-assessment model to convert the subjective probabilities reported in Kriegler et al. (2009) for arriving at corresponding estimates of 'hazard rates'; that is, probability rates that Earth's climate system will be at a tipping point conditional on it not having reached the point until then. The authors have deployed the numbers on perhaps the most sanguine of global climate models, the DICE model constructed by Nordhaus (1994), to find that the social cost of carbon (Chapters 10–12) is far higher than the US$20–40 per tonne of carbon dioxide implied by DICE, as high as US$200 per tonne. The social cost of carbon is the loss in social well-being from a marginal increase in carbon emission. The existence of tipping points (worse, a cascade of tipping points) lying in wait raises the cost of carbon, meaning that it raises the cost of inaction.

Consider then an ecosystem from which goods and services are being drawn in excess of its regenerative rate.[192] It is known that the ecosystem will tip over and collapse if it is continued to be overused, but the state of the ecosystem at which collapse will occur is not known. So long as the ecosystem has not suffered from a regime shift, DM has two options: (a) stick with business-as-usual (i.e. the status quo); or (b) change course of action. Imagine now that DM is able to forecast the impact on the economy if the ecosystem tips over. That forecast will itself be uncertain, but we may assume DM is able to summarise that future in terms of expected social well-being. We assume, reasonably, that if the ecosystem was to suffer from the regime shift, expected social well-being under DM's forecast would be lower than it would be under either of the options (a) and (b). For vividness we imagine that the regime shift would be a catastrophe.

At what point should DM recommend a move from (a) to (b)? If DM ignored the possibility of a catastrophe under (a), her decision would be to stick with (a) longer than she should. Including the possibility of a catastrophe in her deliberation says she should recommend the move from (a) to (b) earlier. That way she would reduce the risk of the catastrophe. Box 5.6 presents a simple model that formalises DM's optimum decision rule.

DM's decision problem found deep expression in early 2020, when the population of a new strain of coronavirus was discovered to be spreading across the globe. Governments were forced to choose between allowing their population to acquire herd immunity (option (a)) and imposing economic lockdown (option (b)). Delaying a move to (b), however, ran the risk of an unacceptable number of additional deaths – the catastrophe.

Box 5.6
When to Stop Business-As-Usual

Let $\underline{u} > 0$ be the constant flow of social well-being under the status quo so long as the ecosystem has not flipped into an unproductive state. The state of the ecosystem is denoted by a scalar number x. The states have been so numbered that the larger is x the *worse* is the state. Without loss of generality we suppose x lies in the interval [0,1] and that currently the state is $x = 0$. For expositional

---

[191] ENSO is the abbreviation of the El Niño-Southern Oscillation. It is caused by an interaction between the atmosphere and the tropical Pacific Ocean and is roughly speaking a periodic variation between below-normal and above-normal sea-surface temperature and dry and wet conditions over the course of a few years.

[192] We are grateful to Martin Rees for his comments on a previous exposition of DM's problem.

Chapter 5: Risk and Uncertainty

ease, the states have also been so numbered that at the rate at which the ecosystem is being exploited under the status quo x moves in tandem with time, at the same rate. That move, which was made in the text, makes the technical aspects of the analysis a lot simpler, for it implies that until the ecosystem suffers from a regime shift, we do not need to distinguish its state from time.

It would be natural to assume not only that the social cost of a regime shift increases with x until it is in effect dead, but also that the probability of a regime shift (Chapter 2) increases with x. Should the ecosystem flip, the drop in social well-being, aggregated over the indefinite future starting from the date at which the flip occurs, is taken to be $m(x)$. We shall think of $m(x)$ as a 'damage function'. It is then natural to assume that $m(x)$ is a non-decreasing function of x. Once the system flips, the economy enters a new regime. We assume also without loss of generality that if DM was to abandon the status quo before the ecosystem flips, the flow of societal well-being would be zero.

We now pose DM's choice problem at $x = 0$. Let the (subjective) probability rate that the ecosystem will undergo a regime shift at some x under the status quo be $\pi(x)$. It is natural to assume that π does not decline with x. Next denote the probability that the regime shift will occur only after the ecosystem passes x by $\Omega(x)$. That means $\Omega(x') = \int_{x'}^{1}[\pi(x)dx$. It also means Ω declines with x.

For notational ease, we assume DM does not discount the future. The expected value of social well-being under the status quo is therefore $\int_0^1 [\underline{u}\Omega(x) - \pi(x)m(x)]dx$. To have a meaningful problem to study, we assume that $\underline{u} > \pi(0)m(0)$. Otherwise DM would abandon the status quo at $x = 0$.

Notice that $\Omega(x)$ decreases with x and goes from 1 at $x = 0$ to 0 as x tends to 1. Under our assumptions that means the function $\underline{u}\Omega(x)$ crosses the function $\pi(x)m(x)$ at a unique value of x. DM's stopping rule is therefore to plan to abandon the status quo at the value of x that satisfies

$$\underline{u}\Omega(x) = \pi(x)m(x)] \tag{B5.6.1}$$

It is simple to check that the stopping rule continues to hold over time, that is, the policy is time consistent.

Let the solution of equation (B5.6.1) be $\underline{x}$. If the ecosystem has not flipped before $\underline{x}$ is reached, social well-being subsequently would by zero by our chosen normalisation. If it flips at $x < \underline{x}$ the loss would be $m(x)$, and DM would be accused of having been overly sanguine. On the other hand, if the flip had not yet occurred at $\underline{x}$, DM would be accused of crying wolf. In either case the accusation would be unwarranted. Figure 5.2 represents DM's problem in the special case we use to illustrate the reasoning.

Figure 5.2 Determining the Stopping Rule $\underline{x}$

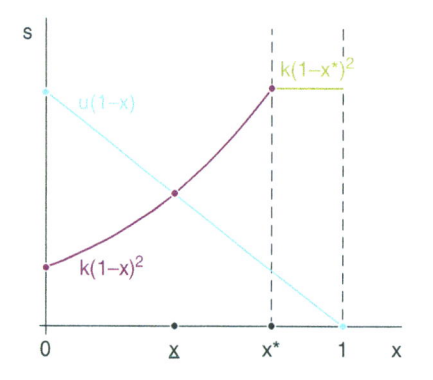

Illustration

Assume that $\pi(x)$ is the uniform distribution in $[0,1]$. That means $\pi(x) = 1$ for all x. We now assume that there is a value of x, say x^*, and a constant $k > 0$ such that

$$m(x) = k/(1-x)^2 \text{ for } 0 \leq x \leq x^*$$
$$m(x) = k/(1-x^*)^2 \text{ for } x^* \leq x \leq 1$$

x^* is to be interpreted as so bad a state that any further deterioration would not increase restoration costs. To have a non-trivial problem, we suppose $k/(1-x^*)^2 > \underline{u} > k$. It is now simple to confirm that DM's optimum policy is to continue with business-as-usual so long as the ecosystem has not tipped over, but to plan to stop when the state of the ecosystem deteriorates to the value of x at which

$$\underline{u}(1-x) = k/(1-x)^2,$$

that is, at

$$\underline{x} = 1 - (k/\underline{u})^{1/3} \tag{B5.6.2}$$

It is clear from equation (B5.6.2) that DM recommends stopping business-as-usual before the ecosystem deteriorates to x^* (Figure 5.2). As noted earlier, the policy is time consistent: So long as the ecosystem hasn't tipped, DM continues to recommend the status quo, until that is, $\underline{x}$ is reached.

5.5 The Value of Keeping Options Open

It is commonly accepted that people should as far as possible keep their options open, that a person should strive for flexibility of choice. The reason usually given is that one never knows what will transpire. Locking oneself into a course of actions is a bad idea because the future is uncertain. What appears to DM today to be the best course of actions may turn out not to have been the best course of actions once she learns more about the biosphere's processes. In previous chapters, we found that ecosystems provide services that are not easily detectable but have been found on close examination to be critical to our well-being. Moreover, switching from one course of actions to another is typically costly. The chosen course may have involved constructing a business centre on what was previously a wetland. As wetland restoration takes time, its destruction can only be reversed at considerable cost. If tomorrow it was discovered that the wetland provided many more valuable services to the community than was known previously, DM would have cause to regret her decision. But if she recognised today that she may discover tomorrow that the wetland was even more valuable than was previously known to be, she would be wary of approving the business centre now. On the other hand, keeping options open also has costs, for it means turning down a proposal to construct a profitable business centre. DM's problem is to find a balance between this contending pair of considerations.

Two features of DM's choice problem are significant. First, destruction of the wetland is known by her to be an act that can only be reversed at considerable cost: restoration takes time and resources and, as noted in Chapter 3, is in any case costlier than conservation. Second, DM knows that with the passage of time she may learn that the wetland is even more valuable than it is judged to be today. It follows that the wetland has a value over and above what DM would impute to it on the basis of what is known today about the services it provides. *Conserving the wetland therefore has an additional worth over destroying it now and restoring it, should it prove to have been more valuable than the business centre.* That additional worth is known as the ecosystem's *option value* (Figure 5.3).[193]

---

[193] Arrow and Fisher (1974) and Henry (1974) applied the idea of option values to provide a new argument in support of conservation science. Arrow (personal communication) attributed the idea to Weisbrod (1964).

Chapter 5: Risk and Uncertainty

Figure 5.3 Option Value

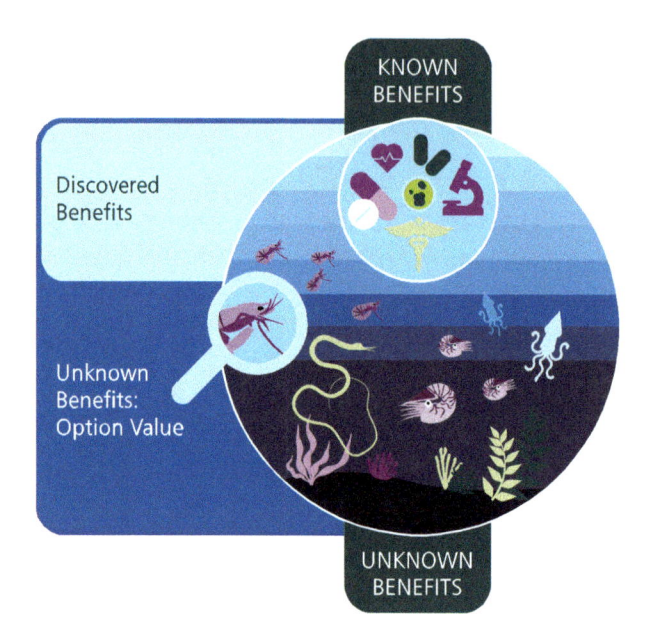

It is the twin presence of uncertainty about the value of the ecosystem she is destroying and the fact that restoration is costlier than conservation (Chapter 3) that makes flexibility an attractive proposition. If DM knows she will learn nothing new about the value of the wetland, she will have no cause to regret her decision, whichever way it goes. Flexibility of choice would have no value to her. Likewise, if wetland restoration was costless, then today's decision would be fully reversible at no cost and the fact the wetland will be lost today would be of no concern to DM; for it could be brought back to life without cost.

Box 5.7 presents an example to demonstrate how option values can be estimated. We will discover that the existence of an option value does *not* depend on whether DM is risk averse. The magnitude of the wetland's option value does depend on her attitude to risk (she would ascribe a great value to it if she was highly risk averse and was far sighted), but that is a different matter. The fact remains that DM would award an option value to the wetland even if she were a risk taker. The reason is that in choosing between waiting for new information and going ahead now, DM is not comparing probability distributions over payoffs, she is instead asking herself whether she should commit to a course of action that is hard to reverse should she wish to reverse it on receiving new information.

The concept of option values is of profound importance in the economics of biodiversity. DM could be a civil servant in her government's Ministry of Planning, tasked with evaluating a project that would involve building a high dam she knows will further fragment a major river system. She could be an official at the World Bank, evaluating a request for a loan to build a system of roads that will penetrate deep into a rainforest to mine the minerals that have been detected there. She could be an analyst in an NGO wanting to estimate the damage to the oceans in permitting a fishing company to deploy radar satellites and drones for tracking schools. She could be an employee of an oil company, instructed because of public protests to evaluate the environmental consequences of installing oil rigs in an estuary that has so far eluded industrial development. And so on. Spatial features of the biosphere matter so much that the processes governing ecosystems in their particularities are only dimly understood even by experts. As noted previously in Chapter 4, the number of species is conservatively estimated to be between 8 and 20 million kinds of organisms with complex cells (known as eukaryotes) and probably a much larger number of archaea and bacteria (known as prokaryotes). Only a little over 1 million species of eukaryotes have been identified so far (Mora et al. 2011). Ecologists despair that the biosphere's book has continued to be written over evolutionary time, and yet we are destroying chunks of it in a matter of decades. Humanity is not merely transforming the biosphere, we are eliminating species at a furious rate. And that is irreversible.

Box 5.7
Option Values

To establish a connection between the narrative in the text and the technical account of choice under uncertainty in Box 5.1, we formalise the option value problem for DM when she evaluates a wetland transformation project.

We assume DM's choice problem involves two periods: $t = 0, 1$. At $t = 0$ the (net) social benefit from constructing a business centre on what is now a wetland is b_0. DM understands there are two possible states of Nature, s_1 and s_2, that the true state will be revealed in period $t = 1$. We are to interpret the state of Nature as the additional knowledge she will acquire about the wetland's ecological worth. If s_1 prevails, the probability of that is π, the social benefit from constructing the business centre will be found to be b_1. If s_2 prevails, the probability of which is $1-\pi$, the social benefit will be found to be b_2.

To have an interesting problem, we assume $b_0 > 0$, $b_1 < 0$, and $b_2 > 0$. DM knows that even if she rejects the proposal now, she will be able to revisit the proposal in the next period. DM is risk neutral.

It proves useful to first consider a scenario where the wetland can be brought back to life without cost at $t = 1$ should it have been destroyed at $t = 0$. That is of course, an absurdity, but it enables us to appreciate the force of irreversibility in decision-making. Without loss of generality let the size of the wetland be 1. If D is taken to mean the extent of the wetland that is destroyed, then D is either 0 or 1.

Let D_0 be the extent of the wetland that is developed at $t = 0$ and by D_1 and D_2 the extent of the wetland that is developed at $t = 1$ depending whether the state of Nature is s_1 or s_2. Finally, let r be the rate of at which DM discounts future social benefits. It follows that the expected present value of the development plan is

$$b_0 D_0 + [\pi b_1 D_1 + (1 - \pi) b_2 D_2]/(1 + r) \qquad (B5.7.1)$$

DM's task is to choose D_0, D_1 and D_2. As her decision is reversible costlessly, she can commit herself to the choice she makes now, for it will make no difference to her choice in the next period. To have a meaningful problem, we assume

$$[\pi b_1 D_1 + (1 - \pi) b_2 D_2] > 0 \qquad (B5.7.2)$$

Equation (B5.7.2) says the expected benefit from the business centre development project is positive. As $b_0 > 0$, it is immediate from (B5.7.1) that if the development is fully and costlessly reversible, DM's decision would be

$$D_0 = 1, D_1 = 0, D_2 = 1 \qquad (B5.7.3)$$

Equation (B5.7.3) says that the business centre should be built now, that it should be retained if it continues to be judged to yield positive benefit at $t = 1$, but that the wetland should be brought back to life at $t = 1$ if it is discovered to have been a lot more valuable than was thought previously. The decision rule is myopic: DM needs only to confirm that $b_0 > 0$ before approving the project.

Irreversibility of the decision at $t = 0$ means that when maximising expected social well-being now (Equation (B5.7.1)), DM must observe the constraints

$$D_0 = 0 \text{ or } 1, \text{ and } D_0 \leq D_1, D_2 \leq 1 \qquad (B5.7.4)$$

Working backward, DM reasons that at $t = 1$ she will choose $D_1 = D_0$ and $D_2 = 1$. Taking her choice at $t = 1$ into account, her choice problem at $t = 0$ reduces to choosing D_0 so as to maximise

$$b_0 D_0 + [\pi b_1 D_0 + (1-\pi) b_2)/(1+r)] = (b_0 + \pi b_1/(1+r))D_0 + (1-\pi)b_2/(1+r)$$

DM would recommend the business centre only if $(b_0 + \pi b_1/(1+r)) > 0$, or in other words, if $b_0 > -\pi b_1/(1+r) > 0$. The condition for accepting the project is stiffer because the project destroys the wetland irretrievably. As expected, the extent of the stiffness depends on the discount rate. If r is large, irreversibility of investment makes little difference. It makes a big difference if r is small.

Notice that the argument establishing that the wetland has an option value does not require DM to be risk neutral. She could be risk averse, or even a risk lover, and still she would find that the wetland has an option value. The reason is that DM is not comparing risks when deciding whether to permit the business centre to be built at $t = 0$. The risk remains the same whether she keeps her options open or forecloses them by recommending that the wetland be destroyed. Her decision rests on the trade-off she makes between flexibility of choice and the cost of waiting.

Box 5.8
Translating Ecological Risks into Economic Risks

How does the risk of ecosystem collapse at the top end of a company's supply chain translate into the company's commercial risks? We study that by deriving the adjustment that firms should make to the value they attribute to ecosystem services in danger of collapse. The analysis correspondingly provides a method for estimating the losses firms suffer by permitting their supply sources to erode; and it provides a method for estimating the losses borne by those who rely on ecosystems that are affected adversely by the neglect by others of correlated ecosystems.

Time is continuous, denoted by $t \geq 0$. Suppose a supply source (e.g., a wetland) of size S, yields P dollars per unit of the asset to a firm. We begin by assuming P is constant. The discount rate the firm applies to future profits is r. We assume $r > 0$. So long as the supply source remains intact, the flow of benefits from it is PS at each moment. The asset would then be worth PS/r to it. We assume however that the ecosystem faces the risk of collapse. Little would be added to what we are after here if, as it would be realistic to assume, the risk is a function of the firm's activities. We assume therefore that the risk is exogenous to the firm.

The case where the risk is characterized by a Poisson process, with a hazard rate h, is trivial, as it means that the value of the supply source to the firm is $PS/(r+h)$. We therefore consider a different scenario. We suppose that the source will collapse at a random date in the next T years. We study the case where the risk is uniform. At $t = 0$, there is a constant probability rate $1/T$ of the supply source collapsing. Viewed from $t = 0$, the probability rate that the source will survive until t, is thus $(T-t)/T$. The *hazard rate* at t is thus $1/(T-t)$, which goes to infinity as t tends to T.[194]

---

[194] We are grateful to Matthew Agarwal for asking how the risk of ecological collapse affects accounting prices.

B5.8.1 A Single Ecosystem

We first apply the construction to calculate the risk-adjusted value of the supply source. As the probability that the asset will exist until t, is $(T-t)/T$, its expected worth to the firm is,

$$PS\left\{\int_0^T [e^{-rt}(T-t)/T]dt\right\} = [PS/r][1-e^{-rT}] - [PS/T]\left[\int_0^T \{te^{-rt}\}dt\right] \tag{B5.8.1}$$

Write the risk adjusted value of S as a function of T as $F(T)$. Integrating the final term on the right-hand side to eq. (B5.8.1) by parts yields:

$$F(T) = [PS/r][(1-(1-e^{-rT})/rT] \tag{B5.8.2}$$

The firm's loss owing to its negligence is,

$$PS/r - [PS/r][(1-(1-e^{-rT})/rT] = PS(1-e^{-rT})/r^2T \tag{B5.8.3}$$

Notice that the risk adjustment the firm should deploy when valuing its supply source is,

$$R(T) = [1-(1-e^{-rT})/rT] \tag{B5.8.4}$$

As $0 < R(T) < 1$, the moral is simple: risk of ecological collapse translates into a risk factor, between 0 and 1, on the ecosystem's value to the firm.

An extension of the model worth considering here involves abandoning the assumption that P is a constant. With the world's rainforests being razed to the ground to make way for cattle ranches, plantations, and mines, we would expect the benefits from S to increase over time relative to our assumed *numeraire*, income. The simplest assumption is that P increases exponentially at, say, the rate $\beta > 0$, that is, $P(t) = P(0)e^{\beta t}$. Were there no danger of collapse, the asset would be infinitely valuable if $r < \beta$; implying the loss owing to the possibility of collapse would be infinite. We therefore assume $r > \beta$. We may then replace r by $(r-\beta)$ in eq. (B5.8.2)–(B5.8.4). That is, the risk adjusted value of the source is larger, the larger is β. Moreover, $P(0)SR/(r-\beta) \to P(0)ST/2$ as $\beta \to r$. That too is exactly what intuition would suggest.

B5.8.2 Coupled Ecosystems

A further extension involves coupled ecosystems. Consider the pair S_1 and S_2, yielding benefit flows P_1 and P_2 per unit of the respective ecosystems. We may think of S_1 as the stock of supply source for firm 1 and S_2 the corresponding stock for firm 2. Assume now that the risk to the productivity of ecosystem S_1 transmits itself to S_2 instantaneously. Thus, the risk that S_2 will collapse has the same uniform distribution as S_1. It then follows from equation (B5.8.3) that the loss to firm 2 is,

$$P_2S_2/r - [P_2S_2/r][(1-(1-e^{-rT})/rT] = P_2S_2(1-e^{-rT})/r^2T \tag{B5.8.5}$$

A simple extension of the coupled system would be a lagged response. Suppose that if ecosystem "S_1" were to collapse at t, the coupled ecosystem "S_2" would remain intact until $t+L$, at which moment it too would collapse. In short, the risk that ecosystem 2 will collapse is a translation by L of the uniform distribution describing the risk that ecosystem 1 will collapse. Write $V(L)$ as the value of S_2 to firm 2 when faced with this inflicted risk. Then

$$V_2(L) = P_2S_2[(1-e^{-rL}) + \left(1-(1-e^{-rT})/rT\right)e^{-rL}]/r \tag{B5.8.6}$$

Chapter 5: Risk and Uncertainty

> Using equation (B5.8.6) it follows that the loss suffered by firm 2 is,
>
> $$P_2 S_2/r - V_2(L) = P_2 S_2[(1 - e^{-rT})/r^2 T]e^{-rL} \qquad (B5.8.7)$$
>
> If the ecosystems were independent of one another, $L = \infty$. In that case the loss would be zero.
>
> A simple extension of the case of two linked ecosystems is one where there is a series of, say, N ecosystems, each coupled directly to the one that precedes it in the numbering. There would now be a cascade of collapses – a domino effect – at uncertain dates.

Annex 5.1 Fat Tails and Unbounded Utilities: St. Petersburg Paradoxes

The axiomatic system Savage (1954) introduced to study choice under uncertainty postulates that the DM can rank all the options available to her even though she does not know the true state of Nature. The state of Nature reveals itself to her only after she chooses. She is therefore uncertain about the consequences of her choice. Savage's axioms on the principles DM should follow was shown by him to imply that her ranking of options can be represented in the form of 'expected utility'; that is, the higher an option is in her ranking, the larger the expected utility of that option. The 'as-if' utility function $u(.)$ in the representation is bounded and the corresponding probability distribution π is subjective (Box 5.1).

It has become customary in the economics of climate change to take a different approach from that of Savage. The practice is to postulate both a utility function u and a probability distribution π with which to express the notion of expected utility, Eu, and to then derive the implications for choice. If the chosen u-function is bounded, there need not be any restriction on the choice of the probability distribution with which to work: As the probabilities sum to 1, Eu is a well-defined function.

What though if the chosen u-function is unbounded? In that case the class of probability distributions has to be suitably restricted if expected utility is to be a meaningful entity. The problem is, Nature provides DM with data that restrict her discretion over the choice of probability distributions. For example, the functional relationship between carbon emissions and the global climate has been found to be mediated by so many factors, that climate scientists fear that unless net global emission rates are reduced to zero by year 2050, the probability that the mean global temperature will exceed 3°C above pre-industrial levels is so large as to imply a distribution that is thick at its upper tail (see below).[195] A thick upper tail means that the chance the biosphere will enter a truly appalling state is non-negligible. If in addition the chosen u-function is unbounded below, it can be that Eu ceases to be a meaningful concept for the problem in hand. It is hard to know in advance what happens when zero meets minus infinity (or for that matter, plus infinity), but in the cases under review here, it means an infinitely awful state of affairs.

To illustrate the problem, suppose utility is a function solely of consumption, c. Consider now a widely used class of u-functions:

$$\begin{aligned} u(c) &= c^{(1-\eta)}/(1-\eta), & \eta &> 0 \\ u(c) &= \log c & &\text{corresponding to } \eta = 1 \end{aligned} \qquad (A5.1.1)$$

---

[195] We are grateful to Ilan Noy of Victoria University, Wellington, for valuable discussions on climate uncertainties.

Chapter 5: Risk and Uncertainty

It is easy to check that if $0 < \eta < 1$, the *u*-function is bounded below but unbounded above; that if $\eta > 1$, the *u*-function is bounded above but unbounded below; and that if $\eta = 0$ it (the log function), the *u*-function is unbounded at both ends.

The *u*-functions in equation (A5.1.1) are easy to work with because they are defined by a single parameter, η. We make use of them in Chapter 11 when developing the concept of intergenerational well-being. It is easy to check that η is the elasticity of marginal utility $[-(du'(c)/dc)/u'(c)]$. As η is a constant in equation (A5.1.1), this class of *u*-functions is called 'iso-elastic'.

Assume now that there are two dates: today and the future. Imagine that *c* is a known, declining function of future global temperature, but that future temperature is unknown. Let $T \,(\geq 0)$ denote the mean global temperature in the future relative to what it is today and let $\pi(T)$ be the density of the probability distribution of *T*. Pindyck (2011) considered the following specification:

$$c = c_0/(1+T), \qquad c_0 > 0 \qquad (A5.1.2)$$
$$\pi(T) = \lambda(1+T)^{-(1+\lambda)}, \qquad \lambda > 0 \qquad (A5.1.3)$$
$$u(c) = c^{(1-\eta)}/(1-\eta), \qquad \eta > 1 \qquad (A5.1.4)$$

In equation (A5.1.2) c_0 is today's consumption. Otherwise the equation requires no comment. But equations (A5.1.3)–(A5.1.4) do. First, $\pi(T)$ is a power function. And because the function declines to zero as *T* goes to infinity at a rate that is far smaller than exponential, statisticians call it 'fat-tailed'. Second, *u(c)* is unbounded below. As noted above, the two features do not go together. To see why, we use equation (B5.1.2) to express *c* in terms of *T* and obtain

$$Eu(c) = {_0\int^\infty} \left\{ [c_0/(1+T)]^{(1-\eta)}/(1-\eta) \right\} \lambda(1+T)^{-(1+\lambda)} dT \qquad (A5.1.5)$$

It is a simple matter to confirm that the integral in equation (A5.1.5) converges if and only if $(\eta-\lambda) < 1$. What happens if $(\eta-\lambda) > 1$? To answer, we note that because $\eta > 1$, $Eu(c)$ in equation (A5.1.5) is negative. It follows that if $(\eta-\lambda) > 1$, $Eu(c)$ diverges to minus infinity, which is another way of saying that DM expects the future to be infinitely awful. That is, a world expecting to face total annihilation.

A sharper result can be obtained if we calculate the expected value of marginal utility $Eu'(c)$ and check what happens to the function as *c* tends to 0 (the infinitely large catastrophe). From equation (A5.1.4) we have $u'(c) = c^{-\eta}$. Routine calculations now show that if $\eta > \lambda$, $Eu'(c)$ tends to (plus) infinity as *c* tends to zero. (As illustrative figures, Pindyck (2011) assumes $\eta = 2$ and $\lambda = 4/3$.)

Imagine now that at a suitable expenditure the economy can avoid global warming altogether, meaning that the investment would ensure that $T = 0$. Then no matter how large the required expenditure, DM would recommend the investment. She would recommend it even if 99.9% of GDP were needed to avoid the risk. This is an unpalatable result, which is why Weitzman (2009) dubbed it the Dismal Theorem and why it has been much discussed in the recent literature on the economics of climate change.[196]

The underlying issues here are not new. In a much neglected paper, Levhari and Srinivasan (1969) studied a fully dynamic model of stochastic economic growth that combined equation (A5.1.4) with a log-normal distribution on the return on investment. The authors showed that if the variance of the log-normal distribution is large relative to the other parameters of the model, the future can be expected to be infinitely awful no matter which consumption policy is chosen. The point is that if the variance on returns is large, consumption drops to within a neighbourhood of zero for an infinite

---

[196] See also Weitzman (2014) for a similar argument in connection with estimating the social cost of carbon.

Chapter 5: Risk and Uncertainty

number of periods no matter how small the neighbourhood is. And that is a world expecting to face total catastrophe.

Why then continue to work with u-functions that are unbounded below, as in equation (A5.1.4), if the distribution DM has to include in her analysis is fat-tailed? The reason is that these extreme examples point to an important truth:

Suppose the chosen u is a bounded function of c. Eu would be well-defined. But if u tends to a large negative limit as c tends to zero, DM's choice would be sensitive to the specifications of the model she uses (Equations (A5.1.3)–(A5.1.4)). Weitzman's purpose in using extreme examples was to show that the welfare economics of climate change is problematic. What DM should recommend turns out to be sensitive to small differences in model specification. Weitzman's analysis says that the recommendation by those many people who take the economics of climate change seriously, that the global tax on greenhouse gas emissions should be in the range of US$40–80 per tonne is in all probability a serious underestimate. But when that is contrasted with the maximum tax rate of US$40 per tonne to be found anywhere, we see how far governments are from taking the economics of the biosphere seriously.[197]

Box A5.1
The St. Petersburg Paradox

That fat-tailed distributions, when combined with unbounded u-functions, yield empirically absurd conclusions, is the stuff of St. Petersburg Paradoxes, attributed usually to the 18th century mathematician, Daniel Bernoulli.[198] The paradoxes involve games of chance and are usually so constructed that the problem is not that expected infinity is unbounded below, but rather that it is unbounded above. The paradoxes can be summarised by means of the simple example of an individual being invited to play the following game of chance:

A fair coin is to be tossed at every stage of the game. The initial stake is £2 and is doubled every time the coin comes up head, but ends at the first time the coin shows tail. The player wins whatever is in the pot on the first time the coin comes up tail. How much would a person be willing to pay to play the game?

The expected value of the pot is

$$EV = £[(1/2)2 + (1/4)4 + (1/8)8 + \ldots] = £(1 + 1 + 1 + \ldots) = \infty \qquad (BA5.1.1)$$

If the criterion the person uses to play the game is EV, then equation (BA5.1.1) says he should be willing to pay an unbounded sum of money to enter the game. That is patently absurd.

Bernoulli argued that people display risk aversion, which means their criterion of choice is not the expected value of the pot but the expected utility from playing the game. Suppose the utility of wealth y is $u(y) = \log y$ and suppose the individual's initial wealth is w. If k is the cost of entering the game, the expected utility the person would enjoy from the game would be

$$Eu = -k + (1/2)\log(w + 2) + (1/4)\log(w + 4) + (1/8)\log(w + 8) + \ldots \qquad (BA5.1.2)$$

---

[197] See Howard and Sterner (2017) for a meta-analysis of estimates in the literature on future global damages from climate change.

[198] See Peterson (2019) for a historical account of the St. Petersburg Paradox, and Arrow and Priebsch (2014) for its connection to the Dismal Theorem.

> It is simple to confirm that Eu in equation (BA5.1.2) is a finite sum. Let k^* be the value of k at which Eu = 0. As k^* is a finite number, the paradox vanishes. For example, if w = £1 million, k^* is a bit under £21; if w = £1,000, k^* is bit under £11; and so on.
>
> Unfortunately, although the paradox vanishes in the case of the present gamble, so long as $u(y)$ = log$y$, it is possible to construct a gamble for which Eu is plus infinity, meaning that the paradox reappears. The result generalises: no matter how risk averse DM happens to be, so long as $u(y)$ is an unbounded function, there is a gamble for which Eu is unbounded. Unbounded u-functions are useful for classroom exercises, but they should be avoided in the economics of the biodiversity and climate change.

Annex 5.2 Catastrophes and Ambiguities

Publications by the IPCC suggest that the effects of our overshoot on the biosphere display themselves more sharply the smaller the region under scrutiny. Some regions lose more than others, and there are regions that may even gain from mild increases in the mean global temperature.[199] The reverse feedback, predicting human adaptation and mitigation in response to global environmental change adds to that uncertainty, perhaps with similar magnification from the global to the local. Nevertheless, it should not be surprising that IPCC has over the years repeatedly raised their estimates of the speed at which the climate system is moving away from its norm over a long past. Members of the Panel have understandably compromised in the direction of caution.

Local ecosystem collapses are catastrophic for communities dependent on them, but in principle policies can be put in place for alleviating human suffering. Emergency aid in the form of food, clothing, shelter and cash are not a substitute of course, but they prevent communities from going under. Possible global catastrophes lying in wait under business-as-usual are a different matter altogether. Strong correlation in the uncertainties people face means there can be little scope for mutual insurance.

The way we have posed the choice problem facing DM here is, however, incomplete. DM's role is to recommend an action (which can also be read as a 'policy'). Each possible action gives rise to a change in the portfolio of assets that her society holds. As shown in Box 5.1, action (a), when combined with a state of the world (s) gives rise to an event. The event is the return on the portfolio DM has recommended. So DM's subjective probability (should she be able to arrive at one!) of s occurring would read as $π(s)$ and the loss in societal well-being should be understood to depend on both a and s, which we may therefore write as $m(a,s)$. In the language of expected utility theory the expected loss in societal well-being would be $π(s)m(a,s)$. However, Annex 5.1 showed that when DM has to include events in geophysical terrains on which no one has even the remotest experience (e.g. mean global temperature 3°C or more than preindustrial times), she would very unlikely be able to arrive at a subjective probability distributions over them. And that says when biospheric collapse is included in the events which DM has to consider, she will want to abandon expected utility theory.

Scepticism toward the expected utility theory Savage (1954) had created is not new, it goes back many years. In a famous paper Ellsberg (1961) pointed to problems with the theory by noting that people treat objectively known risks, such as the chance that a fair coin will land on 'head', differently from unknown uncertainties, such as the chance that a coin with an unknown bias will land on 'head'. Many people are 'ambiguity averse', meaning that they would rather bet on known risks than bet on uncertain prospects

---

[199] Heal and Millner (2014) is an excellent review of uncertainties in forecasts of global climate change.

Chapter 5: Risk and Uncertainty

even if from a subjective probability point of view the two bets are indistinguishable. And there have been many more criticisms levelled against expected utility theory.[200]

What then are the alternatives open to DM? She needs a criterion for choosing from among portfolios whose returns are ambiguous. One decision rule, of potential use to DM when she is advised by experts who differ in their subjective probabilities, is to admit all, provided each distribution is consistent with what is commonly known. The rule DM could adopt would be to rank portfolios according to their lowest expected utility over the set of probability distributions that she has admitted. This is the 'maxmin expected utility' approach of Gilboa and Schmeidler (1989). It combines an extreme form of uncertainty aversion with expected utility theory.[201]

What should DM do if no one is able to arrive at a subjective probability distribution over the returns? Expected utility theory requires of DM to make fine distinctions between events spanning *all* possible patterns of return on each available portfolio of assets. That is to be able to discriminate a lot among events on which she knows little.

One possibility that suggests itself (it is the only one we note here) is to abandon probabilistic reasoning altogether. The idea would be to invoke axioms of choice that are applied to coarser information on uncertain portfolios than is demanded by expected utility theory. Arrow and Hurwicz (1977) proposed a set of axioms that imply what they called the *α-maxmin rule*, which ranks portfolios according to the weighted sum of the best and worst returns. The parameter α, which is a number between 0 and 1, is the weight awarded to the lowest return and $(1-\alpha)$ is the weight awarded to the best outcome. If DM is a pessimist, she would choose an α close to 1.

In practice, people apply their values and the information they have sifted to make decisions in ways that are themselves ambiguous. DM should be expected to be no different. Expected utility theory is an icon around which people can test their reasoned intuitions before choosing even under ambiguity. Our account of DM's use of the theory is no different from this.

---

[200] Kahneman, Slovic and Tversky (1982) and Camerer and Weber (1992) are key references. In the text we are following the exposition in Heal and Millner (2014).

[201] It needs emphasis that decision theorists do not propose decisions rules willy-nilly. Like Savage, they propose axioms DM could find reasonable to adopt when she ranks alternative portfolios with uncertain yields.

Chapter 6
Laws and Norms as Social Institutions

Introduction

The Oxford English Dictionary defines *institution* to be "an established law, custom, usage, practice, organization, or other element in the political or social life of a people". The present chapter unravels their lead, but recasts the concept so as to stress the role of institutions in economic life.[202]

By institutions we mean, very loosely, the *arrangements* that govern collective undertakings. Those arrangements include not only legal entities like the modern firm, but also insurance institutions within village networks, such as the *iddir* in Ethiopia and variants elsewhere in sub-Saharan Africa (Box 6.4). They include bazaars in the Middle East, village networks for saving and credit in Asia and Latin America, the nuclear household in the West, the extended kinship system of claims and obligations in Africa, and elaborate authority structures such as the Mafia. And they include that overarching entity called *government* that exists everywhere in the modern world.

Institutions are defined in part by the rules and authority that structure governing collective undertakings, but also in part by the relationships they have with outsiders. The rules on the factory floor (who is expected to do which task, who has authority over whom, and so on) matter not only to members of the firm, they matter to others too. In many democracies there are laws relating to working conditions in factories. Moreover, environmental regulations, where they are enforced, constrain what firms are able to do with their effluents. In every society there are layers of rules of varied coverage. Some rules come under other rules, many have legal force, while others are at best tacit understandings, founded perhaps on a reading of social norms.

Institutions are overarching entities. People interact with one another *in* institutions. So to understand the structure and function of institutions, we take a possibly unexpected route in this chapter. We ask: *what characteristics must a social engagement have if people are to have informed reasons for trusting one another to fulfil the obligations they undertake*? Taken together, trust in others, confidence in government to deliver and in markets to function well, and the institutional arrangements that enable people to engage with one another for mutual benefit, is called *social capital* – a concept central to the economics of biodiversity.[203]

It is common to regard society as comprising three classes of institutions: households, markets and the state. The economics of climate change is framed with that three-way classification in mind (Nordhaus, 1994, 2007; Stern, 2006). The idea of social capital illuminates a fourth class, comprising *communities* and *civil society*, which play an essential role in enabling households to engage with one another without the direct involvement of either markets or the state. Figure 6.1 provides an overview of society's institutions.

---

[202] This chapter has been adapted from Dasgupta (1988, 2000, 2007).

[203] The literature is vast. Contributions that have direct bearing on the issues studied here include Wildavsky (1987), Coleman (1988), Putnam (1993), Fukuyama (1995), and Granato, Inglehart and Leblang (1996). The way the notion of social capital has been framed here is not quite the same as it has been proposed in the above works, nor is it framed the same way even among them. But ambiguity is not a hindrance in so complex a notion; it is instead a help, for it enables us to avoid a protracted discussion on what social capital means.

Chapter 6: Laws and Norms as Social Institutions

Figure 6.1 Society''s Institutions

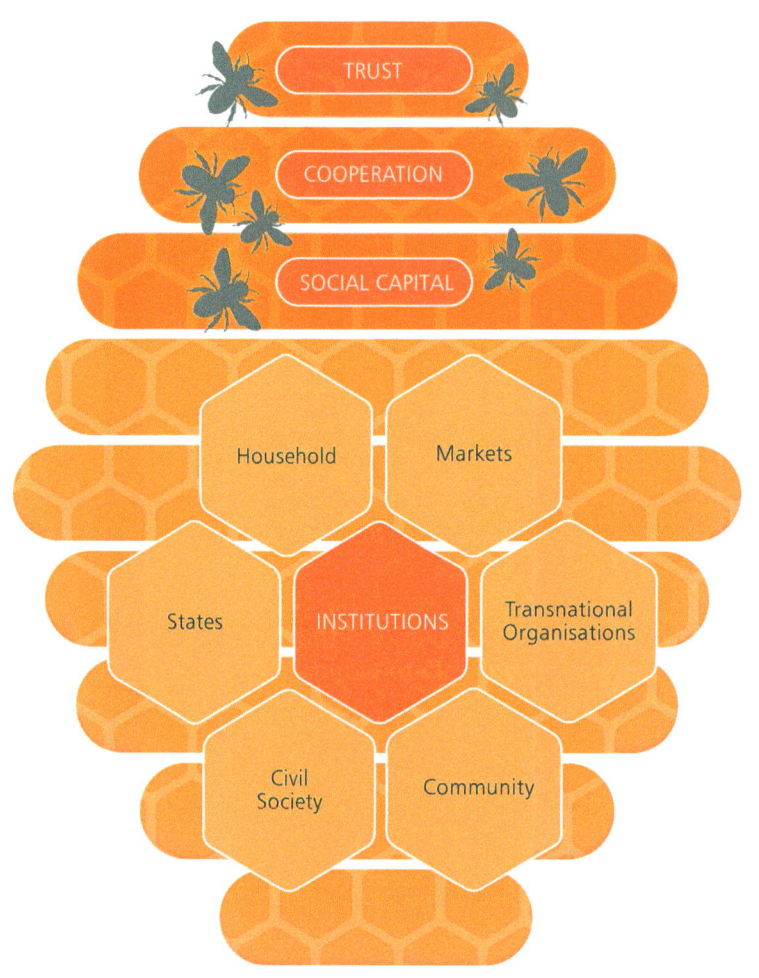

Communities and civil society – the 'third estate', as they are sometimes called – carry slightly different connotations. It is not a travesty to think of 'communities' as institutions that work outside the state (in fact they mostly evolved in traditional societies when there *was* no state, at least not in the contemporary sense) to manage, for example, their local ecosystem (Chapter 8).[204] That can be contrasted with 'civil society', which we may think of as comprising institutions that engage actively with the state, for example, to create a place for services that the state is either unwilling or unable to provide (Chapter 7) and, more generally, to make the state function better (Box 6.4). That can be a cause of friction between the two. Nor are communities and civil society without their own deficiencies. Communities can harbour gross inequities (e.g. between genders or races) and civil society can degenerate into mere special interest groups, even street gangs. Our examples may encourage us to think that communities are rural while civil society is urban. Nothing essential will be lost if we were to imagine them that way, but so as to compress the two together we shall refer to them as *communitarian institutions*.

---

[204] That depends, of course, on how far back one looks. In his classic essay on the world in Homer's songs, Finley (2002) wrote: "The world of Odysseus was split into many communities like Ithaca. Among them, between each community and every other one, the normal relationship was one of hostility, at times passive, in a kind of armed truce, and at times active and bellicose." Finley would seem here to be using 'community' to refer to the social fabric of an entire island kingdom.

The analysis that follows is accompanied by illustrations, taken from empirical studies of various types of communitarian institutions and the purposes they serve. However, societies evolve, so the accounts here may not fit exactly the character of those same institutions today. But that should not detract from the purpose of our illustrations, which is that conservation and restoration of Nature can only be achieved if communities and civil society are invited, even encouraged, to be partners when the programmes are initiated by government and international or national non-governmental organisations (NGOs), and to be active in those programmes where external bodies intend only to finance the programmes. No matter what the character of a conservation or restoration programme happens to be, the engagement of communitarian institutions will prove to be essential.

6.1 Societal Trust and Economic Progress

The effectiveness of an institution depends on the rules governing it and on whether its members obey them. The codes of conduct in the civil service of every country include honesty, but governments differ enormously as to its practice. Transparency International, an NGO, has created a Corruption Perception Index for governments in 180 countries.[205] The index is based on, among other attributes, the perception private firms have acquired on the basis of their experience of the bribes people have to pay officials for any activity they undertake (e.g. to enter and stay in business). The index – which is on a scale of 0 (most highly corrupt) to 100 (most highly 'clean') – is less than 20 in many countries and greater than 80 in only a few countries.

It used to be argued that bribery of public officials helps to raise national income because it lubricates economic transactions. It does so in a corrupt world: if you do not pay up, you do not get to do business. Nevertheless, having to pay bribes raises production costs, and so less is produced. Citizens suffer, because the price they have to pay for products is that much higher.[206]

How do state corruption, ineffectiveness and indifference to the rule of law translate into macroeconomic statistics? They leave their imprint on the rates of return on an economy's capital goods (Chapter 1). Other things equal, a country whose government is corrupt or ineffective, or where the rule of law is not respected, is a country where the productivity of capital goods is lower than that of a country whose government suffers from fewer of those defects (Lambsdorff, 2003).

Economists studying the link between untrustworthiness in public bodies and economic progress have, however, mostly looked to Gross Domestic Product (GDP) growth as a measure of the latter (Mauro, 1995; Mo, 2001). The link between social trust (i.e. responses to variants of such a question as, 'Would you say you can trust people in your society or do you feel you cannot be too careful?') and GDP growth have been also been studied.[207]

There is an intuition which says the level and extent of trust in others is related to confidence in an economy's institutions. But it is hard to establish that empirically, as the geometry of trust can confound matters. Someone could be justifiably wary of local government officials but have confidence in the NGO that serves the community. To confound matters still, it could be that people invest more in their relationships with members of their own community *because* the state is untrustworthy (Dasgupta, 2000).

There is a stronger intuition that says government corruption is bad for Nature conservation. Because much of the world's biodiversity is in the tropics and because a majority of poor countries are in the tropics, even a cursory look at the data suggests that national poverty, state untrustworthiness and biodiversity loss are connected (Annex 6.1). In part, the intuition is that people who are adversely affected by land-use changes involving such activities as mining,

---

[205] https://www.transparency.org/en/cpi/2019   [206] We are focusing here on the instrumental role of honesty, not on its intrinsic virtue.

[207] See for example, Knack and Keefer (1997), Whiteley (2000), Zak and Knack (2001), and Beugelsdijk, de Groot and van Schaik (2004).

construction and deforestation have to protest in concert (individual appeals would be ignored by government), and collective action is never an easy undertaking (Chapter 7). In large part, though, the intuition is built on the fact that capturing such natural assets as forests and sub-soil resources is easy for the state; and that by exporting the products it can earn huge rents without having to invest in either technology or human capital.

But because foreign companies are usually required to bring both technology and human capital with them, there is an intuition that runs the other way, which is that the very existence of a large pool of natural resources in a country encourages the government there to become corrupt. It is hard to avoid the suspicion that the positive relationship that has been found between biodiversity loss and government corruption (Sachs and Warner, 1995, 2001; Leite and Weidmann, 1999) points to a causation that runs both ways.

6.2 The Idea of Trust

Imagine that a group of people have discovered a mutually advantageous course of actions. At the grandest level, it could be that citizens see the benefits of adopting a constitution for their country. At a more local level, the undertaking could be to share the costs and benefits of maintaining a communal resource (irrigation system, grazing field, coastal fishery); constructing a jointly useful asset (restoring a watershed); collaborating in political activity (civic engagement, or lobbying); doing business when the purchase and delivery of goods cannot be synchronised (credit, insurance, wage labour); entering marriage; creating a mutual insurance scheme (Box 6.4); initiating a reciprocal arrangement ('I help you, now that you are in need, with the understanding that you will help me when I am in need'); establishing a rotating savings and credit association (Box 6.4); adopting a convention (send one another Christmas cards); creating a partnership to produce goods for the market; entering into an instantaneous transaction (purchase something across the counter); and so on. Then there are mutually advantageous courses of action that involve being civil to one another. They range from forms of civic behaviour, such as not disfiguring public spaces and obeying the law more generally, to respecting the rights of others.

Imagine next that the parties have agreed to share the benefits and costs in a certain way. Again, at the grandest level the agreement could be a social contract among citizens to observe their constitution; or the agreement could be among nations over proposals on how the demands humanity makes of the biosphere could be reduced to an extent that the Impact Inequality becomes an Impact Equality on a revived biosphere (Chapter 4). Or it could be a tacit agreement to be civil to one another, such as respecting the rights of others to be heard, to get on with their lives, and so forth. Here we will be thinking of agreements over transactions in goods and services. There would be situations where the agreement was based on a take-it-or-leave-it offer one party made to another (as when someone accepts the terms and conditions set by a supermarket when making a purchase there). In other contexts, bargaining may have been involved, such as when someone purchases household fineries at the rural fair in South Asia. The question arises: *Under what circumstances would the parties who have reached agreement trust one another to keep their word*?

Because one's word must be credible if it is to be believed, mere promises would not be enough (witness that we warn others – and ourselves too – not to trust people 'blindly'). If the parties are to trust one another to keep their promise, matters must be so arranged that: (a) at every stage of the agreed course of actions, it would be in the interest of each party to plan to keep his or her word if all others were to plan to keep their word; and (b) at every stage of the agreed course of actions, each party would believe that all others would keep their word. If the two conditions are met, a system of beliefs that the agreement will be kept would be self-confirming.

Notice that condition (b) on its own would not do. Beliefs need to be justified. Condition (a) provides the justification. It offers the basis on which everyone could in principle believe that the agreement will be kept. A course of actions, one per party, satisfying condition (a) is called a *Nash equilibrium*, in honour of the mathematician John Nash – he of *A Beautiful Mind* – who proved that it is not a vacuous concept, demonstrating that the condition can be met in realistic situations.[208] The way we have stated condition (a) is not due to Nash though, but to John Harsanyi, Thomas Schelling and Reinhard Selten, three social scientists who refined the concept of Nash equilibrium so that it could be applied to situations where Nash's own formulation is not adequate.[209]

Notice that condition (a) on its own would not do either. It could be that it is in each person's interest to behave in a self-interested way if everyone believed that everyone else would behave so. In that case non-cooperation is also a Nash equilibrium, meaning that a set of mutual beliefs that the agreement will not be kept would also be self-confirming. Stated somewhat informally, a Nash equilibrium is a course of actions ('strategy', in game theoretic parlance) per party, such that no party would have any reason to deviate from his or her course of actions if all other parties were to pursue their courses of actions. As a general rule, societies harbour more than one Nash equilibrium. Some yield desirable outcomes, others do not. *The fundamental problem every society faces is to create institutions where conditions* (a) *and* (b) *apply to engagements that protect and promote its members' interests*.

Conditions (a) and (b), taken together, require an awful lot of coordination among the parties. In order to probe the question of which Nash equilibrium can be expected to be reached – if a Nash equilibrium is expected to be reached at all – the thing to do is to study human behaviour that are *not* Nash equilibria. The idea is to model the way people form beliefs about the way the world works, the way people behave and the way they revise their beliefs on the basis of what they observe. One can then track the consequences of those patterns of belief formation to check whether the model moves towards a Nash equilibrium over time, or whether it moves about in some fashion or the other but not towards an equilibrium.

This research enterprise has yielded a general conclusion. Suppose the economic environment in a certain place harbours more than one Nash equilibrium. Which equilibrium should be expected to be approached – if the economy approaches an equilibrium at all – will depend on the beliefs that people held at some point in the past. It also depends on the way people have revised their beliefs on the basis of observations since that past date. But this is another way of saying that history matters. The narrative style of social enquiry becomes necessary at this point. Model building, statistical tests on data relating to the models, and historical narratives have to work together synergistically if we are to make progress in understanding our social world.

6.3 The Basis of Trust, 1

Mutual trust is the basis of cooperation. In view of what we have learnt about the multiplicity of Nash equilibria, we are now led to ask what kinds of institution are capable of supporting cooperation. To answer that, it will prove useful to classify the contexts in which the promises people make to one another are credible. There are four cases to consider, involving (1) mutual affection, (2) pro-social disposition, (3) external enforcement, and (4) mutual enforcement. We consider them sequentially.

6.3.1 Mutual Affection

Consider the situation where the people involved care about one another, and it is commonly known that they care about one another. The household is the most obvious example of an institution based on affection. To break a promise we have made to someone we care about is to feel bad. So we try not to

---

[208] *A Beautiful Mind* is the title of a biography of John Nash, by Sylvia Nasar. The book (Nasar, 1998) was made into a 2001 film with the same title.

[209] Binmore (2007) presents an excellent, readable account of their work.

do it. From time to time, though, even household members are tempted to misbehave. As people who live together can observe one another closely, the risk of being caught misbehaving is high. This restrains household members even when the temptation to misbehave is great.

Affection does not extend to more than a few people. Evolutionary biologists have explained why. The household has traditionally been defined as a consumption unit, and that says something about the range of activities it is able to organise. Moreover, as numbers are small, households cannot undertake engagements requiring people of many and varied talents. So households need to find ways to engage with one another if people are to expand the kind of activities they would like to pursue. The problem of trust therefore reappears at the interhousehold level. This leads us to search for other contexts where people can trust one another to keep agreements.

6.3.2 Personal Integrity

One such situation is where people are trustworthy, or where they reciprocate if others have behaved well toward them. Evolutionary psychologists have argued that we are adapted to have a general disposition to reciprocate. Development psychologists have found that pro-social disposition can be formed by communal living, role-modelling, education, and receiving rewards and punishments (be it here or in the afterlife).

We do not have to choose between the two viewpoints; they are not mutually exclusive. Our capacity to have such feelings as shame, guilt, fear, affection, anger, elation, reciprocity, benevolence, jealousy, and our sense of fairness and justice have emerged under selection pressure. Culture helps to shape preferences, expectations and our notion of what constitutes fairness. Those in turn influence behaviour, which are known to differ among societies. But cultural coordinates enable us to identify the situations in which shame, guilt, fear, affection, anger, elation, reciprocity, benevolence and jealousy arise. They do not displace the centrality of those feelings in the human makeup. The thought we are exploring here is that, as adults, we not only have a disposition for such behaviour as paying our dues, helping others at some cost to ourselves and returning a favour, we also ease our hurt by punishing people who have hurt us intentionally, and shun people who break agreements, frown on those who socialise with people who have broken agreements, and so on. By internalising norms of behaviour, a person enables the springs of his actions to include them. In short, they have a disposition to obey the norm, be it personal or social. When they break it, neither guilt nor shame is absent, but frequently the act will have been rationalised by them. Making a promise is a commitment for that person, and it is essential for him that others recognise it to be so. The mutual influence of the sense of citizenship and being members of civil society may be why civics used to be taught at school. Rights and obligations do not always have to be enshrined in the law.[210]

Box 6.1

Civic Virtues

Schama (1987) has drawn a portrait of civic culture among the Dutch burghers of the late 17th century to explain the remarkable prosperity the Dutch enjoyed during the following hundred years. For Schama, the burgher was a citizen first and *homo economicus* second. The author showed that the obligations of civic life conditioned the opportunities of economic prosperity. If any one obsession linked together the burghers' various concerns with family, fortunes of state, power of their empire, the condition of their poor, their standing in history, and the uncertainties of geography, it was the moral ambiguity of their good fortune. Schama's account of Dutch prosperity turns on the way the nation's burghers coped with that ambiguity.

---

[210] Janoski (1998) is a treatise on the subject.

People are trustworthy to varying degrees. When we refrain from breaking the law, it is not always because of a fear of being caught; we refrain from doing it even when we know that no one will ever know. When we resist the temptation to take a closer look at an exotic plant in the forest it is not because others will rebuke us for trampling over fragile Nature, it is because we want to contribute to lessening the tragedy of the commons. The problem is that although pro-social disposition is not foreign to human nature, no society could rely exclusively on it. How is one to tell to what extent someone is trustworthy? If the personal benefits from betraying one's conscience is large enough, almost all of us would betray it. Most people have a price, but it is hard to tell who comes at what price.

People everywhere have tried to establish institutions in which they have an incentive to cooperate. The incentives differ in their details, but they have one thing in common: *those who break agreements without cause suffer punishment*. In contrast to the two sets of circumstances we have just identified, in which the punishment is internal to the person in breach of an agreement, the punishment in the circumstances we next identify is meted out by a source external to the person. The next section studies how that is achieved.[211]

6.4 The Basis of Trust, 2

There are two ways. One is to rely on external enforcement, the other on mutual enforcement. Each gives rise to a particular class of institutions. Depending on the nature of the engagement they would like to enter into, people invoke one or the other. The official term for one is the 'rule of law'; for the other, it is 'social norm'. Casual observation suggests that people in democracies rely heavily on the former, while elsewhere people depend greatly on the latter. Some economists have argued that it is *because* they have been able to depend extensively on the former for centuries that Western democracies are now high income countries.[212]

6.4.1 External Enforcement

One possible way to ensure that the parties trust one another (and themselves!) to keep to their agreement is to have it enforced by an established structure of power and authority. In certain societies, tribal chieftains, village or clan elders, and warlords enforce agreements and rule on disputes. There are also examples where rural communities have created mini republics in certain spheres of life. Village panchayats in India assume that form. The idea there is to elect officials who are then trusted with the power to settle disputes, enforce agreements (they may only be tacit), communicate with higher levels of state authority, and so on. Robert Wade's account of local enforcement of the allocation of benefits and burdens in sharing natural capital in rural South India described such a mechanism in detail (Wade, 1988). Forty-one villages were studied. It was found that downstream villages (those suffering particularly from water shortage) had an elaborate set of rules, enforced by fines, for regulating the use of water from irrigation canals. Most villages had similar arrangements for the use of grazing land. Wade reported that elected village councils ('panchayats') had appointed agents who allocated water among farmers' fields, protected crops from grazing animals, collected levies and imposed fines. In a different setting, Baland and Platteau (1996) have written about 'water masters', who regulated the use of local fisheries in the Niger River delta.

Here we imagine that the external enforcer is the state and that the agreement is drawn up as a legal contract. We include on this list the implicit 'social contract' among citizens not to break the law. However, if contracts are to offer a viable means of doing things, breaches must be verifiable, otherwise

---

[211] Rewarding people for pro-social behaviour is implicit, in that, complying with the agreement is beneficial to the party.

[212] North (1990) read the economic rise of the West in terms of a rising demand for well-defined and secure property rights. The role of authority in ensuring the latter was crucial.

the external enforcer would have nothing to go by if asked to rule on it. To be sure, lawyers make a handsome living precisely because verification is fraught with difficulties.[213] We leave aside the difficulties of verifying breach of contract (Chapters 7 and 8) and note that if the punishment the state imposes for a violation is known to be severe relative to the temptation each party to the agreement faces to violate it, then everyone will be deterred from going that route. If everyone is aware of the force of that deterrence, they will each trust the others not to be opportunistic.

In the modern world, the rules governing transactions in the marketplace are embodied in the law of contracts. A modern firm is a legal entity, as are the financial institutions through which employees are able to accumulate their pension, save for their children's education, and so on. Employees have employment contracts with their firm, the firm complies with environmental regulations at a cost to itself, and so on. The agreements people reach with financial institutions are also legal contracts. When someone goes to the grocery store, the purchases involve the law, which provides protection for both parties (the grocer, in case the cash is counterfeit or the card is void; the purchaser, in case the product turns out on inspection to be substandard). Formal markets, from which people enter and exist when they need to or wish to, are able to function only because there is an elaborate legal structure that enforces the agreements known as 'purchases' and 'sales'. Moreover, it is because the customer, the grocery store's owner and the credit card company are confident that the government has the ability and willingness to enforce contracts that they do business together.

What is the basis of that confidence? After all, the contemporary world has shown that there are states, and there are states. Who, one should ask, keeps an eye on the guardian? One answer – in a functioning democracy – is that the guardians worry about their reputation. A free and inquisitive press, probing NGOs, and active citizens help to sober governments into believing that incompetence or corruption would mean an end to their rule, come the next election. This involves a system of interlocking beliefs about one another's abilities and intentions. The millions of households in a country trust their government (more or less!) to enforce contracts, because they know that government leaders know that not to enforce contracts efficiently would mean being thrown out of office. In their turn, each side of a contract trusts the other not to renege (again, more or less!), because each knows that the other knows that the government can be trusted to enforce contracts, and so on, hopefully in an endless recursion of mutual, self-confirming beliefs. Trust is maintained by the threat of punishment (e.g. a fine, dismissal or incarceration) for anyone who breaks a contract, be it the legal contract (employment contract) or social contract (the contract between the voters and the government to maintain law and order). We are in the realm of beliefs that are held together by their own bootstraps (our earlier condition (b)).

What we have presented is only the sketch of an argument. The complete argument is similar to the one which shows that *social norms* also offer a way to enforce agreements. So we turn to that fourth context in which agreements are kept: the play of social norms. We will see that because societal norms can be a powerful weapon in disciplining external enforcers, they are more fundamental for cooperation than even the law.

6.4.2 Mutual Enforcement in Long-Term Relationships

Social customs differ from rulers' dictates. The motivations they give rise to in us to behave one way rather than another are not the same. Anthropologists have studied the practice of 'gift-exchange' in classical societies (Malinowski, 1926; Sahlins, 1972). The practice involved the exchange of goods people regard as part of their social custom. Modern societies also practise gift-exchange. They take

---

[213] Rough estimates suggest that in the US, for example, gross value added of the legal profession (lawyers, judges, investigators), on people who work in insurance (loss adjusters, insurance agents), and on those in law enforcement (the police) make up more than 4% of GDP (Bureau of Economic Analysis, 2020); and that does not include the defensive measures people take against possible litigations, burglary, and theft.

place on religious occasions and at family celebrations; and they are pervasive in subtle forms in modern societies in such relationships as those to be found between managers and workers (Akerlof, 1982) and landowners and peasants (Rudra, 1982).[214] There are variations of the practice, for it is not restricted to 'exchange'; a unilateral gift would come with an expectation that it will be reciprocated, even among people who are not engaged in a long-term relationship. That would appear to have been as true in ancient, even archaic times as it is today:

"The word 'gift' is not to be misconstrued. It may be stated as a flat rule of both classical and archaic society that no one ever gave anything, whether goods or services or honours, without proper recompense, real or wishful, immediate or years away, to himself or to his kin. The act of giving was, therefore, in an essential sense always the first half of a reciprocal action, the other half of which was a counter-gift." (Finley 2002: 53)

Social norms differ from legal systems in channelling behaviour. Reciprocity is central to the meaning of social norms. Greif (1993) has used norm-guided practices to explain the emergence of institutions that facilitated the growth of trans-border trade in Medieval Europe. He showed that the Maghribi traders during the 11th century in Fustat (now part of Cairo, Egypt) and the Mediterranean acted as a collective to impose sanctions on those of their agents who broke their commercial codes. Elsewhere Greif, Milgrom and Weingast (1994) have provided an account of the rise of merchant guilds in late Medieval Europe. Guilds afforded protection to members against unjustified seizure of their property by city-states. Guilds determined if and when trade embargoes against such states were justified. In their account guilds are not be to viewed as rent-seeking organisations.[215] In a related work, Milgrom, North and Weingast (1990) pointed to the role of merchant courts at the annual fairs in Champagne. The courts helped members to impose sanctions on those who had transgressed on agreements.

Although the law exists today in every country, there are places where people cannot depend on it. It could be that the legal system is corrupt. But even when it is reliable, it could be that the nearest courts are far away and there are no lawyers in sight. When roads are decrepit, villages are enclaves. Much economic life is then shaped outside a formal legal system. Nevertheless, people transact with one another. Saving for funerals in Ethiopia, for example, involves saying, 'I accept the terms and conditions of the *iddir*' (Box 6.4). As there is a lack of developed credit markets where they live, villagers also practise reciprocity to smooth consumption over time. Udry (1990) reported that in the sample of villages in Nigeria he had studied nearly all credit transactions were either between relatives or between households in the same village. No written contracts were involved, nor did the agreements specify the date of repayment or the amount repaid. Social codes were implicitly followed. Fewer than 10% of the loans were in default.

Why would the villagers trust one another? They would if agreements were mutually enforced; that is, *a credible threat by members of a community that stiff sanctions will be imposed on anyone breaking an agreement would deter everyone from breaking it*. This is a common basis for doing business in developing countries. Among the Kofyar farmers in Nigeria, for example, agricultural land is private, but free-range grazing is permitted once the crops have been harvested (Stone, Netting, and Stone, 1990). Kofyar households are engaged in subsistence farming, so labour is not paid a wage. However, in times

---

[214] Reporting an interview he had conducted with a landowner (patron) and a farm worker (client) in West Bengal, India, Rudra (1982: 255) wrote of the relationship between them:

(Landowner): "I may require a labourer to come and help me in the middle of the night, for example, if it has rained and the living quarters and the paddy go down have got flooded. There are no rates for such work to be done at such an hour. Such a service cannot be purchased."

(Farm worker): "I am a poor man and I do not even have enough to eat every day. I may require urgently some money for a funeral in the family. To whom shall I go?"

[215] This should be contrasted with Ogilvie's account of producers' guilds in the duchy of Wurttemberg in the early-modern period (Ogilvie, 1997).

Chapter 6: Laws and Norms as Social Institutions

long gone the Kofyars instituted communal work on individual farms. Although some of this is organised in clubs of 8 to 10 individuals, there are also community-wide work parties. A household that does not provide the required quota of labour without good excuse is fined (as it happens, in jars of beer). If fines are not paid, errant households are punished by being denied communal labour and subjected to social ostracism.

How is mutual enforcement able to support agreements among members of a community? It is all well and good for those with whom a person had reached an agreement to threaten that sanctions will be imposed on him if he breaks it. But why should he believe the threats? He would believe them if sanctions were inscribed into social norms. That may read as simply a rewording designed to make one think that description amounts to explanation, but social norms carry a deep meaning. To uncover it, let us assume an agreement that has been kept by each party is *observable* by all parties. This is a strong assumption, but as with 'verifiability' (Section 6.4.1) it is a useful starting point. Once we draw conclusions from it, we will be able to infer how communities have modified practices in situations where the assumption does not hold even approximately. For example, even if you are unable to observe someone's actions, you may be able to rely on someone else who *can* observe those actions. Trust in that someone is then built on trust in that someone else. That said, anyone who has visited villages in low income countries will have observed that there is often less insistence on privacy there. In tropical villages, cottages are frequently designed and clustered in such fashion that it must be hard for anyone to prevent others from observing what they are about. What is loosely called 'social capital' is the seat of cooperation sustained by the practice of social norms.

By a social norm we mean an accepted rule of behaviour. A rule of behaviour reads like so: 'I will do X if you do Y; I will do Z otherwise'; 'I will do P if Q happens; I will do R otherwise'; and so forth.[216] For a rule of behaviour to *be* a social norm, it must be in the interest of each person to act in accordance with the rule if all others act in accordance with it; that is, the rule should correspond to a Nash equilibrium. To see how social norms work, it is useful to study whether cooperation based on long-term relationships can be sustained between various parties.

Imagine that all parties are far-sighted, that is, they discount future gains and losses at a low rate relative to current gains and losses. That means the private gain to a party from breaking the agreement unilaterally is less than the losses he would suffer if all other parties were to impose sanctions on him. Imposing sanctions could involve, for example, refusing to enter into any transactions with the erring party for some while, shunning the person in other ways for suitable numbers of periods, and so on. But as imposing sanctions is itself not costless, the threat of sanctions must be made credible.

To see how social norms solve the credibility problem, let us call someone a *rule-abider* if the person cooperates with those others who are rule-abiders, but imposes sanctions on those who are not rule-abiders. Although that sounds circular, it is not, because for a rule of behaviour to *be* a social norm requires all parties to start the process of cooperation by keeping their agreement. It would then be possible for any party in any period to determine which party is a rule-abider and which party is not. For example, if ever someone was to break the original agreement, he would be judged to be someone who is not a rule-abider; so, the norm would require all parties to punish the person. Moreover, the norm would require that punishment be inflicted not only on those in violation of the original agreement (first-order violation); but also on those who fail to punish those in violation of the agreement (second-order violation); on those who fail to punish those who fail to punish those in violation of the agreement (third-order violation); and so on, indefinitely. The infinite chain makes the threat of punishment for

---

[216] Game theorists call rules of behaviour 'strategies'. The theorem was proved in its general form by Fudenberg and Maskin (1986). Mailath and Samuelson (2006) contains an excellent, formal explanation for why the theory of repeated games forms the basis of long-term relationships.

errant behaviour credible, because if all others were to conform to the norm, it would not be worth any party's while to violate the norm. Keeping one's agreement would then be mutually enforcing.

All societies appear to have sanctions in place for first-order violations. Second-order sanctions are in play when members of an institution shun any member who continues to do business with someone who has cheated. That sanctions against higher-order violations have not been documented much may be because they are not needed to be built into social norms if it is commonly recognised that people feel a strong emotional urge to punish those who have broken agreements. Anger facilitates cooperation by making the threat of retaliation credible.[217]

Here then is our general finding: *social norms of behaviour are able to sustain a mutually beneficial state of affairs if people care sufficiently about the future benefits of cooperation*. The precise terms and conditions will be expected to vary across time and place, but what is common to them all is that cooperation is mutually enforced, it is not based on external enforcement. Using a numerical example, Box 6.2 illustrates the force of social norms in enabling cooperation.

But there is bad news. Even when cooperation is a possible outcome via the application of social norms, non-cooperation is a possible outcome too. If each party was to *believe* that all others would break the agreement from the start, then each party would break the agreement from the start. Notice that a failure to cooperate could be due simply to a collection of unfortunate, self-confirming beliefs, nothing else. We usually reserve the term *society* for a collective that has managed to reach an outcome that is mutually beneficial to all parties. Social norms are a powerful coordinating device for eliciting cooperation. But in order to be successful each party must believe that all others are abiding by the norm. It is only then that it will be in the interest of each to behave in accordance to the norm.

Then there is further bad news. It is possible for social norms to sustain inequities among the parties bound by them. Worse still, it can be that a social norm sustains a sharing rule which can only be described as exploitation of one group by another. It can be that a norm supports an allocation rule among members of a community in which one group is worse off than it would be if the norm had not been in play. We thus extend the previous proposition by the following: *social norms of behaviour are able to sustain a mutually beneficial state of affairs if people care sufficiently about the future benefits of cooperation; but there are social norms that sustain states of affair where one group exploits another group*. Annex 6.1 provides an example where exploitation masquerades as cooperation.

Box 6.2
The Grim Norm

Consider a pair of individuals, named 1 and 2, who are engaged in a one-shot game. Each can choose one of two options independently of the other (e.g. it may be that they do not reside in the same place). The options are labelled C_1 and D_1 for person 1, and C_2 and D_2 for person 2. Table 6.1 records the incomes the pair would enjoy for each of the four possible pairs of choices. The first quantity within parentheses is person 1's income from the venture, expressed in dollars; the second quantity is person 2's income from the venture. If person 1 (who selects a row) were to choose C_1 and person 2 (who selects a column) were to choose C_2, each would enjoy an income of $50,000. If person 1 were to choose D_1 and person 2 was to choose C_2, the former would enjoy an

---

[217] There is anecdotal evidence of infinite chains of responses demanded by social norms. Tribes in and around the Kakadu National Park, Australia, are known to have practised a form of punishment that involved spearing the thigh muscle of the errant party. If the party obliged to spear an errant party balked at doing so, he in turn would have been speared. If the person obliged to spear the latter miscreant was to balk, he too would have been speared. The oral evidence goes on to say that the chain was unending.

income of $130,000, while the latter would garner only $10,000; if person 1 were to choose C_1 and person 2 were to choose D_2, the incomes garnered by the pair would be reversed.[218] But if person 1 were to choose D_1 and person 2 were to choose D_2 each would enjoy an income of $30,000.

The opportunity open to the two parties is known as the *Prisoner's Dilemma*, a quintessential game in which to discuss the dissonance between individual and group interests.[219] The game demonstrates in a dramatic way the temptation to free ride on others in collective engagements. Even though both parties recognise the benefits of cooperation (it would yield $50,000 to each party, well in excess of the $30,000 each would earn if they did not cooperate), each party has an overwhelming incentive (because the game possesses what game theorists call a 'dominant' strategy) to act opportunistically by defecting from an agreement to cooperate.

Table 6.1 Options – The Grim Norm

		Person 2	
		C_2 (cooperate)	D_2 (defect)
Person 1	C_1 (cooperate)	(50,000, 50,000)	(10,000, 130,000)
	D_1 (defect)	(130,000, 10,000)	(30,000, 30,000)

Ahistorical models can tell us nothing about the play of social norms. It does not make sense even to talk of norms of behaviour unless the past and future are brought in. So, we now imagine that the two parties expect to face the same table of choices and their income consequences period after period, say, annually. Periods are denoted by t, so that $t = 0, 1, 2, \ldots$ *ad infinitum*. At $t = 0$, the pair have the opportunity to enter into a long-term relationship.

This is a far more sophisticated and realistic setting than the previous one-period analysis, for it allows the parties to condition their choice of action at any date on the past behaviour of both parties. That makes cooperation possible, because should a party misbehave, sanctions can be meted out by withdrawing cooperation in the future.

A simple rule of behaviour on which the pair could coordinate would read as follows: *Begin by cooperating and continue cooperating in every period so long as neither party has defected, but defect permanently from the period following the first defection by either party*. Because of the unforgiving nature of the rule (sanctions are to be imposed forever following a breach), it is called 'Grim'.

We want to calculate how far-sighted the parties have to be if Grim is to serve as a social norm. So let $r > 0$ be the rate at which each party discounts his benefits. Because r is constant, calendar time does not matter to the parties: future engagements enshrined in the matrix will look the same to both persons at all future dates. This proves to be very useful in the reasoning that follows.

Suppose person 1 believes that person 2 will play Grim. He would then reason that he could enjoy a gain of $80,000 at $t = 0$ by defecting ($130,000 as against $50,000). But he also realises that if he were to do that it will be in his interest to defect forever from $t = 1$ onward. As r is constant, it means that if it proves to be in person 1's interest to cooperate for some

---

[218] Essentially, opportunistic behaviour on one person's part ruins the other person.

[219] For an interpretation of the game in which 'Prisoner's Dilemma' is an apt name, see Luce and Raiffa (1957).

periods but then defect forever, it would be in his interest to defect forever starting now. This piece of reasoning simplifies matters enormously, because person 1 needs only to compare the two benefit streams – one associated with cooperation forever (V_C), the other associated with defection forever (V_D). They are:

$$V_C = 50{,}000 + 50{,}000/(1 + r) + 50{,}000/(1 + r)^2 + \ldots \qquad (B6.2.1)$$
$$V_D = 130{,}000 + [30{,}000 + 30{,}000/(1 + r) + \ldots]/(1 + r) \qquad (B6.2.2)$$

Obviously, it would be in person 1's interest to play Grim himself if $V_C > V_D$. Comparing equations (B6.2.1) and (B6.2.2) shows that $V_C > V_D$ if $r < 25\%$ a year. The argument is of course symmetric: if person 2 believes that person 1 will play Grim, choosing Grim would be in her interest. We have shown therefore that if $r < 25\%$ a year, Grim can serve as a social norm.

Grim is a useful behavioural rule for study because, as it is wholly unforgiving (a single breach of agreement is followed by permanent sanction), no behavioural rule could serve as a social norm if $r > 25\%$ a year. Even though the norm may seem stylised, Grim has been found in reciprocal relationships (Czakó and Sik, 1988). The authors speculate that it is in force only in those environments in which the parties have access to formal markets as an alternative institution in which to do business. Otherwise there is scant evidence that the grim norm is deployed. Graduated sanctions are in wide use. The first misconduct is met by a small punishment, subsequent ones by a stiffer punishment, persistent ones by a punishment which is stiffer still, and so forth (Ostrom, 1992). Where information is imperfect, a small penalty for the first misconduct could be a warning that others are watching, or it could be that others signal their acknowledgement that the misconduct could have been an error on the part of the offender and that he should try harder next time. And so on.

The parameter r should not be interpreted solely as short-sightedness. It could be that the pair discount future benefits because each believes that there is a good chance they will not have the opportunity to play the game in the future. That may be because people are separately mobile (the argument will not work for collective migration, as in the case of hunter-gatherers); in which case r could be the probability that the two will *not* meet to transact in year t conditional on their having met and transacted until year $t–1$. In that case r is the hazard rate (Chapter 5) and V_C and V_D are expected values of income. We now have an explanation for why people in a mobile society are unlikely to have enough trust in others to enter into ventures without the security of an external enforcer (r would be too large). But an external enforcer would be trustworthy only if there was reason to trust it, which brings us back to the primacy of mutual enforcement and the play of social norms.

6.4.3 Reputation

Cooperation does not have to be based on long-term relationships. Person *A* may know that the agreement they would like to reach with person *B* will last only for a brief period, but if it was to become known that *A* had cheated, no one would cooperate with them in the future, as they would have acquired a bad reputation. That would deter them from betraying *B*'s trust, provided of course that they cared about their future.

If a person is found to be trustworthy in their dealings, they acquire a *reputation* for trustworthiness. That tells us reputation is an asset. In the case of a person, it is part of their human capital; for a firm it enters the bottom line of its balance sheet; for a politician, it enters the mental ledger of voters. Even communities acquire reputations, for being honest traders, skilful fishermen, brave soldiers, admirable scholars, and so on. Sociologists like to say reputation is a social construct. Like trust in the currency notes issued by central banks of countries with a sound reputation for fiscal prudence, reputation for trustworthiness hangs on a system of interlocking beliefs.

Chapter 6: Laws and Norms as Social Institutions

Brand names are codes for reputation. A firm may have changed hands, perhaps several times over the decades, but many are known to have retained their name, to signal that their products are as good as they have always been. Firms worry that customers increasingly want to be satisfied that their products are based on fair trade, that they have not required unsustainable resource use in a low income country, that their packaging is not a persistent pollutant, and so on. Some features of their product, such as packaging, are observable, so customers are able to coordinate their actions by boycotting firms whose products break the code. Other features are hard to observe but are in principle verifiable. But verification can be costly, unless purchasers insist that firms certify that they have complied with fair trade rules, with the requirement of sustainable resource use, and so on. Today such certification frequently appears on the product's wrapper. Information conveyed in that form is inevitably coarse; purchasers may not know what fair trade means, nor what sustainable resource use in a particular context amounts to, and so forth. In order to overcome that, firms could be required to disclose their entire chain of operation, starting at the terms conducted with the suppliers of primary inputs (cocoa beans) to the product on the counter (chocolate bars). If the disclosure is readily available, say, on the firm's website, the firm's claim can be verified (see Chapters 15 and 16 for further discussion on supply chains).

Of course, the certification could be false, or at best vague; but the firm now worries that someone will check. And if it gets around that the firm has shaded its claims, its reputation will suffer, worse, it could get entangled in a legal case. The firm fears that it will be fined and face a boycott from customers. That is the threat of sanctions facing the firm. Box 6.3 formalises this reasoning to establish that reputation is an asset.

Why would a firm wish unilaterally to enter what could appear to be a labyrinth of disclosures? They would do so if customers become increasingly anxious about biodiversity losses in foreign lands. A firm that took customers' increasing concerns seriously and moved first in tying its proverbial hands would gain a reputation for socio-ecological awareness and raise its market share which later movers would find difficult to dislodge (see Chapter 17 for more on how changes in reputation can have financial implications). Game theorists call that a 'first mover advantage'.

Ultimately, it is the force of public insistence on norms of conduct by firms and governments that keep both tolerably clean. Eternal vigilance is the price citizens have to pay not only for their freedom, but for sustainable development too.

Box 6.3
Reputation as an Asset

Consider a firm that has an unsullied reputation for selling a good product, based on fair trade, sustainable resource use, and so on. So long as its reputation remains good, it earns an annual profit of £100 million. Imagine that owners change each year. By ignoring those environmental requirements an owner could earn a profit of £150 million. The temptation to cheat customers is therefore very high. However, customers follow the strategy of buying from the firm *if and only if* its reputation is good. Suppose the rate at which firms' owners discount future profits is r, where $r = 5\%$ a year. The owner at $t=0$ has two options: (i) ignore its commitments and make additional profit of £50 million, but find no buyer at the end of the year; or (ii) respect its commitments, earn £100 million and sell the firm at the end of the year at the firm's capitalised value of $100(1+r)/r$. Therefore, at $t = 0$ the new owner will compare £50 million with £$100(1+r)/r$ million. At $r = 0.05$ a year, $100(1+r)/r = 2.1$ billion. Hence, maintaining the brand name forever is a Nash equilibrium.

6.5 Social Capital as the Basis of Societal Coherence

Using indicators of trust and the exercise of civic norms in a sample of 29 countries from the World Value Survey, Knack and Keefer (1997) found that trust and civic norms are stronger in countries with (i) higher income levels and less inequality of income, (ii) institutions that restrain predatory actions on the part of executives (read, officials), and (iii) homogenous populations with high educational attainment. Zak and Knack (2001) subsequently found a positive relationship between trust and economic growth. The findings, however, are relationships, and so should be seen only as relationships, and causality should not be imputed to them. As our analysis says, the underlying drivers of the relationship are *beliefs*.

But beliefs do not appear out of nowhere. There is a potential role for governments, NGOs and increasingly private firms to help build local institutions through which people are able to identify and take advantage of new opportunities for collective action. Such help involves, among other things, devising clearly defined, fair rules for the allocation of burdens and benefits. That helps to align beliefs. In their study of property rights, migration and local ecosystems in a sample of villages in Northwest India, Chopra and Gulati (1998) found that the effect of conscious efforts to build local institutions had been powerful. Distress migration out of villages was lower where NGOs had been at work than in villages where there had been little attempt to create institutions for managing water and pastureland on a communal basis. Significantly, the authors found that the probability of participation in communal pastureland was higher among villagers who were participating in communal water management schemes than among villagers who were not.

Cooperation enables members to learn about one another's traits – for example, that they are trustworthy so it is frequently remarked that successful cooperation begets further cooperation. In a study of milk producers' cooperatives in South Indian villages, Seabright (1997) found that prior history of cooperative institutions in the community was a positive predictor of a cooperative's success. The author reported that cooperation among producers was more successful in those villages in which members had previously organised communal religious festivals, as opposed to villages in which festivals had been segmented by caste.

How did people who now engage with one another connect in the first place? In traditional societies the answer is simple: mostly they have known one another from birth. People who are engaged in long-term relationships based on social norms, communities for short, know one another, at least indirectly through people they know personally. Each person knows those with whom he shares the local ecosystem. Communities are personal and exclusive. Members have names, personalities and attributes. An outsider's word is not as good as an insider's word. Some call the lubricant that makes for a society *social infrastructure*, others call it *social capital* (Figure 6.2).

Chapter 6: Laws and Norms as Social Institutions

Figure 6.2 Effective and Ineffective Institutions

Box 6.4
The *Iddir* and ROSCAs

A good portion of income risk that rural households in low income countries face is not common among the villagers, but idiosyncratic to the household. The risk of falling prey to illness or death, of damage to stored grain from pests or a leaky roof, or of loss of a draft animal, are incidents that can befall any household.[220] The *iddir* serves as a communitarian insurance arrangement in wide use in Ethiopia (there are variants elsewhere in sub-Saharan Africa). Small landholders who join their local *iddir* make a fixed and regular payment of money into a common pot (the 'premium' is usually smaller for poor women), and withdraw an amount owed to them in the event of a bereavement. No contracts are signed, members know one another. The rules are mutually enforced.[221]

---

[220] See Bardhan and Udry (1999). This is not a self-evident empirical observation. A common mental picture of calamities in low income societies, drawn from newspaper articles or news programmes, is that they befall entire communities at once (drought, flood, and so on).

[221] Aredo (2010) is an excellent reference to the workings of the *iddir*.

> A not entirely dissimilar class of communitarian institutions, known as rotating savings and credit associations (ROSCAs), are widespread in Asia.[222] In the absence of commercial banks, which may in any case be unwilling to lend to poor households, ROSCAs enable people to avoid storing money for a lumpy expenditure. In detail they differ, depending on the terrain, among other things, but there is a common thread running through them. A stylised version would run as follows:
>
> Imagine that each of 12 individuals needs an asset (e.g. a draft animal), which costs US$120. Each person is able to put aside US$10 each month. If they acted alone, each individual would be able to purchase such an asset in 12 months' time. Until then, the accumulating saving would be lying idle under the proverbial mattress. Suppose now they establish a ROSCA, with the agreement that each person would contribute US$10 every month into a common pot for 12 months. The agreement also says that at the end of each month the US$120 that had accumulated in the pot would be awarded by a fair lottery to a member who had not yet won a lottery. At the beginning, all face the same odds (i.e. probability 1/12) that they will be able to purchase the asset in a month's time, the same odds (of 1/11) that those unlucky at the first draw will be able to purchase it in two months' time, and so on. The only person who would not gain *ex post* is the person who lost at each round of the lottery. But they will not have lost either, because they will be able to purchase the asset at the end of the 12th month.
>
> Observers have reported that in the ROSCAs they had studied, every transaction was observed by every ROSCA member at every meeting (Aredo, 2010). That ensured observability – an essential feature of exchanges that involve mutual enforcement.

6.6 Social Capital and Identity

In contrast to norm-based transactions, the hallmark of transactions enforced by the law of contracts is that they can take place among people who do not know one another. In the modern world people are mobile, a pattern of behaviour not unrelated to the fact that we are able to do business even with people we do not know. We all too often do not know the salespeople in the department stores where we shop, nor do they know us. When someone borrows from their bank, the funds made available to them come from unknown sources. Literally billions of transactions take place each day among people who have never met and will never meet. Often, the exchanges take place only once, unlike exchanges based on long-term relationships. Markets are prime examples of institutions offering such opportunities. In contrast to communities, markets are impersonal and inclusive. Witness the oft-used phrase: 'My money is as good as yours.'

And yet, markets cannot function without communities, for as we have seen, mutual enforcement is the basis on which external enforcement (e.g. state enforcement) acquires its force. We have also seen that mutual trust rests on a system of self-confirming beliefs about one another. And that leads to the thought that mutual trust is potentially a fragile state of minds. Like ecosystems, societies need to possess a certain amount of modularity if they are to protect and promote trust. Modularity makes it more likely that a breakdown of trust in one sector of society does not infect the rest but is instead revived by actions taken by some other sector of society.

Despite societal modularity, people could end up not cooperating even if they care a lot about the future benefits of cooperation. To reconfirm that, imagine that each party believes that all other parties will renege on an agreement. It would then be in each one's interest to renege at once, which means of course that there would be no cooperation. Even if people were far-sighted, a *belief* that others will renege can be a cause of a failure to cooperate. No doubt it is only mutual suspicion that ruins the chance to cooperate, but the suspicions would be internally self-consistent. We should conclude that

---

[222] Geertz (1962) is the classic reference. His field study was based in Indonesia.

Chapter 6: Laws and Norms as Social Institutions

even when appropriate institutions are in place to enable people to cooperate, they may not do so. Whether they cooperate depends on mutual beliefs, nothing more.

We now have in hand a tool to explain how a community can skid from cooperation to non- cooperation. Ecological stress – caused, for example, by rising population and prolonged droughts – can result in people fighting over land and natural resources. Political instability – in the extreme, civil war – could in turn be a reason people who previously formed a society become antagonistic toward one another, forming groupings that rely on past relationships for use against others. One group becomes concerned that their source of livelihood will be destroyed or confiscated by another, so they now discount the future benefits of cooperation with the other at a higher rate.

Similarly, if the parties fear that their government is now more than ever bent on destroying communitarian institutions in order to strengthen its own authority, they would discount future benefits and losses at a higher rate. For whatever reason, if the parties discount the future at too high a rate, relationships break down. The points at which those switches occur are *bifurcations*, the points themselves are *tipping points* (Chapter 3). Social norms work only when people coordinate on trust.

Ominously, there are subtler pathways by which societies can tip from a state of mutual trust to one of mutual distrust. We have argued that even if people are far-sighted, non-cooperation is a possible outcome. That tells us that a society could tip over from cooperation to non- cooperation owing merely to a change in beliefs. The tipping may have nothing to do with any discernible change in circumstances; the entire shift in behaviour could be triggered in people's minds. The switch could occur quickly and unexpectedly, which is why it would be impossible to predict and why it would cause surprise and dismay. People who woke up in the morning as friends would discover at noon that they are at war with one another. Of course, in practice there are usually cues to be found. False rumours and propaganda create pathways by which people's beliefs can so alter that they tip a society where people trust one another to one where they do not.

The reverse can happen too, but it takes a lot longer. Rebuilding a community that was previously racked by civil strife involves building trust. Non-cooperation does not require as much coordination as cooperation does. Not to cooperate usually means to withdraw. In order to cooperate, people must not only trust one another to do so, they must also try to coordinate on a social norm that everyone understands. That is why it is a lot easier to destroy a society than to build it.

How does an increase or decrease in cooperation translate into macroeconomic statistics? The numerical example in Box 6.2 captured a salient point, that an increase in cooperation raises wealth by permitting a more efficient allocation of resources. So now consider two communities that are identical in all respects, except that in one community, people have coordinated at a set of mutual beliefs where they trust one another, while people in the other have coordinated at a set of mutual beliefs where they do not trust one another. The difference between the two economies would be reflected in the economy's productivity figures, which would be higher in the community where people trust one another than in the one where they do not. Enjoying greater income, individuals in the former economy are able to put aside more of their income to accumulate capital goods, other things being equal. So, asset accumulation there is higher. Mutual trust would be interpreted from the statistics as a driver of economic growth – Chapter 8 offers empirical evidence of this.

The economics of biodiversity says that societies that have become rich were able to coordinate their mutual trust in activities that helped in the accumulation of produced capital and human capital, but that they did so at the expense of natural capital, often in distant lands (Chapters 4, 14 and 15). Many economies have followed that route to economic development. They have degraded their own natural capital, allegedly as a price for economic growth. The choice of activities on which people coordinate their trust in one another matters enormously. Trust on its own cannot save the biosphere from ruin. The agenda on which trust is to be made to play will prove crucial.

Box 6.5
Social Capital as Societal Coherence

In an early definition, social capital was identified with those " ... features of social organization, such as trust, norms, and networks that can improve the efficiency of society by facilitating coordinated actions." (Putnam, 1993: 167)

There was a reason Putnam spoke of *civil society*, not *communities*, for there are subtle differences between the two. Here we make a sharp (i.e. unreal) distinction by regarding the latter as the seat where social norms of behaviour are deployed because the law is unavailable. In his groundbreaking work, Putnam saw the former as the venue where social networks are able to engage in civic activities as a complement to the state, ideally with the support of the law. The underlying thesis is that network activities help to create trust among members (by, among other ways, enabling members to learn who are trustworthy), which in turn helps people to engage in civic activities. Using Italy as his laboratory, Putnam uncovered contemporary data on memberships of choral societies and football clubs in each of the 20 states in the country. Calling the trust that is created in such networks 'social capital', he found that the networks not only discipline the nation's state governments in their role as suppliers of public services, but also have a long temporal reach. Regions where civic engagement was greater hundreds of years ago enjoy greater levels of civic engagement and better governance today. Putnam identified civil society as the seat of social capital. His finding says that while civic engagement contributes directly to one's well-being (Chapter 11), they are also of instrumental value: accumulation of one type of asset (social capital) improves the quality of another asset (state government).

Putnam distinguished social capital with a horizontal structure from those built on a vertical structure. There is some evidence from elsewhere that, loosely speaking, membership in the former is voluntary (as in choral societies in northern Europe), in contrast to the latter, where membership is hereditary (for example, in the Hindu *jajmani* system of obligations across various layers of society).

It would be a mistake to read Putnam's account of social capital only in instrumental terms, nor do we believe he intended it to be so read. The line separating an instrumental value from an intrinsic value is, in any case, wafer-thin when the instrument advances a value we hold deeply. What is taken to have an intrinsic value could well be an instrument for advancing a more deeply held value. Conversely, what advances a deeply held value could have instrumental advantages.

The human need to belong and participate with others in celebrations of birth and marriage, and to express grief communally when someone in the household dies, would appear to be so deeply rooted as to transcend cultural differences across societies. We have a need to provide our family with material goods and services, but we also have a need to belong and participate in the wider community. That the origins of that enlarged need could be traceable to our personal benefits is of little moment, because they lie in our distant, even evolutionary past. It has, for example, proved to be a puzzle for development economists that the income elasticity of demand for food, even for low income households in Asia and Africa, is no more than 0.7 to 0.8, meaning that of an increase in household income, some 20 to 30% would be expected to be spent on non-food items (celebrations, festivals, funerals, alcohol-laced networking) (Femenia, 2019). This would seem to be true even among very poor, even malnourished people. That people's expressed needs transcend their material requirements only reveals that our well-being depends not only on our engagement with our own selves, but with others too.

The social historian David Hollinger has identified features of someone's *identity* that can be viewed as 'solidarity'. He writes: "To share an identity with other people is to feel solidarity with them; we owe them something special, and we believe we can count on them in ways that we cannot count on the rest of the population." (Hollinger, 2006: 23) By way of illustration Hollinger observes: "Feminism is a solidarity, but womanhood is not." Previously, Sunstein and Ullmann-Margalit (2001) had noted that there are cases where people may not even care for some of the activities their solidarity groups are engaged in (taking part in protest marches, choice of clothing, adopting particular eating habits, and so on), but nevertheless join them so as to express solidarity. Reasonably, the authors named the goods people use to express solidarity, 'solidarity goods'.

The anthropologist Mary Douglas once remarked that to be poor is not to be able to invite one's neighbour home to tea. Social capital has an intrinsic value to each of us. That is why we have chosen here to discuss cooperation and reciprocity in terms of community norms rather than rules in the networks we choose to join.[223]

6.7 The Primacy of Integrity

Ultimately though, neither the rule of law nor the dictates of social norms is enough. The unaccounted-for consequences for others of someone desecrating Nature do not necessarily lie in the immediate future. Nor is the person the sole desecrator, for innumerable acts of desecration, each small in itself, occur all the time, throughout the globe. Large numbers of small acts, taken time after time, have cumulative effects on the biosphere (Chapters 3 and 4). Neither verifiability nor observability can serve to discipline people when the consequences are not traceable to those who are responsible. If we are to make peace with Nature, which in effect means make peace with ourselves, we each will have to serve as both judge and jury. The checks and balances on our actions need to have an internal source, we cannot rely on others to discipline us into using the biosphere only as far as is justifiable. That justification has to be provided by us to our personal selves. And that cannot happen without education. Which is why the Review concludes (Chapter 21) with a plea for a transformation of our education systems towards one where children from an early age are encouraged to try to understand the infinitely beautiful tapestry of processes and forms that is Nature. The realisation that we are embedded in Nature would be a route to an understanding that desecrating Nature is like desecrating our own home. It is only when we appreciate that we are part of Nature and that Nature nurtures us that we will have fewer needs for reviews of the economics of biodiversity.

Box 6.6

Dark Sides of Social Capital

This chapter has explored the virtuous scope of social capital because, or so we argue in later chapters, success in protecting and restoring the biosphere will ultimately depend on whether people are able to act collectively toward that goal. The neighbourhood, and from there through the larger spheres of community and civil society, is the channel citizens, should they be fortunate enough to reside in a benign political climate, have available to them for collective action. A general disposition to abide by agreements, to be truthful, to be able to trust one another, and to act with justice is an essential lubricant of societies. The larger the population over which this disposition is cast by all, the better is the collective outcome.

---

[223] Dasgupta and Goyal (2019) study the notion of social identity when people assume the identity of networks to which they join for the benefits that brings them.

> But in a society where the reach of democracy is weak and trust is confined within groups, the disposition to be honest would be toward members of some particular group (e.g. one's community, or ethnic or religious group), not others. This amounts to group loyalty. One may have been raised to be suspicious of people from other groups, one may have even been encouraged to dupe such others if and when the occasion arose. Society as a whole squanders resources when the disposition for honesty is restricted to certain groups. The forms social capital is able to take in these circumstances are far from benign. They range from street gangs that practise extortion to para-states that sell protection to households and local businesses. The distinction between rebels righting wrongs and extortionists running a business is slippery. Hobsbawm (1959) famously read the practice of banditry since medieval times as rebellion against an oppressive social order, but in a modern classic, Gambetta (1993) recorded interviews with members of the Sicilian Mafia to reveal a business enterprise.
>
> Trust is not a given in any society. How people are able to channel their trust in ways that make for progress is not something that can be written on stone like an edict. We have to discover them continuously.

Annex 6.1 Corruption

The United Nations Convention against Corruption was adopted in 2003. Although there is no explicit reference to 'grand' corruption, the preamble of the Convention expresses concern over "cases of corruption that involve vast quantities of assets, which may constitute a substantial proportion of the resources of States, and that threaten the political stability and sustainable development of those States". The International Anti-Corruption Coordination Centre estimates that every year at least US$20–40 billion of assets are stolen by corrupt leaders and dispersed across the globe. Unlike petty corruption, grand corruption involves the distortion of central functions of government by senior public officials.[224]

Citizens of countries that enjoy large pools of natural resources but where governance is weak suffer from illegal resource extraction and the government corruption that fuels it (Biggs et al. 2017). One by-product of this is often biodiversity loss. A lack of alternative livelihoods serves to abet trade in illegally acquired resources. Conversely, robust democratic institutions can serve as a powerful deterrent to corruption.

A recent report by the UN Environment Programme and Interpol found that globally environmental crime has been growing at 2–3 times the rate of global GDP growth (Nellemann et al. 2016). Grand corruption entrenches poverty and inequality, undermines good business practice and threatens the integrity of financial markets. It also exacerbates societal exclusion. As we see below, it also undermines biodiversity.

An abundance of natural resources can trigger grand corruption, for it creates opportunities for rent-seeking behaviour, especially when legal economic opportunities are scarce. Corruption is a feature of life when officials enjoy a monopoly over decisions over the use of natural resources, and are not accountable (Riley, 1998). Reforms in policy, for example towards greater public control of resources, become susceptible to corruption.

---

[224] One definition of petty corruption is: "Everyday abuse of entrusted power by public officials in their interactions with ordinary citizens, who often are trying to access basic goods or services in places like hospitals, schools, police departments and other agencies." (Transparency International, n.d.)

Chapter 6: Laws and Norms as Social Institutions

Bribery is an example of both petty corruption and grand corruption. The former – known also as petty bribery – involves citizens paying small bribes to public officials to hasten bureaucratic processes or access to public services people are entitled to.[225] Bribery of local officials in developing countries is seen by many as a "worthy, if not required, investment to penetrate otherwise closed markets or sources of natural resources and labour" (Liu, 2018). The distinction between private incentives and the common good is all important here.

High-level government corruption cannot be separated entirely from local-level corruption, for corruption begets corruption. A study on mining found that the activity increased bribery: local officials began insisting on more bribes than previously once mines had begun operation (Kotsadam et al. 2015). Similar studies have found that preferential treatment over concessions and logging practice can be purchased by the payment of bribes to forestry regulators. Where corruption is widespread, government funds earmarked for conservation of natural resources are embezzled. That puts forest resources at a higher risk of overexploitation (Sundström, 2016).

Certain patterns of change in natural resource management can also increase the level of corruption. Exploration for natural resources in remote locations is open to corruption. Deep-sea drilling and mining for coal and gas in high altitudes are examples. These activities are also hard to monitor.

---

[225] Despite its name, the impact of petty bribery should not be thought to be petty. It can be pervasive, affecting citizens in their daily interactions with the state, their standards of living and well-being, as well as having a corrosive impact on growth, business operations, tax revenues, and ultimately the governance and regulatory environment of a country. Petty corruption, in terms of both money and economic distortions, may be as great, if not greater, than grand corruption (World Bank, 1997).

Chapter 7
Human Institutions and Ecological Systems, 1: Unidirectional Externalities and Regulatory Policies

Introduction

In this and the following two chapters, we create a language in which to study the three factors that were identified as comprising humanity's impact on the biosphere (the left-hand side of the Impact Equation in Chapter 4): population (N), the standard of living (y), and the efficiency with which we transform the biosphere's goods and services into the goods and services we produce and consume (α). Our aim is to explain the Impact Inequality.

Processes driving a wedge between our demand for the biosphere's goods and services and its ability to supply them without undergoing decline harbour *externalities*. These are the unaccounted-for consequences for others, including future people, of actions taken by one or more persons. The qualifier 'unaccounted-for' means that the consequences in question follow without prior engagement with those who are affected. Inefficiencies in the production, consumption and exchange of goods and services are an expression of externalities. Public policies can reduce (in some cases, eliminate) externalities. If the policies included redistributive measures, reduction of the externalities would release goods and services while harming no one. That, of course, is what is meant by 'inefficiency'. The colloquial term for externalities is *waste*,[226] as in the case of UK households who have been found to waste 9.5 million tonnes of food and drink per year, 70% of which could have been avoided, at a value of over £19 billion a year (WRAP, 2020).

It is common to read externalities as institutional failure, or more narrowly, as in the economics of climate change, as market failure (Stern, 2006); but that is merely to reword 'externalities'. In this and the following two chapters, we classify externalities and point to ways by which they can be lessened, that is, ways in which agents can be held accountable for their actions. A direct effect of a reduction in externalities is an increase in α. But that in turn has an effect on y, and in the examples we study in Chapter 9, it has an effect on future N as well; the factors embodying human impact (that is, N, y, and α) are not independent of one another.

The way the notion of externalities has been stated here could appear odd, on the grounds that our actions inevitably have consequences for future generations, who by the nature of things cannot engage with us. In fact, we engage with future people constantly, albeit indirectly. Parents care about their children and know that they in turn will care about their children, whose children will in turn care about their children, and so on. By recursion, thoughtful parents take the well-being of their descendants into account when choosing how much to save for their children and

---

[226] That waste almost surely comes allied to externalities is a general theorem and can be found in any microeconomics text. See for example Malinvaud (1972).

Chapter 7: Human Institutions and Ecological Systems, 1: Unidirectional Externalities

invest in them. In part, people do that directly, in part through the state. Engagement with future generations would be imperfect if parents make choices without adequate concern for their children, that is, if they discount the future well-being of their children at overly high rates. Externalities across the generations would be rampant in that case. We ignore that line of analysis in this and the following chapter, as the subject is taken up in Chapter 9. Here and in Chapter 8, we study why reasoned decisions at the individual level can result in collective failure, including why we should not expect even thoughtful parents to engage adequately with *others'* descendants, for example when it comes to emitting carbon, changing land use and damaging coastal zones. Indeed, we should not expect people to engage adequately with others when the effects of their activities are *non-excludable*, that is, it is not possible for people to pick and choose who is affected.

It is customary in economic reasoning to confine attention to externalities that travel in the material world. In fact, externalities are embedded in a larger space. The social world can be as powerful a carrier of externalities as the material environment. A common form in which they appear in the social world is in the way our relationships influence our preferences and wants (Chapter 9). In Chapter 9 we study externalities that shape our preferences and wants.

In this and the following chapter, we study externalities that travel in the material world. Two types of externalities may be contrasted: *unidirectional* and *reciprocal* (Figure 7.1). Unidirectional externalities (the subject of this chapter) are just that, unidirectional, where one agent (or a group of agents) inflicts an unaccounted-for damage or confers an unaccounted-for benefit on another (or others). An example of the former is a company discharging toxic chemicals into waterways; an example of the latter is a neighbour whose garden attracts bees that also pollinate one's own garden. Under reciprocal externalities, each party inflicts an unaccounted-for harm or confers an unaccounted-for benefit on all others in a defined population. An example of the former are the carbon emissions of every household (the population here is the entire world); an example of the latter is a community of beekeepers and apple growers. The usefulness of this seemingly arbitrary two-way partitioning of externalities will become apparent in this and the following two chapters.

Figure 7.1 Reciprocal and Unidirectional Externalities

In this chapter we first unearth the sense of collective failure in circumstances harbouring externalities. We then study unidirectional externalities and the institutional changes that could help to reduce them, if not eliminate them. The focus in Chapter 8 is on the use of common property resources. In Chapter 9 we study fertility and consumption behaviour and the externalities that both motivate our behaviour and are caused by them.

7.1 Property Rights and Wealth Distributions

Human activities involving the biosphere (in other words, *all* human activities) give rise to externalities because *property rights* to large segments of the biosphere are either weakly defined or inadequately enforced. And a common reason for the latter is that Nature is mobile. No one can contain the atmosphere they befoul, the soil they contaminate, the rivers they pollute. Moreover, the harms they cause are *non-excludable* (Chapter 1).

To illustrate, consider that the operating cost of an upstream logging company, which is the company's private cost, typically does not include the cost borne by inhabitants downstream, who now are subject to greater risk from floods (the latter is the externality). Moreover, no one downstream can be excluded from the additional risk (that is non-excludability). The heightened risk is a manifestation of an absence of property rights for downstream inhabitants; they are not compensated for the damage to their property. Chapters 1–5 provided a number of examples of harmful externalities that are a direct cause of biodiversity loss.

The connection between property rights and the distribution of wealth may seem obvious. If one group of people in an economy owns few assets while another group owns a large proportion of assets, the distribution of wealth in the economy is patently very unequal. In the presence of externalities, though, the link between property rights and wealth distributions is sufficiently subtle to go unappreciated. A distinction needs to be made between *polluters' rights* and *pollutees' rights*. Continuing the example above, under the latter, inhabitants downstream have a right to compensation from the logging company; under the former, inhabitants would be obliged to pay the logging company to reduce its activity. Box 7.1 considers an example from international trade that points to a subtle link between externalities and the international distribution of wealth.

By property rights we do not simply mean 'private' property rights, we include community and state property rights. At an extreme end are global property rights, a concept that is implicit in current discussions on climate change. If nations reach an agreement on the maximum they are to emit in the aggregate, then every nation has a right to emit only within the terms and conditions of the agreement. We discuss that case in Chapter 8.

Box 7.1
International Trade, Wealth Transfers, and the Character of Technological Change

Imagine that an upstream logging company exports its timber to a foreign country and does so without compensating inhabitants downstream. The timber is under-priced at the country's border because it does not include the cost borne by downstream inhabitants in the form of a heightened risk from floods. The timber is further under-priced because transportation of the product to the foreign country creates harmful externalities (carbon emissions), and that cost is also not paid for by the timber company. There are thus two elements of cost that are not included in the company's invoice: the cost borne by the inhabitants downstream and the cost inflicted on the world's population via the carbon emissions. The former element is a wealth transfer from downstream inhabitants to the importer; the latter element is a wealth transfer from the world's population to the timber company.

Chapter 7: Human Institutions and Ecological Systems, 1: Unidirectional Externalities

> Aside from the economic waste associated with externalities, this example has a feature that is rarely noted by proponents of free trade. As poor countries depend greatly on the export of primary products (coffee, tea, sugar, timber, fibres, palm oil, minerals), there is a hidden transfer of wealth from them to importing countries, many of which are rich (Chapter 13). If the importing country *is* rich, the wealth transfer is a redistributive insult, especially so because the hidden subsidy is paid by downstream inhabitants in the exporting country, who are often among the poorest communities there (e.g. small farmers and fishermen). The transfer of wealth remains hidden from the national accounts of both countries because national accounts do not record externalities (they are unaccounted-for consequences).[227]
>
> The example also has a general message. Modern consumption patterns, relying as they do on imported primary goods from distant parts of the world, are prone to being under-priced. And they are under-priced at source for reasons similar to the present example. When final products are under-priced, this provides people with an incentive to consume not only too much, but also to consume ecologically damaging goods. Moreover, research and development expenditure is directed towards producing new products and new technologies that are profligate in the use of primary products. This puts further pressure on the biosphere.

7.2 Externalities and Rights

Although the failure to protect property rights has traditionally been at the heart of the theory of externalities, the language of rights sits awkwardly there. We are speaking of fundamental rights, not rights that are assigned to people or organisations because they are instrumental in advancing the well-being of the people involved. But even fundamental rights need to be justified. As elsewhere in the economics of biodiversity, trade-offs have to be weighed if actions are to be judged. Rights short-circuit those complexities.[228]

Rights are peremptory, which is why they are problematic. One way to overcome the problem is to place them in a hierarchy. That was the conclusion Rawls (1972) famously reached when framing his principles of justice. But if note is taken of adverse externalities accompanying a person's actions, it is by no means clear whose rights are to trump. This is why the language of rights sits awkwardly in the economics of biodiversity.[229]

Perhaps the most striking – as well as the most sensitive – sphere where rights would seem to clash is reproduction. The 1994 International Conference on Population and Development reaffirmed the language of rights in the sphere of family planning and reproductive health. The Conference's conclusions read:

"Reproductive rights ... rest on the recognition of the basic right of all couples and individuals to decide freely and responsibly the number, spacing, and timing of their children, and to have

---

[227] Dasgupta (1990) and Chichilnisky (1994) drew attention to this distributive feature of contemporary international trade. Economists had previously pointed to economic exploitation as a driver of trade imbalance between economic centres and their peripheries (e.g. Furtado, 1964). The cause of the imbalance we are identifying here lies in the exploitation of the biosphere.

[228] Moral philosophers question the liberal use of the term 'rights' even in the United Nations' landmark 1948 Declaration of Human Rights. They ask where rights come from. See Blackburn (2003).

[229] In recognisably exceptional circumstances, governments resolve the dilemma by creating a hierarchy of rights. In the context of the COVID-19 pandemic, many governments have insisted that the right not to be infected by others trumps the individual right to do as we please when it comes to wearing face masks and a keeping safe distance from our friends.

information and means to do so, and the right to attain the highest standards of sexual and reproductive health." (UNFPA, 1995).[230]

The qualifier 'responsibly' could be read as requiring couples to take into account the adverse environmental externalities their reproductive decisions may give rise to, but that probably would be a stretched reading. Certainly, writings affirming the UN declaration have interpreted the passage and its intent more narrowly. For example, the fundamental right of individuals "to decide freely and for themselves whether, when, and how many children to have" is central to the vision and goals of Family Planning 2020 (FP2020, 2019). It is also pivotal in the reproductive health indicators of the UN's Sustainable Development Goals (SDGs). In this vision, information and other services pertaining to family planning and reproductive health are rights, as is choosing one's family size. But it is not clear that the two sets of rights have the same force, nor is it clear how they are to be weighed against one another should a choice have to be made at the margin between them.

In a world where the Impact Inequality holds, and holds strongly, it may seem reasonable to insist on the rights of future generations when an appeal is made to curb our impact on the biosphere. Sen (1982), for example, likened persistent pollutants to instruments of oppression: "Lasting pollution is a kind of calculable oppression of the future generation." But if additional births can be expected to contribute further to the discharge of persistent pollutants, why does a couple's reproductive rights trump the rights of future people not to be oppressed? That is the kind of ethical dilemma the language of reproductive rights misses.

That reproductive decisions may involve a clash of rights has not been self-evident to scholars. In a powerful essay that dismissed concerns on over-population, Bauer (1981: 61–64) wrote:

"The comparatively high fertility and large families in many LDCs (less developed countries) should not be regarded as irrational, abnormal, incomprehensible or unexpected. They accord with the tradition of most cultures and with the precepts of religious and political leaders ... Allegations or apprehensions of adverse or even disastrous results of population growth are unfounded. They rest on seriously defective analysis of the determinants of economic performance; they misconceive the conduct of the peoples of LDCs; and they employ criteria of welfare so inappropriate that they register as deterioration changes which are in fact improvements in the conditions of people."

One problem with Bauer's critique is that it gives the impression that societies in past eras were also characterised by large families. While fertility rates were high then, so were mortality rates. In fact, high fertility rates were a rational response to high mortality rates. The contemporary demographic problem in the world's poorest regions is that fertility rates remain high even though mortality rates have fallen considerably (UNPD, 2019b). The main problem with Bauer's critique, however, is that it does not acknowledge reproductive externalities.

Two categories of externalities are involved. One consists of the adverse externalities that are conveyed through the material world because the Impact Inequality today is large (Chapter 4). It is increasingly hard to argue that the vast quantities of produced capital and scientific and technological knowledge we will be bequeathing future people compensate for the vastly diminished biosphere we are leaving behind for them.

The other category consists of the externalities implicit in socially embedded preferences (Chapter 9). Bauer's critique does not acknowledge that individual households may themselves affirm that their reproductive behaviour falls short of what they would ideally favour because they are unable to coordinate their decisions with other households. As in every other field of personal choice, it should be asked whether

---

[230] Moral philosophers would argue that the evaluation of family planning programmes should include the quality of lives that will not be lived on account of the programmes. We avoid those difficult problems by assuming that thoughtful parents reach their fertility desires by taking into account the potential well-being of their offspring and, by recursion, the well-being of their dynasty. On this and related matters in population ethics, see Dasgupta (2019).

Chapter 7: Human Institutions and Ecological Systems, 1: Unidirectional Externalities

a collection of reasoned decisions at the individual level harbour collective failure. This is the central question raised by externalities, and it is particularly apposite in the case of adverse externalities and socially embedded preferences. That family planning services bring many benefits (such as improved health, education, income, and female empowerment) to those who make use of them has been documented repeatedly in recent years (UNFPA, 2019). The focus in this Review on externalities points to the fact that they bring benefits to others as well. Those additional benefits need to be included in the design of social policies. In what follows, we study the tools that can be used for eliminating the externalities that arise from our production and consumption of goods and services. Reproductive externalities raise deeper and more difficult issues, which may be why they have not been studied much in the literature on externalities. We present a few of the difficulties in Chapter 9. The subject remains unsettled.

7.3 Taxing and Subsidising Externalities

A striking feature of well-functioning markets is that they make people responsible to others for what they produce and consume. The qualifier 'well-functioning' is intentional and significant. Well-functioning markets ensure that people pay the social cost of the resources they use, which means that market prices correspond to *accounting prices* (Chapters 1 and 10). Well-functioning markets harbour no externalities because they 'internalise' potential externalities.

Thus, if q_i is the market price of asset *i* and e_i is the value of the externalities generated by the deployment of a marginal unit of *i*, then the asset's accounting price is:

$$p_i = q_i + e_i \tag{7.1}$$

If the market price is nil, as is often the case with global commons (Chapter 8), the asset's accounting price p_i is composed entirely of externalities (Equation (7.1)). When some economists say that their estimate of the social cost of carbon in the atmosphere is US$40 per tonne, they mean that in their view the accounting price of carbon is *minus* US$40 per tonne.

Why not create markets for externalities so as to eliminate them? Although you might expect economists to have explored that avenue first, this is not how they first studied externalities. In his great work on the economics of welfare, Pigou (1920) took the elimination of externalities to be the responsibility of government. He argued that if a factory spews industrial smoke into the atmosphere, the owner should be taxed at a rate equal to the damage suffered by the rest of society from a marginal unit of emission. That way the factory owner would face a price for his output that would include the damage it causes (i.e. market price minus the tax per unit of emission). In this example, the product's market price minus the tax per unit of output is its accounting price. The underlying idea is that because the smoke is a by-product of the factory's output, the tax would make output less profitable. The owner would reduce output, or install smoke scrubbers, or find some other means to reduce pollution. The Pigouvian tax alters his incentives in the right direction and by the right amount. Congestion charges in cities, noise charges at airports, and most famously today, a tax on carbon emissions are further examples of Pigouvian taxes. Box 7.2 provides the formal analysis.

Pigouvian taxes raise the question of what the government should do with the revenue. Suppose, to give an example, the government imposes a carbon tax but spends the revenue on general public services. As the benefits would be spread over the entire population, the tax would be regressive. That is because poor households spend a greater portion of their income than rich households on domestic heating and cooling, and on transportation. But if the revenue was returned on a lump sum basis, say equally to all households as a dividend, the tax-dividend policy would be progressive.

Pigouvian subsidies for beneficial externalities follow the same logic. Consider, for example, a landowner intent on converting a portion of his property into wilderness. Wilderness attracts pollinators, and that benefits neighbouring farmers. If the benefit is unaccounted for, it is a beneficial externality. The Pigouvian route, if we may use that expression, would be for the government to subsidise wilding, at a rate that equals the marginal benefit to farmers.

There is a natural inclination to think that Pigouvian taxes and subsidies are minor fixes for the gigantic set of problems we face on account of the enormous Impact Inequality we have created. A tax here and a subsidy there would certainly prove next-to-useless in closing the gap in the Inequality. But we should be imagining a socio-ecological world where *all* externalities are accounted for by the deployment of Pigouvian taxes and subsidies. Imagine for example that all countries taxed net carbon emissions at US$40 per tonne, or higher. That would be nothing short of a transformative change for our societies. Currently, environmental taxes are very low in OECD countries: only above 3% of GDP for six of the 37 countries (OECD, 2019d), despite the fact that a large proportion of economic activity causes environmental damage, involving externalities. Economies where the accounting prices of ecosystem services were reflected in their market prices would be markedly different.

We will have reasons for abandoning the idea of instituting Pigouvian taxes and subsidies for all externalities (Box 7.3), but they will be deep technical reasons, having to do with the inability of authorities to observe or verify who does what at each instant of time (Chapter 6). Creating protected zones for preserving biodiversity is a natural extension of Pigouvian reasoning, but that involves outright prohibitions, not taxes (Chapters 5, 18–19).

Box 7.2
Pigouvian Taxes and Subsidies

It is simplest to present the logic of Pigouvian taxes and subsidies by returning to the example Pigou himself used – that of a firm whose factory emits toxic fumes. We assume that, other than the harmful externalities the fumes give rise to, the economy does not suffer from any distortion. The model is timeless.

Let p be the price at which the firm sells its product. The cost of producing Q units of output is $C(Q)$, where $C(Q) = 0$, $dC(Q)/dQ > 0$, and $d^2C(Q)/dQ^2 > 0$. In words, production cost is zero if output is nil, and marginal cost of production is positive and increases with the volume of production. In the absence of regulation, the firm's profit function is $[pQ - C(Q)]$, which we write as $\phi(Q)$, as in Figure 7.2. That means the profit maximising output is the value of Q at which the marginal cost of production equals the product's price:

$$p = dC(Q)/dQ \qquad (B7.2.1)$$

(Equation (B7.2.1) can also be expressed as $d\phi(Q)/dQ = 0$).

Let Q be the solution. Q is the value of Q at the peak of the curve $\phi(Q)$.

Production creates toxic fumes as by-product. If the by-product matches output unit for unit, we do not need a separate symbol for it. Denote the damage to residents in the neighbourhood, expressed in monetary terms, by $D(Q)$. Reasonably, we assume that $D(0) = 0$, $dD(Q)/dQ > 0$, and $d^2D(Q)/dQ^2 > 0$.[231] In words, damage is zero if production is nil and the damage caused by a marginal unit of production is positive and increases with output. We now introduce a *regulator*, whose objective is social well-being, V, which in our simple example is the firm's profit minus the damage caused by the firm's production:

$$V(Q) = pQ - C(Q) - D(Q) \qquad (B7.2.2)$$

---

[231] The latter pair of assumptions would not hold if there is a volume of production at which the residents' state of health tips over to an unacceptable level. That would be akin to a threshold (Chapter 3). The Pigouvian tax scheme obviously does not allow the tipping point to be reached. We return to this point below in the text.

Chapter 7: Human Institutions and Ecological Systems, 1: Unidirectional Externalities

To make the idea of Pigouvian environmental taxes transparent, let us imagine that the economy is subject to a 'command and control' system of authority, in that the regulator can instruct the firm on how much to produce. Maximising $V(Q)$ with respect to Q then yields the socially optimal output level of production:

$$p = d[C(Q) + D(Q)]/dQ = dC(Q)/dQ + dD(Q)/dQ \qquad (B7.2.3)$$

Let Q^* be the solution of equation (B7.2.3). The regulator would instruct the factory owner to produce Q^*. It is simple to confirm that $Q^* < \underline{Q}$, which is what intuition demands: output should be reduced, as that is the only way to reduce pollution.

Notice that in deriving equation (B7.2.3) we have shifted the entire burden of agency from the factory owner to the regulator. The Pigouvian solution to the problem of externalities is instead to award agency over the choice of tax rate to the regulator but retain the agency over production decision to the factory owner. Formally, we have a two-stage game in which the regulator makes the first move by imposing a tax rate, followed by the firm having to choose its production level.

To see how the two-stage setting avoids command and control over production decisions, suppose the regulator imposes a tax per unit of pollution, t^*, equal to marginal damage at the socially optimum level of production, $dD(Q^*)/dQ$. Then the firm's profit function, net of tax, would be $pQ - C(Q) - t^*Q$. Profit maximising output would then satisfy the condition that the market price of the product equals the marginal cost of production plus the tax:

$$p = dC(Q)/dQ + t^* = dC(Q)/dQ + dD(Q^*)/dQ \qquad (B7.2.4)$$

It is simple to confirm that the solution of equation (B7.2.4) is Q^*. Ideally, the optimal tax revenue, t^*Q^* would be returned as a lump sum payment to the parties suffering damage. The latter would enjoy a surplus, by the amount $t^*Q^* - \int_0^{Q^*}[D(Q)]dQ > 0$. Notice that the regulator knows in advance what the firm's response would be for any choice of t. That means he can ensure any level of output by a suitable choice of t. To take an example, imagine that the damage caused by the pollution has a tipping point (Chapter 3), say $Q\tilde{\ }$, in that pollution levels in excess of $Q\tilde{\ }$ are deemed to be unacceptable. The damage function in this case would display a steep increase near $Q\tilde{\ }$, in the extreme a large jump at $Q\tilde{\ }$. It follows from equation (B7.2.4), the optimum pollution tax ensures that Q^* is less than $Q\tilde{\ }$.

Pigouvian subsidies for beneficial externalities are obtained by a mere change of signs. Suppose, as in the text below, a landowner converts a portion Q of his land into wilderness. Let $D(Q)$ now denote the benefit to neighbouring farmers from the wilding. Equation (B7.2.4) would now yield the Pigouvian subsidy to the landowner per unit of land wilded.

Figure 7.2 Net Profit Function of Polluting Firm

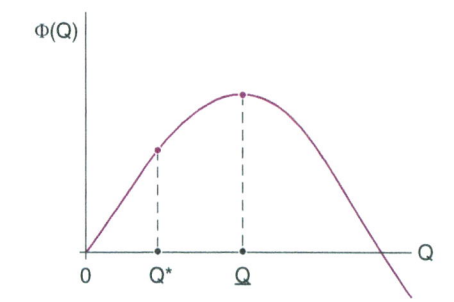

7.4 Quantity Restrictions

There is a problem with the way Pigouvian taxes have been formulated here so far. It presumes that the regulator knows everything he needs to know in order to calculate the correct level of taxes or subsidies. In our example, the regulator is assumed to have an assessment of the (expected) damage caused by the factory's emissions, but he needs to know the firm's production costs too. As a minimum we should imagine that the firm's owner (certainly the firm's manager) knows more about production costs than the regulator.

Box 7.3 shows how the information gap between the two parties can be taken explicitly into account by the regulator. The Pigouvian tax in that case is known among economists as a *second-best* tax.[232] The example in Box 7.3 shows that information gaps among the parties – differences among people in what they know or can know and in what they observe or can observe – is a reason good institutions are neither top-down nor bottom-up, but are polycentric arrangements where information flows every which way. The example also shows, however, that in the face of a possible breakdown of an ecosystem, it can be desirable to move out of a decentralised mode of governance to one where the government imposes *quantity restrictions*. In such cases, it can even be that the regulator finds it necessary to prohibit entry into an ecosystem.

Box 7.3
Rationale for Environmental Regulations

The model analysed in Box 7.2 assumes the regulator knows both the damage function $D(Q)$ and the cost function $C(Q)$. That is altogether too strong a requirement. The former problem can be removed without much fuss. It could be, for example, that the regulator holds reasoned beliefs about the harm factory smoke can inflict on people and that those beliefs can be summarised as a (subjective) probability distribution over an uncertain damage function (Chapter 5). In that case we would interpret $D(Q)$ as the regulator's *expected* damage function.

The problems that arise if the regulator does not know the firm's cost function are less easy to address. It may be reasonable to suppose that the factory owner knows the cost function (or at least knows a lot more about it than the regulator). In contrast to the firm's owner, who would know how to respond to a tax, the regulator would be uncertain of the firm's response. We therefore have a setting in which the participants hold different kinds of information. The regulator's task is to make the best of it.

To formulate the regulator's problem, let $\tilde{\varepsilon}$ be a random variable reflecting his ignorance of the firm's cost function. We may then represent his knowledge of the cost function as $C(Q, \tilde{\varepsilon})$. Imagine now that the regulator wants to maximise the expected value of the V-function in equation (B7.2.2). By assumption the factory owner knows the true value of $\tilde{\varepsilon}$, but the regulator does not. However, the regulator is able to calculate what the owner's response would be in every possible realisation of $\tilde{\varepsilon}$. That is because the regulator knows that if the realisation of $\tilde{\varepsilon}$ is ε, then the owner's profit function is $pQ - C(Q, \varepsilon)$. This gives rise to an interesting problem in incentives.

---

[232] The term 'second-best' misleads because economists call *any* policy that is built on asymmetric information among the relevant actors a second-best. But economists do not claim there to be a hierarchy of information gaps, nor that there is a third-best tax, a fourth-best tax, and so on.

Chapter 7: Human Institutions and Ecological Systems, 1: Unidirectional Externalities

> Suppose, as previously, that there is a tipping point $Q\tilde{\ }$. With a Pigouvian tax, the regulator would be able to ensure that the firm produces less than $Q\tilde{\ }$ only by setting a high tax rate. But from the regulator's point of view (expected value of V), a high tax would lead to an undue contraction of expected output. On the other hand, setting the tax rate too low would run the risk of an output response from the owner that exceeds $Q\tilde{\ }$. A balance has to be struck, and the way to do that is to find a non-linear tax schedule, that is a tax rate per unit output, $t(Q)$, that is not independent of Q. For example, a natural compromise is a Pigouvian tax that comes tied to a condition that output does not exceed a level that is chosen to be well below $Q\tilde{\ }$. That is like a tax schedule in which the tax rate increases sharply with output in the neighbourhood of the tipping point.[233]
>
> If there are several (possibly many) firms polluting the atmosphere, a mixed tax-regulation schedule suggests itself. The idea is to set a maximum for total emissions but then allow a quasi-market system to settle the emission allocation among polluters. A mechanism that has been put into place in some parts of the world for containing global carbon emissions is tradeable emission permits (Asian Development Bank, 2018; European Commission, 2020). The number of permits issued equals the upper limit on aggregate emissions, expressed say in tons of CO_2 per year. Firms may emit only an amount equal to the number of permits it holds. One variant would have the government sell the permits, the sales revenue serving as a tax the government keeps. Another variant would be to have the government issue the permits to firms and allow them to buy or sell permits in an open market. In either case the resulting market prices for permits simulate Pigouvian taxes on global emissions, subject to an upper limit to aggregate emissions.

There is a plethora of tradeable-permits schemes in place today, ranging from hunting and fishing to waste disposal and pollution (Tietenberg, 2003). If the permits are distributed free of charge, the rents are enjoyed by the polluters themselves; if the permits are sold by the authority, the revenue is enjoyed by the authority. Other than that distributional difference, the schemes are the same. Seasonal variations on year-round permits are an obvious modification when the activity is hunting or fishing. Box 7.4 illustrates the application of tradeable permits with the help of a scheme introduced in New Zealand's fisheries.

Prohibitions are frequently used as a device for *restoring* natural capital; for example, populations of species. The 1973 Endangered Species Act in the US is a well-known example (U.S. Fish and Wildlife Service, 1973). The Act instituted a Federal Law for protecting imperilled species. It was designed not only to prevent extinction but also to allow populations to recover to the point where prohibition would not be needed.

A further example is the 1982 moratorium on commercial whaling, established by the International Whaling Commission. Japan, Norway, and Iceland have continued commercial whaling under objection to the moratorium but issue their own quotas. There is overwhelming evidence that the moratorium and quotas have enabled whale populations to escape the 'tragedy of the commons'.

Pigouvian taxes and their extensions are a means for regulating our use of natural capital when that capital has instrumental value. But there are monuments and ecosystems that have cultural or religious significance to communities. Because the objects must not be desecrated, prohibition from any use other than as a cultural or religious object is the only instrument available to the regulator.

---

[233] This result, due to Weitzman (1974), is a simple illustration of 'mechanism design', in which the regulator designs a way to make the best use of differential information among the actors in the economy. The idea is to design a policy that would align the objectives of all the actors with the social objective. That way, every actor would have the right incentives to achieve the common good.

> **Box 7.4**
> **New Zealand's Tradable Permit Scheme for Fisheries**
>
> New Zealand has a tradable permit scheme for fisheries management, with around 100 species and 642 individual stocks under a Quota Management System (QMS). This type of instrument is not unique to New Zealand – other schemes exist, for example, in Canada, USA, Iceland, Australia – however, it is possibly the most extensive and successful of its kind in the world. For example, two international reports, by Kerr, Newell, and Sanchirico (2004), Alder and Pauly (2008) and Holland (2010), assessed New Zealand fisheries as world leaders in the management of marine resources. The objective for managing New Zealand's fisheries as stated in the 'Fisheries 2030' strategy (Ministry of Fisheries, 2009), is to "maximise benefits from the use of fisheries within environmental limits". The tradable permit scheme aims to maintain sustainable fishing yields, prevent overfishing and achieve cost-efficiency. The scheme has enabled New Zealand to increase the number of fish stocks close to, or above, sustainable limits and the value of the fishing industry.
>
> The scheme was introduced in 1986 when there was a widespread perception of overfishing in the industry. An aggregate Total Allowable Catch (TAC), by volume (e.g. tonnes), in a given zone was established for each stock. These were then divided into 100 million quota shares for each fish stock, which henceforth could be openly bought, sold or leased on a Quota Trading Exchange, with aggregation limits in place to ensure against monopolies developing in the industry. Each year the TAC is shared among the various categories of users: commercial fishing, recreational and other uses. Each quota owner receives the rights to an Annual Catch Entitlement. If commercial fishers do not have enough quota allowance for their catch, they face financial penalties, as a disincentive to overfish (Lock and Leslie, 2007).
>
> The QMS is widely viewed as successful in achieving economic and environmental objectives. In terms of environmental objectives, in 2016, of the 150 or so stocks which were assessed, 83% were above the 'soft' limit (deemed overfished), 94% were above their hard limit (biomass is below level at which a stock is deemed to be collapsed and fishery closure should be considered) (Hale and Rude, 2017). There are several success stories for individual fish stocks. For example, the hoki fishery received Marine Stewardship Council Certification in 2001 and was recertified as a sustainably managed fishery for a further five years in 2007 (Fisheries New Zealand, 2020). New Zealand has developed into a major fish exporter, with 95% of New Zealand's commercial catch and 75% of aquaculture production exported, contributing US$1.5 billion to New Zealand's economy (Williams et al. 2017).
>
> Although the QMS is widely seen as a success, they have raised problems. Concentrations in allocation of quotas have led to equity concerns within the industry. For example, as with other tradable permit schemes, the system can provide incentives to focus on quality over quantity ('high-grading') and overruns on catch limits due to bycatch.

7.5 Markets for Externalities

Well-functioning markets do not harbour externalities.[234] Property rights are so extensive that each potential externality is priced in accordance with the consequence to the person who is affected by it. Of course, who is affected by whose actions is itself determined by the prevailing system of property rights. It

---

[234] What we are referring to as 'well-functioning markets' are usually called 'perfectly competitive markets'. Malinvaud (1972) remains the outstanding text on the logic of perfectly competitive markets.

could be, for example, that the body responsible for a wetland, such as the local authority, is required by law to compensate anyone who suffers from the mosquitos that breed there (with the compensation calculated in terms of a price equalling the harm the person suffers from an additional mosquito). In that system, the authority would ideally also have the right to charge anyone who not only does not suffer from the mosquitos but instead enjoys the benefits that pollinators bring to their garden (the price being the benefit to the person from an additional pollinator).[235] But if property rights were allocated the other way round, someone who suffers from mosquitos would have to pay the authority for doing something about the nuisance.

It is a common observation that well-functioning market systems economise on information. The institution requires each individual to know only (i) her own mind and her abilities, (ii) the assets she has a right to, and (iii) the prices of goods and services. She does not need to know what is on other people's minds or their abilities, nor what they are doing. But that economy of information is made possible only by the enormous information burden on the institution itself – markets have to pool all information and embed them in prices. Prices are interpreted as an emergent feature of markets (Adam Smith called that the workings of the 'invisible hand'). You may possess information of a feature of the world that I do not have, but if I can infer that information from your actions correctly, then our collective actions will convey private information through the emerging prices. Box 7.3 showed the implications for resource allocation when that assumption is relaxed.[236]

Nevertheless, imagine that the acquisition of information is so cheap that markets are able to price externalities. Wouldn't *laissez faire* then be superior to an institution where regulators direct some behaviour? Although scholars in the economics of law have argued that *laissez faire* can match even well-chosen government regulations on environmental externalities, we show below that the belief is a stretch. This issue is at the heart of political philosophy, for it has to do with identifying the tasks citizens would assign to the state in a well-ordered society.[237]

7.5.1 Mutualism

In a paper that revived the study of externalities in contemporary economics, Meade (1952) considered a world of beekeepers and fruit growers. Bees pollinate fruit trees while fruit trees provide nectar and pollen to bees; the two populations raise one another's productivities. Ecologists call that kind of population dependence *mutualism*. Meade derived the structure of Pigouvian subsidies that would internalise the externalities, and he showed that beekeepers and fruit growers in the presence of those subsidies would be motivated to increase their stocks to the point where the mutualism was internalised.

In a symposium that questioned the need for a regulator, Cheung (1973) provided evidence that many places have flourishing markets that mediate the mutualism between beekeepers and fruit growers. Beekeepers sell the bees they raise to owners of orchards at what Cheung conjectured was probably the right price. He therefore argued that there is no call for market intervention – so long as property rights are well defined – markets can be relied upon to emerge.

The criticism was misplaced, however. Unlike fruit trees, bees are mobile. It is the essence of Meade's model that in contrast to fruit trees, which are a factor of production in beekeeping, bees provide

---

[235] That right is exercised through property taxes.

[236] For simplicity we are assuming that each person is negligible relative to the size of the economy. That means there is no scope for strategic behaviour. Grossman (1977, 1989) are the classics on prices as carriers of information. He focused on the (realistic) case where prices carry noisy information, but that enabled him to investigate the consequences if prices bring private information into the public domain with unerring accuracy.

[237] The modern classic on the question is Rawls (1972).

a service in which orchards produce fruit. For example, if an orchard owner in Meade's model increases all his inputs by 10% and the population of bees remains the same, the output of fruit in his orchard increases by 10%. But if the population of bees also increases by 10%, his fruit output increases by *more than* 10%. This excess in production is an externality Cheung's competing beekeepers would not be able to internalise. It is easy to confirm that the excess is a feature of the non-linearity in fruit production. The non-linearity arises because bees provide a service in which orchards can thrive.[238]

7.5.2 Prices for Externalities

Public goods are goods or services that are neither rivalrous (access to a public good by any one group of people has no effect on the quantity available to others) nor excludable (no one can be excluded from access to the good). As beneficial externalities are widespread in the production of public goods, markets would be expected to offer very little in the way of incentives for individuals in their private capacity to supply them (the benefit from supplying a public good would be shared with everyone, while the entire cost of production would be borne by the supplier).[239] This is why the common port of call for economists seeking a way out from a potential under-supply of public goods is the government.

In a pioneering work, Lindahl (1958, [1919]) presented the criterion government should use for calculating the quantity of a public good it should produce and the charges it should levy for it.[240] As the benefit people derive from a public good differs from person to person but the quantity enjoyed by all is the same (the latter is the non-excludability condition for public goods), the charge levied for a public good should ideally be person-specific. Someone deriving greater benefit than someone else would be charged more. The government's tax revenue would be the *sum* of those charges.[241]

It did not escape Lindahl's notice that the criterion he had derived for the government's use could be reinterpreted to construct an account of private provisions of public goods. We could imagine for example that private firms supply public goods and set a price for each person equal to what the government would have charged should it have been the supplier. Here again is an argument that there is no need for government to step into an activity that could be performed by the market.

But there is a problem with the argument. It is one thing for government to set differential charges for a public good. That is done all the time, everywhere, but on the basis of people's identifiable characteristics that have little to do with preferences as such. For example, discounts, usually differential discounts, are given in museums and parks for children, senior citizens, residents, and in some places, surcharges for foreign visitors. The authority's motivation is not a direct response to consumer preferences, but rather to the view that museums and parks are 'merit goods'. Private firms cannot be expected to do any better. If firms are to set Lindahl prices for public goods, they would have to set a price for each person that was tailored to the person's preferences. In the extreme, that would require a separate market for each person. Well-functioning markets in contrast require that many people participate in each market, for the reason that consumers need to be price takers if markets are to function well. Arrow (1970), who drew attention to this, pointed out that markets for externalities

---

[238] Production of final goods and services, Y, in the macroeconomic model in Chapter 4* resembles fruit production in Meade's orchards. In our macroeconomic model, the biosphere is the 'atmosphere' in which the global economy is embedded.

[239] Here we are talking of public goods, not 'public bads' (e.g. greenhouse gas emissions). The latter category suffers from the same problem but with the opposite sign (Section 7.3.3), but as we will see in Chapter 8, it harbours additional problems.

[240] See Samuelson (1954) for a concise formulation of Lindahl's construct.

[241] This is in contrast to goods that are rivalrous and excludable (they are called 'private goods'). In a well-functioning market system, everyone faces the same price for a private good and each consumes only to the point where the marginal benefit she derives from the good equals the price. The contrast is sharp: for a private good the price people face in a well-functioning market is the same and the sum of the quantities purchased equals the total amount supplied, whereas in the case of a public good, the charges people pay differ from person to person but they are all supplied the same amount.

Chapter 7: Human Institutions and Ecological Systems, 1: Unidirectional Externalities

would be so thin – in the extreme one person per market – that the 'invisible hand' would be unable to coax competitive prices from market competition.

7.5.3 Harmful Externalities and Non-Linearities

When externalities are harmful there are further problems for the price mechanism. Consider once again the factory emitting smoke. As the emission is a by-product of output, we may assume that the firm's profit increases with emission until reaching a maximum, from which point it declines with emission (Figure 7.2). In contrast, the damage experienced by a household in the neighbourhood increases with emissions.

Imagine now that the household has the option of moving out of the locality should the pollution level be too high. As moving out is costly, it serves as a fallback should the pollution level exceed the cost of changing residence. Let the fallback cost be D and let Q be the level of pollution at which the damage suffered by the household is D.

What would be a competitive market price for the pollution? Without loss of generality let the price be interpreted as positive if the polluter has to pay the pollutee and negative if the payment has to be made the other way around. If the price was positive, the household would demand an unlimited amount of pollution: the household's loss would be D but its gain would be the fortune it would make. But as the firm would be willing only to supply a finite quantity of pollution, the market would not clear.

On the other hand, if the price was either zero or negative, the household's demand for pollution would be zero, whereas the factory owner would want to produce a positive amount. Once again, the market would not clear, meaning that the market mechanism fails to perform at all.[242]

7.5.4 Bargaining Over Externalities: the Coase Theorem

Traditionally, the argument for *laissez faire* did not take prices to be at the heart of market capitalism, it was based instead on the fruits of voluntary exchange among free citizens. In one of the most influential papers ever to have been published in law and economics, Coase (1960) argued that people would be able to conduct mutually beneficial exchanges without the visible hand of the state, even in the presence of externalities, provided they face no constraints on the ability to bargain and contract. If people recognise that there are mutually beneficial agreements that, if carried out, would eliminate an externality, they would have an incentive to bargain their way to make those agreements, perhaps sealing them in the form of legal contracts. It would seem to follow that, even if the price system was to fail, there is a case for *laissez faire*.

The problem with Coase's argument is that it is built on the requirement that every externality is excludable, in the sense that the agent giving rise to it has control over who is or who is not affected by it. To see what goes wrong with non-excludable externalities, consider an economy where every household emits toxic fumes from its kitchen. No household can choose to restrict the emission to any one group of households, meaning that the fumes are a 'public harm'.[243]

Reducing emissions is not costless, so if households acted on their own, they would all enjoy the benefits from a reduction in emissions, but each household's contribution to that reduction would be borne entirely by it. The dilemma facing the households is to find a way to reduce every household's emission. The argument for *laissez faire* in this context is that if households come together to negotiate an

---

[242] The example is taken from Starrett (1972), who produced a general theorem regarding the non-existence of a price system that would internalise harmful externalities.

[243] The example is by no means whimsical. For example, countries and regions in East Asia, South-east Asia and South Asia suffer from the highest levels of air pollution (IQAir, 2020). This environmental tragedy is in part caused by household emissions of soot.

agreement to reduce emissions, in which each household agrees to undertake a specific reduction in exchange for all other households doing the same, a mutually beneficial reduction would be attainable.

There is a problem with this argument. Because participation is voluntary, any given household, call it A, would calculate that if all other households were to negotiate an agreement, it would be better off not to participate. By staying out, A would enjoy fully the benefits of the reduction (that is where the non-excludability of fumes comes in) while not incurring a cost. In short, it pays A to free-ride on others. Because all other households reason in the same way, there will be no pollution-reduction agreement after all. This is the classic argument embodying the 'tragedy of the commons' (Chapter 1) and illustrated by the Prisoners' Dilemma (Box 6.2). Households would thus realise that they have to call upon the government, or some other authority with coercive power, to impose a mechanism by which it will be in the interest of each household to reduce its emissions by an amount they would have liked to have agreed upon but were unable to implement. The environmental regulation in Box 7.3 is an instance of such a mechanism.[244]

7.6 Payment for Ecosystem Services

The idea underlying Pigouvian subsidies has been deployed in recent years as measures to finance the conservation and restoration of Nature (Chapters 18, 19 and 20). Under the name *payment for ecosystem services* (PES), such payments are becoming a familiar tool for conserving and restoring ecosystems and the services they provide.

The underlying idea is simple enough (Section 7.2). For example, the Amazon rainforest is recognised as the Earth's lungs, so if the government in Brazil is convinced that deforesting the Amazon is necessary for economic growth in the country, should the rest of the world not pay an annual fee to Brazil for enabling the forest to recover? If a wetland on my range is a sanctuary for migratory birds, should bird lovers not pay me not to drain it for other use?

As the questions suggest, PES is based on the principle that beneficiaries of ecosystem services should pay to conserve and restore them. The institution translates externalities into financial incentives for local actors so that they provide the ecosystem services enjoyed by the beneficiaries (Chapter 20). PES systems have been much influenced by the fact that landowners manage their property in ways that are not necessarily (perhaps not even usually) consonant with the provision of ecosystem services. Modern agricultural practices are a notable example. Designing PES systems therefore grapple with finding ways in which landowners are to be compensated for providing the services (Engel, Pagiola, and Wunder, 2008).

Institutions that define PES programmes are often thought of as being 'markets', but the expression should be used with care. They are not markets in the sense that was explored in Section 7.3. The payments instituted are more like 'administered prices' based variously on the cost of maintaining the ecosystem service, limiting entry to the ecosystem so as not to have overcrowding, and so on. PES programmes differ in the types and scale of the demand for ecosystem services, the payment source (i.e. who makes the payment), the type of activity for which payment is made, the measure used to judge the quality of the service for which payment is made and of course the size of the payment. The effectiveness of a PES depends crucially on the programme design.[245]

---

[244] Baliga and Maskin (2003) have presented an ingenious scheme for solving the incentive problem in the example in the text but with added realism that the government does not have information about either households' aversion to pollution or the cost it faces for reducing its emission. Box 7.3 set out a vastly simplified case where (i) pollution has a single source (the factory), and (ii) the government maximises the expected value of the losses suffered by all households. Notice that government regulation there also solved a Prisoners' Dilemma that Coasian bargaining would not be able to resolve and for the same reason as in the text above.

[245] See Engel, Pagiola, and Wunder (2008), Jack, Kouskya, and Simsa (2008), and Pattanayak, Wunder, and Ferraro (2010) for reviews of the various PES systems that have been instituted over the years.

Chapter 7: Human Institutions and Ecological Systems, 1: Unidirectional Externalities

As the tropics are rich in biodiversity and many of the world's poorest countries are in the tropics (Chapter 14), it has been a natural thought to build PES programmes around poor communities. A PES system in which the state plays an active role is attractive for wildlife conservation and habitat preservation. Property rights to grasslands, tropical forests, coastal wetlands, mangroves and coral reefs are often ambiguous in developing countries. The state may lay claim to the assets ('public' property being the customary euphemism), but if the terrain is difficult to monitor, inhabitants will continue to reside there and live off its products. Inhabitants are therefore key players. Without their engagement, the ecosystems could not be protected. Certain natural 'wonders' attract a significant number of tourists on a regular basis (one may think of the National Parks in North America or the jungles of Borneo as examples). An obvious thing for the state to do is to tax tourists and use the revenue to pay local inhabitants for protecting their site from poaching and free-riding. Local inhabitants would then have an incentive to develop rules and regulations to protect the site. An alternative would be to hand over the right to charge tourists to the local inhabitants themselves.

Although PES programmes are not markets in the usual sense, they are not Pigouvian schemes either. They resemble Pigouvian subsidies only in the sense that providers of ecosystem services are compensated; the compensation is not necessarily made by government. In one variant, as in the case of the Communal Areas Management Programme for Indigenous Resources (CAMPFIRE) in Zimbabwe (Box 7.5), the state awards inhabitants of an ecosystem (e.g. an open range in the savannah) the right to charge for the services it provides those who are interested (e.g. safari tourists). In another, variant citizens forming an NGO contribute to compensate landowners for providing an ecosystem service (Box 7.6). Then there is a variant where the state pays restoration costs for the benefit of citizens. The now-revived Ganga Action Plan in India to restore the river Ganges is such an example (Box 13.7). In still other variants, a concerned citizen initiates a Trust to which individuals and government contribute so as to restore an ecosystem of value. The urban valley in Wellington known as Zealandia is an example (Chapter 21). And then there are examples where a farmer unilaterally 'wilds' her land to provide ecosystem services such as carbon capture and wildlife experiences, financed in part by government subsidy and in part by tourists.[246] In each of these examples, it is the beneficiary who pays for the ecosystem service.

Box 7.5
Payments for Wildlife Services in Zimbabwe

In the 1960s, wildlife in Zimbabwe was threatened by land-use change. Agricultural expansion was destroying habitats and corridors. Wildlife was the property of the state, so farmers did not benefit from its preservation. Often, the wildlife were pests. The 1975 Parks and Wildlife Act recognised wildlife protection to be a form of land use. The Act granted farmers the right to derive economic benefits from wildlife. CAMPFIRE was initiated in 1989 to aid inhabitants, who were granted authority over the wildlife in their own land (Frost and Bond, 2008). The programme's goal was to enable communities to benefit from ecotourism and safari hunting. Initially, the programme targeted sparsely populated, communally owned areas in the periphery of official conservation areas where wildlife was abundant.

While individual landowners were given responsibility for, and benefits from, wildlife on their land, authority over communal land was delegated to Rural District Councils (RDCs) which were supposed

---

[246] In her best-selling book, *Wilding*, Isabella Tree provides a vivid account of the way she transformed her agricultural farm into a nature reserve (Tree, 2018).

> to devolve rights and benefits further to communities. In due course, village level institutions such Ward Wildlife Committees were established for managing communal lands. It would appear though that these complex institutional relationships introduced economic distortions into the system (Frost and Bond, 2008).
>
> Safari hunting provided the bulk of CAMPFIRE income (Frost and Bond, 2008). The most common source of revenue was the sale of leases by the RDCs to the safari operators who delivered tourism services. The RDCs determined the scale and the location of the safari operation, and in turn paid a part of the revenues collected to local communities (around half of all revenues).
>
> WWF produced a guide on 'Marketing Wildlife Leases' for the RDCs including advice on postal tenders, auction design, contract structure and marketing. WWF (1997) provided evidence that the RDCs were able to raise far more revenue from safari leases by using auctions and competitive tendering than by non-competitive allocation. For example, once competitive auctions were introduced, lease prices in Hurungwe District rose from Z$172,000 to Z$654,000.
>
> Around a fifth of CAMPFIRE revenue was allocated to wildlife management (Frost and Bond, 2008). Several species, such as elephants, increased in number. Unfortunately, financial irregularities were detected by foreign donors, but although foreign contributions to the programme ceased in the early 2000s, CAMPFIRE remains in operation (Tchakatumba et al. 2019).

Today there are literally hundreds of PES schemes round the globe (Salzman et al. 2018). China, Costa Rica, and Mexico, for example, have initiated large-scale programmes in which landowners receive payment for furthering biodiversity conservation, carbon sequestration, landscape amenities and hydrological services. These are multiple objectives, and in practice a PES system may have to be designed to target more than one. That creates problems. A system could aim to reduce fragmentation of an ecosystem but has to check that it does not lead to an increase in disease transmission among populations (Horan, Shogren, and Gramig, 2008); it could aim to improve a habitat for birds even while increasing the watershed's supply of other ecosystem services (Asquith, Vargas, and Wunder, 2008); its purpose could be to increase both carbon sequestration and hydrological services (Pagiola, 2008); it could be designed to reduce local poverty while promoting conservation (Pagiola, Rios, and Arcenas, 2008); and so on. Multiple goals add to design problems.[247]

There are two aspects of special interest in PES systems in low income countries. One is its contribution to poverty reduction, the other to biodiversity conservation. Unfortunately, the findings are mixed on both counts (Zilberman, Lipper and McCarthy, 2008). There are situations where the system would be bad for poverty reduction and distributive justice. Many of the rural poor in low income countries enjoy Nature's services from assets they do not own (Chapter 14). Even though they may be willing to participate in a system of property rights in which *they* are required to pay for ecosystem services provided by landowners (Pagiola, Rios, and Arcenas (2008) reported in their careful study of a silvopastoral project in Nicaragua that they did), it could be that the economically weaker among them are made to pay a disproportionate amount. Some may even become worse off than they were previously.[248] One could argue that in those situations the state should pay the resource owner instead, using funds obtained from general taxation. As Bulte et al. (2008) observe, who should pay depends on the context.

---

[247] Perrings (2014) provides a fine, extended discussion of differences among PES systems.

[248] How this may happen is shown in formal terms in the Annex to Chapter 8.

In a penetrating study of the logic and practice of PES, Pattanayak, Wunder, and Ferraro (2010) noted four pervasive problems with PES for Nature conservation in the tropics. First, problems of moral hazard and adverse selection are prominent. For example, it can be that a landowner is paid to protect an ecosystem service that he would have protected in any case for his own benefit. Second, restricting the use of an ecosystem service may lead to greater ecosystem destruction not covered by the PES (the substitution effect). Third, because payments are based on the improvements to ecosystems, landowners have an incentive to degrade them in anticipation of a PES. Fourth, as noted in Chapter 2, ecosystem functions are not the same as ecosystem services. So even if a PES leads to additional forest conservation, it may not amount to an increase in ecosystem services.

Chapters 18 and 19 explore biodiversity conservation and restoration in greater detail, and Chapter 20 looks at financing that investment in Nature. Here we have studied the logic underlying PES as an institution and the underlying difficulties in designing them.

Box 7.6
Auctions for Providing Temporary Habitat to Migrating Birds

During their annual cycle, migratory birds rely on habitats across vast areas, sometimes spanning countries and continents. However, as habitats for bird populations are being degraded, even destroyed, in many parts of the world, only around a tenth of migratory birds can rely on feeding and breeding in protected areas throughout the year (Reynolds et al. 2017).

The Central Valley of California is a flyway for American migratory birds such as waders (Reynolds et al. 2017; Golet et al. 2018). Unfortunately, over 90% of Californian wetlands have given way to agriculture and urbanisation. Rice fields in Sacramento Valley, however, provide an opportunity to restore temporary habitats for migratory birds. Rice fields are flooded during autumn and winter to decompose rice stubble. However, flooded rice fields provide potential food resources and rest areas for shorebirds. The conservation benefit of these temporary wetlands depends on when and where they appear in order to coincide with migration. For example, although wetlands are reasonably plentiful in the winter, they become inadequate by spring. Using a statistical model and satellite imaging, researchers were able to identify that the greatest additional benefit of wetland enhancement was between February and April. But because fields are used for commercial agriculture, it was not clear whether temporary habitat could be provided at a reasonable cost.

In 2014, The Nature Conservancy (TNC) ran a 'reverse' auction to provide incentives to rice farmers in the Sacramento Valley to flood their fields during bird migration (Reynolds et al. 2017; The Nature Conservancy, 2018). In order to maximise the conservation benefits of field flooding, TNC asked farmers to submit bids at a specified shallow depth for four, six, or eight weeks starting on 1 February. In addition to the duration, farmers were asked to state a flooding area and location as well as a price. TNC received 55 bids covering almost 50 km². In order to select winning bids, TNC did not simply pick the bids with the lowest cost per square kilometre. Instead, TNC used its statistical model in order to 'score' the ecological value of each bid measured in shorebird abundance and surface water availability. The bids which achieved the highest ecological scores per dollar were selected as winning bids. 80% of the bids were winning bids and the average cost of winning bids was 60% lower than that of rejected bids, suggesting that the auction was competitive. The costs of temporary habitat compared favourably to buying at market rates and conserving land in perpetuity in the same areas even at low discount rates.

The auction had a significant impact on bird ecology. Shorebird species richness was more than three times greater and average shorebird density was five times greater in fields that were winners in the auction than in similar non-participant fields (Reynolds et al. 2017).

Chapter 8
Human Institutions and Ecological Systems, 2:
Common Pool Resources

Introduction

In the absence of collective action, the use of the biosphere's goods and services gives rise to an important class of *reciprocal externalities*. The externalities are most powerful when access to a resource base, which may be an entire ecosystem, is unrestricted. Today, the most prominent among *open access resources* are the atmosphere as a sink for gaseous and particle emissions, and the oceans beyond the 200-mile exclusive economic zones (EEZs) of nations.[249] Anthropologists have discovered, however, that ecosystems with small geographical reach, such as village woodlands and ponds, are usually neither private property nor state property, but are instead communal property. In this chapter, we use the conceptual apparatus that was developed in Chapter 6 to provide a sense of the way communities in various parts of the world have tried to manage their local ecosystems. By so doing, we will gain an understanding of the successes and failures of societies to live within their local resource base. Chapter 8* constructs a bare-bones model for studying the relative efficacy of alternative institutional arrangements for managing ecosystems.[250]

8.1 Open Access Resources

Gordon (1954: 135) famously remarked that a resource that is everybody's property is nobody's property. Using the example of fisheries, he showed that the scramble to extract open access resources dissipates all potential rents from them; and he used the dissipation of rents to explain why fishermen working on open waters are (or *were*, when Gordon wrote on the matter) usually poor. Hardin (1968) even more famously called the rent dissipation associated with open access resources the 'tragedy of the commons'. And he described the situation thus:

"Picture a pasture open to all. It is to be expected that each herdsman will try to keep as many cattle as possible on the commons ... As a rational being, each herdsman seeks to maximise his gain. Explicitly or implicitly, more or less consciously, he asks, 'What is the utility to me of adding one more animal to my herd?' ... Adding together the component partial utilities, the rational herdsman concludes that the only sensible course for him to pursue is to add another animal to his herd. And another; and another ... But this is the conclusion reached by each and every rational herdsman sharing the commons. Therein is the tragedy. Each man is locked into a system that compels him to increase his herd without limit – in a world that is limited. Ruin is the destination toward which all men rush, each pursuing his own best interest in a society that believes in the freedom of the commons. Freedom in a commons brings ruin to all." (Hardin, 1968: 1244)

---

[249] As elsewhere in the Review, we use the terms 'resources', 'resource base' and 'ecosystem' interchangeably where there is no risk of confusion.

[250] This chapter draws on Dasgupta (1993, 2000, 2010).

Chapter 8: Human Institutions and Ecological Systems, 2: Common Pool Resources

There is poetic licence here. Animals are not free, so even if the field is open to all at no charge, no herdsman would seek to introduce an unlimited number of animals into it. Even ignoring that licence, the Gordon-Hardin thesis on open access resources does not apply everywhere, nor does it apply at all times. The context matters. When the population of harvesters is small relative to the personal cost of harvesting an open access ecosystem, then other things equal, the ecosystem has no social scarcity value. Even if the community of harvesters were to come together to negotiate a cooperative harvesting policy, they would discover that the rent they should ascribe to the ecosystem is zero. Freedom in the commons in those circumstances does not give rise to externalities. One can go further and say property rights do not matter there. However, when population is large relative to harvesting costs, other things equal, people draw excessively on ecosystems to which access is unrestricted, and that means institutions *do* matter there.[251]

The theorem is instructive because it says that in centuries past, when global population relative to harvesting costs was small, the biosphere in the aggregate was wholly undisturbed by the human presence. Of course, *local* ecosystems are known to have been harvested excessively, even to an extent that communities dependent on them collapsed (Chapters 0 and 6); but that only goes to show that what matters is the size of the community's population relative to the costs involved in harvesting from the local resource base. As was observed in Chapter 4, humanity's demands for goods and services from the biosphere began to exceed the biosphere's ability to supply them on a sustainable basis only recently, perhaps not more than 100 years ago.

Box 8.1
Biosphere as the Common Heritage of Mankind

The idea of global property rights to untapped resources in the biosphere that had not to date been enclosed has had a troubled history. Mutual benefits of cooperation are easily recognisable and implementable in activities such as the use of airwaves and navigation in air and on the seas, and on such matters as units in which to measure time, weights and temperature. If country A refuses country B permission to use its air space, country B can retaliate by refusing A permission to use *its* airspace. And as agreements over measuring time, weight and temperature solve problems of coordination, there is little cause for conflict. Humanity has profited greatly from such agreements.

The UN Convention on the Law of the Sea went a different route by creating EEZs for coastal nations. An EEZ for a coastal nation is a band extending 200 nautical miles from its coastline in which it has exclusive rights to explore and use marine resources, including energy from wind and water. Beyond the EEZs, the oceans are common property of all nations. But nations that have technological or locational advantages over particular geographical terrains do not have the necessary incentives to accept the idea of global property rights to them. Russia has a locational advantage in exploiting sub-soil resources in the Arctic, which is why it rejects the idea of an Arctic Common. Historically, there has usually been a scramble (often a conflict) over natural resources, with the winner taking the most. The Amazon rainforest, for example, has the features of a global public good, but because countries have sovereignty over territories that cover the rainforest, it is not a global common. As it is recognised as the Earth's lungs, it would seem right that all countries pay an annual preservation fee to the nations that have sovereignty over the rainforest. No such agreement appears to be on the horizon.

---

[251] The theorem is stated and proved in Dasgupta, Mitra, and Sorger (2019).

Chapter 8: Human Institutions and Ecological Systems, 2: Common Pool Resources

> That humanity has collective responsibility over the state of the world's oceans was recognised in 1970, when the UN General Assembly issued a Declaration of Principles governing the use of the seabed in areas beyond national jurisdiction (UNGA, 1970). The Declaration accepted that the seabed is a 'common heritage of mankind'. One motivation behind the Declaration was the prospect of mining manganese nodules from the deep seabed. Nothing came of it, in part because the required technology proved too costly.[252]
>
> Not much has come either of the Aichi Biodiversity Targets, agreed in the UN Convention on Biological Diversity (CBD) in 2010. The 194 signatories to the agreement identified 20 conservation goals to safeguard global biodiversity by 2020. Unfortunately, the Aichi Targets were neither integrated with the UN's Sustainable Development Goals (SDGs), nor were they funded adequately. Moreover, in Chapter 4 (Box 4.6) it was shown that the SDGs themselves had not been subjected to an adequate screening for viability.
>
> Demands on the biosphere are further amplified by the common practice among national governments of actively subsidising the use of natural resources. Subsidies on farm acreage and on inputs in agricultural inputs (fertiliser and energy subsidies), are the form most commonly noted, but governments are known to offer fisheries subsidies and favourable terms to themselves and to private companies for logging and extracting minerals. Broadly speaking, government subsidies target natural resources that produce provisioning goods (Chapter 2). Estimates are hard to compile, but global subsidies for energy, agriculture, water, and fisheries are conservatively in excess of US$4–6 trillion annually (Andres et al. 2019; Coady et al. 2019; OECD, 2019b; Sumaila et al. 2019). Rents on the use of those resources are therefore *negative*, meaning that we are encouraged to exploit them even more profligately than open access resources. Environmental subsidies are discussed further in Annex 8.1.

8.2 Common Pool Resources (CPRs)

The size of an ecosystem matters also for another reason. The atmosphere as a sink for pollution embraces all humanity. In contrast, a grazing field is typically contained within the perimeters of a village's jurisdiction. It is a lot easier to institute fines for the overuse of a village grazing field than it is for overuse of the atmosphere as a sink for pollution. Fewer jurisdictions are involved in negotiations in the former case and monitoring people's activities is lower. Taken together, the two features suggest that, as the basis of cooperation, 'mutual enforcement' would be more reliable than 'external enforcement' (Chapter 6).

We are talking here about the role of social norms in sustaining cooperation. Hardin could have used as his example the atmosphere as a sink for carbon emissions; alternatively, he could have used Gordon's observations on marine fisheries. If he had done either he would have nailed his case – poetic licence excluded. Instead he had turned to grazing fields for his illustration. It may be that he had open ranges in mind, but anthropologists, political scientists and environmental activists had observed a different world from the one he had drawn. They had studied village economies, which are confined in space. Working at the fringes of official development economics, they discovered that spatially confined ecosystems such as woodlands and water sources are often common property, but they are regarded as property only of the village community. Access to them by outsiders is not permitted without the community's consent. To contrast spatially confined common-property resources from open access

---

[252] Shackelford (2009) dissects the problems that the idea of a common heritage of mankind has runs into over the decades.

resources, the former are called *common-pool resources,* or CPRs (Ostrom, 2010a). A rich and striking literature has found that communities restrict their own use of CPRs using a wide range of measures, held together by norms of conduct. The norms are effective because community members have an incentive to impose them on one another (Chapters 6 and 7). Hardin had not appreciated that there are CPRs which are not open to all, which is why his analysis came under criticism (Feeny et al. 1990; Ostrom, 1990; Colchester, 1994; Baland and Platteau, 1996).

The literature on CPRs is extensive and rich. Scholars have uncovered socio-ecological details that matter. However, although the literature on CPRs is long on empirical studies, it is somewhat short on the analytical basis of those studies. While details matter, there are features in many of the studies that are common. So, after reporting a sample of the empirical literature, we present a model in Chapter 8* of a grazing field in which we are able to formulate alternative institutions. That institutions matter is a truism, but it is useful to have a formal language in which to understand alternative institutions. The model has been so constructed with that in mind.

8.3 CPRs and the Poor World

The place of CPRs in the lives of the world's rural poor is central to the economics of biodiversity. In an early publication, Falconer and Koppell (1990) recorded the major significance of 'minor' forest products among people with low incomes in the humid forest zone of West Africa. They also recorded the many different ways people there have coped with the decline in access to those products during political upheavals. In a study in the rainforest of Chiapas, Mexico, López-Feldman and Wilen (2008) found that non-timber products there are extracted mainly by the poor. They attributed that in part to the fact that extraction involves search and that the opportunity cost of time (relative to the value of non-timber products to the extractor) is low among those with low incomes.

The Millennium Ecosystem Assessment (2005a-d) and IPBES (2019a) are comprehensive studies on the state of the natural resource base in poor regions. They confirm that the world's poorest people live in especially fragile natural ecosystems. When wetlands, inland and coastal fisheries, woodlands, forests, estuaries, village ponds, aquifers and grazing fields are damaged – owing to agricultural encroachment, nitrogen overload, urban extensions, large dams, resource usurpation by the state, open access or something else – it is the rural poor who suffer most. Frequently, there are no alternative sources of livelihood. In contrast, for eco-tourists or importers of primary products, there is something else, often somewhere else, which means there are alternatives.

As a proportion of total assets, CPRs range widely across ecological zones. In India, they are most prominent in arid regions, mountain regions and unirrigated areas; they are least prominent in humid regions and river valleys (Chopra, Kadekodi, and Murty, 1990; Agarwal and Narain, 1992). In low income countries, CPRs include grazing fields, threshing grounds, swidden fallows, inland and coastal fisheries, rivers and canals, woodlands, forests, and ponds and tanks (or artificial ponds).

Why did communities not permit themselves to divide their local ecosystem and turn them into private properties? One reason would have been the need to pool risks. Woodlands, for example, are spatially heterogeneous ecosystems. Microclimate matters, as does variation in soil quality. In some years, one group of plants bears fruit in one part of a woodland, in other years some other group of plants, in some other part, are fecund. Relative to average output, fluctuations could be presumed to be larger in arid regions, mountain regions and unirrigated areas. If a woodland were to be divided into private parcels, each household would face greater risks than it would under communal ownership and mutual-regulation. The reduction in individual household risks may be small, but as average incomes are very low in Indian villages, household benefits from communal ownership and mutual enforcement of agreements (Chapter 6) could be expected to be large. Communal ownership helps to pool risks.

An immediate corollary is that income inequality is smaller in those locations where CPRs are more prominent. Aggregate incomes are a different matter, though, and it is the arid, mountainous and unirrigated landscapes that are the poorest. However, the dependence on CPRs even within dry regions declines with increasing incomes across households. Theory predicts it and case studies confirm it (Jodha, 1986, 2001; Cavendish, 2000).

Where users are symmetrically placed, distributions would be expected to be symmetric – a subtle matter to devise if the resource is heterogeneous. Rotation of access to the best site is an example of how that can be achieved. It is often practised in coastal fisheries, fuel reserves in forest land, and fodder sites in the grasslands and mountain slopes (Netting, 1981; Baland and Platteau, 1996). Rotation enables users to get a fair share.

It would be possible in principle for the community to parcel out the resource as private property and let households establish a mutual insurance scheme. But that move would jeopardise cooperation in other activities, for at least two reasons. First, cooperation appears to be habit forming (Chapter 6), which means dispensing with cooperation in any one activity could lead to a weakening of cooperation in other activities. Second, cooperation is more robust when sanctions for opportunism in any one venture include exclusions not only from that venture, but also from other collective ventures (Box 6.2). Abandoning cooperation in one field of activity thus reduces the robustness of cooperation in other fields of activity. This fact is an implication of the theory of games. It explains why economic relationships are so frequently tied to one another (Dasgupta, 2007).

Local ecosystems are frequently CPRs also because animals, birds and insects are mobile. Mobility integrates an ecosystem's components. A coastal ecosystem can hardly be split into bits: fish swim. In Chapter 2, it was also noted that the productivity of ecosystems is greater than the sum of the productivities of their spatial parts. Ecosystems therefore have an element of indivisibility to them. But even if it was decreed that no portion could be converted for another use, parcelling ecosystems into private bits would be inefficient because of the externalities that would be created by mobile components. Communal ownership internalises those externalities. Admittedly, private monopoly would avoid the externalities, but it would grant far too much power to one person in the community.[253]

Agricultural land, especially in densely populated areas, is a different matter. Both labour and capital are critical inputs in production. Investment can increase land's productivity enormously. Agricultural land as CPRs would be subject to serious management problems, including those due to the temptations to free-ride on investment costs. The lack of incentives to invest and innovate would lead to stagnation, even decay. The experience with collective farms in what was previously the Soviet Union testifies to that. Those regions of sub-Saharan Africa where land is, or was until recently, held by the kinship were exceptions, but only because land was plentiful in the past and because poor soil quality meant that land had to be kept fallow for extended periods. Of course, it may be that agricultural productivity remained low there *because* land was held by the kinship, not by individuals. As elsewhere in the social sciences, causation typically works in both directions.

Are CPRs important to rural people? In a pioneering study on the significance of CPRs in a sample of semi-arid districts in Central India, Jodha (1986) reported that the proportion of income among rural families that is based directly on CPRs is 15–25%. Cavendish (2000) arrived at even larger estimates than Jodha from a study of villages in Zimbabwe; the proportion of income based directly on CPRs was found to be 35%, the figure for the lowest income quintile being 40%.

Marine fisheries are a major source of food and income among coastal communities. Some 250 million of the world's poorest people depend on coastal CPRs (Berkes et al. 2001). At a country level, some

---

[253] Monopoly of access by sub-groups is a different matter. We discuss that presently.

US$5 billion of products were estimated to have been drawn from CPRs annually by the rural poor in India in the late 20th century (Beck and Nesmith, 2001). Wood-fuel is prominent among CPR products. In the early 2000s, some 2.4 billion of the world's poorest people depended on wood or other biomass fuels for cooking and heating (World Resources Institute, 2005). Such evidence does not, of course, prove that CPRs are well managed, but it does show that rural households have strong incentives to devise arrangements whereby they *are* managed.

CPRs not only supply households with a regular flow of ecosystem services including tangible goods (water, fuelwood, fibres, building material, fruit, honey, fish), they also offer protection against agricultural risks. In a study of households on the margin of the Tapajos National Forest in the Brazilian Amazon, Pattanayak and Sills (2001) found that households made more trips into the forest for non-timber products when times are hard. CPRs are sometimes the only assets to which the otherwise disenfranchised have access. Hecht, Anderson, and May (1988) described the importance of *babassu* products among the landless in Maranho, Brazil. Extraction from those plants offers support to those on the very lowest incomes, especially women. The authors reported that *babassu* products are an important source of cash income in the period between agricultural-crop harvests. Falconer and Koppell (1990) offered a similar picture in the West African forest zone.

Economic theory says that where there is a market for labour, even the casual wage rate of unskilled labourers would be higher in villages with more abundant CPRs (Dasgupta, 1993). There is evidence of this (Barbier, 2005). That said, we are not implying that the poorest among communities feature prominently in community decisions to create the institutions that govern CPRs; we are merely drawing attention to the significance of CPRs. Cooke, Köhlin, and Hyde (2008), for example, found instances of community forestry in low income countries where the importance of wood-biomass among rural households is not appreciated.

Because CPRs are seats of non-market relationships, transactions involving them are usually not mediated by market prices, and so their fate is frequently unreported in national economic accounts. A rich empirical literature identifies community practices over the use of CPRs that vary in their complexity, depending on both geography and history. The practices are guided by social norms, meaning that mutual enforcement is the way rules and regulations are kept.

Examples abound. From his study of communitarian allocation rules over water and grazing land in 41 South Indian villages, Wade (1988) reported that villages downstream had an elaborate set of rules for regulating the use of water from irrigation canals. Fines were imposed on those who violated the rules. Most villages had similar arrangements for the use of grazing land. In an early study of communitarian allocation rules over the use of standing timber in the Swiss Alps, Netting (1981) found village councils marked equivalent shares for community members, who in turn drew lots for the shares. Sanctions were imposed on those who took more firewood than their entitlements. In the summer months, cattle owners were entitled to graze as many animals on the communal alps as they were able to feed from their private supply of hay during the preceding winter. That way the total number of animals was kept roughly in line with the fodder potential of the village meadows. Netting traced the origins of such communitarian arrangements on the Alps to the 15th century.

In a study of the Kuna tribe in Panama, Howe (1986) described an intricate set of sanctions that are imposed on those who violate norms of behaviour designed to protect a common source of fresh water. In a study of sea tenure in northern Brazil, Cordell and McKean (1985) uncovered a system of codes that served to protect the fishery. Violations of the codes were met with a range of sanctions that included both shunning and the sabotaging of fishing equipment.

In the contemporary world, there is a role for government and non-government organisations (NGOs) to help build or rebuild local institutions through which communities could realise the advantages of informed collective management under changing times. Such help would involve, among other things, devising clearly defined rules concerning the allocation of burdens and benefits, rules whose compliance can be observed (and hopefully verified) by the others who are involved.

8.4 Fragility of CPRs

The empirical literature on CPRs is enormous but of uneven quality. It is one thing to record the uses to which CPRs are put by communities, it is a lot harder to determine the extent to which communities in poor societies have successfully managed their CPRs. Any number of confounding factors have to be met if one is to reach convincing evidence of efficiency, not to mention equity, in the use of managed CPRs. That the communities in question are poor makes the task of determining the efficacy of their institutions particularly hard. Enthusiasm for communities as an alternative to markets and the state has also led scholars to claim more than what the evidence is able to shoulder. On occasion, Western scholars have even shown surprise that intricate rules of behaviour could have been devised by poor, often illiterate people. But literacy has little to do with being smart, or even wise, on matters of resource allocation, especially when management errors can mean destitution. However, numeracy does matter, and people even in the poorest communities have been found to be unerringly good in keeping accounts. The issue is not whether community members know, within the knowledge base, what is good for them; the issue is whether the rules of behaviour over the use of their CPRs are efficient and fair.

In some societies, there are easy markers to alert the scholar of inequity in communitarian rules. That women may be excluded from CPRs has been recorded in the study of communal forestry in India (Agarwal, 2001). Entitlements from CPRs have also been known to be based on private holdings, that is, higher income households have been found to enjoy a greater proportion of the benefits. Béteille (1983), for example, drew on examples from India to show that access to CPRs is often restricted to the elite (e.g. caste Hindus). Cavendish (2000) reported that in absolute terms, higher income households in his sample of villages in Zimbabwe took more from CPRs than poor households. In an early review, McKean (1992) noted that benefits from CPRs are frequently captured by the elite.[254] Agarwal and Narain (1989) exposed the same phenomenon in their study of water management practices in the Gangetic plain, as did Bardhan and Dayton-Johnson (2007) in a study of irrigation systems in Mexico and South India.[255]

However, the relative use of CPRs is not uniform across the world. In two large scale studies of household data in India and Nepal, respectively, Bandyopadhyay and Shyamsundar (2004) found that wood-fuel consumption decreases with wealth in India, but increases with wealth in Nepal. Their finding suggests that the availability of cheap substitutes matters. In India, where rural markets are more developed than in Nepal, relatively wealthy households are able to save on labour costs by buying fuel in the marketplace.

A second piece of bad news about CPRs is that they have deteriorated in recent years in many low income countries and regions. Why should that have happened in those places where they had been managed in a sustainable manner previously? There are several reasons:

One stems from deteriorating external circumstances, under which both the private and communal profitability of investment in the resource base decline. Political instability is a general cause. It is, of course, a visible cause of resource degradation, as civil disturbance all too frequently expresses itself by a destruction not only of produced capital, but of natural capital too. But increased uncertainty in communal property rights is a frequent, often hidden cause. People could worry that the state or warlords will assume authority over the CPRs. If the security of a CPR is uncertain, the returns expected

---

[254] See also Bromley (1992) for further wide ranging review of CPR management schemes.

[255] It is even possible that the elite *exploit* others, in the strong sense that the latter are worse off when the CPR is managed by the community than they would have been if the CPR was unmanaged (Chapter 8*). But because cooperation in one activity is usually tied to cooperation in other activities, it would be hard to establish empirically that one group of CPR users is exploiting another group of users.

from collective action are low (Box 6.2). The influence would run the other way too, with growing resource scarcity contributing to political instability, as rival groups battle over resources. The feedback could be positive, exacerbating the problem for a time, thereby reducing expected returns yet further.

The second reason CPRs have deteriorated in many places is rapid population growth. The latter triggers environmental degradation if institutional practices are unable to adapt to the increased pressure on resources. In Cote d'Ivoire, for example, growth in rural population was accompanied by increased deforestation and reduced fallows in the 1990s. Biomass production declined, as did agricultural productivity (López, 1998).

The third reason CPRs have deteriorated in some places is that communal rights were overturned by central fiat. In order to establish its political authority, a number of states in the Sahel, for example, imposed rules that destroyed communal management practices in the forests. Villages ceased to have authority to enforce sanctions on those who broke communitarian rules. But state officials did not have the expertise to manage the commons, and often they were corrupt. Knowledge of the local ecology is held by those who work on the commons. Local participatory democracy offers a mechanism by which that knowledge can inform public policy. Isham, Kaufmann and Pritchett (1997) found evidence from 121 rural water projects (in 49 countries in Africa, Asia and Latin America) that participation by beneficiaries is positively correlated with project performance. Relatedly, Somanathan (1991), Thomson, Feeny, and Oakerson (1992), Isham, Narayan, and Pritchett (1995), and Baland and Platteau (1996), among others, have identified the many ways by which the exercise of state authority damages local institutions and turn CPRs into what are in effect open-access resources.

Even when the state is not predatory, there are problems. In an exceptional study of the quality of forest management in the central Himalayas, Somanathan, Prabhakar, and Mehta (2005, 2009) used satellite images and field surveys on what was a natural experiment in governance to find that crown cover was no less in places that were governed by village councils than it was in areas managed by the state, but that expenditure on governance was an order of magnitude higher in the latter. The authors also reported that relative to unmanaged commons, crown cover in broad leaved forests was higher in village-managed commons but were not appreciably different in pine forests. The authors noted that under the user rules, broad-leaved forests provided more benefits to community members than pine forests. As in any management problem, monitoring costs matter.[256]

Success in managing CPRs by mutual enforcement cannot be guaranteed. Past agreements among members have been known to come under stress from market pressures. One reason is that migration of key members to the city can interfere with the practice of behavioural norms that previously protected the CPRs (Chapter 6).[257] Then there are cases, sadly all too often for forest resources, where the state imposes rules that interfere with communitarian norms, or worse usurps CPRs and makes them state property. That forests provide a multiplicity of goods and services is a reason they are a seat of intense political tension. One group of actors (the state, private companies, plantation owners) would be interested in one class of resources (hardwood, minerals, palm oil), while another group would be interested in another class (honey, water, bark, soft wood). The interests generally cannot play out satisfactorily because they compete against one another. To get at the minerals requires fragmenting the forest massively; to get at the hardwoods or to create plantations would mean destroying the forest; and so on. In the process, communities, often indigenous populations, are made destitute.

---

[256] Baragwanath and Bayi (2020) report a similar finding on a site in the Brazilian rainforest. Deforestation was significantly less in land inside a territory where inhabitants had been given collective property rights than in land outside the border.

[257] This and other reasons for a tension between market and communitarian institutions have been studied by Arnott and Stiglitz (1991).

Democratic movements among stakeholders and pressure from international organisations have encouraged a return to communitarian management systems in some parts of the world. Shyamsundar (2008) is a remarkable synthesis of the findings in nearly 200 articles on the efficacy of a devolution of management responsibilities – from the state to local communities – over the local natural resource base. Her article focuses on wildlife, forestry and irrigation. The balance of evidence appears to be that devolution leads to better resource management, other things being equal. Shyamsundar of course offers a discussion of what those other things are.

The fourth reason CPRs have deteriorated in many places is that cooperation is fragile, dependent as it is on many factors that must work simultaneously in its favour. For example, in the face of growing opportunities for private investment in substitute resources, households are more likely to break agreements that involve reciprocity (Dasgupta, 1993, 2007; Campbell et al. 2001). But when traditional systems of management collapse and are not replaced by adequate institutions, CPRs suffer from neglect. Here are three examples illustrating the phenomenon:

(1) Mukhopadhyay (2008) is a historical study of the transformation of agrarian land in Goa, India, that was earlier owned and regulated by a communitarian institution called the *communidades*. When Goa became a part of India, the government introduced land reforms that gave tenants the right to purchase the land they had worked. Mukhopadhyay does not question the underlying motivation behind land reforms, but notes one unfortunate consequence, which is the breakdown of cooperation among households in maintaining the embankments that had earlier prevented the land from flooding by tidal waters. Over the years, deterioration of the embankments has led to an increase in soil salinity.

(2) In her study of collectively managed irrigation systems in Nepal, Ostrom (1995) accounted for differences in rights and responsibilities among users (who gets how much water and when, who is responsible for which maintenance task of the canal system, and so on) in terms of the fact that some farmers are head-enders, while others are tail-enders. Head-enders have a built-in advantage, in that they can prevent tail-enders from receiving water. On the other hand, head-enders need the tail-enders' labour for repair and maintenance of traditional canal systems, which are composed of headworks made of stone, trees and mud. Ostrom reported that a number of communities in her sample had been given well-meaning aid by donors which installed permanent headworks. What could be better, you may ask. But Ostrom observed that those canal systems that had been improved were frequently in worse repair at the tail end and were delivering less water to tail-enders than previously. Ostrom also reported that water allocation was more equitable in traditional farm management systems than in modern systems managed by external agencies, such as government and foreign donors. She estimated from her sample that agricultural productivity is higher in traditional systems.

Ostrom's explanation for this is that unless it is accompanied by countermeasures, the construction of permanent headworks alters the relative bargaining positions of the head- and tail-enders. Head-enders now do not need the labour of tail-enders to maintain the canal system. So, the new sharing scheme involves even less water for tail-enders. Head-enders gain from the permanent structures, but tail-enders lose disproportionately. This is an example of how well-meaning aid can go wrong if the nature of the institution receiving the aid is not understood by the donor.

(3) Village tanks (or artificial ponds) are one of the oldest sources of irrigation in South Asia. In a study of a group of villages in southern India, Balasubramanian (2008) reports that village tanks have deteriorated over the years owing to a decline in collective investment in their maintenance. That decline has taken place as richer households have invested increasingly in private wells. As low income households depend not only on tank water but also on the fuelwood and fodder that grow around the tanks, construction of private wells has accentuated economic stress among low income people.

> **Box 8.2**
> **Ownership Rights**
>
> Problems arising from badly designed institutions usually have a long history; bad institutions have a tendency to attract opportunism, which begets worse institutions. The current fate of the Brazilian rainforest was probably written decades ago. In a notable paper, Binswanger (1991) argued that in Brazil, the exemption from taxation of virtually all agricultural income, allied to the fact that logging is regarded as proof of land occupancy, provided incentives to the wealthy to acquire forests and then to deforest the land.[258] He argued that the subsidy enjoyed by the private sector has been so large that a reduction in deforestation by the removal of subsidies is in Brazil's own interests, let alone that of the rest of the world. This has implications for international negotiations. The current consensus, at least among farmers and loggers in the Amazon basin, is that Brazil has much to lose from reducing the rate of the deforestation it has been engaged in. If this is true, there would be a case for the rest of the world to subsidise Brazil, as compensation for losses the country would sustain if it were to exercise restraint. Binswanger's account shows that it is by no means clear that the consensus is correct.
>
> Alston, Libecap, and Mueller (1999) took the analysis further. They showed that accelerated deforestation in the Brazilian Amazon, followed by violent conflicts between landowners and squatters, also occurred because of legal inconsistencies between the civil law, which supports the title held by landowners, and the constitutional law, which supports the right of squatters to claim land not in 'beneficial use' (e.g. farming or ranching). Ironically, the latter right reflects the government's stated desire for land reform. The authors have shown that the vagueness of the 'use' criteria, and the uncertainty as to when a landowner's claim to a piece of land or a squatter's counter claim to it is enforced, are together an explosive force.

8.5 Property Rights to Land

Property rights on land have assumed a bewildering variety across regions – land rights in sub-Saharan Africa having traditionally been quite different from those in, say, South Asia. Chance events have undoubtedly played a role in the way patterns of land tenure developed in various parts of the world, but so have economic, demographic and ecological circumstances. Rights to assets which offer multiple services are often complex. Someone may have the right to cultivate a piece of land (in many contexts if they inherited it from their father, in others if they are the one to have cleared it), while others may share the right to the products of the trees growing on this land, while still others may have a concurrent right to graze their animals on the stubble following each harvest. On occasion, the person who has the right to cultivate a piece of land does not have the right to rent it nor to sell it, and on most occasions, he does not have the right to divert water-flows through his land. These last are often group rights (Breslin and Chapin, 1984; Feder and Noronha, 1987).

For clarity, it pays to think of polar cases, which are territorial (or private property) systems, and communal property systems. Even these on occasion can be hard to distinguish. Social groups could assert territorial rights over land and at the same time practise reciprocity over access. This would have much the same effect as controlled communal ownership of all the lands. Matters can also be the other way around. Ownership could be communal across groups, but residence could be confined to given territories. Right of access to resources by one group from the territories of another would have to be

---

[258] Heath and Binswanger (1996) extended their basic argument to neighbouring countries.

monitored to avoid free-riding. But this would look pretty much like private ownership with reciprocity over access (Cashdan, 1983).

Two aspects of spatially spread ecosystems are of vital importance: productivity per unit area and predictability. By *productivity per unit area* we mean here average net primary productivity (NPP) per square mile, which is sometimes also called the ecosystem's 'density'. And by *predictability* we mean the inverse of the variance in density per unit of time, with the allied assumptions that the probability distributions are not overly correlated across spatial groupings of the ecosystem, and not overly correlated over time. Two extreme types of spatially spread ecosystem are then of particular interest. The first is characterised both by high density and high predictability (e.g. river valleys), and the second by low density and low predictability (e.g. semi-arid scrublands and grasslands). Economic theory tells us that communities would tend to institute private property rights over the former category and remain geographically stable. Economic theory also tells us that communities would be dispersed and mobile were they dependent upon the second category of resources. The prevalence of nomadic herdsmen in the Sahel is an instance of this. In any event, we would expect a greater incidence of CPRs in regions where natural capital has low density and low predictability.[259]

Netting (1968) and Feder and Noronha (1987), among a number of other scholars, studied the evolution of land tenure systems in sub-Saharan Africa. Their accounts are consistent with our reasoning. In Africa, land rights were typically held by groups, not by individuals. It is this which is being transformed, a good deal by state fiat (who often claim ownership, as in Ethiopia, Mauritania, Zaire, Zambia, Nigeria and Tanzania), and some by individuals themselves, who break with traditional norms of ownership when land values rise.

Resource density increases with investment and technological improvements, for example terracing, or the introduction of high-yielding varieties of wheat. Predictability can be made to increase at the same time, for example, by the creation of irrigation facilities. The opening of new markets for cash crops also raises resource density. Changing patterns of land tenure often observed in poor countries would seem to be explainable along the lines we have outlined here (Ensminger, 1992).

History tells us that CPRs have declined in importance as economies have developed (North and Thomas, 1973). Ensminger's 1992 study of the privatisation of common grazing lands among the Orma in north east Kenya established that the transformation took place there with the consent of the elders of the tribe. She attributed the move to cheaper transportation and widening markets, making private ownership of land more profitable. However, as the elders were from the stronger families, privatisation accentuated inequality within the tribe.

Box 8.3
Plantations for Palm Oil

Palm oil accounts for one-third of the global consumption of vegetable oil (Statista, 2020). Oil palm is a high yielding crop, producing many more tonnes per hectare than other types (five times the output of rapeseed, six times that of sunflower, and seven times that of soybean). Malaysia and Indonesia produce 85% of the global output of palm oil (Statista, 2020). Expansion of oil palm plantations during 1990 to 2005 occurred there at the expense of forests, to an extent that the lowlands lost more than 50% of their rainforest cover. By some estimates, that resulted in the sharpest declines in biodiversity in any biogeographical area (Edwards et al. 2010).

---

[259] As noted above, this is to some extent confirmed by field studies in villages in the drylands of the Indian sub-continent.

Chapter 8: Human Institutions and Ecological Systems, 2: Common Pool Resources

> Oil palm plantations fragment rainforest. As noted in Chapter 3, fragmentation reduces biodiversity. Sampling bird abundances, species richness and species composition within the Ulu Segama-Malua Forest Reserve and oil palm estates in Sabah, Borneo, Edwards et al. (2010) found that fragments were in fact less valuable for birds than a similar area of contiguous forest. Abundance of imperilled bird species were found to be 60 times lower in fragments, with 1.8 times fewer birds overall.
>
> Oil palm plantations have proliferated in Latin America as well. Their expansion is often a source of social conflict. In Chocœ, Colombia, communities that had been forcibly displaced established Humanitarian Zones, trying to hold on to their lands and livelihoods. Private companies, cultivating palm oil for use as a biofuel, established plantations once the inhabitants had been displaced. That exerted further pressure on the inhabitants to sell their land. In Indonesia too, the arrival of plantations altered local lives, disturbing many of their customs, values and their ability to act as guardians of sites considered sacred in their tradition. In the state of Sarawak in Malaysian Borneo, land disputes took place between indigenous Dayak communities and oil palm companies. Although the courts repeatedly ruled in favour of the Dayak and found the government's limited interpretation of 'native customary rights' to be faulty, the state continued to hand out concessions (Colchester et al. 2007).

8.6 Property Rights and Management: A Schemata

Property rights to an ecosystem – perhaps distinct sets of property rights to the various goods and services produced by the ecosystem – are the rights, restrictions and privileges with regard to its use. Management of an ecosystem, we may call it a resource-base, is a different (though related) matter. Property rights may be well-defined, but a resource could nevertheless be managed badly, owing to disagreements among those holding the rights on how it ought to be used, or to corrupt practices regarding its use, or to a lack of understanding of ecosystem functions, or a combination of all these reasons.[260] Ownership is yet another matter. Someone owns a resource by law but transfers their right of use to another person for a limited period. Ownership does not change, but the right-of-use does.

Private, communitarian and state property rights to resources were discussed in the previous four sections. Table 8.1 presents the classification. We can think of a property rights regime and the associated management structure as defining the institution governing the use of a resource base. In practice, every institution harbours externalities, but some harbour more than others. The relative efficiency of institutions depends on the character of the resource base (e.g. whether its productivity is subject to especial risk, whether the resource base is lumpy, whether its use by someone is observable by other relevant parties); it depends on the rules governing the production and allocation of other goods and services (e.g. credit and insurance); it depends on the extent to which the returns from investment in it can be appropriated by the investor; and so on. The interplay of market and non-market institutions is a crucial determinant of well-being, because the performance of one depends on the performances of the others. Nor do institutions remain fixed, they evolve. The latter is a truism no doubt, but details matter. Ensminger (1992), for example, has narrated how common grazing lands among the Orma in north east Kenya became segmented into private land. She showed that the transformation took place with the consent of the elders of the tribe and attributed this to changing transaction costs brought about by cheaper transportation and widening markets. The elders were, not surprisingly, from the stronger families, and it did not go unnoted by Ensminger that privatisation accentuated inequalities. Institutions can also be created or destroyed by design and legislation. What was once a relatively efficient institution for a resource may cease being so when circumstances change.

---

[260] Ostrom and Schlager (1996) contains an account of the conditions necessary for effective communitarian management.

Table 8.1 Property Rights to Natural Capital

Ownership	Resources
Private	cultivated land, cattle, oil and minerals
Communitarian (CPRs)	grazing land, threshing grounds, ponds, woodlands, mangrove forests, inland and coastal fisheries
Open access	atmosphere, international waters
State	commercial forests, oil and minerals
State as trustee	national parks, cultural monuments

Annex 8.1 Estimating Subsidies

Conventionally, a subsidy is a form of government support to an economic sector (or institution, business or individual) with the aim of promoting an activity that the government considers beneficial to the economy overall and to society at large. The subsidy can be supplied in the form of a monetary payment or other transfer, or through relief of an opportunity cost.

We currently subsidise the use and consumption of provisioning ecosystem goods heavily. Historically, our reliance on the extraction of these services has grown significantly (Chapter 16), given that the goods and services they provide underpin global human activity (fossil fuels, water, fisheries and agriculture). Our current reliance on these goods and services makes reform of subsidies extremely difficult, primarily due to lack of political support to do so.

The estimation of subsidies takes various forms, depending on the mechanism of fiscal transfer, and the impact this can have on actors in the respective sector. This is due to the complexity of, and controversies around, the changes which can be attributed to the subsidy itself. Most recent estimates use differing methodologies across sectors in order to assess the mechanisms by which a subsidy can distort prices and change behaviour. This is by no means uniform across subsidies, the efficacy and efficiency of policy instruments depend on a range of factors including the issue to be addressed, the relevant institutions and the technical limitations and constraints associated with the use of particular policy instruments (OECD, 2013b).

For the sectors considered here, see Table A8.1 for the characteristics of subsidies that have been included in the respective methodologies, and their global estimates[261]. There are other fiscal transfers (explicit and implicit) to these sectors which are not included in these estimates, due to either the technical complexity of measurement, or lack of reporting. Data reported on subsidies remains limited and incomplete, making the estimation of the true cost of subsidies difficult.

Table A8.1 Global Subsidies to Provisioning Ecosystem Goods*

Sector	Subsidy estimate methodology	Estimate per year (year)
Fossil fuels	Assessment of global and regional fossil fuel energy subsidies based on comprehensive country-level estimates for 191 countries. Estimating post-tax subsidies, reflecting the difference between actual consumer fuel prices and prices that would fully reflect supply costs, plus the taxes needed to reflect environmental costs and revenue requirements (Coady et al. 2019).	US$5.2 trillion (2017)

---

[261] Please refer to the source of the estimates for further discussion of the methodologies.

Table A8.1 (cont.)

Sector	Subsidy estimate methodology	Estimate per year (year)
Agriculture	Assessment of agricultural support given in 53 countries. Estimating transfers to individual producers (three-quarters of the total), policy expenditures which have primary agriculture as the main beneficiary (not to individual producers) and budgetary support to consumers of agricultural commodities (depressing domestic prices and distorting the market). Does not include environmental costs (OECD, 2019b).	US$705 billion (2016–18 average)
Water and sanitation	Assessment of subsidies to water and sanitation services globally (excluding India and China due to lack of data). Estimated using a method which aims to maximise both allocative efficiency (recreating competitive market results) and productive efficiency (producing efficient quantities at the lowest possible cost), while allowing each utility to generate sufficient revenue to cover service costs. Does not include infrastructure expansion (tends to be fully subsidised) or environmental costs (Andres et al. 2019).	US$289 – 353 billion (2019)
Fisheries	Assessment of subsidies to the fishing sector globally. Estimates include all direct and indirect transfers from the public sector to the private fishing sector. Capacity enhancing subsidies constitute 63% of the total estimate (those which have the potential to encourage overfishing). Does not include environmental cost of these subsidies (Sumaila et al. 2019).	US$35.4 billion (2018)

*Subsidies to forest activities not included here due to insufficient data.

The estimates in Table A8.1 constitute best estimates of the total amount of subsidies given to the respective sectors. They differ in their methodologies, depending on data availability and the ability to assess the impact of subsidies on particular sectors. Given the differences in methodologies, it may be problematic to aggregate them, however we are able to make a best estimate based on the ranges given, and the fact that they are known to be likely underestimates. We estimate that the total subsidies given to the sectors above lies conservatively in the range of US$5–7 trillion annually (due to the lack of data reported by countries and limited estimation of the true costs of subsidies in terms of damage to natural assets). This is not to say that these subsidies all damage ecosystems (as a proportion would not), but that we subsidise, on a large scale, the activities of these sectors and the consumption of the products they produce, in many cases without knowing the environmental cost of doing so.

Estimates of global subsidies which are environmentally damaging are still being developed. The OECD provides the most recent estimate of subsidies that are known to be harmful to biodiversity, at US$500 billion dollars per year (based on fossil fuel subsidies[262] and government support to agriculture that is potentially environmentally harmful) (OECD, 2019a). It is acknowledged that this is an underestimate. There are difficulties in attributing environmental damage to specific subsidies, given the likelihood of double counting and cross-effects. For example, farming accounts for around 70% of water usage globally, while also contributing to the pollution of the water system through pesticide and fertiliser run-off (Gruère, Ashley, and Cadilhon, 2018), so it is therefore difficult to ascertain what proportion of agricultural or water subsidies exacerbate this.

---

[262] Using an OECD estimate from 2015, different to the IMF estimate used here.

In contrast to subsidies that are environmentally harmful, it is important to note the subsidies allocated to conservation, restoration and sustainable use of the environment. According to the OECD PINE database, biodiversity relevant 'positive' subsidies amount to US$0.89 billion per year (2019 prices, 2012 to 2016 average). The OECD also estimates expenditure of government on support to agriculture that is considered potentially beneficial to biodiversity at €2.6 billion per year in OECD countries[263] (OECD, 2019a). Further, domestic public finance associated with the conservation and sustainable use of biodiversity is estimated to total around US$67.8 billion per year globally (OECD, 2020a). Even accounting for this, subsidies that are harmful to the environment vastly outweigh those which are beneficial. Notably, we subsidise harvesting of the biosphere at a magnitude far higher than that that we spend on its conservation and sustainable use. Chapter 20 provides further information on how finance affects the sustainability of our engagement with Nature.

Further reporting is needed on subsidies and their intended impact to understand fully their true costs and how they can be reformed to reduce environmental damage and increase investment in Nature. Globally, effort is being made to improve reporting on subsidies, as well as efforts to reform harmful subsidies (Chapter 20, Box 20.2).

---

[263] This is a proxy, focusing on two categories of government support – support with environmental constraints, for long-term retirement of resources and specific non-commodity outputs.

Chapter 8*
Management of CPRs: A Formal Model

Introduction

We consider a timeless model of a common property resource (CPR) in which N herdsmen are able to graze their cattle.[264] The model has been constructed deliberately along orthodox economics lines. That enables us to study alternative institutions in a sharp way. For example, the model eschews non-linearities in the processes that drive a CPR's productivity. The model allows us to explore three regulatory regimes for managing CPRs: (i) privatisation of the CPR; (ii) communitarian management; and (iii) state management. In Section 8*.3 we show how features of the socio-ecological world we know can be introduced into the basic model. Only empirical evidence on the significant features of each of the three institutions can tell us how to modify the model. The context will be found to matter. Grazing fields are different from crop fields, and coastal fisheries are different from rain forests. As in all other matters in the socio-ecological world, there is no unambiguous answer.

8*.1 A Timeless World

We begin with a timeless world.[265] Herdsmen are indexed by i ($i = 1, 2, ..., N$). Cattle are private property. The grazing field is taken to be a village pasture. Its size is S. Cattle intermingle while grazing, so on average the animals consume the same amount of grass. If X is the size of the herd in the pasture, total output – of milk – is $H(X, S)$, where H is taken to be constant returns to scale in X and S. Assume $H(0, S) = 0$ for all $S \geq 0$. Assume too that the marginal products of X and S are positive, but diminish with increasing values of X and S, respectively.

As S is fixed and H is constant returns to scale, we may eliminate S by writing $H(X, S) = SH(X/S, 1)$; by letting $S = 1$ without loss of generality; and by defining $F(X) \equiv H(X, 1)$. From the assumptions made on H, it follows that $F(0) = 0$; $F'(X) > 0$; $F''(X) < 0$; and that the average product of cattle ($F(X)/X$) exceeds the marginal product of cattle ($F'(X)$) for all $X \geq 0$.[266] Figure 8*.1 presents the average and marginal product curves.

Herders are interested in the profits they can earn from their cattle. We normalise by choosing the market price of milk to be one. Let the market price of cattle be p (> 0), which will be interpreted as a rental price when we embed the model in time. To have a non-trivial problem to analyse, assume that $F'(0) > p$.

---

[264] The timeless version of the model presented here has been taken from Dasgupta and Heal (1979).

[265] The material here is from Dasgupta and Heal (1979: Chapter 3) and Dasgupta (2008).

[266] $F'(X)$ is the first derivative of F with respect to X and $F''(X)$ is the second derivative.

Chapter 8*: Management of CPRs: A Formal Model

Figure 8*.1 Average and Marginal Productivity Curves.

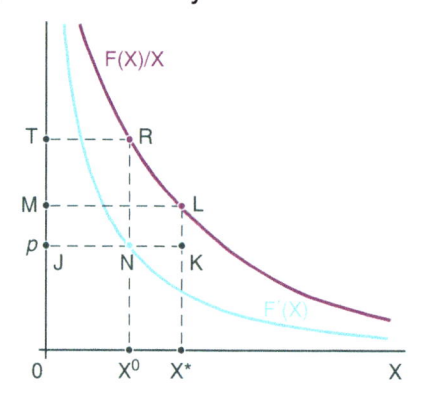

8*.1.1 The Unmanaged CPR

Let us first determine the herd size that would be grazed in an unmanaged CPR. By 'unmanaged' we mean that the field is an open access resource among the villagers. Let x_i be the size of i's herd. For ease of computation, x_i is taken to be a continuous variable. Since cattle intermingle, $x_i F(X)/X$ is i's output of milk. Therefore, i's net profit, π_i, is

$$\pi_i = x_i F(X)/X - px_i \tag{8*.1}$$

We now compute the non-cooperative (Nash) equilibrium of the resulting timeless game. Since the model is symmetric, we should expect it to possess a symmetric equilibrium. (It can be shown that equilibrium in this timeless model is unique.) Consider herder i. If the herd size of each of the other cattlemen is x, equation (8*.1) can be written as

$$\pi_i(x_i, x) = x_i F(x_i + (N-1)x)/(x_i + (N-1)x) - px_i \tag{8*.2}$$

The profit function $\pi_i(x_i, x)$ reflects the crowding externalities each herdsman inflicts on all others in the unmanaged CPR, for π_i is a function not only of x_i, but also of x. Let x^* be the size of each cattleman's herd at a symmetric equilibrium. By definition, x^* would be the value of x_i that maximises $\pi_i(x_i, x^*)$.

To determine x^*, differentiate $\pi_i(x_i, x^*)$ in equation (8*.2) partially with respect to x_i and equate the result to zero. This yields

$$F(x_i + (N-1)x^*)/[x_i + (N-1)x^*] + x_i F'(x_i + (N-1)x^*)/[x_i + (N-1)x^*] - x_i F(x_i + (N-1)x^*)/$$
$$[x_i + (N-1)x^*]^2 = p \tag{8*.3}$$

In a symmetric equilibrium, x_i in equation (8*.3) must equal x^* (Figure 8*.1). Now re-arrange terms to confirm that the aggregate herd size in the CPR, which we write as X^*, satisfies

$$((N-1)/N)F(X^*)/X^* + F'(X^*)/N = p, \qquad X^* = Nx^* \tag{8*.4}$$

Equation (8*.4) says that in equilibrium the price of cattle equals the weighted average of the average product of cattle and the marginal product of cattle, with weights $(N-1)/N$ and $1/N$, respectively (Figure 8*.1). It is easy to confirm that X^* is an increasing function of N.

Notice that aggregate profit for the herders, which we denote by Π^*, is

$$\Pi^* = [F(X^*) - pX^*] > 0 \tag{8*.5}$$

Equation (8*.5) says that N being finite, the profit is not entirely dissipated. In Figure 8*.2, Π^* is the area of the rectangle JKLM.

From equation (8*.5) we conclude that profit per herder is

$$\Pi^*/N = [F(X^*) - pX^*]/N > 0. \qquad (8^*.6)$$

In Figure (8*.2), which depicts the case $N = 2$, the equilibrium pair of profits $(\Pi^*/2, \Pi^*/2)$ is the point A.

Note: It is commonly thought that unmanaged CPRs reflect the Prisoners' Dilemma game that was studied in Chapter 6 (Box 6.2). That is a mistake. As we confirm presently, the equilibrium number of cattle in the CPR exceeds the collective optimum, but x^* is *not* a dominant strategy for the representative herder.

If N is large, the unmanaged CPR approximates an open access resource. To confirm, notice that if N is large, equation (8*.4) becomes

$$F(X^*)/X^* \approx p \qquad (8^*.7)$$

The approximate equation (8*.7) says that profits are dissipated almost entirely.

Figure 8*.2 Profits from CPRs under Cooperation and Non-cooperation

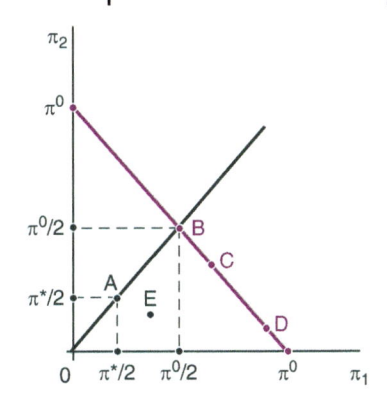

8*.1.2 Regulatory Regimes: Quotas and Taxes

An unmanaged CPR would be unattractive to the herders: they could increase their profits by managing it together. Imagine that cooperation involves negligible transaction costs. What would be a reasonable agreement among the herders? As the model is symmetric, we assume that they agree to maximise aggregate profit and share that profit equally. Maximising aggregate profit $(F(X) - pX)$ yields the condition

$$F'(X) = p \qquad (8^*.8)$$

Equation (8*.8) says that at the optimum the marginal product of cattle equals their price – a familiar result in price theory. Let $X°$ be the solution of equation (8*.8). At the community optimum, aggregate profit, which we denote by $\Pi°$, is

$$\Pi° = F(X°) - F'(X°)X° > \Pi^* > 0 \qquad (8^*.9)$$

In Figure 8*.2, $\Pi°$ is the area of the rectangle JNRT.

Chapter 8*: Management of CPRs: A Formal Model

From equation (8*.9) we conclude that profit per herders is

$$\Pi°/N = (F(X°) - F'(X°)X°)/N \qquad (8*.10)$$

A comparison of equations (8*.4) and (8*.8) shows that $X^* > X°$. In Figure 8*.2, which depicts the case $N = 2$, the pair of profits $(\Pi°/2, \Pi°/2)$ at the community optimum is the point B.

We now study two regulatory regimes for implementing $X°$.

With a *quota restriction*, the herders agree to practise restraint by limiting each to at most $X°/N$ cattle. It is simple to confirm that choosing $X°/N$ is then the dominant strategy for each herder. We conclude that a quota can implement the agreement to limit the aggregate herd size to $X°$.

An alternative would be to impose an *entry tax* on each cow and to share the tax revenue equally as a lump sum income. This is the (Pigouvian) tax solution to the problem. In a tax regime, the herders are free to graze as many cattle as they like but have to pay an entry fee per cow.[267] The optimum tax corrects for the externalities each herder inflicts on all others when introducing a cow in the CPR. If the tax rate is t, the effective price a herder pays is $(p + t)$. The idea therefore is to so choose t that equation (8*.4) reduces to equation (8*.8) and the equilibrium herd size equals $X°$.

Let $t°$ be the optimum tax per head of cattle. On using equations (8*.3)-(8*.4), routine calculations show that

$$t° = (N-1)\{[F(X°)/X° - F'(X°)]/N\} - p \qquad (8*.11)$$

Equation (8*.11) is intuitively obvious. The expression on the right-hand side represents the crowding externalities the representative herder inflicts on the other $(N-1)$ herders. It follows that $t°$ is the rent that ought to be charged for the use of the grazing field each animal brought into the field. That rent is dissipated in an unmanaged field, the larger the number of herders the greater the rent dissipation. If $t°$ was to be imposed as a fee for introducing an animal into the field, each cattleman would limit his herd size to the right number.

Who sets the tax or quota and who enjoys the rents? That is studied next.

8*.2 Mutual Enforcement of Optimum Herd Size

One can imagine two institutional regimes in which optimum quotas or taxes would implement $X°$. In one, the state steps in to claim the CPR as its property and imposes a tax $t°$ on each animal entering the field. What the state does with the tax revenue is a distributional matter of great importance. If it usurps the revenue, the herders earn no profit from their enterprise (after adding the grazing fee to the cost of the animals there would be nothing left), which means they would have been better off if the grazing field had been an unmanaged CPR (profits are positive in the unmanaged CPR, albeit in a dissipated form). If, on the other hand, the state returns the entire tax revenue to the herders on a lump sum basis, it would be as though the field was an optimally managed CPR, fiscally speaking.

Suppose, however, that the field is a CPR. How might the community of herders enforce an agreement to limit the size of each herd to $x°$? In Chapter 6, it was shown that long-term relationships are a venue where mutual enforcement can serve to implement collective agreements. So now imagine that the timeless interaction among the N herders is repeated every period t, where $t = 0, 1, 2, \ldots$ Box 6.2 studied the Grim norm to illustrate how social norms of behaviour are vital for implementing agreements. That same analysis can be deployed to show that so long as the rate at which herders discount their future profits is not too large, an agreement that each herder will limit the size of his herd to $x°$ would be mutually enforcing.

---

[267] The analysis that follows is identical to the one that was used in Box 7.2 to compute Pigouvian taxes to remove a unidirectional externality.

The way to show that is to recall that x* is an equilibrium number of cattle per herder. That is the number each herder would bring into the unmanaged grazing field. From that we may conclude that Grim can be built on x* as a credible threat point. The social norm that each herder abides by would read as: introduce x° animals into the field at t = 0 and continue to introduce x° in every period so long as everyone has introduced at most x° animals each until then; but introduce x* animals into the field forever following the first violation of the agreement by anyone. As noted in Box 6.2, beliefs play a key role here, but the trust that supports the agreement, if achieved, would be rationally held. That is because the hallmark of a social norm is that it is an equilibrium rule of behaviour.

8*.3 Privatising the CPR

An alternative to a regulatory regime involves a change in the property rights to the pasture. Consider privatising the grazing land. The size of the pasture is S. Imagine that S is divided into N equal parts and awarded as private property to the herders. Suppose too that they are able to protect their property rights without cost (e.g. fences are built without cost). What would be the outcome?

Each cattleman owns S/N amount of land after privatisation. If herder i was to introduce x_i cows into his own land, his output would be $H(x_i, S/N)$ and his profit would be

$$\pi_i(x_i) = H(x_i, S/N) - px_i \qquad (8^*.12)$$

Because H is constant returns to scale, $H(x_i, S/N) = H(Nx_i/S, 1)S/N$. Once again, let us normalise by setting $S = 1$. Now $H(Nx_i, 1) = F(Nx_i)$ in our earlier notation. Therefore equation (8*.12) reduces to

$$\pi_i(x_i) = F(Nx_i)/N - px_i \qquad (8^*.13)$$

Notice that unlike equation (8*.2), which represented herder i's profit function in the unmanaged CPR, the profit function in equation (8*.13) harbours no externalities: π_i is solely a function of x_i. Privatisation removes the crowding externalities among cattle. Differentiating π_i with respect x_i and setting the result equal to zero, the profit maximising size of herd is found to be

$$F'(Nx_i) = p \qquad (8^*.14)$$

Comparison of equations (8*.8) and (8*.14) shows that each herder introduces X^*/N cows into his private parcel of land. But this is the cooperative outcome in the case where the pasture is a CPR. Thus, under the assumption that fracturing the grazing field into N parts does not reduce the field's productivity, privatisation also solves the resource allocation problem facing the N herders.

8*.4 Extensions

Working with an exceptionally simple model of a natural asset, in this case a grazing field, we have shown that so long as people are not overly myopic, each of three distinct institutions can, in an artificially designed set of circumstances, sustain an efficient level of usage. In the first, the field is a managed CPR; in the second the field is owned by the state; in the third the field is divided into private lots.

The finding that the three institutions are equivalent works against common intuition and experience, but that is the merit of the model. It is flexible enough to permit modifications in various directions for reflecting the empirical findings reported in Chapter 8. Quantitative models are useful in institutional economics because they point to those parameters that must be estimated if the policy maker is to delineate policy options. Reinterpreting the model allows us to do that. Below we modify, and when necessary reinterpret, the model so as to introduce features of the socio-ecological world that break the equivalence result.

Chapter 8*: Management of CPRs: A Formal Model

(1) *Non-linearities*: It was noted in Chapter 3 that ecosystems lose productivity when they are fragmented. To introduce that feature, recall that in the step leading to equation (8*.12), use was made of the assumption that aggregate production function $H(X,S)$ is constant returns to scale. Fragmentation reduces output per unit area, implying that output from a plot of size $1/N$ would be only a fraction of $H(x_i, S/N)$. It follows that privatisation would not be a substitute for a well-managed CPR, it would be worse.

(2) *Productivity gain from privatisation*: A reverse move enables us to consider cases where privatisation raises productivity because owners are motivated to invest in the ecosystem. Suppose the model is that of agricultural land. Let x_i now be i's labour effort and $F(X)$ as aggregate crop output from the field. Suppose x_i is only partially observable by others. In that case, communitarian management would lose the key to its potency: observability of actions. Consequently, output in a plot of size $1/N$ would be a multiple of $H(x_i, S/N)$: privatisation would raise land productivity.

(3) *State malfeasance*: In Chapter 8, we reported a finding in the Brazilian rainforest that deforestation was significantly less in land inside a territory where inhabitants had been given collective property rights than in land outside the border (Baragwanath and Bayi, 2020). To accommodate this, the model could be modified to read x_i as the labour effort of person i and $F(X)$ as aggregate biomass production. The authors' finding says that when left to the state, $F(X)/x_i$ is a fraction of what it would be under a communitarian management system.

(4) *Inequities in communitarian practices*: The way the model of a managed CPR could be modified so as to accommodate inequities in management practice is trivial, for it involves identifying the elite among the N herder. Suppose they are M in number. We may suppose that they impose a limit on the number of cattle the ($N-M$) others are permitted to introduce into the field. If that limit amounts to $\underline{x}$ per herder, then it must be that $\underline{x} < x^*$ (Equation (8*.4)), otherwise there would be no inequity. The calculation that follows is routine.

(5) *Payment for ecosystem services*: In Chapter 7, it was noted that while payment for ecosystem services schemes (PES) can be so designed as to eliminate adverse externalities, the system can lead to an increase in inequality.[268] To confirm, imagine that the grazing field is owned by a landowner who previously allowed the community of tenants to graze their cattle at no charge. An unmanaged field would be found to be grazed by X^* cattle, whereas a managed field would be found to be grazed by $X°$ Cattle. Aggregate profit for the herders would be Π^* in the former case, and $\Pi°$ in the latter case (Equations (8*.5)–(8*.9)). In the latter case, there would of course be no need for a PES system – use of the field would have to be efficient without the system – but there could be a call for a PES system if the field was unmanaged. So suppose the field is unmanaged to begin with. If the landowner is now encouraged to charge a rent, efficiency would require that he imposes a charge of $\Pi°/X°$ per animal. But in that case the herders would enjoy no profit from the enterprise. The landowner could of course charge a tiny bit less in order to coax his tenants into animal grazing, but the PES system would make the tenants worse off.

(6) *Exploitation masquerading as cooperation*: The matter here is subtle. Recall that aggregate profit in the managed CPR (Equation 8*.9)) and the corresponding figure in the unmanaged CPR (8*.5) are $\Pi°$ and Π^*, where

$$\Pi° = [F(X°) - X°F'(X°)] > \Pi^* = [F(X^*) - pX^*] > 0 \qquad (8^*.15)$$

Profit per herder in the unmanaged CPR is therefore

---

[268] Arriagada et al. (2015), for example found in a study of a PES scheme in Costa Rica no evidence that tenants had become better off when charged for the use of land belonging to the landowner.

Chapter 8*: Management of CPRs: A Formal Model

$$\Pi^*/N > 0 \tag{8*.16}$$

Without loss of generality, imagine that there is a single herder, j, who is discriminated against by all others acting as a group. Suppose now the group insists on a limit, $\underline{\pi}$ on the profit j is permitted to earn, such that

$$0 < \underline{\pi} < \Pi^*/N \tag{8*.17}$$

Inequality (8*.17) says j is so restricted in the number of cattle he is permitted to introduce into the CPR that the maximum profit he would be able to earn, although positive, would be less than the fair share of aggregate profit in the unmanaged CPR. That is exploitation, in the sense that j is worse off in the managed CPR than he would have been had there been no attempt at collectively managing it (Inequality (8*.17)). We now confirm that it is because $\underline{\pi} > 0$ that the group is able to exploit j.

How might the group ensure that j does not earn more than $\underline{\pi}$? It can do that by threatening to swamp the field with cattle – to the point where herding generates no profits – should j not comply with the restriction imposed on him. Formally, imagine that the group threatens to introduce $\underline{X}$ cattle, where $F(\underline{X}) = p\underline{X}$. Herder j would be even worse off if the threat was carried out, because his earnings would then perforce be zero.

The question is whether the threat is credible. It can be made credible by recourse to a social norm that was discussed in Chapter 6, where anyone who refuses to join the rest of the group in carrying out the threat is punished in the same way as j would be punished, and so on recursively, to higher order violations of the norm. It follows that an outcome where j earns $\underline{\pi}$ and the exploiters enjoy the maximum rent they are able to earn is an equilibrium. Of course, in equilibrium the threat does not have to be applied, because it is in j's self-interest to comply (remember, $\underline{\pi} > 0$). The irony is that an unsuspecting visitor could think the community is rich in social capital, but it appears to be rich in social capital only because exploitation is able to masquerade as cooperation .

Chapter 9
Human Institutions and Ecological Systems, 3:
Consumption Practices and Reproductive Behaviour

Introduction

Chapters 7 and 8 studied externalities that travel in the material world. In fact, externalities are embedded in a larger space. The social world can be as powerful a carrier of externalities as the material environment. A common form in which they appear in the social world is in the way our relationships influence our preferences and wants.

In Chapter 6, aspects of human relationships were studied in terms of the underpinnings of laws and social norms. Here we further probe the factors that are at the heart of human relationships. The features of personhood we examine here will be found to be of special significance for policy, for they tell us about the human motivations that drive population numbers (N) and the standard of living (y), both of which appeared in the Impact Inequality (Chapter 4).

The influence of our past choices on our current desires and wants is an externality that affects each of us. Even – perhaps especially – our aspirations are influenced by our own past experiences. The psychological phenomenon may explain the striking finding by Kahneman and Tversky (1979) that our aversion to losing a unit of income far exceeds our sense of gain from an additional unit of income, a feature of our makeup the authors named Loss Aversion.[269]

The effect of someone's past choices on her current well-being is a type of externality that in principle can be internalised by the person herself. If she can anticipate the direct effect of what she is doing now on her well-being in the future, she will want to adjust her choices to take account of that, provided of course she cares about her future well-being. We study that corrective in Chapter 10. Here we are concerned with a class of externalities that are harder to internalise. They arise from those of our desires and wants that are influenced by what others do.

Textbook economics, even graduate textbook economics, takes the human person to be an *egoist*. That same assumption influences the way income-expenditure models that government ministries of economics and finance use to determine their nation's public finances are interpreted.[270] In some aspects of behaviour, we are surely egoistic, but in other aspects we are socially embedded; we look to others when acting. Even in the latter, our motivation is not uniform. In some spheres of life, we are

---

[269] This is closely related to the Endowment Effect (Kahneman, Knetsch, and Thaler, 1990), which is that we are more likely to retain an object than to acquire an identical object. Economists translate that as our 'willingness to accept' an object exceeds our 'willingness to pay' for it. See also Tversky and Kahneman (1992) for a synthesis of these findings.

[270] An account of the vast empirical literature on the consumption and saving behaviour of egoists is in Attanasio and Weber (2010). A special session at the 2011 annual conference of the Royal Economic Society was devoted to revealed preference theory as applied to egoists. See Vermeulen (2012).

competitive; in others, we are conformists. The economics of biodiversity remains seriously incomplete when our innate sociability is not acknowledged. Nor is policy well informed when we neglect our social embeddedness.

In this chapter, we explore the reciprocal externalities that appear in those spheres of life that are driven by our socially embedded preferences. Our account of socially embedded preferences is illustrated by aspects of consumption practices in high income countries and reproductive behaviour in low income countries. The analysis points to patterns of behavioural change that would reduce our impact on the biosphere (N and y in the Impact Inequality) without leading to the extent of loss in personal well-being that models of the egoistic consumer say wouldbe entailed.[271]

9.1 Socially Embedded Consumption Preferences

The term *socially embedded preferences*, or *social preferences* for short, will be used here to identify a person's motivations when their behaviour and practices are influenced by the behaviour and practices of others. It is significant that the pressure we are subjected to when our preferences are socially embedded comes from within ourselves, they are internal pressures. In contrast, the pressures that we experience from the play of social norms arise from external sources. In the latter class of situations, we discipline ourselves, perhaps even unconsciously, to avoid the consequences that would follow if we were to behave otherwise (Chapter 6).

Two broad classes of social preferences have been found to be empirically significant: *competitive* and *conformist* (Figure 9.1). The former, as in the phenomenon of conspicuous consumption – first studied by the sociologist Thorstein Veblen (1925, [1899]) – displays a desire for high social status relative to that of others. The latter embodies a desire to be like others, to not stand out; meaning that we prefer that our choices be close to the choices made by others, rather than be distant from them (fads and fashions in the choice of food, clothing, home appliances, reading, entertainment are only the most striking examples).[272] The former gives rise to a pressure for higher consumption and thus for a profligacy that we would ourselves acknowledge; whereas the latter displays no clear sign as to whether our consumption practices are moderate or profligate.[273]

Conformism can direct us to be profligate, but we would be profligate only because others are profligate. So, it may be difficult empirically to distinguish the two explanations for profligacy – competitiveness and conformism. Nevertheless, conformism is distinct from competitiveness. One reason is that if we were conformists, we would be moderate in our consumption practices if others were moderate in their consumption practices. Each practice could be held together by its own bootstraps, so to speak. In fields of choice where our social preferences are conformist, our demands from the biosphere cannot be predicted *ab initio*.[274] That indeterminacy, as we see below (Section 9.2),

---

[271] The material in this chapter has been adapted from A. Dasgupta and Dasgupta (2017) and Barrett et al. (2020). We are grateful to John Bongaarts, Aisha Dasgupta, Patrick Gerland, Paul Ehrlich and Cosmas Ochieng for their most helpful comments on a previous draft.

[272] Just which population's average behaviour influences our own behaviour is a matter of importance. It could be that people want to engage in communal activities and identify commodities that serve as focal points, or it could be a desire for shared experience (Section 9.2). Trentmann (2016) is a monumental study of the emergence and growth of modern consumption practices since the Early Modern period. Commodities whose demand is in part an expression of the desire to relate to others have been called 'relational goods' (Donati, 2011). In many cultures, religious expenditures are built around them (Iannaccone, 1998). Gilchrist and Sands (2016) provide evidence that the cinema is a relational good. Conformist preferences are discussed also in Sections 9.3 and 9.4 over reproductive behaviour and are given a wider interpretation. In Annex 9.1, it is confirmed that while competitive preferences have the flavour of the Prisoner's Dilemma (Chapter 6), conformism can be exemplified by 'coordination games'.

[273] Socially embedded preferences over commodity categories can in principle assume forms that neutralise sociability, to the point where the choices they elicit cannot be distinguished from the choices made by an egoist (Arrow and Dasgupta, 2009). But as they are knife-edge cases, we ignore them here.

[274] In Chapter 6 we noted a similar indeterminacy in the level of trust in societies. That led us to observe that trust can be fragile.

gives substance to a common observation that consumption patterns in the affluent West is wasteful, worse that it damages biodiversity. It also offers hope that reductions in some forms of consumption may cost far less in psychological terms than they would were the human person an egoist in all manners of consumption.[275]

Figure 9.1 Competitive and Conformist Social Preferences.

To illustrate, imagine as a thought experiment two islands that are identical in all respects, including that consumption choices are driven by the same structure of conformist social preferences. Nothing to distinguish the islands, you might think. But you would be wrong, for it could be that people in one island are moderate while in the other they are profligate. It could also be that inhabitants in the island where moderation is practised enjoy a greater sense of well-being than those in the island where people are profligate. By assumption, though, the islands are identical in their ecology and human motivations. An external observer would be unable to explain why, nevertheless, inhabitants in the two islands behave so differently. In due course, the islands would diverge in other ways also. Their histories would be found to differ. It could even be that the island where moderation is practised enjoys a sustainable state of affairs, whereas in the latter island the Impact Inequality is in play and their future is threatened. But they would have started at the same place (Annex 9.1).

The above thought experiment has been designed deliberately for its artificiality. In fact, of course there would be differences, howsoever small, between any two societies. So we now imagine that some small differences in the two islands veer their practices in the two directions, one island-society toward moderation, the other toward profligacy. If we now add to the experiment the thought that the socio-ecological world is non-linear, experiencing positive feedback in its processes (Chapter 3), those small initial differences would get amplified. Divergence between the islands could be rapid even in historical time.

This is, of course, the stuff of historical studies. The rich, illuminating debate among historians of the origins of the Great Divergence between the West and the rest of the world is not that, in terms of macroeconomic

---

[275] The classic on wasteful consumption is Galbraith (1958). But the author's focus is on private affluence versus public squalor in the US, not the burden modern patterns of consumption place on biodiversity.

Chapter 9: Human Institutions and Ecological Systems, 3

indicators, differences between China and the West (Europe) in, say, year 1500 were large (Chapter 0); quite the contrary, the debate is over the character of their differences that amplified over the following centuries to an extent that it was the West that gave rise to the modern world, not China.

Behaviour directed by social preferences has externalities built into it. The externalities have biospheric implications. That competitive preferences over consumption patterns lead to socially excessive consumption is well known, as evidenced by the term 'consumerism', which carries a pejorative note with it. People acknowledge the force of the 'rat races' they join in search of consumption experiences. However, other than fads and fashions (e.g. Aspers and Godart, 2013), conformist preferences for consumption goods would appear to have been less studied by econometricians.

Although conformist preferences over *reproductive* behaviour can amplify the tragedy of the commons (Section 9.3), they appear infrequently in demography, perhaps because of concerns about state intervention in what are regarded as personal matters. Social preferences over consumption goods, in contrast, are mentioned in the media on occasion, but they have not yet entered policy discourse, perhaps because they too are thought to be in the private domain. But because social preferences give rise to externalities, behaviour responding to them should not be regarded as purely personal matters. Moreover, behaviour based on such preferences can be shifted by social mechanisms that do not involve state directives but instead bottom-up social mechanisms. The latter are likely to be better placed to bring about ecologically desirable changes, which is why they have implications for how we should view policies that directly affect reproductive behaviour (Sections 9.4 and 9.5). Consumption and reproductive externalities in conjunction with environmental externalities (Chapters 7 and 8) lie behind the Impact Inequality of Chapter 4.

Our behaviour over some aspects of consumption are competitive; in others they have been found to be conformist (Douglas and Isherwood, 1979; Trentmann, 2012, 2016; Ng and Lee, 2015); and in others still they appear to resemble those of the egoist, familiar in economics textbooks and official income-expenditure models. In a nearly forgotten work, Duesenberry (1949) applied the idea of competitive social preferences to explain business cycles. He used the term 'demonstration effect' to denote the macroeconomic implications of such preferences. Duesenberry even expanded the notion of social preferences by labelling one's past selves as 'others'. In that reckoning, a person's consumption choice today would be coloured by their own consumption choices in the past (Box 10.6). People commonly use the term 'habit', less commonly 'history dependent aspirations', to describe the phenomenon.[276]

Global output of final goods and services (i.e. global Gross Domestic Product (GDP)) is today round 135 trillion dollars PPP at 2019 prices. The World Bank (2019) reported that the approximately 1.2 billion people on its list of high income countries produce approximately 64 trillion international dollars of output.[277] Assuming for simplicity of calculation that our ecological footprint is proportional to the scale of economic activity, close to 50% of humanity's impact (US$64 trillion/US$135 trillion) on the biosphere can be attributed to some 16% of world population. In contrast, the approximately 670 million people in the World Bank's list of low income countries, which are mostly in sub-Saharan Africa, produce 1% of global output, which translates into about 3% of the global ecological footprint if we use Figure 4.10 for the computation. So, the world's poorest countries cannot even remotely be held responsible for the size of the Impact Inequality today. Table 9.1 contains figures for population and output shares in countries in the four income categories proposed by the World Bank. It also contains figures for total fertility rates (TFR), which are discussed in Section 9.4.

---

[276] The material in this section has been taken from Barrett et al. (2020).

[277] We cite estimates that are even rougher than their published versions, to signal that it is all too common to regard the published figures to be a lot more accurate than they actually are. Throughout we round the numbers. The World Bank (2019) classifies high income countries as those where average income is 52,000 dollars PPP, and upper-middle income, lower-middle income and low income countries as those where average incomes are 17,000 dollars PPP, 6,800 dollars PPP and 2,500 dollars PPP, respectively. See also Table 11.1, where we study a variety of indicators of well-being, including income.

Table 9.1 Income and Population Shares

Countries (by income, dollars PPP)	GDP (2019 prices)	% global GDP	% global population	TFR
High income	62	47	16	1.6
Upper-middle income	48	37	37	1.9
Lower-middle income	19	15	38	2.8
Low income	2	1	9	4.6

Source: World Bank (2019). Figures rounded.

That is the current state of affairs. Output growth in the rest of the world can be expected to make yet further demands on the biosphere. UNPD's median projection of global population for year 2100 is 10.9 billion (UNPD, 2019b). If, as a reasonable aspiration, per capita global output in year 2100 was to be 30,000 international dollars at 2017 prices (which is around the 75th percentile on the distribution of GDP per capita across countries at present), global output at a population size of 10.9 billion would be about 330 trillion international dollars. The UN's Sustainable Development Goals (SDGs) speak of environmental protection, but would not appear to have been influenced by humanity's prospective demands of the biosphere's goods and services, nor of its assimilative capacity.[278] If that demand is to be brought down to sustainable levels, that is, if the reduction is to convert the Impact Inequality (Equation (4.4)) into an Impact Equality, either global output (Ny) will have to decline substantially or technological advances and institutional reforms will have to reduce the environmental impact of high consumption levels (i.e. α will have to be increased). In Sections 9.4 to 9.7, we explore ways in which growth in N could be reduced. But as reducing y as remedy is also typically neglected, we study it first.[279]

9.2 Consumption Practices

To illustrate how an environmentally pernicious category of goods can get established in consumption practices, imagine that there are two categories of consumption goods. Category-1 goods are intensive in the use of the biosphere's goods and services, whereas the production of category-2 goods is intensive in the use of produced and human capital. If the biosphere was freely accessible, category-1 goods would be cheaper than category-2 goods, other things equal. It is then easy to see why people would have been drawn to category-1 goods in the past, and how, owing to the force of socially embedded preferences that initial bias would have triggered further demands for category-1 goods, giving rise to an ever-expanding demand for them. Nor would we expect technological advances to be directed at lessening the production costs of category-2 goods. Quite the contrary, competing entrepreneurs would invest in technologies that economise on factor inputs other than natural capital in the production of category-1 goods. And that would further fuel the global demand for category-1 goods. Moreover, the greater the difference between the market costs of producing goods in the two categories, the bigger would be the social effort required to move consumers away from category-1 goods to category-2 goods. The difference between the two cost structures would be further accentuated by the presence of fixed capital in the sector producing category-1 goods. Retiring

---

[278] In Chapter 4 (Box 4.6), it was shown that the SDGs suffer from more acute quantitative deficiencies.

[279] For simplicity we are identifying y with consumption. Consumption and its impact on the biosphere are also discussed in other chapters. See Chapters 14–16 for discussions on the distribution of y among and within countries, trade, and the increase in our harvesting of provisioning goods. Here, we focus on the socially embedded nature of consumption choices.

machines and equipment before their intended dates can be costly. We are experiencing that phenomenon even as nations try to move away from industries based on fossil fuels.

The current structure of market prices works against our common future; the biosphere is precious but priced cheaply, if it is priced at all. Worse, owing to a wide range of government subsidies, some services come with a negative price. To shift consumption patterns in high income countries, and the aspiring consumption patterns in the rest of the world, away from resource-intensive goods and services will require massive, coordinated actions. In a different context, economists use the term 'big push' to indicate the investment expenditures and collective determination that are required to move an economy from one 'societal equilibrium' to another, superior, one. Chapter 13 studies how decision-makers could locate public policies for implementing the transformative changes that are now needed in the global economy.[280] Chapter 21 outlines options for transformative changes in more detail.

When social preferences dominate consumption choice in a category of goods, the psychological cost to a person of a collective reduction in consumption in that category is likely to be a lot less than what it would be if she were to reduce consumption unilaterally (Annex 9.1). It has been argued, for example, that as between two societies, one in which everyone owned a large house and had little time for exercising or socialising, and another in which everyone owned a modest-sized home but had more time for these other activities, subjective well-being would be higher in the latter society (Frank, 1997). The point made by the author is that the size of a house is a 'positional good'; people care more about the size of their own house relative to that of others than about the absolute size of the house they live in. That reflects competitive social preferences. By limiting house size, resources can be devoted to activities that are known to increase well-being. The argument can be extended, as Frank did, to shifting expenditure away from positional goods towards public goods, including a healthier biosphere.[281]

That consumption is the activity around which human relationships are built and maintained has been found time and again by historians and anthropologists in their studies of traditional societies. Feasts to mark occasions of significance (birth, puberty, marriage, death, harvests and the annual renewal that is spring) are a recurrent theme in their writings. The epic poems of Homer and Vyasa are full of it. There is hardly a book in the *Odyssey* that does not have a line-after-line description of the roasting, carving and communal eating of meat and drinking of wine. And no opportunity is lost in the *Mahabharata* for describing the extravagance with which the king entertains his kinsmen and guests to mark a sacrifice to the gods or to celebrate the establishment of a capital for his kingdom.

Historians of consumerism are aware that the act of eating food communally has been a salient feature of societies everywhere, that feasts have bound societies together since time immemorial. What contemporary historians have sought to understand are the societal changes that led to the democratisation of eating meals together beyond the household unit. The feasts described in epic poetry are enjoyed only by the aristocrats, we are told nothing of how common people celebrated special occasions. The modern historian in contrast is at pains to understand the growth of inns, cafes and restaurants in the general development of societies in the West since the Late Middle Ages and Early Modern times. The changing patterns of consumption (food, clothing, housing) accompanying technological change and consumption practices are reflected in the magnitude of y in the Impact Inequality.

To date, socially embedded preferences have been uncovered mainly by historians, sociologists, and social and behavioural psychologists. What we have from the historians are narratives on the formation of tastes and changes in consumption practices over time (e.g. Trentmann, 2012, 2016); what we have

---

[280] Krugman (1991) used the mechanism based on self-fulfilling expectations we study here to analyse Keynesian unemployment in advanced economies.

[281] The term 'positional' to refer to goods for which people have competitive social preferences is due to Hirsch (1976).

from sociologists are analyses of large-scale market data on contemporary consumption practices (e.g. Warde, 1997; Warde and Martens, 2000); and what we have from social and behavioural psychologists is a rich set of findings from the study of natural and randomised controlled experiments (Section 9.3), and from large-scale surveys on stated happiness and life satisfaction (they are discussed in Chapter 11). The psychological drivers in each of these classes of studies have been identified as residing in social preferences.

9.3 Induced Behavioural Changes

The COVID-19 pandemic offers a natural experiment for studying the force of socially embedded preferences. To combat the virus, draconian restrictions on work, leisure and travel have been imposed by governments throughout the world. Even casual introspection suggests that the world-wide acceptance of the measures is a signal that social preferences have been at work. Bavel et al. (2020) is an early exploration into this.

Socially embedded preferences are able to operate only when behaviour is observable, as the term 'conspicuous consumption' makes clear. It is a lot easier, for example, to encourage people to buy electric cars than it is to persuade people to decarbonise the heating systems in their homes (the latter change remains unobserved by others; Stern, 2014). Appealing to people's sense of identity or social aspirations can have a powerful impact. A campaign in Vietnam aimed at reducing demand for rhino horn among businessmen illustrates this. Rhino horn is a high-status object – horns are sought after as medicinal ingredients, luxury possessions or gifts. Their most prolific users are male members of the business community. The campaign targeted the community's aspirations and beliefs and presented a positive image of self-created success and developed a narrative around 'making your own good fortune' in place of rhino horn. Over the course of the campaign, self-reported rhino horn use declined from 27% in 2014 to just 7% in 2017 (TRAFFIC, 2017; BIT, 2019).[282]

Reliable information about behaviour among peer groups has an impact; the behaviour does not have to be observed directly. A study of farmers' uptake of a payment for ecosystem services (PES) scheme in China's Wolong Nature Reserve found that providing farmers with information about the norms of engagement with the scheme in their neighbourhood group made farmers significantly more likely to re-enrol themselves (Chen et al. 2009). Even information that people around you are changing their behaviour can have an effect. Sparkman and Walton (2017) reported that when cafeteria diners in their study were told that an increasing proportion of people were reducing their meat intake, they were twice as likely to order a meat-free option than if they were given information about population-level behaviour without reference to change.

When people know their behaviour is being observed by their peers, they are more likely to stick with a behaviour they have openly committed themselves to. One example comes from the Northern Republic of Congo, in an area where bushmeat is a common pool resource, and its hunting a major threat to wildlife. In an experiment (Marrocoli et al. 2018), people made individual decisions about the amount of time they spent hunting and farming under different conditions: (i) with no communication between groups, (ii) with communication but with no monitoring system, and (iii) with communication and a self-monitoring system. Communication paired with a self-monitoring system was successful in reducing hunting.

In contrast, there have been few econometric tests for uncovering social preferences from consumer expenditure data, and those that have been carried out reveal weak social effects. The demands made by data on the econometrics were until recently too great relative to the power of econometric theory to

---

[282] Aunger and Curtis (2016) is a powerful essay on effective ways to design studies aiming to uncover the drivers of behavioural change.

identify social effects on individual behaviour.[283] Information campaigns also shape behaviour. One US-based study found that electricity customers receiving information about their consumption relative to that of their neighbours reduced energy consumption by 2% (Allcott, 2011). Yet another randomised field experiment found that social comparison lowered water consumption by over 5% (Ferraro and Price, 2013). That a move to environmentally friendly behaviour would be a gesture toward our descendants is of course, an added bonus. As of now, evidence for the salience of social preferences for the economics of biodiversity comes from historians, sociologists and social psychologists, not from economists.[284]

Box 9.1
Prying Open Behavioural Anomalies

In a well-known body of research, social psychologists have suggested ways to induce change by prying open behavioural anomaly. Because the way options are framed influences what we choose (Tversky and Kahneman, 1981), default options are known to have powerful effects in shifting behaviour. That's called 'nudging' (Thaler and Sunstein, 2008). Reisch and Sunstein (2016) reported that when surveyed, people in most European countries approve of being 'nudged'. It would seem people are aware that the way options are framed matters. This has been recently deployed for the uptake of green energy, for the authors found that even when people say they would like to use a green energy supplier, few do. However, when a green energy supply was made the default, 69% of people in a trial of approximately 40,000 households ended up with one, compared to only 7% when it was it was not the default option (Ebeling and Lotz, 2015). Relatedly, hotels today routinely offer guests the default option of not having their bathroom towels changed.

Nudging people to do one thing rather than another can be achieved even through changing the environment in which people make their decisions. This includes cues in the physical environment, such as size, availability and position of objects or stimuli (Hollands et al. 2017). Social groups experiencing their environment together will be influenced in similar ways and change their behaviour as a group, without the process necessarily being conscious.

Expanding the available set of choices by including seemingly irrelevant ones can affect choice. When more healthy options and more vegetarian meals were available in cafeterias, calorie intake and meat consumption were both reduced (in separate experiments), despite meat and less healthy options still being available (Pechey et al. 2019). Changing portion sizes available also has an impact: when cafeteria portion sizes were reduced, calorie intake decreased significantly, but overall satisfaction with the meal remained much the same (Hollands et al. 2018).

There is evidence that making people associate choices with other things can alter their attractiveness. Using symbols or images can make people more (or less) likely to choose an option, the association of cigarettes with lung disease being a powerful example (Brewer et al. 2016). The positive effect of negative images has also been observed with unhealthy food (Hollands, Prestwich, and Marteau, 2011). Interestingly, positive images have been found to be have little impact on choice, which is why it has been argued that more people could be pushed towards sustainable

---

[283] Regressing individual behaviour against the general practice in society in cross-section data gives rise to what econometricians call an 'endogeneity' problem (see Blume and Durlauf (2001); Brock and Durlauf (2001)). The point is that as society's common practice is made up of individuals' practices, it is not a simple matter to purge the latter's influence on common practice and cipher the influence of the common practice on an individual's choice of what to practise. Natural experiments ease the problem, but they are hard to find; and randomised field experiments are expensive. See Kapteyn et al. (1997), Maurer and Meier (2008) and Blume et al. (2011) for an account of the technical problems that arise in determining the influence of others on one's choices, and also of ways to overcome the problems.

[284] For a study of the economic implications of social preferences that go beyond the competitive-conformist dichotomy, see Dasgupta et al. (2016).

consumption through labelling the environmental harms associated with a product rather than the benefits a product is responsible for (Van Dam and De Jonge, 2015; Scarborough et al. 2015).

Labelling and advertising have an impact on consumer choice. In an effort to combat childhood obesity, Chile introduced a law in 2016 mandating labels on sugary drinks which showed their negative impact on health and restricting child-directed marketing. Data on purchases between 2015 and 2017 showed that after the regulation was implemented, unhealthy drink purchases decreased by 24% per day (Taillie et al. 2020). As with negative image associations on products, negative labelling would seem to have more influence than positive labelling.[285]

9.4 Factors that Slow Fertility Transition

The United Nations Population Division's median projection of world population in year 2100 is 10.9 billion, with a 95% prediction interval of 9.4 to 12.7 billion (UNPD, 2019b). More than three-quarters of the increase from today's 7.7 billion is expected to be in sub-Saharan Africa, where population in 2100 is projected to rise from today's approximately 1.1 billion to 3.8 billion (Figure 9.2), with 95% certainty that it will be somewhere between 3 and 4.8 billion. Comprising around 14% of the world's population today, the region represents only a bit more than 3% of the world economy (World Bank, 2019) and only 6% of the global ecological footprint (Chapter 4, Figure 4.10). So sub-Saharan Africa cannot be held responsible for the global Impact Inequality we face today. However, in a study on the sub-continent facing dire need to raise the material standard of living, Juma (2019) has pointed to the social stresses a slow demographic transition is likely to give rise to in the face of declining biodiversity.

The SDGs are reticent about family planning, and yet it is hard to imagine that they can be met without addressing the subject (Box 4.6 explained why). Abel et al. (2016) have offered a projection for global population in year 2100 in a narrow band of 8.2–8.7 billion. The range lies below the lower bound of UNPD's 95% prediction interval, but the estimates in their paper are based on an assumption that the SDGs will be achieved by year 2030. As that is increasingly unlikely, we turn instead in Box 9.2 to a study published in 2020, and which elicited much interest in the media for its population projections for 2100.

Box 9.2
The World Under Faster Demographic Transition

Vollset et al. (2020) have forecast global population for 2100 based on two assumptions that are notably different from UN Population Division (2019): (i) sharper declines in fertility rates in countries that have already undergone a fertility transition; and (ii) earlier declines in the TFR in low income countries. The authors' forecast of global population in 2100 is 8.79 billion, with a 95% prediction interval of 6.8 to 11.8 billion. Their forecast for sub-Saharan Africa is an increase of approximately 2 billion from today's 1.1 billion, which is fewer by 700 million than the median projection of UN Population Division (2019). As a global population of 8.79 billion in 2100 is lower

---

[285] The behavioural anomalies illustrated in this box are *not* manifestations of socially embedded preferences. They are anomalies, in that they appear to be irrational. What the examples illustrate is that such anomalies can be used to shift choices in directions that every chooser favours. Of course, it may be that what appears as irrational behaviour today has its roots in rational responses to problems our distant ancestors faced; that is, behavioural biases today may be relics of survival strategies in the distant past. For an exploration of this line of thinking for understanding the time-varying discount rates we humans have been found to deploy in our saving behaviour, see Dasgupta and Maskin (2005).

than even the lower bound of 9.4 billion in UNPD's 95% prediction interval, the publication has created much media interest.

In fact, differences between the two sets of projections for global population become significant only after year 2070. Until then, there is less than 5% difference between the two. For example, the difference between the two projections for global population in year 2050 is 211 million. Of a median projection of 9.7 billion in UNPD (2019b), that is negligible.[286] The projections' closeness until 2070 matters far more than their differences for year 2100. Nature responds to the demands we make of it; it does not calculate rates of change in demands, nor rates of change of rates of change in demands. If a few more planetary boundaries are breached by year 2070, differences in population projections for year 2100 will count for little.

Using the same estimates as were reached in Box 4.6, it can be shown that if the Impact Inequality is to be converted into an Impact Equality by year 2050, the efficiency with which natural capital is converted into final goods and services has to grow at an average annual rate of 4.2% until then. That contrasts with the average annual growth rate of 2.7% in α that the global economy has experienced since 1970.

Global population grew sharply in the years following the mid-20th century (Figure 4.2), because the substantial reductions in death rates owing to advances in medicine and public health practices were not matched by reductions in fertility rates. Two other factors that influence a society's population size are its age distribution and fertility rates. The TFR in a population is the average number of live births per woman over her reproductive years (taken to be 15–49).[287] For any given fertility rate, population grows at a faster rate in a society where the age distribution is tilted toward young people. In what follows we use the terms TFR and fertility rates interchangeably.

If a society is at its *replacement fertility rate*, then as the term suggests its population will be stable in the long run. The replacement rate depends on mortality rates, but as rule of thumb it is taken to be 2.1. Table 9.1 shows that fertility rates in richer groups of countries are higher. Today, TFR in sub-Saharan Africa is around 4.7, in contrast to a world average of 2.5. The fertility rate in India has fallen to 2.2, while that in China (at 1.7) is below the replacement rate (UNPD, 2019a). We should remember though that the global ecological footprint depends on *absolute* population size. A population can be stable, but if large it would have a big footprint, other things equal, and could bring the biosphere into disrepair. That is why the replacement fertility rate does not have as significant a place in the idea of 'fertility transition' (the transition from high fertility rates to the replacement rate) in the economics of biodiversity as it does in demography. The economics of biodiversity encourages us to look for transitions that dip below the replacement rate before tending toward it. As Table 9.1 shows, high income and upper-middle income countries, seen as groups, display that feature.[288]

---

[286] We are grateful to Patrick Gerland of the United Nations Population Division for clarifying this for us.

[287] TFR is calculated from current age-specific fertility rates (viz. number of children born last year per 1000 women aged 15–19; the number of children born last year per 1000 women aged 20–24; and so on).

[288] O'Neill et al. (2010) have sketched scenarios of lower global population growth that lead to reductions in greenhouse gas emissions by 16–29%. And yet, the Paris Agreement of December 2015 on climate change made no mention of population.

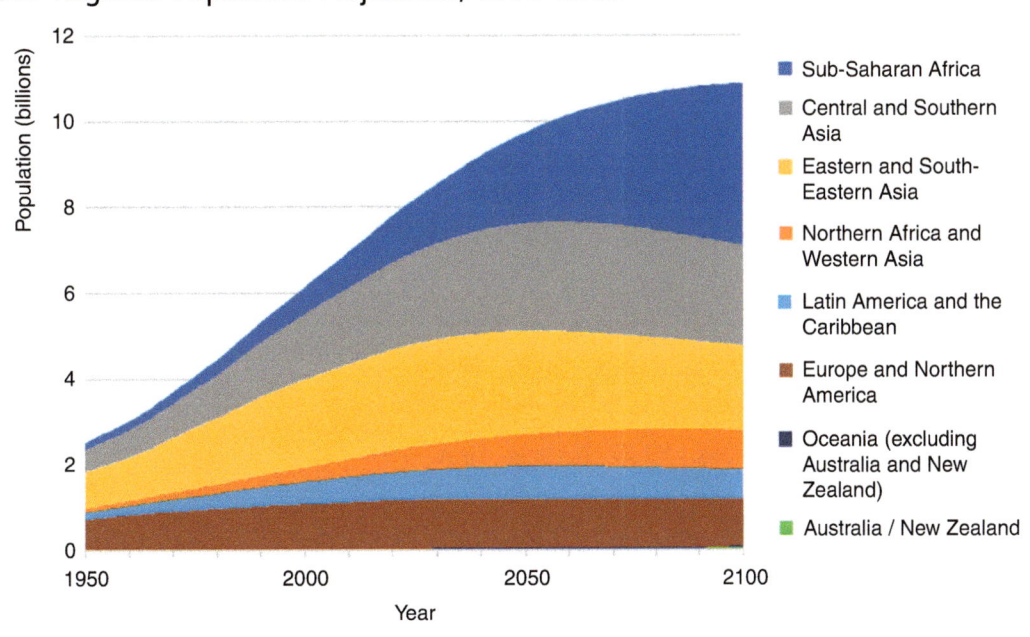

Figure 9.2 Regional Population Projections, 1950–2100

Source: UNPD (2019).

When speaking of average behaviour in large populations there is a danger that differences within them are ignored. Fertility behaviour is not uniform in India. Nor is it uniform in China, where fertility rates are generally higher in those regions in which reproductive restrictions have not been applied by the government. Nevertheless, it can prove useful to speak of average behaviour in regions, especially for macroeconomic reasoning. When demographers speak of fertility rates in sub-Saharan Africa (Bongaarts, 2011; Bongaarts and Casterline, 2013), they recognise that there are differences within the sub-continent, especially between high income urban households and low income rural households, but that there are sharp differences between the average in the sub-continent and elsewhere. Exceptions do not make averages false. Moreover, anthropologists and historians have shown that practices can be so tied to the past that they display inertia. So, it is natural for the social scientist to try to identify institutions and practices that influence fertility rates no matter where. Below we attempt to do just that.

As an organisation for production, consumption and fertility decisions, the 'household' is not as apt a term when applied to sub-Saharan Africa as it is for the Indian sub-continent; for example, in rural areas there is often no common budget for a man and wife (Goody, 1989; Caldwell and Caldwell, 1990). In an early review of fertility intentions, Cochrane and Farid (1989) noted that after distinguishing urban from rural people and literate from illiterate people in sub-Saharan Africa, households in that region had more, and wanted more, children than their counterparts in other less-developed regions. Even young women expressed a desire for an average of 2.6 more children than women in the Middle East, 2.8 more than women in North Africa, and 3.6 to 3.7 more than women in Latin America and Asia. Updated versions of these figures are available, but these data from the mid-1980s are especially striking because the income gap between sub-Saharan Africa and the rest of the developing world was smaller at that time than it is now. The problem for the social scientist is to understand reasons behind the difference.[289]

---

[289] It helps to keep extreme examples in mind. UNPD (2019) reported that Niger's TFR in 2019 was 6.8. Only 15% of women in the age range 15–49 were estimated to use modern methods of contraception (UNPD, 2020). Even though contraceptive use is low in Niger, only 16% of

Sub-Saharan Africa's slower fertility decline has been traced by scholars to many aspects of life that held traditionally and although they are undergoing change, they continue to be different from other regions with low income and low-middle income countries. They include kinship-held agricultural land, the practice of polygamy, lack of access to modern methods of contraception, low education among women, and kinship obligations to share the fruits of effort.[290] Here we examine a pathway that has been less discussed, but that offers an explanation for why the desired number of children has remained substantially higher in sub-Saharan Africa than elsewhere. The account of the pathway combines history with social mechanisms as seen through the lens of economics.[291]

A potential source of reproductive externality is the wedge between the private and social costs of childrearing. The costs borne by parents are lower when childrearing is shared among kin than when households are nuclear. Fosterage within the kinship would appear to remain more prevalent in sub-Saharan Africa than elsewhere. The responsibility for raising children is more diffuse within the kinship group there than elsewhere (Caldwell, 1990; Bledsoe, 1990, 1994). Studies in West Africa revealed that up to half the children were living with their kin at any given time. Nephews and nieces were found to have the same rights of accommodation and support as biological offspring. In that cultural setting, children are seen as a common responsibility, which makes it important that in surveys that seek to identify desired numbers of children it is made clear that the questionnaire refers to biological children. However, fosterage creates a free-rider problem if the parents' share of the benefits from having children exceeds their share of the costs. The corresponding externalities are confined to the kinship. Other things equal, reduction in those externalities would be accompanied by a fall in the demand for children (Dasgupta, 1993).

We do not claim that behaviour is necessarily based on conscious choices. There are demographers who argue that in preliterate societies behaviour was purely norm-based and that basic female education has helped to bring fertility into the calculus of conscious choice, with an attendant reduction in fertility desire. But conscious choice even in modern societies does not preclude the influence of others on what we do (Bledsoe, 1996). More importantly for fertility surveys in low income societies, women selling goods in the marketplace there may be illiterate, but they are never innumerate. We will make use of the distinction presently.

9.5 Socially Embedded Reproductive Behaviour

Fertility behaviour is not only influenced by private desires and wants; it is also shaped by societal mores, which is subsumed under social preferences. Anthropologists have reported that in parts of sub-Saharan Africa women acknowledge that they are able to acquire social status through reproductive success (Bledsoe, 1994). That attitude toward reproduction, which has been called 'children as wealth' (Guyer and Belinga, 1995), has a competitive edge to it.

However, as it is not known how prevalent that phenomenon is in the sub-continent, we focus on conformist preferences.

---

women said they did not wish to get pregnant and were not using contraception. They also said their desired number of children was 9.5. Income per capita in Niger is round 930 international dollars, which is only a bit above the World Bank's absolute poverty line of 1.90 international dollars a day. The country's population in 2020 was estimated to be 24 million and is projected to rise to 66 million in 2050 (UNPD, 2019).

[290] See for example Fortes (1978), Bongaarts, Frank, and Lesthaeghe (1984), Bledsoe and Pison (1994), Guyer and Belinga (1995), Kohler, Behrman, and Watkins (2001), and Bongaarts and Casterline (2013).

[291] But changes have been observed, certainly in East Africa. Fostering is declining, land registration is increasing, and urbanisation is eroding pro-natal institutions. Whether those changes will markedly influence fertility rates in the near future is unclear. And delay matters to Africa's prospects (Figure 9.2). We are grateful to John Cleland for helpful discussions on this.

We say reproductive behaviour is conformist when the family size a household desires stays close to the average family size in the community or, more broadly, in the world households come into contact with (A. Dasgupta and Dasgupta, 2017). As with conformist consumption practices, conformism in fertility preferences can give rise to more than one possible outcome. So long as all others aim at large families, no household will wish to deviate from the practice; if, however, all other households were to restrict their fertility, every household would wish to restrict its fertility. A society can thus get embedded in a self-sustaining mode of behaviour characterised by high fertility and stagnant living standards, even when there is another potentially self-sustaining mode of behaviour that is characterised by low fertility and rising living standards which is preferred by all.

Studies on fertility have pointed to choices that are guided in part by attention to others. In her highly original work on demographic change in Western Europe over the period 1870 to 1960, Watkins (1990) showed that differences in fertility and nuptiality within each country declined. She also found that in 1870, before the large-scale declines in marital fertility had begun in most areas of Western Europe, demographic behaviour differed considerably within countries. Differences among provinces within a country were high even while differences within provinces were low. Spatial behavioural clumps suggest the important influence of local communities on behaviour. In 1960, differences within each country were considerably less than in 1870. Watkins explained this by increases in the geographical reach national governments achieved over the 90 years in question. The growth of national languages could have been the medium through which reproductive behaviour was able to spread.

Watkins' study was historical, as were the studies Montgomery and Casterline (1998) used to distinguish pathways by which reproductive practices diffuse within a society. A similar analysis was conducted by Basu and Amin (2000) on the experience in West Bengal, India, where fertility rates declined in the early 1970s, ahead of the northern states of India and neighbouring Bangladesh. The authors attributed it to historical and cultural factors in the state that combined to promote interaction between the elite and the general public.

A feature of historical studies of the diffusion of behaviour across space and time is that they do not necessarily identify the behavioural fundamentals (or 'drivers') on which the diffusion process is built, but could be seen to differ from one another in regard to the transmission mechanism. The drivers could be knowledge acquisition, or they could be pure mimicry, or what Cleland and Wilson (1987) called 'ideation', or the advent of modernity, or the desire to belong to one's (possibly expanding) peer group, or the force of celebrity culture, and so on. These fundamentals are not unrelated, but they are not the same. It could be that people observe successful behaviour and copy it, or that the language in which newspapers are read spreads, or that people discuss and debate among themselves, and so forth.[292]

The notion of conformism we are adopting here is built on the common structure of all such diffusion processes. Studying the common structure offers the advantage that we are able to analyse the resting (i.e. equilibrium) points of a wide variety of diffusion processes without having to identify the processes themselves. The account presumes that fertility preferences are socially embedded, but it does not specify the reasons households are influenced by the behaviour of others. Being analytical, the account is able to entertain counterfactuals. It allows us to ask how a household's behaviour would change if other's behaviour was to change.

Conformist preferences are depicted in a stylised form in Figure 9.3. The curve ABCDE is a household's desired number of children, plotted against the average number of children per household (the

---

[292] Diffusion processes had been studied long before, in connection with technology adoption. In a classic paper, Griliches (1957) conducted an empirical study of the spread of hybrid corn in the US. In his model, farmers observed the successful adoption of new varieties before adopting them themselves. The process gave rise to the now-familiar logistic pattern of adoption. On the effect on reproductive behaviour of the diffusion of ideas brought about by family planning programmes, see Babalola, Folda, and Babayaro (2008) and Krenn et al. (2014).

horizontal axis) in the community. The curve is upward sloping and intersects the 45° line OF at three points, B, C and D, each of which is a social equilibrium. B and D are stable while C is unstable. In the figure, which is purely illustrative, every household desires d children if all other households have d children each; and b, if each among all others have b. (The same reasoning holds at C, where the number of desired children is c. But as C is unstable, we ignore it.[293]) Imagine now that every household prefers the outcome in which all households have b children each to the one in which all have d children each. As having either b or d children are both stable outcomes, a fertility rate of d would be just as tenacious as a fertility rate of b.[294]

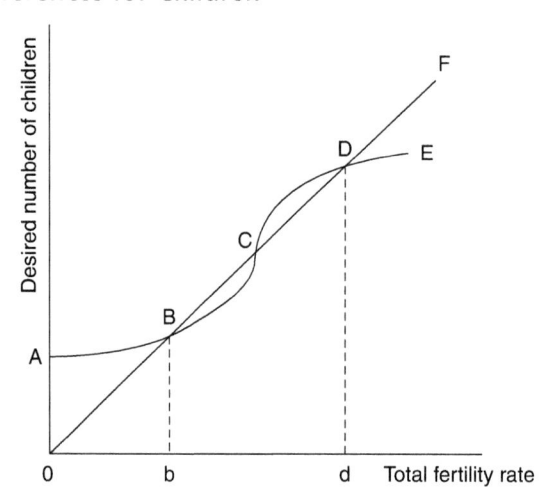

Figure 9.3 Conformist Preferences for Children

Source: A. Dasgupta and Dasgupta (2017). *Permission to reproduce from John Wiley and Sons, Inc.* Note: The curve ABCDE shows a representative woman's desired number of children as a function of the average number of births per woman in the population. B and D are stable equilibria; C is unstable.

Policies can be used to shift the curve ABCDE (and therefore the equilibrium points b, c, d) as well, thereby to influence the beliefs on the basis of which households act. Family planning programmes involving the participation of communities are a way to shift behaviour (see below). Fertility transitions can be interpreted as disequilibrium phenomena (Dasgupta, 2002), where practices change slowly in response to gradual changes in the social environment, until a tipping point is reached from which society shifts rapidly to a new stable equilibrium, say from high fertility to low fertility.

The common structure of diffusion processes that we are studying proves useful also for interpreting statistical regularities between actual fertility (TFR) and wanted fertility (WTFR). Pritchett (1994), for example, regressed TFR on WTFR in 43 countries in Asia, Africa and Latin America and found that about 90% of cross-country differences in TFR are associated with differences in WTFR. He also found that excess fertility (TFR − WTFR) was not systematically related to actual TFR, nor an important determinant of it. The author concluded that high fertility is due entirely to the strong desire for children.

The study has been found to be wanting (Lam, 2011). Nevertheless, imagine the finding had been robust. That fertility preferences are conformist tells us we should expect a correlation between TFR and WTFR, but it also warns us not to attribute causality to the relationship. For it would be as true to say

---

[293] To see why C is unstable, notice that if the average family size exceeded c, the representative household would desire to increase its size, and would do so as long as its desired family size exceeded the community average. The reverse argument would come into play if the population-wide family size was smaller than c. This same basic argument can be used to shown that b and d are stable family sizes.

[294] Annex 9.1 provides a numerical version of this possibility in the form of a simple social environment, known as a 'coordination game'.

fertility rates in those countries where they are high are high because people have a strong desire for children as it would be to say that people there have a strong desire to have children because fertility rates are high (A. Dasgupta and Dasgupta, 2017).

The theoretical observation that reproductive behaviour is guided by conformist preferences, more particularly that the drivers of such behaviour can give rise to any one among a multiplicity of reproductive outcomes (Dasgupta, 1993), was given additional support in a study of contraceptive use in rural Kenya, which found that in communities having dense social networks and a poorly developed market economy, a woman would be unlikely to use contraceptive methods if contraception use in her network was low, whereas she would be likely to use such methods if contraception use in her network was high (Kohler, Behrman, and Watkins, 2001). Community discussions on the benefits of smaller family size can coordinate households to act upon lowered fertility targets. Collective deliberation – at the national level 'national conversation' – is the democratic way for not only sharing information but also coordinating decisions.

That reproductive behaviour is guided by conformist preferences was also given support in an analysis of a natural experiment which found that state-level fertility rates declined in step following staggered introductions of cable TV across Indian states in the 1980s (Jensen and Oster, 2009). Further support has been provided in an analysis of contraceptive uptake in Bangladesh (Munshi and Myaux, 2006). The study concerned women living in the same community but belonging to different religious groups. After controlling for individual differences in education, age, wealth and the like, the study found that a woman's choice to use contraception depended strongly on the predominant choice made by other women in her religious group and was unaffected by the predominant choice made by women belonging to the other group.

Nobles, Frankenberg, and Thomas (2015) analysed data bearing on fertility responses in Indonesia to the 2004 tsunami in the Indian Ocean. The tsunami killed large numbers of residents in some communities but caused no deaths in neighbouring communities. The authors found that there was a sharp increase in fertility in the aggregate following the disaster; mothers who had lost children were more likely to bear additional children. Interestingly, the authors also found that women without children before the tsunami reproduced earlier in communities where tsunami related mortality rates were higher.

Persistent reproductive practices that go counter to present-day interests may of course have had a rationale in the past. But even when circumstances have changed, a society can remain stuck in a mode of behaviour characterised by high fertility (d) even when there is an alternative that is characterised by low fertility (b), which would be preferred by all. Newspapers, radio, television and the internet communicate information about other lifestyles. The media can be a vehicle by which conformism becomes based on the behaviour of a wider population than the local community, which disrupts existing practices. Education of women makes a break with tradition, partly because it delays marriage, partly because it leads to better birth spacing, and partly also because it offers a way toward economic security for the woman in an urban milieu.[295]

9.6 Importance of Investment in Family Planning and Reproductive Health

The transition from d to b children per household requires neither coercion nor taxation, nor even education. Conformist reproductive behaviour can be shifted by changing expectations about others'

---

[295] Farooq, Ekanem, and Ojelade (1987) is an early study of the phenomenon in West Africa. Lutz, Butz, and KC (2014) is a collection of essays on the effectiveness of education, especially women's education, in changing fertility behaviour.

fertility choices. Family planning programmes can be so designed as to encourage members of communities to share information about modern methods of contraception and discuss the advantages of smaller families (Box 9.2).

Currently the EU to Africa Official Development Assistance (ODA) towards 'family planning and population services' is less than 1%. The World Bank also attaches low priority to it. Moreover, developing countries themselves relegate family planning to minor government departments. The introduction of family planning-related media messages is credited in part for the rise in contraceptive demand and use in countries where governments have taken family planning seriously. Between 2000 and 2019, total demand for family planning (which is the sum of contraceptive use and unmet need) among women of reproductive age rose from 48% to 61% in Malawi, and from 27% to 44% in Rwanda (UNPD, 2020). In contrast, little or no change in demand was observed in those countries, such as Nigeria and the Democratic Republic of Congo, where family planning programmes remain weak.

Over the years, female education has been seen by development experts as the surest route to women's empowerment, including, for example, choice over birth spacing. All governments recognise the importance of women's education for empowering women. And yet even today nearly 30% of women between 15–24 years of age in low income countries are illiterate (World Bank, 2019). Family planning programmes, in contrast, are affordable by governments even in low income countries, they offer an easy and effective route for governments to empower women, and yet they remain low on the development agenda. It is a paradox.[296]

By providing access to subsidised contraceptive commodities and services, family planning programmes were successful in accelerating fertility declines in Asia and Latin America in the 1960s and 1980s. Cleland et al. (2006) estimated that promotion of family planning in countries with high birth rates had the potential at that time to avert more than 30% of all maternal deaths and nearly 10% of childhood deaths. The rationale for vigorously expanding the content and reach of such programmes today also lies in the more than 217 million women in developing, mainly low income countries who have reported they want to prevent pregnancy but are not using modern contraception. Among them over 150 million use no method of contraception and nearly 66 million rely on traditional methods (UNPD, 2020). The Guttmacher Institute (2020) estimated that there are nearly 111 million pregnancies in low and middle income countries annually that are unintended. The institute also estimated that if all unmet need for modern contraception were satisfied in developing countries, there would be a 68% decline in unintended pregnancies, amounting to around 35 million a year. Meeting unmet need for contraception would reduce pregnancy-related deaths by 70,000. Many unintended pregnancies end in abortion, a significant proportion of which are performed under unsafe conditions.

In addition to reducing unintended pregnancies, contraceptive use among women enhances their own health and that of their children by spacing births and providing greater opportunity for education. The Guttmacher Institute (2020) reported the estimate that more than 2.5 million babies in low and middle income countries die each year in the first month of life. Access to modern family planning is a means for women to have greater control over their lives and for improving the chance of having healthy babies. If the benefits of modern family planning are high, the costs are low. By one estimate (Guttmacher Institute, 2020), expanding and improving services to meet women's needs for modern contraception in developing countries would cost under US$2 a year per person. As routes to

---

[296] The significance of family planning programmes is reflected in the demographic histories of Bangladesh and Pakistan. In 1970 (i.e. before the partition of Pakistan into two states), the TFR was high in both Bangladesh and Pakistan – 6.9 in Bangladesh and 6.6 in Pakistan – but by 2000 it had dropped to 3.2 in Bangladesh but was 5.0 in Pakistan (A. Dasgupta, 2021).

fertility transitions, investment in community-based family planning programmes should now be regarded as essential.[297]

The questions directed at women to elicit their fertility desires are not framed in a way that acknowledges social preferences. The questionnaires do not ask women what their fertility desires would be were fertility choices of their peer group to be different from what they are currently. Figures for reported desire for family size are therefore very likely to be underestimates. We study the complexity of eliciting desired family size next.

9.7 Unmet Need, Desired Family Size, and the UN's Sustainable Development Goals[298]

UNFPA (1995) took it that family planning and reproductive health policies should address *unmet need*, meaning that they should be made to serve women aged 15 to 49 who are seeking to stop or delay childbearing but are not using modern methods of contraception (Bradley et al. 2012; Alkema et al. 2013). Although the idea of unmet need could appear straightforward, it has in practice been interpreted in different ways over the years. It is currently measured using more than 15 survey questions, including questions on contraceptive use, fertility intentions, pregnancies, postpartum amenorrhea, sexual activity, birth history and menstruation. Women's reported fertility intentions are inferred from such questions as: 'Now, I have some questions about the future. Would you like to have a(nother) child or would you prefer not to have (any more) children?' This is followed by a question about how long the woman wants to wait if she responded to the previous question that she does want a(nother) child.[299]

Unmet need is recognised by experts to be large. It is hard to measure, but there are good reasons to believe it is underestimated. Unmet need is determined from respondents' expressed want for children, taken together with responses to other questions. But in matters of life and death, needs assume an independent status; they even serve as the basis on which commodity rights are founded. The philosopher David Wiggins has argued that a statement of the form 'person A needs commodity X' is tantamount to a challenge to imagine an alternative future in which A escapes harm without X (Wiggins, 1987: 22). Expressed wants or desires for children, used to calculate unmet need for family planning, may not adequately convey her true need for family planning; that is, for her own best interests. To infer needs solely from wants is therefore to undervalue the significance of family planning. Moreover, none of the survey questions is conditioned on the behaviour of others.

Closely related to wants is the notion of desired family size, which is obtained from answers to the following question: 'If you could go back to the time when you did not have any children and could choose exactly the number of children to have in your whole life, how many would that be?' The WTFR is calculated by first dividing the number of observed births into those that occurred before and after the desired family size is reached (the former are considered as wanted, the latter unwanted). WTFR is then obtained with the same procedure as the one used in calculating TFRs (that is, from age-specific fertility rates), but only wanted births are included in the numerator of these rates.

There are dangers of biases in responses to the question at the basis of desired family size, but the requirement that family planning programmes have quantitative estimates of it is clear enough. Notice

---

[297] Of the many publications that have reported the benefits family planning services provide to poor households, see especially Cleland et al. (2006), Bongaarts (2011, 2016), Miller and Babiarz (2016) and Guttmacher Institute (2020). Box 9.3 provides an account of the Matlab (Bangladesh) experiment in family planning and reproductive health, 1977 to 1996.

[298] This and the following section are based on A. Dasgupta and Dasgupta (2017).

[299] Casterline and Sinding (2000) discuss ways in which the measure of unmet need can be used to inform family planning policies.

though that women are not asked what their desire would be if the prevailing fertility practices of others were different. In fact, there is no mention of the prevailing fertility rate. Since respondents are not invited to disclose their conditional desires, they very likely disclose their desired family size on the assumption that fertility will remain at its prevailing rate. A direct way to discover socially embedded preferences would be to reconstruct the questionnaires by asking a series of conditional questions, which we collapse here for convenience into one: 'If you could go back to the time when you did not have any children and could choose exactly the number of children to have in your whole life, how many would that be, assuming everyone else in your community had n children over their whole life?'

The survey could pose the conditional question in an ascending order of n, say from 0 to 10. We may now imagine that in Fig. 9.2 imagines the answers to $n = b = 2$; $n = c = 4$; and $n = d = 5$ are, respectively, 2, 4 and 5. It also imagines that answers to the questions in which $n = 0, 1, 3, 6–10$, respectively, differ from 0, 1, 3, 6–10; which is why the latter numbers are not social equilibria. No doubt responding to a string of conditional questions would tax respondents, but not to ask them is to misread fertility desires.[300]

Fabic et al. (2015) defined total demand for modern contraception to be the number of women who want to delay or limit childbearing (i.e. the sum of contraceptive users and women with unmet need). The role of family planning, the authors argued, is to supply that demand. The suggestion is that the success of family planning should be measured by the ratio of family planning users to the total demand. The United Nations has adopted this measure in SDG No. 3.7.1. It is known as 'demand for family planning satisfied with modern contraceptive methods', or 'demand satisfied' for short. Formally, if X is the number of women aged 15–49 who are users of modern contraceptives, Y is the number of women with unmet need, and Z is total demand for modern contraception, then $Z = X + Y$ and the UN's 'demand satisfied' is $X/Z = X/(X + Y)$.

Reproductive rights are at the heart of X/Z, which is its attraction. The indicator reflects voluntarism, rights and equity, informed choice, and the imperative of satisfying individuals' and couples' own choices with regard to the timing and number of children. But there are problems. The use of demand satisfied as the measure of success could create perverse incentives among programme managers. A programme's performance would improve if more women were to declare that they want to get pregnant. As long as women want many children, Y (unmet need) remains small, and therefore Z (total demand) is only marginally greater than X (the number of modern contraceptive users). The country scores well on the indicator demand satisfied and appears not to need further family planning programme effort. The success could mask a situation where contraceptive use is low and stagnant and high fertility rates persist. Moreover, fertility preferences, which contribute to the measurement of Y, are themselves influenced by the behaviour of others. Y could therefore be small in a society that harbours another outcome in which Y is large.[301]

If family planning programmes were strengthened sufficiently to meet unmet need everywhere in Africa, population there would be some 1 billion smaller in 2100 than is currently projected (Bongaarts, 2016). That in itself would be a substantial gain for people in the region. Family planning is undervalued where it is needed most. Greater investment in the service, bringing it into alliance with other social programmes, could be expected to reduce the population projections further (Box 9.2). Recognition that fertility practices are socially embedded and that family planning and reproductive health

---

[300] Because people's preferences differ, we should expect the responses to differ but discover that each individual's preferred number of children is an increasing function of n. That finding would reveal socially embedded preferences. Although the SDGs include universal access to sexual and reproductive health and emphasise reproductive rights, the family planning indicator that is advanced (SDG Indicator 3.7.1) is focused on the satisfaction of the individual woman's expressed desires; it does not say anything about the value in increases in the demand for family planning that would be prompted by community engagement.

[301] We are grateful to Aisha Dasgupta for drawing our attention to this.

programmes can be designed in ways that not only do not offend traditional mores but are welcomed by women to whom they are addressed, should allay fears that any talk of fertility reductions in poor countries is a tacit nod at command and control. Social embeddedness also offers hope to people everywhere that the environmental demands of the 1.2 billion people in high income countries can be reduced by the people themselves at little personal cost. For if relative consumption matters (Section 9.1), a reduction in consumption among all should not be expected to prove too costly to people. The population–consumption–environment nexus is one area of the human experience where the cost of necessary social change is probably much less than is feared.

Box 9.3
The Matlab Experiment, 1977 to 1996

The International Centre for Diarrhoeal Disease Research (Dhaka), with technical suggestions from Population Council, initiated an experiment in 1977 to test the benefits of family planning and reproductive health programmes.[302] A control area was chosen in Matlab, a sub-district in Bangladesh, where people received the same limited family planning and reproductive health services as the rest of the country. An experimental area in Matlab was provided with free services and supplies, home visits by trained female family planning workers, and a comprehensive media communication outreach to husbands and village and religious leaders to address potential familial and social objections. The results were impressive. Use of contraceptives jumped from 5% to 33% among married women of reproductive age (Bongaarts, 2016). A difference of about 1.5 births per woman between the experimental and control areas was observed until 1990, and a smaller difference continued until 1996 when the experiment ended. Among the long-term consequences of this difference in fertility rates were better educated children in the experimental area, larger households assets, greater use of preventive health services, greater birth-spacing and lower child mortality. Studying household data from the experimental area, Joshi and Schultz (2013) have reported that the decline in the TFR continued beyond 1996. The Matlab experiment has shown that family planning programmes can succeed in highly traditional societies. Its finding is consistent with the experience of countries that have adopted comprehensive family planning programmes, such as Iran in 1989 and Rwanda in the mid-2000s. Iran's TFR declined at a remarkable pace – from 5.6 in the late 1980s to 2.6 a decade later. In Rwanda, the TFR dropped from 6.1 in 2005 to 4.6 in 2010, and the proportion of married or in union women using contraceptives rose from 17% to 52%. Family planning programmes, when designed with the community in mind, are able to elicit the latent demand among women by providing supplies that match the demand.

Annex 9.1 Socially Embedded Preferences for Consumption: Formulation

Consider a large population in a timeless economy.[303] To avoid unnecessary notation, individuals are assumed to be identical. There are two goods, preferences over one of which relative to the other are socially embedded. The other good is a composite commodity which we call income (or wealth, in the timeless economy they are the same).

---

[302] The material here has been taken from Bongaarts (2016). For an early assessment of the Matlab experiment, see Caldwell and Caldwell (1992).

[303] Socially embedded preferences in an intertemporal setting are studied in Chapter 10.

Chapter 9: Human Institutions and Ecological Systems, 3

Let y denote an individual's income, c his consumption, and c^* the average consumption level in the population. An individual's preferences are assumed to be in the form of a quasi-linear well-being function v that reads

$$v(c,c^*,y) = u(c,c^*) + y, \qquad \partial u/\partial c > 0 \tag{A9.1}$$

An individual's sociality is reflected in the presence of c^* in the u-function, which we assume is concave in c and c^*. We assume as well that $\partial u/\partial c$ declines with increasing c (diminishing marginal well-being to increases in one's own consumption).

Let q be the price of the consumption good. (We may imagine the commodity is imported.) As individuals are identical, the condition that market equilibrium must satisfy is

$$\partial u(c,c^*)/\partial c = q, \text{ and } c = c^* \tag{A9.2}$$

In contrast, the socially efficient expenditure on consumption satisfies

$$\partial u(c,c^*)/\partial c + \partial u(c,c^*)/\partial c^* = q, \text{ and } c = c^* \tag{A9.3}$$

The externality arising from socially embedded preferences is reflected in the difference between the expressions in the left-hand side of equations (A9.2) and (A9.3). The equations match the directive of equation (7.1) in Chapter 7 that the accounting price of a commodity is its market price plus (or minus) the externality its use gives rise to. For we may re-write equation (A9.3) as

$$\partial u(c,c^*)/\partial c = q - \partial u(c,c^*)/\partial c^* = p \tag{A9.4}$$

In equation (A9.4), $\partial u(c,c^*)/\partial c^*$ is the corrective tax (or subsidy) and p is the accounting price of the consumption good. If individuals were to face p as the price of the consumption, they would consume the socially optimum quantity.

Competitive and conformist preferences are formulated next.

Competitive Preferences

We say that v represents competitive social preferences if

$$\partial u/\partial c^* < 0 \tag{A9.5}$$

Condition (A9.5) says that, other things equal, an increase in the average consumption reduces his sense of well-being. Thus while consumption matters to the person (Equation (9.1.1)), relative consumption matters too.

Condition (A9.5) implies that there is excessive consumption in market equilibrium (Equation (A9.2)). The resource waste accompanying competitive preferences resembles outcomes in the Prisoner's Dilemma game (Box 6.2).

It proves useful to add further structure to the u-function:[304]

$$\text{For all } c, c^*, \text{ and for all } \lambda > 1, u(\lambda c, \lambda c^*) > u(c,c^*) \tag{A9.6}$$

$$\text{For all } c \text{ and } c^*, \partial u/\partial c^* = \gamma \partial u/\partial c, \quad -1 < \gamma < 0 \tag{A9.7}$$

Blanchflower and Oswald (2004) estimated from the General Social Surveys of the US that an individual's sense of well-being would increase if everyone's income was to increase proportionately by $\lambda > 1$, but only by 2/3 as much if only his consumption was to increase by the same proportion. Formally that means

---

[304] The additional structure is taken from Arrow and Dasgupta (2009).

$$u(\lambda c,\lambda c^*) - u(c,c^*) = (2/3)[u(\lambda c,c^*) - u(c,c^*)], \text{for all } c,c^*, \text{all } \lambda > 1 \quad (A9.8)$$

Equation (A9.8) implies that $-1 < \gamma < 0$.

Conformist Preferences

Conformist social preferences point to a different aspect of human sensibilities. In a large population c^* is regarded by an individual as a parameter. If q is the market price of the consumption good, and individual would select a level of consumption c that satisfies

$$\partial u(c,c^*)/\partial c = q \quad (A9.9)$$

This gives rise a demand function $\underline{c}(c^*,q)$. In a market equilibrium

$$\underline{c}(c^*,q) = c^* \quad (A9.10)$$

Equation (A9.10) can have multiple solutions, which can be ranked in terms of individual well-being v. Figure 9.3 illustrates this in the context of fertility behaviour.

In contrast to market equilibrium, the social optimum (Equation (A9.3)) is given by the condition

$$\partial u(c,c^*)/\partial c + \partial u(c,c^*)/dc^* = q, \qquad c = c^* \quad (A9.11)$$

Thus

$$\partial u(c,c^*)/\partial c = q - \partial u(c,c^*)/dc^*, \qquad c = c^* \quad (A9.12)$$

The accounting price of the consumption good is

$$p = q - \partial u(c,c^*)/dc^*, \qquad c = c^* \quad (A9.13)$$

The optimum tax is $-\partial u(c,c^*)/dc^*$. Whether it is positive or negative depends on the sign of $\partial u(c,c^*)/dc^*$ at the optimum.

It is simplest to illustrate the multiplicity of market equilibria with the help of a social environment known as a *coordination game*.

Consider two individuals (1 and 2), each of whom can choose between moderation (M) and profligacy (P). The game is represented by the matrix below, where person 1 chooses a row and person 2 chooses a column. Their welfare outcomes are represented by the numbers in parentheses, the first number in each of the four pairs representing the welfare v of person 1.

Imagine now that the two choose independently of each other. It is easy to check that both (M,M) and (P,P) are equilibrium outcomes, but the former is the one both would prefer. In a 2x2 game it may appear easy to imagine that the pair could coordinate their choices to achieve moderation. In a large economy with differing preferences, coordination can be so hard that it requires a coordinator .

Table A9.1 Two-Person Game with Conformist Preferences

Person 1	Person 2	M (moderation)	P (profligacy)
M (moderation)		(100, 000)	(20, 20)
P (profligacy)		(20, 20)	(40, 40)

Chapter 10
Well-Being Across the Generations

Introduction

What should we mean by human well-being across the generations? How ought we to take the interests of people in the distant future into account when making our own decisions? In which normative language should we deliberate over the rate at which our society invests for the future? What considerations should inform the way a society balances its investment portfolio among produced capital, human capital and natural capital? What role do social capital and other forms of enabling assets, such as financial institutions, play in protecting and promoting well-being across the generations? What should the balance be between private and public investment in the overall investment that a generation makes for the future?

For many people today, there is a simple answer: "Let individuals in the market-place decide."

That is not an outlandish thought. The idea would be that people have a right to judge for themselves how best to reach answers to the questions. People care about their children and know that their children in turn will care about their grandchildren, that their grandchildren in turn will care about their great-grandchildren, and so on. By recursive reasoning thoughtful parents would care about all their descendants. So, when parents reflect on how much they should bequeath to their children, they are guided by all their descendants' well-being. Voluntary participation in free and fair markets would enable parents to express that extended care, while relying on their own resources. Or so it could be argued.

One problem with the reasoning (there are other problems as well) is that even if every household is able to internalise the well-being of its own descendants, no household would be expected to take into account the positive externalities it confers, nor the negative externalities it inflicts on *other* households and *their* descendants. Because much of the biosphere is an open-access resource, market prices are imperfect signals of resource scarcity; in many cases they mislead hugely (Chapters 7 and 8). Even if people coordinate their choices among themselves before acting on the basis of their socially embedded preferences (Chapter 9), their choices would be incongruent with the common good of all households because of pervasive externalities that travel through the material world (Chapters 7 and 8).

So, we now imagine that the person asking the questions views the world with a wider lens than she does when shopping in the supermarket. There are others in her society, and she wants to include them in her deliberations. We could imagine she raises the questions as a citizen, *qua* citizen; but she could be someone who has to think about well-being across the generations in her role as civil servant, Member of Parliament, delegate at the Conference of Parties on the United Nations Convention on Biological Diversity (CBD), and so on. In any of these roles, she feels she must view the world from others' perspectives also, not just her own. We call her the *social evaluator*.

As biodiversity loss is a global phenomenon, we imagine the social evaluator adopts a global view; she is engaged in global *asset management*. That means we could equally refer to her as the *citizen investor*.[305] But she is not a Utopian. The social evaluator/citizen investor accepts that many features of the social world (such as what households do with their assets) are not for her to dictate excepting through public policies; she knows they are, rightly, outside her control.

---

[305] The normative reasoning we develop here is not restricted to the globe in its entirety. The social evaluator could be doing the thinking for a nation or even a smaller socio-political unit.

Chapter 10: Well-Being Across the Generations

There are many constraints, some institutional, others socio-ecological, that she will respect. But she allows herself free rein to explore alternative socio-ecological futures that are consistent with those constraints. In the economist's jargon, she is in a *second-best* world. But she needs an ethical framework for balancing the competing claims of the generations. This chapter studies the ways moral philosophers and welfare economists have suggested the competing claims should be considered.

In Sections 10.1 to 10.4, we sketch three styles of ethical reasoning that are distinct but nonetheless commend the same structural form for the idea of intergenerational well-being. The common form says that well-being across the generations is *the (possibly discounted) sum of the well-beings of all who are here today and all who will ever be born*.[306] We give an account of each style of reasoning, in particular what each says about the rates at which the well-being of future generations ought to be weighed against our own well-being (time rates of discount). We then deploy (Sections 10.5, 10.6 and 10.7) that common formulation to develop such notions as the rates the social evaluator would recommend for weighing future benefits and costs of alternative courses of action today (consumption rates of discount). The social evaluator knows that the term-structure of those weights are fundamental in the asset management problem people face.[307]

Our use of the term 'well-being' will be kept loose here. Griffin (1986) contains a measured book-length analysis of the concept in its many guises, but he also develops his preferred interpretation. Briefly, he thinks of personal well-being as a measure of the extent to which one's informed desires are realised. He also discusses measurement problems.

The qualifier 'informed' is meant to bear ethical weight. Here we keep Griffin's conception in mind and interpret the fulfilment of informed desires as a flourishing life. The formulation of well-being across the generations reviewed here is, however, not tied to that notion of well- being. Moreover, Griffin's account can itself be broadened to include the health and well-being of all sentient creatures, still more broadly the health of the processes that govern Nature (weathering of rocks, circulation of the oceans, material cycles and so on). Even if we were to so broaden the ethics, it would be a conception that works through features of the human experience that make life go well; it would not endow Nature an independent source of value. In Chapter 12 we touch on the latter, very complicated, set of issues.

Chapter 11 will explore the notion with a different slant and review a literature that studies well-being by asking people to reflect on what makes their lives go well. We will discover that its findings, coarse though they are in comparison to the explorations of moral philosophers, are congruent with them.

10.1 Classical Utilitarianism

In his statement of Utilitarianism, Sidgwick (1907: 409–11) wrote:

"By Utilitarianism is ... meant the ethical theory, that the conduct under which, under any given circumstances, is objectively right, is that which will produce the greatest amount of happiness on the whole; that is, taking into account all whose happiness is affected by the conduct. ... We shall understand then, that by Greatest Happiness is meant the greatest possible surplus of pleasure over pain, the pain being conceived as balanced against an equal amount of pleasure, so that the two contrasted amounts annihilate each other for purposes of ethical calculation."

---

[306] The expressions 'well-being across the generations' and 'intergenerational well-being' are used synonymously.

[307] The account that follows of the three styles of ethical reasoning is taken from Dasgupta and Heal (1979: Chapter 9).

Chapter 10: Well-Being Across the Generations

A page later, Sidgwick is explicit about the population that is to be covered by Utilitarianism:

"It seems . . . clear that the time at which a man exists cannot affect the value of his happiness from a universal point of view; and that the interests of posterity must concern a Utilitarian as much as those of his contemporaries, except in so far as the effect of his actions on posterity – *and even the existence of human beings to be affected* – must necessarily be more uncertain."

Well-being is taken here to be a measurable, scalar quantity. To formalise the theory, we simplify the exposition by conflating time and generations and denote both by t. One interpretation of the move would be that an individual's lifetime well-being is a weighted sum of his well-beings over his life. This runs against the view that a person's life is not the sum of its parts, not even a weighted sum of its parts (for example, memory is missing from our account). Our move, however artificial, is meant to make the exposition simple and to keep it in line with the entire literature not only on the economics of climate change, but also more generally on environmental and resource economics.[308]

Time t is discrete and takes the values 0 (the present), 1, 2, . . . and so on. The social evaluator views humanity's prospects at $t = 0$. There is a risk that humanity will become extinct due to a global catastrophe.[309] That risk has two sources. One is independent of human actions, and the other is traceable to our own actions. Among the former is the prediction that over the next millions of years, the Earth will be scorched by an expanding sun; among the latter are losses in biodiversity that carbon emissions, land-use changes, deforestation and pollution are causing. One can presume the citizen investor is anxious to identify policies that would eliminate the latter sources of risk (Chapter 5). We may even presume that she has read this Review! But it does mean the risk of extinction is in part endogenous in her analysis.[310]

For a given course of human activities over time, let $\Omega(t)$ be the subjective probability the social evaluator attributes to humanity surviving *beyond t*, and by $N(t)$ her population projection for t conditional on humanity surviving until then. She would be especially interested in locating policies which, if they were undertaken, would mean that $\Omega(t) = 1$ for many generations. But because she knows that there are extraneous reasons life on Earth will become extinct in the very long run, $\Omega(t)$ is vanishingly small for t large enough.[311]

If $u(t)$ is per capita well-being at t, conditional on humanity surviving until t, intergenerational well-being in Sidgwick's Utilitarianism would be

$$V(0) = N(0)u(0) + \ldots + \Omega(t)N(t)u(t) + \ldots = {}_{t=0}\Sigma^{\infty} [\Omega(t)N(t)u(t)] \qquad (10.1)$$

Equation (10.1) is the expected value of *well-being across the generations*. The percentage rate of decline in $\Omega(t)$ is the (subjective) probability rate of extinction in period t conditional on humanity having survived until period t–1. It is known as the *hazard rate*.[312]

---

[308] An integrated life would mean that lifetime well-being is a non-linear function of momentary well-beings. The problem is to find an adequate expression for the non-linearities. Introducing memory is a relatively straightforward matter (today's well-being is a recursive function of past experiences), but it is not clear how more complex features of an integrated life should be introduced. We do not know of any attempt by economists or philosophers to address these deep problems.

[309] A catastrophe that causes human extinction need not be a one-shot event. Our formulation below includes protracted processes.

[310] Of course, $\Omega(0) = 1$ and $N(0)$ is a given number. $N(t)$ is bounded above.

[311] Rees (2003) speculated that the probability rate of a global catastrophe, including in the extreme human extinction, may be as high as 0.7% a year. If the extinction rate is governed by a Poisson process, it means there is a 50% chance that humanity will not survive beyond another 100 years (i.e. if t were measured in years, $\Omega(100)$ would equal 0.5). Ord (2020) is more optimistic; he reckons the probability that future possibilities for humanity will be seriously compromised in the next 100 years by a catastrophe is no more than 1 in 6.

[312] Yaari (1965) used equation (10.1) to study ideal markets for annuity that people could use to save for an uncertain life span.

Chapter 10: Well-Being Across the Generations

Rawls (1972: 184–7) interpreted the reasoning behind Utilitarianism – he called Sidgwick's version Classical Utilitarianism – in these words:

"Something is right, a social system say, when an ideally rational and impartial spectator would approve of it from a general point of view should he possess all relevant knowledge of the circumstances. Thus he imagines himself in the place of each person in turn, and when he has done this for everyone, the strength of his approval is determined by the balance of satisfactions to which he has sympathetically responded. When he has made the rounds of all the affected parties, so to speak, his approval expresses the total result. Sympathetically imagined pains cancel out sympathetically imagined pleasures, and the final intensity of approval corresponds to the net sum of positive feeling."

The 'net sum' Rawls speaks of is the 'equally-weighted sum'. Classical Utilitarianism is famous for being unconcerned with the distribution of well-being among people. Distribution of goods and services is a different matter, and the theory is sensitive to that. Edgeworth (1881) developed the idea that additional increments to a person's well-being are not independent of his standard of living, in particular the higher the living standard a person enjoys, the smaller the gain in well-being from a small increment in his living standard would be. This is the condition of 'diminishing marginal utility', an assumption that is routinely made in the economics of climate change (Box 10.4), both for normative and for broad empirical reasons. A further notable feature of Utilitarianism is that it makes a direct connection between the ethical case for equality among people in the distribution of the standard of living and for being averse to facing risk in one's living standard (Chapter 5). Both are reflected in diminishing marginal utility.

Conditional on humanity surviving until date t, how would the ideally rational and impartial spectator view the future if he was to place himself at t? Let $V(t)$ denote the expected value of well-being across the generations from the vantage point of t. Applying Bayes' theorem for updating probabilities, equation (10.1) yields

$$V(t) = {}_{s=t}\Sigma^{\infty}\{[\Omega(s)N(s)u(s)]/\Omega(t)\} \tag{10.2}$$

What loss would a Utilitarian social evaluator attribute to humanity suffering the ultimate catastrophe – extinction – at t? Equation (10.2) says the loss would be $V(t)$, as $V(t)$ would be the expected value of foregone well-being across the generations. It is important to recognise that $V(t)$ does not measure the value of *actual* lives lost; it is not an aggregate measure of the value of life. Instead, it is the value of *potential* lives lost, which is a very different ethical object. The value of a statistical life (VSL), a concept central to the meaning and measurement of human capital is built on $V(t)$. This is explained in Box 10.2. We will make use of the notion in Chapter 13 (Annex 13.1) when developing the idea of inclusive wealth.

Box 10.1
Weights and Measures

Savage's axioms of rational choice under uncertainty yield a u-function that is unique up to positive affine transformations (Box 5.1). Decision theorists say that such a u-function is 'cardinal'. Our social evaluator has to contend with a deeper ethical problem than Savage's decision maker, for she has also to reflect on what makes life go well and what does not, which means she has to distinguish a life that is good from a life that is not good. A Utilitarian would say that for two people to suffer from a life that is not good is twice as not-good as one person suffering it. Our decision-maker needs to add an axiom to the set proposed by Savage so as to be able to make that additional judgement.

Making judgments on where to draw a line between a good life and a life that is not good is commonplace. When the World Bank proclaims that 1.90 international dollars a day is an absolute poverty line, it is meant that a person is (absolutely) poor if her income is below 1.90 international dollars a day but is not if her income is above it. The UN's Sustainable Development Goals pay special attention to poverty elimination using an absolute poverty line as a quantitative indicator (Goal 1 being elimination of poverty; Chapter 4). In short, the sign of u in equation (10.1) matters.

We say the u-function is *strongly cardinal* if it is unique up to *proportional transformations*. That says if u is an admissible measure of per capita well-being and if β is a positive number, then βu is an equally valid measure of per capita well-being.[313] Empirical estimates of well-being as constructed from responses to large-scale polls on happiness and life satisfaction (Chapter 11) assume that individual well-beings are strongly cardinal.

It is then possible to deploy strong cardinality of well-being for empirical studies of absolute poverty. If the World Bank's poverty line of 1.90 international dollars per day is used for the purpose, then u would be deemed to be zero for anyone whose daily income is 1.90 international dollars; u would be negative for someone below the poverty line, it would be positive for someone above the poverty line.

It is an explicit condition in equation (10.1) that the ideally rational and impartial spectator is able to compare individual well-beings. Our social evaluator recognises that. Personal well-beings are said to be *fully comparable* if, when she chooses to measure someone's well-being by a scale in which u is replaced by βu, where β is positive, she is required to multiply the well-beings of all other individuals by that same β. Multiplying all the $u(t)$s in equation (10.1) by β, however, has no ethical significance, as the social evaluator's ordering of policies remains unaffected.

Strong cardinality and full comparability are familiar notions. The weights of objects (measured, say, in a vacuum-sealed flask at ground level in a given latitude) are fully comparable. Suppose we find x to be a heavier object than y. If that is to be a meaningful finding, the units in which they are measured must be the same; it's no good measuring x in ounces and y in grams. Reference to grams and ounces tells us that we can say a lot more than merely that x is heavier than y: we can say how much heavier x is proportionately than y. The reason we can is that if x is found to be twice as heavy as y using one system of units (ounces), it will be found to be twice as heavy as y using any other system of units (grams). And that's because an ounce is proportional to a gram. We can move from one system of units to another with impunity so long as the corresponding transformations (grams to ounces) are applied consistently.

The ideally rational and impartial spectator has a universal point of view. He is able to peer into the future from any period in time should humanity exist at that time. Equation (10.2) reflects his reasoning. The way the Utilitarian social evaluator applies that reasoning is as follows:

She takes as her database the wide array of assets generation-0 has inherited from its predecessors. Given that inheritance, generation-0 has available to it a set of possible socio-ecological futures, the possible futures being conditional of course on there *being* a future. Call that set of possible futures, $\Xi(0)$. We are to imagine that the social evaluator selects the possible future in $\Xi(0)$ that maximises $V(0)$. That simultaneously yields investment decisions which add to, or subtract from, the assets generation-0 had inherited, and which in turn determines the possible socio-ecological futures that would be open to generation-1, should generation-1 exist.

Next, the social evaluator places herself in period-1, should generation-1 exist. She takes as her database the wide array of assets generation-1 has inherited from generation-0. Given that inheritance, generation-1 has available to it a set of possible socio-ecological futures, the possible futures being conditional of

---

[313] In fact, as can be easily checked, it would make no sense to define u to be per capita well-being unless the u-function was strongly cardinal.

Chapter 10: Well-Being Across the Generations

course on there *being* a future. Call that set of possible futures, $\Xi(1)$. We are to imagine that the social evaluator selects the possible future in $\Xi(1)$ that maximises $V(1)$. That simultaneously yields investment decisions, which add to, or subtract from, the assets generation-1 had inherited, which in turn determines the possible socio-ecological futures that would be open to generation-2, should generation-2 exist.

And so on – the social evaluator applies the reasoning sequentially, for all t, on behalf of all potential generations. In so doing, she makes the Utilitarian viewpoints of all generations congruent with one another. Her exercise informs each generation to choose its activities and leave behind assets that can sustain the subsequent sequence of socio-ecological futures that it deems to be right on Utilitarian grounds, aware that succeeding generations will choose in accordance with what it had planned for them, conditionally of course. In modern game theoretic parlance, the Utilitarian optimum is a non-cooperative equilibrium among the generations. It follows that there is no need for an intergenerational 'contract'.

This last is not a trivial argument, for it means that even though it is not possible for the generations to devise a binding agreement among themselves, that is of no moment. We noted the force of this reasoning in our discussion on social norms (Chapter 6). The reasoning will be applied for determining the optimum rate of saving in stylised models of the global economy (Annex 10.1).

Utilitarianism involves two related notions: (i) happiness, whose numerical measure is also called *utility*; and (ii) summation as the required operation for combining individual utilities to express *social well-being*. Sidgwick (1907: 119–50) contains three chapters on empirical hedonism, where the sense in which 'happiness' is used is a lot more considered than is suggested in the frequent criticism that Utilitarianism views humans to be mere pleasure machines. That is why we are replacing the term 'happiness' with 'well-being', but are otherwise in this section retaining the teleological conception underlying (Classical) Utilitarianism. For now, we imagine, for simplicity, that the social evaluator is concerned only with human well-being. Subsequently, we broaden her understanding of the good and the right.

Utilitarianism has had a rough time in the hands of modern philosophers. Sen and Williams (1982) dismissed applications of Utilitarianism to public policy by calling such attempts "Government House Utilitarianism".[314] Our social evaluator, however, is not seeking a normative guide for personal decisions, she is seeking a guide in her role as *citizen*. As a citizen investor she needs ethical guidance that is not a prop for paid officials to act in nepotistic – never mind predatory ways – but is instead impartial over people's needs and sensitivities. The ideally rational and impartial spectator may well be the person she wants to consult.

Box 10.2
The Value of a Statistical Life

The expected value of well-being across the generations (Equation (10.1)) gives rise, if only via several steps, to an important concept in the economics of health: the *value of a statistical life* (VSL). The concept is relevant for the allocation of resources by government between a population's health and other assets, and it is relevant for the medical profession when it deliberates on how scarce treatments should be allocated among competing needs. It is central also to the measurement of human capital accumulation (Chapter 13).

There is a risk of fatality no matter what we choose to do, but the risks differ. Crossing a road involves a risk, as does climbing the staircase in our home, but they are unlikely to be the same. The risk a window cleaner faces on a skyscraper differs from the risk he faces when cleaning the windows of

---

[314] Nordhaus (2007) used the same term to dismiss the approach taken by Stern (2006), who argued in favour of low rates with which to discount future economic losses from climate change (Box 10.4).

a bungalow – the former is likely to be greater than the latter. How should one infer the cost individuals themselves impute to an increase in the risk of their own death? To get a feel for the question, consider the following thought experiment:

There are 7 million people, each of whom is willing to pay up to US$1 to prevent an increase in the risk of their own death by 1/7,000,000. Imagine also that the increases are independent across individuals. The law of large numbers says the expected number of additional deaths that would occur if the risk was not eliminated is almost surely 1. We may then say the total amount the population is willing to pay so as to prevent the death of one random person is US$7 million. The figure of US$7 million is the VSL.

The example is only a start, for VSL would be expected to depend on age and state of health, among many other factors. That it does has been confirmed from market data on wage differentials among workers facing differences in fatality risks (controlling for other factors), it has been confirmed from differences in the amount of life insurance people purchase (again, controlling for other factors), and so on. Sadly, but not unexpectedly, income is also a factor. Poor people cannot afford not to take risks the rich would avoid.[315] This is reflected in national averages. VSL in the UK lies in the range US$7–13 million (Social Value UK, 2016), whereas in India, Shanmugam and Madheswaran (2011) have estimated it to be a bit over US$1 million. Aldy and Viscusi (2008) conducted a meta-analysis of international data to estimate that the percentage increase in VSL associated with a percentage increase in income is in the range 0.5 to 0.6.

For assigning quantitative values to increases in age-specific life expectancy, institutions such as the World Health Organization (WHO) have modified VSL by constructing the values of (statistical) *life-years*. WHO has also experimented with ways to adjust values of life years for the quality of life; hence 'quality adjusted life years' (QALYs) and even 'disability adjusted life years' (DALYs). Health economics obliges the social evaluator to give quantitative expression to trade-offs among individual well-beings that most people find repugnant even to contemplate. We citizens should feel relieved that it is *she* who has to do the thinking, not we.

10.2 Utilitarian Reasoning behind the Veil of Ignorance

An alternative conception of intergenerational well-being (although he would not have used the term) was offered by Rawls (1972), who moved away from the teleological viewpoint that shapes Utilitarianism and looked instead for principles of justice. Rawls took the principles to be those that free and rational individuals concerned to further their own interests would agree should govern the basic structure of their society if they had to choose them from behind a veil of ignorance; that is, ignorance of their own abilities, of their psychological propensities, and of their status and position in society of which they are to be members. The position of equality among the choosing parties in this thought experiment was named by Rawls *the original position*. On justice among the generations, Rawls (1972: 287–8) wrote:

"The (choosing) parties do not know which generation they belong to ... They have no way of telling whether it is poor or relatively wealthy, largely agricultural or already industrialized ... The veil of ignorance is complete in these respects ... Since no one knows to which generation he belongs, the question is viewed from the standpoint of each and a fair accommodation is expressed by the principle adopted. All generations are virtually represented in the original position, since the same principle would always be represented."

---

[315] Viscusi and Aldy (2003) is the classic reference on global estimates of VSL. Viscusi (2018) is a book-length treatment of the subject.

Chapter 10: Well-Being Across the Generations

Rawls' conception of justice is built on choice under uncertainty. However, the veil of ignorance he imagined is so thick that the world presents itself ambiguously (Chapter 5, Annex 5.2).[316] Which may be why Rawls advanced *maxi-min* as the decision rule the chooser would adopt behind the veil. Today we could use the axiomatics of the decision rule "α-max-min" in Arrow and Hurwicz (1977) to understand the reasoning our social evaluator might deploy if she pursued Rawls' line of thought (Chapter 5, Annex 5.2).

Rawls abandoned the maxi-min rule when sketching the principle of just saving. The reasoning he deployed to advance his saving principle is, however, very problematic (Arrow, 1973; Dasgupta, 1974; Solow, 1974a). So we avoid the route he took and adopt an extension of a line of argument put forward by Harsanyi (1955), who also regarded choice under uncertainty to be at the core of the social evaluator's mode of reasoning. Central to Harsanyi's ethical theory is the notion of 'impersonality' (Harsanyi, 1955: 316):

"[A]n individual's preferences satisfy his requirement of impersonality if they indicate what social situation he would choose if he did not know what his personal position would be in the new situation chosen (and in any of its alternatives) but rather had an equal chance of obtaining any of the social positions."

Applying the notion of impersonality to her problem, the social evaluator imagines she could be a member of any generation with equal probability, conditional on the world surviving until then.[317] If she were to choose in accordance with Savage's axioms, her ranking of alternative saving policies would correspond to the ranking implied by expected utility across the generations (Box 5.1). In this interpretation, choice under uncertainty requires the social evaluator to interpret well-being across the generations as the expected value of well-beings (Equation (10.1)). But she adopts the criterion by applying the restriction of equi-probability, she does not arrive at it from an assessment of the ideally rational and impartial spectator of Classical Utilitarianism. Harsanyi's Utilitarianism comes via a different route.[318]

10.3 Discounting Future Generations

The risk of human extinction has two sources. One is independent of human actions, the other is traceable to our own actions. Restoring natural habitats increases biodiversity. That in turn raises $\Omega(t)$. The value of an increase in $\Omega(t)$ would be reflected in a corresponding increase in intergenerational well-being $V(0)$, provided of course that well-being is positive. Here we adopt the view that so long as there is room for reducing the risk of human extinction, the citizen investor will recommend that it be taken, which is to say she eliminates from consideration all socio-ecological futures in $\Xi(0)$ that add to the risk of extinction beyond what is unavoidable. She is then left with a truncated set of socio-ecological futures, in which the only risk of extinction is exogenous to her choice. Call that set $\Xi^*(0)$. Creating a map of $\Xi^*(0)$ involves the use of a macroeconomic model of the economy, in the manner of the one presented in Chapter 4*.

Utilitarianism does not have the machinery for her to limit her attention to $\Xi^*(0)$. One can imagine a world where the well-being of the present generation and that of generations in the near future can be increased greatly by plundering Nature, even if by so doing it raises the likelihood that there will be no human life on Earth beyond the near future. That would appear to be the path humanity has been

---

[316] The thickness of the veil is warranted because Rawls looked for principles of justice that would govern the basic structure of society, not over such details as whether an industry should be nationalised or whether pollution should be controlled by taxes or quantity restrictions.

[317] In making this move, we are conflating Rawls' 'veil of ignorance' with Harsanyi's conception of 'impersonality'. But we are doing that *only* for the problem in hand. The two theories are otherwise not the same. On this, see Dasgupta (2019).

[318] Mathematically, 'equi-probability' in Harsanyi's conception serves the same role as 'impartiality' does in Classical Utilitarianism. Equal probabilities over an indefinite deterministic future is of course meaningless, but by equi-probability we mean equal probabilities conditional on humanity surviving.

following since we entered the Anthropocene. The social evaluator needs a wider ethical conception to foreclose those options. She can do that by appealing to an imperative that human life is sacred. That would be to appeal to a two-tiered system of ethics, in which a deontological imperative overrides Utilitarian considerations. The former limits the social evaluator to $\Xi^*(0)$, while the latter directs her to deploy the Utilitarian mode of reasoning to judge the relative merits of the socio-ecological futures in $\Xi^*(0)$.

Deontological constraints that serve to limit the space in which Utilitarian reasoning reigns is routine in the public sphere (it is commonplace in the private sphere too). Constitutional prohibitions on capital punishment, for example, are adopted by countries where the matter is subsequently not open to negotiation by Utilitarian considerations. Protection of sacred groves, more generally sacred landscapes, in societies that respect deeply held beliefs about the sanctity of particular ecosystems respond to deontological directives (Chapter 12). They impose constraints on what can be permitted.

We are talking here of deontological constraints restricting the social evaluator's set of choices to $\Xi^*(0)$, not constraints that are of help in carrying out Utilitarian recommendations. The latter was explored in Chapter 7 (Box 7.3). There we found that limitations on what a regulator can verify (i.e. moral hazard and adverse selection) can limit the effectiveness of Pigouvian externality taxes, that quantity restrictions are in many cases a better set of instruments. The reasoning there was entirely Utilitarian. Here we are talking of deontological constraints. In Chapter 12, we explore and affirm philosophical reasoning that demands that, as a minimum, the social evaluator restricts her attention to $\Xi^*(0)$ and, by recursion, to $\Xi^*(t)$ for all t.

But that move leads to a new problem:

Suppose the social evaluator has good reasons to think that the likelihood of human extinction for reasons exogenous to human activities is negligible for the foreseeable future. Formally, that would be to set $\Omega(t) = 1$ for a long stretch of time. So, Utilitarianism directs the social evaluator to weigh the well-being of generations equally for that long stretch of time.

To award the same weight to the well-being of all future generations relative to the weight awarded to the present generation in a deterministic world is taken by many moral philosophers and economists to be incontrovertible.[319] Sidgwick (1907) took the directive to be axiomatic. In his famous paper on optimum national saving, Ramsey (1928: 543) regarded the practice of discounting future well-beings in a deterministic world to be ethically indefensible and thought it "arises merely from the weakness of the imagination". In a book that laid the foundations of the modern theory of economic growth, Harrod (1948: 40) wrote that discounting future generations is "a polite expression of rapacity and the conquest of reason by passion". Strong words, and to some people Ramsey's and Harrod's strictures read like Sunday pronouncements. Solow (1974b: 9) expressed the feeling exactly when he wrote, "In solemn conclave assembled, so to speak, we ought to act as if the [discount rate on future well-beings] were zero."

Parfit (1984) applied sophisticated reasoning in support of the Sidgwick-Ramsey-Harrod view on discounting for time and the generations. That view has influenced one strand of the welfare economics of global climate change (Broome, 1992, 2004, 2012; Cline, 1992; Stern, 2006). The problem is, whether the social evaluator should discount the well-beings of future generations is not that simple a matter. Consider the following ethical tension she could face:

(1) Low levels of well-being among generations in the distant future would not be seen to be a bad thing by her if future well-beings were discounted at a positive rate. It could then be that, by applying positive discount rates the present generation finds it justifiable to destroy biodiversity so as to raise the material basis of its own well-being. But if it were to do that, generations in the distant future would inhabit a near-barren Earth. That suggests the social evaluator should *not* discount future well-beings.

---

[319] Ramsey (1928), Sidgwick (1907), Mirrlees (1967), Rawls (1972), and Parfit (1984) are among the most prominent.

Chapter 10: Well-Being Across the Generations

(2) In a world where the return on judiciously chosen investment is positive (Chapter 2), not to discount future well-beings could mean that the present generation and those immediately following it should sacrifice themselves by saving at enormously high rates for the many, many generations that will follow. But if that was carried out, the demands of intergenerational equity would not be met: early generations would be required to suffer from privation for the benefit of future generations, who would be progressively wealthier. That suggests the social evaluator *ought* to discount future well-beings at well-chosen positive rates.

In a remarkable body of work, Koopmans (1960, 1965, 1967, 1972, 1977) showed that to insist on awarding the same weight to the well-beings of all generations in a world where well-chosen investment has a positive yield leads to ethical incoherence. Seedlings grow to become trees, fisheries can recover if left alone, and with but little expenditure the health of wetlands have been known to improve. Judiciously chosen investment has a positive return, meaning that a unit of consumption foregone today can generate more than a unit of consumption in the future. The productivity of judiciously chosen investment creates an asymmetry between the present and future. Box 10.3 presents an example that shows there can be deep *in*equity in the distribution of the standard of living *ex post* if equity is insisted upon *ex ante*. Koopmans' work reveals that the asymmetry created by the productivity of well-chosen investment is so built into the arrow of time, that symmetric treatment of well-beings across the generations can lead to incoherence in ethical thinking.

Box 10.3
Zero Discounting and the Never-Ending Potlatch

Policies that commend themselves in a world that will not be subject to the risk of extinction for a long stretch of time can be expected to be similar to policies that commend themselves in a world that is imagined to be everlasting. In order to show the problems that can arise if the generations were awarded the same ethical weight, we study an everlasting world. Assume also that population is constant. Imagine now that output over time is a steady stream of a completely perishable good. In this economy, generation-0 has an investment opportunity in which a unit of consumption foregone yields a perpetual stream of r units of consumption from $t = 1$ onward. The rate of return on the investment, which is r, is positive. (The example is taken from Arrow (1999), who in turn based it on Koopmans (1960, 1965).)

Not to discount future well-beings would imply that the present value of returns is infinite. That is because the infinite sum, $r+r+\ldots$, is unbounded. Thus generation-0 incurs a finite loss for a unit of consumption sacrificed, but the investment yields an infinite gain to future generations no matter how small r happens to be. It is obvious the investment should be undertaken, meaning that consumption should be reduced by generation-0.

Now suppose a similar investment project is also available. The same reasoning says the new project should also be accepted, further reducing current consumption. And so on, until current consumption is reduced to near-zero. Thus, *any* consumption sacrifice by generation-0 (short of 100% of available consumption) is good. Many would regard that to be an unacceptable burden on the present generation.

The reasoning goes further. Generation-0, having sacrificed nearly everything for future generations, is followed by generation-1. Imagine that it too is faced with a string of investment projects of the same kind. An identical reasoning would come into play for members of generation-1, who also would be required to consume at a near-zero level. And so on, down the generations. But that means every generation is required to live in penury for the sake of a future that is always just beyond. Koopmans (1965, 1967) likened this consequence of Utilitarian reasoning to a 'never-ending potlatch', which of

> course cannot be an optimum policy. But that only shows that Utilitarianism is incoherent under the circumstances of the model: there *is* no optimum policy.[320]
>
> The argument extends to models with durable capital goods. There are classroom models in which an optimum saving policy under Classical Utilitarianism exists, but which requires each generation, even if the early generations are very poor, to save nearly 100% of GDP so as to accumulate wealth for the generations that are to follow. That is shown in Box 10.6. That is not a never-ending potlatch (despite the high saving rate, consumption grows with time and the generations), but it comes pretty close to it.
>
> An infinite horizon in a deterministic world could seem altogether too artificial a construct. So imagine instead that there is a risk of extinction, and that the hazard rate our social evaluator believes drives the extinction process is zero for T periods, following which it is positive. Expected well-being is well-defined, implying that there is no incoherence in Harsanyi's version of Utilitarianism. All potential generations are awarded equal weight, so we are in the realm of an ethics that dates back to Sidgwick, Pigou, and Parfit. However, applying the same example as above, it is easy to confirm that if T is large, the social evaluator will award near- zero consumption to early generations in order that later generations, should they appear, enjoy enormously high consumption levels. To insist on awarding equal weight to all potential generations is to require the social evaluator to commend brutal inequality.

10.4 Intuitionism and Pragmatism

The social evaluator is now faced with an ethical dilemma: on deontological grounds she has eliminated all socio-ecological futures that add to the risk of extinction from natural causes, only to find herself possibly having to commend courses of action that leave the present generation in penury.

Koopmans (1972) provided a set of ethical axioms for which well-being across the generations in a *deterministic* world assumes the form in equation (11.1), but in which $\Omega(t) < 1$ for all t nevertheless. Because Koopmans' model economy does not allow the possibility of human extinction, $\Omega(t)$ in his rendering of equation (10.1) represents *pure well-being discounting*. The final example in Box 10.3 showed that in certain circumstances the social evaluator could find it ethically necessary to deploy a time discount rate over and above the hazard rate of extinction.

The lesson we should draw from Koopmans' work is this:

It is foolhardy to regard any ethical judgement as sacrosanct, as one can never know in advance what it may run up against. A more judicious tactic would be to play off one set of ethical assumptions against another in not-implausible worlds, check their implications, and then appeal to our normative senses before arguing over policy. The idea is to reach a balance among our intuitions on the reading of both facts and ethical directives. That lesson is the hallmark of Intuitionism. Rawls (1972), as always, gives a sympathetic airing to ethical reasonings not his own. And here he is on Intuitionism (page 34):

"I shall think of intuitionism ... as the doctrine that there is an irreducible family of first principles which have to be weighed against one another by asking ourselves which balance, in our considered judgment, is the most just. ... Intuitionist theories ... have two features: first, they consist of a plurality of first principles which may conflict with one another to give contrary directives in particular types of cases;

---

[320] The flavour of the argument remains if the horizon is long but finite. The present value of a *T*-period consumption flow, *r*, without time discounting is *Tr*. If *T* is large, Total Utilitarianism demands an attenuated form of consumption postponement: An optimum exists, but the overwhelming bulk of consumption is to be enjoyed by the last few generations. In the limit, if $T \to \infty$, the never-ending potlatch emerges, and there is no optimum policy.

and second, they include no explicit method, no priority rules, for weighing these principles against one another: we are simply to strike a balance by intuition, by what seems to us most nearly right."

That there can be an irreducible family of first principles is doubtful. Even as citizens, certainly in daily life, we tend to muddle along, guided by weak directives that have worked for us in the past but which we are ready to chuck if they do not lead to outcomes that feel right. We grow up under some directives but are willing to revise them in later years because of further reflection and experience. It may be that the revisions converge in time, so that we are led to acknowledge a set of first principles; but even that can be doubted. All ethical principles are tentative in this conception and even as citizens we are seen to be ready to shift gear if things do not pan out in ways that make life go well. This looser ethical structure is called Pragmatism.[321]

10.5 Discounting in Arbitrary Futures

We take Pragmatism to be our guide and use Koopmans' insights to interpret the weight $\Omega(t)$ to reflect a combination of extinction risks and the desire for equity among the generations. Imagine now that the social evaluator places herself in period t-1 and that the well-being of generation-$(t-1)$ is reduced by one unit. Equation (10.1) says that other things equal the potential well-being of generation-t, which is $N(t)u(t)$, must increase by $\Omega(t)/\Omega(t-1)$ units if $V(0)$ is to be unaffected. Because $\Omega(t)/\Omega(t-1)$ is the rate at which the potential well-beings of the two generations can be substituted for one another without affecting $V(0)$, $\Omega(t)$ is called the *well-being discount factor* at t. It should be noted that $\Omega(t) \geq \Omega(t-1)$.

Define the *well-being rate of discount* $\delta(t)$ as the percentage rate of decline in $\Omega(t)$ between t-1 and t. Thus

$$\delta(t) = -[\Omega(t) - \Omega(t-1)]/\Omega(t-1) \geq 0 \qquad (10.3)$$

Equations (10.1) and (10.3) taken together say that, other things equal, the rate at which the potential well-beings of generations t–1 and t can be substituted without affecting $V(0)$ is $(1+\delta(t))$. But that is another way of saying that $(1+\delta(t))$ is the rate a marginal unit of well-being in t ought to be discounted by the social evaluator from the vantage point of period $(t-1)$. By recursion it means that the factor at which the social evaluator would discount a marginal unit of well-being at t from her perspective in period $t = 0$ is the product $(1 + \delta(1)) \times \ldots \times (1 + \delta(t))$.[322]

The reasoning just carried out can be extended to identify the rates at which the *determinants* of well-being can be substituted across generations without affecting $V(0)$. As will be shown below those rates are of supreme importance for defining the accounting prices of capital goods (Chapter 1); they create the link between the social worth of goods and services across time.

It is customary in economics to call the determinants of well-being, *consumption*. It is also customary in models of economic growth to imagine that consumption can be so aggregated that it can be expressed as a scalar. We have followed that practice in this Review so far (recall especially Chapters 4* and 5); we continue to do so now.[323]

In our formulation, the well-being of a generation is the product of population size and average well-being. The expression is thus neutral as regard the distribution of well-being *within* a generation. Cline (1992), Nordhaus (1994, 2007), and Stern, (2006) have gone even further. So as to focus attention on the distribution of goods and services *across* generations, they have bypassed concerns

---

[321] Pragmatism has a rich history, recounted in Misak (2016). Our interpretation of the force of Koopmans' work has been influenced by a modern rendering of the normative basis of Pragmatism in Shamik Dasgupta (2020), to whom we are grateful for helpful discussions.

[322] As noted previously, $\delta(t)$ is the hazard rate t in Classical Utilitarianism.

[323] In Box 10.7, the framework is extended to include the socially embedded conceptions of well-being that were discussed in Chapter 9.

even over the distribution of *consumption* within generations. We follow that route here, because not to do so would only mean additional notation.[324]

Assuming then that all individuals in a generation consume their average consumption level, equation (10.1) can be expressed as

$$V(0) = N(0)u(c(0)) + \ldots + \Omega(t)N(t)u(c(t)) + \ldots \tag{10.4}$$

Let $C(t)$ denote total consumption in period t. That means $c_t(t) = C_t(t)/N(t)$, of which $N(t)$ is a demographic projection. A *consumption stream* can be expressed as a sequence $\{C(0), \ldots, C(t), \ldots\}$. Let the set of consumption streams available for consideration be $\Re(0)$.[325] Each consumption stream in $\Re(0)$ gives rise to a well-being stream in a set that was denoted in Section 10.4 as $\Xi^*(0)$.

Consider an arbitrary consumption stream $\{\underline{C}(0), \ldots, \underline{C}(t), \ldots\}$. Projection of per capita consumption in period t is then $\underline{C}(t)/N(t)$. It will be convenient to call it a *reference stream*. However, describing a future exclusively in terms of consumption across the generations hides an enormous amount of information. Even to identify it requires the social evaluator to draw a mental picture of the circumstances that bring it about. She has to make a projection of not only population, technological and ecological possibilities in each period, but also the institutions that provide incentives to people to consume and save in such ways as to bring about the consumption stream she is considering. There is an inevitable circularity in reasoning here, but it is the kind of reasoning people carry out all the time. When making future projections on any matter, we are not satisfied until the various factors that are likely to bring them about 'hang together', and that requires us to consider what would happen if things were to pan out differently. This involves considering counterfactual worlds. If we did not try to do that, projections into the future would amount to no more than wishful thinking. If projections have any significance, they have to be justified by argument and evidence.

Consider now a thought experiment in which the aggregate consumption in the reference stream is reduced in period $t-1$ by one small unit. Suppose now that, other things equal, the amount of consumption that would have to be added to $\underline{C}(t)$ in order that $V(0)$ is unaffected is $1+\rho(t)$. We call $\rho(t)$ the *consumption discount rate* at t and $1/[1+\rho(t)]$ the *consumption discount factor* at t.[326]

Any mention of 'discount rates', and one thinks of positive numbers. But should $\rho(t)$ necessarily be positive?

There are two reasons $\rho(t)$ could be positive. First, it may be that $\delta(t)$ is positive. Second, if $\underline{c}(t)$ exceeds $\underline{c}(t-1)$ in the reference stream, an extra unit of consumption for someone in period-t would be less valuable than an extra unit for someone in period $t-1$ because well-being increases with consumption at a diminishing rate. Thus, rising consumption provides a second justification for discounting marginal gains and losses in future consumption at a positive rate.

The pair of considerations tells us that $\rho(t)$ depends on (i) the rate at which well-being in period t is discounted in $t-1$; (ii) the rate at which consumption per capita is projected to change over the two periods (this is where the reference stream comes into play); and (iii) the rate at which marginal

---

[324] To include the possibility of intra-generational inequality in consumption, let j be an index of persons in each generation, which means j at time t assumes the values 1 to $N(t)$. If $c_j(t)$ is the consumption level of the individual labelled j at time t, then the well-being of generation-t would be $\sum_{j=1}^{N(t)}[u(c_j(t)]$. If u is an increasing function of $c_j(t)$ – i.e, $du(c)/dc > 0$ – but marginal well-being diminishes as $c_j(t)$ – i.e. $d^2u(c)/dc^2 < 0$ – then $V(0)$ is averse to inequality within generations and the social evaluator has an ethical language in which to consider intra-generational equality. In the text we express $du(c)/dc$ by the usual notation, $u'(c)$, and $d^2u(c)/dc^2$ by the equally usual notation, $u''(c)$. The assumptions on the u-function can be traced to Edgeworth (1881).

[325] Economists call $\Re(0)$ the *feasible* set of consumption paths.

[326] $(1+\rho(t))^{-1}$ is the value our social evaluator imputes to a marginal unit of consumption at t from the vantage of period $t-1$. By recursion it means that the product $[(1+\rho(1))\times\ldots\times(1+\rho(t))]^{-1}$ is the value she imputes to a marginal unit of consumption at t from the vantage of period 0.

well-being $u'(c)$ diminishes with greater consumption (i.e. the elasticity of marginal well-being with respect to consumption per capita).

Several questions arise: How is $\rho(t)$ to be derived? Should $\rho(t)$ be constant over time or would it depend on time? Does $\rho(t)$ reflect the 'opportunity cost' of capital; if so, how should the social evaluator first determine what that cost is? Can $\rho(t)$ be inferred from 'market observables', such as risk-free interest rates on government bonds? Is $\rho(t)$ inevitably positive or are there circumstances when it is negative? And how should $\rho(t)$ be modified when the future is not only uncertain, but the uncertainties are perhaps correlated over time?

Denote the elasticity of marginal well-being by $\eta(c)$. It is a positive number because marginal well-being $u'(c)$ is assumed to decline with increasing consumption per capita, c. Let $g(\underline{c}(t))$ denote the percentage rate of change in per capita consumption in the reference stream between period $(t\text{-}1)$ and period t, and let $\eta(\underline{c})$ be the elasticity of marginal well-being at $\underline{c}$. In Box 10.4 it is shown that

$$1 + \rho(t) \simeq (1 + \delta(t))(1 + g(\underline{c}(t)))^{\eta(\underline{c}(t))} \tag{10.5}$$

In theoretical work on the economics of climate change, it has frequently been assumed that time is continuous (i.e. the length of a period is infinitesimal). In that case, as is shown in Box 10.4, Equation (10.5) can be expressed as an exact equation:

$$\rho(t) = \delta(t) + \eta(\underline{c}(t))g(\underline{c}(t)) \tag{10.6}$$

Koopmans' ethical axioms on the ordering of consumption streams in $\Re(0)$ could appear to imply that intergenerational well-being is additively separable over each generation's *well-being* (Equation (10.4)); but care is needed. The axioms he proposed (Koopmans, 1972) do not yield a numerical value of either δ or η. We have called η the elasticity of marginal well-being with respect to consumption, but the η that is yielded by his axioms also reflects a conception of equity in consumption among the generations. And we have already noted that δ reflects not only the risk of extinction, but a conception of equity in well-being across the generations. Indeterminacy of δ and η is the hallmark of Koopmans' Pragmatism. Suppose, for example, we are convinced the elasticity of marginal well-being can be estimated from data on consumer expenditure. Koopmans' axioms would permit the social evaluator nevertheless to transform the estimates before placing them in the formula for intergenerational well-being (Equation (10.4)). The axioms he proposed do not yield an "irreducible set of first principles", to use Rawls' phrasing of Intuitionism. Pragmatism encourages the social evaluator to conduct numerical experiments with alternative values of δ and η before commending policies. Box 10.4 shows how that experimentation can be practised.[327]

Box 10.4
Consumption Discount Rates (CDRs): Basics

B10.4.1 CDRs in Arbitrary Reference Paths

Let $\underline{c}(t)$ be per capita consumption in period t on the consumption path that the social evaluator chooses to study. If $g(\underline{c}(t-1))$ is the percentage rate of change in $\underline{c}$ between periods $t\text{-}1$ and t, then

$$\underline{c}(t)/\underline{c}(t-1) = 1 + g(\underline{c}(t-1)) \tag{B10.4.1}$$

Define the elasticity of marginal well-being with respect to per capita consumption as

$$\eta(c) = -cu''(c)/u''(c) > 0 \tag{B10.4.2}$$

---

[327] Ramsey (1928) conducted such experimentation in his classic paper on the optimum rate of national saving.

Consider now a small perturbation to the reference consumption path in periods (t-1) and t, leaving per capita consumption unperturbed in all other periods. Let the perturbations be denoted as $\Delta \underline{c}(t-1)$ and $\Delta \underline{c}(t)$.[328] If the perturbation has no effect on V(0), we have from equation (11.4) that

$$\Omega(t-1)u'(\underline{c}(t-1))\Delta \underline{c}(t-1) + \Omega(t)u'(\underline{c}(t))\Delta \underline{c}(t) = 0 \tag{B10.4.3}$$

From the definition of the CDR, $\rho(t)$,

$$\rho(t) = -[\Delta \underline{c}(t)/\Delta \underline{c}(t-1)] - 1 \tag{B10.4.4}$$

Moreover, equation (10.3) is

$$\delta(t) = -[\Omega(t) - \Omega(t-1)]/\Omega(t-1) \geq 0 \tag{B10.4.5}$$

To derive equation (B10.4.6) write, as in the text, $g(c(t)) = [c(t) - c(t-1)]/c(t-1)$. Now consider the pair of variations $\Delta \underline{c}(t-1)$ and $\Delta c(t)$ on the reference consumption path $\{\underline{c}(0), \ldots, \underline{c}(t), \ldots\}$ that leave V(0) in equation (10.4) unaffected. Then equation (B10.4.3) reads

$$u'(\underline{c}(t-1))\Delta \underline{c}(t-1) + [u'(\underline{c}(t))\Delta \underline{c}(t)]/(1+\delta(t)) = 0 \tag{B10.4.6}$$

Using equations (B10.4.1)-(B10.4.6) we obtain the approximate equation

$$1 + \rho(t) \simeq (1+\delta(t))(1+g(\underline{c}(t)))^{\eta(\underline{c}(t))} \tag{B10.4.7}$$

Taking logarithms of both sides of equation (B10.4.7) yields

$$\log(1+\rho(t)) \simeq \log(1+\delta(t)) + \eta(\underline{c}(t))\log(1+g(\underline{c}(t))) \tag{B10.4.8}$$

Suppose now, as is realistic, that $\delta(t)$ and $g(\underline{c}(t))$ are both small. Because $\log(1+x) \simeq x$,

Equation (B10.4.8) reduces to

$$\rho(t) \simeq \delta(t) + \eta(\underline{c}(t))g(\underline{c}(t)) \tag{B10.4.9}$$

In many exercises, it proves convenient to assume that time is continuous, not discrete. So we make the length of the period that separates points in time vanishingly short. In that case the approximate equation (B10.4.9) becomes exact:

$$\rho(t) = \delta(t) + \eta(\underline{c}(t))g(\underline{c}(t)) \tag{B10.4.10}$$

In the special case where δ is independent of t, η is independent of the $\underline{c}(t)$, and $g(\underline{c}(t))$, the CDR ρ is constant.[329]

B10.4.2 CDRs in the Economics of Climate Change

Economists writing on global climate change (Cline, 1992; Nordhaus, 1994; Stern, 2006) have used equation (B10.4.10) to identify optimum policies for correcting climate externalities. The authors assumed that δ is independent of t and η is independent of the reference consumption path. As noted in Chapter 5 (Box 5.1), the family of u-functions for which η is a positive constant has the form

---

[328] The two-period perturbation to the reference path is the most elementary investment project it is possible to imagine. Investment projects more generally are perturbations of consumption at more than two periods (dams can last more than 100 years!). If $\Delta C(t)$ is a small perturbation to $C(t)$, then $\{\Delta C(0), \ldots, \Delta C(t), \ldots\}$ represents a small investment project. Some elements in the sequence are negative (investment outlays), others are positive (gains in consumption). If the project is to be recommended for acceptance, it must entail $\Delta V(0) > 0$.

[329] It will be noticed that in Savage's theory of decision under uncertainty (Chapter 5), η is the measure of relative risk aversion. That shows there is a deep connection between risk aversion and inequality aversion. Harsanyi (1955), whose work was reported in Section 11.3, pointed to the connection. The classic on this equivalence is Atkinson (1970), who provided an explicit demonstration of it.

Chapter 10: Well-Being Across the Generations

$$u(c) = c^{(1-\eta)}/(1-\eta), \qquad \eta \neq 1,$$
$$u(c) = \log c, \qquad \eta = 1. \qquad \text{(B10.4.11)}$$

The larger is η, the greater is the curvature of u-function. Notice also that $u(c)$ is bounded above but unbounded below if $\eta > 1$, whereas $u(c)$ is bounded below but unbounded above if $\eta < 1$. The logarithmic function is of course unbounded at both 'ends'. The authors' most-preferred specifications for δ and η were:

Cline: $\delta = 0$ and $\eta = 1.5$

Nordhaus: $\delta = 3\%$ a year and $\eta = 1$

Stern: $\delta = 0.1\%$ a year and $\eta = 1$

For his base case, Stern (2006) assumed that $g(\underline{c}(t)) = 1.3\%$ a year under business as usual (i.e. the reference path). Using this figure in equation (B10.4.10) yields:

Cline: $\rho = 1.95\%$ a year

Nordhaus: $\rho = 4.3\%$ a year

Stern: $\rho = 1.4\%$ a year

Relative to the present (i.e. $t = 0$), a marginal unit of consumption is awarded half the weight if it is to appear 51 years in the future for Stern, 36 years in the future for Cline, and 17 years in the future for Nordhaus. Differences in their choice of CDRs explain to an extent disagreements among the authors on the urgency over global climate change. Nordhaus' 4.3% a year may not seem very different from Stern's 1.4% a year but is in fact a lot higher when it is put to work on the economics of the long run. Just how much higher can be seen from the fact that the present-value of a small loss in consumption 100 years from now if discounted at 4.3% a year is *17* times smaller than the present-value of that same consumption loss if the discount rate used is 1.4% a year. The moral is banal: If the time horizon is long, even small differences in CDRs can mean large differences in the message cost-benefit analysis gives us. The reason Cline (1992) and Stern (2006) have recommended that the world spend substantial sums today to tame climate change, while Nordhaus (1994) has recommended a far more gradualist investment policy can be traced in part to the difference in their choice of δ.

Drupp et al. (2018) reported findings of a survey in which economists were asked to say what rates they thought would be appropriate for discounting investment projects. While 30% of respondents recommended Stern's 1.4% or lower, only 9% recommended Nordhaus' 4.5% or higher, with 61% forming the middle ground between the two. The authors also reported that there is much less disagreement among economists than we are led to believe from the literature on discounting. Ninety per cent of economists find a discount rate of 1–3% acceptable for long-run public projects.

10.6 Directives on Discounting

Equation (10.6) offers a number of morals for the social evaluator:

(1) Because well-being is not proportional to consumption, the two should be regarded as different objects. If well-being is chosen by the social evaluator as the unit of account, $\delta(t)$ is the rate she should use for discounting the future deployment of resources, whereas if consumption is chosen by her to be the unit of account, it is $\rho(t)$ that she ought to use for discounting the future deployment of resources.

Chapter 10: Well-Being Across the Generations

(2) ρ(t) is not a primary ethical object, it has to be derived from her conception of intergenerational well-being (Equation (10.4)) and the reference stream round which she is conducting her thought experiment. CDRs cannot be plucked from the air.

(3) δ(t), η(c̲) and the forecast g(c̲(t)) together determine ρ(t). The latter increases with δ and g(c̲(t)), respectively, and increases with η *if and only if* g(c̲(t)) > 0. We have highlighted the qualifier 'if and only if' for a good reason. In studies of long-run economic development, it has become customary to confine attention to future projections in which consumption increases indefinitely. Although the tradition has been maintained in the economics of climate change, the previous chapters have shown why it has to be discarded in the economics of biodiversity.

(4) If g(c̲(t)) > 0, δ and η play similar roles in the determination of ρ(t): a higher value of either parameter would reflect a greater aversion toward intergenerational consumption inequality. But if g(c̲(t)) < 0, δ and η contain diametrically opposite directives: a larger figure for η would lower ρ, which would act as a corrective to consumption inequality across the generations. In contrast, a larger figure for δ would raise ρ, and that would imply an ethical preference for even greater inequality in consumption across the generations.

(5) Unless consumption is constant, discount rates depend on the unit of account: ρ(t) = δ if and only if g(c̲(t)) = 0. Even though well-being and consumption are not the same object their respective discount rates are the same in a stationary economy.

(6) Just as increasing consumption provides a reason CDRs should be positive, declining consumption would provide a reason they could be *negative*. To illustrate, suppose it is forecast that c̲(t) will decline at a constant rate of 1% a year (i.e. g(c̲(t)) = −0.01) from what it is today. Assume δ = 0 and η = 2. In that case ρ(t) = −0.02 per year.

That CDRs are not independent of the reference stream is not a trivial observation. In a Focus article in *The Economist* (26 June 1999: 128), the author uncovered a disturbing tendency of compound interest to make large figures in the distant future look very small today:

"Suppose a long-term discount rate of 7 percent (after inflation) is used ... Suppose also that ... benefits (from a project) arrive 200 years from now ... If global GDP grows by 3 percent a year during those two centuries, the value of the world's output in 2200 will be US$8 quadrillion (a 16-figure number). But in present-value terms, that stupendous sum would be worth just US$10 billion. In other words, it would not make sense for the world to spend any more than US$10 billion (under US$2 a person) today on a measure that would prevent the loss of the planet's entire output 200 years from now."

The argument is of course pure rubbish. Humanity would cease to exist if world output was to be zero. Where the author gets things wrong, after posing the problem, is to assume that the rates to be used for discounting future income losses are independent of the economic forecast that says there will be income losses. The final sentence in the passage points to the Ultimate Perturbation (zero world output in year 2200). The perturbation would involve a sharp decline in output just around year 2200. CDRs would be *hugely* negative in that brief period. So, where does the 7%-a-year discount rate come from? Discounting future incomes or consumption produces paradoxes only when it is not recognised that, as discount rates are features of the reference consumption stream around which investment projects are defined, they cannot be plucked from thin air.

(7) Because much of the biosphere is an open access resource, private rates of return on investment could be positive for a period of time, even while ρ(t) is negative. It was noted previously (Chapter 8) that

Chapter 10: Well-Being Across the Generations

individuals have an incentive to rush and privatise open access resources.[330] That rush would be read as *impatience* by someone unaware that much of the biosphere is an open access resource. Economists, however, should not be unaware. It is a fundamental mistake to infer δ by estimating it from consumer behaviour in a world where much of the biosphere remains free to all.[331] The common practice among contemporary macroeconomists of explaining macroeconomic statistics by assuming that an economy can be likened to a *single* dynasty should be abandoned. It certainly has no place in the economics of biodiversity.

10.7 Social Rates of Return on Investment

CDRs have been defined on arbitrary reference streams. But the task before the social evaluator is to identify the optimum consumption stream. Formally, her problem is to find the consumption stream (we may suppose in advance that it is unique) in $\Re(0)$ that maximises $V(0)$. To identify the optimum the social evaluator needs to relate the consumption rate of discount to the productivity of investment. That fundamental relationship is studied next.

10.7.1 The First Best

Consider once again an arbitrary reference stream in $\Re(0)$. We write it as $\{\underline{C}(0), \ldots, \underline{C}(t), \ldots\}$. Suppose now, as a further thought experiment, that the social evaluator considers reducing consumption in period t-1 by one small unit. That reduction would amount to a saving, and her intention is to find a project in which to put the saving to work. To break down a complicated problem into a series of simple problems, she considers projects that last two periods, the first being the investment period, the second being the period it produces output. The social evaluator therefore needs to estimate the additional consumption that would be made available in the second period if project output was consumed entirely.[332] We call a two-period project an 'elementary project'. For example, the project could be to fell one fewer tree (the export revenue from which would have been consumed in the reference stream), and to fell the tree in the following period for export and use the revenue for consumption purposes.

Let $1 + \gamma(t)$ be the addition to consumption in period t that would be made available by the elementary project. (In the example $1 + \gamma(t)$ would represent the size of the tree in period t for potential export.) We say $\gamma(t)$ is the *yield* – alternatively *own rate of return* – on the investment (Chapter 1). It measures the project's productivity as measured in consumption units.

It should be evident that, as with CDRs, a project's yield depends not only on the reference stream but also on the nature of the project. A reference stream in which ecosystems are in a degraded state offers fewer options than one in which ecosystems have remained healthy; and it may be that the yield from a project that expands a road system is lower than a project that restores a tropical rainforest.

The social evaluator now seeks a project with the *maximum* yield in period t. Let us write that maximum as $r(t)$. Thus $\gamma(t) \leq r(t)$ for all possible elementary projects starting in period t-1. To distinguish projects with the maximum yield from all other elementary projects we shall call $r(t)$ the *social rate of return* on investment. It is a simple matter to confirm that the optimum consumption stream must be the reference stream along which the CDR equals the social rate of return on investment in each period (Box 10.5). The necessary condition for optimality is thus

---

[330] For a proof of this claim in a fully dynamic economy, see Dasgupta, Mitra, and Sorger (2019).

[331] The mistake is common. See for example, Nordhaus (2007).

[332] It is assumed that consumption in all other periods is to remain the same as in the reference stream.

$$\rho(t) = r(t), \qquad \text{all } t \geq 0 \qquad (10.7)$$

Equation (10.7) is the fundamental principle of optimum saving. We confirm below that the equation is an arbitrage condition (Chapter 1).

To express the equality of the consumption rate of discount and the consumption rate of return in its most familiar form, assume time is continuous. On using equations (10.6) and (10.7), we then have

$$\delta(t) + \eta(\underline{c}(t))g(\underline{c}(t)) = r(t) \qquad (10.8)$$

Equation (10.8), due in this form to Ramsey (1928), is routine in the economics of climate change and, more generally, in growth and development economics. It is known as the Ramsey Rule. The Rule is a necessary condition for optimality. It can be used by the social evaluator to estimate the optimum level of investment in the economy at each moment in time. For reasons that will become clear presently, the optimum we have just characterised is known as the *First Best*. Annex 1 to this chapter works with a simple model economy to make the required calculations.

We have arrived at equation (10.8) through a sequence of thought experiments involving elementary, 'two-period' projects. The equation itself provides a condition the optimum consumption stream must necessarily satisfy. It bears stressing therefore that an analysis that involves a sequence of marginal changes should not be mistaken for marginal change. The consumption stream the social evaluator projects into the future under prevailing and projected institutions and policies (i.e. under 'business as usual') is most likely to be very far from what she judges on using equation (10.8) to be the optimum consumption stream. The leap from business as usual to where citizens would like their society to be can be huge. In our own world the Impact Inequality is so large that the institutional changes that will be required are enormous if we are to convert the inequality into an equality, and with the biosphere brought back into a healthy state.

10.7.2 Second-Best Optima

Institutional failure in general, and externalities in particular, is the prime reason economies fail to attain consumption and investment mixes that satisfy the Ramsey Rule. As the massive biodiversity loss we are witnessing today speaks to that, we study the implications of detrimental externalities whose presence directs investment away from Nature conservation and restoration toward investment in produced and human capital. It would be unrealistic to imagine a society being able to leap from a distorted economy to the First Best. Some distortions can be removed with relative ease if attention is paid to them, but many will be too rigid to change at once. So our social evaluator's task is to maximise $V(0)$ within a constrained set of possible socio-ecological futures. The feasible set of consumption streams in which she is to limit her search is therefore a subset of $\mathfrak{R}(0)$. Call that set $\mathfrak{R}^*(0)$. Thus $\mathfrak{R}^*(0) \subseteq \mathfrak{R}(0)$. The optimum in $\mathfrak{R}^*(0)$ is known as a *Second Best*.

Box 10.5

Optimum Saving Principle

Well-being and consumption are not the same good. Let well-being be the unit of account (i.e. *numeraire*). The *accounting price of consumption* in period t is the social worth of a marginal unit of consumption relative to well-being: $u'(\underline{c}(t))$. As $\eta(\underline{c}(t))g(\underline{c}(t))$ is the percentage rate of change in the accounting price of consumption, it is the 'capital gains' on holding consumption between periods $t-1$ and t. (If $g(\underline{c}(t)) < 0$, the 'gains' are losses.) Moreover, because $1+\delta(t)$ is the rate at which well-being is rate at which well-being can be traded off between periods $t-1$ and t without affecting $V(0)$, $\delta(t)$ can be read as the *own rate of return* on well-being. By its definition $r(t)$ is the own rate of return (or yield) on consumption. Equation (10.8) is thus the arbitrage condition for efficient portfolios (Chapter 1).

Chapter 10: Well-Being Across the Generations

To prove the optimality condition (Equation (10.7)), suppose $\rho(t) < r(t)$ along the reference path. Investment projects are perturbations to reference paths. Consider now the simplest possible investment project it is possible to undertake in the reference path $\underline{C}(t)$: Reduce consumption in period t-1 by one unit and consume the maximum output the project can supply in t, while keeping consumption in all other periods the same as in the reference path. That maximum supply is, by definition, $1 + r(t)$. So, the social value in period t-1 of an increase in expected consumption in period t from the project is $[1 + r(t)]/1 + \rho(t)] > 1$. As the project's investment outlay was 1 unit of foregone consumption, $V(0)$ increases. The project passes the social cost-benefit test: its *present value* (PV) is positive. The additional saving in period t-1 should be undertaken. The project should be recommended by the social evaluator.

Undertaking the additional saving in t-1 increases $g(\underline{c}(t))$ and does not reduce η. That means $\rho(t)$ increases slightly. If there are diminishing social returns on investment with greater investment, $r(t)$ declines marginally, and the two together reduce the gap between $\rho(t)$ and $r(t)$. The social evaluator now conducts social cost-benefit analysis of a sequence of investment projects, accepting those whose PV is positive and rejecting those whose PV is negative, until the gap closes to zero. Each step in the sequence can be thought of as an exercise in social cost-benefit analysis. A reverse sequence of project analyses can be conducted if in the reference path $\rho(t) > r(t)$.

The procedure that has just been described is known as the 'hill-climbing method'. The method is an algorithm. Social cost-benefit analysis can be interpreted as steps in an algorithm that solves the social evaluator's optimisation problem.[333]

There is a caveat. As has been demonstrated here, equality of $\rho(t)$ and $r(t)$ is a necessary condition for optimality. To locate the reference path that maximises $V(0)$ requires further conditions. There are two reasons the algorithm we have described to understand the role social cost-benefit analysis plays in the search for the optimum distribution of well-being across the generations is not sufficient. The first reason is easy to demonstrate by considering the following example:

Suppose r, δ, and η are constants. Then equations (10.6) and (10.7) can be expressed as

$$d\underline{c}(t)/dt = [(r - \delta)/\eta]\underline{c}(t) \quad \text{(B10.5.1)}$$

It is easy to see what the problem is. The equation identifies the optimum rate of change in consumption per capita, but it does not say what the initial consumption level should be. The social evaluator needs a further condition to identify the socially optimum $\underline{c}(0)$.

Integrate equation (B10.5.1) to obtain

$$\underline{c}(t) = \underline{c}(0)\exp[(r - \delta)/\eta] \quad \text{(B10.5.2)}$$

There is an infinity of $\underline{c}(0)$'s for which equation (B10.5.2) holds. In Box 10.6, it will be confirmed that there is a maximum in that range. All initial levels of consumption per capita less than the maximum are inefficient, resembling the indefinite consumption potlatch described in Box 10.3. If the initial level of consumption per capita is chosen higher than the maximum, the economy runs out of capital assets in finite time, and that is the ultimate catastrophe.

The other reason the optimality condition in Equation (10.6) is not sufficient has to do with the non-linearities in the structure of the optimisation exercise. Non-linearities are present in the set of feasible socio-ecological futures, $\Re(0)$. They display themselves in ecological processes (Chapter 3) and they arise in the presence of harmful externalities (Chapter 7).[334]

---

[333] This interpretation of social cost-benefit analysis was emphasised in Dasgupta, Marglin, and Sen (1972).

[334] Key publications illustrating the two, respectively, are Brock and Starrett (2003) and Starrett (1972).

> The presence of non-linearities means that the social evaluator's maximisation problem may have more than one 'hill', that is, the hill-climbing algorithm that defines marginal social-cost benefit analysis could lead to a local peak, at which there is no further marginal project worth undertaking, but the destination may nevertheless not be the global peak. Which is why the economics of biodiversity requires the social evaluator to scan beyond the horizon, to check if there are better configurations of society's capital goods far away from where society is now. Marginal cost-benefit analysis is of no use for that exercise.

10.8 Accounting Prices

If we are to take a single insight for the economics of biodiversity from previous chapters, it is that Nature is mobile and is in large measure also silent and invisible (Chapter 0). These features give rise to the pervasiveness of detrimental externalities in our use of the biosphere. The social evaluator has to accept that there are externalities that simply will not be eliminated. She recognises also that there are financial limits on amounts that can be expected to be available for protecting and promoting the biosphere. In formal terms that is to recognise the inevitability of underinvestment in Nature. It is against that underinvestment she fashions a second-best economy.

The term 'accounting price' has been used informally in previous chapters. We now have the machinery to define it formally. To preserve neutrality among goods and service in the definition of accounting prices, let well-being be the unit of account.

Consider an arbitrary reference stream in $\Re^*(0)$. The accounting price of a commodity in period t on the path is the change in intergenerational well-being V(t) that would occur if a unit more of the commodity was to be made available at no cost at t.

What we have defined is the *spot* (accounting) price of the commodity. Its accounting price in period 0 is the *present value* of its spot price. Let $p_i(t)$ be the spot price of commodity i. To determine its present value, $p_i(t)$ is discounted at well-being discount rates $\delta(t)$. In practical work the unit of account (*numeraire*) is a commodity, not well-being. Suppose consumption was chosen as *numeraire*. In that case $p_i(t)$ would be the period t spot accounting price commodity i relative to consumption. Its present value would be calculated by discounting $p_i(t)$ at the consumption rates of discount $\rho(t)$. Environmental and resource economists have studied the influence of *substitution* possibilities among produced capital and natural capital on the structure of accounting prices (Dasgupta and Heal, 1979; Drupp, 2018; Traeger 2011; Baumgärtner et al. 2015).[335] In Chapter 4*, we laid stress on the influence of *complementarities* among the biosphere's regulating and maintenance services on accounting prices. *The latter have a deeper effect because they set bounds on economic possibilities no matter how substitutable the factors of production happen to be.*

Notice that accounting prices can be defined on *any* consumption stream in $\Re^*(0)$. But the task before the social evaluator is to estimate the accounting prices that would correspond to the optimum consumption stream in $\Re^*(0)$. In Box 10.5, it was shown that accounting prices in arbitrary consumption streams can be used iteratively to identify the optimum consumption stream in $\Re^*(0)$. The procedure outlined there displays the link between social cost-benefit analysis of investment projects and locating the optimum level of investment in an economy.

The social evaluator's task is to find the consumption stream in $\Re^*(0)$ that maximises V(0). But $\Re^*(0)$ is circumscribed by institutional constraints. The social evaluator would ideally like to break through some

---

[335] The landmark exposition of the structure of accounting prices remains Koopmans (1957).

Chapter 10: Well-Being Across the Generations

of those constraints, but that may not be feasible. In the presence of institutional rigidities, consumption and investment would not be perfectly substitutable for one another at the margin.

Along the optimum $u'(\underline{c}(t))$ is the accounting price of consumption relative to well-being, but because there is underinvestment in the economy, it is *less than* the accounting price of investment.[336] Our focus here is the underinvestment in Nature. If $p_n(t)$ is the accounting price of natural capital, $u'(\underline{c}(t)) < p_n(t)$ on a second-best. At the margin investment in natural capital is not the same commodity as consumption, nor is it the same commodity as investment in produced and human capital. An example of a second-best accounting system is provided in Annex 10.1.

The classics on ways to estimate second-best accounting prices in dynamic economies are Marglin (1963a,b). The constraints Marglin considered are reflected in an inability of government to channel first-best investment levels in produced capital.[337] In contrast, the second-best we are interested in here arises because investment in natural capital is perforce so limited as to be below its first-best level. *In such an economy, the consumption rate of discount is less than the social opportunity cost of investment in natural capital.* Despite its salience, this directive has to the best of our knowledge been ignored in the integrated assessment models that in large measure define the economics of climate change.

Box 10.6
Accounting Prices in a Dynamic Economy

We work with the framework that was introduced in Box 10.5 but now specify it in the way integrated assessment models have made familiar in the economics of climate change. Time is denoted by t and is taken to be continuous. The stock of capital good i at time t is $k_i(t)$, and so on for all other time-dependent variables.

Accounting prices should ideally be estimated from four items of information:

(i) A descriptive model of production possibilities in the economy (as in Chapter 4*)

(ii) An account of the institutions and practices that shape people's choices

(iii) The size and distribution of the economy's capital goods

(iv) A conception of intergenerational well-being.

Together (i)-(iii) are the basis for estimating the changes that take place to the allocation of goods and services if an additional unit of a capital good is made available free of charge; (iv) is the coin on the basis of which the changes are evaluated.

Generalising the model in Chapter 4*, let $k(t)$ denote the vector of capital goods at date t. The social evaluator can make use of her understanding of (i)-(iii) to make an economic forecast for all dates $s \geq t$. For policy analysis, she also needs to have a sense of what the forecast would be if the composition of assets had been otherwise. She needs to have that sense because investment projects involve the reallocation of capital goods.

Let $c(s)$ be the potential standard of living at time s and $u(c(s))$ the potential flow of well-being at s. Using her forecast, the social evaluator can express $c(s)$ as $c(s,k(t))$. Intergenerational well-being at t is

---

[336] This is demonstrated formally in Annex 10.2.

[337] Marglin's approach was applied by Little and Mirrlees (1969, 1974) and Dasgupta, Marglin, and Sen (1972) in their work on social cost-benefit analysis in labour-surplus economies.

> $$V(t) = {}_t\!\int^{\infty}[u(c(s, k(t)))]e^{-\delta(s-t)}ds, \qquad \delta > 0 \qquad (B10.6.1)$$
>
> From equation (B10.6.1), it follows that $V(t)$ is an implicit function of $k(t)$. So we may write the equation as
>
> $$V(k(t)) = {}_t\!\int^{\infty}[u(c(s, k(t)))]e^{-\delta(s-t)}ds, \qquad \delta > 0 \qquad (B10.6.2)$$
>
> The accounting prices of capital goods follow directly from equation (B10.6.2). Writing by $p_i(t)$ the accounting price in well-being units of asset i, we have
>
> $$p_i(t) = \partial V(k(t))/\partial k_i(t) = {}_t\!\int^{\infty}[u'(c(s))\,\partial c(s, k(t))/\partial k_i(s)]e^{-\delta(s-t)}ds \qquad (B10.6.3)$$
>
> In Chapter 13, we will find that the expression for intergenerational well-being on the left-hand side of equation (B10.6.2) is useful for identifying those reference consumption streams that represent 'sustainable development', whereas the expression on the right-hand side is useful for conducting 'policy analysis'. We will also find that when a policy involves choice of an investment project, the criterion of choice that ought to be adopted is the net present value (PV) of the project's social benefits.

10.9 Should Environmental Projects be Evaluated Using Lower Discount Rates?

The rates our social evaluator will want to use to discount future benefits and costs depend on her choice of unit of account, or *numeraire*. That was directive (a) in Section 10.6. Economists often use the neutral term *social discount rates* when they have not specified the *numeraire*. That practice alone shows that price changes matter. The significance of relative prices was already noted in Chapter 1 (Box 1.3), where we proved that efficient portfolios satisfy 'arbitrage conditions' (Equation (B1.3)).[338]

Social discount rates are related to one another by the same logic. That was shown in the derivation of Equation (10.6). We now extend it to all pairs of goods. Let $\rho_i(t)$ be the social discount rate at t if i is *numeraire* and let $\rho_j(t)$ be the social discount rate if j is *numeraire*. If $p_{ij}(t)$ is the accounting price of j relative to i, then

$$\rho_j(t) + [dp_{ij}(t)/dt]/p_{ij}(t) = \rho_i(t) \qquad (10.9)$$

Equation (10.9) is simple to prove.[339] The equation implies, among other things, that if $p_{ij}(t)$ is forecast to increase at t, then the social discount rate if j is *numeraire* should be less than the social discount rate if i is *numeraire*. If we now deploy the Impact Inequality in our global economic model of Chapter 4* and forecast that the global economy will continue to accumulate produced capital and decumulate natural capital for a while, we would conclude that the accounting price of natural capital relative to

---

[338] To recall, arbitrage conditions say that if $ri(t)$ and $rj(t)$ are the rates of return on assets i and j and $p_{ij}(t)$ is the price of j relative to i, then $r_j(t) + [dp_{ij}(t)/dt]/p_{ij}(t) = r_i(t)$.

[339] Imagine the social evaluator is offered the following two options at t: (i) convert one unit of i into $1/p_{ij}(t)$ units of j, and receive $[1 + \rho_j(t)]\Delta/p_{ij}(t)$ units of j a brief interval Δ later; and (ii) forego one unit of i at t and receive $[1 + \rho_i(t)]\Delta$ units of i at $(t+\Delta)$, which she can then convert into j at the price $p_{ij}(t+\Delta)$. If the social evaluator is to be indifferent between the options, the two options should reward her with the same quantity of i at $(t+\Delta)$. In that case the accounting prices $\rho_i(t)$, $\rho_j(t)$, and $p_{ij}(t)$ must be related to one another by the condition

$$\{[p_{ij}(t+\Delta) - p_{ij}(t)]/\Delta\} + (1 + \rho_j(t))p_{ij}(t) = (1 + \rho_i(t))p_{ij}(t+\Delta)$$

Taking the limit $\Delta \to 0$ yields equation (10.9).

produced capital will increase for that while. Equation (10.9) then says social discount rates if produced capital is chosen as *numeraire* would be higher than the corresponding rates if instead natural capital is chosen as *numeraire*. That in turn means if produced capital is *numeraire* future benefits and costs ought to record the rise in value of natural capital; but if instead natural capital is *numeraire*, future benefits and costs should record the decline in the value of produced capital.

It is tempting to conclude that 'green' projects ought to be evaluated using preferential discount rates. Formally that would be correct (Hoel and Sterner, 2007). But in practice, that would be so cumbersome as to lead inevitably to errors. A piece of equipment, when deployed in a green project, would have to be recorded at a lower accounting price than when deployed on a building site. And that move would have to be made for every piece of produced capital.

All investment projects involve a mix of produced capital, human capital and natural capital. Even projects that are designed to conserve biodiversity require produced capital and human capital (how else would the projects be protected against intruders?). Tempting though it is to conservationists, it would be unsafe to apply preferential discount rates even to nature-conservation projects, for there is always a risk the social evaluator will overlook to correct for capital gains and losses of some goods and services. The far safer practice would be to choose a *numeraire*, apply it to all projects, and measure benefits and costs on the basis of one set of accounting prices.

10.10 The Idea of Investment

To invest in a capital good is to increase it beyond what it would be if there was to be no investment in it. We are talking of *net* investment, that is, investment net of depreciation. Common-sense notions of investment, however, carry with them a sense of robust activism. When a government invests in roads, the picture drawn is of bulldozers levelling the ground and tarmac being laid by people in hard hats. But the notion of capital goods we are using here extends beyond produced capital to include human capital and natural capital. That training people to be teachers is investing in human capital is simple enough. To leave a forest unmolested may not sound much like investment, but it is an investment. It enables the forest to grow. And to allow a fishery to re-stock is to invest in the fishery, and so on.

The examples of elementary projects in the previous section could suggest that investment amounts to deferred consumption, but the matter is subtler. Providing additional food to undernourished people via, say, food guarantee schemes not only increases their current well-being, it enables them also to be more productive in the future and to live longer. Because their human capital increases, the additional food intake should count also as investment. Note though that food intake by the well-nourished does not alter their nutritional status, which means the intake is consumption, not investment. By net investment in a capital good asset we should mean the accounting value of the change in its stock.

In deriving equation (10.8), there is no mention of the character of the elementary project. It could have been investment in a network of roads (produced capital), to primary education (human capital), to ecosystem restoration or to Nature conservation (natural capital). To determine the character of optimum investment requires the social evaluator to dig deeper and conduct *social cost-benefit analysis* of alternative investment projects. In Chapter 13, it is shown that social cost-benefit analysis is an essential tool for managing an economy's assets.

Let $k_i(t)$ be the projected stock of the ith asset on a reference consumption stream in $\Re^*(0)$. Then $p_i(t)k_i(t)$ is the accounting value of the stock of asset *i*. The sum of the accounting values of an economy's stock of assets is known as *inclusive wealth*, the qualifier 'inclusive' is there to remind us that an economy's wealth consists not only of produced and human capital, but natural capital too. Chapter 13 develops the role of inclusive wealth in the economics of biodiversity. It is demonstrated there that the criterion our social evaluator ought to use for managing the economy's assets is inclusive wealth. Social cost-benefit analysis is a calculus for maximising an economy's inclusive wealth. Inclusive wealth is a *stock*.[340] In Chapter 13 it is shown that an economy's inclusive wealth is at a maximum at the optimum in $\Re^*(0)$. It is also shown there the iterative exercises described in Box 10.5 for locating the optimum consumption stream in $\Re^*(0)$ involve the use of inclusive wealth as the coin with which to conduct social cost-benefit analysis of investment projects. Economic progress, it will be argued, should be growth in inclusive wealth, not GDP.

GDP, which is the market value of the final goods and services an economy produces, is a *flow*. The rogue word in gross domestic product is 'gross', as the index does not include the depreciation of capital goods. Which is why it is possible for GDP to grow for a period while inclusive wealth declines. Even if produced capital and human capital were to grow in magnitude, inclusive wealth would decline if natural capital were to decline in quality, or quantity, at a high enough rate. But if inclusive wealth was to continue to decline, GDP would eventually have to decline. You cannot degrade the biosphere indefinitely and expect living standards to rise continually. We discuss this further in Chapter 13, where the concept of inclusive wealth is studied.

Box 10.7
Discounting When Preferences are Endogenous

The language of optimum asset management does not change when economic models allow for richer conceptions of human well-being. In Chapter 11, we review what is becoming increasingly clear about the sources of well-being. There is a lot more than consumption, narrowly defined, that makes for a flourishing life. To show how widening the notion of well-being influences discount rates, we consider first socially embedded preferences. The effect of past consumption on current well-being is considered next. For simplicity of notation, time is assumed to be continuous.

The expression for intergenerational well-being is the continuous-time version of equation (10.1):

$$V(0) = {_0}\int^{\infty}[e^{-t}u(c(t))]dt \qquad (B10.7.1)$$

B10.7.1 Socially Embedded Preferences

As in Chapter 9, we suppose that well-being is a function not only of personal consumption c, but also of the average consumption level c^* in the economy. Thus $u = u(c,c^*)$. There is no need to specify whether consumption is competitive or conformist.[341]

We now have to extend the notion of rates of discount and rates of return by focusing on *output*, not consumption. And we need to distinguish the proportions of output devoted to different forms of consumption and investment. For the simple model we are considering here, the accounting price of output devoted to consumption includes not only the contribution output makes to individual consumption but also to average consumption.

Write $\partial u/\partial c$ as u_c and $\partial u/\partial c^*$ as u_{c^*}. In well-being units, the accounting price of output devoted to consumption is then

$$p(t) = u_c(c(t),c^*(t)) + u_{c^*}(c(t),c^*(t)), \qquad c(t) = c^*(t) \qquad (B10.7.2)$$

It is now a simple matter to confirm that the rate at which the output that is devoted to consumption ought to be discounted is $\{\delta - [dp(t)/dt]/p(t)\}$. If $r(t)$ is the social rate of return, the condition for optimality is

---

[340] Formally, stocks of capital goods are state variables in dynamic models.

[341] The analysis presented here is taken from Arrow and Dasgupta (2009).

Chapter 10: Well-Being Across the Generations

$$\delta - [dp(t)/dt]/p(t) = r(t) \tag{B10.7.3}$$

In a large economy, the individual person will regard c^* as a parameter. Denoting economic variables in the market with the lower-case m, the market equilibrium price of consumption relative to well-being would be

$$q(t) = u_c(c_m(t), c_m^*(t)), \qquad c_m(t) = c_m^*(t) \tag{B10.7.4}$$

If $r_m(t)$ is the market rate of return on investment, then the dynamical trajectory for $q(t)$ would be

$$\delta - [dq(t)/dt]/q(t) = r_m(t) \qquad c_m(t) = c_m^*(t) \tag{B10.7.5}$$

Routine computations show that if social preferences are competitive (Chapter 9), $p(t) > q(t)$, meaning that there should be an externality tax per unit of consumption, amounting to $u_{c^*}(c(t), c^*(t))$, so that consumption in market equilibrium corresponds to socially optimum consumption.

B10.7.2 The Effect of the Past

The simplest way to model the effect of past consumption on current well-being is to think of past consumption as 'adaptive memory'. The contribution of past consumption to memory of that consumption decays at an exponential rate λ, which means the farther into the past one peers the smaller is its effect on the present. Define $z(t)/\lambda$ as the discounted integral of consumption until t. Thus

$$z(t)/\lambda = \int_{-\infty}^{t} [c(s)e^{\lambda(s-t)}]ds, \qquad \lambda > 0 \tag{B10.7.6}$$

z is the return on past consumption.[342] Intergenerational well-being is $V(0)$ in equation (B10.7.1).

From equation (B10.7.6) we obtain

$$dz(t)/dt = \lambda[c(t) - z(t)] \tag{B10.7.7}$$

We now suppose that current well-being is a function of current consumption c and the index of past consumption z, that is $u(t) = u(c(t), z(t))$. Features that would be natural to impose on the u-function are $u_c(c,z) > 0$, $u_z(c,z) < 0$, $u_c(c,z) + u_z(c,z) > 0$, and concavity. That $u_z(c,z) < 0$ means memory is a burden; an additional unit of consumption today can be expected to reduce well-being tomorrow, other things equal. However, the externality travels across time and can be fully internalised by each individual person.[343]

Let $p(t)$ be the accounting price of consumption in well-being numeraire and let $\psi(t)$ be the accounting price of $z(t)$. Then

$$p(t) = u_c(c,z) + \lambda\psi(t) \tag{B10.7.8}$$

As $u_z(c,z) < 0$, $\psi(t) < 0$: memory inflicts a negative externality on current well-being. If $r(t)$ is the social rate of return, $p(t)$ and $\psi(t)$ satisfy the pair of arbitrage equations

$$\delta - [dp(t)/dt]/p(t) = r(t) \tag{B10.7.9}$$

$$\delta - [d\psi(t)/dt]/\psi(t) = -\lambda + u_z/\psi(t) \tag{B10.7.10}$$

Ryder and Heal (1973) have provided a complete characterisation of the optimum consumption policy.

---

[342] If c has been constant since time immemorial, $z(t) = c$.

[343] The model here is taken from Ryder and Heal (1973).

10.11 Population Ethics

If population is exogenously given, the u-function in our formulation of well-being across the generations (i.e., the additive form of eq. (10.1) in sect. (10.1)) may be read as being cardinal, meaning, unique up to positive affine transformations (Box 10.1). The sign of u would not matter in decision making. That's the standard assumption in classical decision theory. However, when we assume population numbers to be endogenous, the u-function in the additive form of eq. (10.1) is required to be strongly cardinal, meaning, unique up to positive linear transformation (Box 10.1). In that case well-being *levels* have operational significance; the sign of u matters in decision making. We confirm this by means of an exercise in optimum population in Box (10.8) below.[344]

Consumption and investment decisions today influence future population numbers. Likewise, projections of future populations have a bearing on current consumption and investment decisions. Optimum choices in the face of this mutual influence were studied formally by Dasgupta (1969) in a simple model of production. As in Sect. (10.1), the normative basis of choice was taken there to be Classical Utilitarianism. It was shown that if output is subject to constant returns to scale in produced capital and population numbers (human capital), the optimum policy leads in the long run to unbounded population and unbounded GDP. But it was also shown that if in addition to produced capital and human capital, output requires a third factor of production, fixed in size (land), economic possibilities are altogether different: optimum consumption, investment, and population policies lead in the long run to a stationary state, in which population and output are both finite. The biosphere isn't, of course, a fixed factor (it is self-regenerative but can be traduced through overuse); however, the optimum policy would have the same properties if the constraining factor of production were taken to be a bounded self-regenerative asset. Technical progress would overcome some of the biosphere's constraints, but as we noted in Ch. 4, it cannot lead to indefinite growth even if human ingenuity were boundless.

10.11.1 Zero Well-Being

Implicit in the formulation here (and the more general formulation in Sect. (10.1); eq. (10.1))) is the idea that non-existence should be calibrated as $u = 0$. Sidgwick (1907: 124–125) instead spoke of 'neutral feeling' when alluding to $u = 0$:

"If pleasure ... can be arranged in a scale, as greater or less in some finite degree, we are led to the assumption of a hedonistic zero, or perfectly neutral feeling, as a point from which the positive quantity of pleasures may be measured ... For pain must be reckoned as the negative quantity of pleasure, to be balanced against and subtracted from the positive in estimating happiness ... "

Zero well-being is a fundamental notion in ethics. Sidgwick's reference to neutral feeling invites us to assess life from the inside. But the idea of neutral feeling and the corresponding idea of a neutral life also point, even if several steps removed, to a comparison of life with non-existence. That latter exercise requires calibrating well-being in terms of something outside our experience, which is why the move could be regarded as questionable. Yet, when in deep despair people have been known to say they would rather not have been born, an utterance that doesn't sound incomprehensible. Comparison of life with non-existence is an unavoidable exercise in ethics if for no other reason than that reproduction is never a certainty. Suppose a couple understands there is a 90% chance of producing a happy child and a 10% chance they will be unable to conceive. Neutral feeling in Sidgwick's sense cannot cover the latter event in the couple's reasoning because the intended child does not exist there.

---

[344] This and the following section have been adapted from the central essay, "Birth and Death" (Kenneth Arrow Lectures), in Dasgupta (2019).

Nagel (1979: 2) famously suggested that death is not an unimaginable condition of the living person: " ... the value of life and its contents does not attach to mere organic survival: almost everyone would be indifferent (other things equal) between immediate death and immediate coma followed by death twenty years later without reawakening." Nagel also suggested that death is a mere blank, and that it can have no value whatever, positive or negative, which suggests that he thinks the blank cannot be used as a benchmark against which other states of affair are compared. But that thought cuts against the suggestion that one can imagine non-existence by imagining being in a coma for the rest of one's life. It would seem, for Nagel non-existence is the real blank, being totally unconscious for the rest of one's life is a simulation of that blank.

We are thinking of someone's life in its entirety, not her life at a moment in time. The u-function in eq. (10.1) is calibrated by using Nagel's 'blank' as a point of reference. We are attaching the number 0 to that point. There is no loss in generality doing that, because we would merely recalibrate the u-function if we were to choose the number to be other than 0. Zero well-being is therefore the measure of a person's life that, taken as a whole, goes neither well nor not-well for her. In view of the additive structure of well-being across the generations that the Pragmatist social evaluator has been assumed here to have adopted (eq. (10.1)), $u = 0$ is also the level of well-being at which, in her judgement, an additional life adds no further value to the world that contains it. Call the living standard at which $u = 0$, *well-being subsistence*. Using the notation in eq. (10.1), we would call c^* well-being subsistence if $u(c^*) = 0$ and would say that c^* is the living standard at which life is neither good nor not good.

It could be thought that $u = 0$ is the point of indifference between dying and continuing to live, or the point of indifference between life and death. Below we uncover the reason that interpretation is misconceived. The reason also steers us away from the thought that in the contemporary world well-being subsistence can be identified as the standard of living the World Bank proposed as representing extreme poverty – to wit, 1.90 dollars-a-day. In Box 10.1 we suggested that the *idea* of a poverty line could be used as a reference for well-being subsistence. Whether that should be 1.90 dollars per day, however, is a different matter. The concerned citizen will inevitably bring her experiences and social expectations to bear for identifying what well-being subsistence conveys to people in her society. If she is in an affluent culture, she would regard a life at 1.90 dollars per day a bad state of affair for the individual living it. She would identify well-being subsistence as being higher than the World Bank's poverty line.

Determining well-being subsistence will always prove contentious, for it involves a deep and difficult value judgment. But the exercise cannot be avoided. Cultural experiences colour the exercise. Among other things, they tell us well-being subsistence is not independent of the living standard of others in one's society and beyond (as in the socially embedded preferences of Ch. 9); and that means in the contemporary world well-being subsistence is higher, possibly a lot higher, than it was in the past.

10.11.2 Death

In a moving discourse on the place of autonomy and responsibility in personal well-being, Williams (1993: 50–102) drew attention to an aspect of personal responsibility that starts not from what others may demand of someone, but from what that someone demands of himself. Williams reminded readers that Sophocles had reported that Ajax, being slighted by the award of Achilles' arms to Odysseus, had intended to kill the leaders of the Greek army. To prevent the massacre, Athena made Ajax mad. It is significant that Ajax's condition didn't affect his purposes; rather, it altered his perception. Thinking that he was killing Odysseus and the others, Ajax slaughtered the army's flocks of sheep and cattle. In Sophocles' account the despair arising from the shame Ajax felt on awakening left no option open to him but to take his own life. And Williams observes (p. 76) that when Ajax says he must go, " ... he means that he must go: period."

Sidgwick in contrast offered a view of life that is at odds with Sophocles' account. In the chapter that introduces Utilitarianism to his readers, Sidgwick (1907: 414-415) wrote:

" ... I shall assume that, for human beings generally, life on the average yields a positive balance of pleasure over pain. This has been denied by thoughtful persons: but the denial seems to me clearly opposed to the common experience of mankind, as expressed in their commonly accepted principles of action. The great majority of men, in the great majority of conditions under which human life is lived, certainly act as if death were one of the worst of evils, for themselves and for those whom they love ... "

Nagel (1979) concluded that if death is an evil, it is the loss of life that is objectionable. The conclusion is incontrovertible to the secular mind, but there are three circumstances of death that should be distinguished, and they don't point in the same direction. There is death that comes naturally to one in the fullness of time; there is death that comes not from one's hands before one's time; and there is death that is brought on one by one's own deliberate action. Nagel contrasted the first two but didn't speak to the third. And it is the latter that should make us pause before accepting Sidgwick's conclusion that life, all in all, is a positive good for most people.

Religious prohibition, fear of the process of dying (the possibility of suffering pain, the feeling of isolation), the thought that one would be betraying family and friends, and the deep resistance to the idea of taking one's own life that has been built into us through selection pressure would cause someone even in deep misery to balk. It may even be that no matter what life throws at us we adjust to it, if only to make it possible to carry on. But the acid test for Sidgwick's inference that "life on the average yields a positive balance of pleasure over pain" is to ask ourselves whether we shouldn't pause before creating a person and imagine the kind of life that is likely to be in store for the potential child. The desire to procreate springs from our deep emotional needs, and our direct motivation to reproduce can be traced to a wide variety of reasons, but here we are concerned only with the life of the prospective child.

In his speculations on population ethics, Parfit (1984) studied the asymptotic properties of the iso-quants on the positive quadrant of the function Nu and noticed that they converge to $u = 0$ as N tends to infinity. He saw that as a repugnant feature of Classical Utilitarianism, read it as a Repugnant Conclusion (Parfit, 1984: 388), and interpreted it thus:

"For any possible population of at least ten billion people, all with a very high quality of life, there must be some much larger imaginable population whose existence, if other things are equal, would be better, even though its members have lives that are barely worth living."

The play on words in the passage should baffle the reader. We are being asked to consider a figure for world population that, as we noted in chapter 9, will in all probability be reached by the end of this century and a figure that is most unlikely to be *sustainable* at reasonable material comfort (Ch. 4); we are then made to imagine an Earth where, because of population pressure, people scramble for resources in order to eke out an existence, having *lives that are barely worth living*. But someone whose life is barely worth living doesn't enjoy a life of positive quality; she suffers from a life that is not only not good (as experienced by her) but is positively bad. In the contemporary world over a half billion people are malnourished and prone regularly to illness and disease, many of whom are also debt ridden, but who survive and tenaciously display that their lives are worth living by the fact that they persist in wishing to live. If you were to say that you would not wish the circumstances they endure on anyone, we wouldn't take you to mean their lives aren't worth living; we would take you to be saying that their circumstances are so bad that you wouldn't wish them on even your worst enemy, that something

Chapter 10: Well-Being Across the Generations

ought to be done to improve their lives, that if someone were to disregard the countervailing needs you and your household may have, you wouldn't want to *create* children facing those circumstances.[345]

Death *relieved* the intolerable pain Ajax experienced on awakening from the madness Athena had inflicted on him. Ajax knew it would, which is why he chose it. It was better for him that he paid the price of death than that he carried on. The inference Sidgwick drew from the fact that death is generally thought to be one of the worst evils, that life on the average yields a positive balance of pleasure over pain, is altogether unfounded, and it is hard to imagine how so profound and careful a thinker could have made such an elementary arithmetical error. That death is a horror to most people doesn't imply that life is on balance pleasurable. On the contrary, the greater is the horror that taking one's own life poses to someone (betrayal of one's family and friends, revelation of one's misery to others when one wants it to remain undisclosed even after death), the *more* he would be willing to carry on in a state of misery. To illustrate Sidgwick's error, imagine that in the units chosen to measure u, the horror of suicide for someone is −300. The person would choose to continue to live so long life offered her a value exceeding −300; and that could be as low as −299.99.[346]

One way to interpret life in the range between the point at which a person takes his life and well-being subsistence (as judged by him) is to view it as *bearable* but not good. The person would not contemplate suicide, but could wish he hadn't been born, that he didn't have to go through his life's experiences. Zero well-being is then the transition point from the bearable to the good.[347] That is why estimating well-being subsistence from people's behaviour or responses to questionnaires would be a mistake.

Box 10.8
Optimum Population

In Box (10.5) we studied the conditions that an optimum consumption policy must satisfy. To study the conditions that an optimum population policy must satisfy, we consider a timeless version of the model in Dasgupta (1969). To focus on the choice of population size, we hold fixed all other factors of production.[348] Let N denote population size and output an increasing function, $F(N)$. As all other factors of production are being held fixed, we assume $F''(N) < 0$.[349] It follows that average product exceeds marginal product, that is, $F(N)/N > F'(N)$.

As previously, let u denote per capita well-being and assume marginal well-being is a declining function of consumption c (i.e., u is strictly concave). Utilitarianism would therefore advocate equal treatment of all. If C denotes aggregate consumption, the objective would be to choose N to maximize $Nu(C/N)$. In a stationary state $C = F(N)$. Write $c = C/N = F(N)/N$. We suppose that

---

[345] Variations on the theme of Parfit's Repugnant Conclusion, built not only on the positive quadrant of the iso-bars of Nu but also on points to the east of $N = 0$ where $u \leq 0$, have been conducted, admiringly, by large numbers of population ethicists. We have no explanation for why none of the authors paused to ask what $u = 0$ might mean. Cowie (2020) is an excellent review of that literature and an account of writings that have found the Parfitian literature to be based on a misreading of the neutral life. Parfit's interpretation of $u = 0$ hadn't changed over the years. On the Repugnant Conclusion, he wrote recently, "Compared with the existence of many people who would all have some very high quality of life, there is some much larger number of people whose existence would be better, even though these people would all have lives that are barely worth living." Parfit (2016: 110).

[346] We do not know whether Parfit's misreading of $u = 0$ had its origins in Sidgwick's error.

[347] We owe this interpretation to Robert Solow.

[348] It is not necessarily being assumed that they are at their optimum levels.

[349] In what follows, $F''(N) = d^2F(N)/dN^2$ and $F'(N) = dF(N)/dN$.

optimum *N* is large enough to allow us to regard N to be a continuous variable. It follows that optimum *N* must satisfy the condition,

$$u(c) = u'(c)[F(N)/N - F'(N)] > 0 \tag{B10.8.1}$$

Equation (B10.8.1) is easy to interpret. When an additional person is created, at the average consumption level *c*, the gain in total well-being is *u(c)*. But there is a loss, which is the value of the net reduction in consumption for others. That net reduction is the difference between what the additional person consumes and what he contributes to production: $F(N)/N - F'(N)$. Eq. (B10.8.1) says that at the Utilitarian optimum, the gain equals the loss. The equation also says that the optimum living standard *c* exceeds well-being subsistence.

It pays to parametrize both the well-being and production functions.[350] So, we assume:

$$u(c) = B - c^{-\eta}, \quad B > 0, \eta > 0 \tag{B10.8.2}$$
$$F(N) = N^\alpha, \quad 0 < \alpha < 1 \tag{B10.8.3}$$

Ramsey (1928) called *B*, the least upper bound of *u*, Bliss. Let *c** denote well-being subsistence and *c°* the optimum per capita consumption rate. From eq. (B10.8.2) we have $c^* = (1/B)^{1/\eta}$. Moreover, from eq. (B10.8.3) we have $F(N)/N - F'(N) = (1-\alpha)N^{-(1-\alpha)}$. Using these simplifications, eq. (10.8.1) reduces to

$$c° = \left[\left(1 + \eta(1-\alpha)\right)/B\right]^{1/\eta} < c^*(1+\eta)^{1/\eta} < ec^* \approx 2.74c^* \tag{B10.8.4}$$

Eq. (B10.8.4) says optimum consumption is not much above well-being subsistence. For $\eta = 1$ and $\alpha = 0.5$, to take an example, $c° = 1.5c^*$. Moreover, $c° \to c^*$ as $\eta \to \infty$. It would seem Utilitarianism is 'pro-natalist,' the ethics commends large populations, where people experience lives that are not far above well-being subsistence. But that is not what Parfit (1984) had in mind when commenting on the iso-bars of the function *Nu*. In arriving at his Repugnant Conclusion, he simply misinterpreted well-being subsistence.

10.12 The Repugnance of Existential Risks

Death is inevitable for every person. While each death is a loss, we mostly accept its inevitability. The possibility of human extinction in contrast is so alien to our sensibilities, that we do not yet have an accepted vocabulary in which to deliberate it. We realise that the loss would be all the lives that could have existed but would now not be lived; the unsettled vocabulary is over the way we should measure that loss.

The conception of well-being across the generations we have developed here (eq. (10.2)) says that the loss that would occur if humanity were to cease to exist at *t* would be *V(t)*, which is the (possibly) time discounted expected well-being of all the generations that would otherwise exist. Exogenous risks of extinction, such as the kind of event that led to the fifth extinction some 66 million years ago, are probabilistically inevitable; we are not responsible for them. In contrast, we *would* be responsible for the risks of extinction that would be avoided if only we were to change our behaviour. That is why extinction risks that our actions give rise to have greater moral gravity than those over which we have no control.

In arriving at an expression for well-being across the generations, we expunged all risks of humanity's extinction that could arise from human action. The risks of extinction that remained in the formulation

---

[350] The following exercise is taken from Dasgupta (1969).

are exogenous. Actions that endanger humanity's existence were prohibited on deontological grounds. But that was a pedagogical move. Climate change and biodiversity loss heighten the risk of extinction. Emitting even a billion tonnes of carbon into the atmosphere today raises the risk of extinction, even if ever so slightly. The rate at which our activities are causing species to become extinct are contributing to the risks of human extinction. Our concerned citizen will urge decision makers to find ways to reduce those risks but will acknowledge that they cannot be eliminated, which means she will interpret the risk of extinction embedded in eq. (10.1) – more generally, eq. (10.2) – as a combination of exogenous and endogenous risks. The problem is, advocating vigorous moves to reduce the endogenous risks because of the deontological considerations that arise from eq. (10.2) are too impersonal. The concerned citizen will seek a reason that is closer to home.

In a deep meditation on the significance of a possible nuclear holocaust in which humanity suffers extinction, Jonathan Schell distinguished two types of death:

"It is of the essence of the human condition that we are born, live for a while, and then die . . . But although the untimely death of everyone in the world would in itself be an unimaginably huge loss, it would bring with it a separate, distinct loss that would be in a sense even huger – the cancellation of all future generations of human beings." (Schell, 1982: 114–115)

Schell's book was originally published as a three-part essay in *The New Yorker* in 1981, at a time the Cold War had created an especial chill. Schell was a writer, not a professional philosopher, but he made not one false move in philosophical reasoning in the crucial middle chapter, Second Death. Utilitarianism measures the loss from the Second Death in terms of the well-being of all who would not exist on account of human extinction. The loss, if it were to occur at t, would be measured as $V(t)$ in eq. (10.2). Schell however made a different move, which could be read as internalising well-being across the generations within each of us. He wrote of the loss each of us alive today would suffer if we were to discover that there will be no one after we are gone and located that loss *not* to any attachment that we may have to humanity writ large, but to a devaluation of our own lives. And he used the artist and his art to make the point:

"There is no doubt that art, which breaks into the crusted and hardened patterns of thought and feeling in the present as though it were the prow of the future, is in radically altered circumstances if the future is placed in doubt. The ground on which the artist stands when he turns to his work has grown unsteady beneath his feet." (Schell, 1982: 163)

Schell spoke of the artist, but he could have made the same case for all who create ideas and objects. Future people add value to the creators' lives by making their creations durable. Here the fact of a general assumption that people desire to have children is significant. An artist may regard his work to be far more important than parenting, but he is helped by the presumption that there will be future generations to bestow durability to his work.

The examples Schell pointed to were works of art and discoveries in the sciences. Those creations are public goods, and most people don't have the talent to produce them. Confining attention to public goods is not only limiting, but it also raises an ethical dilemma: Suppose we all were indifferent to having children and stared only at the prospective costs of raising them. We would then free-ride, and the artist would be mistaken in his assumption that there will be future people to give durability to his work.

Nevertheless, the direction Schell was pointing to is exactly right. Public goods aren't the only objects of ethical significance. Our values and practices are significant too. Many are private, even confined to the family, and it is important to us that they are passed down the generations. Procreation is a means of making one's values and practices durable. We imbue our children with values we cherish and teach them the practices we believe are right not merely because we think it is good for them, but also because we desire to see our values and practices survive. Those values and practices are not public goods. On the contrary, we cherish them *because* they are intimate. They are stories we tell our children about their grandparents' foibles, of our own joys, sorrows, and discomfiture, and we instruct them on the family rituals we ourselves inherited from our parents. Our descendants do something supremely important for us: they add value to our lives that our own mortality would otherwise deprive them of. That is the reason we would not practise reproductive free riding even if we found reproduction to be personally costly.[351]

The springs that motivate humankind to assume parenthood are deep and abiding. Their genetic basis explains the motivation but doesn't justify it. Justification is to be found elsewhere. Our children provide us with a means of self-transcendence, the widest avenue open to us of living *through* time, not merely *in* time. Mortality threatens to render the achievements of our life transitory, and this threat is removed by procreation. The ability to leave descendants enables us to invest in projects that will not cease to have value once we are gone, projects that justify life rather than merely serve it. Alexander Herzen's remark, that human development is a kind of chronological unfairness because those who live later profit from the labour of their predecessors without paying the same price; and Kant's view that it is disconcerting that earlier generations should carry their burdens only for the sake of the later ones, that only the last should have the good fortune to dwell in the completed building – or in other words, that we can do something for posterity but it can do nothing for us – are a reflection of an extreme form of alienation.[352]

The motivation we are identifying here transmutes from the individual to the collective. Every generation is a trustee of the wide range of assets – be they cultural or moral, produced, or natural – it has inherited from the past. Looking backward, it acknowledges an implicit understanding with the previous generation, of receiving the capital in return for its transmission, modified suitably in the light of changing circumstances and increasing knowledge. Looking forward, it offers an implicit proposal to the next generation, of bequeathing its stocks of assets that they in turn may be modified suitably by it and then passed on to the following generation. This perspective is not at odds with the conception of well-being across the generations in eq. (10.2). In our account of population ethics in a world moving through time, generation-*t* would be moved to internalize the potential well-being of its descendants, expressed in $V(t)$. Our descendants are not us, but they are not outside us either.[353]

Schell's reflections point also to the intrinsic value of Nature. It's a mistake to seek justification for the preservation of ecological diversity, or more narrowly the protection of species, solely on instrumental grounds; that is on grounds that we know they are useful to us or may prove useful to our descendants. Such arguments have a role, but they are not sufficient.[354] Nor can the argument rely on the welfare of the members of such species (it does not account for the special role that species preservation plays in the argument), or on the 'rights' of animals. A full justification bases itself as well on how we see ourselves, on what our rational desires are. In examining our values and thus our lives, we are led to ask whether the destruction of an entire species-habitat for some immediate gratification is something that

---

[351] In a wide-ranging and moving essay on death and the afterlife, Scheffler (2013) has also observed that our own lives would be diminished if there were to be no future people.

[352] Rawls (1972: 291) has a characteristically profound criticism of Kant's perspective.

[353] The idea of stewardship would appear to be common among different cultures. On African conceptions of intergenerational ethics, Behrens (2012: 189) writes: "African thought does not limit moral consideration to only the current generation. It conceives of a web of life that transcends generations, and of the environment as a resource shared by different generations. This entails a direct moral obligation to preserve the environment for future persons, since it is a communal good. Africans also expect that the current generation should develop an attitude of gratitude towards their predecessors for having preserved the environment on their behalf. This virtue of gratefulness ought to be realized by the current generation seeking to reciprocate by preserving the environment for future generations, in turn." We are grateful to S.J. Beard for this reference.

[354] We discuss the intrinsic value, even sacredness, of Nature more fully in Ch. 12.

we can live with comfortably. The idea of intergenerational exchange is embedded in the perspective of eternity, but the intellectual source of such exchange is a far cry from the conception that balked Herzen in his effort to locate mutually beneficial terms of trade. The mistake is to see procreation and ecological preservation as matters of personal and political morality. They are at least as much a matter of personal and political ethics.

Annex 10.1 A Simple Exercise in Optimum Saving

Equation (10.7) says that future consumption costs and benefits should be discounted at rates that equal to consumption rate of return $r(t)$. As $r(t)$ measures the increase in consumption that would be foregone in period t if an additional unit of consumption was not saved in period t-1, it is also known as the 'opportunity cost of capital'. In what follows, we use a simple model economy to derive first- and second-best optimum saving for the future. The model is a grotesque caricature of economic possibilities in our world, but it has the merit of containing extreme cases that mimic real possibilities.

The economy consists of a single non-deteriorating commodity whose output can be consumed or invested for the future. Population is constant, output (GDP) is proportional to the stock of the commodity, and well-being is iso-elastic in consumption ($\eta > 1$). Writing f for output (GDP) and k for the stock of capital,

$$f(k) = rk, \qquad r > 0 \qquad (A10.1.1)$$

$$u(c) = -c^{-(\eta-1)}, \qquad \eta > 1 \qquad (A10.1.2)$$

From equation (A10.1.1), it follows that $df/dk = r$, which means the own rate of return on investment (Chapter 1) is constant.

For computational ease we suppose time is continuous. Because output can be either consumed or invested, the income-expenditure equality can be written as

$$dk(t)/dt = rk(t) - c(t) \qquad (A10.1.3)$$

A10.1.1 First-Best Optima

Write $\mu = (r-\delta)/\eta$. Applying equations (A10.1.1), (A10.1.2) and (A10.1.3) to equation (10.8) reduces the optimality condition to

$$dc(t)/dt = [(r - \delta)/\eta]c(t) = \mu c(t) \qquad (A10.1.4)$$

Equation (A10.1.4) says that if $r < \delta$, $c(t)$ declines to 0 at an exponential rate. We consider the analytically more interesting case, $r > \delta$, which says the consumption rate of return r exceeds the rate at which time is discounted, δ. That in turn means $\mu > 0$. Integrating equation (A10.1.4) yields

$$c(t) = c(0)e^{\mu t} \qquad (A10.1.5)$$

Equation (A10.1.5) says $c(t)$ grows exponentially at the rate μ. We reconfirm a point that was made previously, that although the equation gives us the rate at which consumption ought to grow, it doesn't reveal the level of consumption that ought to be set initially, $c(0)$. That's the indeterminacy mentioned in the text.

The simplest way to determine the optimum initial consumption, $c^*(0)$, is to observe from equation (A10.1.3) that if $c^*(t)$ grows at the constant rate μ, so should $k(t)$ be required to grow at that rate. The reason is that if the growth rate of $k(t)$ was to be less than μ, capital would be eaten into, which means the stock would be exhausted in finite time. The economy would then cease to exist ($V(0)$ would be minus infinity if the consumption path was to be thus). If, on the other hand, the growth rate of $k(t)$ was to exceed μ, there would be over-accumulation of capital, in the sense that consumption would be lower at every date than it needs be. The situation would resemble one where the social evaluator recommends that a part of the initial capital stock $k(0)$ should be thrown away, to be followed by a saving policy that satisfies equation (A10.1.4).

Exponential growth in our linear economy (Equation (A10.1.5)) says that the saving rate should be constant. Let s be the proportion of output (GDP) that is invested at each instant. Then equation (A.10.1.3) can be re-written as

$$dk(t)/dt = srk(t) \tag{A10.1.6}$$

Equation (A10.1.6) says that intended saving equals intended investment. Integrating it yields

$$k(t) = k(0)e^{srt} \tag{A10.1.7}$$

But we are insisting that both $k(t)$ and $c(t)$ should grow at the same rate. Equations (A10.1.5) and (A10.1.7) therefore imply

$$\mu = (r - \delta)/\eta = sr \tag{A10.1.8}$$

The saving rate in equation (A10.1.8) is the optimum. We write it as s^*. Thus

$$s^* = (r - \delta)/\eta r < 1 \tag{A10.1.9}$$

Equations (A10.1.7), (A10.1.8) and (A10.1.9) tell us that the optimum rate of growth of consumption, g^*, is

$$g^* = (r - \delta)/\eta > 0 \tag{A10.1.10}$$

Thus the optimum consumption path is

$$c^*(t) = (1 - s^*)k(0)e^{s^*rt} \tag{A10.1.11}$$

Equation (A10.1.9) offers as elegant a simplified answer as there could be to the question that opened this chapter.

If $\delta = 0$, equation (A10.1.9) reduces to $s^* = 1/\eta$. As an example, suppose $\eta = 1.1$ Then $s^* \simeq 0.9$, which means that a bit more than 90% of GDP ought to be saved for the future, no matter how poor the economy is to start with. The closer is η to 1 from above, the closer is s^* saving rate to unity. This brings us back full circle to the paradoxes of infinity discussed in Box 10.3, and the St. Petersburg Paradoxes in Annex 5.1.

A10.1.2 Second-Best Optima

If we are to squeeze anything out of this linear economy to inform the economics of biodiversity, we borrow from the insights that were gained in previous chapters of the pervasiveness of detrimental externalities in our use of the biosphere. There are environmental externalities that simply cannot be eliminated. Measures to protect and promote the biosphere in certain ways that are technically and ecologically feasible are not institutionally feasible. In formal terms that is to recognise that there is inevitably to be an underinvestment in Nature.

To capture the implications of that idea in a model that bears no resemblance to the world we inhabit, we now suppose that for whatever reasons, the savings rate is less than s^*, say $\underline{s}$. By equation (A10.1.7) this gives rise to the reference path in which the stock of capital rises at the exponential rate $\underline{s}r$. Thus

$$\underline{k}(t) = k(0)e^{\underline{s}rt} < k(0)e^{s^*rt} \tag{A10.1.12}$$

Chapter 10: Well-Being Across the Generations

That means

$$\underline{c}(t) = (1-\underline{s})k(t) = (1-\underline{s})k(0)e^{\underline{s}rt} \qquad (A10.1.13)$$

Comparison of equations (A10.1.11) and (A10.1.13) shows that $\underline{c}(t) > c^*(t)$ for an initial period of time, followed by $\underline{c}(t) < c^*(t)$ for all subsequent time. This is intuitively obvious. A lower rate of investment in capital leads to higher consumption levels initially, but the lower rate of accumulation means that eventually consumption will be lower. The crossover date T is given by the equation

$$(1-\underline{s})k(0)e^{\underline{s}rT} = (1-s^*)k(0)e^{s^*rT}$$

This is a grotesquely stylised version of the Impact Inequality and its aftermath.

A10.1.3 Accounting Price of Capital

To determine the accounting price of capital, use equation (A10.1.13) to note first that at date t intergenerational well-being in the second-best optimum

$$V(t) = {}_{m=t}\int^{\infty}[u((1-\underline{s})k(t)e^{\underline{s}r(m-t)})]dm \qquad (A10.1.14)$$

Thus intergenerational well-being at t is a function of the stock of capital at t. The accounting price of capital at t, which we write as $p_k(t)$ is defined as

$$p_k(t) = \partial V(t)/\partial k(t) \qquad (A10.1.15)$$

An easy calculation using equations (A10.1.2) and (A10.1.14) confirms that

$$p_k(t) = u'(\underline{c}(t)) \text{ if } \underline{s} = s^* \qquad (A10.1.16)$$

but that

$$p_k(t) > u'(\underline{c}(t)) \text{ if } \underline{s} < s^* \qquad (A10.1.17)$$

Equation (A10.1.16) says that investment and consumption have the same accounting price relative to well-being, while equation (A10.1.17) says that if there is underinvestment in the reference path, the accounting price of investment exceeds that of consumption. This result generalises all the way to include models with finite bounds and many forms of capital.

The moral is this: underinvestment in natural capital relative to other forms of capital carries with it the implication that its accounting price exceeds that of other forms of capital. Cost-benefit analysis of conservation and restoration projects should include a premium on them.

Annex 10.2 Uncertainty and Declining Discount Rates

How are CDRs affected by uncertainty in socio-ecological futures? There is an intuition, arising from a *precautionary motive* that says CDRs should be lowered to adjust for uncertainty about the future, other things equal of course (Levhari and Srinivasan, 1969; Gollier, 2008; Arrow et al. 2014; Cropper, 2014). So imagine that the growth rate in consumption per capita, g, is uncertain and is subject to shocks that obey independent and identical normal distributions, with mean μ_g and a variance σ_g^2. Being a normal distribution, the growth rate could be either above or below μ_g in a symmetric way. Then it can be shown (Levhari and Srinivasan, 1969; Gollier, 2008) that

$$\rho(t) = \rho = \delta + \eta\mu_g - 0.5\eta^2\sigma_g^2 \qquad (A10.2.1)$$

The social evaluator therefore pretends that the world is deterministic, that μ_g is the sure rate of growth, but then introduces a precautionary motive to discount future consumption. If that adjustment is made, the optimum policy in the pretended world is the same as the optimum policy in the world facing the uncertainty. This is called 'certainty equivalence'. The third term on the right-hand side of equation (A10.3.1) reflects the precautionary motive. It gives the social evaluator a reason for putting aside more for the future. It does so by reducing the CDR under certainty by $0.5\eta^2\sigma_g^2$.

In the economics of biodiversity, we should be especially interested in forecasts in which $\mu_g = 0$. Equation (A10.2.1) then reduces to

$$\rho(t) = \rho = \delta - 0.5\eta^2\sigma_g^2 \qquad (A10.2.2)$$

If the variance is large, $\rho < 0$.

A10.2.1 Declining Discount Rates

The UK government (HM Treasury, 2018) uses a declining term structure of CDRs for use in public sector projects (Table A10.2.1). For example, the annual rate for discounting benefits and costs in year 10 is recommended by the UK government to be 3.5%, but a rate of 2.5% is recommended for use for discounting benefits and costs in year 70, falling to 1% in year 300. The Green Book further recommends presenting projects both using the standard discount rate, and one where the pure time preference = 0 when evaluating long-term projects. This is also included in the table below.

Table A10.2.1 Declining Discount Rates in the Green Book

	0–30	31–75	76–125	126–200	201–300	301+
Green Book discount rate	3.50%	3.00%	2.50%	2.00%	1.50%	1.00%
Reduced discount rate (pure time preference=0)	3.0%	2.57%	2.14%	1.71%	1.29%	0.86%

Source: HM Treasury (2018)

A10.2.2 What is the Rationale for using Declining CDRs?

What is the rationale for using declining CDRs? One rationale (there are others) would be that shocks to output are positively correlated over time. Positive correlation means that if the economy has experienced a growth rate less than its expected value one year, the chance that it will be less than the expected value the following year will be greater than 50%. Of course, if the economy were to experience a growth rate in excess of its expected value one year, the chance that it will be in excess of the expected value the following year will also be greater than 50%. But because the social evaluator is risk averse ($\eta > 0$), she will worry more about the downside risk. Positive correlation over time in the shocks the economy is subject to makes future consumption riskier, increasing the strength of the precautionary effect in equation (A10.2.1) as the social evaluator peers further into the future.

The above analysis assumes that the nature of the stochastic consumption-growth process can be adequately characterised by econometric models that are estimated using historical data. Suppose instead the social evaluator is *uncertain* about the expected rate of growth in consumption μ_g. It can be shown (Gollier, 2007) that the precautionary motive leads to a declining CDR. Figure A10.2.1 (taken from Arrow et al. 2014) demonstrates the CDR as a function of time for the case where $\delta = 0, \eta = 2$, and $\sigma_g^2 = 3.6\%$; and where the expected rate of growth of consumption is equal to an annual 1% and 3% with equal probability.

Figure A10.2.1 Certainty-Equivalent Discount Rate Assuming Consumption Growth is a Sequence of Independent Variables with Uncertain Mean

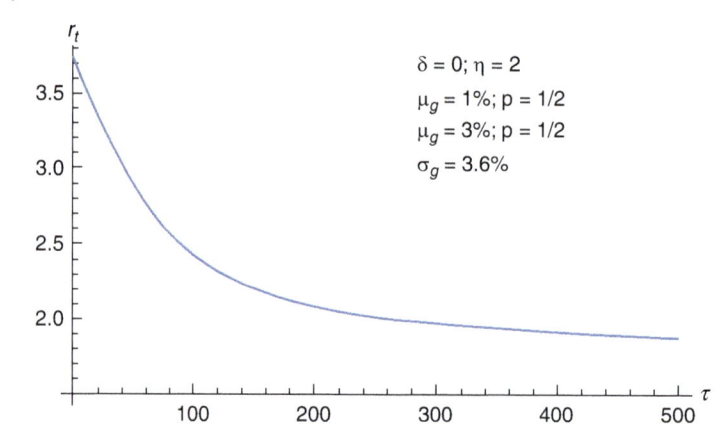

$\delta = 0; \eta = 2$
$\mu_g = 1\%; p = 1/2$
$\mu_g = 3\%; p = 1/2$
$\sigma_g = 3.6\%$

Source: Reproduced from Arrow et al. (2014), based on Gollier (2008). *Permission to reproduce from Oxford University Press.*

Annex 10.3 Three Tiers of Ethical Reasoning

"It is not a trivial question Socrates said: what we are talking about is how one should live." Thus begins Bernard Williams' classic study on how things were in moral philosophy towards the end of the 20th century (Williams, 1985). The generality of *one* in the question already stakes a claim; the implication is that something relevant or useful can be said to anyone. The *we* more than suggests that the agent herself is invited to think through the question. There is also the idea that one must think about *a whole life*, not just a slice of it, from every aspect and all the way down. And to anyone reading this Review, the question includes not only the person's engagement with other people, but also a recognition that she is embedded in Nature.

Economics, in the form we are constructing it here for studying our place in the biosphere, draws its inspiration from Socrates' question, but so addresses it as to make it useable by public decision makers. This Review however is meant not only for public decision makers, but also, more importantly, for the *concerned citizen*, someone whom we have been referring to as the 'social evaluator'. And she will want a reasoning that moves from the individual to the collective, and from there to decisions on the supply of public goods, such as election of public servants whose choices are to be directed at advancing the interests of the collective. Three tiers of ethical reasoning are involved here. Socrates' question subsumes all three.[355]

A10.3.1 Individual and Social Well-Being Functions: The First Two Tiers

We began this chapter by regarding an individual's well-being to be a measure of the extent to which her informed desires are fulfilled. Informed desires are those that, all things considered, she can *justify* to herself. That alone tells us we are in the ethical realm. But it is only the first tier of ethical reasoning, for the citizen will want to extend her sympathy to others when deliberating on decisions that involve the public sphere. And those others include future people, conditional on there being people in the future. That involves the second tier of ethical reasoning.

---

[355] We are most grateful to Eric Maskin, discussions with whom over many years has shaped the material in this section (see Dasgupta and Maskin, 2004, 2008, 2020).

We noted that in various versions of consequentialist thinking, or what we have been calling here Utilitarianism (sec. 10.1–10.4), extended sympathy involves the citizen placing herself successively in everyone's shoes, as it were; for as Atticus Finch in *To Kill a Mockingbird* famously advised his daughter Scout, "You never really understand a person until you consider things from his point of view – until you climb into his skin and walk around it". The point of the thought experiment is that it would not only enable the citizen to measure, howsoever approximately, each person's well-being in every state of affair, but also to have an inbuilt mechanism for comparing well-beings. She would then combine the individual well-being functions she has constructed to arrive at her conception of social well-being, that is, her *social well-being function*. We call the ranking – or *ordering* – of states of affairs that corresponds to the concerned citizen's social well-being function her *social preferences*.[356] In this chapter we have found that three distinct styles of ethical reasoning yield, under certain assumptions, the same functional form of social well-being, namely, the discounted *sum* of individual well-beings (eq. (10.1)).

Social well-being as we are conceiving it here is not a Platonic object, rather, it is a personal evaluation of the concerned citizen. No two citizens can be expected to construct the same social well-being function, not even if both subscribe to a common ethics, for example, utilitarianism. For one thing, their measurement of individual well-beings would be expected to differ; for another, the way they compare individual well-beings to arrive at the discounted sum of individual well-beings would differ too. The entire exercise involves value judgments.

A10.3.2 Voting Rules: The Third Tier

That should not matter in a well-ordered society; for in such a place citizens agree to disagree (see the discussion on Pragmatism in sec. 10.4). Which is why, in a democratic society citizens need to find a procedure by which they can express their social preferences for reaching collective decisions, such as over the supply of public goods. A prominent instance where citizens are called upon to express their social preferences is the election of public servants, for example members of the legislature, whose responsibility is, in effect, to choose policies that speak to citizens' conceptions of social well-being. We call the way citizens' social preferences are combined to reach a collective decision, a *voting rule*. That involves the third tier of ethical reasoning. To illustrate, we consider situations where citizens cast their votes among candidates for a public office.

Political candidates represent the policies they advocate, such as resources devoted to investment in Nature, and policies give rise to consequences. For the Utilitarian citizen the chain linking policies to their consequences is a mapping from political candidates to the (uncertain) realisation of individual well-beings. Citizens in a well-ordered society cast their votes in line with their conception of social well-being, not their private interests, which is what children used to be taught in civics classes. When screening candidates, the citizen therefore looks for how closely a candidate's proposed policies meet her conception of social well-being. That means citizens can be expected to disagree over candidates (the 'ballot') not only because their social well-being functions differ, but also because their confidence in the candidates' abilities and aptitudes differ. Disagreements can be massive.

In its sharpest form a voting rule selects a winner from the voters' expressed social preferences over the candidates on ballot. It has become customary in welfare economics to call such a voting rule a *social choice function*. A more general formulation would be for a voting rule to arrive at an ordering over *all* candidates from the voters' expressed social preferences over them. For historical reasons it is now

---

[356] Harsanyi (1955) called them 'ethical preferences.' In the text we use the terms 'social preferences' and 'social preference orderings' interchangeably.

customary to call such a voting rule a *social welfare function*. Fortunately, the problem of collective choice can be studied entirely in either formulation (Reny, 2001).[357]

Social choice functions are meant to do the job voting rules are designed to perform, which is to pick a winner.[358] The advantage in defining voting rules instead as social welfare functions is that no fresh election would be required should the winning candidate be unable to assume office; the rule would simply declare the second ranked candidate to be the winner.

In a great masterpiece of 20th century social thought, Kenneth Arrow introduced the idea that *democratic voting rules* should not be installed without being screened for whether they embody democratic values (Arrow, 1951). In the following section we present Arrow's famous Impossibility Theorem, that there is no voting rule satisfying a set of seemingly mild ethical properties. We then argue that Arrow's result is overly pessimistic. To illustrate that, we develop the problem of collective choice in terms of social choice functions and show that *majority rule* has deep ethical significance. We then discuss the merits of the familiar *rank-order rule*. They differ in their normative properties, but our conclusion will be that majority rule and the rank-order rule offer the most democratic of rules for collective choice.

A10.3.3 Democratic Voting Rules and Arrow's Theorem

A typical voting rule in the West requires voters to register only their most preferred candidate among those on the ballot. For example, in the rules governing election to UK's parliament – *plurality rule*, or 'first-past-the-post' in common parlance – the candidate receiving the highest percentage of votes is declared winner even if the vote falls short of the 50% threshold. Arrow recognised that such a system ignores voters' social preferences over their less-favoured candidates and took it to be obvious that voters should be required to rank all candidates on the ballot.

That may seem a tiresome, technical requirement, but it matters hugely, because among other shortcomings, the first-past-the-post system allows, even encourages, 'spoiler candidates' to undermine democracy. To see how, imagine that in an election in which there are three candidates, x, y, z, 47% of the electorate vote for x, 48% vote for y, and 5% vote for z. Under the first-past-the-post rule y would be declared the winner. But suppose each of the 5% of voters supporting z, had they been asked, would have declared a preference for candidate x over candidate y (because, say, candidate z espouses an extreme form of policies espoused by x and is thus even farther from y in his political views). That preference won't of course be registered under plurality, but as 52% of the electorate favour candidate x over candidate y, the rule can scarcely be called democratic. Candidate z spoils the democratic mandate of x by taking advantage of a defective electoral system and hands over the election, perhaps unwittingly or perhaps owing to hubris, to candidate y. Instances of spoiler candidates overturning democracy would appear to have occurred in recent years in Presidential elections in the US and France (Dasgupta and Maskin, 2004).[359]

Arrow's work can be read as drawing a distinction between voting rules, which involves the third tier of ethical reasoning, and the considerations guiding the citizen on whom to (more accurately, what to) vote for, which are reached through the second tier of ethical reasoning. His monograph was titled 'Social Choice and Individual *Values*, not 'Individual *Preferences*'. Ethical considerations over

---

[357] The terms we are deploying here can be confusing. 'Social *well-being* functions,' as we have defined them here, reflect citizens' social preferences (the second tier of ethical reasoning); whereas 'social *welfare* functions', as defined in the established literature, are what we are calling voting rules (the third tier of ethical reasoning). The confusion has arisen because, as we see below, it has been customary to conflate the second and third tiers into one.

[358] We allow for ties, that is, we are not demanding that there be a unique winning candidate.

[359] The requirement that voters rank all candidates does not, on its own, rule out spoiler candidates syphoning votes from more favoured candidates. Voting rules must be so constructed as to rule them out. See below in the text.

voting rules are thus different from the ones citizens will wish to entertain for articulating their extended sympathies, such as for example the additive form of social well-being in eq. (10.1). Arrow disallowed voters to register the intensity of their preferences over the candidates on ballot because otherwise voters would be tempted to inflate their feelings about them, that is, they would vote tactically. Riots outside polling stations would be recurrent expressions of that. Nor did he allow interpersonal comparisons of citizens' social well-being functions at the polling station, for that would lead quickly to social unrest. Arrow proposed instead a set of axioms that are widely thought to be democratic and are based *only* on voters' rankings over the candidates on ballot and asked whether there is a voting rule satisfying them that selects a winning candidate no matter what voters' rankings over the candidates happen to be. In what follows, we assume for simplicity of exposition that citizens' social preference orderings are strict, meaning that no voter is indifferent between any pair of candidates.[360]

There are several versions of Arrow's axioms. Here we consider a version for deployment on voting rules that are social choice functions:

Consensus Principle: If every voter ranks candidate x over candidate y, then candidate y will not be elected.

Anonymity: All citizens' votes should count equally; that is, who the voter is should not determine her influence on the election.

Neutrality: The voting rule should not favour one candidate over another.

Ordinality: Who the winner is should depend *only* on citizens' rankings of candidates (i.e., intensities of social preferences or some other cardinal features of those preferences should not count).

Independence of Irrelevant Alternatives (IIA): If candidate x is the winner among a set of candidates (the "ballot"), then x should still be the winner if the ballot were to be reduced by dropping any losing ("irrelevant") candidate.

The first four conditions are so transparent as to require no discussion; they may be viewed as minimal features of democracy. It is the fifth condition (IIA) that has elicited much disquiet among welfare economists, mostly because well-established voting rules such as the rank-order (or Borda) rule (see below) do not satisfy it. One can argue though that IIA is an appealing requirement because, as can be readily confirmed, it among other things, prevents spoiler candidates and vote splitting from tarnishing the democratic process.[361]

Arrow's remarkable finding was that there is *no* voting rule satisfying all five requirements when the number of candidates is three or more, the number of voters is three or more, and the *domain of social preferences is unrestricted* (i.e., the voting rule should select a winner no matter what the voters' rankings of candidates happen to be). In the process of narrowing down the list of voting rules by requiring them to satisfy democratic values, he found that there were none left!

An example of Arrow's Impossibility Theorem had been discussed by the 18th century thinker, the Marquis de Condorcet, in his dissection of *majority rule*. Condorcet asked us to consider three voters,

---

[360] Our reading of the place of Arrow's theorem in citizens' ethical reasoning differs from established readings (e.g., the excellent text by Gaertner, 2009), in that it has been common to regard the theorem as being directed at what we are calling the second tier of ethical reasoning. So, for example, his formulation has been criticised for, among other things, disallowing interpersonal comparisons of social preferences. It has even been suggested that he had been swayed by the then prevailing popularity of Logical Positivism. Arrow (1951) himself didn't help matters by calling the aggregator of voters' social preferences, 'social welfare functions.' We avoid arguing over the correct interpretation of his theorem. Our intention here has been only to establish the presence of three tiers and to show that Arrow's axioms should be read as being directed at the third tier.

[361] In an important recent paper, Maskin (2021) has shown that IIA can be weakened without losing its force on these counts.

Chapter 10: Well-Being Across the Generations

numbered 1, 2, 3, who are to vote on three candidates, x, y, z. Majority rule declares a candidate to be the winner if and only if she beats all others in head-to-head contests by a majority of votes. Notice that the rule satisfies Arrow's latter four axioms. Imagine now that voter 1 ranks candidate x over candidate y and candidate y over candidate z; that voter 2 ranks candidate y over candidate z and candidate z over candidate x; and that voter 3 ranks candidate z over candidate x and candidate x over candidate y. Condorcet noted that in such a case x beats y in a head-to-head contest because two voters, 1 and 3 (a majority) rank x over y; that y beats z in a head-to-head contest because two voters, 1 and 2 (who also form a majority) rank y over z. As x beats y and y beats z, we may think that x beats z, but we would be wrong, because two voters, 2 and 3 (yet another majority)– rank z over x, meaning that candidate z beats candidate x in a head-to-head contest. We have a contradiction, for we are left with a cycle – called a *Condorcet cycle*: x beats y, who beats z, who beats x, who beats y, . . . ad infinitum. As the ruling is intransitive, the voting rule is incoherent.

Arrow showed that in an unrestricted domain of social preference orderings, there are profiles of orderings for which at least one of the axioms is necessarily violated. One may take Arrow's finding to be saying that democracy works only when citizens share something like a common ethical culture.[362] To illustrate, suppose voter 1 in Condorcet's example ranks (candidate) x over y and y over z; voter 2 ranks x over z and z over y; and voter 3 ranks z over y and y over x. In this example the rankings don't clash in the way they do in the Condorcet example: there is an agreement among voters, that x is not worst. It is easy to see that in this situation majority rule would declare x the winner, because x beats both y and z in head-to-head contests.

Another set of circumstances where majority rule identifies a winning candidate is where candidates can be labelled in a (politically) left-right manner. Imagine now that voters' social preferences differ, in that no two voters agree on who is the best candidate, and that social preferences are single peaked; that is, they tail off on either side of each citizen's most preferred candidate. Citizens in that world disagree on politics but share a political culture: they all view candidates in a left-right manner. In a famous work on political arithmetic, Black (1948) showed that under majority rule the median voter's most preferred candidate would be the winner.[363]

We have so far imagined that citizens in the polling booth declare their ranking of candidates in accordance with their social preferences. Formally, we have assumed that *sincere voting* is a Nash equilibrium (Ch. 6). But if sincere voting is not an equilibrium, even upstanding citizens would be tempted to vote tactically. That would prove costly to citizens, as they would now wish to forecast how others are likely to vote; and that in turn would require them to try to discover other citizens' social preferences, which is a tall order. For democracy to work well, voting should require only private information, that is, it should require the citizen to know her own mind, not that of all citizens. The problem is then to design voting rules under which citizens do not have an incentive to misrepresent their social preferences. We call a voting rule with that property *strategy-proof*.

It is intuitively clear that if social preferences are sufficiently aligned with one another, citizens would vote sincerely. For example, in the example we have just studied, there is no room for tactical voting. Under majority rule candidate x, who is the favourite of both voters 1 and 2, would be elected. And there is nothing voter 3 can do to influence the outcome, meaning that she also has nothing to gain from misrepresenting her social preferences. But what if social preferences are highly non-aligned?

---

[362] Sen (1966) identified a formal way of articulating what a common ethical culture means for majority rule to work well.

[363] The 'median' voter would exist if the voting population were large in numbers. Otherwise, an appropriate modification of the word 'median' would be required to identify the voter whose most preferred candidate is the winner.

Once again, negative findings rear their head. Gibbard (1973) and Satterthwaite (1975) proved that there is *no* strategy-proof voting rule satisfying *anonymity, consensus*, and *decisiveness* when the number of candidates is three or more and the domain of social preferences is unrestricted.[364]

These are depressing findings. So, in what follows we explore a few positive results that have been obtained in the search for democratic voting rules. To do that, it helps to assume the constituency to be large.[365] Given the context – citizens electing members of the legislature – the assumption is reasonable. In the UK, for example, the average constituency today is composed of approximately 72,000 citizens with a right to vote in parliamentary elections, which for the kind of calculations the findings on voting rules we report here, can be thought of as being infinitely large. But we assume the constituency to be large for a technical reason, which is that the proportion of voters expressing any particular social preference ordering over candidates can be taken in principle to assume any figure from 0 to 1. That means profiles of social preference orderings, one per voter, that lead to ties are very unlikely to be the outcome of elections. And that helps to establish the findings on the specific voting rules we study below.

A10.3.4 Majority Rule

In today's social sensibilities majority rule has an intuitive appeal unmatched by any other collective rule for political expression. That may be because when the number of candidates is two, any number of voting rules yield the same outcome. Even if voters are asked only to record their most favoured candidate, their lower ranked candidate is revealed by default, which is why, for example, the plurality rule reduces to the majority rule. So does *runoff voting* (as in France's Presidential elections), under which, like plurality, voters cast a vote for their most preferred candidate, but unlike plurality a candidate wins only if he receives a majority of votes, failing which the two highest vote getters go to a runoff. When the number of candidates is two, there is no need for a runoff: the candidate with the higher number of votes in the first-round wins.

Kenneth May (1952) crystalized the appeal of majority rule by showing that despite the plethora of voting rules that are equivalent to majority rule in terms of outcome when the number of candidates is two, majority rule is the only voting rule that satisfies a set of appealing requirements.

Notice that when the number of candidates is two, there is no place for axiom IIR, because there is no 'irrelevant' candidate. May asked us instead to consider the following requirement:

Positive responsiveness (PR): If candidate x rises relative to candidate y in some voter's social preferences, then (i) x does not fall relative to y in the ranking of candidates under the voting rule, and (ii) if x and y were previously tied in the ranking of candidates under the voting rule, x is now strictly above y.[366]

May showed that when the number of candidates is two, majority rule is the only voting rule that satisfies *decisiveness* (i.e., the rule identifies a unique winner), *anonymity, neutrality*, and *positive responsiveness* on an unrestricted domain of social preferences.

When the number of candidates is two, there is no scope under majority rule for tactical voting on the part of citizens: majority rule is non-manipulable. But the idea of studying voting rules when the number of candidates is two is appealing also because one can break up a voting rule when the number of candidates is more than two by applying the rule on pairs of candidates sequentially. The procedure is of course still vulnerable to Arrow's Impossibility Theorem and the Gibbard-Satterthwaite Theorem, but

---

[364] Reny (2001) presents a unified framework for proving the Arrow and the Gibbard-Satterthwaite Impossibility Theorems.

[365] Formally, it is to assume that there is a continuum of voters.

[366] In his original formulation of the Impossibility Theorem, Arrow (1951) postulated 'positive responsiveness' (PR) instead of 'unanimity'. It can be shown that unrestricted domain, PR, and IIA imply unanimity.

Chapter 10: Well-Being Across the Generations

Condorcet cycles and tactical voting express themselves in an interesting form: the *order* in which the pairwise applications are conducted matters. For example, in the case we studied to exhibit Condorcet cycles, suppose we begin the contest by pitting x against y. As noted previously, x beats y; and when winner x is pitted against z, it is z who wins. But suppose instead that in the first-round y was pitted against z. In that case y would win; but against candidate x, y would lose, meaning that x becomes the winner. If a voting rule is transitive, the order would not matter. Riker (1986) found that legislators in US Congress and UK Parliament have manipulated the order in which amendments are made to bring about the outcome they desired. These are examples of agenda setting.

Arrow's theorem has bite when voters' social preferences are markedly non-aligned. Dasgupta and Maskin (2008) showed that the restrictions on social preferences over candidates under which majority rule yields a winner are fewer than the restrictions demanded by any other voting rule satisfying Arrow's axioms. Formally, it is to say that if a voting rule satisfying Arrow's axioms yields a winner in any profile of social preference orderings, then so does majority rule, and that there exists at least one profile of social preference orderings for which the voting rule in question does not yield a winner but for which majority rule does. That is the sense in which majority rule could be said to be the most robust among all democratic voting rules satisfying Arrow's axioms. Which is another reason majoritarianism is an appealing voting rule.

Majority rule has a further appealing feature. Dasgupta and Maskin (2020) have shown that the consensus principle, anonymity, neutrality, decisiveness (i.e., the rule identifies a unique winner), IIA, and strategy-proofness *uniquely* characterise majority rule on any domain of social preferences for which there exists a voting rule satisfying these axioms. The authors showed moreover that majority rule satisfies the six axioms on any restricted domain of social preference rankings without a Condorcet cycle. Their characterisation theorem complements the one by May (1952).

Notice that if, as we are assuming here, the electorate is large, there is little-to-no incentive for a citizen to vote tactically; that's because her vote is one among many. The requirement that a voting rule should be strategy-proof has bite in a large polity when coalitions can form. In their formulation Dasgupta and Maskin took strategy-proofness to include tactical voting by coalitions and showed that their characterisation of majority rule holds even when coalitions are limited to arbitrarily small size.

A10.3.5 Rank-Order Rule

That majority rule does not yield a result when voters' social preference orderings are hugely non-aligned could rankle the democrat. Which is why political theorists have explored the normative basis of other voting rules. Of the many voting rules on offer, the *rank-order rule* has received much attention among social choice theorists. The rule is named after Condorcet's arch intellectual rival, Jean-Charles de Borda. Under the Borda rule, citizens rank all candidates on the ballot. If there are N candidates, each voter assigns N points to her most favourite candidate, N-1 points to her next favourite, N-2 to the next in her ranking, and so on. The winner is the candidate with the highest total.

It is immediate that the Borda rule yields a winning candidate no matter how non-aligned voters' social preferences happen to be. The question arises as to which of Arrow's five axioms are violated by the rule. It is simple to check that the rule satisfies the first four of Arrow's axioms. However, the following, textbook example, shows that it violates the fifth: IIA:

There are 4 candidates w, x, y, z, and 17 voters. Imagine that of the voters, 6 rank w over x over z over y; 5 rank z over x over y over w; 4 rank y over z over x over w; and 2 rank x over y over z over w. It follows that the Borda scores for w, x, y, and z are, respectively, 35, 49, 38, and 48. So, x is the winner. Now suppose one of the losing candidates, y, drops out of the ballot. The Borda scores for the remaining candidates w, x, and z would then be 29, 32, and 37, implying that now z is the winner. That is a violation of IIA.

Maskin (2021) has shown that one does not need IIA to rule out spoiler candidates, that it can be replaced by a weaker condition serving the same purpose. He has also provided an axiomatic foundation for the Borda rule. Notice that applying numerical scores to candidates in accordance with their place in a voter's social preferences amounts to admitting a form of preference intensities to those preferences, the scores being numerical measures of intensities. That suggests the voters could resort to insincere voting. The great attraction of majority rule is that the winning candidates beats all other candidates on a head-to-head contest. Its weakness is that it does not necessarily yield a winner.
A compromise choice would be to apply majority rule first to see whether there is a winner. If the rule leads to intransitivity, deploy the Borda rule to pick a winner.

Chapter 11
The Content of Well-Being: Empirics

Introduction

In Chapter 10 it was shown that despite wide differences in their foundations, three prominent styles of ethical theory interpret well-being across the generations to be the discounted sum of individual well-beings. We interpreted personal well-being to be the extent to which one's informed desires are realised, and assumed it is a function of the individual's standard of living. Realisation of informed desires applies to the cognitive component of happiness (some call it 'contentment'), but not to the affective component (which can be called the 'hedonic level of affect') although it could have a bearing on affect. So, we now dig deeper into the content of well-being.[367]

In the formal models that were developed in previous chapters, the living standard was represented by the quantity of an all-purpose commodity to which the average person in society has access. We called the all-purpose commodity a consumption good. But the presumption that the sole factor in well-being is consumption may seem otiose. It is widely held instead that personal and social engagements are central to one's sense of well-being; they shape the way one lives. But all engagements require goods and services; so, we may think of engagements as production activities, in which goods and services are inputs. Partaking of a meal (alone or in the company of others) is an engagement, but food items are the inputs that make the meal possible; friendship involves investment in time, which is a scarce resource; a walk in the countryside requires that there *be* a countryside; and so on.

Goods and services come with various characteristics, and for each individual there is (given the resources at his command) a best way of obtaining the commodities that are best suited for the activities that are from his point of view the best for him to engage in. Of course, the best way to obtain goods and services with those characteristics involves further engagements – getting a job, making contacts, urging government to supply public goods – but the regress is circular, meaning that it is closed. The economist's presumption that the only things people care about are goods and services should be seen as a filtered expression of people's projects and purposes. Commodities have instrumental value; they do not necessarily have intrinsic worth. Irrespective of what a person values, he will be found to value commodities and thence consumption, broadly construed. The well-being function *u* in the previous chapter should be interpreted in that light. Although for the main part we used an aggregate measure of consumption to denote an individual's projects and purposes, it was shown how the framework could be extended if we were to include consumption's variety, for example, other people's consumption.

Over the millennia, two lines of enquiry into the content of well-being have given rise to a gigantic, frequently contentious literature, spread over many systems of thought. One strand looks for objective measures (*beings* and *doings*), the other looks for subjective measures (*attitudes* and *feelings*). And there is a long philosophical tradition, dating back at least to Aristotle, that sees the two as being so entangled with each other that on many issues it proves near-impossible to distinguish them (it is, for example, not self-evident whether one should respond to the question "Are you well?" by describing one's physical or mental state).[368] Philosophers and social scientists differ as well over how the measures

---

[367] Professor Ruut Veenhoven (personal correspondence) has suggested that contentment depends on the satisfaction of wants and the hedonic level of affect depends on the gratification of needs.

[368] Using pithy examples, the philosopher Hilary Putnam (2004) demonstrated the entanglement of even facts and values.

Chapter 11: The Content of Well-Being: Empirics

are to be interpreted, for example, whether they should be taken to *determine* well-being or to *constitute* well-being. Here we sketch a line of thought that goes back at least to Aristotle's *eudaimonia*, which today reads "doing well and living well".

In his *Nicomachean Ethics*, Aristotle placed great weight on 'doing' and learning from it: "Anything that we have to learn to do, we learn from the actual doing of it: people become builders by building and instrumentalists by playing instruments" (Thomson, 1976: 63). Rawls (1972: 424–3) christened "learning by doing" the Aristotelian Principle, and interpreted it to be saying "other things equal, human beings enjoy the exercise of their realised capacities (their innate or trained abilities), and the enjoyment increases the more the capacity is realised, or the greater its complexity."

If a pair of concepts are entangled, it must be that they *are* a pair of concepts, that they are not the same. In this chapter, we first study objective measures of well-being that are today routine in discussions on international development; we then report on subjective measures that have been studied by social scientists from responses to questionnaires. We will find that the two lines of empirical enquiry point to similar features of life that make it go well. We will also find that the features are not at variance with one another. They even reinforce one another. The one objective measure that stands out as separate is income: within wealthy countries the share of the variance in subjective well-being measures that can be explained by household disposable (equivalised) income inequality has been found to be negligible. Health, marital status and employment status are far more important.

11.1 Objective Measures of Well-Being

The World Bank has created a four-way classification of countries, based on income per capita: high income countries, at an average of US$52,000 PPP; high-middle income countries, at US$17,500 PPP; low-middle income countries, at US$6,800 PPP; and low income countries, at an average of US$2,500 PPP.[369] Table 11.1 presents a snapshot of the quality of life today in countries grouped under this classification. The additional measures in the table are the total fertility rate (TFR), the under-5 mortality rate (i.e. proportion of children who die before reaching their fifth birthday), life expectancy at birth, the youth literacy rate, civil and political liberties, and government corruption. Many would regard them as *objective* measures of well-being. Some would say they are measures of the quality of life (Nussbaum and Sen, 1993).

Being a snapshot, Table 11.1 does not distinguish income from wealth; income could even be taken to be a fixed proportion of wealth. Rawls (1972), who took a broader view of income and wealth, argued that they represent one category of goods whose distribution forms the content of justice. The World Bank classification takes a narrower view. It sees an individual's income as her command over consumption goods in a year, it does not distinguish between private and public goods. That is one reason Table 11.1 includes other measures of well-being.

What is the basis of their inclusion? Reproduction is one of the deepest expressions of human wants; no account of well-being can bypass it. The global TFR today is just under 2.5. However, whether a society's TFR necessarily reflects desired fertility on the part of women is a different matter, which is why we discussed it at length in Chapter 9 and report it across countries in the four income levels in Table 11.1.

---

[369] See World Bank (2019). By income is meant Gross National Income (GNI), which for our purposes can be read as Gross National Product (GDP). Above we have presented rounded the figures that the World Bank uses to base their classification (Table 11.1).

Table 11.1 Social Statistics by Income Level (2018)

	High income	Upper-middle income	Lower-middle income	Low income
Population (millions)	1,231	2,838	2,872	651
GDP per capita PPP (2017 prices)	49,286	16,414	6,368	2,421
Total fertility rate*	1.6	1.9	2.8	4.6
Under-5 mortality rate (per 1000)	5	13	49	68
Life expectancy at birth (years)	81	75	68	63
Youth literacy (% of people ages 15–24)	100	98.3	89.7	75.6
Civil liberties (1 highest – 7 lowest)	2	5	3	6
Political rights (1 highest – 7 lowest)	2	5	4	5
Government corruption (100 least – 1 most)	68	38	36	25

Source: Rows 1–6: World Bank (2019); Rows 7–8: Freedom House (2019); Row 9: Transparency International (2019). Note: Income groups calculated using World Bank data.

Although the under-5 mortality rate is closely related to life expectancy at birth (the early years of one's life are especially hazardous for diseases), we have presented them both so as to give a sense of their numbers in the contemporary world. Together with literacy, they are aspects of human capital. We have selected *youth* literacy because it reflects the outcome of recent government policies toward human capital formation.

Civil and political liberties together are a category of public goods that are central to liberal theories of justice. They are also prominent in contemporary discussions everywhere on human rights. Civil and political liberties are not the same, but we should expect them to be strongly correlated (citizens in a country where political liberties are highly restricted are unlikely to enjoy many civil liberties). Dasgupta (1993: 115; Table 5.3) reported a rank correlation coefficient between political and civil liberties of nearly 0.8 in what in the late 1980s were low-middle income and low income countries.[370]

If civil and political liberties offer an essential space for households and civil society to function, government corruption hinders daily economic life (Chapter 6, Annex 6.1). Although we should expect the two to be related, they are not the same. This is one reason Table 11.1 is even more revealing than would have been expected. The correspondence between per capita income and the other measures that reflect well-being is striking: *The average person in countries that are richer lives longer, is better educated, enjoys greater political and civil liberties, and is less hampered by government corruption; moreover,* as noted in Chapter 9, *TFRs are higher than the global average in countries where family planning resources open to women are especially limited*. Study any one of the measures of well-being in a country, and you will get a sense of what the other measures are likely to record, at least in comparison to those in other countries. That points to an important statistical truth: *the various aspects of well-being are on average not in conflict, citizens do not face a cruel choice among them*.

---

[370] Estimates of civil liberties are published regularly by Freedom House. Their index, running from a score of 1 to 7 (7 being the worst), is based on such measurable features of life as freedom from extended police detention without charge, freedom to practise any religion, freedom to publish books and to read them, the right to seek information and to teach ideas, freedom from political press censorship, freedom of movement within one's own country, freedom from police searches of homes without warrants, the right of women to equality (with men) of movement and physical protection and of access to occupations, freedom of radio and television broadcasts from state control, freedom from torture or coercion by the state. Unsurprisingly, political and civil liberties are highly correlated; across all countries the correlation coefficient is around 0.9.

Dasgupta (1993: Chapter 5) reported the above finding from a study of data from what in the 1980s were low-middle income and low income countries. As each country in the sample represented a political unit, the author did not weight countries by population size (India and China were awarded the same weight as Mauritius and Mali).[371] Moreover, the finding is statistical, meaning that exceptions do not override it. Correlation is, to be sure, not causation, but the evidence points to a thesis that was explored in Chapters 0 and 6, that small differences in the quality of life can accumulate over time, each feature of life influencing the others, step by step, to create large differences in them over time. Path dependence in socio-ecological processes is another face of mutual causation.

11.2 Measuring Well-Being by Asking People

In all cultures that have left a written record, sages and philosophers have sought to understand and advise on the character of a life well-lived. There is, however, another way to identify what makes life go well or badly; it is to ask people. The General Social Survey in the US asks respondents to report their emotional state: "Taken all together, how would you say things are these days? Would you say you are very happy, pretty happy, or not too happy?" In contrast, the European Social Survey asks respondents to evaluate their lives on a 11-point scale (0–10): "All things considered, how satisfied are you with your life as a whole these days?" A sharper measure of a person's emotional state is to ask her to recall the emotions she experienced the previous day, whether they were pleasant or unpleasant ('affect').

Life satisfaction, happiness, and affect, do not reflect the same aspect of personal well-being. The first is an assessment of one's life, and the latter two reflect mood (a person could be sad or downcast owing to a bereavement but nevertheless satisfied with her life as a whole).[372]

The distinction between cross-section studies (e.g. comparisons of responses to questionnaires across countries in a given year) and time-series studies (e.g. responses to a questionnaire in a country over a period of time) is also useful to bear in mind. Using findings in cross-section data to reject contrary findings in time-series is unhelpful. Some of the controversy in the empirical literature on subjective well-being can be traced to that.[373]

Let us look at each of the three features more closely.

11.2.1 Happiness

Methods for eliciting subjective well-being that are the most widely discussed today involve large-scale surveys in which people are asked to record their feelings. Responses are often awarded numbers, as for example to the question in surveys on happiness conducted by researchers in Erasmus University, Rotterdam: "Taking all things together, would you say you are: very happy (weight 4), quite happy (3), not very happy (2), not at all happy (1)?"

The question is far less nuanced than is customary in philosophical discourses on the notion of well-being, but the approach is in the Utilitarian tradition, where well-being is taken to be a subjective, but measurable, entity. Bentham (1789) had taken well-being to be happiness and understood someone's happiness to be the sum of the pleasures and pains she experiences. Edgeworth (1881) suggested that the unit of personal well-being could be taken to be the smallest perceptible change for the better.[374]

---

[371] The statistical analysis in that publication did not include TFR or government corruption.

[372] Kahneman and Riis (2005) is an excellent introduction to this complex literature.

[373] Layard, Mayraz, and Nickell (2010) provide a lucid critique of this in connection with the question whether relative income matters to people.

[374] The idea has been extended by Y.-K. Ng (1975).

Hedonic psychologists have gone a lot farther. They have not only studied the distribution of subjective well-being in individual countries (where the data may be thought to reflect a degree of homogeneity of culture), they have also made international comparisons. And they have used the data to identify features of life that matter most to people. Not surprisingly, there are problems of interpretation, for the term 'happiness' and its equivalents in other languages may not convey the same meaning in all cultures. Nevertheless, respondents appear to understand the word and are able to respond to the questionnaires. There is also widespread evidence that happiness fosters physical health and thus extends life (Veenhoven, 2010).

Happiness figures report mood, whereas social scientists are concerned as well with life evaluation.

11.2.2 Life Satisfaction

'Life satisfaction' is a measure of the latter. In the Gallup World Poll (GWP), respondents (comprising fresh annual samples of people aged 15 or over in each of more than 150 countries) are asked to evaluate the quality of their lives on a 11-point scale, running from 0 to 10. The rungs 0 and 10 are, respectively, the worst possible and the best possible life for them. Respondents are asked to reflect on their life and anchor themselves on a rung on the ladder. The 11-rung ladder was proposed by Cantril (1965) and is now called the *Cantril ladder scale*. Because the questions ask people to assess their feelings, interpreting the responses poses problems. They are retrospective judgements, which may be presumed to have been arrived at only when asked and is influenced by the respondents' current mood and ability to remember.

There are further problems. Some people are by nature measured in the way they express their feelings; others are more extreme. Translated into the GWP, the former group would be taken to confine themselves to a narrower set of rungs than the latter group. In a wide-ranging essay on the measurement of subjective well-being, Kahneman and Krueger (2006) proposed a way to counter differences in personal scales by recording only the most extreme feelings an individual experiences in each episode that he has encountered during the study. Subjective well-being can then be made comparable across people on the basis of those extremes. The proposal amounts to regarding subjective well-being to be an ordinal, level-comparable measure.[375] On the other hand (Kahneman and Krueger do not discuss the possibility), it could be that some people are not so much measured in their utterances as compared to others as they are to experiencing a more limited range of intensities in feeling. If that were so (and Kahneman and Krueger do note that a person's subjective evaluation of his well-being is to a significant extent a personality trait), no re-scaling of subjective well-being across people would be needed.

11.2.3 Affect Balance

Emotions can be recorded in terms of 'affect', which covers enjoyment, laughter, worry, sadness, depression and anger. The idea is to get a sense of a person's emotional state by asking her to recall her emotions as she experienced them during the previous day. The GWP includes responses to questions relating to positive affect (experiencing pleasant emotions and moods) and negative affect (experiencing unpleasant, distressing moods) as well as constructing from the responses 'affect balance', which in a quantified form is the algebraic sum of positive and negative affect.

Responses to surveys on happiness and life satisfaction are based on personal judgement and memory. The latter is notoriously defective, the former is known frequently to be unreliable. The Experience Sampling Method has been developed to avoid both pitfalls. But it is awkward and involves additional costs. Participants carry a handheld computer that instructs them several times in a day to record not only the

---

[375] In a somewhat different context Sen (1970) and Hammond (1976) used ordinal, level-comparable measures of personal well-being to explain the appeal of 'max-min' as a measure of social well-being in policy analysis.

activities they were engaged in and the people they were interacting with immediately before being prompted, but also their current subjective experience (feeling angry, happy, tired, impatient and so on – in other words, affect balance). This approach is particularly useful in identifying the types of activities and their duration that make daily life go well. Investigators have used the method also to uncover links between emotional state and health. The *Cantril ladder* measures the cognitive component of happiness, while the Experience Sampling Method, Affect Balance and the Day Reconstruction Method (Box 11.1) tap the affective component.[376]

Box 11.1

Day Reconstruction Method

Responses to surveys on happiness and life satisfaction are based on personal judgement and memory. The latter is notoriously defective, the former is known frequently to be unreliable. The Day Reconstruction Method, reported in Kahneman et al. (2004), was developed to avoid both pitfalls. Respondents are asked first to complete a diary in which episodes that occurred in the preceding day are recorded. They are then asked to describe each episode (noting its location, time and duration, the people involved, and (from a list provided in the survey) the activities that took place). The idea is to ascertain how respondents felt during each episode. In order to do that, respondents are asked to express the intensity of their feelings in regard to a selected set of markers (or 'affective dimensions'), such as happy, worried, and angry. Intensities run on a scale from 0 (read, "not at all") to 6 ("very much"). In the authors' study involving 909 working women in Texas, participants experienced the highest net affects when they were engaged in leisure activities and the lowest when at work (be it paid or personal work).

Traditional economic models in which people choose their work and leisure balance were built in anticipation of such findings as these. Social contact during each episode was found to be associated with high positive emotions, a feature of the life that has been stressed repeatedly in work on social capital (e.g. Helliwell and Putnam, 2004; see Chapter 6). That is the sense in which there has been a coming together of the various approaches scholars have pursued to unravel the content of personal well-being.

11.3 Intranational and International Comparisons of Well-Being

In an analysis of GWPs from 2006 through mid-2011, Helliwell and Wang (2013) drew a quantitative picture of both the average level of life satisfaction and the distribution of life satisfaction among people in the more than 150 countries represented in the pooled data. Figure 11.1 reproduces their findings.

---

[376] We are grateful to Rutt Veenhoven for discussions on this. A rich, extensive literature has analysed survey data on subjective well-being. Diener, Helliwell, and Kahneman (2010) and Helliwell, Layard, and Sachs (2013) are key references, on which we have drawn here. Inglehart and Welzel (2005) is a landmark work on both subjective and objective measures of well-being. Even when they are attempting to estimate the same aspect of well-being (life evaluation, say), the surveys are not designed in the same way, which is why their findings do not always point in the same direction and have given rise to a number of puzzles and disagreements. See Clark, Frijters, and Shields (2008), Deaton (2008), Deaton and Stone (2013), and Stevenson and Wolfers (2013) for reflections on those puzzles. Sidgwick (1907: 123–150) anticipated several problems of interpretation in empirical hedonism. See also, Stiglitz, Sen, and Fitoussi (2009), OECD (2013a), and Adler and Fleurbaey (2016) on the various ways one can identify and measure human well-being.

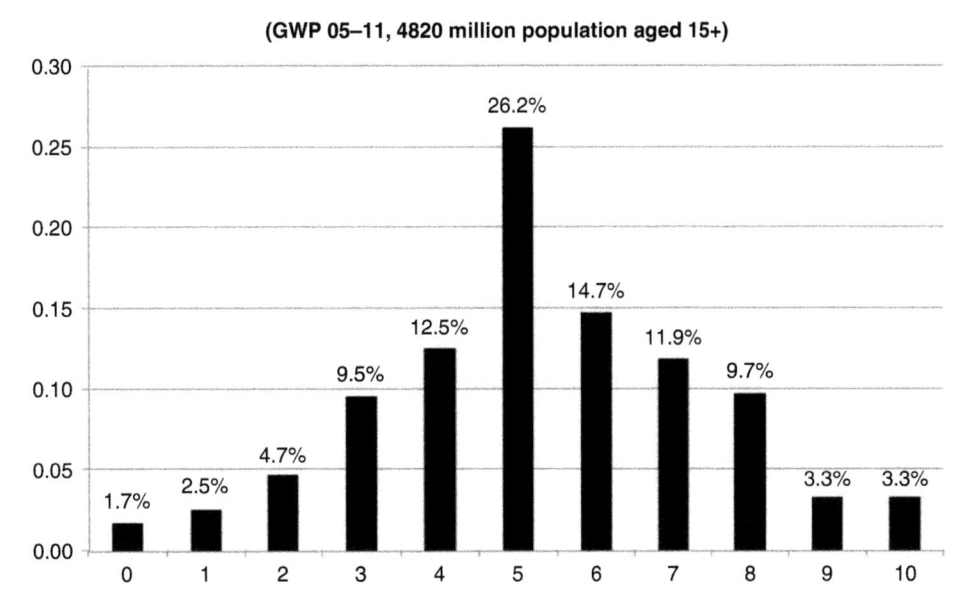

Figure 11.1 World Distribution of Cantril Ladder

(GWP 05–11, 4820 million population aged 15+)

Score	Percentage
0	1.7%
1	2.5%
2	4.7%
3	9.5%
4	12.5%
5	26.2%
6	14.7%
7	11.9%
8	9.7%
9	3.3%
10	3.3%

Source: Helliwell and Wang (2013).

The height of each of the 11 bars represents the proportion of the sample populations who gave that score as their life satisfaction (the table records global averages).[377] The histogram is unimodal, with a maximum at the mid-point, 5 (more than 25% of respondents gave their answers as 5). The bars taper quickly on both sides of the mid-point: 12.5% of people evaluated their lives as 4, while 14.7% evaluated them as 6, the percentages dropping rapidly to the tails (1.7% evaluated their lives at 0, while 3.3% evaluated them at 10). The authors noted that respondents in every country in the sample gave life evaluations that covered the entire range of scores (0 to 10). We are to conclude that no matter how rich a country is, it harbours people who rate their lives at the lowest rung (perhaps because they are relatively poor); and that at the other end, no matter how poor a country happens to be, it harbours people who rate their lives at the highest rung (perhaps because they are themselves rich). The authors noted that in contrast to low income countries, health, personal security, job satisfaction, ambient air quality, civic engagement and personal relationships matter more in high income countries than income. The latter two point to social engagement, which matters to people everywhere.

Helliwell and Wang (2013) also reported that average figures for life satisfaction differ across the world's regions, with a difference in the modal value of three points between North America/ Australia/New Zealand (29% of responses make up the modal value, 8) and South-East Asia (35% of responses make up the modal value, 5).[378] *Rich people on average say they have better lives than what poor people on average say about their lives.* This is true not only across countries but also more generally within countries. The data confirm that people in low income countries, when compared to people in high income countries, are on average less happy, respond with a lower evaluation of life and display lower affect balance.

Responses to what matters to people make for interesting reading. The Gallup Millennium World Survey asked respondents to identify two things that matter most in life. Good health and a happy family life appeared in more than 40% of responses, followed (a bit over 20%) by a job. Interestingly, the standard of living appeared in only 13% of responses (Veenhoven, 2010).

---

[377] The total adult population in the more than 150 countries from which the samples were drawn was about 4.8 billion.

[378] Helliwell and Wang (2013), Tables 2.2–1 and 2.2–4.

There is no presumption in any of the surveys on subjective well-being that the factors influencing the quality of life are independent of one another. We have noted previously that both theory and evidence point to mutual determination. Over time each factor influences, and is in turn influenced by, the others. The features of life pointed at by the authors were assumed by those who designed the surveys to matter, they assumed nothing more. Pollsters were not seeking to identify the causal mechanisms in the social world that give rise to wide differences in the extent to which people are able to make their lives go well.

Box 11.2
Relative Income Matters

Does economic growth raise happiness? Clark, Frijters, and Shields (2008) studied data on happiness from the General Social Survey and on life satisfaction from the Eurobarometer Survey. The authors found that since 1973 average life satisfaction in a number of European countries had remained approximately constant, even though real income per head in each country in the sample had increased sharply. If we are to take reported happiness or life satisfaction at their face value, it would seem that self-evaluation levels run flat through time as rich countries grow richer. The finding, noted first by Easterlin (1974), is known widely as the Easterlin Paradox. The Paradox can be read as two conclusions:

Conclusion 1 – At a point in time within any society, richer people are on average happier than poorer people (a cross-sectional finding).

Conclusion 2 – Over time within many countries, the population does not on average become happier when the country's income rises (a time-series finding).

That relative incomes matter to people, the subject of Chapter 9, has been documented elsewhere as well. The European Social Survey (ESS) asked people: "How important is it for you to compare your income with other people's income?" Those who said income comparisons were more important were on average less satisfied with their lives. Respondents were also asked: "Whose income would you be most likely to compare your own with?" The most important group mentioned was "colleagues". Perceived relative income would appear to have a large effect on life satisfaction (Layard, Clark, and Senik, 2012). It is a reflection of competitive preferences (Chapter 9).[379]

Conclusion 2 in the Easterlin Paradox, however, is contentious. In an analysis of data from representative national surveys undertaken from 1981 to 2007, Inglehart et al. (2008) reported that happiness rose in 45 of the 52 countries for which substantial time-series data were available and that the 45 countries in question include the United States and most European countries. Alpizar, Carlsson, and Johansson-Stenman (2005) in a contrasting, small-scale survey-experimental work in Costa Rica, found both absolute and relative consumptions to be salient. Individual well-being would seem to depend on *both* absolute and relative consumption. The ESS findings says that personal well-being increases with one's own income but, other things equal, declines with a rise in others' income.[380]

Layard (2011: 31–32) reported that an analysis of a set of global surveys on happiness and their relationship with household incomes revealed that in countries where per capita income was in excess of US$20,000 a year, additional income was not statistically related to greater reported

---

[379] Competitive preferences found a telling expression in a remark attributed to Gary Feldman of Stamford, Connecticut, one of the richest towns in the US: "I might be in the top one percent, but I feel I am in the bottom of the people I know." (*The Guardian*, 16 February, 2013).

[380] In the language of Chapter 9, the finding says that $\partial u_i(c_i,\underline{c})/\partial c_i > 0$ and $\partial u_i(c_i,\underline{c})/\partial \underline{c} < 0$ – where u_i is person i's well-being, c_i is i's consumption level and $\underline{c}$ is the average consumption level of others.

> happiness. The finding can be read as saying that, at least in that sample, which was carried out in the 1990s, income beyond US$20,000 a year contributed little to personal well-being (i.e. $u_i(c_i,\underline{c})$ remains constant if setting $c_i = \underline{c}$, both c_i and $\underline{c}$ were to increase at the same rate beyond a figure of US$20,000 year). Layard's finding from his sample should be contrasted with that in Blanchflower and Oswald (2004), which we reported in the Annex 9.1 (Equation (A9.8)).

11.4 Measurement and Interpersonal Comparisons of Subjective Well-Being

The questions posed in the surveys were not designed to serve the purposes of economic evaluation (Chapter 13), at least not directly. The surveys, even when they are confined to, say, Bangladesh cannot tell directly whether reviving the degraded Sundarbans is worth the cost, nor even when confined to India whether constructing a coal-fired power plant to bring electricity to villagers in India is a better option than spending a bit more cash to make solar-powered voltaic cells available to them. The surveys cannot tell, because they do not ask respondents to conjecture what their subjective well-being would have been had circumstances they have encountered been otherwise (that is, the surveys do not allow us to infer what a person's self-assessed well-being would be in alternative social states – counterfactuals have no place in them). The surveys' purposes have been to measure subjective well-being and uncover what matters to people. However, one can perform statistical tests on large-scale survey data to calculate the contribution various aspects of life make to happiness or life satisfaction, and have that information serve the purpose of economic evaluation. As a move away from unsubstantiated speculation on what make lives go well, the surveys have proved invaluable.

But there are problems of interpretation. Scholars analysing the data have said little about the degrees of freedom they believe are inherent in subjective measures of well-being. For example, negative values are not offered for consideration in the GWP, which suggests designers of the poll cannot imagine that life can be wretched (Chapter 10). One can argue of course that the exact ladder does not matter, that even if the unit of subjective well-being is taken to be fixed (say, the smallest perceptible change for the better is taken to be a unit), the level is a matter of choice. If that were so, the poll could equally have been designed to ask respondents to assess their lives on a 11-point scale that runs from −4 to +6 (subtracting 4 from each of the previous rungs). The problem is, we do not know whether the responses would have been the same in the rescaled version. It may be that numbers and their sign have psychological significance. If they do, the distribution of responses would have been different.

Suppose though that the responses would have been the same. The Helliwell-Wang analysis of their pooled dataset (Figure 11.1) would still have required all respondents to have been offered the *same* ladder. Otherwise computing the distribution of life satisfactions in a population would not have been possible. Imagine for example that a unit of life-satisfaction is also regarded a matter of choice, not just its level. That would be tantamount to assuming that subjective well-being is cardinal, perhaps even strongly cardinal (Chapters 5 and 10). But if it *is* cardinal, the same measurement system must be used on all respondents if their responses are to be comparable. To see this formally, suppose u_i is the self-assessed measure of life satisfaction for person i in the GWP under the 0–10 ladder. If, as in the way temperature is measured, the index of life satisfaction is cardinal, the poll could equally have been based on a system of measurement that recorded i's life satisfaction as $\alpha_i u_i + \beta_i$, where α_i is a positive number and β_i is a number of either sign. But for someone to compute average life satisfaction and the distribution of life satisfaction at a point in time, α_i and β_i would have had to be independent of i. That is another way of saying that the index of life satisfaction would have had to be fully comparable across people.

Chapter 11: The Content of Well-Being: Empirics

Now consider the other extreme assumption, that numbers have such psychological significance for people that their responses would depend on the ladder they were offered. In that case the interpretation of responses to a survey would be anchored to the ladder on which the survey had been built.

International comparisons of average happiness are not uncommon today. The King of Bhutan has asserted that citizens' happiness is the right goal of government. Today that tiny kingdom is cited in the media as a place where people on average are happier than people elsewhere. Such comparisons require that happiness is strongly cardinal and fully comparable across people. Average happiness is then the ratio of the sum of individual happiness to the population over which that sum had been computed.

Box 11.3

Are International Comparisons of Subjective Well-Being Meaningful?

Veenhoven (2010) observed that people do not appear to have a problem reporting their assessment of life. Responses to questions on overall happiness are typically prompt everywhere. Moreover, the data show that affective experience dominates the overall evaluation of life.

The latter should perhaps not cause surprise. Evolutionary psychologists have suggested that mood may serve as a meta-signal of how we are doing as a whole. Veenhoven explores the thought that happiness is grounded in the gratification of basic needs. As basic needs are universal, there may be a common ground for interpreting responses to life-evaluation questionnaires.

Diener, Helliwell, and Kahneman (2010) reported cross-country variations in affect balance during the day preceding the polls. In the GWPs, the correlation between national averages of affect balance and life satisfaction was found to be 0.58. Veenhoven (2010) reported the higher correlation coefficient of 0.70 in a set of studies that have linked self-rated overall happiness and average affect. Reported emotions and life satisfaction seem to tell a broadly consistent story about the level of well-being across nations and cultures. The data confirm that people in low income countries, when compared to people in high income countries, are on average less happy, respond with a lower evaluation of life and display lower affect balance. Table 11.2 identifies a set of societal characteristics whose differences across countries explain 75% of the differences in average happiness in nations. Wealth, however, is dominant (column 2): controlling for wealth, differences in the other characteristics explain a markedly small percentage of differences in average happiness. The author observes that the societal conditions that make people happy are not always the conditions they subscribe to. For instance, average happiness is lower in nations where women are discriminated against, even when the practice is widely approved. Likewise, corruption reduces happiness even in societies where favouritism is seen as a moral obligation (Chapter 6).[381]

---

[381] The striking figure in Table 11.2 is the −0.33 for economic equality, when wealth is controlled for. Rutt Veenhoven (personal communication) has suggested the possibility that the balance of positive and negative effects of income inequality depends on the level of economic development. In developed economies there are small negative correlations, while in developing nations the correlations tend to be positive.

Table 11.2 Average Happiness by Societal Characteristics in 136 Countries

		Zero order	Wealth controlled
Wealth			
	Income per capita	+0.79	
Freedom			
	Economic freedom	+0.62	+0.11
	Political freedom	+0.50	+0.07
Peace			
		+0.39	+0.15
Justice			
	Corruption	−0.77	−0.14
	Rule of law	+0.70	+0.06
Equality			
	Income equality	+0.27	−0.33
	Gender equality	+0.67	+0.19
Education			
	School enrolment	+0.57	+0.12
	Intelligence	+0.63	+0.21

Source: Veenhoven (2010: 337).

11.5 Determinants of Well-Being

Analysing a pooled data set that included replies to questions on the circumstances and background of respondents, Helliwell and Wang (2013) estimated that inter-regional differences in five broad features of life explain 95% of most of differences in average life satisfaction:

(1) Income. On average rich people are happier than poor people, not only when the average person in a high income country is compared to the average person in a low income country, but also when the comparison is made between rich people and poor people within countries

(2) Health

(3) Personal relationships (having friends to count on in times of need)

(4) Having a sense of freedom to make life choices

(5) Absence of corruption in the public sphere

Among the other determinants of well-being are personal traits, such as disposition, which can be a reflection of the person's culture.[382] The authors also noted that health, personal security, job

---

[382] There is evidence too that religion has a positive influence on subjective well-being (Inglehart, 2010). See also Inglehart et al. (2008) for a global study of human well-being based on reported happiness over the period 1981–2007. The authors identify freedom of choice as a significant determinant of personal well-being.

Chapter 11: The Content of Well-Being: Empirics

satisfaction, ambient air quality, civic engagement and personal relationships matter more than income to people in rich countries than they do in poor countries. There is thus support for the idea that not only do people everywhere entertain trade-offs among the factors affecting their lives, but that income loses its relative appeal as people become richer.

In an earlier publication, Helliwell (2003) examined three rounds of the World Values Survey, which asked respondents (90,000 adults in 46 countries) to score their level of life satisfaction on a scale of 1 to 10 and features of life that influenced subjective well-being. Seven factors were found to be especially important (see also Helliwell and Huang, 2010): (i) wealth, (ii) family relationships (e.g. whether divorced or separated), (iii) work (e.g. whether employed in a secure job or unemployed, and conditions in the workplace if employed), (iv) social capital (the extent to which one can trust others), (v) health, (vi) personal freedom, and (vii) personal values (whether the respondent, living in a religious country, is religious).

These findings have been justifiably influential. OECD (2013a) has constructed an aggregate index of national well-being for 40 countries, built on an array of life's circumstances: (i) income and financial wealth; (ii) housing; (iii) job (earnings, job security, employment rate); (iv) health; (v) education (including its returns); (vi) work-leisure balance; (vii) community (quality of social support system); (viii) quality of the (natural) environment; (ix) governance (involvement in democratic practices); (x) safety (murder and assault rates); (xi) life satisfaction.

The index of each of the eleven items is constructed out of up to three indicators. The classification is not unproblematic. Factors (i)-(vi) are directed at personal circumstances and experiences and (vii)-(x) point to the socio-ecological environment in which the individual functions, but (xi) points to the object of the exercise; it is not a factor that contributes to life satisfaction. Nevertheless, the important point remains: the features of life that one would have thought matter, do matter. The findings we have reported here do not say whether the features are complements or substitutes, but cross-country data suggest they can all improve more or less in tandem with one another.[383]

These findings have been re-confirmed in an exceptional paper by Benjamin et al. (2014), who have developed a method for constructing well-being functions (the u-functions of Chapter 10) from surveys in which respondents are asked to report the extent to which the aspects of life that matter to them are being realised. The authors applied their method by asking more than 4,500 people in the US and constructed a well-being function based on their responses. The factors included family, health, security, values, freedom, happiness and life satisfaction. Their work demonstrates the precise sense in which the idea that well-being is a function of consumption is merely a reduced form of the notion that well-being depends on our engagements.

Box 11.4
New Zealand's Living Standards Framework

New Zealand's Treasury has implemented an ambitious new Living Standards Framework (LSF). The framework aims to strengthen four types of assets: natural capital, human capital, physical/produced capital, and the enabling assets of financial and social capital. These categories of assets are then combined with 12 metrics of well-being: civic engagement and governance, cultural identity, environment, health, housing, income, jobs, knowledge and skills, time use, safety, social

---

[383] But there are exceptions, even statistical exceptions. Veenhoven (2010) reports that correlation between happiness and education ranges from −0.08 to +0.27 in various studies.

connectivity, and subjective well-being. The LSF is still in development, with several natural capital indicators yet to be developed. In 2018, the country's statistical agency began compiling these accounts following the UN System of Environmental-Economic Accounting (SEEA).[384]

The New Zealand LSF can be used for conducting policy analysis to evaluate alternative policy interventions. LSF contains an accompanying guidance on ways to quantify well-being impacts from policy interventions through an updated cost-benefit analysis tool. It provides a link to the 12 well-being metrics and the four asset domains in the LSF. Government departments and agencies are required to describe the impact of proposals across various areas of well-being and wealth and the impact to economic risk and resilience in Impact Assessment Statements. LSF commends using measures of both the 'means' (an inclusive measure of wealth) and 'ends' (well-being) in policy analysis.

On 30 May 2019, the New Zealand Treasury put its LSF and policy analysis methodology to work, with its first Well-being Budget. The well-being approach shifted the focus of government priorities around intergenerational well-being outcomes: (i) supporting mental health with a focus on under 24 year-olds; (ii) improving child well-being; (iii) supporting Māori and Pacific incomes, skills and opportunities; (iv) supporting digital innovation, and social and economic opportunities; (v) improving the productivity of businesses and regions; and (vi) the transition to a sustainable and low-emissions economy.

11.6 Nature and Well-Being

There is a new entry in national and international book-keeping on well-being. It reflects an increasing awareness among governments and international agencies that modern life has grown distant from Nature and that it should be a cause for public concern, that macroeconomic policy should be more sensitive to the place of biodiversity in our everyday life. The UN Population Division (2019c) notes that 55% of the world's population live in urban areas, a proportion that is projected to increase to 70% by 2050. More than 50% now live a distance from a landscape that is rich in biodiversity. Outside summer months, adults in high income countries spend less than 10% a day outdoors (Diffey, 2011). With growing urbanisation, it should not be surprising that there is today concern among psychologists that detachment from the natural world lowers our sense of well-being, adversely affecting even our mental health (Maas et al. 2009).

In their admirable survey of a growing literature on the role played by our direct experiences with biodiversity in personal well-being, Capaldi et al. (2015) distinguish two aspects of those experiences: contact with Nature and connectedness with Nature. The former could even involve interaction with the natural world via indoor plants or from exposure to virtual representations of Nature such as photographs or paintings of natural landscapes. Contact need not be regular; it can be brief and intermittent.

The latter points to the extent to which people internalise the experiences they have with Nature. It has been found, for example, that individuals who are more connected with Nature spend more time outdoors, and that contact with Nature often increases momentary feelings of connectedness, although the causal link could be go either way (Capaldi et al. 2015: 2). Thus, if contact with the natural world is a means to furthering personal well-being, connectedness with Nature is an aspect of well-being itself.

Humans cannot survive without the rich biodiversity of microbiota that exist on our skin, in our gut, in our urogenital tracts and in our airways. They influence our susceptibility to disease and play an

---

[384] See Karacaoglu (2020) for a more comprehensive account of the LSF.

important role in our health with changing environmental conditions.[385] We are ecosystems ourselves, with exchanges taking place constantly between bacteria, viruses, fungi, Archaea and protozoa in the natural environment and those in our bodies. We need contact with sources of diverse microbiota. Loss of biodiversity can and often does have an impact on our health (e.g. weaker resistance to disease). Studies suggest that macro-biodiversity (e.g. plants and trees) in urban environments is associated with environmental microbe diversity and in turn with a healthy human microbiome, known to be linked to a wide range of health outcomes. One recent exploratory study used post-mortem human microbiome assessments (n=48) and data on 'green remediation' in Detroit, US, and found suggestions of a 'healthier' microbiome amongst individuals residing in locations with green infrastructure interventions (Pearson et al. 2019). Increased urbanisation and less exposure to the natural environment has contributed to a rise in dysfunctional immune responses and non-communicable diseases such as asthma and type 1 diabetes, which in turn have been traced to reduced exposure to a biodiverse microbiota.

Three psychological theories have been invoked to explain why contact with Nature and connectedness with Nature heighten our sense of well-being: biophilia, attention restoration and stress reduction.[386] The first looks back to our evolutionary past. Urban living is a recent phenomenon. To our hunter-gatherer ancestors an understanding of Nature's rhythms was necessary for survival. Connectedness is an expression of that understanding. Biophilia (Wilson, 1984; Kellert and Wilson, 1993) says the need to make contact with the natural world is an innate trait in humans.

The second, attention-restoration theory, distinguishes directed attention (used for executive functions, involving prolonged concentration) and involuntary attention (which is effortless yet demanding). The former is a limited resource, meaning that a person becomes fatigued after an extended use and may lead to negative emotional states, for example, irritability and reduced cognitive performance. The natural environment would appear to be a restorative; it provides a way to engage our involuntary attention, attracting rich stimuli and furthering our emotional functioning (Berman, Jonides, and Kaplan, 2008).

The third, stress-reduction theory, holds that exposure to certain non-threatening natural environments that were beneficial over evolutionary time elicits a variety of stress-reducing psychophysiological responses. There is evidence that contact with Nature can reduce pulse rates, reduce cortisol levels and improve immune functioning.

Psychologists would appear to be on firmer ground when reporting the role contact with Nature plays in our sense of well-being. The influence on our well-being of connectedness with Nature is less assured empirically, at least as of now. The reason may be that connectedness is more difficult to achieve than making contact with the natural world. So, most studies have looked for the influence of contact on hedonic well-being.[387] A meta-analysis of 32 randomised-control studies with over 2000 participants has revealed that contact with Nature results in moderate but significant increases in positive affect as well as in small but significant decreases in negative affect (McMahan and Estes, 2015). Exercises in green spaces have been found also to improve hedonic well-being.

There is evidence too that repeated contact with Nature contributes not only to long term hedonic well-being, but to life satisfaction as well. A series of large-scale European studies based on data from

---

[385] It can be argued that the organisms residing in our bodies in symbiotic relationships with us are part of our own make-up. This is discussed in Chapter 12.

[386] We have adopted the classification in Capaldi et al. (2015), which also cites references to these findings.

[387] There is a fourth, broader model of the human person, named by Kaplan and Kaplan (2009) the 'reasonable-person model', which focuses on how environmental factors influence cognition and behaviour, and thus well-being, by supporting human informational needs. This has been less explored in the nature/well-being literature.

national surveys has found that living in an area with more green space is associated with less mental distress than otherwise. A longitudinal study covering over 10,000 UK residents found that living in a greener urban space was associated with greater life satisfaction (White et al. 2013).

The latter study points to a relationship between contact with Nature and long-term well-being. There is evidence that exposure to biodiversity even increases one's sense of autonomy – the sense of being in command of oneself. Mayer et al. (2009), for example, report that in their study participants who were randomly assigned to take a walk in a green space reported reduced public self-awareness. Research from the literature on outdoor education and experiential learning also points to the positive effect Nature-immersion experiences have on the sense of autonomy and other measures of psychological well-being (self-esteem, self-regulation, social competency) as well as objective measures such as vitality.

In a consensus statement, Bratman et al. (2019) include on their list of benefits of Nature experiences positive social cohesion, the ability to carry out life tasks, reductions in mental distress, improvements in cognitive functions, memory, attention, creativity and children's performance at school. The authors also note that in addition to improvements in psychological well-being, Nature also helps to reduce the risks of poor mental health and ameliorate the burdens of existing illness.[388]

Research has also found that the amount of green space in one's neighbourhood correlates positively with social ties and with pro-social activities with neighbours (see Chapter 6 on social capital). There are, to be sure, confounding factors in these studies, and they demonstrate correlations, causation is a lot harder to determine. Current research in the neurosciences on the links between contact and connection with the natural world and personal well-being is beginning to offer a physiological explanation for why it is that local biodiversity loss is detrimental to human well-being.

Box 11.5
The Well-Gardened Mind

Gardens have played a part in human societies for millennia, from Persian gardens that may have existed as early as 4000 BCE to temple gardens in India, China and Japan, and monastic gardens in Europe. The word 'paradise' correlates in Persian with an 'enclosed green space'. In today's more democratic sensibilities, we do not see paradise in an enclosed space, we see it rather in an unbroken green landscape. But the role that gardens play in recuperation grew in prominence in 18th century Europe, where mental hospitals were increasing built in the countryside to aid recovery (Stuart-Smith, 2020).

Evidence suggests that green spaces enable social contact and thus may be associated with perceptions of greater social belonging. Lower percentages of green space in the living environment have been found to be associated with higher likelihood of people reporting a feeling of loneliness and shortage of social support (Hartig et al. 2014). While all social groups have been found to benefit from green space, socio-economically deprived and disadvantaged populations appear to benefit more than average from greener living environments. Access to good-quality natural environments, especially in urban areas, can also reduce socio-economic inequalities in health (Mitchell, Norman, and Mullin, 2015; Wheeler et al. 2015).

In the contemporary world, we rely on medication to treat mental health, but there has been a resurgence of interest in Nature, more particularly biodiversity, as a source of treatment for mental

---

[388] The consensus statement is significant because the authors ranged across the natural, social and health sciences.

health and well-being. Horticultural therapy is increasingly available today (Stuart-Smith, 2020). Gardening has been found to boost mood and self-esteem and to alleviate depression and anxiety (Gonzalez et al. 2010). A randomised control trial comparing gardening to Cognitive Behaviour Therapy found the two to be equally effective in reducing stress disorders (Stigsdotter et al. 2018). Gardening programmes have been used in prison rehabilitation programmes. Riker's Island, one of the biggest penal colonies in the world, runs a green prison programme that enables prisoners to grow and care for plants and offers internships following their release in parks and gardens. For those who attend what is now known as the Green-House Programme, only 10% of those paroled returned to prison within three years compared to California's average recidivism rate of nearly 65% over the same period (van der Linden, 2015).

Chapter 12
Valuing Biodiversity

Introduction

Ecosystems are capital goods. Biodiversity is a characteristic of ecosystems. In the terminology introduced in Chapter 1, it is an *enabling asset*.[389] Biodiversity loss is not the same as environmental pollution. Air pollution would be bad even if it did not emanate from activities that harm biodiversity; we dislike pollution if only because it is bad for human health. In contrast to biodiversity, pollutants are a form of capital *stock* (but with a negative accounting price; Chapter 2).

In Chapter 2, we studied the sense in which ecosystems that are rich in biodiversity are productive. So one can argue that biodiversity's value is derived from the productivity it confers on ecosystems. Against that is the view that biodiversity also has a value independent of its effects on the biosphere's ability to produce goods and services for us. Bearing that in mind, six sources of biodiversity's value may be distinguished, although inevitably, they blend into one another:

One is human existence itself. Extreme forms of biodiversity loss kill (toxic pollution in water; loss of vegetation cover upstream causing landslides downstream; degraded mangroves ceasing to protect villages against storms). So one measure of biodiversity loss is an expression of the human lives that are lost in consequence.

A second source of value is biodiversity's direct contribution to human health. Our lives are better when our health is better. Pollution in water causes physical illness, and an absence of green space in our lives affects our mental health (Chapter 11). Pandemics are, by definition, a runaway breach of biodiversity. Moreover, natural products form a sizeable proportion of the pharmaceuticals we use to maintain and regain our health. If health is an end, then biodiversity is a means to that end.

Closely related to these (purists would say it is indistinguishable from them) is a third source of value. It rests simply on the fact that biodiversity is a source of enjoyment; Nature has an *amenity value*. Ecotourism is an expression of that.

A fourth source of value arises from biodiversity's role in making available the wide range of Nature's goods and services on which we depend (Chapter 2). Economists call that the *use value* of Nature. This source of value draws on biodiversity as a means *par excellence* to human ends and is the aspect of Nature most commonly discussed in environmental and resource economics. The Annex to this chapter and Annex 13.1 contain examples of how natural capital accounts are created from estimates of their use value.

A fifth source of value in biodiversity – we usually ascribe it to species – lies in its very existence. We may never get to meet gorillas in the jungles of Rwanda, we have no direct experience of their presence, but nevertheless may feel gorillas ought to be protected. Economists call that its *existence value* (Freeman, 2003).

The sixth source of value is so closely related to the fifth that aspects of it are very much like the fifth – is sufficiently complex to be traceable to related two considerations. It can be that we value Nature because it is sacred to us (Section 12.6), or it can be that we value Nature because we recognise it has

---

[389] The rationale for distinguishing capital goods from enabling assets is developed more fully in Chapter 13.

value independent of whether it means something to us, for example, that it has moral worth (Section 12.7). Taken together, we may call it Nature's *intrinsic value*.

Economists have mostly studied the value of biodiversity indirectly. They have done that by valuing items of natural capital, for example, ecosystems (e.g. fisheries), specific species (e.g. endangered ones) and natural resources (e.g. ground water; Box 13.4). That may be an acknowledgement that biodiversity is an enabling asset, not a capital good.[390]

The value of an item of natural capital is its accounting price (Chapter 10). Rules governing the dynamics of accounting prices are developed in Chapter 13*. Those rules are none other than the arbitrage conditions we introduced in Chapter 1. This chapter provides an outline of the methods economists have deployed for estimating accounting prices of natural capital. It also provides illustrations of the way the methods have been put to work.

Reflecting as they do a balance between the possible and the desirable, accounting prices are not 'objective' quantities. It was noted in Chapter 10 that consumption discount factors, which are the present values of accounting prices of future goods and services, cannot be estimated in the way the dimensions of a table can be measured. For all practical purposes people can agree on the dimensions of a table, for example, that it is 4 m long and 2.5 m wide. Not so on consumption discount factors. If the citizen (previously we referred to her in turn as the 'social evaluator' and the 'citizen investor') is a 'pragmatist' (Chapter 10), she will at best be able to offer a range of values with which she feels comfortable. Others will no doubt want to know how she arrived at that range, and may arrive at ranges of their own. The same is true for accounting prices. Discussion with oneself and with others is essential if the socio-ecological world is to be navigated wisely. It is far better that such discussion takes place and differences in understanding are aired than citizens throwing up their hands and permitting economists to place a value of zero to natural capital when deliberating policy. So long as we have a grammar in which the biosphere appears as the seat of our activities, it is better to be vaguely right than precisely wrong.[391] To claim that there is an 'economic argument' on the one hand and an 'environmental argument' on the other is to misunderstand the nature of economics.

In democratic societies, differences in people's assessments of accounting prices influence the way policies requiring their use are selected. We may think of candidates for political office as bearers and carriers of contingent policies. It is then no exaggeration to say that citizens express their differences over accounting prices by casting votes on their favoured political candidates. This caveat should be borne in mind when mention is made of 'estimating' accounting prices.

12.1 Estimating Accounting Prices: General Observations

Chapter 10 (Box 10.6) noted that the social evaluator needs four items of information if she is to estimate accounting prices:

(1) A descriptive model of production possibilities in the economy (as in Chapter 4*)

(2) An account of the institutions and practices that shape people's choices

(3) The size and distribution of the economy's capital goods

(4) A conception of intergenerational well-being.

---

[390] There are exceptions of course (see Hanley and Perrings, 2019). Christie et al. (2006) reported estimates based on the views expressed by the public in the UK on whether a given policy would protect rare or common species, whether the species are well-known and whether the policy would restore habitats or whether it would create new habitats.

[391] The aphorism appeared in Shove (1942), who attributed it to Wildon Carr.

Chapter 12: Valuing Biodiversity

Taken together, items (1)-(3) form the basis for projecting changes that are likely to occur in resource allocations, now and in the future, if an additional unit of a capital good is made available free of charge today. Item (4) is the coin on the basis of which those changes are evaluated. Estimates of the social cost of carbon are based on methods that deploy all four items on the above list (e.g. Stern, 2006, 2015).

There is a Utopian world in which government and community engagements serve as a backdrop against which individuals and groups create a system of production, consumption and exchange in which payments for goods and services equal their accounting prices. In this scenario, many of the transactions are imagined to take place in markets that are subject to the discipline of the law and social norms. In those happy circumstances, the social evaluator would not need to estimate accounting prices, she would only have to look up the prices people face. Economists call that state of affairs a 'first-best' (Chapter 10).

In imperfect economies (Chapters 7-9), market prices are reasonable approximations of the accounting prices of some goods and services, while for others they are not. Equation 7.1, which says the accounting price (p) of a capital good is the algebraic sum of its market price (q) and the externality (e) associated with its production and consumption, is key. If, as in the case of public goods, there are no market prices to use even as a first step, the entire burden falls on estimating the externality. The theoretically valid procedure for estimating the externalities is to put the definition of accounting prices directly to use, as in the above list (items (1)-(4)). In Chapter 13*, the global model in Chapter 4* is used to show how in principle that can be done.

To get an intuitive feel for the effects of social and ecological constraints on accounting prices, consider the first item (1) on the above list. Imagine there is a nearly depleted resource that has no near substitutes in either production or consumption. Then, other things equal, its accounting price would be high. On the other hand, if the resource does have substitutes, and in large quantities, its accounting price would be low. Consider next item (3). If a resource is highly productive, then other things equal its accounting price would be high. To illustrate item (4), imagine that special weight is placed by society on species preservation and a particular species is endangered. Then its accounting price will be high. The preservation order would not speak of the species' accounting price, it would simply be a command, the order being that the species must be protected.

To illustrate (2), imagine that the social institutions in a country are so ineffective that a class of capital goods is wasted (Chapters 7 and 8). Then, other things equal, their accounting prices would be low. As this could seem paradoxical, it requires an explanation.

If a commodity is wasted and the prevailing institutions are unlikely to change, an additional good will find little use because that too would be wasted. Of course, if the prevailing institutions are dysfunctional, there is a strong case for improving them or establishing better ones (Chapters 7-9). Similarly, if current policies are wrong-headed, they ought to be revised. If the project being evaluated is an institutional reform under which the resources in question would be better used, their accounting prices will not be low (Chapter 13). Accounting prices do a lot of work for us in summarising information about an economy.

If the good is not a 'good', but is instead a 'bad' (sulphur emission), its accounting price is negative (Chapter 2). But even if a good is a 'good', its accounting price could be negative if the economy's institutions are dysfunctional; offering an extra unit of the good to someone could be a bad thing. The use of fossil fuels provides an example. Their use by motorists yields private benefits to them, but they lead to collective losses (enhanced greenhouse effect). If emissions are untaxed, private benefits would be impervious to the collective damage caused by the combustion of fossil fuels. Imagine now that the collective damage arising from a small increase in the use of fossil fuels exceeds private benefits. An

Chapter 12: Valuing Biodiversity

additional unit of fossil fuels awarded to a motorist will then reduce social well-being. The accounting price of fossil fuels in a motorist's vehicle would be negative.

In practice, estimating accounting prices involves various degrees of short cuts, ranging from the *ceteris paribus* assumption on all features of the economy other than the resource (as in applications of contingent valuation methods; Section 12.2), to models of the entire economy (as in estimates of the value of natural capital as an aggregate; Chapter 13). The starting point of most short cuts has been people's *willingness to pay* for an item of natural capital. The thought here is that (i) a person's willingness to pay measures the item's contribution to his well-being and (ii) social well-being is an aggregate of individual well-being (Chapter 10).

There are two ways to determine people's willingness to pay. One is to ask people ('stated preference') and the other is to infer it from what they do ('revealed preference').

12.2 Stated Preference for Public Goods

Public goods offer the motivation for a popular valuation method, in which representative samples of people are asked hypothetical questions about their willingness to pay for such measures as protecting a species from extinction and improving the quality of drinking water and ambient air. The idea is to sum the individual responses so as to arrive at an estimate of the willingness-to-pay of the entire population. An alternative is to ask people how much they would be willing to accept as compensation for permitting it to deteriorate from its present state to some other, specified, state. The idea here, again, is to sum the individual responses so as to arrive at an estimate of the sum of the willingness-to-accept by the entire population. The two alternative methods, taken together is called the *contingent-valuation method* (CVM).[392]

CVM is attractive because it appeals to our democratic instinct, that people should be asked for their opinion on matters that may be of concern to them. The method is attractive too because in principle it can reveal not only an amenity's value, but also respondents' sense of a species' existence value – perhaps even its intrinsic value. It is known, for example, that people derive satisfaction from the mere knowledge that the large primates in the forests of Uganda exist, they do not feel they have to view them. CVM is in principle capable of coaxing such information from respondents. However, people have been found to be willing to pay more to prevent an amenity from deteriorating to some specified level or a species from becoming extinct if they are informed that others will contribute as well than if they assumed that others were not contributing anything (a possible sign that preferences are socially embedded; Chapter 9). The underlying *ceteris paribus* assumption has frequently been overly strong in questionnaires.

Despite the widespread use of CVM in its original form, there are problems internal to the method. We discuss a few in Box 12.1.[393]

---

[392] Contingent valuation in its original form, in which willingness to pay or willingness to accept is elicited for a change in environmental quality that is identical across all survey participants, is not used much anymore. It has been superseded by choice experiments, in which the attributes of an environmental change are randomly varied across participants. One reason for the increased popularity of choice experiments is that they are less subject to some (but not all) of the criticisms of CVM (Johnston et al. 2017). In the text, we are using 'CVM' as shorthand for a broader set of stated preference methods that includes choice experiments. We are grateful to Jeffrey Vincent for helpful discussions on this.

[393] See Hausman (2012) for an uncompromising critique of CVM. Kling, Phaneuf, and Zhao (2012) is a reflection on the uses to which CVM can be put fruitfully. In the Introduction to an exceptional collection of essays estimating accounting prices of natural capital in South Asia, Haque, Murty and Shyamsundar (2011) explain why CVM has limited use in measuring the value of natural capital in low income countries.

> **Box 12.1**
> **Contingent Valuation Methods: Problems**
>
> CVM has unravelled interesting features about people's stated values. On species preservation, respondents have been known to provide answers that are independent of the scale of operation. Willingness-to-pay has been discovered to be the same regardless of whether, say, 100 or 10,000 cranes are saved. This is an illustration of the 'embedding problem'.
>
> It is as though people are willing to put aside so much for preserving populations from extinction regardless of the size of the population or even the number of species being preserved. More problematically, respondents have been known to give different answers to the same question when it is put differently.[394] This means investigators can elicit the kind of answers they would like to hear by framing the questionnaire in suitable ways. As CVM is not infrequently used in environmental litigations, the method has proved to be controversial. Vast sums of money can be involved. The court cases have not been the seemliest of occasions for the practice of economics.[395]
>
> The application of CVM ought to have been more circumspect. It is all well and good to ask people for their opinion, but if they are uninformed of the functional attributes of the resources about which their opinion is sought, the point of the exercise is questionable. It is not un-common that we say, 'I don't know enough, tell me more that is known about the matter before I can give you a reasoned response.'
>
> In democracies, citizens delegate responsibilities in the public domain to elected representatives. We are not uncomfortable that political decisions sometimes go against what would have been our collective 'will' had we been asked to vote directly on those matters.[396] Limits on the scope of CVM ought to be guided by this fact. CVM would be less than appropriate for valuing biodiversity in tropical rainforests. Normal precautions should have been taken before huge chunks of those forests were converted into crop land, cattle ranches and plantations.[397] But the prevailing institutions were not appropriate for the practice of the principles; the structure of incentives was wrong (Chapter 8). And they continue to be wrong. Altering institutions is ultimately a political choice, and that implies preserving biodiversity is also a political choice.

12.3 Revealed Preference for Amenities

Market prices are usually thought to be the institution *par excellence* where individual preferences are revealed. If you are found purchasing a commodity in the marketplace, it would seem reasonable to conclude that you are willing to pay at least the amount you have paid. Differences in the quality of a marketed good can thus be inferred from differences in prices. For example, the commercial value of residential land has been found to depend on the landscape surrounding it (e.g. the view from the house). A property's *hedonic price* is a measure of its quality. The method is also used to infer the value of a statistical life (Box 10.2). Other things equal, differences in wages in jobs involving different levels of risk offer a way to infer the willingness to accept risks.

---

[394] This is called the 'framing problem' Tversky and Kahneman (1988).

[395] A notable example is the Exxon Valdez case, where the state of Alaska sued Exxon after one of its oil tankers crashed off the state's coast. For an assessment of the strengths and weaknesses of CVM, see the report on the NOAA Panel on Contingent Valuation (co-chaired by K.J. Arrow and R.M. Solow) in Arrow et al. (1993).

[396] In the UK, capital punishment is often cited as an example.

[397] Vincent et al. (2014) is an application of CVM to the valuation of rainforests (in peninsular Malaysia) in which the authors so composed their questionnaires, that they were able to unravel to an extent respondents' understanding of the value of different features of their forests.

Chapter 12: Valuing Biodiversity

This line of thought has been extended to determine quality differences in amenities even when they do not have markets (Freeman, 2003). Sites possessing amenity value are often public goods (national parks, beaches). Entry fees limit congestion, but usually they are a small fraction of the cost tourists incur for visiting the sites. That offers a method for estimating the willingness to pay for the amenity: *the cost of travel*. The method has been used widely to study tourism.[398]

Box 12.2
Valuing Tourism

In 2014, travel and tourism generated some US$7.6 trillion, 10% of global GDP (at 2014 prices; see WTTC, 2015). Ecotourism is a small, though significant proportion of it. A 2015 study found that globally, natural protected areas received 8 billion visitors a year, generating up to an estimated US$600 billion, making it approximately 8% of the travel and tourism market at the time (Balmford et al. 2015).

Mukhopadhyay et al. (2020) have estimated that the *consumer surplus* people enjoy annually by visiting coastal and marine ecosystems in India adds up to the strikingly high figure of 53 billion dollars PPP (about 1% of the country's GDP), of which two-thirds are enjoyed by foreign tourists. To the best of our knowledge, there have been few studies of whether tourists increase the Impact Inequality. In a carefully crafted global study, Hunter and Shaw (2005) found that, unfortunately, they probably do. Therein lies another irony. In the absence of tourism, local inhabitants may have little incentive to protect an amenity, but tourism is itself a cause of wider environmental damage, for example, additional carbon emissions.

12.4 Productivity as Accounting Price

Methods based on stated and revealed preference do not make direct use of item (1) on the list we constructed for estimating accounting prices, which is why neither method on its own is adequate for estimating the use value of ecosystems. In a pioneering work, Barbier (1994) explored the use value of tropical wetlands by identifying the services they provide. To do that, the author had to study the dynamics of tropical wetlands. The idea, which we developed in Chapters 2 and 3, was to view a wetland as a *production system*. Item (1) on our list is prominent in this way of thinking. Either stated or revealed preference could be used to value the inputs and outputs of the production unit (item (4) on our list; see Box 12.3), but that would come only after item (1) had been put to work.[399]

As illustration, consider that fisheries have use value and are marketed. The present value of profits a fishery is expected to generate is its market value. But if the fishery also discharges pollutants into neighbouring assets, the market value would be an overestimate of its accounting value (Equation (7.1)).

Shrimp farms offer a telling example. Individual farms are small in acreage and closely spaced. Farms damage one another by spreading diseases and causing eutrophication of the lagoons feeding water to them. Fish farms also damage neighbouring crop farms by discharging saline water. The former is a crowding externality (Chapter 8*), while the latter is a unidirectional externality (Chapter 7).
Box 12.3, based on Umamaheswari et al. (2011) and Rohitha (2011), gives examples of the two forms of negative externality.

---

[398] In a valuation study of public parks in northern Pakistan, Khan (2011) observed that as entry was free, the local government did not have the resources to maintain them. This would appear to be a common problem in low income countries.

[399] Vincent (2011) contains an excellent account of the practical steps involved in estimating the value of ecological production units. Valuation exercises based on production structures were made familiar previously in water management studies. Brown and McGuire (1967) is a key reference.

> **Box 12.3**
> **Social Costs of Shrimp Farming**
>
> Umamaheswari et al. (2011) studied the effect of shrimp farms on neighbouring paddy fields by contrasting the case of farms in a pair of neighbouring villages in the delta of the Kaveri River at the border of the state of Tamil Nadu, India. Water salinity is measured by electrical conductivity (EC). Normalised, water for which EC < 1 is suitable for crop cultivation. The EC level in paddy fields in the two villages differed greatly. In one village, where the shrimp farms were not adjacent to paddy fields, EC ranged from 0.02–2.13, whereas in the other, where the farms were adjacent to the paddy fields, EC ranged from 4.95 to 15.90. The authors estimated a production function for paddy cultivation, in which the inputs included water quality and the mitigating measures farmers deploy to counter water salinity. Controlling for those factors, paddy output per hectare was found in the former village to be higher by more than 50% of the output in the latter village.
>
> Rohitha (2011) estimated crowding externalities across shrimp farms in the Dutch Canal wetland system in Sri Lanka. A spurt of uncoordinated development there had resulted in the establishment of 1,300 shrimp farms in a land area of only 3,750 hectares. Because individual shrimp farmers did not exercise collective restraint in the amount of feed they used, nor treat the farm waste before discharging it, the wetland lagoon had suffered increasingly from eutrophication (Chapter 8*). This had caused a decline in output of the farms' aggregate output (the farms draw raw water from the canal system). It had also caused a decline in the lagoon's fish harvest. The author estimated a production function for individual shrimp farms, in which water quality is an input. The annual profits that would have accrued to farmers from discharging clean water was found to be 7–10% higher than they currently enjoyed. The author also estimated that resuscitation of the Dutch Canal would be economically profitable.

12.5 Human Health

Pollution affects human health. Ambient air and water pollution are the two dominant sources of damage to health in low income countries. Most studies are localised – to a city, even a neighbourhood. The World Health Organization (WHO) guidelines on air quality suggest a target for small particulates (PM_{10}) of 50 micrograms per cubic metre of air (50 µg/m3). Air pollution in excess of that is injurious to health. According to recent figures on air quality, the annual average PM_{10} concentration in Lahore, Pakistan in 2010 was 198 µg/m³ while in Dhaka, Bangladesh in 2015, it was 146 µg/m³ (WHO, 2018). More recently, between 2016 and 2017, it has been a staggeringly high 273 µg/m³ in Delhi, India (The Indian Express, 2016).

Industry and transportation are the major sources of air pollution in cities. In a study of air pollution in Kanpur, an industrial city in India, Gupta (2011) estimated that morbidity costs amount to about 1% of income there. Bogahawatte and Herath (2011) report a similar estimate for Puttalam district, Sri Lanka. The figures are small, but are an underestimate of the cost of pollution; mortality was not included.[400] In a previous work, Gupta (2008) had found that if morbidity alone is taken to be the damage, the cost of reducing pollution in Kanpur by a quarter would exceed the benefit, but if mortality was included, the benefits would swamp the costs.[401]

---

[400] Lost output and medical expenditure are the common measures of health damage.

[401] The value of a statistical life is the coin with which mortality costs are measured. As noted in Chapter 10 (Box 10.2), VSL is a large number everywhere. Increases in human capital assume enormous importance relative to accumulation of produced capital when expectancy of life is included. We confirm this in Chapter 13 (Annex 13.1).

Chapter 12: Valuing Biodiversity

The presence of organic waste is the major source of water pollution. Parikh, Parikh and Raghu Ram (1999) estimated that about 1 million children died annually in India from waterborne diseases. If those deaths had been valued at the prevailing value of statistical lives (say, US$500,000; Box 10.2), those early deaths would have been found to add up to a staggering half a trillion US dollars annually; India's GDP in 2000 was about half a trillion US dollars. That would not mean that waterborne diseases should have been eliminated at once and entirely, for that would have required expenditure so huge that little would have been left for other expenditures, which in turn would have led to a larger number of deaths from other causes. As always in social cost-benefit analysis, counterfactuals have to be reckoned with.

Then there are the pathogens that devastate nations, even the entire globe. COVID-19 is a reminder that such viruses, when unleashed, devastate lives. Biodiversity loss allied to movement of people and goods can be expected to increase the frequency of epidemics, even pandemics, much as the frequency of extreme weather events are being caused by the growing concentration of greenhouse gases in the atmosphere.

These are examples of the health costs of pollution. On the flip side is biodiversity as a source of health benefits. Box 12.4 looks at the range of pharmaceutical products that are obtained from natural products. It also points to the conflicts that have arisen over misappropriation of property rights of tropical countries that are home to the natural products.

Box 12.4
Pharmaceuticals, Traditional Medicines, and Social Returns on R&D

Biodiversity has been a source of medicines for thousands of years. A huge proportion of our approved antibiotics, antivirals and antiparasitic drugs are derived from natural products. We also rely on biodiversity for new sources of compounds. Between 1981 and 2014, for example, 686 of the 1,328 new therapeutic agents approved for pharmaceutical drugs were based on compounds from natural resources (Newman and Cragg, 2016).

Plants have always been a source of new drugs, but only a small fraction have been studied for potential so far. All living groups of species have the potential to contain compounds that act pharmaceutically; marine and microbial realms are especially thought to hold vast untapped potential. Although we know of no estimates, the option value of biodiversity (Chapter 5), even when restricted to its value as potential pharmaceutical products, could be of the same order of magnitude as the known medicinal value of natural products.

In reverse, pharmaceutical drugs can have direct impacts on biodiversity (Depledge, 2011). Synthetic hormones and hormonally active compounds are of concern, as are endocrine disrupting compounds. Even very low levels of these compounds can have significant effects on our ecosystems. For example, ethinylestradiol (EE2), found in human contraceptive pills, can have dramatic effects on wildlife (Schwindt et al. 2014). In one study in Canada, the introduction of EE2 into a lake led to the collapse of the fish population (Kidd et al. 2007). Similarly, veterinary compounds such as diclofenac can cause damage to ecosystems. Vultures are well-known casualties.

Millions of people rely on traditional medicine. For poor households in low income countries they may be the only available treatment. The medicines are sourced from the wild. Degradation of ecosystems and the accompanying biodiversity loss reduce the source of traditional medicine. Demand for herbal medicines is, however, rising. In 2016, the Chinese government announced their aim to integrate traditional Chinese medicine into their healthcare system by 2020 (Willis, 2017). Sales of herbal medicines globally amount to US$60 billion (Tilburt and Kaptchuk, 2008).

> There is an uneasy relationship between traditional medicines, practitioners and the potential for pharmaceutical drugs. Local, indigenous knowledge is valuable for the communities that are served by health practitioners, but this information is also of potential value to the pharmaceutical industry (WIPO, 2015). That wealth is transferred from low to high income countries because patents are not protected is yet another impediment to sustainable global development.
>
> Tensions are present even within countries between citizens and private companies. Production of knowledge has beneficial spillovers (in Chapter 4*, open science and publicly available technology were denoted by the factor *A* in the aggregate production function). In a survey of studies that had compared private rates of return on investment in research and development to their social rates return, Griliches (1987a,b) noted that the latter figures were more than twice the former, returns *ex post* being as high as 100%. In a study that applied a different method of analysis, Jones and Williams (1998) estimated that social rates of return on investment in research and development in high income countries were as a minimum 30%, while the corresponding private rates were in the range 7–14%. The authors estimated that socially optimum investment levels, as a share of GDP, are two to four times the actual investment. There is little reason to think that investment in natural products as a source of new drugs is any lower. Moreover, if the option value of species in the wilderness in their possible role as pharmaceutical products were to be included (Chapter 5), the social rates of return on *protecting* them would be even higher.

The move from valuation studies of individual resources, even ecosystems, to aggregate measures of natural capital involves a huge leap. Inevitably, one loses details even while many items of capital are missed entirely. On the other hand, aggregate measures find a place in the balance sheet of national economies, which is why they are necessary. The UK's natural capital accounts are among the first such accounts to be published (Bright, Connors and Grice, 2019), as are estimates of China's Gross Ecosystem Product (Annex 12.1). There are also published estimates of movements over time of the inclusive wealth of nations (e.g. UNU-IHDP and UNEP, 2014; Managi and Kumar, 2018). Each of these exercises should be seen as probes. Accounting values of natural capital do not adequately include the health benefits a less degraded Nature confers on us. They are even farther away from being able to include the benefits to our mental health that green spaces confer on us. And if we consider the intrinsic value of Nature, we begin to appreciate how far estimates of movements of the value of natural capital are from what we should be after.

In Chapter 11, we reviewed studies that have found a link between our connection, even contact, with Nature and our mental health. The significance of this finding is obvious in a world that is becoming increasingly urban. These are early days for including quantitative estimates of that loss in the accounting prices of green space. But in principle that can be done.

12.6 Nature's Existence Value and Intrinsic Worth: Sacredness

One may doubt, however, that hard-nosed cost-benefit analysis could be the right language in which to express all our values. Aspects of Nature require care and consideration in other ways. Many people, perhaps in all societies, locate the sacred in Nature. And the sacred is not negotiable, unless we rationalise by imagining it to be incorruptible (Box 2.8). We compromise – there are always contending needs and wants – but we do not put a price on the sacred. Our urge is simply to protect it. In Benin, for example, legislation has even enshrined the sanctity of sacred groves into law. Many social practices of communities there rely on materials and resources drawn directly from forests; leaves, animals, water and stones. These areas are considered sacred because they have traditionally been seen to be inhabited by deities or spirits. The sites also serve as spaces for rituals. Hunting is prohibited there, as is setting fires to the groves. Traditionally, custody of sacred forests was entrusted to members of a certain lineage.

Activities such as logging for timber or gathering plants for food and medicine were regulated by local communities (Houngbo, 2019).

Benin has approximately 2,940 sacred groves – covering an area of 18,360 hectares, which accounts for 0.2% of the national territory. Most of the sacred groves themselves are small; 70% of the total number are less than 1 hectare in extent, while 18% cover an area between 1 and 5 hectares, and only 12% are larger than 5 hectares (Soury, 2007).

Forests are, however, disappearing in Benin. In 1990, they covered 50% of land; today the figure is less than 40% (FAO, 2015a). Deforestation in Benin has its roots in common with other regions; agricultural extensions, hunting, grazing and timber extraction. In 2012, the Benin government introduced a decree (Interministerial Order No. 0121) that gave legal recognition to sacred forests. They are protected areas. The law recognises sacred forests and sites where gods, spirits and ancestors reside. It protects the areas through local management, by integrating traditional custodians and community authorities.

Many today would regard an awareness of the sacred to encompass a sense of awe and wonder, of self-transcendence, through which we locate ourselves within the landscape around us and imagine what lies beyond. That sense is not confined to what cosmopolitans call 'traditional cultures'. A number of the most sublime songs of the poet Rabindranath Tagore have roots in the Vedanta, but they are detached from the rituals that had been given expression in Vedic times. They invoke the transcendent, but they are not tied to religion. That sense of spirituality is often experienced today not in isolation, but communally, such as among bird-watchers, hikers, cyclists, surfers, divers and anglers (Grove-White, 1992). Gould et al. (2015) reported findings from interviews they conducted in Hawaii and British Columbia to identify the cultural and ethical values respondents associated with ecosystems. The authors used the responses to devise a protocol that can be used to elicit the intrinsic values people ascribe to Nature. The historian Simon Schama (1995) has argued that it is a mistake to think that Western cultures have abandoned the spiritual aspects of Nature – that they have abandoned the myths that were created around Her. He showed that the transcendent has been expressed repeatedly in art and architecture.

The sense of transcendence that Nature invokes may still exist everywhere, but as our Review has shown, it has taken a severe beating in modern times. Contending needs and wants have displaced our feeling that by protecting the landscape we protect ourselves. The ability of rich societies to desecrate the landscape elsewhere has helped to accommodate those conflicting feelings, and the discipline of economics has increasingly aided that accommodation. Economic justification in rich countries is all too often a euphemism for commercial justification; and in poor countries the term 'economic development' is routinely used to justify blatantly unworthy investments.

That need not have been. As this Review has tried to show, correct economic reasoning is grounded on our values. Biodiversity does not only have instrumental value, it also has existence value – even an intrinsic worth. These senses are enriched when we recognise that we are embedded in Nature. To detach Nature from economics is to imply that we consider ourselves to be external to Her. The fault is not in economics; it lies in the way we have chosen to practise it.

12.7 Nature's Intrinsic Worth: Moral Standing

The sacredness of Nature points to an existence value, even intrinsic worth, that we impart to it. Does Nature also have moral standing? The UK Environmental Law Association (UKELA) has published studies that face the question from a legal point of view. They have proposed 'wild law' and 'earth jurisprudence' – concepts that give legal standing to non-human parts of the biosphere, such as plants, animals and ecosystems. These ideas go beyond traditional notions of conservation, in which Nature is protected because humans want it to be, in providing legal rights and interests to Nature so that it can be represented in its own right in courts of law.

Chapter 12: Valuing Biodiversity

In a 2009 study, the UKELA found instances of wild law in many countries around the world, including New Zealand, India, Ecuador, the United States and Colombia (Warren, Filgueira, and Mason, 2009). These legal terms are clearly founded on moral thinking about the nature and value of ecosystems, but are primarily concerned with legal processes. Here we explore the moral status of ecosystems.[402]

To assess whether ecosystems have moral standing, it is useful to start by comparing them with the entity most universally accepted as being morally significant: ourselves. In doing so, however, it is important to distinguish between *persons* and *human beings*, where it is typically the notion of personhood that provides the basis of our moral standing.[403] This has a long and varied history and can be found in different contexts, albeit often with different terminologies. The distinction is important to a number of philosophical schools, even if it is handled in different ways. Oversimplifying hugely, one might say that Buddhism denies the existence of the person, Confucianism sees persons as socially constructed, Christianity links the person to the soul, Kantianism takes it to be the basis for moral responsibility, existentialism regards it to exist outside of the causal confines of the universe, and much of contemporary consequentialism sees the person as being the source of important axiological distinctions.[404]

Among the many views that accept the distinction between human beings and persons, we can draw a broad distinction between those who think that personhood is an innate and intrinsic property of all humans and those who think it is acquired, typically at the point where we develop grammatic language and a social identity. The notion of personhood relates to two aspects of a person in particular:

(1) Being rational, in the sense of identifying and responding to reasons for acting in a certain way or making a certain choice (rather than doing so automatically)

(2) Identifying a self, the extended portion of space and time that is 'mine' and that is the metaphysical ground for these reasons.

The question as to whether it is appropriate to impart a notion of personhood to ecological systems is thus a question about whether there is some aspect of these systems that is at least partially analogous to the quality of personhood as possessed by human beings, and is thus worthy of our moral respect. There are at least three ways in which we might answer this question in the affirmative. We can call these arguments from 'extension', from 'qualitative similarity' and from 'teleology'.

The argument from extension involves the thought that human personhood, and all the rights and interests inherent in it, should be extended to cover the ecological systems we inhabit; they are really part of us and should be respected as such. A key aspect of personhood is identifying a self that contains this person. Normally this self is identified by the limits of the human body and the span of our lives. However, this turns out to be a very fuzzy sort of boundary to draw. Our teeth are a part of us and have moral importance as a result of being so. However, should they become detached (as children's teeth naturally do) they would seem to lose some, but perhaps not all, of this significance. Second, many of us are now cyborgs, in that we are intimately bound up with technologies that we also incorporate into our selves, from artificial heart valves to robotic arms. Third, we naturally extend the self in some respects to cover personal property, especially where this has more than monetary value to the person

---

[402] This section, exploring ideas that are rarely discussed in the economics of biodiversity, has been put together from notes prepared by Simon Beard of the Centre for the Study of Existential Risk, University of Cambridge. We asked Dr. Beard to convey to us the contemporary philosopher's first cut into the question of whether Nature has a moral standing. He prepared his notes only to suggest *how* we may think on the matter, not what we should think.

[403] As one common thought experiment puts this point, 'You wouldn't deny the moral standing of Superman simply because he came from the planet Krypton'.

[404] Dennett (1993: Chapter 13) and Parfit (2012) contain illuminating discussions on personhood.

Chapter 12: Valuing Biodiversity

who possesses it. Finally, our bodies are not, in fact, just one organism, but an entire microbiome in which non-human cells (on our skin, in our guts and just about everywhere between) outnumber the human cells by quite a margin.

It is natural for us to adopt a notion of ourselves that is sufficiently permissive to encompass these kinds of oddities and boundary cases. However, there is an argument to be made that we significantly underestimate just how embedded our human bodies are in Nature (Chapter 4), and that such broader notions of the self would actually let in far more than we realise. We may correctly judge that, because none of us could digest food without the bacteria in our gut, we should treat those bacteria as part of us. So why should we not just as equally judge that because we could not breathe without the plants in our garden and elsewhere that we should also treat these as part of us? Effectively, what such an argument does is to make the case that we should internalise all of those ecosystem services that are essential to human survival, and not just the tiny percentage of these that take place within the confines of our bodies. This kind of argument is quite powerful in its implications for how we should treat natural ecosystems (i.e. no less favourably than we would ourselves), but ultimately all it leaves us with is all of the same human persons we began with, only this time these persons are understood to overlap and extend to cover the entire biosphere they inhabit.

The second argument, from qualitative similarity, is very different and starts with the ecosystems themselves. This argument attempts to show that the functioning of individual ecosystems is so sophisticated, in terms of their complex adaptability, that they create or possess a kind of personhood directly analogous to that of human beings. It is important to note that this takes place at the level of the ecosystem, rather than of the individual plants and animals that make it up. The reason for this is that there are certain hard limits on the sophistication of plants and animals that do not possess a grammatical language. They can learn things in a 'Skinnerian' fashion, choosing to do what leads to rewards and avoid what leads to harms, but because they cannot conceptualise the world they cannot imagine or respond to things they have not yet experienced. The most they could possibly achieve is inference by induction, not the kind of creative reasoning that is involved in creating a self and responding to reasons. Most of the development of plants and animals therefore happens at the level of the population, rather than the individual, in which multiple variations are tried out simultaneously and the most successful ones selected, allowing the population to respond adaptively via the process of evolution. However, evolution does not affect individuals, so its genius is quite separate from what individual animals are capable of.

Population evolution does not take place in a vacuum, and the driving force behind most evolution is the exploitation of niches within a complex ecosystem. So, the sophistication of what can be achieved by the combination of individual learning and population-based evolutionary selection is best understood at the level of the entire ecosystem, a group of species simultaneously evolving to exploit complementary niches (Chapters 2 and 3). Ecosystems thus exhibit far greater adaptability in response to environmental changes than individuals do, making them a stronger candidate for this kind of personhood. At the level of biomes (such as forests, deserts, tundra and the like) this makes them very resilient (it was recently discovered, for instance, that rainforests were able to survive even at the south pole until relatively recently). However, at the level of the entire earth system, there is an even higher degree of adaption and resilience, to the point where it survives (and *eventually* repopulates) even terrible events like the end-Permian 'great dying' that killed 95% of all life on earth.

At those higher levels it is, therefore, not unreasonable to talk about ecosystems or the biosphere exhibiting the same kind of properties that persons have, both in terms of having a defined self with clear interests (systemic stability and resilience in the face of a general trend towards increasing entropy) and being able to detect and respond to reasons relating to these interests (environmental changes that threaten this stability). This argument is related to the concept of Gaia (Lovelock, 1995; which was alluded to in Chapter 2), a way of understanding global changes as representing complex and purposeful adaptation of change. It is doubtful that such a way of thinking could ever be proven or disproven scientifically. However, it makes sense at a human level, and provides a basis for granting these systems the kind of moral standing we grant to another human being.[405]

The third kind of argument, from teleology, takes as its starting point a view about personhood and the self that would invalidate the above argument, namely a kind of hard realism about consciousness. In this view, there is just something about humans that makes us conscious and gives us free will, and this is irreducible to any of our physical properties and so cannot be understood as extending to other physical systems based on their being analogous to us. As such, there is nothing that we share with other complex adaptive systems like forests, or even the entire biosphere. In an interesting but highly controversial work, the philosopher Thomas Nagel (2012) defends this view, but then goes on to argue that it implies that consciousness must exist outside of the laws of physics. In arguing this, Nagel does not try to support any other kind of account of what consciousness is or how it arrives, such as dualism or religious accounts of the soul, but he does suggest that these accounts are correct in looking for something teleological in Nature, a purpose or driving force that exists outside of the laws of physics and has caused the universe to be one in which consciousness arises in us. This teleological force gives credence to a kind of natural law, whereby ecosystems and the biosphere are worthy of respect, not because they are persons in themselves, but because, like human beings, they exist for the purpose of bringing consciousness into being and supporting its existence.

This argument is thus similar to the argument from extension, in that it is ultimately our personhood that matters, rather than that of ecosystems themselves. However, unlike that account, the value we impart to ecosystems by our existence is not merely derived from these ecosystems being important to us, but rather their being necessary components of the universe that brought us into being. This value does not extend to cover every single plant or animal, because individual death and suffering are also intrinsic parts of this universe, but it does imply that there is something important about how the world works, why humans arose when and how we did, which parts of the earth continue to sustain us (not only physically, but culturally, emotionally and even spiritually), and how this might extend into the future.

To the economist, these three kinds of arguments may appear strange and muddy, and yet to avoid them would be to leave the economics of biodiversity diminished. Of the three, it is the second argument, about the qualitative nature of ecosystems, that flows most readily from the Review. However, although the three arguments are very different, even contradictory, each points to a direction of thought worth pursuing; for asking people to disclose the value they place on Nature is only the first step towards an understanding of the full value of Nature, including its moral worth.

Annex 12.1 Valuing Ecosystem Services in China: Gross Ecosystem Product

In the late 1990s, China suffered a series of natural disasters that were exacerbated by changes in ecosystems, wrought by resource extraction and pollution (Bryan et al. 2018; Lu et al. 2019). In 1997, water extraction for human use exacerbated drought along the Yellow River (Shiau, Feng, and Nadarajah, 2007). Shortly after, in 1998, the River Yangtze flooded, intensified by deforestation upstream. The floods killed 3,600 people, inundated 5 million hectares of crop land and cost

---

[405] One might respond that ecosystems cannot be persons like us because it is not possible to know how it is like to *be* a forest? However, all this argument says is that ecosystems exhibit properties that are analogous to personhood, not that they are persons like us. Furthermore, when mathematicians have tried to model human consciousness with any kind of precision, they have found that the same models would apply to a wide variety of other systems, suggesting that even though it may seem incredible to us, there may really be something it is like to be a forest and we are just not in a very good position to imagine it.

Chapter 12: Valuing Biodiversity

US$36 billion (Ye and Glantz, 2005). In 2000, another natural disaster hit; overgrazing and desertification led to dust storms in northern China that covered Beijing seven times in one month (Wang et al. 2004). Costs were estimated at US$2.2 billion (Ai and Polenske, 2008).

Prompted by these disasters, a number of large-scale policies have been implemented to conserve and restore ecosystems and biodiversity and lessen the adverse impacts of economic activity on ecosystems (Bryan et al. 2018). In 2018, the People's Republic of China took the striking step of enshrining the concept of 'ecological civilisation' in its constitution, which emphasises the need for people to engage with Nature in ways that allow people to live well and within the bounds of the biosphere (Hanson, 2019). There is recognition that this requires sectoral reforms, spatial planning, technological innovation, ecosystem conservation and restoration and regulation. There has also been significant investment in cross-regional payments for ecosystem service schemes (PES), known in China as 'eco-compensation' programmes (Leshan et al. 2017, 2018). The Sloping Land Conversion Programme is one of the largest PES initiatives in the world. By the end of 2012, the total area of afforested land was a little under 30 million hectares, with total investment of around 440 billion yuan (approximately US$66 billion at current exchange rates), of which approximately 325 billion yuan was paid directly to 32 million households in 25 provinces (Liu and Lan, 2015). The programme aimed to reduce soil erosion, deforestation and flood risk by restoring forests and grasslands.[406] Evidence suggests the conversion of land through this payments programme, from cropland to forest and grassland, has sequestered significant amounts of carbon, reduced soil erosion into the Yangtze and Yellow Rivers, and reduced flood risk (Song et al. 2014; Gutiérrez Rodríguez et al. 2016).

A12.1.1 Gross Ecosystem Product

Gross Ecosystem Product (GEP) has been developed as a measure of the value of flows of ecosystem services in China, aggregated as a single monetary estimate (Ouyang et al. 2020). The methodology uses spatially explicit modelling to predict flows of ecosystem services and applies various valuation methods to estimate their value. These value estimates are aggregated into one overall monetary measure of the contribution of ecosystems: its GEP. A sub-set of ecosystem services are included, focusing on those that are clearly important and are feasible to model and value with available data and based on current scientific understanding.

China's National Ecosystem Assessment assessed the status and trends in terrestrial ecosystems, ecosystem quality and ecosystem services between 2000 and 2010 (Ouyang et al. 2016). Seven ecosystem services were mapped for China's land area: food production, carbon sequestration, soil retention, sandstorm prevention, water retention, flood mitigation and provision of habitat for biodiversity. These services are supplied by China's ecosystems, which include forests, wetlands, croplands and grasslands. The ecosystem service assessment was undertaken using spatially explicit, integrated models that link ecosystem conditions and processes to ecosystem services, based on 'ecological production functions' (Kareiva et al. 2011; Sharp et al. 2018). The regulating services of soil retention, sandstorm prevention and water retention were measured through reductions of rates of soil erosion, wind erosion and storm run-off respectively, using models that link ecosystem conditions and processes to the provision of ecosystem services. The assessment drew on data from satellite images, soil and hydrological measurement, 114,500 field surveys, biodiversity records and surveys of desertification (Ouyang et al. 2019). The assessment identified areas that have experienced an increase or decrease in ecosystem services over the period 2000 to 2010, as shown in Figure A12.1.

---

[406] The goal of alleviating poverty was also added subsequently.

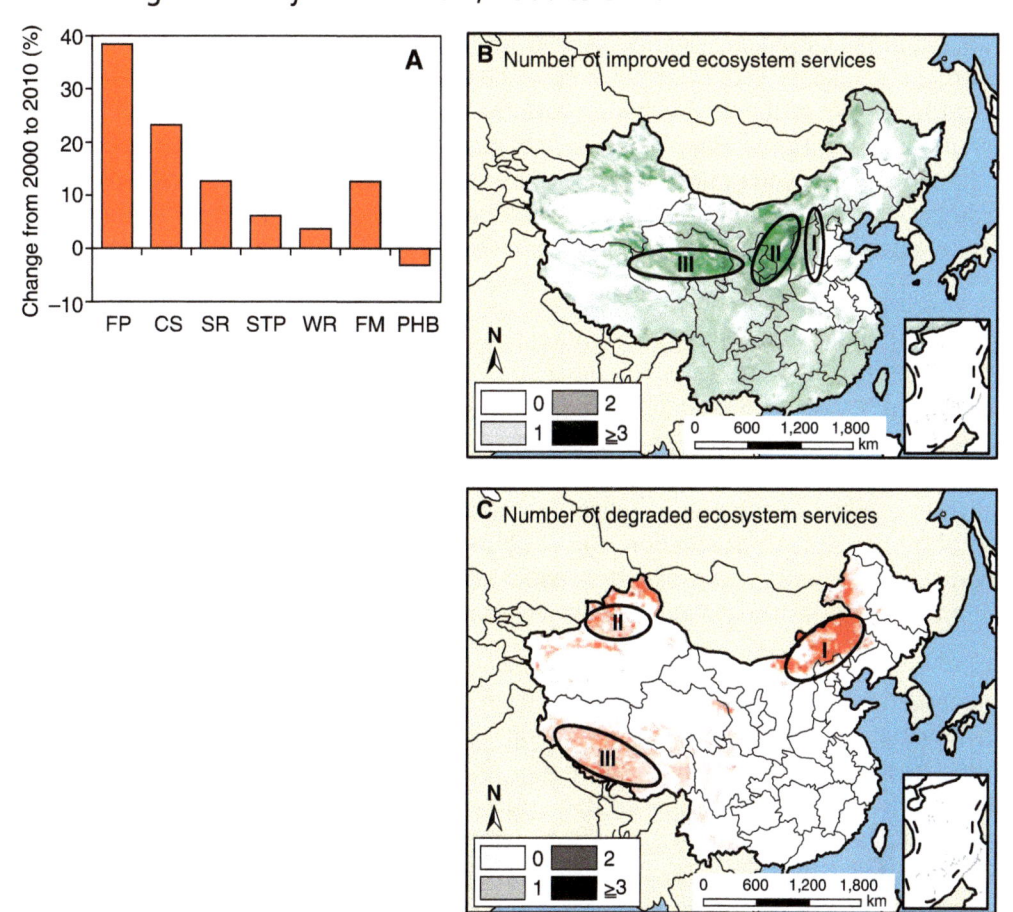

Figure A12.1 Changes in Ecosystem Services, 2000 to 2010

Source: Ouyang et al. (2016). *Permission to reproduce from Springer Nature.*

Drawing on data from this national ecosystem assessment and other sources, GEP measures have been estimated by using spatially explicit integrated ecological-economic modelling to predict the flow of ecosystem services and then applying economic valuation methods to estimate the value of those services (Ouyang et al. 2020).[407] The value of the flows of ecosystem services are then combined to provide a single aggregate monetary estimate. A range of valuation techniques are used to estimate the values. Market prices were used where markets exist, mostly for provisioning goods, such as agricultural production and animal husbandry, based on published data on production and prices. The provision of regulating services was estimated using the InVEST model suite – Integrated Valuation of Ecosystem Services and Trade-offs (Sharp et al. 2018) and data from government sources. Market and non-market valuation methods were used to estimate monetary values for the provision of those services.

---

[407] GEP follows broadly the framework and definition of accounting value used in the United Nations System of Environmental-Economic Accounting (SEEA) and the SEEA Experimental Ecosystem Accounting (EEA) (United Nations et al. 2013, 2014).

A12.1.2 Valuation of Ecosystem Services – the case of Qinghai

GEP has been piloted in Qinghai, a province in the north-west of China (Ouyang et al. 2020). The GEP measure includes a sub-set of ecosystem services, but a sizeable number. Provisioning goods valued were agricultural crops, animal husbandry, fisheries, forestry and nursery production in Qinghai, and water supply originating from Qinghai. Regulating services included soil retention, sandstorm prevention, flood mitigation, air purification, water purification and carbon sequestration. The cultural service of eco-tourism was also included. The full valuation results are provided in Table A12.4. For brevity, here we give an overview of some useful, illustrative examples of valuation of provisioning, regulating and cultural services.

As noted previously, provisioning goods often have a market price, which makes valuation relatively easy. Ecosystems in Qinghai provide various agricultural products (grains, vegetables, fruit etc.), animal husbandry products (meat, dairy, eggs, honey, wool, cashmere etc.), fishery products (freshwater fish), forest products (timber, walnuts etc.) and nursery products (flower plants and seedlings). The value of these ecosystem services was calculated as the product of the total biophysical amount of the products harvested (tons of traditional Chinese medicinal herbs harvested) multiplied by their market price. One complication is to adjust the values to include only the portion due to inputs from nature, and subtract the contribution of other inputs to production, such as human labour and machinery (e.g. labour to pick the medicinal herbs). In the Qinghai example, these could not always be accounted for due to a lack of data on costs and input for some sectors, leading to overestimates in the value of some ecosystem goods.

A harder provisioning good to measure is the supply of water. The problem is not only the monetary valuation – market prices, albeit possibly distorted, exist for hydropower, agricultural crops that are irrigated with water, domestic water use and so on – but also to understand the hydrology for approximating how, where and when water flows. Qinghai is an important water tower of Asia, situated at the source of the Mekong, Yangtze and Yellow Rivers. Water from Qinghai is used in downstream provinces, for agricultural irrigation, hydropower, industrial production and domestic use. In their GEP estimates, Ouyang et al. valued all four types of use of water supply. The way that water travels downstream requires approximations. The authors estimated the fraction of surface water in each downstream province that originates in Qinghai, adjusting the amount of water flowing out of Qinghai into the Yangtze and Yellow Rivers, based on the length of the river between where it leaves Qinghai and enters into another province.

Taking downstream hydropower production as an example, the authors calculated the fraction of hydropower electricity generated at each dam downstream attributable to water resources from Qinghai. Valuation was then based on market prices – the market price of industrial water, domestic water, crop production and so forth. Continuing with the hydropower example, the value of hydropower was estimated as the product of the amount of electric power generation attributable to water resources from Qinghai and the market price of electricity. Again, the value of other human-produced inputs, such as dams and machinery needed to convert hydropower to electricity were not included due to data constraints, leading to overestimates of the value of this ecosystem service.

Regulating services are often overlooked as they are harder to estimate. Laudably, the authors undertook valuation of a large number of regulating services. The use of ecological production functions and spatial models (in this case, the InVEST model suite) makes estimation of regulating services feasible to tackle, albeit still not easy, as illustrated in this case study. For brevity, we look here at the regulating service that was found to be most important in Qinghai – sandstorm prevention (17.1% of the total value of GEP). For further information, Ouyang et al. (2020) includes full details of the approach taken to accounting and valuation of the other regulating services – carbon sequestration, soil retention, air purification, water purification and flood mitigation. Like water supply, sandstorm prevention is also a service produced by ecosystems in Qinghai that provide benefits to people living outside the province.

Just as water supply primarily benefits people living downstream, sandstorm prevention primarily benefits people living downwind. Ouyang et al. (2020) measure sandstorm prevention as the difference between wind erosion without vegetation cover and wind erosion under the current land cover pattern. They used the Revised Wind Erosion Equation (RWEQ) model to estimate sandstorm prevention, which combines empirical and process modelling, and has been tested in the field. The model was embedded in a Geographic Information System, providing spatial functionality, which is often needed for modelling of regulating services. The RWEQ estimates sand (and soil) loss as a function of several factors: weather, soil erodibility, soil crust, surface roughness and vegetation cover. This allows estimation of the transport of sand by wind without vegetation cover and with the current land pattern. The ecosystem service of sandstorm prevention is the difference.

To value the benefits of sandstorm prevention, Ouyang et al. (2020) estimate the benefits as reduction in the health costs of people living downwind of vegetation that prevents windborne sand and dust. The accounting price is estimated as the reduction in health costs from reduced exposure to wind-borne sand and dust, which is found by comparing the difference between potential exposure to windborne sand and dust with no vegetation and the actual exposure. A variety of assumptions must be made – for example, the size of the exposed population, prevailing winds, the proportion of the population that will become sick due to exposure to sand and dust, the health cost per person who becomes ill from that exposure, the days of exposure per year and so on. This example of valuation of a regulating service makes clear the importance of the underlying data and modelling assumptions, and the underlying science (anemology, hydrology, ecology) to understand how a specified change in land or water use or condition affects the provision of an ecosystem service.

It is perhaps unsurprising that water-related ecosystem services were found to constitute nearly two-thirds of the value of GEP for Qinghai, given they benefit downstream communities by providing a clean, reliable water supply. In 2000, GEP was higher than GDP (¥81.5 billion for GEP compared with ¥26 billion for GDP). This makes intuitive sense given GEP includes measures of the value of non-marketed ecosystem services and Qinghai provides water supply to downstream regions.

Cultural services are notoriously hard to value in monetary terms, but tourism and recreation can be relatively tractable. In Qinghai, eco-tourism was valued using information on the number of tourist visits and on-site surveys of visitors to three frequently visited eco-tourism sites: Beishan Forest Park, Kanbula National Forest Park, and Qinghai Lake. Data on the number of tourists in the province was collected in 2000 and 2015. Questionnaire-based surveys (462 respondents) were conducted at the three scenic locations to assess expenditure per trip to visit. Valuation was based on travel expenditure,[408] including a number of elements of cost, such as expenses at the recreation site (entrance fees etc.), travel expenses (tickets, fuel etc.), accommodation costs and the cost of time spent by the tourist to travel to the site (based on the visitor's salary). The findings are displayed in Table A12.1.

Table A12.1 Ecotourism Monetary Value in Qinghai (2015) (billion yuan)

Travel cost	Time cost	Total visitor cost
18.1	3.5	21.6

Source: Ouyang et al. (2020).

---

[408] The valuation was based on a zonal consumer cost model, a relatively simple approach to the travel cost method, which is described in further detail in the Supplementary Information to Ouyang et al. (2020). The authors acknowledge the limitations of this expenditure approach in approximating welfare benefits from outdoor recreation and eco-tourism, pursuing the travel cost method for its relative tractability.

Chapter 12: Valuing Biodiversity

Looking across estimates of the values of all ecosystem services, the authors found that large scale investment in restoration in the province led to an increase in ecosystem service values of 127.5% between 2000 and 2015 (Ouyang et al. 2020). This is encouraging in terms of justification for restoration and PES, of which there are many schemes in the province (Table A12.2).

Table A12.2 Eco-compensation Programmes in Qinghai 2010 to 2015 (billion yuan)

Eco-compensation programme	Compensation payments (billion yuan)
Sloping Land Conversion Program	4.234
Compensation for Ecological Benefits of Public Welfare Forest	3.649
Natural Forest Conservation Program	1.798
Three-North Shelter Forest Program	0.557
Wetland Ecological Compensation Program	0.218
Grassland Ecological Protection Subsidy Policy	9.736
Return Grazing Land to Grassland	2.871
Ecological Financial Transfer for Key Ecological Function Areas	9.721
Ecosystem Restoration of Qinghai Lake Basin Program	0.450
Sanjiangyuan Ecosystem Protection and Restoration Project	12.584
Qilianshan Ecosystem Protection and Ecological Construction	0.300
Total	45.819

Source: Ouyang et al. (2020). Supplementary Information Table S-10.

A12.1.3 Policy Use

GEP is a flow, not a stock. It therefore cannot be used to assess sustainability – it would be possible to increase the value of the flow of ecosystem services for a limited period, by running down stocks of natural assets. Alongside GEP, efforts are underway in China to track changes in the stock of natural assets measured mostly in biophysical rather than monetary terms to account for the depreciation (or appreciation) of those assets (Ouyang, Xu, and Xiao, 2017).

China's National Ecosystem Assessment enabled identification of areas that are important for ecosystem service provision (Ouyang et al. 2016). Sixty-three areas were identified as critical for provision of ecosystem services covered in the assessment and designated as 'key ecological function zones'. These areas provide a high proportion of important ecosystem services for China (Table A12.3) and have been made a priority for conservation and spatial planning. They cover 4.74 million km2, 49.4% of China's terrestrial land area, including forests, grasslands and watersheds. They were designated as 'restricted development zones' in the National Regional Development Strategy of China (UNEP, 2016; Ouyang et al. 2019) to conserve them as sources of important ecosystem services.

Table A12.3 Proportion of Ecosystem Services provided by Key Ecological Function Zones

Ecosystem service	Percent of ecosystem services provided
Carbon sequestration services	78%
Soil conservation services	75%
Sandstorm prevention services	61%
Water resource conservation service	61%
Flood mitigation services	60%
Natural habitat for biodiversity	68%

Source: Ouyang et al. (2019).

GEP is now being piloted in other provinces and at municipal and county levels in China to further develop the measure. Ouyang et al. (2020) document uses of GEP data among a variety of government agencies, including the Ministry of Finance, National Development and Reform Commission, the Ministry of Agriculture and the Ministry of Housing, as well as the Ministries one would expect – of Ecology and Environment, and Natural Resources. The combination of ecological information on stocks of national assets – through China's National Ecosystem Assessment and related ecological metrics – as well as estimates of the value of ecosystem services through GEP measures – are being used to guide investments in conservation and restoration of natural assets and to inform large-scale payments ('eco-compensation programmes') that compensate regions that provide ecosystem services (Ouyang et al. 2019, 2020). The government now requires consideration of ecological benefits, as measured by GEP, in the evaluation criteria of local governments' performance, which could create real accountability among officials for how they affect ecosystem services (Ouyang et al. 2020). However, given it focuses on ecosystem service flows, not stocks, GEP alone should not be used as a sustainability index.

Table A12.4 GEP Accounting in Qinghai (2000 to 2015)

(*Prices in billion yuan)

Types of services	Category of ecosystem services	Accounting items	2000 Biophysical quantity	2000 Monetary value*	2015 % of total value	2015 Biophysical quantity	2015 Monetary value*	2015 % of total value	2000–2015 (2015 constant price in billion yuan) Amount of change	2000–2015 (2015 constant price in billion yuan) % change	2000–2015 (Current price in billion yuan) Amount of change	2000–2015 (Current price in billion yuan) % change	Valuation method
Provisioning / material goods	Production of ecosystem goods	Agricultural crop production (x10³ t)	1652.1	1.0	1.2	3091.2	5.6	3.0	4.2	310.6	4.6	482.1	Market prices
		Animal husbandry production (x10³ t)	458.7	1.1	1.4	724	5.8	3.1	4.2	266.4	4.7	419.4	
		Fishery production (x10³ t)	1.2	0.01	0.01	10.6	0.3	0.1	0.3	2351.5	0.3	3375.0	
		Forestry production (x10³ t)	19.5	0.2	0.2	10.4	0.7	0.4	0.5	247.1	0.6	392.1	
		Plant nursery production (x10⁹ plants)	0.3	0.2	0.2	11	0.7	0.4	0.5	190.8	0.6	312.2	
	Total ecosystem goods			2.5	3.0		13.1	7.1	9.7	284.1	10.7	444.5	
	Water supply	Water use in downstream agricultural irrigation (x10⁹ m³)		11.8	14.5		15.0	8.1	−1.5	−9.3	3.2	26.8	Market prices for water
		Water use in households (x10⁹ m³)		4.3	6.5		13.8	7.4	6.4	86.5	8.5	160.4	
		Water use in industry (x 10⁹ m³)		19.4	23.8		29.2	15.8	2.2	8.1	9.8	50.5	
		Hydropower production (x 10⁹ kwh)	21.3	11.3	13.9	92	48.8	26.3	37.5	331.6	37.5	331.6	Market prices for electricity
	Total water supply			47.8	58.7		106.7	57.6	44.5	71.6	58.9	123.3	

(*Prices in billion yuan)		2000				2015				2000–2015 (2015 constant price in billion yuan)		2000–2015 (Current price in billion yuan)		Valuation method
Types of services	Category of ecosystem services	Accounting items	Biophysical quantity	Monetary value*	% of total value	Biophysical quantity	Monetary value*	% of total value		Amount of change	% change	Amount of change	% change	
Regulating services	Flood mitigation	Flood mitigation (x10^9 m^3)	0.07	0.02	0.03	0.07	0.03	0.02		0.001	2.3	0.01	45.0	Avoided water storage costs
	Soil retention and nonpoint pollution prevention	Retained soil (x10^9 t)	0.4	4.8	5.9	0.4	7.0	3.8		0.13	1.9	2.1	44.5	Avoided treatment costs
		Retained N (x10^3 t)	9.8	0.01	0.01	10	0.02	0.01		0.0003	1.9	0.01	103.9	
		Retained P (x10^3 t)	0.7	0.002	0.002	0.7	0.002	0.001		0.00004	2	0.00004	2	
	Water purification (wetland)	COD purification (x10^3 t)	33.2	0.02	0.03	104.3	0.1	0.1		0.1	214	0.1	528.0	
		NH-N purification (x10^3 t)	3.5	0.003	0.004	10	0.02	0.01		0.01	186.8	0.01	473.6	
		TP purification (x10^3 t)	-	-	-	0.9	0.003	0.001		-	-	-	-	
	Air purification	SO$_2$ purification	32.0	0.02	0.02	150.8	0.2	0.1		0.1	370.9	0.2	841.8	Avoided air filtration costs
		NO$_x$ purification	-	-	-	117.9	0.1	0.1		-	-	-	-	
		Dust purification	105.5	0.02	0.02	246	0.04	0.02		0.02	133.3	0.02	133.3	
	Sandstorm prevention	Sand retention (x10^9 t)	0.3	21.4	26.2	0.5	31.7	17.1		1.5	4.9	10.3	48.2	Avoided health costs
	Carbon sequestration	Carbon sequestration (x10^9 t)	0.01	2	2.4	0.02	4.7	2.5		1.9	67.4	2.7	137.3	Afforestation cost
	Total regulating services		28.3	34.7		43.9	23.7	3.9		9.8	15.6	55.3		
Cultural/ non-material services	Ecotourism	Tourists (x10^6 3.2 persons)	3	3	3.7	23.2	21.6	11.7		17.4	408.8	18.6	621.3	Travel expenditures
Overall total				81.5	100		185.4	100		75.5	68.8		127.5	

Source: Ouyang et al. (2019)

Chapter 13
Sustainability Assessment and Policy Analysis

Introduction

Humanity's future will be shaped by the portfolio of assets we inherit and choose to pass on, and by the balance we strike between the portfolio and the size of our population. Assets are durable objects, producing streams of services. Their durability enables us to save them for our own future, offer them as gifts to others, exchange them for other goods and services, and bequeath them to our children. Durability does not mean everlasting. Assets depreciate, but unlike services they are not fleeting. Perhaps because financial capital has figured prominently in economists' writings, the qualifier 'capital' is sometimes added to assets, as in 'capital assets'. Assets acquire their value from the services they provide over their remaining life (Chapter 1).

This chapter constructs a two-way classification of assets and identifies an inclusive measure of wealth based on them. It then establishes a series of propositions that reveal a deep connection between inclusive wealth and intergenerational well-being. We prove that any change in one is mirrored in a corresponding change in the other. The correspondence is named the *wealth/ well-being equivalence theorem*. We show that the theorem offers a way to judge whether the development path an economy is following meets the requirement of sustainability. The theorem is also shown to be the right criterion for evaluating public policy. Inclusive wealth is the coin with which economic progress or its absence should be measured. Private companies prepare balance sheets. The theorem shows why countries should now prepare wealth accounts.

13.1 Capital Goods

In common parlance, to say an object is an 'asset' is to convey the idea that it has positive worth; assets are taken to be goods. But one virtue of moving from the literary to the formal is that it allows us to extend the use of concepts so as to bring disparate objects under a common framework. It was noted in Chapter 2 that assets contribute not only directly to our well-being, but also indirectly, as sinks for pollution. It was shown that pollution should be viewed as depreciation of assets. The task is to estimate the depreciation. For example, economists have estimated the damage an additional tonne of carbon in the atmosphere is likely to inflict over time on agricultural production, coastal habitats, human health, and so on. It is an accounting price with a negative sign, so it is known as the 'social cost of carbon'. As noted previously, estimates differ widely, depending on the assumptions made. There are outliers as low as US$10 per tonne and as high as US$200 per tonne. The median value of the social cost of carbon used by regulatory agencies in the US is $40 per tonne. The figures are underestimates, because – for lack of quantitative estimates – none includes the negative impact global climate change will have on biodiversity, a link that has been studied extensively in Lovejoy and Hannah (2019).

Assets are often called 'capital goods'. But economists have typically confined the use of the term to assets that are material (i.e. tangible) and alienable (i.e. whose ownership is transferable) – a piece of furniture, an orchard, a lathe, and so on. One reason for the latter restriction has been the desire to work with assets that can be measured and compared with one another. It is not enough to say that houses can be measured in physical units (floor space, say), they need to be compared with other capital goods, such as cars. We need a common unit. Valuing assets is a way to do that (Chapter 12). Goods that are both material

Chapter 13: Sustainability Assessment and Policy Analysis

and alienable can be exchanged in markets. As a possibly crude, first approximation, market prices offer a measure of their accounting prices. But many natural assets fall outside the market system (Chapters 7–9). That explains why environmental and resource economists, more generally ecological economists, have tried to find ways to estimate accounting prices (Chapter 12). It explains particularly why estimates of the social cost of carbon appear prominently in the economics of climate change.

Of the myriad assets that make up the biosphere, land is perhaps the most familiar. It is often claimed that the market price of a hectare of agricultural land is a good approximation of its social value as a piece of land, but it could be that the land would have a greater social value if it were allowed to return to its original state as a forest. Valuing a watershed is harder, but again, it is manageable. Among the services it offers neighbouring communities is purifying water. One way to value that service is to estimate the cost of purifying that same flow of water by other means, for example, a water purification plant. As was noted in Chapter 2 (Box 2.7), studies have found that 'Nature-based' solutions for ecosystem restoration are often able to offer higher returns on investment than technological solutions involving investment in produced capital.[409] Of course, a watershed provides many other services, such as offering a habitat for biodiversity, whose value is not easily measurable. So, we would have an underestimate of its accounting price. Then there are objects in Nature that communities would rightly refuse to value (sacred groves, for instance; Chapter 12). Even when they are recognised to be biased, estimates of accounting prices can be useful. If a cost-benefit study recommends a moratorium on harvesting a particular species even when its accounting price is known to be in excess of the price used in the study, we can be sanguine that a moratorium is the right policy. In Annex 13.2 we provide such an example. That measurability is perforce incomplete is no argument for avoiding measurement.

What about a person's knowledge, skills, reputation, and state of health? They are non-alienable (knowledge, skills and reputation are also intangible). Nevertheless, contemporary economists include them on the list of capital goods by calling them 'human capital'. Economists use market wages and salaries to estimate the value of human capital to the individual possessing it.

The term human capital reminds us that assets can be ends, they can be means to ends, or they can be both. Reading is a pleasurable activity, but it is also necessary in a job that requires literacy. Similarly, a person's health is both a desired end for him and a means to employment, meaning that health should be a component of human capital. These examples suggest that valuing human capital on the basis of their market prices is to underestimate their worth. As noted above, working with biased estimates can nevertheless be revealing. If restoration of a wetland trumps investment in a sewage treatment plant when the wetland is valued only for the service it provides in purifying water, it would surely trump the plant even more convincingly if the other services that are provided by it were included (carbon storage, habitat for pollinators, and so on).

13.2 Enabling Assets

Once you include knowledge, skills, reputation, and health in the category of capital goods, it is hard to know where to stop. Should you not include institutions, such as the state, communities and the market system, and publicly available knowledge systems such as calculus? They too are assets. There is then the temptation to go for broke and speak of institutional capital, financial capital, knowledge capital, and cultural capital. Today some people even refer to religious capital. In Chapter 6, we built the idea of institutions on the broader notion of mutual trust and the social capital that embodies it. It may then seem reasonable to call all durable goods capital goods. As we see below, there are severe disadvantages in doing that.

---

[409] Chichilnisky and Heal (1998) provided an example of that by comparing the costs of restoring the Catskill watershed in the state of New York to the construction cost of a water purification plant for supplying potable water to New York City.

Although there are many ways of classifying assets, only one has proved useful in empirical work (we offered a sketch of the classification in Chapter 1). It is one thing to recognise that a durable object has worth; it is another thing to measure that worth (try, for example, to compare the value to a nation of 'good governance' with the real estate value of its capital city). The requirement that a durable good be measurable if it is to be called a 'capital good' was at the root of the complaints Arrow (2000) and Solow (2000) made of attempts to regard social capital on par with buildings, roads, labour and land. We take heed of their caution in the Review by creating a three-way partition of *capital goods*: *produced capital* (buildings, roads, ports, machines, instruments), *human capital* (population size, health, education, reputation, knowledge and skills) and *natural capital* (ecosystems, sub-soil resources). What remains – institutions and practices, more generally, social capital, and publicly available knowledge – we will call *enabling assets*, because they confer value to the three classes of capital goods by facilitating their use.[410]

Knowledge in the sciences, technologies, and the arts and humanities – they are all enabling assets – are created and acquired, but they are created and acquired by people (human capital) in combination with produced capital (books, laboratories and equipment) and natural capital (raw material). Of course, institutions are also created by people, as is mutual trust, which is the glue that holds communities together (Chapter 6). We study the real economy in the Review. Financial capital facilitates exchange (among people and across time), so it too is an enabling asset (Chapters 20). In our classification, there are thus three categories of capital goods and a wide range of enabling assets. In Chapter 2, a parallel was drawn between biodiversity in ecosystems and what we are now calling enabling assets in social systems. Well-functioning institutions lubricate societies and enable them to thrive. Enabling assets are not always usefully measurable, but that does not matter, for they enable human societies to function healthily; and these functions can be measured.[411]

Management of a society's portfolio of assets involves *policy analysis* and *sustainability assessment*. We call the two generically as *economic evaluation*. This chapter explains the distinction between the two and shows how they should be undertaken. Our classification of assets into capital goods and enabling assets proves useful in developing a quantitative language for the exercise. Because assets are central to our well-being, we should expect to find a numerical index for them which is useful in economic evaluation. Below we show not only that there is such an index, but also identify it as an inclusive measure of *wealth*. For simplicity of exposition, we consider national economies, closed to trade.[412]

Box 13.1
The UK Government's Green Book

The UK government's guidance for conducting policy appraisal, known as the 'Green Book', was updated in 2018 to include considerations of natural capital (HM Treasury, 2018). It advises that for any policy that is likely to have significant impacts on the natural environment, changes to stocks of natural capital should be considered, as well as possible ecological tipping points. This means in

---

[410] Here we are following the classification in Dasgupta and Mäler (2000). Of the three categories of capital goods, produced capital has the oldest pedigree, being the substance of Classical Political Economy of the 18th and 19th centuries. (Land of various qualities served as an additional, indestructible factor of production.) The paper that placed human capital in the category of capital goods and created the modern literature on human capital is Schultz (1961). A huge contemporary literature in ecology and environmental economics has introduced natural capital in economic reasoning.

[411] There is an extensive literature using cross-country data to show that enabling assets, such as social capital, contribute to economic performance. The literature has mostly studied their contribution to Gross Domestic Product (GDP) growth. See, for example, Jones and Romer (2010), who find that the quality of institutions matters.

[412] Introducing international trade requires minor modifications to the terms used to define variables (e.g. Gross National Product (GNP) rather than GDP, domestic versus foreign assets, and so on).

Chapter 13: Sustainability Assessment and Policy Analysis

addition to possible losses in ecosystem services from a specific policy intervention, the consequent degradation of renewable assets such as losses in woodland or fishery stocks should be assessed. The Green Book also says natural capital stock levels should be systematically measured and monitored.[413] Interestingly, it recommends that the cumulative effect of multiple investment decisions upon natural capital should also be taken into account, including whether the ecosystems that are likely to be affected are near their tipping point.[414] It also includes provisions for using subjective well-being among policy objectives (Chapter 11).

The Natural Capital Committee (2020) found that across a sample of UK government impact assessments there was little evidence of the new updates in the Green Book being used to inform environmental policies. But then, these are early days – the revisions date back only to 2018.

Biodiversity is particularly difficult to value. Recognising this, some recommend the use of 'look-up' values, as is common with carbon emission costs, to allow for initial or incremental assessments of biodiversity impacts. Others (Bateman et al. 2013; Bateman and Mace, 2020) have proposed using biodiversity as an unvalued constraint on decision-making. These methods are second-best substitutes for more detailed assessments where impacts are large (Chapters 5, 7, 12).

13.3 The Idea of Inclusive Wealth

The accounting value of an economy's stock of capital goods is its *inclusive wealth*. The qualifier signals that the notion of wealth adopted here differs from the one in common use in two ways: (i) accounting prices are not necessarily market prices; and (ii) in addition to produced capital, wealth includes human capital and natural capital. As accounting prices measure the social worth of goods and services, the inclusive wealth of a nation is the social worth of its capital goods. As noted previously, by 'social worth' we mean not only the worth to people who are alive at that date, but also to future people.[415] In Chapter 10, we named the person engaged in economic evaluation the *social evaluator*. She evaluated options on behalf of society. To do that, she needs a criterion with which to compare options. Inclusive wealth will be shown to be the criterion she needs. As comparison of options is at the heart of asset management, we also call her the *Citizen Investor*.

To establish the salience of inclusive wealth in economic evaluation, it helps to study accounting prices more closely than we have so far. The inclusion of human capital in inclusive wealth says, for example, that personal characteristics of the individual with access to a piece of capital good matter. A piano in the possession of someone who can play the piano has a higher accounting price than it would have were it to be in the possession of someone innocent of music, other things equal. An individual's well-being is shaped by the extent to which her projects and purposes are realised. They in turn are rooted in her engagements, both with her own self and with others (Chapters 7–11), which means accounting prices of capital goods are person-specific. It also means that the degree of fairness in the distribution of well-being influences accounting prices. A practical way to record that influence is to apply distributional weights to personal wealth. Inclusive wealth is then a weighted sum of different forms of personal wealth, the weights that are awarded being larger the smaller is someone's wealth.[416] So as to avoid repetition, we will sometimes drop the qualifier 'inclusive' from inclusive wealth.

---

[413] Section 6.48: 45.

[414] Section A2.3: 61.

[415] For expositional ease, we also eschew risk and uncertainty. Chapters 5 and 10 provided the tools to use for measuring wealth when the future is uncertain. The crudest approximation would be to interpret economic variables by their expected values.

[416] Dasgupta, Marglin, and Sen (1972) developed the use of distributional weights in their work on social cost-benefit analysis of investment projects in developing countries.

An economy's institutions and practices endow its capital goods with their social worth, which is why we are calling them enabling assets. That means a society does not have to rely on accumulating capital goods in order to increase inclusive wealth. It could raise inclusive wealth by simply bringing about such changes to its institutions and practices that create greater trust among people. Those changes would express themselves through an altered set of accounting prices for the same portfolio of capital goods; a writing desk, for example, has a higher accounting price in someone's study than in a war zone. An economy can become wealthier simply by improving the quality of its enabling assets.

Accounting prices can be defined only when a socio-ecological future has been specified. The set of possible socio-ecological futures in period t was denoted in Chapter 10 by $\Re^*(t)$. $\Re^*(t)$ is the set of all future possibilities in a second-best world. For concreteness, socio-ecological futures were called consumption streams. Our social evaluator's task is to compare the social worth of the consumption streams in $\Re^*(t)$. She considers an arbitrary consumption stream, a *reference stream*, and estimates accounting prices of goods and services on it. Inclusive wealth in any period on the reference stream is the accounting value of the economy's stock of capital goods.

13.4 Inclusive Wealth and Intergenerational Well-Being

Why should we be interested in inclusive wealth? The reason is this:

Assets in their totality are the factors that determine intergenerational well-being. Moreover, accounting prices of capital goods measure the (marginal) contribution they make to intergenerational well-being (Chapter 10). We should therefore expect an intimate connection between inclusive wealth and intergenerational well-being. In Box 13.2, we demonstrate:

Wealth/Well-Being Equivalence Theorem – A small perturbation to an economy increases well-being across generations if and only if it is associated with an increase in inclusive wealth.

The theorem says that although inclusive wealth and intergenerational well-being are not the same entity, they move in step with each other; there is perfect correspondence between the two. Wealth and well-being are two sides of the same coin.[417]

We are talking of ends and means here. Despite the equivalence theorem, ends (enhancing well-being across generations) are the right starting place for economic evaluation. Ends are the right starting point because they are antecedent to means. One can articulate ends even without asking whether they can be realised, but it makes no sense to talk of means if the ends they are meant to advance are not articulated first. The wealth/well-being equivalence theorem does not deny the antecedence of ends; what the theorem says is that if the means to a set of ends have been identified, it does not in principle make any difference whether we examine the extent to which the ends have been (or are likely to be) furthered by a change to an economy or whether we estimate the degree to which the means to those ends have been (or are likely to be) bolstered by that change: the two point in the same direction. The wealth/well-being equivalence theorem draws attention to the fact that no matter what conception of ends the Citizen Investor may adopt, the source of the means to those ends lies in a society's capital goods. Their accounting prices serve to tie them to the ends. The theorem says, for example, that weapons and human capital deployed to abduct citizens from their homes in the middle of the night are to be awarded very large, possibly unboundedly large, negative accounting prices. Likewise, steel put to use in making ploughs

---

[417] Wealth is the dynamic counterpart of income. The welfare significance of national income was explored by Hicks (1940), Samuelson (1961), Mirrlees (1969) and Sen (1976), among many others. As the authors confined themselves to perturbations of stationary states, their findings have no empirical import. Only Samuelson addressed the problems a dynamic economy poses for the national accountant. In the final page of his article, Samuelson speculated that something like a wealth index is needed for economic evaluation but provided no argument. The wealth/well-being equivalence theorem furnishes the argument.

Chapter 13: Sustainability Assessment and Policy Analysis

differs from steel used to manufacture guns. Accounting prices of capital goods depend on their location and the use to which they are put.

The equivalence between inclusive wealth and intergenerational well-being holds as tightly in a society where the ends are far from being met owing to misallocation of the means or unjustified usurpation of the means by the powerful, as it would in a society where they are met as far as is possible under the prevailing scarcities of the means. The equivalence theorem is utterly wide in its reach.

That inclusive wealth is equivalent to well-being across generations is not an empirical law, it is an analytical proposition. Being an equivalence relationship, it does not say whether a society is doing well or badly, whether it is well governed or badly governed. But both theory and experience say that it is commonly easier to measure the means to the ends than it is to measure the ends themselves, which is why in empirical work we are drawn to the means.

There is a subtler reason for evaluating policies and assessing the progress or regress of economies in terms of the means to our ends. It is a mistake to suppose we come armed with our ends. For the most part, they are inchoate in our minds. Studying means clarifies the trade-offs involved in promoting our ends. The interplay of the 'is' and the 'ought', to use that well-worn distinction, helps us to better understand our ends.[418]

Box 13.2
Wealth and Well-Being: The Formal Connection

B13.2.1 Theory

In Chapter 10, a broad version of Utilitarianism was used to show that social well-being in any period t (we called it intergenerational well-being previously to emphasise the dynamic setting) has the form (Equation (10.2))

$$V(t) = {}_{s=t}\Sigma^{\infty}\{[\Omega(s)N(s)u(s)]/\Omega(t)\} \tag{B13.2.1}$$

In equation (B13.2.1), $N(s)$ is population size in period s, $u(s)$ is average well-being in the population in period s and $\Omega(s)/\Omega(t)$ is the probability that the economy will survive until period s conditional on it having survived until period t.

The wealth/well-being equivalence theorem holds for *any* socio-ecological future in $\Re(t)$. But to state and prove the theorem, we have to peer more closely into $V(t)$.

Human capital is embodied in people. Each individual person possesses a unique stock of human capital. For any period, s we let j denote persons, which means j takes numerical values from 1 to $N(s)$. Let $c_j(s)$ be consumption by j in period s. $u(c_j(s))$ is then j's well-being in s. The well-being of generation s can then be expressed as ${}_{j=1}\Sigma^{N(s)}[u(c_j(s))]$, correspondingly $V(t)$ as

$$V(t) = {}_{s=t}\Sigma^{\infty}\{{}_{j=1}\Sigma^{N(S)}[\Omega(s)u(c_j(s))]\}/\Omega(t) \tag{B13.2.2}$$

The expression for $V(t)$ says that the human capital of every individual taken together sums to total human capital.

In Chapter 10, it was shown that for any future projections of an economy's enabling assets, social well-being in period t is a function of the economy's stock of capital goods in the period. For the sake

---

[418] The remarks in Chapter 10 on 'pragmatism' are relevant here.

of evenness in our treatment of capital goods, we choose $V(t)$ to be the numeraire in period t. All accounting prices are relative to the numeraire. In order to present the equivalence between intergenerational well-being and inclusive wealth in the most economical way, we suppress the index for time. For that reason, we use the time-neutral term 'social well-being' to represent well-being across the generations.

Capital goods are numbered by the index i, which takes the values 1,2. We note again that i ranges over all items of produced capital, human capital and natural capital. Let $K_i(t)$ be the stock of capital good i in period t and $p_i(t)$ be i's accounting price in well-being units. The accounting value of i is $p_i(t)K_i(t)$. Inclusive wealth is defined as

$$W(t) = {}_i\Sigma[p_i(t)K_i(t)] \tag{B13.2.3}$$

Imagine now that the economy experiences a small perturbation. Denote the perturbation by Δ. Two extreme classes of perturbations are to be noted (intermediate cases follow readily as a mixture of the two). One involves perturbing the economy's enabling assets while holding the quantities of capital goods fixed, the other involves perturbing the deployment of capital goods while keeping the enabling assets as they are. The former implies that the perturbation to inclusive wealth is

$$\Delta W(t) = {}_i\Sigma[K_i(t)\Delta p_i(t)] \tag{B13.2.4}$$

The latter implies that the perturbation to inclusive wealth is

$$\Delta W(t) = {}_i\Sigma[p_i(t)\Delta K_i(t)] \tag{B13.2.5}$$

Here we focus on the latter type of perturbation.

Denote by $\mathbf{K}$ the vector of capital goods; that is, $\mathbf{K} = \{K_1, \ldots, K_i,\}$. We know from Chapters 10 and 12 that social well-being can be expressed as $V(\mathbf{K})$. The structure and character of the economy's enabling assets are taken to be implicit in the V-function.

The perturbation to the stock of capital goods causes $V(t)$ to change by

$$\Delta V(\mathbf{K}(t)) = {}_i\Sigma\{[\partial V(\mathbf{K}(t))/\partial K_i(t)]\Delta K_i(t)\} \tag{B13.2.6}$$

By definition of accounting prices (Chapter 10),

$$p_i(t) = \partial V(\mathbf{K}(t))/\partial K_i(t) \tag{B13.2.7}$$

Using equation (B13.2.7) in equation (B13.2.6) yields

$$\Delta V(\mathbf{K}(t)) = {}_i\Sigma[p_i(t)\Delta K_i(t)] \tag{B13.2.8}$$

Equation (B13.2.3) and (B13.2.8) together imply that the change in social well-being resulting from the small perturbation to the economy equals the corresponding change in the economy's inclusive wealth *at unchanged accounting prices*:

$$\Delta V(\mathbf{K}(t)) = {}_i\Sigma[p_i(t)\Delta K_i(t)] = \Delta W(t) \tag{B13.2.9}$$

Equation (B13.2.9) is the wealth/well-being equivalence theorem.

A similar argument leads to the wealth/well-being equivalence theorem when the perturbation involves a small change in the structure and character of the economy's enabling assets (Equation (B13.2.4)).

Chapter 13: Sustainability Assessment and Policy Analysis

> The theorem holds for small perturbations. It can be extended to cover large perturbations by summing the small perturbations on both sides of equation (B13.2.9). The sum incorporates changes to the set of accounting prices that must be taken into account when the perturbation is large.
>
> ### B13.2.2 Simplifications in Practical Work
>
> $V(t)$ is an expression for aggregate well-being across the generations. In most discussion on public policy, attention is paid today to the well-being of the *average* person. In practical applications of the wealth/well-being equivalence theorem, economists have presented their findings in terms of *inclusive wealth per capita* (see below). The move requires justification.
>
> From equation (B13.2.2), well-being of the average person across the generations is found to be
>
> $$V(t)/{}_{s=t}\Sigma^{\infty}[\Omega(s)N(s)] = {}_{s=t}\Sigma^{\infty}\{{}_{j=1}\Sigma^{N(s)}[\Omega(s)u(c_j(s))]\}/\{\Omega(t){}_{s=t}\Sigma^{\infty}[\Omega(s)N(s)]\} \qquad (B13.2.10)$$
>
> Cross-national studies are hampered by a lack of data on the distribution of well-being. The above average can therefore be simplified if we replace the sum of individual well-being in a period (s) by the product of the population number ($N(s)$) and the average well-being in that period ($u(s) = {}_{j=1}\Sigma^{N(s)}[u_j(s)]/N(s)$). Formally, the average person is taken to be the person that matters, his consumption $c(s)$ being the generation's average, that is, $c(s) = {}_{j=1}\Sigma^{N(s)}[c_j(s)]/N(s)$. In that case, equation (B13.2.10) reduces to
>
> $$V(t)/{}_{s=t}\Sigma^{\infty}[\Omega(s)N(s)] = {}_{s=t}\Sigma^{\infty}\{\Omega(s)N(s)u(c(s))]\}/\{\Omega(t){}_{s=t}\Sigma^{\infty}[\Omega(s)N(s)]\} \qquad (B13.2.11)$$
>
> Write average well-being across the generations in t as $V_{av}(t)$ and inclusive wealth per capita as $W_{av}(t) = {}_i\Sigma[p_i(t)K_i(t)]/N(t)$. Arrow, Dasgupta, and Mäler (2003) showed that if population numbers change at a constant rate over the period studied, then $V_{av}(t)$ and $W_{av}(t)$ enjoy the same equivalence relationship as $V(t)$ and $W(t)$ do in the wealth/well-being equivalence theorem. This version of the theorem has been the basis of applied work on sustainable development (see Arrow et al. (2012, 2013; UNU-IHDP and UNEP (2012, 2014); and Managi and Kumar (2018). We summarise their findings in Box 13.6 and Annex 13.1. Yamaguchi (2018) has studied the differences between conceptualising intergenerational well-being in its total and its population-average forms.

13.5 Inclusive Wealth and the Substitutability of Capital Goods

An economy's inclusive wealth is a weighted sum of the stocks of all the capital goods it possesses (Equation (B13.1.1)). The weights are their respective accounting prices. The index is a linear function of the stocks. That feature may give the impression (e.g. Daly et al. 2007) that the wealth/well-being equivalence theorem is built implicitly on the assumption that the various capital goods are perfect substitutes for one another in production.

The impression is mistaken. Nowhere in the derivation of the wealth/well-being equivalence theorem was there any mention of the substitutability of one capital good for another in production or consumption, for example, whether a piece of produced capital is a substitute for a piece of natural capital. But as equation (B13.1.5) makes clear, accounting prices are themselves functions of the stocks of capital goods. The extent to which various capital goods substitute for one another in production is reflected in the structure of accounting prices. And there may be little-to-no substitution possibilities between key forms of natural capital and produced capital, or for that matter any other form of capital. Imagine, for example, that a tropical forest has been so degraded that it has come perilously close to its tipping point (Chapters 3 and 5). Crossing it would lead to massive climate change. The accounting value of the forest as it entered the uncertain zone would be so large that any further diminution of its stock would lower inclusive wealth dramatically.[419]

13.6 Six Questions

The meaning of economic evaluation (i.e. policy analysis and sustainability assessment) may be intuitively clear, but to nail it down we list six questions the social evaluator frequently asks about the economy she is interested in. The questions are:

(1) How is the economy doing?

(2) How has it been doing in recent years?

(3) What would be our forecast of the future if policies and institutions evolve in the way we expect them to evolve?[420]

(4) How is the economy likely to perform under alternative policies?

(5) Which policies should we support?

(6) What would be an ideal set of policies?

The questions prompt the social evaluator to make *two* types of comparisons (Chapter 1):

Questions (1)–(3) require that she assesses whether an economy is on a path of *sustainable development*; that is, she evaluates a change that has been or is likely to be experienced by an economy *as it moves through* time. We call that evaluation exercise *sustainability assessment*. An example of the kind of question she asks is: "Are the prospects people have for improving their lives and the lives of their descendants better now than they were a decade ago?"

Questions (4)–(6), on the other hand, prompt the social evaluator to engage in *policy analysis*. There she evaluates the economic change that would be brought about by a proposed shift in policy *at a point in time*. An example of the kind of question she asks in policy analysis is: "Is the wetland restoration project being proposed likely to increase well-being?"

Question (3) is the socio-ecological forecast round which both sustainability assessment and policy analysis are conducted.

The wealth/well-being equivalence theorem spoke to the equivalence of inclusive wealth and intergenerational well-being in economic evaluations of perturbations experienced by an economy. We now put flesh to those perturbations by first studying sustainability assessment (Sections 13.7–13.9) and then policy analysis (Sections 13.10–13.11). There we also extend the theorem to include large perturbations.

Inclusive Wealth and Sustainability Assessment

13.7 SDGs and the Idea of Sustainable Development

In September 2015, the United Nations General Assembly agreed on an agenda for sustainable development in member countries. Nations committed themselves to meeting 17 Sustainable

---

[419] This and other common misconceptions of the wealth/well-being equivalence theorem are discussed in Arrow et al. (2007) and Arrow et al. (2013).

[420] In Chapter 10, where the focus was on a broad conception of consumption, we named such a socio-ecological future the 'reference consumption stream'.

Chapter 13: Sustainability Assessment and Policy Analysis

Development Goals (SDGs) (see attached chart) by year 2030. Reasonably, the quantitative targets to be met do not distinguish ends from means. Thus, Goals 1–6 should be taken to be ends, whereas Goals 7, 9 and 17 are means to ends, and Goal 8 reflects both ends and means. The SDGs involve 169 socio-economic targets. To measure progress in meeting those targets, it was proposed to track more than 240 socio-economic indicators over the coming years (United Nations, 2015).[421]

Figure 13.1 17 Sustainable Development Goals (SDGs) Adopted by All United Nations Member States in 2015

Source: United Nations (2015).

International agreement on the SDGs was a remarkable, even noble, achievement, for the Goals unpick necessary features of lives that can in principle be well-lived. But there is a problem. The Goals are not accompanied by an analysis examining whether once they are achieved they are sustainable.

The Brundtland Commission (1987) defined sustainable development as "development that meets the needs of the present without compromising the ability of future generations to meet their own needs". The requirement is that, relative to their respective demographic bases, each generation should bequeath to its successor at least as large a productive base as it had inherited from its predecessor. If it were to do so, economic possibilities facing the successor would be no worse than those the generation faced when inheriting the productive base from its predecessor.

The wealth/well-being equivalence theorem suggests that by an economy's productive base we should mean its inclusive wealth. We now show that the suggestion follows directly from the theorem.

---

[421] COVID-19 will no doubt cause the United Nations to revisit the SDGs. That the target date for the Goals, 2030, was unfeasible even before the pandemic struck was shown in Chapter 4 (Box 4.6).

13.8 Economic Progress as Growth in Inclusive Wealth

Although the Brundtland Commission defined sustainable development in terms of an economy's productive base as it moved through time, the intention was to base the idea on the sustainability of intergenerational well-being. For, without the latter as the object of interest (the ends), there is no reason why a society should bother about its productive base (the means). So, we shall say an economy enjoys sustainable development over a period of time if intergenerational well-being does not decline over the period. Whereas the Brundtland Commission spoke of sustainable development in terms of the means, we are defining it in terms of the ends. The wealth/well-being equivalence theorem assures us that the two ways of speaking amount to the same thing. To confirm that, let us interpret the perturbation to which the economy is subjected as the passage of time. We then have (Box 13.3 provides the proof):

Sustainable Development Theorem – Intergenerational well-being increases over a period of time if and only if inclusive wealth increases over that same period of time.

The theorem implies that inclusive wealth is the right measure of intergenerational well-being, not GDP nor any of the other measures that have been suggested in recent years, such as the United Nations' Human Development Index (HDI).[422] Private companies produce annual balance sheets. A country's inclusive wealth accounts would be its national economy's balance sheet. To date, only a few countries have begun preparing balance sheets.

Box 13.3

Sustainable Development as Growth in Inclusive Wealth

To prove the Sustainable Development Theorem, it is simplest to assume that time is a continuous variable. The stock of capital good i at time t is $K_i(t)$. Define the small perturbation to the economy to be a brief interval of time $[t, t+\Delta]$. We may then express equation (B13.2.4) as

$$V(\mathbf{K}(t+\Delta)) - V(\mathbf{K}(t)) = {}_i\Sigma\{[\partial V(\mathbf{K}(t))/\partial K_i(t)] (\Delta K_i(t)/\Delta t)\Delta t \qquad \text{(B13.3.1)}$$

Divide both sides of equation (B13.3.1) by Δt and recall equation (B13.2.5). Taking Δt to zero then yields

$$dV(\mathbf{K}(t))/dt = {}_i\Sigma p_i(t)dK_i(t)/dt \qquad \text{(B13.3.2)}$$

Equation (B13.3.2) is the formalisation of the Sustainable Development Theorem and its two corollaries in continuous time. The equation holds so long as the passage of time is brief. The argument can be extended to cover non-negligible periods of time by summing the small perturbations in both sides of equation (B13.3.2). The sum incorporates changes to the set of accounting prices that must be taken into account when the period being covered is not brief.

---

[422] The UN's HDI (UNDP, 1990) is a linear aggregate of GDP per capita, life expectancy at birth (human capital) and literacy (human capital). The two equivalence theorems were established in a general setting by Dasgupta and Mäler (2000) and Arrow, Dasgupta, and Mäler (2003a-b). Dasgupta (2004) is a book-length treatment of the idea of sustainable development and the role that the wealth/well-being equivalence theorem plays in both policy appraisal and sustainability assessment. Agliardi (2011) extended the theorem by introducing a stationary stochastic process driving consumption. Arrow et al. (2004) studied the connections between sustainable development and optimum development. Dasgupta (2014) and Irwin, Gopalakrishnan, and Randall (2016) are non-technical accounts of the wealth/well-being equivalence theorem and its extensions. Arrow et al. (2012, 2013) applied the theorem by estimating movements in inclusive wealth in five countries (Brazil, China, India, US and Venezuela) over the period 1995 to 2000. Yamaguchi, Sato, and Ueta (2016) applied the theorem to a region in Japan. UNU-IHDP and UNEP (2012, 2014) and Managi and Kumar (2018) contain pioneering estimates of the wealth of more than 120 countries, akin to estimates the World Bank provides annually of the GDP of nations. Tomlinson (2018) has applied the theorem to the recent economic history of Nigeria. The theorem has also been used to motivate the development of methods for estimating accounting prices.

The Sustainable Development Theorem can also be stated in terms of inclusive investment. In common parlance as well as in the language of accountancy, investment in a capital good net of depreciation equals the rate at which the capital good grows.[423] We then have a corollary that asserts the equivalence between sustainable development and positive inclusive investment:

Corollary 1 (Sustainable Development as Positive Investment) – Intergenerational well-being increases over a period of time if and only if net inclusive investment is positive over that same period.

Box 13.4

Composition of Inclusive Wealth

Fenichel et al. (2016) have estimated that groundwater withdrawal from the Kansas High Plains Aquifer during 1996 to 2005 amounted to an annual loss in wealth of the state of Kansas, USA, by some US$110 million (2005 prices). The authors noted that the figure is substantially more than the state's annual investment in schools during the period. The loss in wealth were not recorded in Kansas' economic statistics.

To illustrate the Sustainable Development Theorem more generally, suppose an economy in a given year has invested, in accounting prices, US$40 billion in produced capital, spent US$20 billion on salaries for employees in education, and depleted and degraded its natural capital by US$70 billion. The economy's System of National Accounts (SNA) would record the US$40 billion as investment ('gross capital formation'), the US$20 billion as consumption, and, if the market prices of natural capital were zero, remain silent on the US$70 billion (accounting dollars) of loss in stocks of natural capital. The theorem says that in contrast the US$20 billion expenditure on education is=investment in human capital and the US$70 billion is depreciation of natural capital. It would also say that owing to the depreciation of natural capital, the economy's wealth will have declined over the year by US$10 billion; and that is before taking note of the depreciation of produced capital and human capital. The moral we should draw is that development was unsustainable that year.

The Sustainable Development Theorem can also be cast in terms of familiar national accounts. Inclusive investment net of depreciation is equal to Net Domestic Product (NDP) minus aggregate consumption. We then have:

Corollary 2 (Sustainable Development and NDP) – Intergenerational well-being increases over a period of time if and only if aggregate consumption is less than GDP minus depreciation of capital goods.

The intuitive reason behind Corollary 2 is that if aggregate consumption is less than NDP over a period of time, net investment in productive capacity is positive in that same period of time. That in turn means that the productive capacity of the economy expands during the period. Below we show that Corollaries 1 and 2 are the foundations of policy analysis as well.

The criterion for sustainable development over a period of time is not that the economy's NDP increases over the period, rather, it is that consumption does not exceed NDP. Expansion of the productive base points to net inclusive investment as being the right index with which to judge the progress or regress of economies.

The move from national income accounts to national balance sheets would be far reaching. National income accounts are built on *flows*, whereas national balance sheets would report *stocks*. Flows do not

---

[423] Negative net investment would amount to decumulation of the capital good.

have the future built into them; stocks do. Our engagement with our descendants involves the transmission of stocks to them, not flows (other than as returns on stocks). The concentration on flows has proved to be an impediment for the economics of biodiversity. Box 13.4 provides an example.

Box 13.5

Trade, Externalities, and Wealth Transfers

The Sustainable Development Theorem tells us also to curb our enthusiasm for free trade in a distorted world. To illustrate why, it proves useful to return to the example we studied in Chapter 7 (Box 7.1). Imagine that timber concessions have been awarded in an upstream forest of a low income country by its government so as to raise export revenue. As forests stabilise both soil and water flow and are a habitat for birds and insects (these are 'regulatory' and 'provisioning' goods; Chapter 2), deforestation erodes soil and increases water runoff downstream, and reduces pollination and pest control in nearby farms. If the law recognises the rights of those who suffer damage from deforestation, the timber company would be required to compensate downstream farmers. But compensation will not be forthcoming unless the government insists, because Coase's theorem (Chapter 7) does not apply. Problems are compounded because damages are not uniform across farms, their geography matters. Moreover, downstream farmers may not even realise that the decline in their farms' productivity is traceable to logging upstream. In those circumstances, the timber company's operating cost would be less than the social cost of deforestation (the latter, at least as a first approximation, would be the firm's logging costs and the damage suffered by all who are adversely affected). So the export would contain an implicit subsidy (the 'externality'), paid for by people downstream. And we have not included forest inhabitants, who now live under even more straightened circumstances or may have had to leave their home. *The subsidy is hidden from public scrutiny, but it is real and amounts to a transfer of wealth from the exporting to the importing country.* Ironically, some of the poorest people in the exporting country would be subsidising the incomes of the average importer in what could well be a rich country. Compensation to displaced inhabitants and downstream farmers, paid for by a tax on the harvesting of timber, would be the appropriate policy. Trade is discussed further in Chapter 15.

13.9 Total Factor Productivity Growth

A standard method for accounting for technological change in empirical work in growth and development economics is to construct a macroeconomic model of the economy and estimate the percentage rate of change in total factor productivity (TFP). Solow (1957) famously named TFP growth the 'residual'.[424]

Contemporary estimates of the residual should be treated with scepticism. They are based on growth models that do not include natural capital as factors of production. If the rate at which natural capital is degraded was to increase over a period of time, growth rates of TFP obtained from regressions would be overestimates. The reason is that the regressions would read environmental degradation as increases in knowledge and improvements in institutions. Worse still, the greater was the under-coverage of natural capital, the larger would be the bias in the estimate of TFP growth. By plundering Earth, TFP could be raised by as much as we like.

---

[424] Arrow et al. (2012) estimated (Annex 13.1) that growth in inclusive wealth per capita in India was, in the period 1995–2000, in large measure a consequence of TFP growth.

Chapter 13: Sustainability Assessment and Policy Analysis

Theory guides and helps to shape empirical research. These are early days in the preparation of inclusive wealth accounts. But it is sobering to realise though that 60 to 70 years ago estimates of national incomes were subject to uncertainties of a magnitude people are minded to think no longer exists in current estimates. The fact is, of course, that we take contemporary estimates of national incomes too much at face value. Official estimates, most especially in low income countries, are reticent to speak of the proportion of incomes that go unrecorded. Estimates of transactions falling outside the market system or operating within a black-market system suggest that the errors in official estimates of national incomes are substantial.

In recent years, there have been attempts to track the progress or regress of nations using the wealth/well-being equivalence theorem. Arrow et al. (2012, 2013) have tried to track changes in inclusive wealth per capita in five countries (Brazil, China, India, US and Venezuela), selected on the basis of their stage of development and of their resource base) over the period 1995 to 2000. Annex 13.1 contains an account of their findings for India.

Managi and Kumar (2018) have estimated changes in inclusive wealth per capita in 140 countries over the period 1990 to 2014. We reported their work briefly in Chapter 4. Box 13.6 presents a summary of their findings when averaged across all countries on their list.

Box 13.6
Global Inclusive Wealth Change 1992 to 2014

In their *Inclusive Wealth Report 2018*, Managi and Kumar (2018) tracked inclusive wealth per head over the period 1990 to 2014 in 140 countries. The value of produced capital (PC) was obtained from official national accounts. Data limitations meant that natural capital (NC) was limited to minerals and fossil fuels, agricultural land, forests as sources of timber and fisheries. Market prices were used to value them as well. The accounting value of human capital (HC) was estimated by using the approximations in Arrow et al. (2012) for both education and health. Adjustments were then made to the values of the three classes of capital goods by including population growth, the social cost of carbon emissions (at US$50 per tonne), capital gains on oil, and official estimates of technological progress. The aggregate of those estimates yielded the measure of inclusive wealth per capita.

Figure 13.2 depicts time series of global per capita accounting values of the three classes of capital goods over the period 1992 to 2014. Figure 13.2(a) depicts time series of global accounting values of per capita produced capital, per capita natural capital, and per capita human capital,[425] respectively. Globally, produced capital per head doubled and human capital per head increased by around 30%,[426] but the value of the stock of natural capital per head declined by nearly 40%.

When adjustment for capital gains on oil (for oil-importing countries the corresponding term was a negative contribution to inclusive wealth; for oil-exporting countries the contribution was positive) and technological progress are included, global inclusive wealth per head was found (Figure 13.2(b)) to have grown by about 44% (implying an annual rate of increase of 1.8%), in contrast to GDP per capita, which grew by nearly 80% (i.e. an annual increase of 4%).

Strikingly, only 84 out of the 140 countries were found to have experienced non-declining inclusive wealth per head during the period in question. Moreover, in most countries (and they included both developed and developing economies) the share of the accounting value of produced capital per

---

[425] Note *education approach* estimates for human capital are referred to here and differ to estimates using the *frontier approach* shown in Chapter 4. See Managi and Kumar (2018).

[426] Note this is around 13% when using the *frontier approach* (Chapter 4). See Managi and Kumar (2018).

head in global inclusive wealth per head, and to a lesser extent that of human capital per head, increased, even while the share of the accounting value of natural capital per head declined (Figure 13.2(c)).

Figure 13.2 Growth rates of Inclusive Wealth Per Capita and its Components (Panels a and b) and Global Aggregate Wealth Composition Change (Panel c)

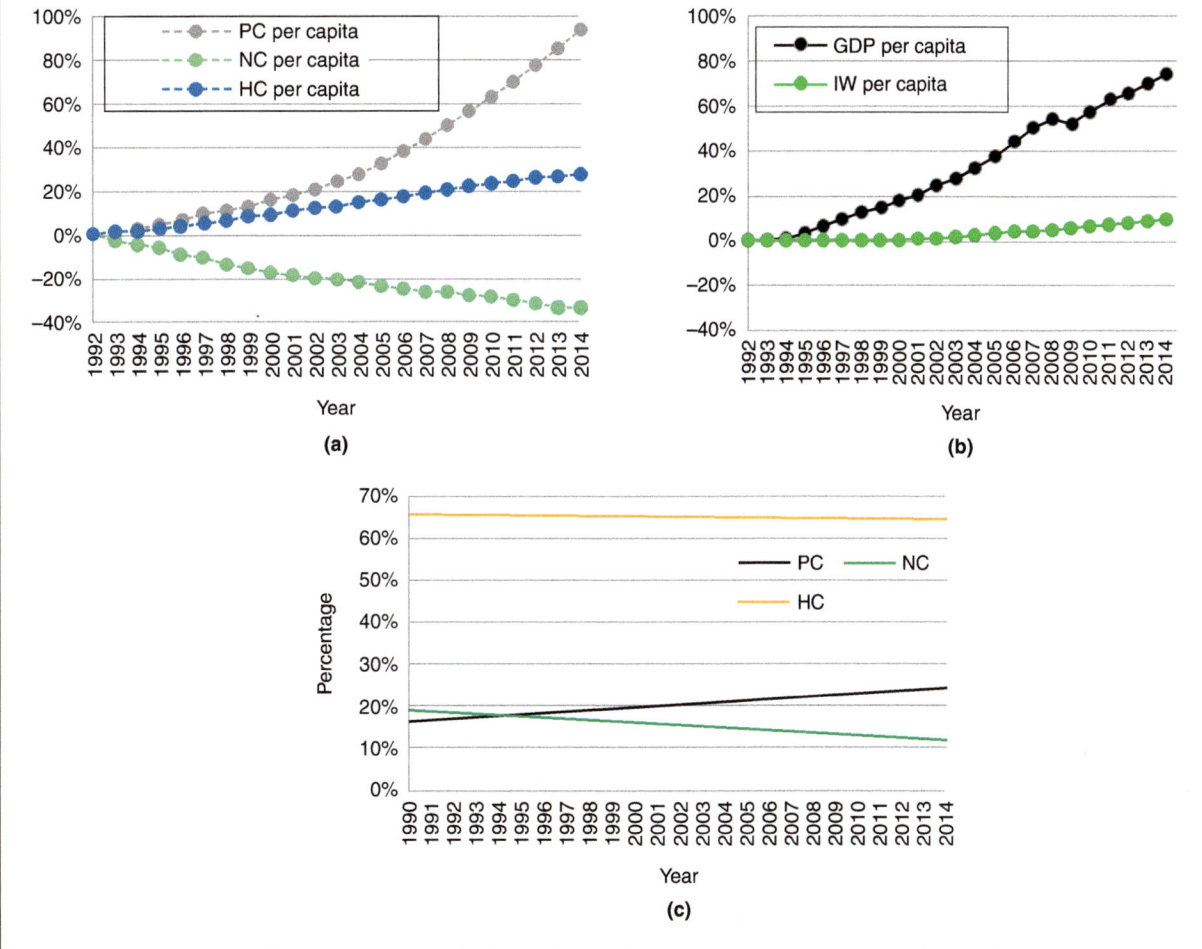

Source: Managi and Kumar (2018) and Yamaguchi, Islam and Managi (2019). Permission to reproduce from Springer Nature.
Note: IW, PC, NC and HC stand for inclusive wealth, produced capital, natural capital and human capital, respectively.

Tight theoretical reasoning is useful to economic accountants because it reminds them of the purpose behind their activity. In practice, empirical corners have to be cut. An understanding of the tight theory helps accountants to justify the corners they choose to cut. We should not expect national accountants to produce wealth accounts in the form the wealth/well-being equivalence theorem requires of them. Wealth estimates will probably be displayed on a sectoral basis, as national income accounts are. The latter are compiled from sectoral accounts such as manufacturing, agriculture, mining, and so on. In a similar vein, we would expect wealth accounts to display changes to stocks of specific forms of natural capital, human capital and produced capital. Moreover, there are Nature's objects and sites of cultural significance that resist being valued and placed in comparison to marketed goods. Societies record their presence and allocate funds to preserve and restore them. They fall outside the scope of national accounts.

Chapter 13: Sustainability Assessment and Policy Analysis

13.10 Growth in GDP and Inclusive Wealth

Is GDP growth compatible with sustainable development? Specifically, are there policies governments today can pursue that will enable GDP to grow even as the inclusive wealth of their economies increases?

GDP growth is, in principle, compatible with sustainable development. There are economic models in which natural resources are needed in production but in which GDP and consumption can both grow indefinitely even as inclusive wealth increases indefinitely. The models demonstrate that GDP growth per se is not an obstacle to sustainable development. Dasgupta and Heal (1979: Chapter 10), for example, studied production and consumption possibilities in a model economy in which produced capital and an *exhaustible* resource are the sole factors of production. Population was assumed to be constant.[427] Being an exhaustible resource (oil and natural gas were taken to be the prototypical examples), depreciation is a key feature of the model: what is extracted for the purposes of production simply disappears in the process of production (the model did not include carbon emissions). The authors showed that in their model economy there are policies under which GDP, consumption, and inclusive wealth grow towards infinity. The stock of natural capital (the depletable resource) is driven in the long run to zero of course, but substitution possibilities in the model economy are taken to be so large that accumulation of produced capital (which also tend to infinity) can not only compensate for vanishingly small quantities of natural capital in production but can permit unbounded growth in output and consumption.

In this Review, we have shown why such models are not to be believed even remotely. The question with which we began this section can only be answered within the context of complete macroeconomic models of the long run, in which natural capital plays an essential role – from source to sink. In Chapter 4*, such a model was constructed as a prototype of the kind governments and international organisations should now construct.

The qualification is all important in today's global economy. The findings reported by Arrow et al. (2012) – see Annex 13.1 and Managi and Kumar (2018) in Box 13.6 – suggest that in recent years, both GDP per capita and inclusive wealth per capita have grown not only in a number of countries, but globally as well. But the authors of both studies acknowledged that lack of quantitative information meant the accounting prices of natural capital they deployed were in all probability serious underestimates. Nor did the authors build their studies on the backs of a formal model. In large measure, market prices were used for such natural resources as land and forests. Their studies are reconnaissances and represent a first cut into what is an extremely difficult set of empirical problems; they are not a guide to what is possible in today's world.

Until a model with features of the prototype in Chapter 4* is estimated and fed with data concerning current stocks of produced, human and natural capital, we cannot answer the question with which we began. What economists will have to check using such a model is whether GDP growth can be maintained for an extended period even while *politically acceptable* consumption levels remain below GDP net of depreciation of all capital goods. The evidence presented in Chapter 4 casts doubt that such a possibility exists for the global economy today, for humanity's ecological footprint is far in excess of the biosphere's ability to meet it on a sustainable basis. There is evidence also that we have left the safety zones defining two of the nine planetary boundaries identified in Rockström et al. (2009) and are perilously close to leaving the safety zone of two further boundaries. Protection and restoration of the biosphere should now be a priority; otherwise the global footprint will continue to increase. The metaphor that best reflects the present situation is 'putting out fires before all else'. Whether the redirection of consumption and investment that is now required is compatible with global GDP growth in the immediate future is an empirical matter.

---

[427] We offered a formal sketch of the model in Chapter 4*, Section 4*3.

Earlier in the Review (Chapter 4), we defined the efficiency parameter, α, in the Impact Inequality relative to the index of human activity per person per unit of time, y_i. For reasons we explained in that chapter, estimating y_i poses significant measurement problems, and so we used a measure of the standard of living, income per capita per unit of time, as a proxy. If an entirely different pattern of consumption and investment were introduced by a society, the underlying activity per person in economic unit i and income, y_i would be entirely different. It follows that human activities that make little demand on the biosphere could in principle be associated with high yi and large and increasing (though not unboundedly increasing!) α.

For the reasons shown in this chapter, inclusive wealth is the right measure of sustainable prosperity, not GDP growth. A focus on GDP leads to a focus on immediate consumption and immediate (gross) investment, with no attention to impacts on the biosphere. Some readers may nonetheless ask whether increasing inclusive wealth and closing the Impact Inequality are together possible even as economic growth (that is, GDP growth) takes place in the short to medium term. In Chapter 4*, we identified limits to GDP growth in future, due to the efficiency parameter α being bounded in how much it can grow. As noted above, this α is defined relative to the index of activity, or the makeup of consumption and investment. If the composition of consumption and investment were to change, the relevant α would be different and the relevant limit to growth would change too. This means it could be possible for GDP to grow without an increase in demands on the biosphere in the medium-term *if* it is possible to change the structure sufficiently. To understand if it is possible in practice for the make-up of the economy to lead to sustainable growth, though, brings us back to using inclusive wealth to assess the sustainability of various paths. Indeed, asking how an emphasis on inclusive wealth affects GDP growth is not the right question. Instead, we should ask how we can keep aggregate consumption below Net Domestic Product (that is another version of the Wealth/Well-Being Equivalence Theorem), and how changes to the composition of consumption and *net* investment should change to make that happen. Economics and finance ministries need to reorient the questions they ask in developing policy. The model in Chapter 4* and its welfare counterpart, Chapter 13*, provide what we hope will prove to be a useful foundation.

Inclusive Wealth and Policy Analysis

13.11 Net Present Values

The wealth/well-being equivalence theorem also provides a criterion for policy analysis:

Inclusive Wealth Criterion – At any given date, a policy change increases intergenerational well-being if and only if it raises inclusive wealth.

It will prove instructive to develop the foundations of policy analysis by studying one sub-species of public choice, namely, choice of investment projects.[428] Feasibility reports of investment projects typically present quantitative estimates of goods and services that are to be shifted from alternative activities (e.g. consumption) to capital accumulation for a period (a project's 'gestation period'). The reports then provide forecasts of positive benefits in subsequent years of the project's life. Acceptance of a project amounts to a small reallocation of goods and services between alternative activities. The re-allocation is small if the project happens to be small relative to the size of the economy. We are then in the realm of a much-studied subject: *social cost-benefit analysis*.[429]

---

[428] Extension to policies that perturb the economy's enabling assets follows along the lines of the analysis in Box 13.2.

[429] Consumption projects, in contrast would envisage an increase in consumption in initial years of the project followed by a reduction in consumption in later years. The difference is only a matter of the assignment of signs to elements of the perturbation.

Chapter 13: Sustainability Assessment and Policy Analysis

Consider as an example a proposal to build a social housing project on an unused piece of wetland. The project's feasibility report contains estimates, expressed in dollars perhaps, of the investment involved (draining the land, constructing the buildings and ancillary structures). The report estimates that construction requires so many labour hours in each year of the investment phase, so many types and quantities of machines and equipment and intermediate goods, and so on. The report also estimates the expenditure incurred by various parties and the transfer of resources that the project envisages among them during the project's life (government, taxpayers, people who will occupy the homes). If the public are environmentally minded, they will insist the report also itemises the services the wetland currently provides to the local community (e.g. filtering water) and a description of the animal and bird populations that make it their habitat. Their disappearance will be seen as a loss. All that information, and it is usually very detailed, is in the form of flows of goods and services. The problem is to appraise the project.

It will not do to simply ask whether the project will enhance intergenerational well-being, for that would be to re-ask whether the project should be accepted. The idea is to put the project data to use by applying accounting prices to the items. Decades ago, welfare economists proved that the project can be evaluated by estimating the present-value of the flow of net social profits it gives rise to. By *Net Present Value* (NPV), we mean a weighted sum of the flow of the project's net social profits. The weights are the consumption discount factors (Chapter 10), that transport (social) benefits and (social) costs in the future to the present. It can be shown that the project should be accepted if its NPV is estimated to be positive (unless, that is, a close variant of the project with an even higher NPV is identified) but rejected if it is found to be negative. That is known as the *NPV-criterion*.[430] The criterion also confirms that the accounting price of a capital good is the present-value of the flow of social benefits that would be enjoyed if the economy were to be provided with an additional unit of the good.

The NPV of a project summarises the social value of the reallocation of the economy's capital goods that is brought about by the project. If the NPV is positive, the reallocation raises intergenerational well-being; if it is negative, the reallocation reduces intergenerational well-being. The NPV-criterion thus leads us back to the arbitrage conditions for portfolio management. In Chapters 1 and 10, it was argued that efficient portfolios necessarily satisfy arbitrage conditions. The NPV-criterion is another way of expressing those conditions.

Arbitrage conditions compare small changes to the composition of portfolios. Likewise, the NPV criterion holds only for small projects, but small only in relation to the overall economy for which a project represents a change. A billion-dollar project in a trillion-dollar economy is small. That is why the NPV criterion offers a powerful language for policy analysis. This is illustrated in a bold and interesting exercise in social cost-benefit analysis (Markandya and Murty, 2000, 2004) of a proposal put forward by the government of India in 1985 to restore the Ganges River (Box 13.7).

Box 13.7
Restoring the Ganges

The Ganges (the river's Sanskritic name is Ganga) originates high in the Himalayas in a glacier near the Chinese border, and comes to an end 1,500 miles further east in the tidal bores of the Bay of Bengal. In between the two ends is the Ganga Basin, which is more than 860,000 km^2 of tributaries,

---

[430] Prest and Turvey (1965) is an early survey of the then existing literature on social cost-benefit analysis. Little and Mirrlees (1969, 1974), Arrow and Kurz (1970), and Dasgupta, Marglin, and Sen (1972) are book-length treatments of the theory and of ways to make use of the social NPV-criterion in project evaluation.

canals, tanks, dams and tube wells supplying water for irrigation, industry, and domestic use to more than 500 million people. There are about 52 cities (including Delhi, Varanasi, Kanpur, and Kolkata), 48 towns and thousands of villages in the basin. Nearly all the sewage (mostly untreated) from these populations goes directly into the river, totalling in the early years of this century some 1.4 billion litres per day, along with a further 260 million litres of industrial waste, runoff from the more than 6 million tons of fertilisers and more than 9,000 tons of pesticides used in agriculture, and large quantities of solid waste, animal carcasses, and human corpses (the latter being released into the river for spiritual rebirth). Unsurprisingly the Ganga basin has been unable to assimilate so ferocious an assault on its integrity. By the 1980s, over 600 km of the river were effectively dead zones (Wohl, 2011). In 1985, the Ganga Action Plan (GAP) was launched, one of the largest single attempts to clean up a polluted river anywhere in the world. The main objective was to raise the river water quality to 'bathing standard'. GAP included numerous water cleaning schemes such as sanitation projects, river-front development and diversion of sewage. The programme entailed investment costs of Rs.7,657 million (or £205 million) and operating costs of Rs.480 million (£6.4 million) (at 1995 prices and exchange rates).

Capital expenditure in the first phase of the programme was to be met by the central government. Maintenance and operating costs were to be shared by central and provincial government in a federalist institutional structure, with co-operation between the riparian communities, who paid for damages incurred downstream in the form of sewerage charges, and local governments across provinces. Water quality modelling showed improvements in water quality (measured in terms of dissolved oxygen and biochemical oxygen demand (BOD)) everywhere, albeit quite small improvements in some places. A cleaner Ganges would provide recreation and tourism benefits, a variety of non-use benefits to people who do not use the river directly but would gain welfare from knowledge that the river is clean and species like the Ganges dolphin are protected. The project would also bring health benefits to households living along the river, agricultural benefits to farmers and improved fish stocks.

To estimate the economic benefits of the programme, Markandya and Murty (2000) surveyed 2,000 households across 10 cities in India and asked respondents to evaluate their willingness to pay for a range of water quality scenarios (measured by BOD levels), using a Contingent Valuation methodology. They estimated willingness to pay for users and non-users of the Ganges for different levels of water quality across a range of demographic, economic and social characteristics. With estimates for the flow of costs and benefits over the duration of the programme, NPVs of the benefits of cleaning the Ganges were calculated at both market and accounting prices (Table 13.1).

Table 13.1 Estimates of Net Present Value of Benefits of Cleaning the Ganges for Various Beneficiaries at Shadow Prices and with Income Distribution Effects (Rs. Million at 1995 to 1996 prices)

	At shadow prices	With income distribution effects $\varepsilon = 1.75$	$\varepsilon = 2.0$
Users	29.11	2.79	1.98
Non-users	6,871.03	439.74	295.45
Farmers	574.93	1,709.84	1,997.88
Health beneficiaries	826.93	2,549.29	2,873.58
Fishermen	N/A	N/A	N/A

Table 13.1 (cont.)

	With income distribution effects		
Unskilled labour	1,919.42	5,708.36	6,670.00
Industrial units	−1,504.59	−144.44	−102.31
Government	−4,569.32	−4,569.32	−4,569.32
Net Present Value	4,147.51	5,696.26	7,167.26
Benefit-Cost Ratio	1.68	2.21	2.53

Source: Markandya and Murty (2000). Note: Rate of discount is taken as 10%. The cost to the government is the present value of costs incurred up to the year 1996–1997.

With the benefits that could be quantified, at a social discount rate of 10% (an unusually high figure) and accounting prices[431], the authors found that the Ganges cleaning programme would have delivered an NPV of around Rs.4,150 million, an Internal Rate of Return of 15.4% and a Benefit-Cost Ratio (BCR) of 1.68. Positive net benefits would have accrued to a range of groups (e.g. Ganges River users, non-users, farmers, and unskilled labourers). If weights are added to adjust for income distribution effects, the BCR rises to 2.21 and NPV to approximately Rs.5,700 million.

As benefits would have accrued to many different groups, a range of institutional structures and instruments could be designed to sustain a cleaning programme. For instance, a cleaning programme could be designed in line with the 'polluter pays' principle, for example through a water charge per kilo litre collected as a tax, or as part of a water tariff from households and industries. It could be designed in line with the 'user pays' principle with government or charity involvement to collect a tax or charge from users.

Unfortunately, the GAP was abandoned. A plan to revive the Ganga has been put forward in recent years. The Markandya-Murty study showed that even by the narrow notion of benefits the authors took in their study, reviving the river would bring enormous benefits.

13.12 Inclusive Wealth and the Present Value Criterion

Investment projects reallocate an economy's stocks of capital goods from one set of uses to another. The present value of a project's flow of net social benefits (dollars per year) has the dimensions of stock (dollars, period). So, it should be no surprise that a project's NPV is the change in inclusive wealth brought about by the project. It means the wealth/well-being equivalence theorem is another way of expressing the criterion for project evaluation that was developed decades ago by applied welfare economists. This is stated as (the proof is given in Box 13.8):

The Wealth/NPV Equivalence Theorem The NPV of social benefits from a project is positive if and only if the project gives rise to an increase in inclusive wealth.

The theorem reaffirms that portfolio managers intent on maximising wealth funds will wish to hold those portfolios that satisfy the arbitrage conditions.

---

[431] Accounting prices added a 40% premium for capital and 50% reduction in the market wage rate in comparison to market prices.

> ## Box 13.8
> ## The Equivalence of Wealth and Present Values
>
> We assumed in Chapter 10 that in order to conduct economic evaluation the Citizen Investor chooses a reference consumption stream. In period t, the set of available consumption streams was denoted as $\Re^*(t)$. It should be apparent to her that the reference stream also contains future projections of the economy's stocks of capital (Chapter 4*). Thus, consumption by each person in any future period s is a function of the vector of capital stocks in period t. We write that functional relationship as $c_i(s, \boldsymbol{K}(t))$.
>
> We use the formulation of intergenerational well-being in Box 13.2, but for notational convenience assume time is continuous. Thus
>
> $$V(\boldsymbol{K}(t)) = \int_t^\infty \{\Omega(s)[\Sigma_{j=1}^{N(s)}(u(c_j(s, \boldsymbol{K}(t))))]e^{-\delta(s-t)}ds\}/\Omega(t), \quad \delta > 0 \qquad (B13.8.1)$$
>
> Consider now a proposal for a small investment project, to start at t. The project would perturb the economy by displacing capital assets from their previous deployment to the project. On using equation (B13.8.1), the proposed perturbation can be expressed as
>
> $$\Delta V(\boldsymbol{K}(t)) = \Delta[\int_t^\infty \{\Omega(s)[\Sigma_{j=1}^{N(s)}(u(c_j(s, \boldsymbol{K}(t))))] e^{-\delta(s-t)}\}ds/\Omega(t) \qquad (B13.8.2)$$
>
> Equation (B13.8.2) can in turn be expressed as
>
> $$_i\Sigma[\partial V(\boldsymbol{K}(t))/\partial K_i(t)]\Delta K_i(t) = \int_t^\infty \{\Omega(s)[\Sigma_{j=1}^{N(s)}(\Delta u(c_j(s, \boldsymbol{K}(t))))] e^{-\delta(s-t)}ds\}/\Omega(t) \qquad (B13.8.3)$$
>
> Because Δ is small,
>
> $$\Delta u(c_j(s, \boldsymbol{K}(t)))/\Delta t = {}_i\Sigma\{_{j=1}\Sigma^{N(s)}[(\partial u/\partial c_j)\partial c_j/\partial K_i(t)]\Delta K_i(t)/\Delta t$$
>
> Moreover,
>
> $$\partial V(\boldsymbol{K}(t))/\partial K_i(t) = p_i(t) \qquad (B13.8.4)$$
>
> So, equation (B13.8.3) can be approximated to read
>
> $$_i\Sigma[p_i(t)dK_i(t)]/dt =$$
> $$_i\Sigma\{\int_t^\infty [\Omega(s)[_{j=1}\Sigma^{N(s)}(\partial u/\partial c_j(s))\partial c_j(s)/\partial K_i(t)]dK_i(t)/dt]e^{-\delta(s-t)}ds\}/\Omega(t) \qquad (B13.8.5)$$
>
> The left-hand side of equation (B13.8.5) is the contribution of the project to inclusive wealth; that is, it measures the project's contribution to inclusive investment. The right-hand side of the equation is the NPV of the project's flow of social benefits. The latter is the received criterion in social cost-benefit analysis: if the project's NPV is positive, it is worth undertaking, but not otherwise. The left-hand side of equation (B13.8.4) says that the NPV criterion could equally be taken to be the project's contribution to inclusive wealth.

The NPV-criterion for choice of investment projects takes us away from the demand commonly made that macroeconomic policies should be so chosen as to enhance GDP. A macroeconomics that sees growth and distribution of GDP as the aim of public policy is inconsistent with the wealth/well-being

Chapter 13: Sustainability Assessment and Policy Analysis

equivalence theorem and its various corollaries. We have no explanation for how and why that incongruity has remained in economics.[432]

The wealth/well-being equivalence theorem and its various extensions tell us that by 'economic growth' we should mean growth in wealth, not growth in GDP even when corrected for population and the distribution of incomes.[433] Similarly, the theorem says that by poverty we should mean a paucity of wealth. The theorem tells us to measure a society's *prosperity* by its inclusive wealth. Estimating stocks is no doubt hard work, but it should not be avoided. The rogue word in Gross Domestic Product is 'gross'; the measure does not account for depreciation of capital goods. However, growth in NDP does not imply sustainable development. For development to be sustainable consumption must not exceed NDP. That means 'green GDP' is a misnomer.

Box 13.9

Appraisal of the Restoration of the Exmoor Mires

Mires are ecosystems that accumulate peat, such as blanket bogs, valley bogs and fens. 13% of blanket bog globally is in the UK. It is the most common type of mire in Exmoor – blanket bog and peat cover the central moorland. But it has been dried out over centuries of moorland reclamation, drainage and cutting of peat for domestic use. In consequence, it has lost much of its unique biodiversity and ability to provide ecosystem services, such as water filtration, carbon storage and sequestration, and opportunities for recreation.

In response, the Exmoor Mires Partnership was created to restore the Exmoor Mires peatland. The project aimed to "restore degraded Exmoor peatlands on a landscape scale and to promote the regeneration of moorland bog vegetation".

Bright (2017) compiled restoration accounts to evaluate whether the restoration project was a beneficial investment. The accounts were compiled once the project had already been initiated, it was not an ex-ante analysis. Restoration accounts are a subset of UK's natural capital accounts (Bright, Connors, and Grice, 2019). One should not expect the desired end state of an ecosystem from restoration to be the same as its earlier, undisturbed reference state. Furthermore, the costs of restoration may not be the same as the costs of maintaining the current quantity or quality of the ecosystem. Restoration accounts provide additional information on the benefits and costs of achieving a specific restoration target, which then help determine whether the project is worthwhile. Based on the desired biophysical state of the restored ecosystem, the ecosystem accounts can be used to estimate the resulting physical flows of ecosystem services and their monetary value (additional to those ecosystems currently provided, without restoration). The monetary value of the stock of the restored ecosystem assets can be estimated based on the discounted sum of these monetary ecosystem service value flows (Figure 13.3). These benefits can be compared with the costs of restoration to estimate the NPV or benefit-cost ratio for project evaluation.

---

[432] Nevertheless, GDP has a strong intuitive appeal. Annex 13.3 discusses that appeal. We argue there that as an economic indicator GDP has many virtues, but that as the wealth/well-being equivalence theorem shows, it should not be used to judge the progress or regress of economies.

[433] The definition of inclusive wealth includes a recognition of population change and disparities among individuals.

Figure 13.3 Baseline Natural Capital Accounts and Restoration Cost Accounts

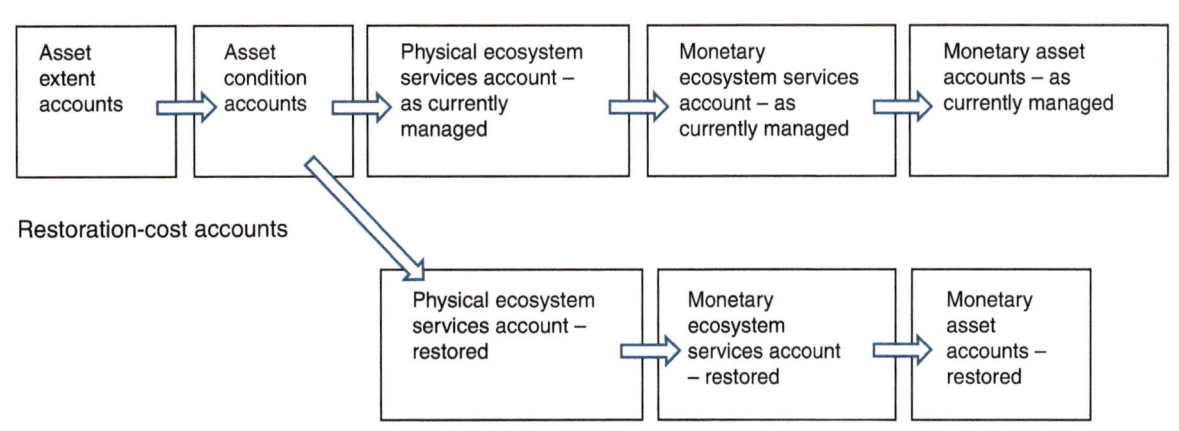

Ecosystem accounts

Asset extent accounts ⇨ Asset condition accounts ⇨ Physical ecosystem services account – as currently managed ⇨ Monetary ecosystem services account – as currently managed ⇨ Monetary asset accounts – as currently managed

Restoration-cost accounts

Physical ecosystem services account – restored ⇨ Monetary ecosystem services account – restored ⇨ Monetary asset accounts – restored

Source: Bright, Connors, and Grice (2019).

The findings of these accounts are described in Table 13.2, which shows the extent of the Mires with different peat condition (pristine, near natural, modified, recovering, drained and actively eroding). In 2006, at the outset, the total project area of 2,573 hectares was classified as 'drained' and 65 hectares as recovering. By 2016, the closing year of the accounts, restoration was underway, with 1,537 hectares moving to an improved condition (65 hectares from drained recovery to modified status, and 1,472 hectares from drained to recovering status). 1,101 hectares remained drained. The impact of the project came from changing the conditions of the peatlands, rather than their extent.

Table 13.2 Exmoor Mires: extent and condition account

Peat Condition Indicators	1	2	3	4	5	6	Total (hectares)
	Pristine	Near natural	Modified	Drained: recovering	Drained	Actively eroding	
Status 2006	0	0	0	65	2,573	0	2,638
Regeneration gains	0	0	65	1,472	0	0	1,537
Regeneration reductions	0	0	0	-65	-1,472	0	-1,537
Status 2016	0	0	65	1,472	1,101	0	2,638

Source: Bright (2017).

The improved peatland condition was found to increase the provision of ecosystem services such as grazing space, water quality, flood protection and opportunities for recreation. Physical and monetised restoration accounts were compared with accounts that represented a scenario without the restoration project for identifying the additional benefits. These benefits were compared with the costs of restoration. The discount rate, in accordance with the recommendations in the Treasury Green Book 2016, was taken to be 3.5% a year for the first 30 years, falling in steps to 1% a year beyond 300 years.[434] The project was evaluated with a 100-year horizon. The discounted benefits of the services provided by the restoration project was found to be £50.15 million. The discounted costs of the project was £5.63 million. Together they meant an NPV of £44.52 million and a benefit-cost ratio of 8.9.

---

[434] The logic of time varying social discount rates was discussed in Chapter 10.

13.13 Lower Discount Rates for Environmental Projects?

Recall that the rate of return on an asset is the algebraic sum of its yield and the capital gains it enjoys on the commodity selected as numeraire. Recall also that arbitrage conditions describe the equality of rates of return (adjusted for risk) on all assets in the Citizen Investor's portfolio (Chapter 1). If the economy is in a stationary state, spot prices are constant over time, meaning that there is no capital gains term in the arbitrage conditions. But if the economy is not in a stationary state, spot prices change over time. It follows that unless the economy is in a stationary state, the system of discount rates our Citizen Investor will wish to use will depend on the commodity she chooses as numeraire. In a consistent accounting system, there can only be *one* numeraire, which is to say that only *relative* prices matter, absolute prices have no operational significance.[435]

This has far-reaching implications. Suppose, for example, the Citizen Investor believes that produced capital will continue to be accumulated in the global economy even while natural capital continues to depreciate. In that case, the accounting price of natural capital relative to consumption will be expected to rise over time. That would be capital gains on natural capital. For concreteness, imagine that the Citizen Investor chooses consumption as numeraire and the discount rate she uses to appraise projects is 3% a year. Suppose next that she expects the accounting price of natural capital relative to consumption to increase at 1% a year. She now considers an investment project that is expected to increase the flow of the biosphere's goods and services by B, at a cost in terms of foregone consumption of E every year. Viewed from the present, the project's net benefit in year t would read as $[B(1.01)^t - E]/(1.03)^t$. The expression could be interpreted as saying that as against a discount rate of 3% a year for the project's running costs, the project's gross benefits should be discounted at the lower rate of 2%. On the other hand, if the Citizen Investor chooses natural capital as numeraire, the social discount rate would be 2% a year, but the project's running costs would be found to be *declining* at an annual rate of 1% a year (the capital losses suffered by produced capital). Now the project's net benefit in year t would read as $[B - E/(1.01)^t]/(1.02)^t$. Since the figures are proportional to one another, project appraisal would be unaffected by a switch in numeraire.

The exercise we have just performed should not be interpreted as saying that environmental *projects* should be appraised on the basis of a lower discount rate. For suppose there is a competing project that would produce a consumer good but would incur a recurrent cost involving the same set of inputs as the environmental project. If the Citizen Investor was to deploy two systems of accounts, there would be confusion. The recurring costs would have to be valued differently in the two projects. That could of course be done, but in practice it would lead to errors.

Box 13.10

The Plight of Slow Growth Forests

The literature on the economics of climate change has shown that policy analysis can be sensitive to the choice of discount rates (Stern, 2006; Nordhaus, 2007; see also Chapter 10, Box 10.4). In Chapter 10, we noted that if the rate of return on an investment is expected to fall short of the rate used to discount future benefits and costs, the investment would not be undertaken. Slow growth forests (more specifically, trees) are a casualty to this reasoning. Suppose the regeneration rate of forest biomass (S) has the form made familiar in Chapter 3 (Box 3.2; Equation (B3.2.2)):

---

[435] The point arises if only in a slightly different form when people compare weights. An ear of corn could serve as numeraire, but if it is measured in grams in one place and in ounces in another place, people would have difficulty communicating, unless of course they reach an agreement over the conversion factor to be used from one system of weights to the other.

$$dS(t)/dt = rS(t)[1 - S(t)/\underline{S}][(S(t) - L)/\underline{S}]$$

The forest's intrinsic growth rate is r. It is simple to confirm that if the discount rate chosen by the forest owner exceeds r, he will choose to deplete it entirely. The irony is that slow growing trees are often giants (redwood, teak, oak), meaning that their $\underline{S}$ is large. In contrast, consider fast growing forests with lower maximum sustainable yield. If the discount rates in use (e.g. market rates) lie between the intrinsic growth rates of the two classes of forests, the vegetation that survives would be devoid of slow growing giants. The Citizen Investor would no doubt avoid this eventuality. That's because she would find it natural to work with low discount rates (Chapter 10).

13.14 Optimum Development

Investment projects have been defined here as perturbations to socio-ecological futures. As those projects are part of more general macroeconomic policies (taxes, subsidies, bank rates – more generally the lowering of economic distortions), we should think beyond investment projects and speak of 'policy change'. Although the NPV-criterion is valid only for small shifts in policy, it remains vital in the search for the optimum composition of consumption and investment even if they are far from where the economy happens to be currently. To illustrate the way optimum policies can be found, we consider the global economy.

Recall that the starting point in policy analysis is the construction of a macroeconomic model, as in Chapter 4*. The model provides a map of alternative socio-ecological futures. We once again borrow from Chapter 10 and summarise socio-ecological futures in terms of consumption streams. The set of all feasible consumption streams was denoted there by $\mathfrak{R}^*(0)$. $\mathfrak{R}^*(0)$ is limited by institutional and technological constraints, meaning that the optimum stream in $\mathfrak{R}^*(0)$ is only a second-best.

Socio-ecological futures are assumed to be evaluated by our social evaluator/Citizen Investor. In view of the magnitude of the Impact Inequality in the contemporary global economy (Chapter 4), she knows that the best future is far from the one that will unfold under what people often call 'business as usual'. The problem before her is to identify not only the optimum stream in $\mathfrak{R}^*(0)$, but also a set of policies that would implement the optimum. We now describe an algorithm that enables her to do that in certain circumstances. It involves conducting a sequence of hypothetical exercises in social cost-benefit analysis.

Imagine that the social evaluator starts with an arbitrary reference consumption stream. For example, the reference stream could be the socio-ecological future under business as usual. She next estimates the accounting prices for goods and services in the reference stream. She then considers a policy shift that gives rise to a small change in the reference stream with which she began her exercise. Using the accounting prices, she had estimated, the social evaluator evaluates the NPV of the change. She rejects the change if its NPV is negative but accept it if its NPV is positive.

With the acceptance of a proposed change, the reference stream the social evaluator now evaluates (the second round in the iteration) is slightly different from the reference stream in the initial round. She next estimates accounting prices on the new reference stream. They will be found to be slightly different from the accounting prices in the first round. The social evaluator now repeats the previous process.

And so on. At each step in the algorithm, those incremental policy changes whose NPVs are negative are rejected, whereas a change whose NPV is positive is accepted. The algorithm thus weaves its way through alternative socio-ecological futures. The process continues until the social evaluator finds a consumption stream that cannot be bettered by the NPV-criterion.

Chapter 13: Sustainability Assessment and Policy Analysis

The algorithm just described is known as a 'hill-climbing procedure': At each iteration, a project is approved only if its NPV is positive. The idea is to climb until well-being across the generations is at a peak. Radner (1963) and Malinvaud (1967) identified a class of economic environments (i.e. $\Re^*(0)$s) in which the algorithm just described works; that is, it leads the social evaluator to a peak.[436] As the economic environments they studied contain a single peak, the algorithm leads the social evaluator to the unique optimum. At the optimum, the mix of capital goods and their accounting prices satisfies the arbitrage conditions described in Chapter 1.

Arbitrage conditions are established by considering marginal changes to portfolios. There is a risk that they are thought to commend only small changes. To think so would be a mistake. The hill-climbing method is an algorithm. At each step, the process involves evaluating small changes. But there are many steps in the process. So the resting point of the algorithm – the optimum consumption stream – may be far from the reference stream at the first round of the algorithm. The social evaluator is able to identify the optimum mix of capital goods only by repeated use of the NPV-criterion. At the optimum, the NPV-criterion amounts to the arbitrage conditions.[437]

Unfortunately, non-linearities in ecological processes (Chapter 3) violate the conditions under which the hill-climbing procedure is able to locate the social optimum. In the presence of non-linearities, intergenerational well-being has multiple peaks in $\Re^*(0)$. The algorithm leads the social evaluator to a local peak, whereas the search for the optimum is a search for the global peak (Figure 13.4).

Although the wealth/NPV equivalence theorem says that inclusive wealth is maximised at the global peak, finding the global peak requires the social evaluator to scan the economy in its entirety. Macroeconomic models, such as the one studied in Chapter 4*, enable her to do that scanning. Box 13.11 explains why the NPV-criterion on its own is inadequate for identifying the transformative changes that are now needed in our relationship with the biosphere.

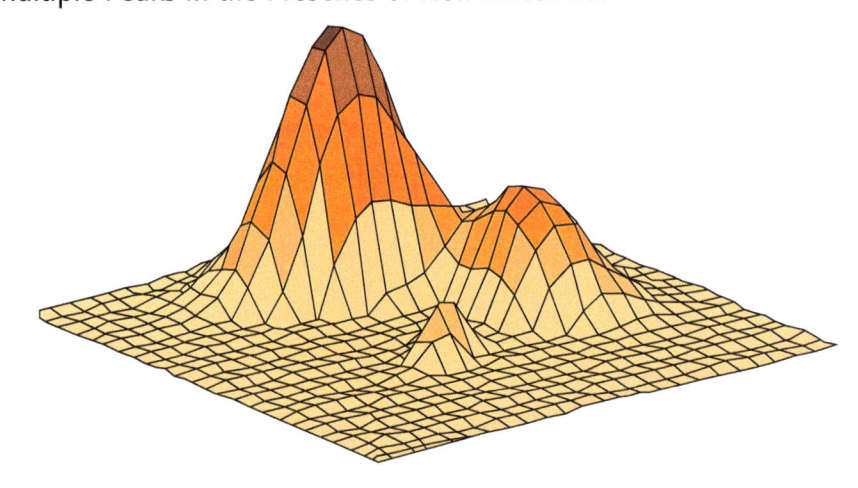

Figure 13.4 Multiple Peaks in the Presence of Non-Linearities

Source: Aguilar-Rodríguez, Payne and Wagner (2017). *Permission to reproduce from Springer Nature.*

---

[436] The idea to mimic the efficiency of ideal markets in national planning by using accounting prices has its origins in the works of economists Oscar Lange and Abba Lerner (Lerner, 1934); Lange (1936, 1937). Heal (1969) proposed a different algorithm that is also effective.

[437] Annex 13.2 reports a study by Spence (1974), in which the author applied the theory sketched here to determine the optimum harvesting of blue whales, a species that came perilously to extinction in the later 1960s. Spence showed that a moratorium on whale hunting was in the commercial interest of the International Whaling Commission. A moratorium was a far cry from the rate at which whales has been hunted in the previous century.

Chapter 13: Sustainability Assessment and Policy Analysis

> **Box 13.11**
>
> **Project Complementarities and Lumpy Investments**
>
> The size of the Impact Inequality today is large. Marginal changes to global activities would not close the gap between the demands that humanity makes on the biosphere and the ability of the biosphere to meet those demands on a sustainable basis. Previous chapters have shown why large changes to the mix of global consumption and investment activities are now needed if the gap is to be closed. Emerging from COVID-19 will offer our battered economies an opportunity to consider transformative changes. Application of the NPV- criterion bit by bit, from one small project to another, is of little-to-no use for identifying optimal changes to an economy.
>
> Why? The reason is a combination of two factors: (i) when it is applied iteratively, the NPV-criterion directs the economy only to a local optimum; and (ii) ecological processes are non-linear. We observed earlier (Chapters 3–4 and 4*) that those non-linearities give rise to complementarities among ecosystem services. None on its own could be a viable service. Which is why the structure of accounting prices in a global economy that has for decades accumulated produced capital and degraded the biosphere will not signal that we should make a transformative move toward protecting and repairing it. If green investment projects make their appearance before the social evaluator one by one, in small bits, none on its own is likely to make the cut, and only special pleading will get them safely past national planning commissions and ministries of industry and transport. Previous chapters traced the biosphere's non-linearities in part to mutualism among communities of organisms and to Nature's mobility. The non-linearities were illustrated in Chapter 3 by the fate of ecosystems upon repeated fragmentation. Nature's interconnectedness is the reason 'green projects' complement one another. Projects that extend an entire network of roads by eating into Nature bit by bit can be expected to trump the alternative of keeping Nature safe from destruction. That is because evaluating conservation and restoration projects singly does not record their complementarities.
>
> When, however, an investment that is designed to protect and restore a set of interconnected ecosystems is pitched against an investment of comparable size in produced capital, the matter is different. Today there is talk in many circles that once we are past the COVID-19 crisis, the global economy should make a shift toward reviving our ailing biosphere, for if nothing else, it would offer huge employment opportunities. In a report on differential labour intensities in various sectors, Levy, Brandon, and Studart (2020) have found that projects involving Nature restoration create, on average, over 10 times more jobs than investments of a comparable size in fossil-based industries. That green investment is labour-intensive is, of course, an attraction, but it is an *added* attraction. The primary reason for embarking on a transformative change in our investment activities is, however, that not to do so will increase the size of the Impact Inequality. Social cost-benefit analysis of green projects, taken singly, will not signal that.

13.15 Production and Consumption Targets

In practice, shortcuts have to be found to the way we have posed society's optimisation problem in this chapter. In international negotiations over global climate change, the approach has been to agree on an upper limit to the mean global temperature (2°C above that prevailing before the industrial revolution) and then to search for the optimum means to ensuring that the limit is not breached. The general idea in this form of policy analysis is to agree on a goal and to then find the optimum means to achieving that goal. As we noted in Chapter 5, the existence of tipping points is a reason for setting

Chapter 13: Sustainability Assessment and Policy Analysis

targets. The Aichi Targets to achieve 2020 goals that were adopted in 2010 by the Parties to the Convention on Biological Diversity are an example.[438]

Allocating budgets across government departments has a similar flavour, but for a different reason. They are based on precedence and are the starting point of ministerial deliberations. Budgets are apportioned among departments and the task of each department is to find the most beneficial way to spend its budget. A department that is guaranteed, say, 0.7% of a nation's GDP could decide how to spend it by evaluating projects. Generally, competing government departments would negotiate their budgets over a common pool of funds. That involves a previous round of policy analysis, where departments justify their budgetary demands. But no matter what shortcut is found to be necessary for optimising the allocation of investment, the reasoning involves the wealth/well-being equivalence theorem.

13.16 Internal Rate of Return

The early literature on cost-benefit analysis of investment projects identified a different criterion. It asked the social evaluator to estimate a project's *internal rate of return*, which is the constant discount rate that, if applied to the flow of social benefits and costs of the project, yields an NPV equal to 0. It is simple to show that if the project incurs costs during an initial period of time and provides positive benefits for the remainder of its life, its internal rate of return is unique.

The internal rate of return is an attractive criterion for use in project evaluation because it seemingly bypasses the need to identify social discount rates and, thus, the need to engage in ethical reasoning (Chapter 10). But the gain is illusory. The social evaluator could no doubt set herself the task of ranking projects that come to her desk in terms of their internal rates of return, but she will need to furnish herself with a cut-off rate below which she is to reject projects. The social rate of discount supplies her with that cut-off rate. The moral is banal: there is no way to bypass normative reasoning in policy analysis.

These remarks bring us back full circle to the idea we advanced in Chapter 1 that the economics of biodiversity is about portfolio management. All decision-making units in the world are engaged in managing assets, be they small households in poverty-stricken villages in a poor country; be they companies in rich countries purchasing primary products abroad and selling the finished products domestically; be they governments concerned about the smog that engulfs their capital city; or be they the United Nations grappling with not only global climate change but also the entangled problems that have arisen from large-scale biodiversity loss. In the remainder of this Review, we study a number of extensions to this that serve to inform decision- makers of the measures they may wish to explore to reduce the Impact Inequality, and perhaps even bring the biosphere back to health.

Annex 13.1 Economic Growth and Sustainable Development

Empirical work is forced to cut theoretical corners. But even for the most hard-boiled empiricist the wealth/well-being equivalence theorem is of use. If national income statisticians were to remain unaware of it, they would not know which corners they would be obliged to cut when preparing their nation's balance sheet.

Arrow et al. (2012, 2013) made an early attempt at applying the theorem to data. The publications should be viewed as reconnaissance exercises. You know they have got it wrong, but you also know they are in the right territory.

---

[438] Weitzman (1970) presents an algorithm that searches for an optimum that is subject to production targets.

Chapter 13: Sustainability Assessment and Policy Analysis

The authors estimated the change in wealth per capita over the period 1995 to 2000 in Brazil, China, India, United States and Venezuela. The choice of countries was in part designed to reflect different stages of economic development and in part to focus on particular resource bases. Because of an absence of data, the authors did not study wealth inequality within countries. Here we summarise the steps they took to enquire whether economic development in India was sustained during the five years in question. Details can be found in their paper.

A13.1.1 Wealth in India: Estimates

A13.1 provides estimates of wealth per capita in 1995 and its growth during the following five years. Columns (1)-(2) provide estimates of stocks per capita for 1995 and 2000, respectively, for three categories of assets: produced capital (row (1)); human capital, divided into education and health (rows (2)-(3)); and natural capital (row (4)).

The value of produced capital in 1995, amounting to US$1,530 per head, was calculated from government publications on past capital investments. The implicit assumption was that prices used by the government to record expenditures are reasonable approximations of accounting prices. Using the methods summarised in Klenow and Rodríguez-Clare (2005), the value of education per person (US$6,420) was estimated on the basis of a functional relationship between wage differences and differences in levels of education.

No data are currently available for calculating the contribution of health to labour productivity. For that reason, the authors studied longevity only. Its accounting price was estimated from the value of a statistical life (VSL). VSL in India in 2000 was estimated to be approximately US$1.26 million. Arrow et al. (2012) estimated the value of a statistical life-year and used that to value the increase in life expectancy between 1995 and 2000 (row (3), column (3)).

Four categories of natural capital were included in the study: forests (valued for their timber), oil and minerals, land, and carbon concentration in the atmosphere. For the sake of transparency, atmospheric carbon has been excluded from columns (1) and (2) but included in the estimate of the change in wealth over the 5-year period.

The value of land was taken from Deininger and Byerlee (2011). Using market prices for timber and oil and minerals, the accounting value of natural capital in 1995 was estimated to be US$2,300 per person (row (4), column (1)). Because of the lack of relevant data, the figure does not include the value of all the many ecological services that forests provide. Moreover, ecosystems such as fisheries, wetlands, mangroves, and water bodies are missing from Table A13.1. That means US$2,300 is an underestimate, in all probability seriously so. Adding the figures, wealth per capita in 1995 was found to be approximately US$1,270,000 (row (5), column (1)).

Population in India grew at an average annual rate of 1.74%. Column (3) records changes in per capita capital stocks over the period in question; and column (4) presents the corresponding annual rates of change. The former is embellished by two factors. First, India is a net importer of oil, whose real price rose during the period. The capital losses owing to that increase amounted to wealth reduction in India, which was calculated to be US$140 per person (row (5), column (3)). Second, during 1995 to 2000 global carbon emissions into the atmosphere was over 35 billion tons. At levels of concentration in 1995 (380 parts per million in 1995), carbon was taken to be a global pollutant. The theory of public goods says that the loss to India over the period would have been global emissions times the accounting price of carbon specific to India. In their base case, Arrow et al. (2012) took the global accounting price to be *minus* US$50 per ton. Using the estimates of Nordhaus and Boyer (2000), the loss to India per ton of carbon emissions was taken to be 5% of the global shadow price, which is minus US$2.50. This amounted to a loss per person of US$90 (row (6), column (3)).

The Economics of Biodiversity: The Dasgupta Review

Row (7) records the change in wealth per capita in India over the period 1995 to 2000. It translates to 0.20% a year, a figure so near to zero as to be alarming. However, the estimate does not include improvements in knowledge and institutions. Arrow et al. (2012) modelled the latter as 'enabling assets' and interpreted improvements in them as growth in Total Factor Productivity (TFP), which in India has been estimated to be 1.84% a year (row (8)). Based on a formula the authors derived for including the residual in wealth calculations, row (9) records the annual rate of growth of wealth per head in India during 1995 to 2000 as having been 2.01%.

A13.1.2 Wealth in India: Commentary

The composition of wealth in Table A13.1 does not have direct implications for policy. A mere study of the relative magnitudes of the different forms of wealth would not tell us their relative importance. Suppose, for example, that the value of asset *i* swamps all other forms of capital, by a factor of 1,000. We noted in Chapter 13 that this does not mean investment ought to be directed at further increases in *i*, for we do not know the costs involved in doing so. Only social cost-benefit analysis, using the same accounting prices as are estimated for sustainability assessment, would tell the social evaluator which investment projects are socially desirable. Taken at face value, Table A13.1 reveals a number of interesting characteristics of India's economic development during the final years of the 20th century. It is as well to highlight the most striking:

(1) Of the three types of capital goods comprising measured wealth, produced capital is the smallest. Even though the value of natural capital in both years is in all likelihood a serious underestimate, it was considerably greater in 1995 than produced capital.

(2) The rapid growth of produced capital (7.30% a year), as against a 0.15% annual rate of decline of natural capital meant that by 2000 their stocks were pretty much the same.

(3) In 1995, human capital in the form of education was over four times that of produced capital. But the ratio declined over the five-year period owing to slower growth in education.

(4) Health swamps all other forms of wealth. That it is some two orders of magnitude larger than all other forms of wealth combined in what was in 1995 a low income country is unquestionably the most striking result of the study. That the finding is a cause for surprise is, however, no reason for dismissing it. Health has been much discussed in the development literature but has not often been valued within the same normative theory as produced capital and education. Health dominates because of the high figure for VSL reported in the empirical literature. Longevity matters to people everywhere and matters greatly. In democratic societies that should count.[439]

(5) Growth in wealth per capita in India has been in great measure a consequence of TFP growth (the 'residual'). But contemporary estimates of the residual should be treated with the scepticism, because they are based on models that do not include natural capital as factors of production. We noted this in Chapter 13. If the rate at which natural capital is degraded were to increase over a period of time, TFP growth obtained from regressions based on those models would be overestimates. Worse still, the greater is the under-coverage of natural capital, the greater is the bias in the estimate of TFP growth. By plundering Earth, TFP could be raised by as much as the authorities like.

The value of natural capital in Table A13.1 is probably a serious underestimate. When national accounts are better prepared, health and natural capital will in all probability be found to be much the most significant component of the wealth of nations. That is also why official ignorance of the state of an economy's stock of natural capital assets should now be a priority for governments. For example, in

---

[439] Becker, Philipson, and Soares (2005) included longevity increase in estimates of the growth of income per head to show that the economic performance of developing countries in recent decades was considerably superior to that of rich countries.

a review of the empirical literature on forest services (carbon storage, ecotourism, hydrological flows, pollination, health, and non-timber forest products), Ferraro et al. (2012) have found little that can be used reliably in wealth estimates. But even if figures for natural resource stocks were available, the deep problem of imputing values to them would remain. Market prices may be hard facts, but accounting prices are often soft (Chapter 12). The issue is not merely one of uncertainty about the role natural capital plays in production and consumption possibility, it is also that people differ in their ethical values. The sensitivity of wealth estimates to accounting prices should become routine exercise in national accounts. It may be that in the foreseeable future, wealth estimates are best left disaggregated and presented in terms of their components, say on a sectoral basis; at the very least as in rows (1)-(4) in Table A13.1, but preferably in a more disaggregate form (e.g. the accounting value of forests, ground water levels, mangroves, fisheries) and that too within bands; they should not be presented as precise figures. Accounting prices are too fragile to support point estimates.

Table A13.1 Per Capita Wealth and its Growth in India, 1995 to 2000

	(1) 1995 stock (2000 US$)	(2) 2000 stock (2000 US$)	(3) Change (1995–2000)	(4) Growth rate (% per year)
(1) Reproducible capital	1,530	2,180	650	7.31
(2) Human capital, 1 (education)	6,420	7,440	1,020	2.99
(3) Human capital, 2 (health)	1,259,320	1,268,710	9,390	0.15
(4) Natural capital	2,300	2,280	−20	−0.16
(5) Oil (net capital gains)			−140	
(6) Carbon damage			−90	
(7) TOTAL	1,269,570	1,280,380	10,810	0.17
(8) TFP				1.84
(9) Wealth *per capita*				2.01

Source: Arrow et al. (2012), Table 5 (modified). Note: Figures rounded.

That people may never agree on the wealth of nations is, however, no reason for abandoning inclusive wealth as the object of interest in sustainability assessment. Our ignorance of the economic worth of natural capital remains the greatest barrier to an understanding of economic development. Until that ignorance is lifted, policy analysis will remain crippled and sustainability will continue to be a notion we admire but cannot put into operation.

Annex 13.2 Saving the Blue Whale

The blue whale is a marine mammal. Up to 30 metres in length, with a maximum recorded weight of about 175,000 kg it is the largest animal on the planet, possibly the largest that has ever existed. Sub-species of blue whale breed in the Southern Ocean, the North Atlantic, and in the Indian and South Pacific Oceans. The species were abundant until the beginning of the 20th century, when their population could be assumed to have been at carrying capacity. IUCN (2008a) estimates that before whaling began, the largest population was in the Southern Ocean, numbering perhaps 240,000. But with advances in technologies to kill what were in effect open access resources (Chapter 8), blue whales were hunted close to extinction by the late 1950s. Some estimates suggest their population had become as low as 1,640 (Spence, 1974).

Chapter 13: Sustainability Assessment and Policy Analysis

Whale products include meat for consumption and oil for manufacturing soaps, cosmetics, and perfumes. As there are a number of near-substitutes, demand for whale products can be taken to be fairly elastic, a matter of importance to the studies we report here.

The Golden Age of whaling was the decade 1928 to 1938, when catches were as high as 26,000 blue whales a year. Stocks dropped and harvesting costs (which include search costs) rose correspondingly to such levels that by the early 1960s annual catch was small (about 1,450 in 1958) in comparison with previous levels. As concerns arose that blue whales would become extinct unless whalers were prohibited from exercising freedom on the oceans, the International Whaling Commission imposed a ban on whaling in 1966. As was noted in Chapter 7, an alternative policy would have been a tax on catch. The tax would have served as the rents that the international community could have earned from an asset that is among the resources dubbed in 1958 at the First UN Convention on the Law of the Sea to be the Common Heritage of Mankind. However, as was noted in Chapters 5 and 7, prohibitions can be shown to be superior to taxes as policy instruments when a resource is threatened with extinction and the point at which extinction occurs is not known with certainty. In an interesting and important paper, Spence (1974) showed that a moratorium on whale hunting was the right instrument even if there was no uncertainty about the species' recovery rate.[440]

To get a sense of the problem facing the International Whaling Commission, we may think of the Commission to be representing a monopoly, interested in profits from whale products. Harvesting whales is a production activity. The activity is harder the smaller is the whale population (search costs are higher).[441] We assume that if X is the number of boats deployed in whaling and S is the whale population, the harvest rate R can be expressed in the form of the function

$$R = XS^\alpha \qquad 0 < \alpha < 1 \qquad (A13.2.1)$$

Let t denote date. We assume for simplicity that the price of a whale's products, P, and the rental price of a boat, Q, are constant over time. It follows that the industry's profit at a date t, which is assumed to be continuous, can be expressed as

$$\Pi(t) = [P - Q/S^\alpha(t)]R(t) \qquad (A13.2.2)$$

The regeneration rate of the population is assumed, as in Chapter 3, Box 3.3, equation (B3.3.3), to be

$$dS(t)/dt = G(S(t)) - R(t) \qquad (A13.2.3)$$

and

$$G(S(t)) = rS(t)[1 - S(t)/K(t)][(S(t) - A)/K(t)], A, r, K, (K - A) > 0 \qquad (A13.2.4)$$

Let the long run commercial interest rate on savings be a constant $\rho > 0$. Then the NPV of profits from harvesting at the rate $R(t)$ would be

$$\Pi(S(0)) = \int_0^\infty \{[P - Q/S^\alpha(t)]\, R(t)\} e^{-\rho t} dt \qquad (A13.2.5)$$

We imagine 1966 to be the year when the Commission's optimisation problem is posed. That means $t = 0$ in 1966. In the early 1960s, blue whales were endangered (population size approximately 1,500), but it was thought, rightly as it turned out, that recovery was possible so long as harvesting was restrained. So, we assume that $S(0) > A$. The problem facing the whaling industry was to choose $R(t)$ for all t so as to maximise Π subject to the regeneration equation (A13.2.4) and the initial whale population $S(0)$. That is

---

[440] Spence and Starrett (1975) identified a more general set of circumstances than in Spence's specification, under which a moratorium is the right policy.

[441] We assume, with Spence (1974), that we can ignore various sub-species of blue whales and talk simply of blue whales.

Chapter 13: Sustainability Assessment and Policy Analysis

the Commission's asset management problem (Chapters 1 and 13). We take it that $r > \rho$ (otherwise the Commission would not have found it profitable to preserve blue whales).

Notice that the present value of profits in equation (A13.2.5) is proportional to R. But if the whale population were allowed to increase, harvesting costs would decline (Equation (A13.2.1)). These twin features of the maximisation problem suggest that if $S(0)$ is small, a temporary moratorium on whale hunting is optimum policy, as it would allow the whale population to recover at the fastest possible rate.

How long should the moratorium last? The answer depends on the target population size that is implied by profit maximisation. And that, in turn, depends on the various parameters in equations (A13.2.2)–(A13.2.4); that is, P, Q, r, ρ, and so on. Spence (1974) showed that to be the case. He showed that under a wide range of plausible parameter values, it was most profitable for the international whaling industry to agree on a moratorium until the desired long-run whale population was reached, and for the industry to then harvest blue whales at a rate equal to the population's (optimum) sustainable yield. Let S^* be the target. To see how to calculate S^*, we have to dig deeper into the structure of the problem facing the Commission.

From equation (A13.2.3), we know at a stationary state $G(S) = R$. If T is the date at which the most rapid approach path to S^* is reached, we may write equation (A13.2.5) as

$$\Pi(S^*) = \int_T^\infty \{[P - Q/S^{*\alpha}] G(S^*)\} e^{-\rho(t-T)} dt \tag{A13.2.6}$$

Or, writing $\Pi(S^*)$ as the stationary NPV of industry profits,

$$\Pi(S^*) = [P - Q/S^{*\alpha}] G(S^*)/\rho \tag{A13.2.7}$$

Blue whales can survive only in the oceans. So we need to distinguish between the accounting price of blue whales in their habitat and the accounting price of whales once they are harvested.[442] The latter is P per whale. Denote the former by M and the maximised value of Π by Π^*. By definition (Chapter 10), the accounting price of a blue whale *in situ* is, from equation (A13.2.7),

$$M(S) = \partial \Pi^*(S)/\partial S \tag{A13.2.8}$$

We now recall from Chapters 1 and 2 that the portfolio problem here requires that the rates of return on the two assets in the International Whaling Commission's exercise be equal: blue whales *in situ* and the financial asset that yields a rate of return of ρ. At a stationary state the own rates of return (i.e. yields) are equalised.

In order to determine the own rate of return on the stock of blue whales, the Commission had to calculate the profitability of departing from the stationary state by harvesting one unit less at some date and then enjoying a steady stream of profits in perpetuity, other things being held fixed (Chapter 10). The return from that move can be decomposed into two parts. First, there is the stock's yield in perpetuity (in terms of a larger future stock) at the rate $dG(S)/dS$. Second, there would be a future gain in profits arising from the fact that harvest costs would be slightly lower in perpetuity. On using equation (A13.2.2), that gain per unit of time is $\alpha Q G(S) S^{-(1+\alpha)}/M(S)$. Once the whale population reaches the optimum stationary state S^*, the arbitrage condition (Chapter 1) becomes,

$$\rho = dG(S)/dS + \alpha Q G(S) S^{-(1+\alpha)}/M(S) \tag{A13.2.9}$$

It is a routine matter to check that we can determine S^* from equations (A13.2.7)–(A13.2.9).

---

[442] The difference could be viewed as a transport cost.

Chapter 13: Sustainability Assessment and Policy Analysis

So we have one equation (Equation (A13.2.9)) and one unknown (S^*). The problem is, the equation may have multiple solutions.[443] Some are local minima of $\Pi^*(S)$, others are local maxima. Note though that the optimal policy is not necessarily the one that aims for the S at which $\Pi^*(S)$ is the global maximum. The reason is that profits are zero during the approach to the optimum stationary stock. If the highest peak, as in equation (A13.2.3), is sufficiently far away from the initial stock, $S(0)$, it may be better to aim for a solution to equation (A13.2.7) that is only a local maximum of $\Pi^*(S)$ but is closer to S(0). The industry loses by enjoying the long run stationary profit flow than could have been achieved but gains in getting to the lower flow of profits at an earlier date. This reasoning allows the Commission to identify the ideal target stock, S^*.

Spence (1974) conducted sensitivity analysis by considering alternative sets of values for the parameters in the optimisation problem and found that the moratorium should have been between 10 and 20 years. So, saving the whales was optimum from the standpoint of commercial profits.

But there is a problem with the entire exercise. Π^* is the value of blue whales in terms of the profits the whaling industry can earn from selling whale products. It can be argued though that even at a minimum, blue whales have a value *in situ* as a carbon store. The latter value reflects a global benefit, not commercial profit. Recently, Chami et al. (2019) reported work conducted at the International Monetary Fund, which has found that at an accounting price of US$70 for a ton of carbon, the social value of an average blue whale would be US$6 million. P in contrast is even today at best US$90,000. It follows that if we were to impute a direct value to whales as a store of carbon and not just as source of whale products, the accounting price of a whale *in situ* would be greater than M. Replacing M in equation (A13.2.6) by that higher value would mean that the long run stationary population of whales would be higher.

Even this extension could appear unethical to people. Many would insist blue whales possess an intrinsic value, let alone an existence value many people would regard them to have. If we add those values to the commercial value that Spence considered, the long-run stationary population humanity should strive to attain for blue whales would be higher still. Which is why Spence's formulation, repugnant though it would appear to many people, is of enormous value. It provides us with a biased estimate of a whale's worth, and this is useful information.

His analysis recommended preservation of blue whales solely on commercial grounds. But if preservation is the right policy when the accounting price of blue whales is estimated from their market price only after they have been slaughtered, the recommendation would be reinforced if their intrinsic worth, as stock, were added. That is the message we should take away from Spence's exercise.

Annex 13.3 The Significance of GDP

We have confirmed that economic evaluation involves wealth comparisons, over time and in the choice of policies. Economic progress should be read as growth in inclusive wealth. Nevertheless, in national economic accounts the measure most commonly taken to correspond to well-being is GDP, which is the market value of the flow of final goods and services in a country in a given year. The index is a measure of economic activity. In any given year, if Z is aggregate private consumption (i.e. total consumer spending), I is gross investment in produced capital, G is the sum of government expenditures, and Ex and Im are the market value of exports and imports, respectively, then

$$GDP = Z + I + G + Ex - Im \qquad (A13.3.1)$$

---

[443] Spence and Starrett (1975) has a fine explanation for why we should expect the stationarity state of the arbitrage conditions in this problem not to be unique.

In a market economy, GDP is also the sum of domestic wages, salaries, profits, interests, rents and government income net of taxes (the output has to reach *somebody*'s hands!). That means GDP and gross domestic income are two sides of the same coin.

If the economy is closed to trade with the outside world, $Ex = Im = 0$, and equation (A13.3.1) reduces to

$$GDP = Z + I + G \tag{A13.3.2}$$

Equation (A13.3.2) is a reasonable approximation for economies that trade little and is exact for the world economy, which has no outside world to trade with.

GDP should not be used to evaluate long run prospects. Among other things, GDP ignores the depreciation of capital goods. The measure was created and designed for a different purpose from economic evaluation. Estimating the magnitude of economic activity became necessary in the Great Depression of the 1930s, when some 25% of working-age people in Europe and the US were recorded as being unemployed and a corresponding proportion of factories and resources lay idle (Kuznets, 1941). In the years following the Second World War, when reconstruction of Europe and the Far East and economic development in what were previously European colonies became a matter for economic policy, GDP assumed its role as the measure of long-term economic progress. It is hard to tell why that transfer of GDP's purpose from the short to the long run came about.[444]

But even as you go beyond GDP, you find yourself returning to it. The index remains essential in short-run macroeconomics. It allows economists to estimate the gap between the economy's potential output and actual output and is useful also for studying household and corporate behaviour. As public goods and services are typically supplied by government (local or national), the government requires funds to finance them. If the resources are to be obtained from taxes, there has to be sufficient income in the economy *to* tax. So, economics and finance ministers are drawn to GDP for revenue. As a criterion for evaluating short run economic policy, GDP has served admirably, but ignoring capital depreciation is indefensible in economic evaluation involving the long run.

Here is a sectoral illustration of what goes wrong. Repetto et al. (1989) and Vincent, Ali and Chang estimated the decline in forest cover in Indonesia and Malaysia, respectively. They found that when the decline is included, national accounts look different: net domestic saving rates turn out to have been some 20–30% lower than recorded saving rates. In their work on the depreciation of natural resources in Costa Rica, Solórzano et al. (1991) found that the depreciation of three resources – forests, soil, and fisheries – amounted to about 10% of GDP and over a third of domestic saving.

That GDP can mislead badly when it is put to use in sustainability assessment and policy analysis is now well known, which is why the measure has a hard time these days among the general public and non-governmental organisations. Hardly a month goes by before another publication bearing the title, 'Beyond GDP', makes an appearance. Nevertheless, GDP growth is likely to remain a measure of economic progress in official eyes. Here are two reasons why:

A13.3.1 GDP and Global Competition

GDP is the market value of final goods and services. Those goods and services can be deployed so as to gain advantage in the international sphere. Never mind that a country may be enjoying a large GDP while at the same time depleting its natural resources, ruining the environment, perhaps even running

---

[444] Lewis (1988) offers an account of the historical roots of modern development theory, but does not question that GDP is the object of normative interest. Coyle (2014) contains a history of the evolution of the ways in which GDP has been measured. She explains why seemingly small changes in the way GDP is measured can and have brought about large changes in estimates.

rough-shod over indigenous populations for access to land and minerals; GDP can be (and is) used by governments as a strategic weapon in a world where nations compete against one another for economic and political influence. Not only does a nation's status in the world rise if it enjoys high rates of growth of GDP, a large GDP enables a nation to tilt the terms of trade with the rest of the world to its advantage. History is replete with examples that demonstrate the strategic advantages of GDP growth.

A13.3.2 GDP and Employment

There is another systemic problem in modern industrial societies that makes GDP pivotal. Economists have yet to discover ways to manage the macro-economy in which GDP is delinked from recorded employment. That GDP needs to keep rising if employment is not to decline is a view that appears to be shared by economists and decision makers, be they Keynesians or otherwise. Politicians and media commentators become deeply anxious when spending on consumer goods shows signs of stalling. We are encouraged to think that to consume is to contribute to the social good. It is more than an irony that short-run macroeconomic reasoning is wholly at odds with accumulation of inclusive wealth.

Chapter 13*
Accounting Prices and Inclusive Wealth

Introduction

Here we derive equations for optimum allocations of goods and services in the global economic model of Chapter 4*. The equations are dynamic, and represent rules for determining the optimum mix of consumption and investment in capital goods. The criterion for optimality adopted here is the present value of well-being across the generations (Chapter 10). For vividness, we may imagine that the social evaluator, who was introduced in Chapter 10, performs the optimisation exercise and derives the movement of accounting prices of goods and services over time. The procedure is at the heart of a decentralised economy guided by the state on matters where the price system performs badly. We confirm that the dynamical equations that govern the movement of accounting prices are none other than the arbitrage conditions of Chapter 1 (Box 1.3). That is why in Chapter 10 the social evaluator was also referred to as the citizen investor.

We illustrate the mode of analysis for arriving at arbitrage conditions by studying a fully optimising economy, or, what in Chapter 10 was called a 'first-best'. It is assumed that institutions are in place for supporting the optimum programme (e.g. for eliminating externalities). The arbitrage conditions express the way accounting prices can be used in those institutions to sustain the optimum, at each moment in time and across time. The formula for inclusive wealth in the economy is derived. From the wealth/well-being equivalence theorem (Chapter 1) we know that arbitrage conditions also reflect the rules to be followed by the social evaluator if the optimality criterion she adopts is the maximisation of inclusive wealth. Those same rules govern social cost-benefit analysis. The present chapter thus establishes a claim with which we began (Chapter 1) – that the economics of biodiversity is the study of portfolio management.

It cannot be emphasised strongly enough that the *structure* of the arbitrage conditions we derive is independent of whether the economy is fully optimum or it is plagued by imperfections. Optimisation of resource allocation in imperfect economies, no matter how limited are the gains to be enjoyed from optimising the allocation of resources, would yield arbitrage conditions having the *same* structure: *Resources would be so allocated that the rates of return on all capital goods (adjusted for risk) are equated; moreover, the rate of return on each capital good would be the sum of its yield and the capital gains/losses on holding it*. Accounting prices, however, do depend on the economy's imperfections, as does the way the social evaluator should measure yields. We study a fully optimising economy only because otherwise we would have to specify the institutional constraints that characterise the imperfections.[445]

Notation:

Y: real global GDP (aggregate output)

A: stock of publicly available knowledge

S: stock of natural capital

---

[445] The classic on the optimum resource allocation in a dynamic imperfect economy (i.e. a 'second-best' economy; Chapter 10) is Arrow and Kurz (1970). Among other things, the authors derived accounting prices that supported the arbitrage conditions in an economy where, because of a lack of full flexibility in raising taxes, the government is unable to supply the first-best (Chapter 10) quantity of public goods.

Chapter 13*: Accounting Prices and Inclusive Wealth

K: stock of produced capital

H: stock of human capital

R: extraction rate of Nature's goods and services for production

G: regeneration rate of natural capital

    N: global population size

    h: human capital per capita ($h = H/N$)

    J: long-run global population size ($J = J(h)$; $dJ(h)/dh < 0$)

C: aggregate consumption

    I_K: gross investment in produced capital

    I_H: investment in human capital

    I_A: public investment in research and development

    I: gross aggregate investment ($= I_K + I_H + I_A$)

    $\alpha_Z(A)$: efficiency parameter

13*.1 The Model

The stock/flow equations characterising the global economy constructed in Chapter 4* were six in number: equations (4*.2), (4*.4), (4*.7), and (4*.10)–(4*.12). We first restate them here for convenience ((13*.1)–(13*.6)).

Time $t \geq 0$ is continuous. Aggregate output at t is

$$Y(t) = A(t)S^\beta(t)K^a(t)H^b(t)R^{(1-a-b)}(t), \qquad \beta > 0;\ a, b, (1-a-b) > 0 \qquad (13^*.1)$$

The dynamical stock/flow equations are

$$dS(t)/dt = G(S(t)) - R(t) - Y(t)/\alpha_Z(A), \qquad \alpha^* > \alpha_Z(A) > 0 \qquad (13^*.2)$$

$$dK(t)/dt = Y(t) - C(t) - I_H(t) - I_A(t) - \lambda K(t) \qquad (13^*.3)$$

$$dN(t)/dt = N(t)[J(h(t)) - N(t)], \qquad dJ/dh < 0 \qquad (13^*.4)$$

$$dA(t)/dt = I_A(t) \qquad (13^*.5)$$

$$dH(t)/dt = N(t)dh(t)/dt + h(t)dN(t)/dt = I_H(t) \qquad (13^*.6)$$

For notational simplicity we have assumed in equation (13*.4) that only human capital per person affects J, the latter being the size the human population gravitates to in the long run. Thus we write $J = J(h)$. Human capital includes not only education as conventionally understood, but also knowledge of modern family planning services and access to them (Chapter 9).

The efficiency parameter α_Z is assumed to be an increasing function of publicly available knowledge A, but for reasons explained in Chapter 4 (Section 4.7), it is assumed that $\alpha_Z \to \alpha^*$ as $A \to \infty$.

13*.2 The Optimisation Problem

Let $c(t)$ be consumption per capita at t. Thus $c(t) = C(t)/N(t)$. Nature is taken to have non-use values over and above use value. For simplicity of notation, we amalgamate all non-use values (Chapter 12) and write the well-being function as $u(c,S)$.[446] It is assumed that (i) u is strongly cardinal and concave (Chapters 5, 10); (ii) $u(c,S) > 0$ for some c and S; (iii) $u \to -\infty$ as $c \to 0$; (iv) $u_c, u_S > 0$.[447] Subsequently, we study the structure of accounting prices if people care more about the quality of their environment as their standard of living rises.[448] Formally, that would mean

$$\partial(u_S/u_c)/\partial c > 0 \tag{13*.7}$$

Social well-being at t is $N(t)u(c(t),S(t))$ and the criterion the social evaluator adopts at $t = 0$ for comparing alternative socio-ecological futures is

$$V(0) = \int_0^\infty [u(c(t),S(t))]e^{-\delta t}dt, \quad \delta > 0 \tag{13*.8}$$

The stocks of capital goods at $t = 0$ are a legacy of the past. They are $A(0)$, $K(0)$, $S(0)$, $N(0)$, and $h(0)$, The social evaluator only studies those socio-ecological futures that contain the biosphere in the stability regime it is in at the initial date. That requirement would follow from her recognition that to cross a planetary boundary would be a catastrophe.[449]

Because $H(t) = h(t)N(t)$, it helps to use equation (13*.4) and re-express equation (13*.6) as

$$dh(t)/dt = I_H(t)/N(t) - h(t)[J(h(t)) - N(t)] \tag{13*.9}$$

The social evaluator's problem is to choose the socio-ecological future that maximises $V(0)$ in equation (13*.8) subject to the dynamical equations (13*.1)-(13*.5) and (13*.9), with $A(0)$, $K(0)$, $S(0)$, $N(0)$, and $h(0)$ serving as initial conditions. Every socio-ecological future can be attained by suitable choice of the four *control variables* $C(t)$, $I_H(t)$, $I_A(t)$, and $R(t)$. The capital stocks $A(t)$, $S(t)$, $K(t)$, $N(t)$, and $h(t)$, are called *state variables*.

It will prove helpful visually to write $K^a H^b R^{(1-a-b)}$ as $F(K,H,R)$. That way, we will be able to express the marginal products of the three forms of capital goods as F_K, F_H, and F_R. Equation (13*.1) can thus be expressed as

$$Y(t) = A(t)S^\beta(t)F(K(t),H(t),R(t)) \tag{13*.1a}$$

13*.3 Arbitrage Conditions and the Structure of Accounting Prices

So as to remain neutral among the various forms of capital goods, we choose social well-being $Nu(c,S)$ as the unit of account (*numeraire*). Let p_A, p_S, p_K, p_h, and p_N denote the accounting prices of A, S, K, h, and N.[450] Each is a function of time. At an optimum, the former four accounting prices are related by the following conditions:

$$\begin{aligned} u_{c(t)} = p_K(t) &= p_h(t)/N(t) = p_A(t) \\ &= p_S(t)[1 + A(t)S^\beta(t)F_{R(t)}/\alpha_Z]/[A(t)S^\beta(t)F_{R(t)}] \end{aligned} \tag{13*.10}$$

---

[446] We could include health, an aspect of human capital, as contributing directly to well-being u. The u-function could also be enlarged to reflect socially embedded preferences (Chapter 9). But those extensions would merely add to the notation.

[447] u_c and u_S stand for $\partial u(c,S)/\partial c$ and $\partial u(c,S)/\partial S$, respectively.

[448] This issue was given careful attention by Heal (1998) and Hoel and Sterner (2007).

[449] In order to make planetary boundaries vivid, $G(S)$ was taken from Chapter 4* (Equation (4*5)) to be of the form $G(S) = rS(1-S/\underline{S})[(S-L)/\underline{S}]$, where $L, r, \underline{S} > 0$. The parameter L represents the critical boundary. In the text here, we suppose that $G(S)$ increases with S for small S, attains a maximum (the maximum sustainable yield) at some value of S and declines thereafter and is negative at large enough values of S. G is thus a bounded function of S.

[450] The subscript to the accounting prices are just that, they are there to distinguish between the capital goods. Subscripts to other dependent variables such as $F(K,H,R)$ and $G(S)$ are (partial) derivatives with respect to the variable represented in the subscript. For example, $F_K(K,H,R) = \partial F/\partial K$. See footnote 419.

Chapter 13*: Accounting Prices and Inclusive Wealth

The first three equalities in equation (13*.10) are simple to understand. Global accounts of the division of output into consumption and various forms of investment (Equation (13*.3)) say that a unit less of consumption at t can be invested on an equivalent basis in any of produced capital, human capital, and knowledge. Equation (13*.10) says that at the margin an optimum deployment of consumption and the three forms of investments contribute equally to intergenerational well-being $V(0)$. The social evaluator would thus be indifferent between the option to consume an additional unit of output or to invest the additional unit of output in any of the three capital goods. Their accounting prices must therefore be the same. The presence of $N(t)$ in the third expression in the equation is explained by the decision to represent human capital in per capita terms (Equation (13*.9)).

The fourth equality is only slightly complicated. When an additional unit of natural capital is withdrawn for use in production, it raises output, whose accounting value is $p_K(t)AS^\beta(t)F_R(t)$. But there is a cost associated with the withdrawal, consisting of the value of a unit less of natural capital, which is $p_S(t)$, plus the value of the damage caused by the waste from the additional output, which is $p_S[1+A(t)S^\beta(t)F_{R(t)}/\alpha_Z]$. The fourth equality says that at an optimum the benefit equals the cost.

There is no corresponding equation for the accounting price of human numbers, N. *That is because the model does not permit population to be controlled directly.* Future population size can be influenced by investment in human capital h, but h itself is a stock of capital. There is, however, an arbitrage condition relating N with the other capital stocks. That is derived below.

The *numeraire* is intergenerational well-being $N(t)u(c(t),S(t))$. We now follow the reasoning and notation in Chapter 10. The date at which the optimisation exercise is being undertaken by the social evaluator is $t = 0$. Consider produced capital first. Its accounting price at time t is $p_K(t)$. That's the *spot price*. As the well-being discount rate is δ, the *present-value accounting price* of produced capital is $e^{-\delta t}p_K(t)$. Using the first part of equation (13*.10) it follows that the *consumption discount rate* $\rho(t)$ is the percentage rate of decline of $e^{-\delta t}p_K(t)$. At an optimum, $\rho(t)$ equals the *own rate of return* $r_K(t)$ on produced capital. The latter is produced capital's *yield* (Chapter 1), and it equals the marginal product of produced capital, $A(t)S^\beta(t)F_{K(t)}$, less the rate of depreciation of produced capital, λ, less the degradation of natural capital caused by the waste resulting from production, $p_S(t)A(t)S^\beta(t)F_{K(t)}/[p_K(t)\alpha_Z]$. Equality of $\rho(t)$ and $r_K(t)$ can thus be expressed as

$$\delta - [dp_K(t)/dt]/p_K(t) = A(t)S^\beta(t)F_{K(t)} - \lambda - p_S(t)A(t)S^\beta(t)F_{K(t)}/p_K(t)\alpha_Z \qquad (13^*.11)$$

Re-express equation (13*.11) as

$$\delta = A(t)S^\beta(t)F_{K(t)} - \lambda - p_S(t)A(t)S^\beta(t)F_{K(t)}/[p_K(t)\alpha_Z] + [dp_K(t)/dt]/p_K(t) \qquad (13^*.12)$$

Equation (13*.12) is the *arbitrage condition* (Chapter 1) between well-being and produced capital.[451]

Let $r_S(t)$ be the own rate of return on investment in natural capital. It consists of four terms: (i) its marginal contribution to the flow of well-being, $N(t)u_S(t)/p_S(t)$; (ii) its marginal contribution to output, $\beta p_K(t)A(t)S^{(\beta-1)}(t)F(K(t),H(t),R(t))/p_S(t)$; (iii) natural capital's net regeneration rate, $G(S(t))$; (iv) the marginal damage caused by the waste created by the additional output, $\beta A(t) S^{(\beta-1)}(t)F(K(t),H(t),R(t))/\alpha_Z$. The own rate of return is the algebraic sum of the four terms. Equality between $\rho(t)$ and $r_S(t)$ can thus be expressed as

---

[451] Arbitrage conditions between capital goods are known in mathematics as Euler-Lagrange equations. They can be arrived at by using standard variational techniques in optimisation theory. In this Review, we have been deriving the conditions intuitively, but the intuition has followed the mathematics closely.

… Chapter 13*: Accounting Prices and Inclusive Wealth

$$\delta - [dp_S(t)/dt]/p_S(t) =$$
$$N(t)u_S(t)/p_S(t) + \beta[(p_K(t)/p_S(t)) - 1/\alpha_Z]A(t)S^{(\beta-1)}(t)F(K(t),H(t),R(t)) + G_S(t) \quad (13^*.13)$$

Equation (13*.13) can now be re-expressed as

$$\delta = N(t)u_S(t)/p_S(t) + \beta[(p_K(t)/p_S(t)) - 1/\alpha_Z]A(t)S^{(\beta-1)}(t)F(K(t),H(t),R(t))$$
$$+ G_S(t) + [dp_S(t)/dt]/p_S(t) \quad (13^*.14)$$

Equation (13*.14) is the arbitrage condition between well-being and natural capital.

Consider next capital goods A, h, and N. The arbitrage conditions between each of them and well-being are

$$\delta = [S^\beta(t)F(K(t),H(t),R(t))]\{p_K(t) - p_S(t)[(1-(d\alpha_Z/dA(t))/\alpha_Z]/\alpha_Z\}/p_A(t)$$
$$+ [dp_A(t)/dt]/p_A(t) \quad (13^*.15)$$

$$\delta = N(t)A(t)S^\beta(t)F_{H(t)}[p_K(t) - p_S(t)/\alpha_Z]/p_h(t) - [J(h(t)) + h(t)dJ(h(t))/dh(t)]$$
$$+ p_N(t)N(t)[dJ(h(t))/dh(t)]/p_h(t) + [dp_h(t)/dt]/p_h(t) \quad (13^*.16)$$

$$\delta = u(c(t),S(t))/p_N(t) + p_h(t)h(t)/p_N(t)$$
$$+ J(h(t))-2N(t) + [dp_N(t)/dt]/p_N(t) \quad (13^*.17)$$

As the reasoning behind the expressions for the rates of return on A, h and N on the right-hand side of equations (13*.12), (13*.14), and (13*.15)-(13*.17) is similar to the one we pursued for K and S, we do not rehearse them here. But equation (13*.17) stands out. It is the only equation that makes use of the fact that the u-function is strongly cardinal. The sign of u matters.

It is interesting that the optimum programme in an economy that has depleted its natural capital greatly may involve an initial period of belt-tightening, when well-being is maintained at a negative level. $R(t)$ is kept low, meaning that $Y(t)$ is kept low. That way natural capital is able to regenerate, which in turn will enable the economy in the long run to sustainably maintain a positive level of well-being.[452]

13*.4 Arbitrage Condition Between Produced Capital and Natural Capital

The five arbitrage conditions can be converted into arbitrage conditions that relate the five rates of return on capital goods directly to one another. The procedure for doing that was explained in Box 1.3 of Chapter 1 (see equation (B1.3)). As there are five capital goods in the model here, there are 10 possible arbitrage conditions (4+3+2+1).[453]

To illustrate them, we confine ourselves to the arbitrage condition between produced capital and natural capital. Let the accounting price of natural capital relative to the accounting price of produced capital be denoted as $p_{KS}(t)$. Equating the right-hand sides of equations (13*.12) and (13*.14) then yields

---

[452] That this is a possible course of events was shown in Dasgupta (1969).

[453] As shown in Chapter 10, the arbitrage conditions on their own are not enough, for they do not on their own say what the accounting prices ought to be at $t = 0$. The social evaluator needs a condition that points to the optimum long run state of affairs. That is known as the 'transversality condition'. We bypass it here and assume that the social evaluator has mastered Mirrlees (1967) who showed how accounting prices at the initial date can be computed.

Chapter 13*: Accounting Prices and Inclusive Wealth

$$A(t)S^{\beta}(t)F_{K(t)} - \lambda - p_{KS}(t)A(t)S^{\beta}(t)F_{K(t)}/\alpha_Z =$$
$$N(t)u_S(t)/p_{KS}(t) + \beta[1/p_{KS}(t) - 1/\alpha_Z]A(t)S^{(\beta-1)}(t)F(K(t),H(t),R(t))$$
$$+ G_S(t) + [dp_{KS}(t)/dt]/p_{KS}(t) \qquad (13^*.18)$$

Equation (13*.18) says that the yield on produced capital equals the yield on natural capital plus the accounting capital gains on natural capital relative to produced capital.

Centuries ago, natural capital globally was plentiful while produced capital was small in quantity. Even unguided intuition says the yield on natural capital would have been a lot smaller than the yield on produced capital. Equation (13*.18) would then say that the capital gains on natural capital relative to produced capital would have been positive.

Today matters are different. The evidence collated in Chapter 4 showed that the stock of natural capital relative to produced capital is perilously low, we are even in danger of crossing further planetary boundaries. If the social evaluator was charged with optimising the global allocation of resources today for now and the future, she would find the yield on natural capital higher than that on produced capital. Box 2.3 produced evidence of that; the yield on primary producers was found to be at least 19% a year, far in excess of yields on produced capital. The contemporary global economy, however, suffers from an enormous range of negative externalities arising from the fact that much of natural capital remains free, while a number of services are enjoyed at a negative price. If anything, market prices of natural capital relative to produced capital are declining, directly at odds with the arbitrage condition (13*.18).

13*.5 Inclusive Wealth and the Long Run

Inclusive wealth in the optimising global economy, expressed in well-being *numeraire* is

$$W(t) = p_K(t)K(t) + p_A(t)A(t) + p_h(t)h(t) + p_S(t)S(t) + p_N(t)N(t) \qquad (13^*.19)$$

In the long run, the optimum can be shown to converge to a stationary state. Stocks in a stationary state remain constant over time, as do accounting prices, which means capital gains are zero in each of the arbitrage conditions. Inclusive wealth remains constant in a stationary state.

Interestingly, the stationary solution of the combined set of stock-flow equations (Equations (13*.1)–(13*.6)) and arbitrage conditions (Equations (13*.11)–(13*.17)) –12 in number – is not unique. Appealing to the findings of Kurz (1968), we know that if capital stocks have intrinsic value (i.e. u is a function of S), there is a multiplicity of stationary states that satisfy the 12 equations. It can be shown that the stationary state to which the optimising economy converges in the long run depends on the size of the capital stocks the economy had inherited at $t = 0$. In short, history matters.

13*.6 Counterfactual Futures

It is a truism that policy decisions should be evidence based. But evidence drawn from bad models can mislead hugely. In Ch. 4 we showed that reliance on models that don't view the human economy as being embedded in Nature have had catastrophic consequences. In what follows we formalise the idea underlying *economic forecasts* by using the descriptive side of the dynamic model in section 13*.3, which was in turn the model presented in Ch. 4*.

Intertemporal economic models are systems of dynamic equations. The solution to the equations of a given model is a mapping from the set of all possible initial capital stocks into the set of all future economic possibilities. Call the mapping a 'resource allocation mechanism', or RAM (Dasgupta and Maler, 2000). For example, in an optimizing economy, RAM translates initial capital stocks into optimum economic paths. RAM reflects the economic analyst's understanding of the evolution of technology and social institutions and is the basis on which forecasters project future economic outcomes.

Imagine that an economy is perturbed by the injection of an additional unit of an asset at date t. The resulting change in well-being across the generations is the asset's accounting price at t (Ch. 10, sec. 10.8). As the impact of the perturbation would ripple through the economy's future, an asset's accounting price reflects not only ethical values but also projections of the economy into the future with and without that injection. Formally, if $S(t)$ is the stock of an asset and $V(t)$ is intergenerational well-being, then the asset's accounting price $p_S(t)$ is

$$p_S(t) = \partial V(t)/\partial S(t) \tag{13*.20}$$

Equation (13*.20) is at the heart of both sustainability assessment and policy analysis. Assessing sustainability involves measuring the changes an economy undergoes with the passage of time. In contrast, policy analysis involves measuring the perturbation created by a policy shift at a moment in time. Both exercises involve comparing intergenerational well-being $V(t)$ before and after the economy experiences the perturbation.

Imagine, for example, a national government providing incentives to shift capital goods around from one use to another, its intention being to allow deteriorating ecosystems to recover. The economic possibilities in the future under the new policy will be different from those that would have loomed were the policies to remain as they are ('business as usual'). Thus, evaluating counterfactual future histories is at the heart of policy analysis. Formally, policy analysis requires the social evaluator to ask what the future socio-ecological possibilities would be like if the stocks of capital the economy has inherited were to have been a different mix. To answer that requires her to have a socio-ecological model to work with; for without a model, she wouldn't know how the world would be like were the economy to be perturbed by the policy change.

Part II – Extensions

Chapter 14
Distribution and Sustainability

Introduction

Chapter 10 looked at the distribution of well-being across time and the generations. This chapter examines the distribution in space rather than time, of components of the Impact Equation introduced in Chapter 4. That is, it breaks down the Impact Equation – a *global* condition of sustainability – to examine how both our demands on the biosphere and the biosphere's supply are distributed geographically and among income groups.

On the demand side, there is significant variation in human activity (y), population (N) and efficiency in our engagement with the biosphere (α), both within and between countries. On the supply side, the stock of the biosphere (S) is spatially varied – there exists an inherent heterogeneity in ecosystems and natural resources throughout the world – and the intensity of depletion varies between and within countries and regions.

As noted in Chapter 4, when demands for an ecosystem's goods and services exceed its ability to supply them on a sustainable basis, the stock depreciates. To maintain our natural assets intact, demands on local ecosystems should match their regenerative rates. Demands have outstripped supply in some ecosystems and geographies but not in others, further affecting the distribution of natural assets. Trade breaks the link between demand and supply at smaller scales and is examined in Chapter 15.

14.1 Global Variation in Demand and Supply

Ecological footprint accounting has underscored the current variation in both humanity's demand and the biosphere's supply across the world. Ecological footprint accounting is an approach used to estimate the ecological impact of lifestyles and consumption patterns, including the quantity of food, goods and services residents consume, the natural resources they use, and the carbon dioxide emitted to provide these goods and services (Wackernagel and Beyers, 2019). Ecological footprint estimates for countries reveal significant differences: countries in the northern hemisphere have larger footprints per person than countries in the southern hemisphere (Lin et al. 2018; Wackernagel and Beyers, 2019).

Recent estimates have highlighted that for over 70% of countries, their ecological footprint (as estimated by how much biologically productive land and water is required to produce all the resources the country consumes and to absorb the waste it generates) is greater than their biocapacity (as measured by the ecosystems' capacity to produce biological materials used by people and to absorb waste), leading to what is known as an 'ecological deficit' (Figure 14.1) (Lin et al. 2018; York University Ecological Footprint Initiative and Global Footprint Network, 2020).[454]

---

[454] The minority of countries that are relatively more sustainable (as measured by a comparison of biocapacity and ecological footprint) tend to be in forested regions, such as the tropics and boreal latitudes.

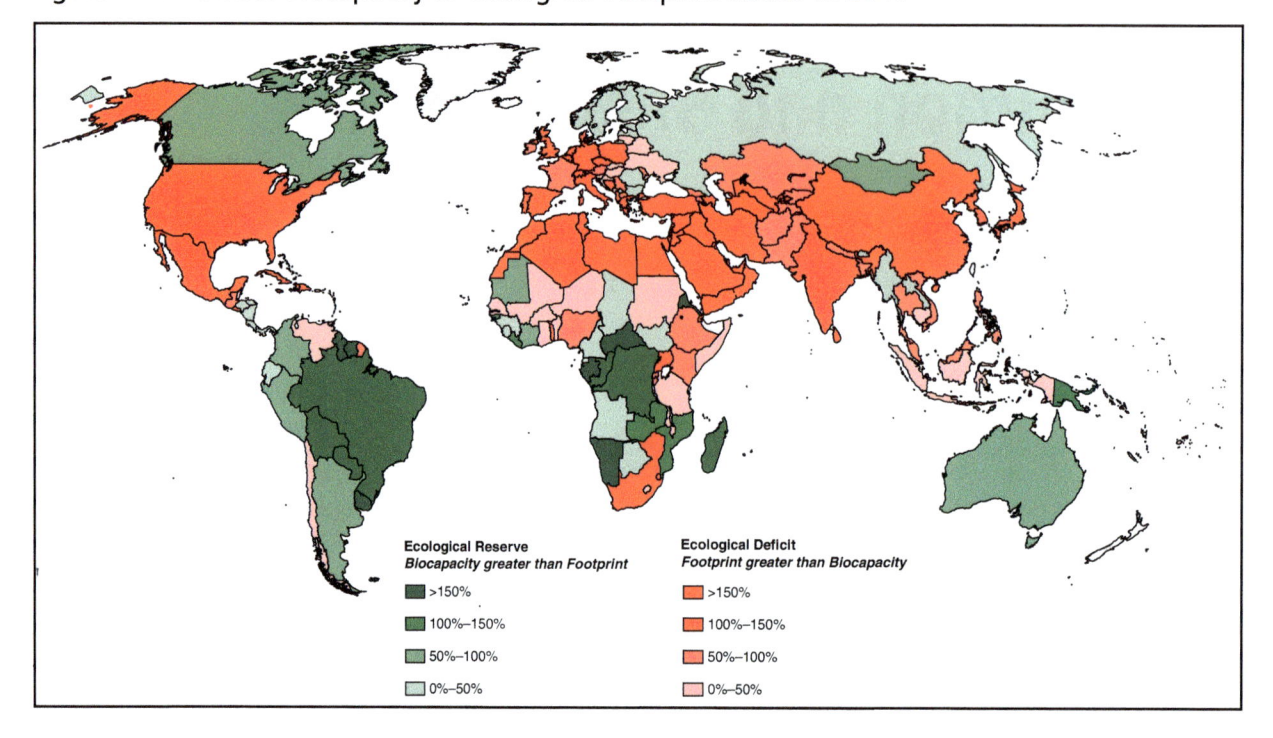

Figure 14.1 Global Biocapacity to Ecological Footprint Ratios in 2017

Source: York University Ecological Footprint Initiative and Global Footprint Network (2020).

14.2 Distribution of Humanity's Demands

The extent of the variation in both the level and growth of income across countries has already been discussed in Chapter 4. For example, income per capita in high income economies is more than 10 times higher, on average, than in low income economies. High income economies accounted for nearly half of PPP-based global GDP in 2017, while upper-middle income economies produced just over a third, and lower-middle and low income economies accounted for 16% and 1% of the global total, respectively (World Bank, 2020d). Given the relationship between income and consumption, these differing levels of income indicate that human economic activity varies significantly between countries.

There is evidence that human activity within countries also varies significantly in relation to income and wealth, with consequences for demands on (and thus condition of) natural assets (Berthe and Elie, 2015; Cushing et al. 2015).[455] For example, Barrett et al. (2020) suggest that around 50% of humanity's impact on the biosphere can be attributed to only around 16% of the world's population.[456] Chancel and Piketty (2015) found that around half of global carbon emissions could be attributed to the richest 10% of people globally, and that the majority of these individuals live in advanced economies. In contrast, the activities of the bottom 50% of the global population in terms of income produces only around 10% of global emissions. Gössling and Humpe (2020) found that in 2018, only 1% of the world's population likely accounted for more than half of the total emissions from passenger air travel.

---

[455] There is evidence that income inequality can hinder protection of natural assets, through influencing the actions of groups or institutions, shaping cooperation in sustaining local commons, or through market concentration of businesses (Hamann et al. 2018).

[456] In a similar analysis, Teixidó-Figueras et al. (2016) suggest that the world's top 10% of income-earners are responsible for between 25% and 43% of environmental impact. In contrast, the authors estimate that the world's bottom 10% of income-earners exert only around 3% to 5% of environmental impact.

While redistributions of incomes and wealth could temper some *extremities* in terms of human activities (and therefore reduce impacts on the biosphere), ecological footprints may be a concave rather than convex function of income at all levels (Chapter 4). Evidence presented in Chapter 4 suggests, worryingly, that egalitarian redistributions of incomes could lead to a larger global ecological footprint, *other things equal*. In other words, shifting around income would ultimately not reduce our ecological footprint unless it was accompanied by a change in the composition of human activity to make it less harmful to the biosphere. Therefore, changes to consumption and production patterns, or improvements in efficiency, are required *alongside* redistributions of income and wealth (Chapters 15 and 16).

Different consumption and production patterns have different impacts on the biosphere (e.g. WWF, 2018; Wackernagel and Beyers, 2019). The extent of the impact for any consumption activity depends on a range of factors, including: the type and amount of natural resources used in production; the distance between origin of production and destination of consumption; and the mode, route and duration of transportation (Wiedmann et al. 2006). Supply chains play an important role in determining the extent of the ecological impact of consumption activities (Chapter 15).

One important aspect of individuals' demand on the biosphere is their diet (Box 14.1). Aspects of diets contribute in different ways to major drivers of biodiversity loss, and diet varies significantly between countries.[457] While the average calorie intake per capita has risen globally over recent decades,[458] high income countries have the highest intake levels, and low income countries the lowest, particularly those in sub-Saharan Africa.[459] A key driver of this differential across countries is meat consumption. The share of calories derived from animal protein tends to increase with income level, and meat consumption is highest in advanced economies (FAO, 2017).

Diets in low income countries tend to be characterised by a high energy share from cereals, roots and tubers. As well as consumption of meat and other animal products, consumption of fruits and vegetables is also higher in high income countries (FAO, 2017). It is estimated that, in order to feed the global population with the average national diet for parts of north and south America, Europe and Oceania, more than 100% of the Earth's total land would be needed (Alexander et al. 2016).[460] Activities associated with the production or harvesting of food from ecosystems (as well as other provisioning goods such as fibre and energy) have been found to have significant impacts on biodiversity and ecosystems' ability to provide

---

Box 14.1
Environmental Impact of Different Diets

In a comprehensive global study, Poore and Nemecek (2018) assessed the environmental impact of food commodities via five environmental indicators: land use, freshwater withdrawals, greenhouse gas emissions, eutrophying emissions (such as phosphates) and acidifying emissions (such as sulphates). They assessed a total of 38,700 commercial farms in 119 countries and 40 products representing approximately 90% of calories consumed globally (Poore and Nemecek, 2018).

---

[457] The average global diet has contained increasing amounts of refined carbohydrates, added sugars, fats and animal products, and decreasing amounts of vegetables, coarse grains, fruits, complex carbohydrates and fibre (Popkin, Adair, and Ng, 2012; Keats and Wiggins, 2014; Khoury et al. 2014; Tilman and Clark, 2014).

[458] Calorie intake has been found to increase with income, as does dietary diversity (FAO, 2017).

[459] Disparities in food consumption are stark: around 860 million people still suffer severe food insecurity, of which 48% are in Africa (particularly sub-Saharan Africa) and 45% are in Latin America (World Food Programme, 2017).

[460] If the average North American diet was adopted globally, 178% of current agricultural land would be required to feed the world's population, whereas if the average Indian diet was adopted, only 55% of current agricultural land would be required (Alexander et al. 2016).

Chapter 14: Distribution and Sustainability

> The study revealed huge differences in the greenhouse gas emissions of different foods, and therefore different diets. Meat and other animal products generally have higher footprints than plant-based products, and for most foods, the majority of greenhouse gas emissions result from land-use change, which is also the primary driver of terrestrial biodiversity loss.
>
> Land-use change and the farm stage of the supply chain together make up more than 80% of the footprint for most food products (Poore and Nemecek, 2018). Diets rich in animal products have much higher footprints than those rich in plant products: greenhouse gas emissions from plant-based food are 10–50 times less than those of animal products. On average for all products, 25% of producers were responsible for 53% of each product's environmental impacts (Poore and Nemecek, 2018). For freshwater extractions, the skew was particularly pronounced: a mere 5% of the world's food is responsible for approximately 40% of the freshwater impact (Poore and Nemecek, 2018).

regulating services (see Chapter 16 on the extraction of provisioning goods and its consequences for regulating services).

Areas where human activity per capita is the highest tend to be the areas where a minority of the global population resides. For example, high income economies were home to just 17% of the world's population in 2017, but accounted for 49% of global GDP (World Bank, 2020d). The vast majority – around 75% – of the global population live in countries in which the average GDP per capita is below the global average (World Bank, 2020d). While population dynamics vary significantly in countries and regions, there is significantly less global population dispersion latitudinally.[461] It has been estimated that around 90% of the global population lives in the Northern Hemisphere, with the remainder residing in the Southern Hemisphere, and that around half of the world's humans reside within the area between 20°N and 40°N of the equator (Kummu and Varis, 2011). The main reason for this relative concentration is that many factors determining suitable living conditions for humans depend on the distance to the equator. For example, terrestrial temperatures and precipitation patterns depend strongly on latitude (though other factors, such as elevation, distance to the sea and wind patterns, also have influence).

In addition to the aggregate number of people, population density – the average number of people per unit of area – has been found to play a role in determining aggregate demands on the biosphere (Teixidó-Figueras et al. 2016). Population density influences geographic patterns of material, energy and land use. In modern economies, cities and metropolitan areas are centres of economic activity and therefore centres of human demand. Despite covering only around 3% of the Earth's land, it has been estimated that the world's cities consume 60–80% of manufactured energy and are responsible for 70% of carbon emissions (Fong et al. 2014).

While urban areas account for the bulk of human activity, they are also home to most of the world's population. For example, more than half of the world's population now live in urban areas – increasingly in highly dense cities. In most high income countries, more than 80% of the population live in urban areas (UNPD, 2019c).[462] In many low to lower-middle income countries, the majority still live in rural areas (UNPD, 2019c).[463] The density and proximity of the population and human activities in urban areas means

---

[461] Kummu and Varis (2011) suggest that natural resource demands are most pressing between 5°N and 50°N.

[462] In most upper-middle income countries – such as those in Eastern Europe, East Asia, North and Southern Africa, and South America – between 50% and 80% of people live in urban areas.

[463] For most of human history, the majority of people lived in small communities. Over the past few centuries – and particularly in recent decades – this has shifted dramatically. Contemporary urbanisation differs from historical patterns of urban growth in scale, rate, location, form and function (Seto, Sánchez-Rodríguez, and Fragkias, 2010).

that *per capita* consumption isrelatively lower than sparsely populated areas (Box 14.2) (Krausmann et al. 2008; Krausmann et al. 2009).[464]

Box 14.2
Urbanisation and Nature

Urban areas and their inhabitants are net importers of ecosystem goods and services and big consumers of natural resources to maintain the lifestyle and economic activities of the inhabitants. Cities are also a key source of waste and emissions. Globally, cities are growing twice as fast in terms of the land they cover as they are in terms of population and are expected to quadruple in area by 2050 (Angel et al. 2011). This growth in size threatens surrounding ecosystems which can exacerbate risks to urban populations. It is estimated that the growth of urban areas could threaten 290,000km2 of natural habitat (an area the size of Italy or the Philippines), and 40% of strictly protected areas will be within 50 km of a city by 2030 (McDonald et al. 2018).

Many people living in urban poverty are highly vulnerable to natural disasters and the degradation of ecosystems. Nearly 60% of cities with at least 500,000 inhabitants are at high risk of six natural disasters: floods, droughts, earthquakes, landslides, cyclones and volcanic eruptions (UNPD, 2018). For example, in 2005, unprecedented floods in Mumbai, India, killed over 1,400 people and caused US$2 billion in damages (Ranger et al. 2011). Further analysis of potential effects of floods showed that poorer households were disproportionately more likely to be exposed to the impacts of flooding in the city, and risks to richer households were virtually absent (Hallegatte et al. 2016).

Well-designed cities offer potential solutions for sustainable living taking advantage of the density and proximity of the population and businesses. Urban planners are increasingly looking to Nature to help improve quality-of-life for their inhabitants and ameliorate such problems as climate change and poor health as well are improving resource efficiency. Cities are hugely important for our society and economy, ensuring they are sustainable by increasing resource efficiency, reducing land take, reducing emissions and waste and incorporating green and blue infrastructure offers huge opportunities socially, economically and environmentally.

Population density can also directly impact the stock of natural assets and the provision of ecosystem services. There is some evidence increasing urban population density has led to declining ecological integrity at local scales, although the magnitude of this effect varies (Chace and Walsh, 2006; McDonnell and Hahs, 2009; Fuller and Irvine, 2013). Urbanisation has also been found to have an impact on Nature via land-use (Kalnay and Cai, 2003; Wu, Zhang, and Shen, 2011), which then affects natural habitats, as cities expand, onto either agricultural land or other already converted land (McDonald et al. 2018).

Urban growth has not only reduced the extent of natural habitat but has been found to degrade and fragment remaining habitats in certain areas leading to adverse consequences for biodiversity (McDonald et al. 2018). In a study assessing the consequences of compact and sprawling urban growth patterns on bird distributions throughout the city of Brisbane in Australia, Sushinsky et al. (2013) found that urban growth of any type reduces bird distributions overall, but more compact development of cities substantially slows these reductions (Sushinsky et al. 2013). Other studies have suggested that

---

[464] In addition to population, other demographic factors play a role in determining the extent of demands on the biosphere. These include gender, urbanisation, age profile, and household size, which can also influence impact on the biosphere (O'Neill et al. 2010; Zagheni 2011; Rosa and Dietz 2012; Lugauer et al. 2014; Bruins et al. 2017).

Chapter 14: Distribution and Sustainability

habitat fragmentation[465] in urban environments can result in lasting changes to the physical environment (for example, light and temperature) (Gonzalez, Mouquet, and Loreau, 2009; Pereira, Navarro, and Martins, 2012), which lead to the degradation of ecosystem function and declines in provisioning goods (Ziter, 2016). More efficient use of land in urban areas has been identified as being critical in minimising humanity's demands on the biosphere (Nelson et al. 2010).

Advances in technologies have the potential to increase efficiency globally, but their implementation to date has varied significantly across countries (see Chapter 16 for more details on technology to improve our efficiency in food production). Moreover, as pointed out in Chapter 7, the failure of market prices to reflect accounting prices means that incentives for the use of technology usually do not create incentives to be efficient with Nature's services in production. Food losses and waste claim a sizeable proportion of agricultural output – it is estimated that around a third of all food produced globally is lost or wasted – and reducing them would lessen the need for production increases (FAO, 2017). Some innovations – such as precision agriculture, vertical farming and cellular agriculture – have the potential to improve efficiency by reducing agriculture's contribution to damaging natural assets and pressures on biodiversity loss (see Chapter 16 for more on these technologies). But precision agriculture technologies, for example, are mainly implemented in high income countries (Keskin and Sekerli 2016; Bramley and Ouzman 2018; Lowenberg-DeBoer and Erickson, 2019).

The implementation of technologies varies between countries. One barrier to implementation is the issue of insecure property rights (see Chapter 7); variation in property rights also partly explains differences in efficiency between countries. The enforcement of property rights is unequal around the world, with people in low income and middle-income countries generally having lower security (Property Rights Alliance, 2019). Without sufficient security, producers have reduced incentive to make significant investments in improving the sustainability of their operations (Wang et al. 2016; The Food and Land Use Coalition, 2019). For example, tenure security has been shown to be a particular problem for indigenous people and local communities, who manage some of the most biodiverse areas on Earth (Rights and Resources Initiative, 2015; Garnett et al. 2018).

14.3 Distribution of the Biosphere's Supply

The distribution of the world's natural assets – and their associated biodiversity – is uneven across the Earth. For example, 17.3% of the Earth's land surface maintains 77% of all endemic plant species, 43% of vertebrates and 80% of all threatened amphibians, in 35 biodiversity hotspots (Mittermeier et al. 2011). Similarly, biodiversity is clustered in the sea: just 0.012% of the oceans have between 45% and 54% of restricted-range species, in 10 coral reef areas (Roberts et al. 2002). The distribution of the stock of natural assets, along with their quantity and quality (as discussed in Chapter 2, Annex 2.3), determines the availability of many ecosystem services to people. For some services, proximity to the stock is essential; for example, urban cooling by street trees is very localised. In contrast, mitigation of climate change from carbon storage in biomass can occur anywhere, and benefits the whole of society globally, including people living far away.

In a global study, Dickson et al. (2014) found that marine ecosystems and biodiversity[466] were concentrated in south-east Asia and along coastlines, especially the west coasts of South America, Africa and Europe (Figure 14.2). Terrestrial ecosystems, such as tropical and boreal forests, are concentrated in the equatorial regions and parts of Canada and Russia. Freshwater resources (i.e. renewable water

---

[465] Habitat fragmentation may be defined as a discontinuity in the spatial distribution of natural resources and ecosystem conditions. Fragmentation can affect the survival, reproduction and mobility of multiple interacting species (McDonald et al. 2018).

[466] Key assets included were freshwater resources, soil quality for plant growth, terrestrial organic carbon, terrestrial biodiversity, marine biodiversity and marine fish stocks.

resources replenished annually including lakes) are unevenly distributed throughout the world, with large quantities in Greenland, the west coast of North America, much of South America, the Congo basin, and in large parts of south and south-east Asia. Global terrestrial organic carbon stocks are high in tropical and boreal forest regions: stocks in the tropics are predominantly found in vegetation, and stocks in the boreal regions are predominantly in soils (Dickson et al. 2014). On land, biodiversity is richest in the tropics,[467] and marine biodiversity is highest along the coasts in the Western Pacific.

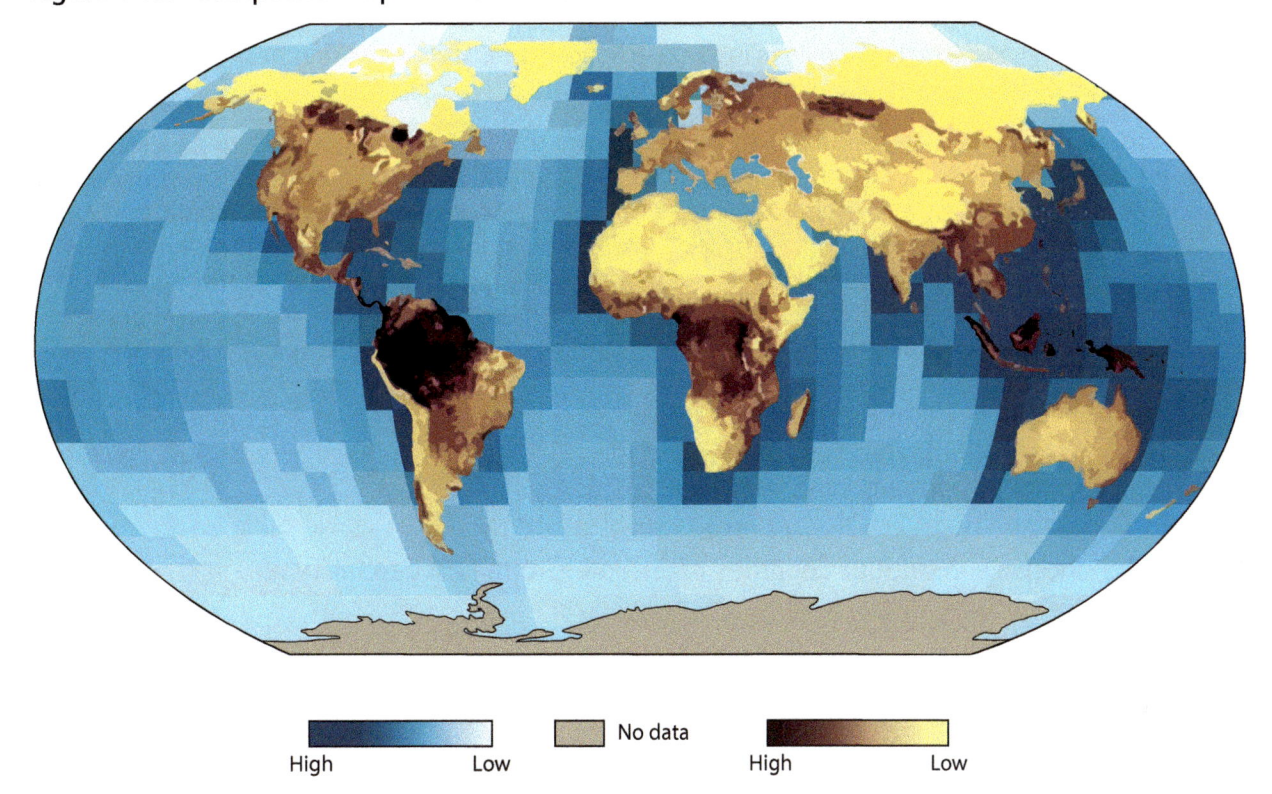

Figure 14.2 Composite Map of Natural Assets across the World

Source: Dickson et al. (2014).

Closer inspection of diversity in the tropics reveals further variation across different regions. Comparison of numbers of indigenous vascular plant species in Africa, Latin America and south-east Asia showed that south-east Asia was proportionately the richest in plant diversity for its size (Raven et al. 2020). However, Latin America is estimated to have a third or more of existing vascular plant species. In the Afrotropical Region, the island of Madagascar makes an important contribution to diversity for its size in all the taxa studied except for butterflies (Raven et al. 2020). The uneven distribution of the world's natural assets is the result of both natural variations in natural processes and properties (such as rainfall, temperature, elevation, soil and geography), and variation in the drivers and intensity of change and depletion.

The spatial configuration of different ecosystems influences their forms and functions as well as the ecosystem services they provide.[468] The provision of services such as crop pollination and flood mitigation depends on specific ecosystems located in a specific configuration in relation to human communities and

---

[467] Extensive areas of largely intact biodiversity have been identified in tropical rainforests close to the equator (Dickson et al. 2014).

[468] Devising models to show where Nature matters most to people has been challenging (Rieb et al. 2017). For services for which the spatial configuration matters (unlike carbon sequestration for climate regulation, which provides global benefits regardless of where it is located), models need to be able to work at fine scales to link the spatial distributions of natural assets with flows of ecosystem services in relation to people.

Chapter 14: Distribution and Sustainability

economic activities, for people to be able to benefit. Other benefits, such as the contribution Nature can make to climate regulation and mitigation, are global, such that everyone benefits regardless of their location. Brauman et al. (2019) highlighted the importance of the location of natural assets and availability of ecosystem services. The authors showed that the impacts of declining ecosystem services vary with geography and among people. For example, storm damage affects coastal communities where there has been a loss of mangroves more than people living inland (Brauman et al. 2019). As noted in Chapter 11, other studies have highlighted the importance of local greenspace for a wide range of health and well-being outcomes (for example, Twohig-Bennett and Jones, 2018). In England, a 10-year survey of over 450,000 people between 2009 and 2019 showed the importance of local greenspace: an increasing number of people travel less than a mile to local greenspaces, with visits increasing from 1.2 to 1.7 billion over the length of the survey (Natural England, 2019).

The extent of geographical variation in natural assets, and the variation in their depletion, means that there are significant differences in the domestic supply of natural assets across countries, and thus the proportion that natural assets currently constitute of countries' Inclusive Wealth (see Chapters 13 and 13* for more on Inclusive Wealth).[469] Many low income countries still possess much of their endowment of natural capital (defined as renewable and non-renewable natural resources). UN Environment Programme (UNEP) data highlighted that, over the past few decades, natural capital accounted for 42% of wealth among low income countries (it was the largest single component of wealth), but only 14% of the wealth of an average high income country (Figure 14.3) (Managi and Kumar, 2018). While different countries have differing levels of natural capital, the extent of degradation across countries has been vast. Out of the 140 countries sampled, almost 90% have seen their natural capital per capita decline from 1990 to 2014.

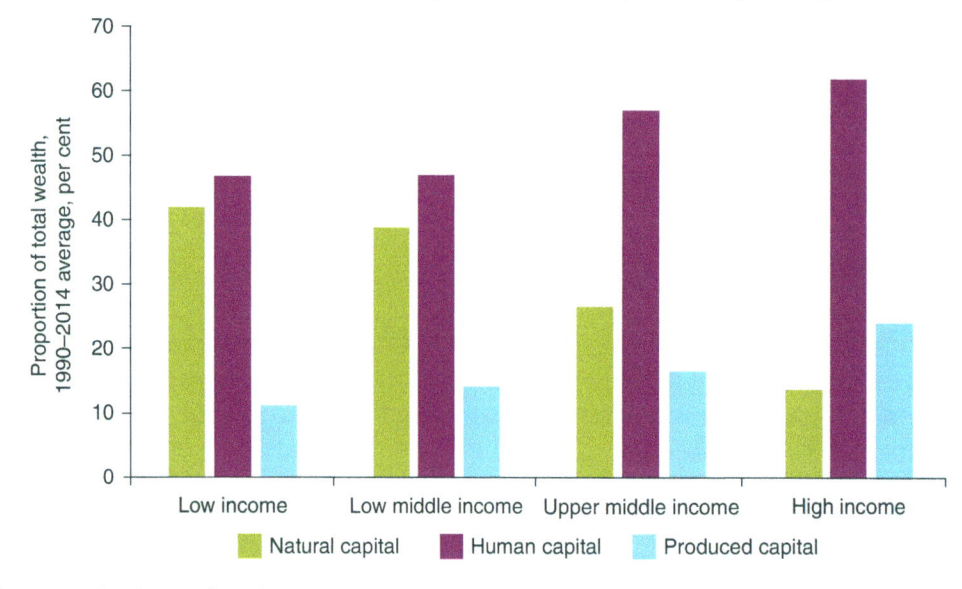

Figure 14.3 Inclusive Wealth of Countries by Income Group and Capital Type

Source: Based on Managi and Kumar (2018) and Review calculations.

Chapter 4 highlighted the extent of the depletion in both the quantity and quality of the stock of natural assets at the global level. Land and sea-use change has been identified as the most important direct

---

[469] The domestic supply of natural assets in a country also affects exposure to risks from degradation of local ecosystems that produce services that matter locally – continuing the earlier example, mangrove destruction results in increased risk from storm surge among coastal communities (see Chapters 5 and 17).

anthropogenic driver, followed by direct exploitation, climate change, pollution and invasive non-native species (Ichii et al. 2019). But depletion is not uniform across geographical regions, as the underlying drivers of change vary in intensity (Ichii et al. 2019; IPBES, 2019a; WWF, 2020a). For example, in Africa, the impact of direct exploitation has been found to exceed that of land- and sea-use change. In the Americas, these two drivers have a similar impact, while in Europe, Central Asia, and Asia and the Pacific, land and sea-use change has been found to be the key driver of change in the state of Nature (Ichii et al. 2019).

The extent of regional variation in ecosystem depletion has also been highlighted, for example, by the Living Planet Index, which tracks the abundance of almost 21,000 populations of mammals, birds, fish, reptiles and amphibians around the world (Chapter 2, Annex 2.3). It shows an average 68% decrease in population sizes between 1970 and 2016 (WWF, 2020a), but the decline varied significantly between regions, and was far greater in Latin America and the Caribbean (94%) than in any other region (Figure 14.4).

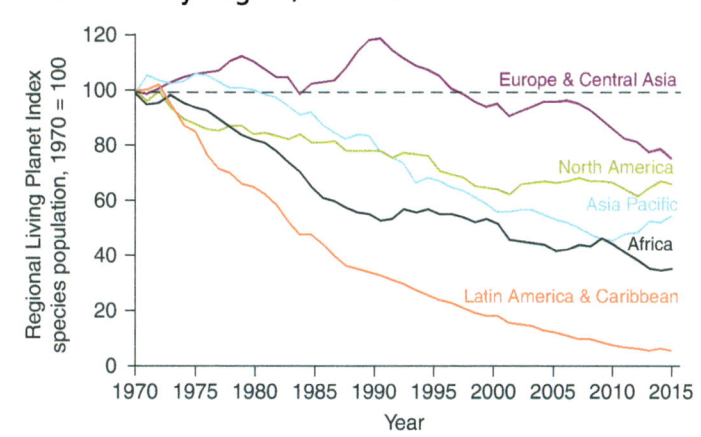

Figure 14.4 Living Planet Index by Region, 1970 to 2016

Source: WWF (2020a).

Cumulative pressures on the ocean are not equally spread. A third of fish stocks are overharvested (FAO, 2018a) and vast areas are considered dead zones (Breitburg et al. 2018). A study of the cumulative impacts of human pressures found that 59% of the ocean had experienced increases in impacts. The largest increases in cumulative impacts were in parts of the Black Sea, tropical Atlantic Ocean, temperate north-west Pacific Ocean, and sub-tropical Indian, Atlantic and Pacific Oceans. 85% of coastal countries experienced increasing cumulative impacts in their coastal zones. Climate change stressors are driving most of the change (Halpern et al. 2019). Efforts to improve fisheries management have resulted in some recoveries and alterations of the distribution of these changes (Duarte et al. 2020a).

The impacts of climate change on terrestrial ecosystems also differ across ecosystems and geographies. The risk of drought is increasing; particularly vulnerable regions are Central and South America, the Mediterranean Basin, South Africa and South Australia (IPCC, 2014). Climate change is also influencing distributional shifts in ecological communities and geographical ranges of species, with regional and local impacts (Pecl et al. 2017). Polar regions are particularly vulnerable, and both Greenland and Antarctic ice sheets are losing mass (Hock et al. 2019). Arctic surface air temperatures are increasing at twice the global rate, which is contributing to the ongoing greening of the Arctic (Swann et al. 2010; IPCC, 2014). This, in turn, is changing ecological communities and reducing the abundance of lichen – a critical winter forage for reindeer. This reduction affects reindeer herding by local communities, with implications for the local economy (Blok et al. 2011; Pecl et al. 2017).

Chapter 14: Distribution and Sustainability

14.4 Interactions Between the Biosphere and Societal Inequalities

Discussions of socioeconomic issues, such as poverty and inequality, often do not consider the role the biosphere plays in the lives of people living in poverty, as discussed in Chapters 1 and 8 (Dasgupta, 2013).[470] But there are several interactions and feedback loops through which the biosphere can exacerbate societal inequalities (Hamann et al. 2018). Evidence shows that ecosystem services can make important contributions to poverty alleviation, but inclusion of ecological indicators is not standard practice in poverty assessments (Schaafsma and Fisher, 2016). This suggests ecosystem management and poverty alleviation policies should not be independent (Schaafsma and Fisher, 2016).

Chapter 8 discussed the extent to which poorer people depend relatively more on common pool resources than richer members of the community, who benefit more from them in absolute terms (Schaafsma et al. 2014). It has been estimated that, globally, ecosystem services and other non-marketed goods make up between 50% and 90% of the total source of livelihoods among rural and forest-dwelling poor households – the so-called 'GDP of the poor' (Kumar, 2010).[471] This contrasts starkly with GDP estimates of which, on average globally, agriculture, forestry and fisheries account for only around 4% of countries' output (World Bank, 2020a).

There is increasing evidence that adverse changes in the biosphere can both create and exacerbate inequalities within societies, and even between entire countries (Box 14.3). Ecosystem change affects who benefits from ecosystem services and who bears the costs of losses in those services, and when these benefits and costs are incurred (Rieb et al. 2017). Inequalities can be created or exacerbated either through shocks or extreme events, such as storms, floods, droughts, epidemics and wildfires (Fothergill and Peek, 2004; Yamamura, 2015), or through gradual environmental change, such as climate change (Mendelsohn, Dinar, and Williams 2006; Diffenbaugh and Burke, 2019). People with lower incomes are often more vulnerable to extreme environmental shocks than others, as they are more exposed to the impacts (Cutter, 2006).

A study in Sri Lanka found that low income households suffer greater losses from floods and droughts than households with higher income (De Silva and Kawasaki, 2018). Gradual changes, such as the impacts of climate change on the distributions of marine species and fisheries, can affect livelihoods where people have limited capacity to adapt, potentially reinforcing economic hardship (Stoll, Fuller, and Crona, 2017). Chaplin-Kramer et al. (2019) found that globally, by 2050, as many as 4.5 billion people could be exposed to higher levels of freshwater pollution than today, and 5 billion may be subject to insufficient pollination for their crops – both services are affected by proximity to the service and the area over which the service is supplied (such as a watershed for hydrological services).

The environmental justice literature has highlighted the extent to which low environmental quality and high environmental hazard risks are typically found in minority and economically disadvantaged communities (for example, see Walker, 2012). This literature has focused on pollutants that degrade natural ecosystems and directly harm human health, such as pollutants that affect air quality (Mitchell and Dorling, 2003; Mitchell, Norman, and Mullin, 2015). Issues of gender and income inequality can also undermine sustainability (Chan et al. 2019).

---

[470] This section focuses on inequality in space, but inequality may occur across both space and time, as noted in Chapter 10.

[471] The 'GDP of the poor' constitutes the proportion of GDP that can be attributed to the rural and forest-dependent poor directly from main natural-resource-dependent sectors (agriculture, forestry and fisheries). It is thereafter adjusted to add the value of ecosystem services and the value of natural products not recorded in GDP statistics (non-market prices of goods used for subsistence).

For example, in India and Nepal, there is strong evidence of the importance of including women in forest management groups for better resource governance and conservation outcomes (Leisher et al. 2016).

Researchers are now beginning to assess the social distribution of natural ecosystems as 'goods', and to quantify their benefits. Some examples include access to urban parks (Xiao et al. 2017), water in urban environments (Raymond et al. 2016), highly diverse ecosystems (Davis et al. 2012), tranquil places (Mitchell and Norman, 2012) and Nature more broadly (Mullin et al. 2018). Decision support tools exist that model the provision of multiple services, from stock through to beneficiaries, but improving our understanding of how ecosystem services affect human well-being has been identified as a crucial research frontier (Rieb et al. 2017). Researchers have begun to disaggregate beneficiaries into groups, and identify populations that are vulnerable to ecosystem service degradation (Daw et al. 2011). Understanding the relationship between Nature and the benefits we derive from it in spatial terms offers significant opportunities to identify areas to improve benefit gaps, aligning Nature and society's needs much better.

Box 14.3
The Distributional Effects of Ecosystem Collapse

The impact of collapse in certain ecosystems – that is the possibility of crossing tipping points as discussed in Chapter 3 – is likely, to be unequal in its impact across countries and income groups due to differences in dependence on natural assets and ecosystem services. This is illustrated by Johnson et al. (forthcoming), who use a linked economy environment model (GTAP-InVEST) to assess the potential combined impacts of three regime shifts over the course of a 10-year period (2021 to 2030).

The authors assess the impact of 'collapses' in three ecosystems – defined as a 90% reduction in the flow of ecosystem service value – on real economic activity across countries. The three 'collapses' that make up the Biodiversity and Ecosystem Services (BES) 'collapse' scenario are in tropical forests, wild pollinators and marine fisheries. Collapses in these ecosystems result in changes to land-use, climate, marine fishing activity and second-round effects, which are in turn used to estimate changes in ecosystem services which affect real economic activity. Impacts are propagated across countries through trade and sectoral interlinkages in the model.

Under the BES scenario, the fall in the level of global real GDP is a little over 2% (−2.4%) by 2030 (with these losses expected to be permanent).[472] But beneath this global aggregate, the analysis suggests that the impacts vary significantly across regions, both by income group and geographically. Low and lower-middle income countries experience disproportionately large GDP contractions (−10.1% and −7.2%) compared to upper-middle and high income countries (−3.7% and −0.8%) (Figure 14.5). This heterogeneity in impacts is due, in part, to these countries relying relatively more on ecosystem services for real economic activities (particularly forestry ecosystem services).

Sub-Saharan Africa and South Asia are disproportionately affected (−9.8% and −6.4% of GDP, respectively) by the 'collapses' (Figure 14.6), with hits of 20% or more to the level of GDP in countries such as Bangladesh, Democratic Republic of Congo, Indonesia, Madagascar, Pakistan and Ethiopia. These findings support the case that taking actions to conserve and restore natural assets is pro-poor, as such actions would reduce the risk of ecosystem collapse, which if realised, would have uneven distributional implications.

---

[472] It should be noted that these estimates are conservative for several reasons. For example, the underlying models do not include all forms of ecosystem services for example.

Chapter 14: Distribution and Sustainability

Figure 14.5 Change in 2030 Real GDP Under BES Collapse Scenario Relative to Baseline, by Country Income Group

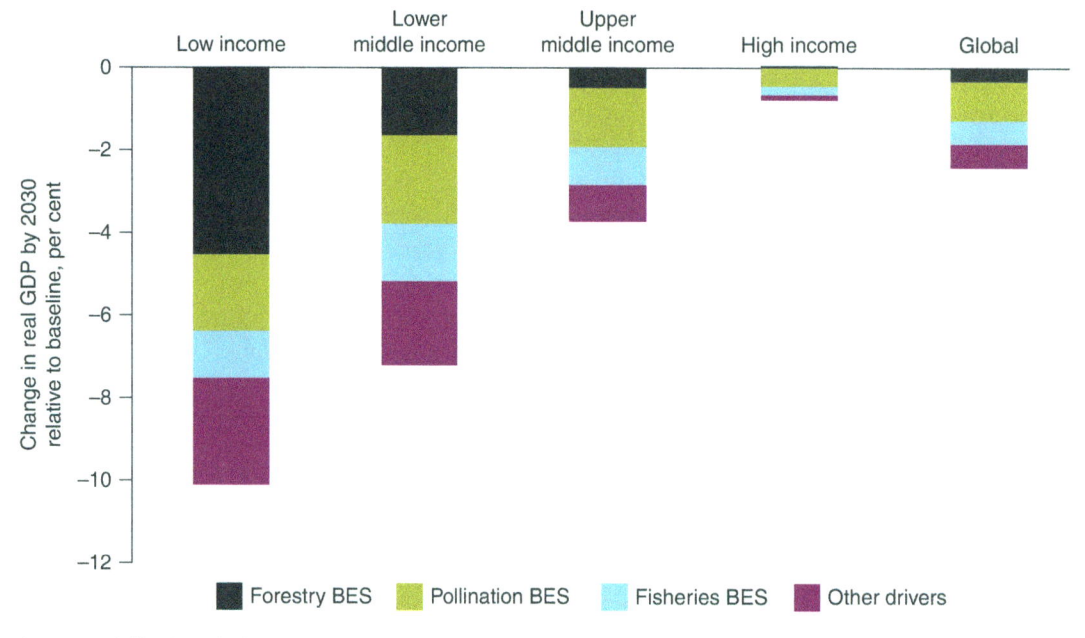

Source: Johnson et al. (forthcoming).

Figure 14.6 Change in 2030 Real GDP Under BES Collapse Scenario Relative to Baseline, by Geographic Region

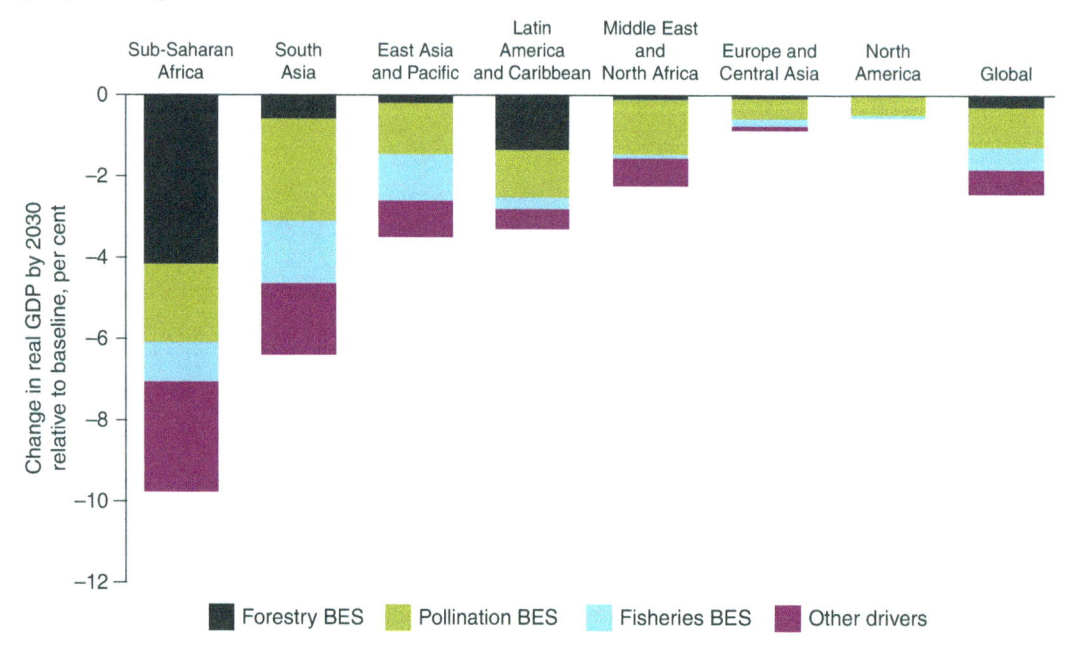

Source: Johnson et al. (forthcoming).

Chapter 15
Trade and the Biosphere

Introduction

Trade allows goods and services to move great distances from where they are produced to where they are consumed. It has allowed a decoupling of local endowments of natural capital, and production and consumption of goods and services that depend on those assets. As noted in Chapter 4 and 4*, achieving Impact *Equality* is a condition for *global* sustainability. But sustainable management of ecosystems at smaller scales remains of key concern. Imbalances between demands on ecosystems and their sustainable supply also occur, affected by where goods and services are consumed and produced, and flows of those goods and services, enabled by domestic and international trade. For example, it is unlikely, indeed near impossible, that a city would be able to meet the needs of its citizens with the ecosystems within its administrative boundary, but this is not necessary if the city can benefit from ecosystem services provided elsewhere. Trade does not change people's needs, but it can allow local or country-level inequalities in Nature's supply and our demands, to be sustained over time.[473]

15.1 Trade and the Impact Equation

Increased economic activity resulting from trade expansion increases human demands on the biosphere and thus depletes natural capital. In addition, trade liberalisation may lead to specialisation by some countries in activities that can significantly deplete local ecosystems, such as extractive or pollution-intensive sectors, if environmental policy stringency differs between countries.

Conversely, trade and trade liberalisation can also facilitate the spread and adoption of technologies that enable more efficient production processes, which may reduce demands on the biosphere. Open markets can improve access to new technologies that make local production processes more efficient, by diminishing the use of inputs such as energy, water, and environmentally harmful substances (see Chapter 16 for examples of these technologies in the food sector). However, this positive effect of trade through enhancing efficiency – through improving productivity in both production and waste – is conditional on market prices and policies reflecting accounting prices, as noted in Chapter 7. In addition, specific elements of trade (for example, wildlife tourism) are supported by the presence of certain species, as well as markets for goods which protect habitats that are recognised as globally valuable (for example, products such as food, rubber and wood harvested at sustainable rates from tropical rainforests).

The impacts of trade and trade liberalisation on the biosphere tend to be discussed in the context of three main channels: scale, composition and technique (Grossman and Krueger, 1991; WTO, 2011). The relative strength of each channel ultimately determines the net impact of trade, both locally – in the importing and exporting countries – and globally.

15.1.1 Scale

The scale effect relates to the impact of trade on the overall level of economic activity and ramifications this has in terms of the depletion of natural capital, through the extraction of resources and emissions of waste and pollution. The presumption is that openness to trade, through the exploitation of

---

[473] We are extremely grateful to James Vause at the UN Environment Programme World Conservation Monitoring Centre for his paper which formed the basis of this chapter.

Chapter 15: Trade and the Biosphere

comparative advantage, increases economic activity and hence human use of the biosphere. An increase in the scale of economic activity and biosphere use will widen the Impact Inequality, all else equal.[474]

Human activity across countries has increased at an unprecedented rate over the past 70 years (Chapters 0 and 4). This is partly through integration of countries into a global economic system (Baldwin and Martin, 1999).[475] Greater trade openness and trade volumes between countries have directly supported increased human activity globally (Frankel and Romer, 1999; Alcalá and Ciccone, 2004), and thereby increased our global ecological footprint. The World Trade Openness Index – defined as the ratio of total trade to GDP – has more than doubled in half a century, with trade having increased more than proportionately with global GDP (Figure 15.1). While global trade has been part of the world economy for thousands of years, its importance was modest until the beginning of the 19th century, as the sum of worldwide exports and imports never exceeded 10% of global output before 1820 (Estevadeordal, Frantz, and Taylor, 2003).

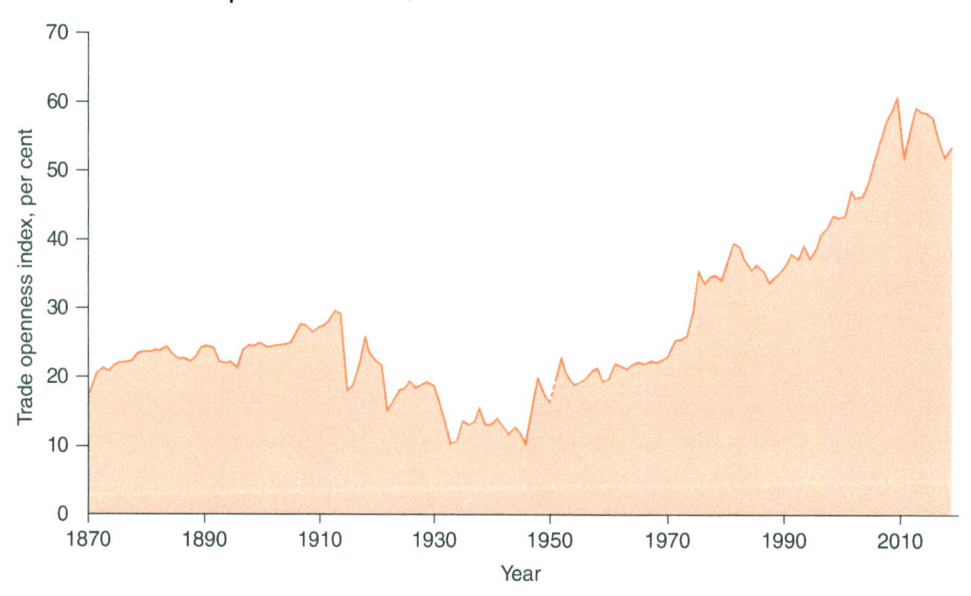

Figure 15.1 World Trade Openness Index, 1870 to 2017

Source: Our World in Data (2018), Feenstra, Inklaar and Timmer (2015) and Klasing and Milionis (2014). Note: The dotted break in line represents the adjoining of different data sources.

15.1.2 Composition

The composition of trade in countries varies, as trade liberalisation changes the mix of a country's production towards those products where it has a comparative advantage. In the classic theory of comparative advantage (Ricardo, 1817), trade causes countries to specialise in the goods that they can produce most efficiently, and thus to become more efficient in their use of factors of

---

[474] The impact of trade is not necessarily detrimental at smaller scales. As one example, a densely populated country with a large service sector may disproportionately import goods, while exporting services that have a smaller ecological footprint than goods. This country's citizens will be highly dependent on foreign-based natural capital.

[475] This process of integration – often termed 'globalisation' – has led to increased access to goods, services, investments, and technologies throughout the world, and has resulted in unprecedented growth in trade between countries. It is generally agreed that there have been two waves of globalisation: roughly 1820 to 1914, and 1960 to present (Baldwin and Martin, 1999); we may be on the cusp of a third wave (Baldwin, 2016).

production.[476] This will influence the composition of trade (i.e. what types of goods and services are produced and consumption), and therefore is a key factor on the overall impact of trade on the biosphere. The composition channel means trade influences the *forms* of production and consumption throughout the world, altering both the extent and location of impacts of production on ecosystems. However, as the theory of comparative advantage requires all goods and services to have perfect markets (i.e. accounting prices equal market prices), the *economics of biodiversity* is obliged to study the implications of trade in goods and services on a case-by-case basis.

As noted in Chapter 14, different types of consumption and production have divergent impacts on the biosphere (WWF, 2018; Wackernagel and Beyers, 2019). The composition effect describes how trade influences patterns of production (and consumption) around the world. The impact of a consumption activity depends on a range of factors, including: the type and amount of natural resources used in production; the distance between origin (production) and destination (consumption); and the mode, route and duration of transportation (Wiedmann et al. 2006). Supply chains affect the extent of the ecological impact of consumption activities. These chains – in which goods and services are produced, transformed and transferred to consumers in various countries – involve a series of complex interactions between producers, traders, manufacturers and consumers. Supply chains are the link between the direct drivers of environmental change (such as land-use change and climate change) and the resulting impacts (such as species loss) and the actors driving these activities, ultimately ending with consumption (WWF and RSPB, 2017; Balvanera and Pfaff, 2019). The choices of many actors along the supply chain will influence how and where a product is made, and therefore its impact on the biosphere. Box 16.1 in Chapter 16 looks at issues of food loss and waste along supply chains.

15.1.3 Technique

Trade can improve access to new techniques and technologies that make local production processes more efficient, through improving access to better technologies and the adoption of new production methods. International technological collaborations have increased significantly over the past three decades: the number of countries involved increased from 79 in 1991 to 125 in 2007, and the number of connections between countries involved roughly quadrupled (De Prato and Nepelski, 2014). Transnational patent networks have increased; from 1994 to 2017, a large number of patents were transferred between Asia, North America and Europe, and a strikingly small number of patents were transferred to African countries, as some countries were able to act as 'gatekeepers', influencing the transfer of technologies between pairs of other countries (Yang et al. 2019).

The ability of trade to influence technique and technology transfer depends on a range of factors, including standards and regulations, the potential to invest in higher-input or mechanised systems connected to foreign direct investment, and financial and technical support available to help countries access global markets. Trade has been found to increase production efficiency[477] (Coe, Helpman, and Hoffmaister, 1997, 2009; Edwards, 1998; Hufbauer and Lu, 2016; Alcalá and Ciccone, 2004). This said, conventional focus has been to minimise costs in terms of labour and produced capital (Chapters 4 and 4*) and not natural capital given insufficient accounting of natural capital in the production process (Chapter 13*) and pricing distortions (Chapter 7).

In terms of the Impact Inequality, these techniques can in principle increase the efficiency with which the biosphere's goods and services are converted into GDP (α_x) and the extent the biosphere is

---

[476] The composition channel thus influences α (the efficiency with which the biosphere's goods are converted into GDP and the disposal of waste), if, for example, production moves to locations that will better manage waste associated with producing specific goods or commodities.

[477] See Chapter 4* for a discussion on why α in the Impact Equation does not equate to total factor productivity (TFP).

transformed by global waste products (α_z), thus reducing demands on the biosphere overall. Assuming there was sufficient accounting of natural capital, trade could be associated with an increase in α_x and α_z – that is the more effective use of natural resources in the production process in relation to the impact on the biosphere. Technique will also affect both the stock of the biosphere (S) and its regenerative rate (G), depending on how the techniques of production promulgated through trade affect biosphere processes. For example, extension of intensive agriculture may result in changes in G, as ecosystems may become less suited to the delivery of multiple benefits, and less stable over time (see Chapters 16 and 19 for more on this).

15.2 Trade Expansion and Pressures on the Biosphere

The increased scale of global trade that has taken place over multiple decades has directly contributed to pressures on the biosphere. There has been increased extraction of provisioning goods, such as agricultural and wildlife (see Chapter 18 for more on the wildlife trade). Trade in agricultural products has at least trebled since the turn of the 21st century (FAO, 2018b; Bellmann, Lee, and Hepburn, 2019), and trade in threatened species – live or as derived products – has also quadrupled over the period 1975 to 2014 (Harfoot et al. 2018). Increased trade has also led to alterations in natural assets themselves, both on land and at sea, so that the output of goods (such as those derived from mining or farming) can be increased to meet increasing demands. For example, agriculture has significantly altered vast landscapes, through altering land for the rearing of livestock and production of crops. As covered in more detail in Chapter 16, land-use change is a major driver of terrestrial species loss, and the majority of cropland areas brought into production between 1986 and 2009 were for exported crops (Kastner, Erb, and Haberl, 2014).

Increased production, enabled by global trade, also contributes to pollution, including greenhouse gas emissions causing climate change (Poore and Nemecek, 2018). Trade has also been associated with the spread of invasive non-native species, pests, and pathogens. Spread can occur directly, via the trade in live plants or animals; as a result of organisms being carried alongside a product; or as a result of the process of trade itself (for example, via ballast water in commercial shipping). There is a strong positive relationship between the degree of international trade in countries and the number of invasive species present (Westphal et al. 2008). Chapter 19 covers invasive non-native species in more detail, including their costs and approaches for their eradication.

The magnitude of trade-related pressures on the biosphere has been assessed, by isolating the impact of trade on global resource use. For example, Plank et al. (2018) show that, between 1990 and 2010, trade expansion resulted in increased resource use (including biomass resources) against a baseline of no trade expansion. Lenzen et al. (2012) developed a holistic (economy-wide) approach focused on 7,000 'critically endangered', 'endangered' and 'vulnerable' species. They related production in different industrial sectors and countries to different anthropogenic threats (for example, pollution and hunting, as identified for each species in the IUCN Red List).[478] Globally, they found that around 30% of species' threats were due to international trade (Lenzen et al. 2012). Box 15.1 discusses the relationship between the extraction of natural resources and trade further.

---

[478] By using a multi-region input-output model to track the financial flows of goods through trade, Lenzen et al. (2012) were able to highlight species' threats 'footprints' around the world that were associated with individual countries' patterns of consumption.

Box 15.1
The Global Resources Outlook

The Global Resources Outlook – developed by the UN Environment Programme (UNEP) International Resource Panel – evaluates the use of renewable and non-renewable natural resources as they move through our economies and societies (IRP, 2019). It presents a story of accelerating natural resource extraction since the start of the millennium. The Global Resources Outlook attributes more than 90% of biodiversity loss to the extraction and processing of materials, fuels and food; within this, 80% of land-use-related biodiversity loss is attributed to biomass extraction (IRP, 2019). Trends in extraction of provisioning goods and their future prospects are discussed in more detail in Chapter 16.

Looking at the extraction of biomass, fossil fuels, metals and non-metallic minerals, and their trade over time, reveals that their trade has been increasing more quickly than overall production. Although biomass extraction seems to increase relatively slowly, global biomass demand increased from 9.1 to 24.1 billion tons between 1970 and 2017, an average annual increase of 2.1% – above the annual increase in the world's human population (1.6%; IRP, 2019).

These data suggest that any reductions in demand for extraction due to increased efficiency have been neutralised by the impact of parallel increases in the main drivers of resource extraction: human population and affluence. At a global level, material productivity (as measured by US$ / kg material used) increased from 1970 to 2000, but it has declined since. Data on the physical trade balance (relating to the actual volume of material trade) show Europe as a consistent net importer of material, with Asia and the Pacific increasing net imports rapidly. Net exporters of materials, especially Latin America and the Caribbean, West Asia and Eastern Europe/Central Asia have all seen increases in net exports over the same time period.

Figure 15.2 Global Material Extraction, in Four Main Material Categories, 1970 to 2017

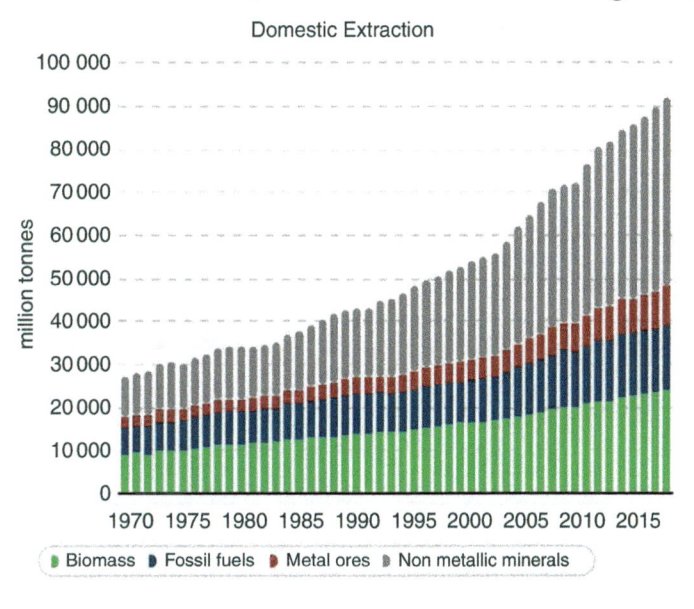

Source: IRP (2019).

Chapter 15: Trade and the Biosphere

Figure 15.3 Global Trade in Materials, in Four Main Material Categories, 1970 to 2017

Physical Trade (million tonnes), stacked bars by category: Biomass, Fossil fuels, Metal ores, Non metallic minerals. Total rises from roughly 2,500 million tonnes in 1970 to approximately 11,500 million tonnes in 2017.

Source: IRP (2019).

Figure 15.4 Drivers of Domestic Extraction, 2000 to 2016

Net Change % DE:
- Africa: 59%
- Asia + Pacific: 120%
- EECCA: 62%
- Europe: 4%
- Latin America + Caribbean: 55%
- North America: −11%
- West Asia: 78%
- World: 66%

Drivers (Population | Affluence | Technological coefficient):
- Africa: −30% | 52% | 37%
- Asia + Pacific: 26% | 85% | 9%
- EECCA: −16% | 3% | 75%
- Europe: −21% | 7% | 18%
- Latin America + Caribbean: 25% | 26% | 4%
- North America: −39% | 13% | 14%
- West Asia: −13% | 63% | 28%
- World: 25% | 31% | 10%

Source: IRP (2019).

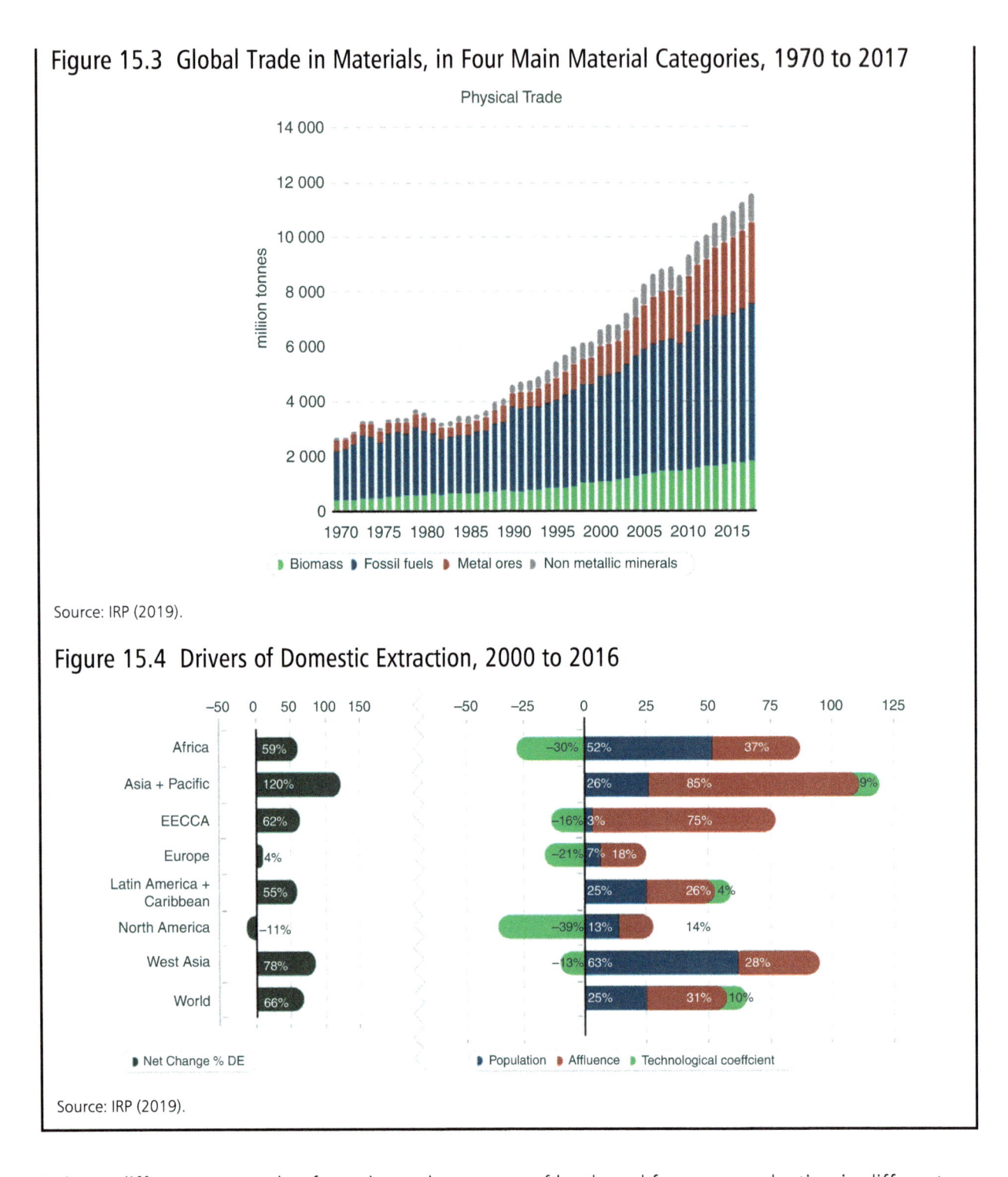

Using a different approach – focussing only on areas of land used for crop production in different countries and understanding the impacts of converting habitats for agricultural production – Chaudhary et al. (2016) estimated that 17% of species losses due to agricultural production resulted from trade. Globally, they estimate that 70% of agricultural land-use impacts on species losses are accounted for by just 13 crops. Tropical crops, such as sugarcane, palm oil, coconut, cassava, rubber and coffee, have a disproportionate impact, and are responsible for 23% of global impacts on biodiversity while using up just 10% of crop land (Chaudhary et al. 2016). If forestry, pasture and urban areas are added to the analysis, the share of species losses attributable to land use for export production increases to 25% (Chaudhary and Brooks, 2019).

Other studies of drivers of impacts over time, assessing the role of trade alongside others as the cause of land-use changes, have broadly consistent results. Pendrill et al. (2019) found that 29%-39% of the increase in carbon emissions associated with deforestation was driven by international trade, which itself was mainly driven by agricultural extension to meet demand for beef and oilseeds. Using satellite imagery to analyse the drivers of deforestation, Curtis et al. (2018) found that 27% of global forest loss was permanent, and was for commodity production; despite increasing corporate commitments over the study period, the rate of commodity-driven deforestation did not decline.

Defries et al. (2010) assessed demographic, agricultural and other economic factors as potential drivers of deforestation in Asia, Latin American and Africa. The most powerful predictors of forest loss were urban population growth and agricultural exports (measured as agricultural trade per capita).[479] Lambin and Meyfroidt (2011), using case studies to explore the complex drivers of land-use and land-cover change, found that people's responses to economic opportunities, mediated by institutional factors, were the key drivers of change. A further explanation of the importance of trade is provided by Meyfroidt et al. (2013), who showed that, while the majority of global agricultural and forestry production remains destined for domestic markets, international trade is far more volatile than domestic trade, and therefore plays an outsized role in determining changes in land use.

The links between trade and terrestrial biodiversity are mirrored by similar connections in the marine environment. For example, in relating species' threat hotspots to global supply chains, Moran and Kanemoto (2017) identified hotspots of pressure associated with consumption in the United States of species harvested off the Caribbean coast of Costa Rica and Nicaragua, and with consumption in the European Union of species from the islands around Madagascar, Réunion, Mauritius and the Seychelles. The dominant hotspot, however, is Southeast Asia, facing threats from fishing, pollution and aquaculture and absorbing pressure from consumption in the USA, the European Union and Japan. Crona et al. (2016) highlight the specific role global trade can play in dampening any price signals that might otherwise reach consumers connected to marine ecosystem degradation by both failing to reflect the true accounting prices of ecosystem services (Chapter 7), and allowing impacts on individual fisheries to be compensated through sourcing similar products from a different source area or aquaculture.

Trade expansion over recent decades has allowed higher-income countries to 'off-shore' the adverse impacts of their consumption on ecosystems and biodiversity, through trade in commodities, goods and services with lower-income countries. More than 50% of the biodiversity loss associated with consumption in developed economies is estimated to occur outside their territorial boundaries (Wilting et al. 2017), and exports from developing economies have been found to be more ecologically intensive than those from advanced economies (Moran et al. 2013).[480] Countries with larger ecological footprints generally have lower domestic levels of environmental degradation, while countries with lower ecological footprints experience higher exported levels of environmental degradation (Jorgenson and Rice, 2005; Jorgenson and Kick, 2006). The phenomenon of

---

[479] Rural population growth and income per capita or national income growth (measured by GDP) were not significant predictors of forest loss at a country level (Defries et al. 2010). A published comment on this paper highlighted that the results seemed to be dominant in Asia and Latin America, but less clear in Africa (Fisher, 2010).

[480] Verones et al. (2017) suggest that low income countries with large human populations can reduce their biodiversity footprint through trade, if they have high domestic demand and high levels of endemism. Similarly, where species' vulnerability is higher in higher-income countries, transferring pressures to countries where species are (currently) less vulnerable can reduce net biodiversity loss (depending on how biodiversity is measured).

Chapter 15: Trade and the Biosphere

'off-shoring' environmental degradation appears to exacerbate global disparities, notably in carbon emissions and resource use contexts (Wiedmann et al. 2020). Box 15.2 discusses a recent study into the global ecological footprint of the UK.

Box 15.2
Exploring the UK's Global Ecological Footprint

WWF and the Royal Society for the Protection of Birds (RSPB) found that, between 2016 and 2018, the average land area required to supply UK demand for beef and leather, cocoa, palm oil, pulp and paper, rubber, soy and timber was 21.3 million ha – 88% of the total land area of the UK (WWF and RSPB, 2020). 28% of this footprint is assessed to be at high or very high risk of driving continued deforestation, conversion of natural ecosystems and/or human rights abuses in the countries where the supply chain operates.

The study found that – although the UK has less than 1% of the world's population and around 2% of global GDP – UK consumption is responsible for 9% of the global cocoa land-use footprint, 5% of the global palm oil footprint and 5% of the global pulp and paper footprint. For these commodities (and soy), between 63% and 89% of the UK's land-use footprint is in countries that are at high or very high risk of driving continued deforestation, conversion of natural ecosystems and/or human rights abuses (WWF and RSPB, 2020).

Production of traded commodities has been found to contribute to the major drivers of biodiversity loss in exporting countries. Lenzen et al. (2012) suggest that, although there are also growing domestic markets in developing countries for products such as coffee and chocolate, through their consumption of the products of developing countries, individuals in advanced economies have contributed to increased threats to species in the product's country of origin. Increasing pressure on biodiversity in developing countries resulting from trade is explored further in Box 15.3.

Box 15.3
Trade Induced Biodiversity Loss Across Regions

Seeking to explain the drivers of biodiversity loss from agriculture and forestry over time, Marques et al. (2019) looked at the relationships between biodiversity (as measured by bird species richness) and population growth, economic development and technological progress in 2000 and 2011.

The impact of population and GDP growth driving demand for agricultural and forestry products outweighed the impact of reductions in the land-use impact per unit of GDP, resulting in a net 3–7% increase in biodiversity loss over the period.[481] A multi-regional input-output analysis, used to quantify consumption drivers, allowed regions where biodiversity impacts occur (left-hand side of each panel in Figure 15.5) to be mapped to regions where consumption is driving that impact (right-hand side of each panel in Figure 15.5). The authors found that in 2011, 33% of biodiversity impacts in Central and southern America, and 26% of impacts in Africa, were driven by consumption in other regions of the world.

---

[481] It should be noted that this is different to related findings in IRP (2019), and therefore there is evidence either way.

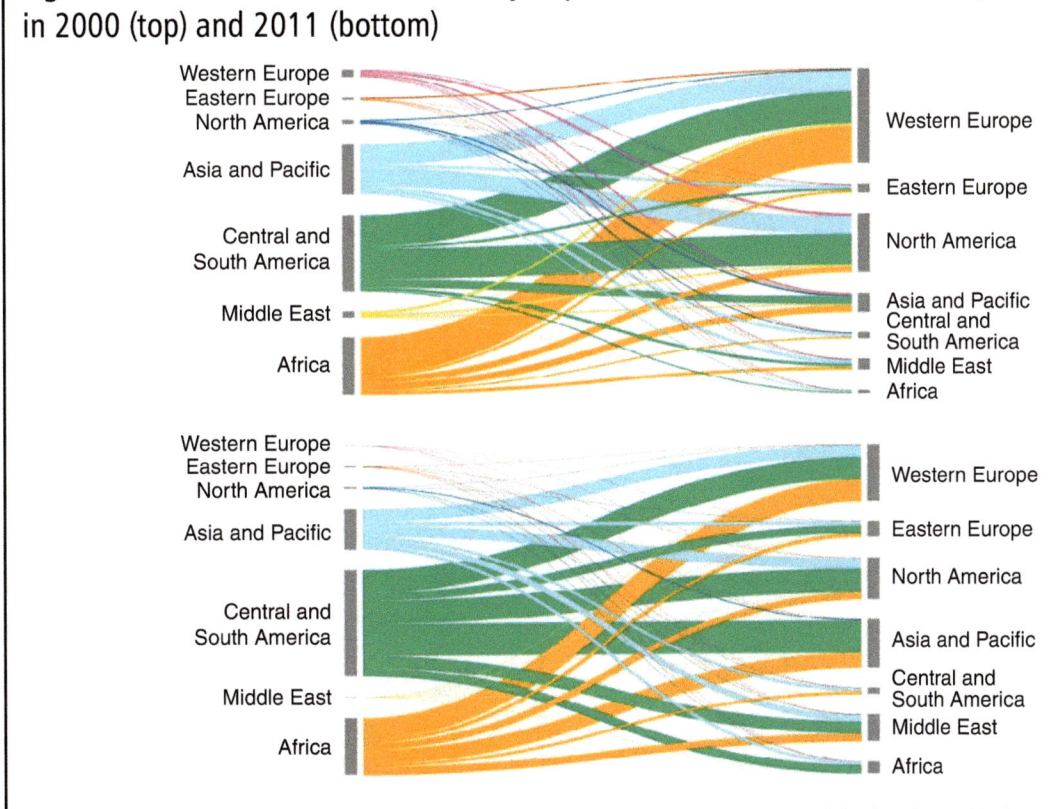

Figure 15.5 Links between Biodiversity Impacts of Production and Consumption in 2000 (top) and 2011 (bottom)

Source: Marques et al. (2019). *Permission to reproduce from Springer Nature.* Note: On the left of each panel is the region where biodiversity impacts occur; on the right is the region where consumption is driving the impacts. The width of the flows between the regions represents the magnitude of the impacts. The visualised impacts represent 22% and 25% of the yearly global totals, respectively, for biodiversity in 2000 and 2011.

Some of the consequences of damage to the biosphere from trade highlighted in the previous section, such as land-use change and sea-use change, affect natural assets that provide global benefits; therefore, the changes are of global significance and concern (see Chapters 7 and 8 for more on institutions for managing global public goods). More locally, trade can affect ecosystems and people's livelihoods. For example, a study in India, Brazil and Indonesia showed that ecosystem services derived from natural assets for free made up between 47% and 89% of the real value of the consumption bundles of nearly 500 million rural people (Kumar, 2010). These benefits are placed at risk when habitats are converted to meet demands enabled by trade without considering the ecosystem services that flow to local people (see Chapter 8 for more on this). For example, Chaves et al. (2020) – by linking trade with deforestation, and deforestation with changes in mosquito communities and connected to this, human exposure to malaria – estimate that about 20% of malaria risk in deforestation hotspots is driven by international trade in deforestation-implicated export commodities, such as timber, wood products, tobacco, cocoa, coffee and cotton.[482]

Global trade is able, to an extent, to mitigate the impacts of any possible ecosystem collapse (as discussed in Chapter 5), at least for a while. For instance, trading markets could allow buyers, especially

---

[482] The authors highlight that disturbances in natural forest landscapes can lead to changes in mosquito communities, causing an increase in the abundance of vector species, and subsequently the risk of human exposure to vectors and thus to malaria, in areas where the landscape is conducive to vector–human contact.

those who are better able to cope with any associated additional costs of materials, to find alternative sources for the goods they want (catalysing further extensions in output elsewhere) (Crona et al. 2016). But such opportunities are unlikely to be as readily available for the people living in ecosystems degraded through the impacts of production. As discussed in Chapter 14, vulnerable groups in society often feel the greatest negative effects of land degradation, and often experience them first and most directly.

15.3 Enhancing Trade Practices and Policies to Support Sustainability

The evidence provided here suggests that trade may be widening the Impact Inequality. Indeed, at present, there is no global mechanism to ensure that the changes catalysed by trade are compatible with moving towards sustainability. Trade practices and policies can support a shift to sustainability, through actions which span international boundaries and influence what products are traded, and how much of each product is traded. The notion that less trade in certain goods – those which are most harmful to the biosphere – may improve overall outcomes for society has been highlighted by Kohn and Capen (2002).[483] As Chapter 6 discussed, the effectiveness of institutions and interactions between layers of governance[484] determine the ultimate efficacy of any actions; those discussed below should be viewed in this context.

15.3.1 Border Adjustment Taxes and Non-Discriminatory Regulations

Where countries take domestic action to improve the sustainability of production, they often cite the risk that trade undermines such standards, as their producers cannot compete with producers in other jurisdictions who are subject to different rules. In climate-change policy, this has prompted discussions of 'border carbon adjustments', which would effectively place higher tariffs on imports that do not comply with tighter domestic emissions standards.

As Chapter 7 highlighted, there are limits to the extent to which measures to eliminate externalities, such as Pigouvian taxes, can correct for externalities, but they can make some difference if widely applied and correctly designed. Border adjustment taxes – taxes on goods and services based on where they are consumed rather than where they are produced – related to environmental issues face both technical and political challenges. As Polasky et al. (2004) discuss, setting the appropriate rate for a domestic tax on land-use conversion that reflects the marginal damage from biodiversity loss requires knowledge of the marginal utility of biodiversity protection, and of both economic and ecological impacts.

To be effective, any tax would need to vary by ecosystem (or habitat type as a proxy) and level of degradation, and over time. To set a tax rate at the border which attributes a cost to lower biodiversity standards would require a deep understanding of the ecological impact of different goods and services. If a border adjustment tax was introduced unilaterally that was not well designed, countries may be accused of disguised protectionism, which could lead to retaliation. Bellmann, Lee, and Hepburn (2019) highlight alternatives to border adjustment taxes, such as taxes imposed equally on imports and domestically produced products (e.g. on salt or sugar content of food). In principle, these taxes are not discriminatory (and hence, less likely to be challenged as protectionist by other countries), but, like in

---

[483] The authors model the optimum level of trade, where trade affects biodiversity through commercialisation of habitat, pollution and species invasions. The analysis showed that trade was lower once external impacts on biodiversity were taken into account, and that the more damaging a production process was for biodiversity, the more trade was reduced.

[484] Brondizio et al (2009) highlight that governance needs to evolve as distant areas become more functionally interdependent through trade and investment.

other measures, the challenge of determining the cost of the externality remains.[485] Box 15.4 presents an example of competition and the adoption of environmental policies resulting in retaliation by another country.

Box 15.4
The US Shrimp/Turtle Dispute

In 1996, the US introduced a ban on imports of shrimp from countries that did not have a certified regulatory programme to protect sea turtles that was similar to the programme in the US (UNEP/WTO, 2018). The intention was to avoid accidental capture of sea turtles by requiring all exporters to the US to use 'turtle excluder devices' to allow turtles to escape from shrimp fishing nets.

The US claimed that the General Agreement on Tariffs and Trade (GATT) allowed the ban, via exceptions in Article *XX* (which lays out specific instances in which World Trade Organization (WTO) members are exempt from GATT rules, in some cases relating to environmental protection). Other countries claimed that the law was a disguised restriction on free trade, and challenged the measure via the WTO's dispute resolution process (UNEP/WTO, 2018).

The ban was judged as contrary to WTO rules, because it was arbitrary and unjustifiably discriminatory, but the WTO stated that it was not issuing a decision suggesting that WTO members could not adopt measures to protect endangered species. The rule was amended in 1999 to allow countries to apply their own policy solutions, provided that they were as effective in protecting sea turtles as the US's approaches. This amendment ensured that the ban on shrimp coming from countries with inadequate protection regimes complied with WTO rules (UNEP/WTO, 2018).

15.3.2 Trade Agreements and Liberalising Trade

Alongside raising standards and enforcement of environment regulation, Regional Trade Agreements have, since 2000, increasingly included requirements to reaffirm commitments and take action to meet internationally agreed goals under multilateral environmental agreements, including the Convention on International Trade in Endangered Species (CITES) (Box 15.5) and the Convention on Biological Diversity (CBD) (Karousakis and Yamaguchi, 2020). 'Sustainability' provisions in Regional Trade Agreements tend to be conditions upon which the reduction of other barriers to trade are contingent.

Adequate enforcement of rules (as highlighted in Chapter 7) is one of the main problems identified in using trade agreements to achieve sustainability in connection with trade. As Kettunen et al. (2020) highlight, there is a compliance gap in free trade agreements. Effective implementation requires effective monitoring processes, clear goals and mechanisms to address non-compliance (Chapter 6 provides detailed discussion of these requirements of effective institutions). Connected to setting clear goals, Kettunen et al. (2020) suggest it is not yet possible for the assessment of environmental impacts linked specifically to EU Free Trade Agreements to be undertaken in a comprehensive and robust way. At present, it seems there is limited capacity both to understand and to limit the impacts of trade agreements on the biosphere.

---

[485] In contrast to the 'stick' approach provided by border adjustment taxes and other taxes, 'carrot' approaches have been suggested, such as sustainable commodity import guarantees, which are financial instruments to create incentives for sustainable sourcing (Global Resources Initiative, 2020).

> **Box 15.5**
> **The Convention on International Trade in Endangered Species**
>
> The Convention on International Trade in Endangered Species (CITES) agreement between governments aims to ensure that international trade in specimens of wild animals and plants does not threaten their survival. Though it does not cover all species traded, CITES accords varying degrees of protection to more than 37,000 species of animals and plants, whether they are traded live or as 'products' (e.g. fur coats or dried herbs). CITES works by agreeing restrictions on the levels of trade, ranging from trade that is only permitted in exceptional circumstances (for example, as part of captive breeding programmes) for species threatened with extinction, to trade that is restricted (allowed by permit only) in certain geographical areas.
>
> The Convention recognises that, where trade in wild animals and plants crosses borders between countries, the effort to regulate it requires international cooperation to safeguard certain species from over-exploitation. But Lenzen et al. (2012) argue that there is little difference between direct trade in species or specimens and trade in commodities the production of which puts the same species at risk. For example, Mexico's spider monkey *Ateles geoffroyi* is listed in Appendix II of CITES and therefore has some protection from the direct threat of international trade, but its long-term future would be more secure if its habitat was protected from the encroachment of coffee plantations. They therefore argue that restrictions in the trade of 'biodiversity-implicated' commodities should be considered.[486] Chapter 18 provides further discussion on wildlife trade.

In their 2018 report, the UNEP and WTO highlighted the potential benefit of liberalising trade in environmental goods (UNEP/WTO, 2018). Under an Environmental Goods Agreement, WTO members would negotiate reductions in tariffs on a wide range of goods that reduce pressures on the biosphere, including solar cooking stoves, devices to prevent bycatch of aquatic mammals in fishing nets, and instruments to monitor environmental quality. The agreement could make technologies that reduce, minimise or monitor harmful impacts on ecosystems more competitive, and help them to infiltrate new markets. For example, goods include energy- efficient LED bulbs, solar panels, solar cook stoves, air and water filters, machines for recycling solid waste, floating barriers to contain oil spills, and devices to prevent turtles and aquatic mammals such as dolphins from being trapped in fishing nets (UNEP/WTO, 2018).

It has been highlighted that more frictionless trade has the potential to reduce waste (and thereby increase α_z) particularly food waste. Leclère et al. (2020) highlight the role smoother trade can play in mitigating food waste, which is critical in the context of reducing pressures for habitat conversion. Bellmann, Lee, and Hepburn (2019) also highlights the benefits of facilitating smoother trade in fruit and vegetables, to reduce waste. The authors highlight that, due to the perishable nature of many fruit and vegetable products, measures aimed at easing transit over borders by reducing unnecessary bureaucracy and waiting times could improve their availability, reduce costs, reduce waste and improve food quality and safety for consumers. (Box 16.1 in Chapter 16 provides further evidence on food loss and waste in supply chains.)

15.3.3 Supply Chain Policy Development

Recent initiatives aimed at reducing the biodiversity impacts of international trade are increasingly focussed on driving changes in production processes, through standards and other mechanisms

---

[486] Restricting international trade of 'biodiversity-implicated' commodities is challenging in terms of the data needed to identify geographically specific threats to species associated with the production of certain commodities and to prevent goods from these areas entering large-scale, relatively generic, supply chains.

that are linked closely to international trade, to transfer signals along supply chains. For example, commitments have been made to remove deforestation from supply chains, including the New York Declaration on Forests (to halve deforestation by 2020 and end it by 2030) and the Amsterdam Declaration (an initiative to achieve deforestation-free supply chains by 2020). However, there is little evidence that the goals of these declarations have been or will be met (TRASE and Forest 500, 2019).

Green et al. (2019) explore the inconsistency between political commitments to reducing forest loss and the delivery of these ambitions. They show that most of the top 10 countries driving biodiversity loss through deforestation in Brazil have signed one or both of the New York Declaration on Forests and the Amsterdam Declaration. Most progress has been made in areas where specific and large-scale or sector-wide actions have been taken, for example, both the government-led moratorium on new concessions for palm oil and timber plantations and logging activity on peatland in Indonesia and the business-led moratorium on soy sourced from land deforested after 2006 in the Amazon in Brazil have reported to have seen reduced deforestation in target regions, although there is evidence of displacement of deforestation to other regions (NYDF Assessment Partners, 2019).

The ability to monitor and regulate supply chains is limited by their complexity and opaque nature. Both Green et al. (2019) and Heron et al. (2018) highlight the complexity of supply chains as an impediment to their successful management to reduce impacts on ecosystems. Green et al. (2019) focus on soy production in the Cerrado in Brazil; through detailed geographically specific examination of trade connections they identify groups of actors who are disproportionately driving biodiversity loss. They suggest that making supply chain connections more transparent will help actors within the supply chain understand where they drive impacts and how they can reduce these. Heron et al. (2018) also looked at soy, recognising that the way in which soy is embedded within different final products (especially meat products) shields retailers from direct connection with any externalities associated with soy production, and so limits their incentive to push sustainability standards. The bulk trade in soy, and the involvement of large businesses in the distribution and processing of soy, also limits the ability of retailers to press for sustainability standards, due to difficulties in segregating and tracing supply.

As Chapter 6 discussed, increased transparency improves our ability to monitor actions taken to achieve sustainability commitments over time, and to hold those who make those commitments to account. Alternative approaches to reduce the impact of traded goods targeting different points in the supply chains include due-diligence approaches, which place the responsibility on importers to prove certain environmental impacts have not arisen in the production of the goods they intend to use or market; and the application and demonstration of adherence to international production standards. In the latter case, the responsibility to demonstrate compliance lies with the producer or exporter.

One group of trade standards related to the biosphere are sanitary and phytosanitary standards (SPS). SPS require that certain conditions are met to ensure the protection of humans, animals and plants from diseases, pests, and contaminants. For example, SPS are used to limit pesticide residues that are permissible on food, which may bring wider environmental benefits from reduced pesticide use and exposure at the farm level (Box 15.6 discusses SPS in relation to invasive non-native species). The WTO has an SPS agreement and encourages WTO member countries to base SPS measures on international standards and guidelines in order to protect ecosystems and biodiversity associated with agricultural production (WTO, 2020).

The Standards and Trade Development Facility supports lower-income countries in adhering to international standards, which allows them to engage in international trade, as, for example, goods imported into the EU must meet EU SPS, so exporters unable to meet the standards are unable to access the market.[487] The Standards and Trade Development Facility has worked to help reduce trade barriers by providing support to enable the uptake of international wood-packing standards in Botswana, Cameroon, Kenya and Mozambique. This enabled broader regional trade, reduced the spread of wood-boring insects outside their native range, and promoted the repair and recycling of wood packaging to reduce pressure on forests (UNEP/ WTO, 2018).

Box 15.6
Trade Standards and Invasive Non-Native Species

Sanitary and phytosanitary standards (SPS) play a role in the control of invasive non-native species, as the WTO Agreement on the Application of Sanitary and Phytosanitary Measures includes in its definition of sanitary and phytosanitary measures "any measure applied to prevent or limit other damage within the territory of the Member from the entry, establishment or spread of pests". Tackling invasive species is particularly well-aligned for invasive plant species, as the International Plant Protection Convention (the international standards-setting body in relation to plants) has a direct interest in addressing invasive plant species issues, unlike the World Organisation for Animal Health (the corresponding international standards-setting body for animal health) which does not have the equivalent mandate (Standards and Trade Development Facility, 2013).

Relying on SPS is likely to be insufficient to tackle invasive species for a number of reasons (Perrings et al. 2010). First, because regulation of international markets beyond those admitted under the WTO Agreement on the Application of Sanitary and Phytosanitary Measures is required. Second, SPS highlight the control of invasive species as an international public good, giving a strong rationale for broader international cooperation, especially to support low income countries, both to finance control efforts and to build capacity for managing invasive species risks. Efforts to prevent species invasions are only as good as the benefits offered by the least effective state in a particular region (i.e. there is a 'weakest link' effect). Third, there may be a need for an international body responsible for monitoring trends in invasive species, carrying out risk assessments, and helping to coordinate international responses to invasive species threats. Clearly, preventing movements of invasive non-native species is beyond what can be delivered through the regulation of trade alone.

'Sustainability standards' are not limited to SPS, and are not necessarily mandated at national or international levels. Voluntary sustainability standards are considered to be private schemes and are therefore outside the remit of the WTO. The UN Forum on Sustainability Standards mapped 241 voluntary sustainability standards in 80 sectors and 180 countries. Sustainable Development Goal 15 (SDG 15 – relating to life on land) was found to be the third most frequently targeted when the various standards were assessed in relation to their contribution to the SDGs (UNFSS, 2018). In spite of the proliferation of standards, sustainably certified production, offering some contribution to sustainability, remains niche in the agriculture sector as a whole. Certification levels are highest for tropical commodity crops (UNEP/WTO, 2018). For coffee, roughly 25% of production is certified and for cocoa, tea and palm oil between 10% and 20% of production is certified (UNEP/WTO, 2018). While coverage of, and the requirements of, sustainability certifications vary, there has been a recent upward trend in the adoption of voluntary sustainability standards (Box 15.7). For example, certification for marine fisheries has increased over the past two decades (Balvanera and Pfaff, 2019; FAO, 2020c).

---

[487] In 2014, the UN Conference on Trade and Development (UNCTAD) estimated that lower-income countries' agricultural exports to the EU were about US$3 billion lower (14%) than they would be if the EU had not implemented SPS (Murina and Nicita, 2017).

Box 15.7
Integration of Environmental Issues into Trade Standards and Agreements

Consumer demand for sustainably and ethically sourced ingredients and the changing regulatory and policy landscape (e.g. the evolution of the SDGs, the Nagoya Protocol and other international instruments) have been identified as opportunities for increasing the sustainability of trade (UNCTAD, 2019). Figure 15.6 provides indicators of these changes. For example, the land area used for a range of highly traded crops grown under certified sustainability standards increased by 59% in 2013 to 2017 and by 18% in 2016 to 2017 (Willer al. 2019).

Figure 15.6 Selected Products Certified by Sustainability Standards, 2008 to 2017

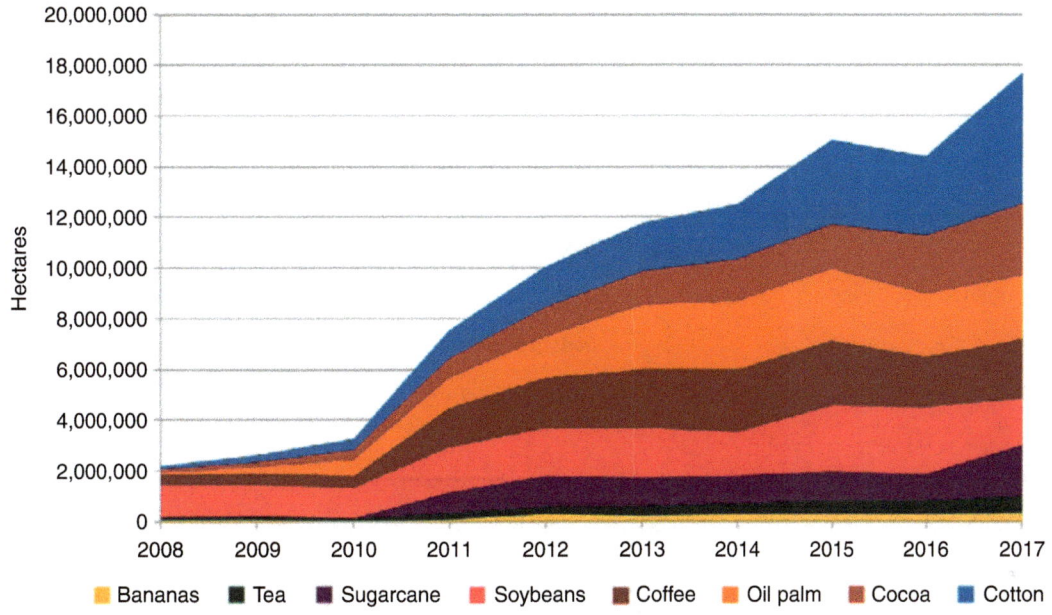

Source: Willer et al. (2019). Note: The products are sorted by area. For the purpose of the figure, it is assumed that a maximum amount of multiple certification is occurring within each commodity and the minimum possible VSS-compliant area is shown. This corresponds to the standards with the largest compliant area operating within a given sector.

Likewise, country commitments to address sustainability issues in trade seem to be increasing, with both the number of Regional Trade Agreements featuring environmental provisions increasing over time (Figure 15.7). Financial commitments to environmental objectives as part of aid expenditure dispersed to support trade has also increased (and accounts for approximately a third of total spend) (UNEP/WTO, 2018). Regional Trade Agreements tend to focus on relatively similar environmental issues, such as the level and enforcement of domestic environmental laws; they may, for example, feature commitments to 'improve', 'adopt', 'harmonise', 'effectively apply', 'not waive' or 'not relax' environmental laws (UNEP/ WTO, 2018).

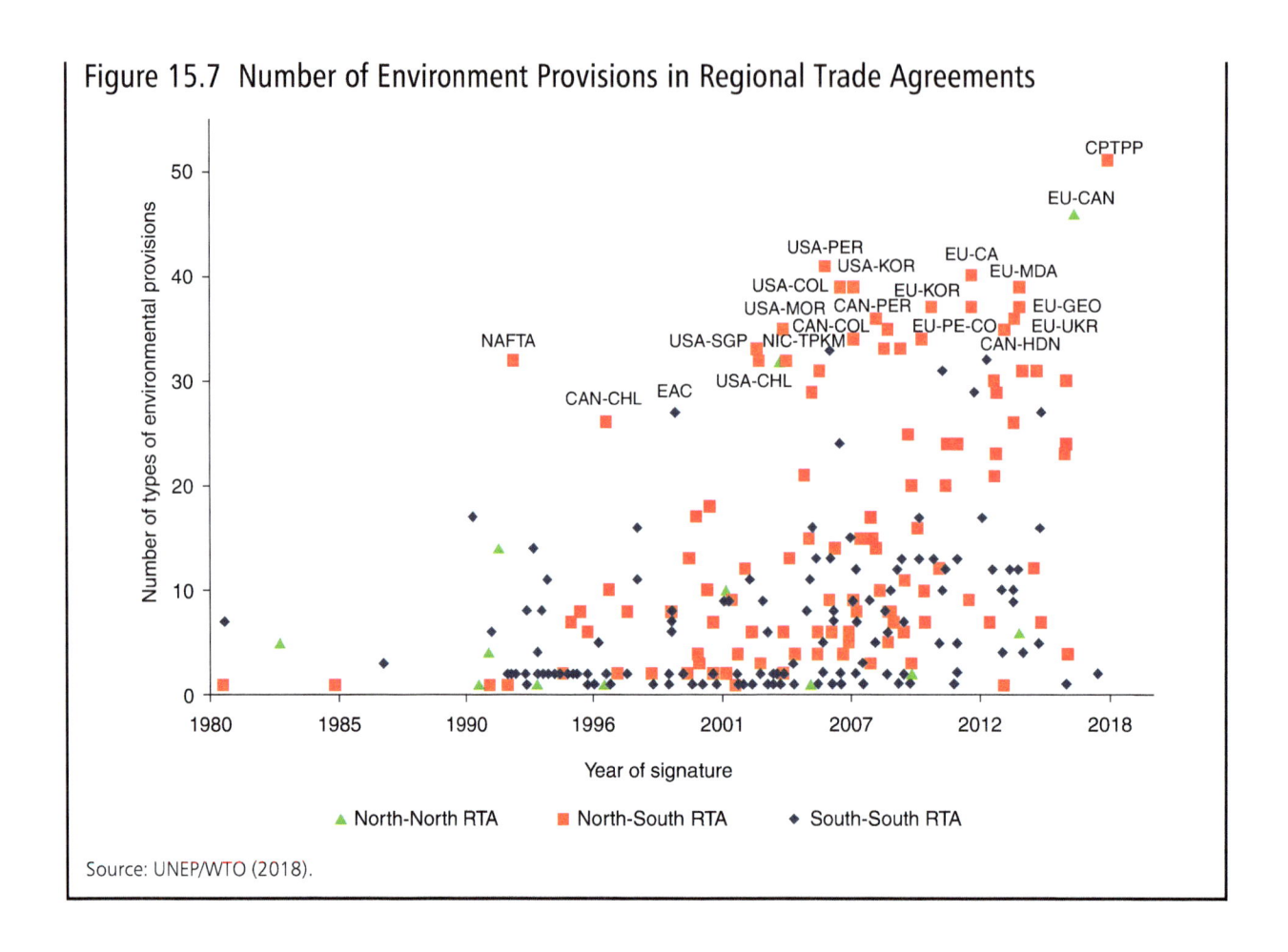

Chapter 15: Trade and the Biosphere

Figure 15.7 Number of Environment Provisions in Regional Trade Agreements

[Scatter plot showing number of types of environmental provisions (y-axis, 0-50) against year of signature (x-axis, 1980-2018). Data points classified as North-North RTA (green triangles), North-South RTA (red squares), and South-South RTA (black diamonds). Notable labelled agreements include NAFTA, CAN-CHL, EAC, USA-SGP, USA-MOR, USA-COL, USA-PER, USA-KOR, USA-CHL, NIC-TPKM, CAN-COL, CAN-PER, EU-KOR, EU-CA, EU-PE-CO, EU-MDA, EU-GEO, EU-UKR, CAN-HDN, EU-CAN, and CPTPP.]

Source: UNEP/WTO (2018).

There is some evidence of limits to the extent to which biodiversity is captured in voluntary sustainability standards for commodity crops, since, particularly in the tropics, they are focussed on conservation areas and actions within farms, rather than at a larger scale (Edwards and Laurance, 2012; Tscharntke et al. 2015). However, there is other evidence to suggest scope for voluntary sustainability standards to reduce adverse impacts on the biosphere. For example, Smith et al. (2019) show that global compliance with the leading voluntary sustainability standard for sugarcane, Bonsucro, would reduce irrigation water use by 65%, greenhouse gas emissions from cultivation by 51%, and nutrient loading by 34%. Certification schemes can exclude smaller producers who cannot afford to collect data on their production impacts (Clark and Martínez, 2016). Standards developed for large-scale producers or high-income consumer preferences may be difficult to apply in other contexts (Foley and McCay, 2014). More generally, the existence of weak and ineffective schemes can undermine the credibility of the practice in the long term (Mori Junior, Franks, and Ali, 2016). There have been some innovative initiatives to scale up certified production (Box 15.8 includes an example).

Box 15.8

Verified Sourcing Areas

In response to the demand for deforestation-free supply chains, the Sustainable Trade Initiative has proposed the development of verified sourcing areas, aimed at sourcing large volumes of commodities produced in line with sustainability commitments at competitive prices. The approach seeks to certify large areas rather than individual production units, and by doing so improve coordination and production standards on a broad scale, delivering economies of scale in certification, and potentially reducing displacement of environmental impacts (IDH, 2020b).

Coalitions of local actors – farmers, producers, government and civil society – are required to come together to develop action plans with measurable sustainability commitments, allowing various outputs from the whole area to be certified if the commitments are met. It is envisaged that plans could be developed with financial support from committed buyers, given the costs of bringing stakeholders together. Verified sourcing areas are currently being piloted in the Mato Grosso region of Brazil and in Aceh Tamiang in Indonesia (IDH, 2020a).

15.3.4 Aiding More Sustainable Trade

To improve outcomes of trade for biodiversity, Karousakis and Yamaguchi (2020) suggest that Official Development Assistance (ODA)[488], including the Aid for Trade initiative, can help countries overcome barriers to enter markets for sustainable agriculture, forestry and fisheries, via sustainable certification schemes. The Aid for Trade initiative was launched by WTO members in 2005. It aims to support developing countries to acquire the skills, supply capacity and infrastructure they need to benefit from international trade.

Aid for Trade accounts for roughly 30% of ODA, and almost 40% of Aid for Trade investments included an explicit environmental objective by 2014, predominantly supporting renewable energy, low-carbon transport and sustainable agriculture (UNEP, 2020a). Aid for Trade can support the development of trade policy and regulation, but productive capacity building and the development of economic infrastructure is also important. Karousakis and Yamaguchi (2020) suggest that increased development assistance can also support capacity for enforcement and the application of property rights, including intellectual property, to ensure that developing countries are able to protect their natural assets and benefit from them.

---

[488] For more on ODA, see Chapter 20.

Chapter 16
Demand for Provisioning Goods and Its Consequences

Introduction

As formalised in the global model in Chapter 4*, harvesting of provisioning goods has consequences for our global stock of natural capital, and its ability to provide regulating and maintenance services on which economies and societies also depend. Building on Chapters 2 and 4, this chapter examines our current extraction of provisioning goods and expectations for our future demand, and the consequences for regulating and maintenance services.

16.1 Current Harvest of Provisioning Goods and Future Prospects

16.1.1 Terrestrial Food Production

As noted in Chapter 4, human economic activity per person and population have both risen rapidly since the middle of the 20th century. The trends in indicators of terrestrial biodiversity and the biosphere described in Chapters 2 and 4 have been significantly affected by land-use change for commodity production; this is the primary driver of terrestrial biodiversity loss. A large proportion of land-use change over the past few centuries (accelerating since 1900) has been for the expansion of crop and pastureland. Figure 16.1 shows changes in global land use since 1700: growth in land for food production at the expense of natural habitat is clear (Ramankutty et al. 2018). Newbold and colleagues (2015) assessed the consequences of land-use change for several measures of biodiversity, across a wide range of taxa; in the worst affected habitats, which faced pressures from combinations of plantation, cropland, pasture, infrastructure and urban land expansion, species richness was reduced by an average of 77% and total abundance by 40% (Newbold et al. 2015). Figure 16.2 maps changes in species richness due to land-use change globally.

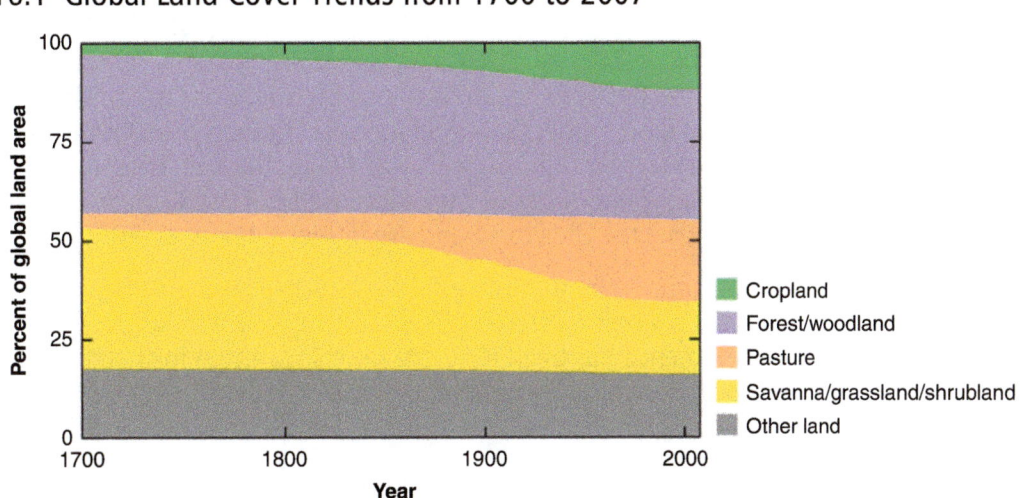

Figure 16.1 Global Land Cover Trends from 1700 to 2007

Source: Ramankutty et al. (2018). *Permission to reproduce from Annual Reviews, Inc.*

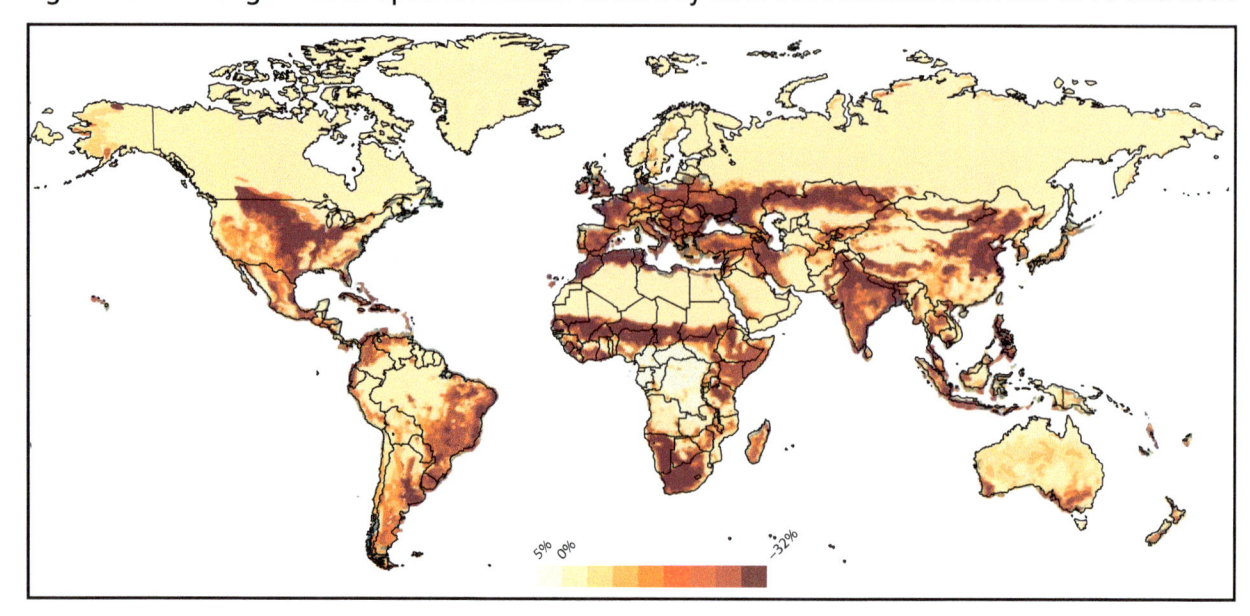

Figure 16.2 Change in Local Species Richness Caused by Land Use Pressures Between 1500 and 2000

Source: Newbold et al (2015). *Permission to reproduce from Springer Nature.*

Over a third of Earth's land surface is currently used for crops and pasture (Foley et al. 2005; Ramankutty et al. 2018; Balvanera and Pfaff, 2019). Most new cropland has replaced forests, and most new pastureland has replaced grasslands, savannahs, and shrublands (Figure 16.1) (Ramankutty et al. 2018).[489] Box 4.1 in Chapter 4 explained the consequences of forest conversion for species loss.

Between 1850 and 1950, agriculture expanded rapidly in North America and the former Soviet Union (Ramankutty and Foley, 1999). In the past 70 years, this expansion has shifted to the tropics. Gibbs et al. (2010) found that land used for agriculture increased by over 100 million hectares between 1980 and 2000 across the tropics, with half of this land directly converting tropical forests. Most of this conversion was, and continues to be, for the production of traded commodities (Pendrill et al. 2019).[490] Cattle pasture contributed to the largest agricultural land expansion in South and Central America: increasing by approximately 42 million hectares. African cropland area has increased by 50% in East Africa, and 25% in West Africa, mainly at the expense of shrubland. In south-east Asia tree plantations occupy the largest share of agricultural land, which rose by 7 million hectares in the last 20 years of the 20th century. By the 1990s, oil palm was responsible for more than 80% of the expansion in tree plantations (Gibbs et al. 2010). Habitat converted to agricultural land has suffered severe biodiversity loss across all measured taxa (Newbold et al. 2015).

Land used for livestock feed contributes to the footprint of animal products (particularly beef). Globally, 35% of crop production is for animal food, rather than food for direct human consumption (Foley et al. 2011). Land is also used to produce food for our pets (cats and dogs): the first assessment of the environmental 'paw' print of pet food found that its production used 0.8–1.2% of global agricultural land (equivalent to approximately twice the UK's total land area) (Alexander et al. 2020).

Of the commodities produced through this land conversion, beef has by far the greatest impact on land use and other environmental inputs per unit of protein. It has been estimated that approximately 71% of rainforest conversion in South America has been for cattle ranching, and a further 14% for commercial cropping including soya for animal feed (Godfray et al. 2018).

---

[489] There are notable exceptions including North American prairies which have been replaced by cropland, and substantial Latin American deforestation which has historically been, and remains, for livestock grazing (Ramankutty et al. 2018).

[490] The relationship between trade and the biosphere was examined in Chapter 15.

Future Prospects

As our global population grows, sustainable food production will pose ever increasing problems. Dietary change (notably increases in meat consumption) and increasing use of land for biofuel will put further pressure on the biosphere. Studies have suggested that unless there are dramatic changes in food waste and consumption patterns, agricultural production would need roughly to double by 2100 (FAO, 2009; Royal Society, 2009; Foley et al. 2011; EAT-Lancet Commission, 2019).

IPCC's Special Report on Emissions Scenarios (SRES) storylines have been combined with FAO data to project future food demand to 2100 (IPCC, 2000; Bodirsky et al. 2015). The SRES scenarios can be represented on a two-by-two grid with economic versus ecological values, orthogonal to increasing globalisation versus increasing regionalisation (Figure 16.3). Scenarios A1 and A2 represent a future characterised by materialistic attitudes. By contrast, scenarios B1 and B2 describe futures where societies have stronger environmental awareness. In A1 and B1, economies are globalised, with greater cultural and social interactions among regions, whereas in A2 and B2 economies are more self-reliant, and look for local solutions to economic and social problems (Figure 16.3) (Bodirsky et al. 2015).

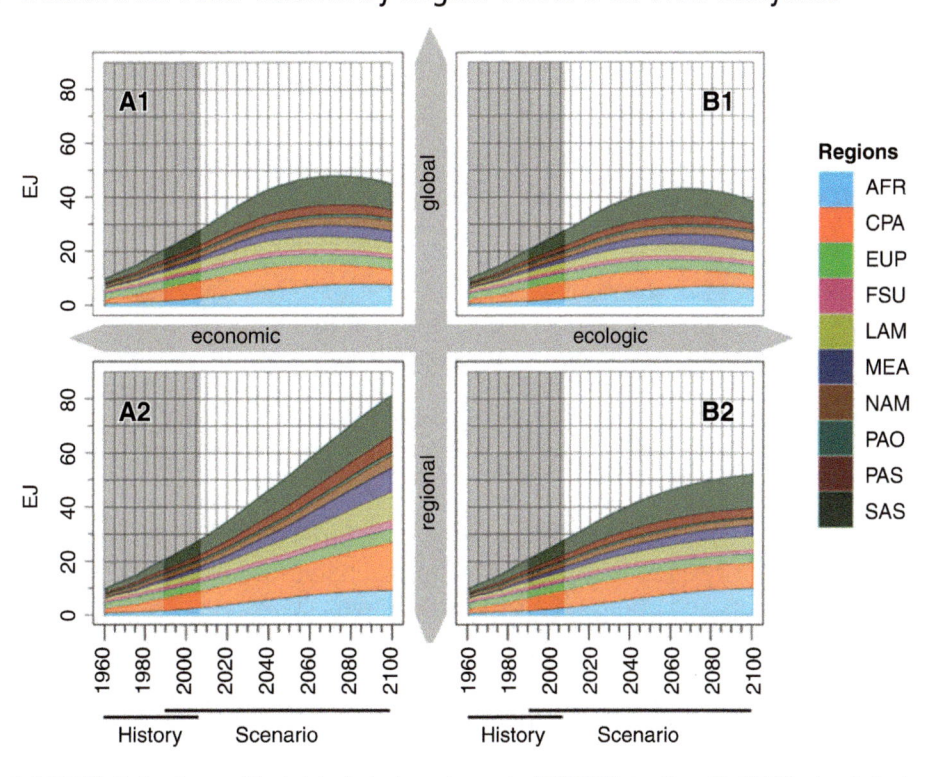

Figure 16.3 Demand for Total Calories by Region Under IPCC SRES Storylines

Source: Bodirsky et al. (2015). Notes: Demand for total calories by region under IPCC SRES storylines. EJ: 1018 joule a-1. Region key: AFR: Africa, CPA: Centrally Planned Asia (including China); EUR: Europe (including Turkey); FSU: Former Soviet Union; LAM: Latin America; MEA: Middle East; NAM: North America; PAO: Pacific OECD (Australia, Japan and New Zealand); PAS: Pacific Asia, SAS: South Asia (including India).

During the first half of the 21st century, there is a strong increase in demand in all scenarios. Moving beyond 2050, scenarios start to diverge: we see continuous growth (A2; economic values, and increasing regionalisation), stagnation (B2; ecological values, and increasing regionalisation), or a slight decrease (A1; economic values and globalisation, and B1; ecological values and globalisation) (Figure 16.3). Population growth differs in these scenarios: global population grows by 65% between 1990 and 2050 in scenarios A1 and B1, by 78% in scenario B2, and 114% in scenario A2. Demand per capita also increases in all scenarios. Specifically to provisioning goods, demand for animal products

Chapter 16: Demand for Provisioning Goods and Its Consequences

increases in all scenarios, outpacing the average increase in food consumption in each case (Bodirsky et al. 2015). This analysis illustrates the scale of the challenge: even when ecological values become globally more important to citizens and decision makers, more food, including more animal products, will still be needed to satisfy demand.

As well as increasing consumption and population, climate change will put pressure on production systems, and climate change and production will continue to put combined pressure on ecosystems. Projections of the future impact of climate change on crop yields are increasingly negative, with close to 20% of major crops projected to lose potentially 50–100% of their yields by the beginning of the next century due to changing temperatures and weather patterns (Figure 16.4) (FAO, 2016). Lower yields are likely to lead to increased pressure on land use for food, increases in food prices, and a rise in the number of malnourished people globally (Wheeler and von Braun, 2013; Nelson et al. 2014; Rosenzweig et al. 2014).

Figure 16.4 Projected Changes in Crop Yields for All Locations Worldwide Due to Climate Change

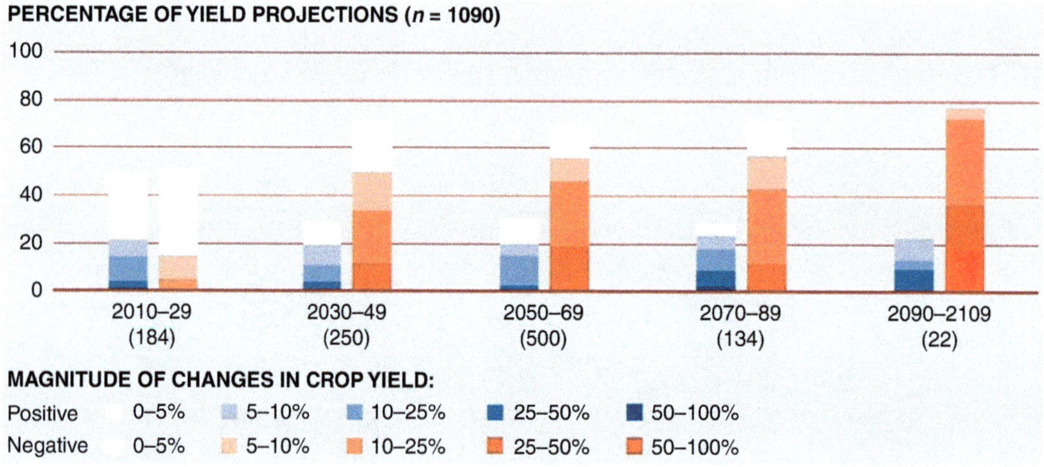

PERCENTAGE OF YIELD PROJECTIONS (n = 1090)

Bar chart showing yield projection percentages across time periods: 2010–29 (184), 2030–49 (250), 2050–69 (500), 2070–89 (134), 2090–2109 (22).

MAGNITUDE OF CHANGES IN CROP YIELD:
Positive: 0–5%, 5–10%, 10–25%, 25–50%, 50–100%
Negative: 0–5%, 5–10%, 10–25%, 25–50%, 50–100%

Source: FAO (2016). Note: Numbers of estimates of change in crop yield are shown in brackets.

16.1.2 Fibre and Biofuels

A rapidly growing share of global agricultural land is used to grow crops for products other than food, such as biofuels and fibre for textiles. In 2008, 4% of global harvested biomass was used for material (including textiles and timber) and energy (Carus and Dammer, 2013).

In 1995, more than 132 million hectares of arable land were used to grow biomass for non-food products. This figure increased to more than 178 million hectares in 2010, a growth of 35% in only 15 years. Non-food agricultural areas thus account for approximately 13% of overall global agricultural land area (Poore and Nemecek, 2018). With 81.8 million hectares, representing a share of 46% of global non-food cropland in 2010, the Asia-Pacific region was the most significant producer of feedstocks for non-food commodities (Bruckner et al. 2019). Looking at the final destination of non-food products, the European Union (EU) is the most significant consumer of global non-food cropland (Bruckner et al. 2019).

Fibre for textiles is one of the biggest occupiers of non-food cropland: cotton production occupies approximately 2.5% of global agricultural land (FAO, 2015b). The large area under cotton cultivation makes it one of the most significant crops in terms of land use, after edible grains and soybeans. While the global area devoted to cotton production has remained relatively constant over the last three decades, regional changes have occurred: the area has shrunk by about 40–50% in Brazil and the USA, but increased in Australia, Africa, and South Asia.

During the same period, yields of cotton have increased by almost 100% (FAO, 2015b). As well as having a significant land footprint, cotton production has disproportionately high agricultural inputs: although it only uses 2.5% of agricultural land, it consumes 16% of global insecticide use – more than any other single crop (FAOSTAT). The biodiversity impacts of chemical agricultural inputs are severe: as well as threatening insect species including valuable pollinators, agrochemicals also reduce soil biodiversity and disrupt biological processes occurring in soil (see Box 2.5 on soil biodiversity and the services it provides, and Niinimäki et al. 2020).

Future Prospects

One of the most important future changes to non-food cropland will be the expansion of bioenergy crops. IPCC scenarios suggest that global production of biofuels will need to grow rapidly in order to implement large-scale negative emissions technologies such as Bioenergy Carbon Capture and Storage (BECCS). The interaction of this with expansion of food cropland and climate change itself will have consequences for terrestrial biodiversity. Hof and colleagues combined land use and climate change projections to investigate their potential impact on terrestrial vertebrate biodiversity under high and low emissions scenarios (Representative Concentration Pathways (RCP)) 6.0 and 2.6 respectively in Figure 16.5). The total impacts on vertebrate diversity are similar in both high and low emissions scenarios, with the impact of land-use change being greater in the low emissions scenario (as a result of bioenergy cropland expansion (the yellow areas in Figure 16.5)), and the impact of climate change being greater in the high emissions scenario (the red areas in Figure 16.5). Many areas will be threatened by combinations of pressures: for example in the high emissions scenario, mammals in parts of central Africa will be at risk from pastureland expansion and climate change together (Hof et al. 2018).

Figure 16.5 Global Distribution of Climate Change and Land Use Threats to Vertebrate Diversity Under Emissions Scenarios for 2080

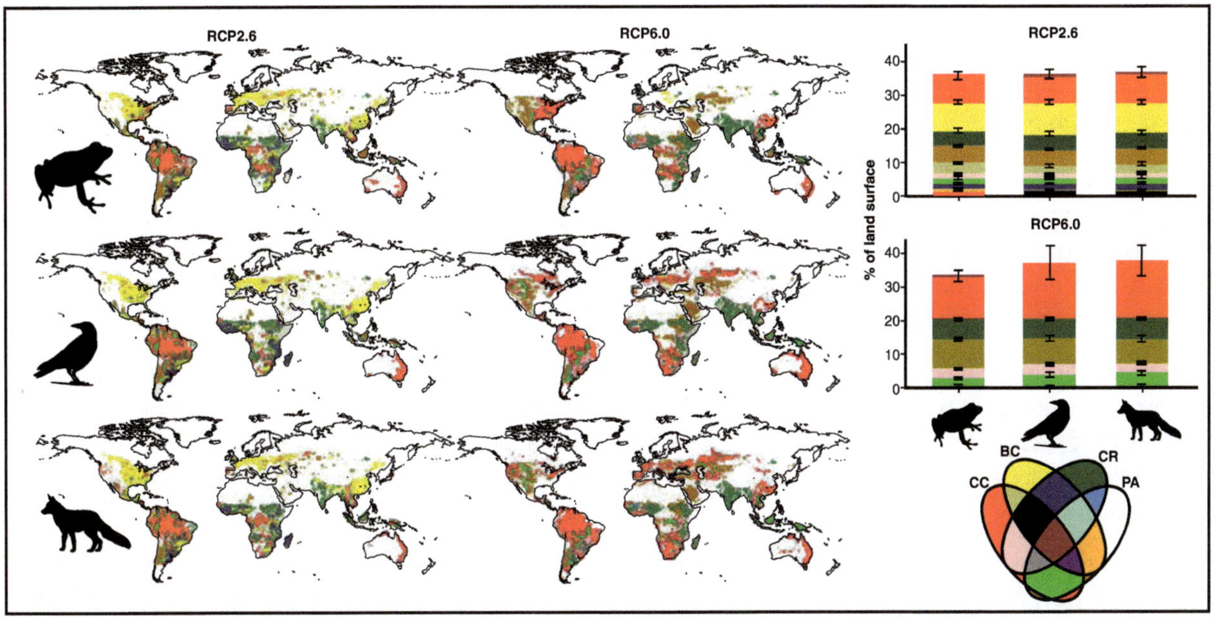

Source: Hof et al. (2018). Note: Global distribution of climate change and land-use threats to vertebrate (amphibian, bird, and mammal) diversity under low (RCP2.6) and high (RCP6.0) emissions scenarios for 2080. BC: bioenergy cropland; CC: climate change; CR: non-bioenergy cropland; PA: pasture.

16.1.3 Timber

After conversion of land for agriculture, logging is the biggest threat to species on the IUCN Red List (Maxwell et al. 2016), and conversion of natural forest to timber plantation is one of the main drivers of extinction and biodiversity loss (Newbold et al. 2015, and discussed in Chapter 4). Forests currently cover 31% of the Earth's terrestrial area, and many people depend on them for food and livelihoods in part or entirely (FAO, 2020b). Between 1990 and 2015, global total forest area fell from 4.28 billion to 3.99 billion hectares, while over the same period the area of planted forest increased from 167.5 to 277.9 million hectares (FAO, 2015a; Payn et al. 2015). The global harvest of roundwood in 2018 was estimated to be just under 4 billion cubic metres, of which approximately 50% was used in industry and 50% was for fuelwood (FAO, 2020a). Harvests of industrial roundwood have fallen in high income countries, but are increasing in middle income countries (Balvanera and Pfaff, 2019). Forest biomass generates energy as solid, liquid, and gaseous fuels, accounting for 14% of the global energy mix in 2014 (International Energy Agency, 2019).

Forest management usually aims to maximise timber production (FAO, 2015a). High-intensity forest management results in high merchantable timber yields, but in many cases contributes to carbon loss, biodiversity loss, and overall forest degradation, with loss of other ecosystem services, particularly regulating and maintenance services including water quality and climate regulation (Trumbore, Brando, and Hartmann, 2015; Schulze, Malek, and Verburg, 2019). Conventional management practices usually result in forest that is low in overall species and tree age diversity. Alternative management systems aim for more diversity in tree species, age, and structure (Puettmann et al. 2015). The impact on biodiversity of managing forest for timber production varies significantly by management type, geographic context, and affected taxonomic group (Chaudhary et al. 2016).

Future Prospects

A 2019 study projecting global planted forest area through to 2070 under several of the IPCC's Shared Socioeconomic Pathways (SSPs) concluded that it would increase in both high, low, and middle income countries in all SSPs. Increases range from a 46% increase from 2015 levels under SSP3 (which represents a relatively poor and unequal world), to a 66% increase under SSP5 (representing a relatively wealthier world with higher equality) (Nepal et al. 2019). This increase is due to increased demand for timber and energy wood globally due to current consumption trends continuing (as discussed in Chapter 4).

16.1.4 Water[491]

Only around 3% of the world's water is freshwater – fit for human consumption. Freshwater ecosystems only make up 1% of the world's freshwater resources; of the remaining 99%, 69% is perennially frozen, and 30% is groundwater (Carpenter, Stanley, and Vander Zanden 2011). They are threatened by human activity and environmental change (including climate change, fragmentation, nitrogen deposition, shifts in precipitation and changing runoff patterns).

The World Economic Forum's annual Global Risk Report in 2016 ranked water crises as the global risk of highest concern over the decade to follow (World Economic Forum, 2016). Climate change and increasing pressure on the global food system (discussed above) will intensify challenges with water availability. Scarcity is already a real problem: estimates suggest that approximately 30% of the global population do not use safe drinking water, and lack access to basic sanitation (WHO and UNICEF, 2017).

Globally, the abstraction, use and pollution of fresh water by humans for agriculture, industry, drinking and other purposes has increased hugely over the course of the last century, from just over 500 billion

---

[491] We are grateful to Ed Barbier for providing material for this section and have drawn from his 2019 book *The Water Paradox*.

m³ to approximately 4 trillion m³ globally per year. Water use continues to increase in order to sustain the growing demand for food and other uses (Figure 16.6) (Burek et al. 2016).

At the start of the 21st century, agricultural production accounted for 92% of humans' water footprint, approximately 77% of which is used in non-irrigated agricultural systems (Mekonnen and Hoekstra, 2010; Hoekstra and Mekonnen, 2012). Abstraction of water for agricultural use has had huge impacts on fresh-water resources globally; for example, the Aral Sea, a vast lake in Central Asia with a surface area of 68,000km² in 1960, was extensively used for irrigation and had dried up completely by 2014. In addition to the quantity used, loading water with nutrients, pesticides and livestock antibiotics negatively affects its quality, biodiversity and human health (Ramankutty et al. 2018).

Water abstraction affects wetlands, rivers, and lakes. These freshwater ecosystems are already experiencing the most severe biodiversity loss: the Living Planet Index for freshwater has declined by 83% since 1970 (WWF, 2018). As noted in Chapter 3, dams controlling the flow of freshwater for human consumption (as well as hydropower) cause large-scale fragmentation of freshwater ecosystems, and associated damage to the life within them (Nilsson et al. 2005; Barbarossa et al. 2020).

Physical features of water have implications for how it functions as an asset. It is viewed as a renewable resource (much available freshwater is in lakes, rivers, streams and other water bodies, which frequently renew through local water cycles). It is highly mobile (it flows, it evaporates, it transpires, making it difficult to measure and capture and meaning that availability fluctuates), and other substances – including sewage and toxic waste – can easily dissolve in it. One consequence of these features is difficulty in enforcing exclusive property rights over water. There is also a high cost of storing, transporting and distributing water, meaning it is susceptible to scale economies – consequently, water is generally supplied publicly or through regulation of a single large private investor (which could be described as a monopoly).

A disturbing trend is that water-scarce low and middle income countries are becoming increasing exporters of water through agricultural and other trade (trade is discussed further in Chapter 15). For example, Egypt exports scarce water through cotton (a water-intensive crop, as mentioned above), fruit and vegetables to countries including the US and Germany. The problem is not trade itself, but the externalities and institutional failures which enable the export of water-intensive commodities from already water-stressed regions. Agricultural trade is increasingly dependent on non-renewable groundwater for irrigation: groundwater depletion increased by 22% in the decade to 2010 (Dalin et al. 2017). Global policy in many cases is increasing water scarcity through regulating, subsidising, and distorting water prices.

The 'price' paid for water has little to do with its scarcity. According to the economist Michael Hanemann, the price consumers pay for water reflects:

"[A]t best, its physical supply cost, and not its scarcity value. Users pay for the capital and operating costs of the water supply infrastructure . . . . Thus, in places where water is cheap, this is almost always because the infrastructure is inexpensive, or the water is being subsidised, rather than because the water per se is especially abundant." (Hanemann, 2006)

Governments around the world typically subsidise water, contributing to its underpricing. Other classes of subsidy exacerbate this: agricultural subsidies contribute to the over-use of water: subsidising agricultural production leads to more production than is necessary, and thus drives overuse of agricultural inputs including irrigation water (for more detail on the global scale of agricultural and other subsidies see Annex 8.1).

Policy (contributing to underpricing) and institutional and governance failures compound water scarcity by encouraging its wasteful use. Most of the world's existing water abstraction regimes and their governance arrangements evolved during times of comparative water abundance, and are therefore not equipped to handle water scarcity (Young, 2014). There is a mismatch between

Chapter 16: Demand for Provisioning Goods and Its Consequences

governance and institutions, and management needs: they have been slow to adapt to changing conditions, including climate change (Olmstead, 2014). To make governance as effective as possible, it needs to occur at the right level (Chapter 7 discussed polycentric governance). For water, this is a problem, as resources often cross boundaries between established governance units. It is widely recognised that the river basin (or other body of surface water) is the appropriate unit of governance for water management, however, different uses of water are typically governed at different levels and by different institutions. Irrigation, industrial, and municipal uses are administered through separate government agencies, and can be covered by different administrative units, which makes integrated river basin management a challenge.

Future Prospects

The International Institute for Applied Systems Analysis published the Water Futures and Solutions Initiative, modelling global water demand and supply under three IPCC scenarios (combining the IPCC's SSPs and RCPs): 'Sustainability' (resulting in low challenges with respect to sustainability, mitigation and adaptation), 'Middle of the Road' (intermediate challenges, could be seen as 'business as usual'), and 'Regional Rivalry' (high challenges to sustainability agendas). Under the 'Middle of the Road' scenario, the share of water used for agriculture will decrease from its current level of 92% to 59% in the 2050s. Asia remains the largest water user for all sectors, especially agriculture. Domestic demand will increase rapidly in sub-Saharan countries, driven by socio-economic and population growth (Figure 16.6). These predicted increases in water use will mean that strategies to improve efficiency (such as improved irrigation) need to be implemented to lessen damage to water-dependent ecosystems (Burek et al. 2016).

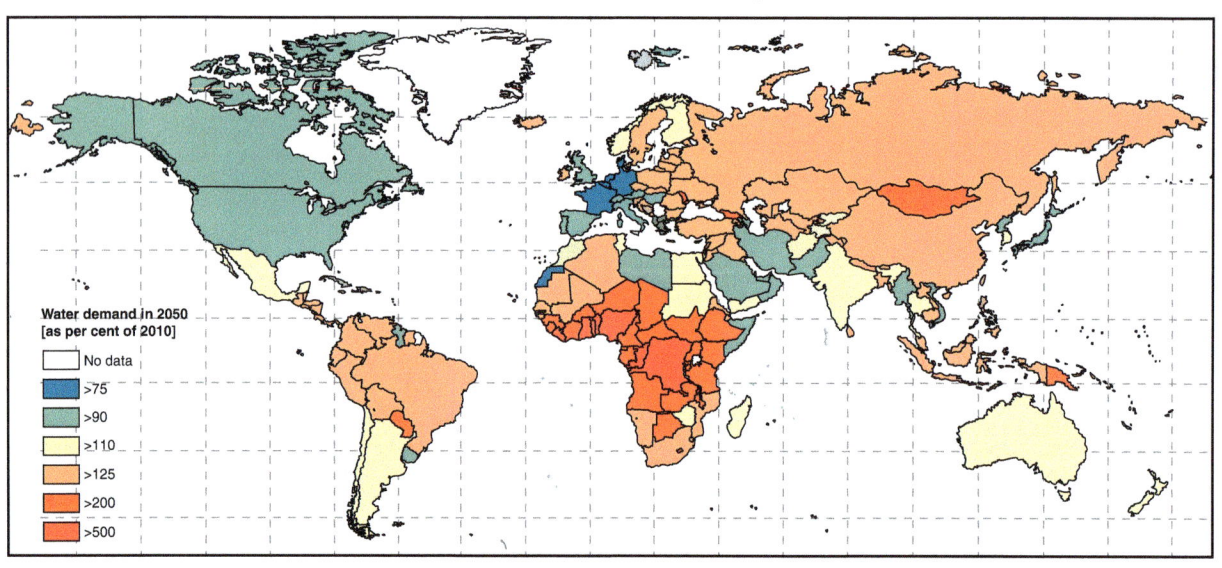

Figure 16.6 Projected Global Water Demand in 2050, Compared to Demand in 2010

Source: Burek et al. (2016) ('Middle of the Road' scenario).

16.1.5 Fisheries and Aquaculture

Since 1950, global fishing has increased steeply; Figure 16.7 shows increases in global fishery catch and aquaculture production. Over the same period, catch per unit effort (CPUE) has steeply declined. The contribution of global aquaculture to world fish production has increased massively in the last two decades: from 26% in 2000 to 46% in 2016–18 (FAO, 2020c)[492].

---

[492] Trends in aquaculture are also having an important impact on marine and coastal biodiversity, however, these trends are not discussed here.

Rousseau and colleagues reconstructed the number of vessels and their engine power, and the CPUE of the global fishing fleet of industrial and artisanal fisheries since 1950 (Rousseau et al. 2019). While fish stocks have stabilised in Northern Europe, North America (where the size of fleets has been reduced) and Oceania (as a result of improved fishery management), fishing power has increased recently in other areas, such as south-east Asia (Rousseau et al. 2019). If these trends continue, a million more fishing vessels could appear in the global fishing fleet in the next few decades.

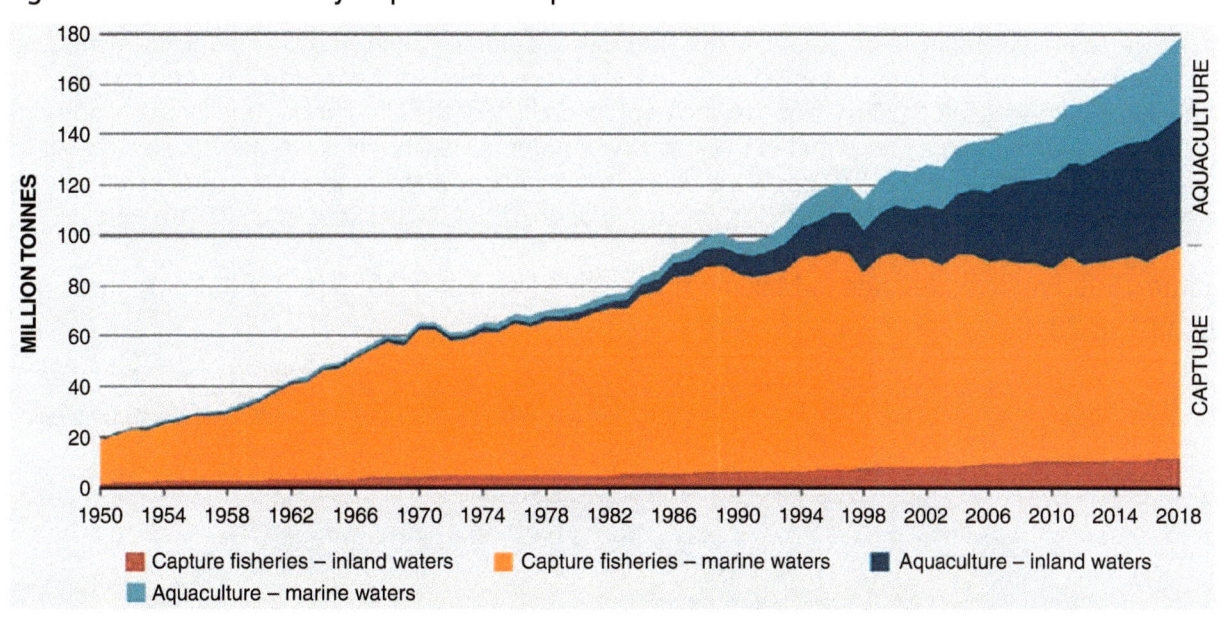

Figure 16.7 Global Fishery Capture and Aquaculture Production Since 1950

Source: (FAO, 2020c). Note: Excludes aquatic mammals, crocodiles, alligators and caimans, seaweeds and other aquatic plants.

Analysis of nearly 5,000 fisheries, covering 78% of global fish catch, found that though approximately a third of fisheries were in good biological condition, the world's median fishery was overfished, and with overfishing continuing (Costello et al. 2016). Overfishing is a significant threat to capture fishery sustainability and marine biodiversity (Costello et al. 2019). There are significant barriers to reducing the power of the global fishing fleet; the capture power of small-scale fisheries (making up a significant proportion of total global fisheries) matches that of industrial counterparts in many places, and these are usually not accounted for in national fish capture data. An estimated 23% of global fish catch is from unassessed fisheries (Costello et al. 2012). In some parts of the world, small-scale fisheries are operating in the absence of governance (Gilman, Passfield, and Nakamura, 2014). Without these fleets being recognised properly, decisions are based on an incomplete understanding of the pressures on marine ecosystems (Rogers et al. 2020). An estimated US$22.2 billion was spent on capacity-enhancing fishery subsidies in 2018, a large proportion of the total estimated US$35.4 billion spent on all fishery subsidies (Sumaila et al. 2019).[493] The World Trade Organization has been tasked with eliminating capacity-enhancing subsidies, which would significantly benefit marine ecosystems (Sumaila et al. 2019).

Future Prospects

Managing fisheries below their maximum sustainable yield avoids fish population crashes and maximises both ecosystem health and long-term food production (as illustrated in Box 3.3 in Chapter 3 on multiple stability regimes in fisheries). This requires reducing fishing pressure and allowing

---

[493] See annex to Chapter 8 for more estimates of subsidies globally.

Chapter 16: Demand for Provisioning Goods and Its Consequences

overfished stocks to rebuild. Costello and colleagues suggest that reducing overfishing and overall catch, using catches more efficiently (despite improvements, bycatch is still a significant problem (WWF, 2020b)), and raising production from underfished stocks could raise global fish capture by 20% above current levels, and over 40% above projected future capture with continued current fishing pressures (Costello et al. 2019).

Climate change will alter the availability of marine food all over the world. In some places, this will cause increases in fish availability, while in others, availability will decrease. Figure 16.8 shows projected changes in fish catch globally, comparing 2001–2010 to predicted catch in 2051–2060 under an IPCC scenario for moderate to high warming (IPCC, 2014). Most decreases in potential catch will occur along coastlines and in the tropics, while most increases in potential catch will occur further from land in the open ocean – as shown in Figure 16.8 (IPCC, 2014). People who depend on fish for their livelihoods and large proportions of their nutrition are more concentrated in areas where potential catch is likely to decrease as a result of climate change (Selig et al. 2019). Coastal decreases in fish catch availability are located in the same places as high economic and nutritional dependence on the ocean (IPCC, 2014; Selig et al. 2019).

Figure 16.8 Predicted Change in Maximum Available Fish Catch 2051 to 2060, Compared to 2001 to 2010

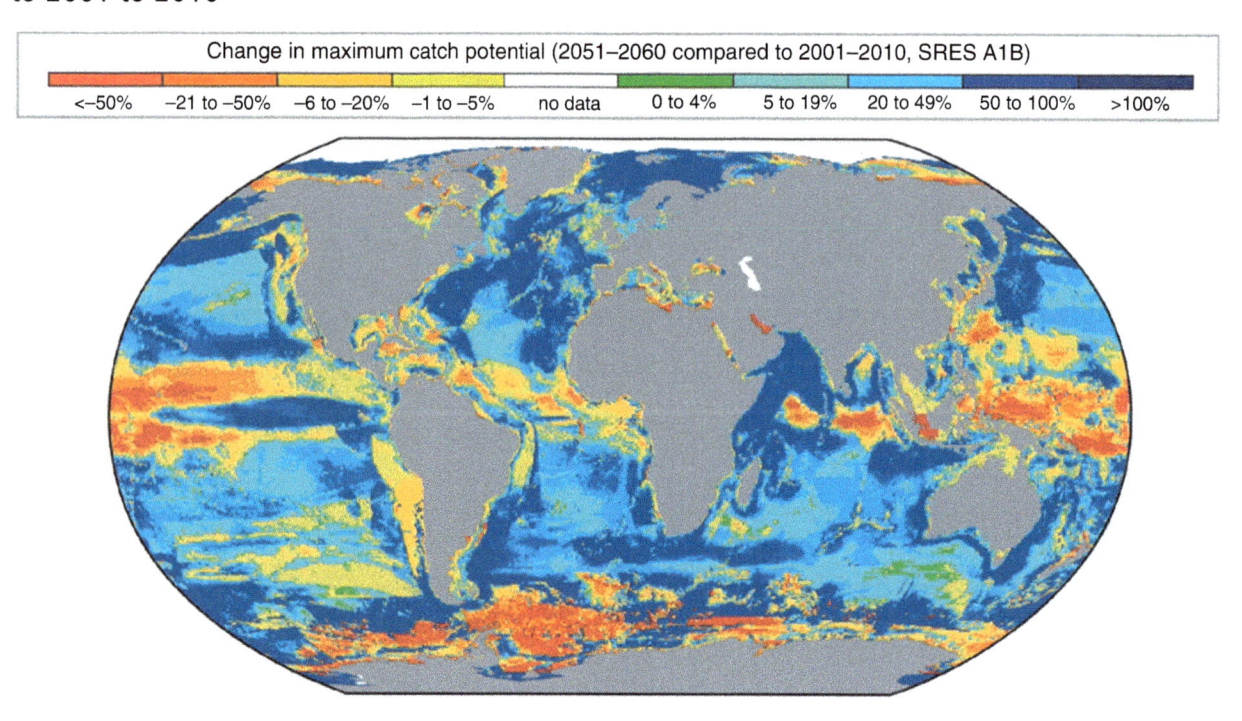

Source: IPCC (2014).

16.2 Trade-Offs Between Provisioning and Regulating Services

Building on evidence provided in Chapters 2 and 4, we examine further the impact of the huge expansion in provisioning goods documented in this chapter on the capacity of ecosystems to provide regulating and maintenance services without which production, even life, would not be possible.

The mobility of Nature means that the impacts on regulating services of our extraction of provisioning goods can be felt at far removed locations. For example, the large-scale use of fertiliser for improving agricultural production has consequences for water quality that occur over a much wider spatial scale

(Tilman et al. 2002). The bluegum plantations in the Nilgiri Plateau in southern India, which produce paper pulp and tannin, have reduced water yield from affected catchments by as much as 23% (Samraj et al. 1988).

Increased extraction of provisioning goods now comes at the expense of future provision of those same services, as well as regulating and maintenance services. In Australia, woody vegetation was cleared for agricultural production in the early 20th century. This vegetation was in fact providing an important regulating service: it maintained groundwater at deep enough levels that salts were not carried up to the topsoil. When the vegetation was removed, the water table moved upwards towards the surface, bringing with it high levels of salt which made the land unsuitable for conventional agriculture (Greiner and Cacho, 2001; Briggs and Taws, 2003).

Chapter 3 examined non-linear ecosystem dynamics; these features make it difficult to project the potential consequences of our increasing extraction of provisioning goods, and to assess the reversibility of resulting declines, in particular of regulating and maintenance services (see also Cavender-Bares et al. 2015). Uncertainty caused by lack of knowledge of the biosphere's processes, and the consequences of this for decision-making, are discussed in Chapter 5. Changes in pollination over the last 50 years provide an illuminating example of large effects across time and space. Due to habitat conversion for production, use of agricultural inputs to maximise production, and invasive species there has been a decrease in pollinator diversity in most global regions. Loss of diversity is irreversible, and will impact future pollination (Potts et al. 2016). Benefits of pollination are discussed in Chapter 2, here we focus on the results of pollinator declines caused in large part by increased consumption of provisioning goods. The honeybee (*Apis mellifera*) makes the most significant contribution to agricultural pollination. The US has seen a 59% loss of honey bee colonies between 1947 and 2005 (National Research Council, 2007), while 25% of colonies were lost from Europe between 1985 and 2005 (Potts et al. 2010). In parts of the UK, honeybee numbers have fallen by 60% since 1980 (Biesmeijer et al. 2006). Almost all wild colonies have been eradicated from Europe and the US due to the invasive parasitic mite *Varroa destructor*, leaving only those kept by beekeepers. Although the highest volume crops (such as rice and wheat) are wind-pollinated, a large proportion of fruit crops are vulnerable to pollinator declines.

The cultivation of pollinator-dependent crops has steadily increased since the 1960s. A large proportion of this increase in the face of declining pollinator numbers can be explained by use of commercial pollinators or hand pollination (though this is rare). However, over the past 50 years, yields of crops with greater pollinator dependence have increased at a lower rate and become more variable than crops that are less pollinator dependent, suggesting that pollination services are compromised by pollinator decline (Garibaldi et al. 2011). Human-made alternatives to wild pollination might partly replace the pollination role of wild animals for some crops, but they cannot replace pollination of wild plants, nor the cultural value of pollinator species (Garibaldi et al. 2013). Declines in pollinator wild species cause declines in species depending on wild plants for food, as well as decreasing numbers of the plants themselves, with far-reaching consequences for ecosystems (Balvanera and Pfaff, 2019).

In aquatic ecosystems, fish play important roles in regulating trophic structure, on which the stability and resilience of the system depends. Overfishing of top predators in the Black Sea in the late 20th century led their populations to crash in the 1970s, resulting in increased numbers of planktivorous fish at lower trophic levels. When top predators were eliminated, the system shifted from having four trophic levels, to having three: zooplanktivores (top level), whose population rose; zooplankton, whose population fell; and phytoplankton, whose population rose. This gap at the top of the food web left space for an alien carnivore species to become the top predator: populations of the comb jelly *Mnemiopsis leidyi* exploded. The comb jelly hunted in a completely different way: predatory fish feed selectively, but ctenophores (such as *M. leidyi*) feed unselectively on their prey, leading to an even larger disruption of the food web. Fishing continued but focused on lower trophic levels, causing another population crash – this time of planktivorous fish – in the 1990s (Daskalov et al. 2007).

Chapter 16: Demand for Provisioning Goods and Its Consequences

Fish can create and maintain their own habitats: when populations decline due to overfishing, habitat for the remaining members of the population declines in quality (Holmlund and Hammer, 1999). Salmon deposit their eggs in self-created riverbed gullies, and cover them in a layer of gravel. This spawning process disturbs the sediment, removing aquatic macrophytes and fine sediment particles (Field-Dodgson, 1987). The process probably also displaces invertebrates and makes them available to fish higher in the water column as food (Bilby et al. 1998). Over years of repeated spawning in the same site, the riverbed becomes modified; dunes in the sediment form, and the spaces between them provides habitat for juvenile salmon (Field-Dodgson, 1987). When populations decline, the structure of this habitat is degraded, making survival of large numbers of juvenile salmon less likely.

16.2.1 Balancing Delivery of Provisioning and Regulating Services

Several approaches attempt to reach an optimum state which enhances provisioning and regulating services simultaneously. Studies have examined population-level responses of large numbers of species to increasing agricultural yields, from undamaged habitats to high-yielding production systems. They have concluded that most species suffer least from any type of food production if that production is at its highest sustainable yield, and as much remaining area as possible is used for habitat protection or restoration (see Chapter 19 for more on sustainable production systems in terms of their implication for conservation and restoration) (Phalan et al. 2011; Balmford, Green, and Phalan, 2015). This is also the case for carbon storage (Williams et al. 2018).

Researchers have looked at how to increase yields in sustainable ways. Foley and colleagues analysed areas with potential to increase crop yields, and areas with potential to grow more crops for direct human consumption; closing yield and calorie gaps (Foley et al. 2011). They found the calorie gap could be closed by shifting production of 16 major crops[494] to 100% human food, instead of the current mix of production for livestock feed and human food. This would have consequences for meat production, and may necessitate a shift towards more plant-centred diets.

In addition to changing the balance of crops produced for food and for feed, closing yield gaps could go some way to address increasing need without further expanding agricultural land. Bringing yields of the same 16 major crops to within 95% of their full potential yield would increase global production by 58% (Foley et al. 2011). In theory, these land-use efficiency improvements could largely be accomplished by improving nutrient and water supplies to crops in low-yielding regions. However, closing yield gaps without increasing environmentally damaging agricultural inputs will require improving crop genetics, and adopting aspects of organic and precision agriculture. As this analysis focused on 16 major global staples, it did not address the need to diversify crop production for increased human health and production system resilience.

Land sparing is the idea that food production will cause minimal harm to biodiversity if it uses minimal land to produce as efficiently as possible, and 'spares' large areas for Nature. Folberth and colleagues argue that closing yield gaps by spatially optimising fertiliser inputs and allocating 16 major crops (plus other row crops, vegetables and fruits, and non-food crops)[495] across global cropland could grow food for the world's population on as little as 50% of current cropland areas (Folberth et al. 2020). It is interesting to imagine what could be achieved if maximum land sparing was implemented globally, while bearing in mind that it may not be the best approach for both Nature and people. Folberth and colleagues compared a maximum land sparing scenario with a targeted land sparing scenario, which

---

[494] The 16 crops are: barley, cassava, groundnut, maize, millet, potato, oil palm, rapeseed, rice, rye, sorghum, soybean, sugarbeet, sugarcane, sunflower and wheat.

[495] Folberth et al. cover the same crops as Foley et al. (2011), with the addition of cotton, and the subtraction of rye. Folberth et al. also include other row crops, vegetables and fruits, and other non-food crops

Chapter 16: Demand for Provisioning Goods and Its Consequences

abandons cropland located in biodiversity hotspots and uniformly releases an additional 20% cropland all over the world. They found that the difference between maximum and targeted land sparing was approximately 10% of current cropland area, but the differences in biodiversity outcomes of maximum rather than targeted land sparing were minimal (Figure 16.9).

Figure 16.9 Cropland Area Needed to Grow 16 Major Crops With Existing Cropland and Under Two Land Sparing Scenarios

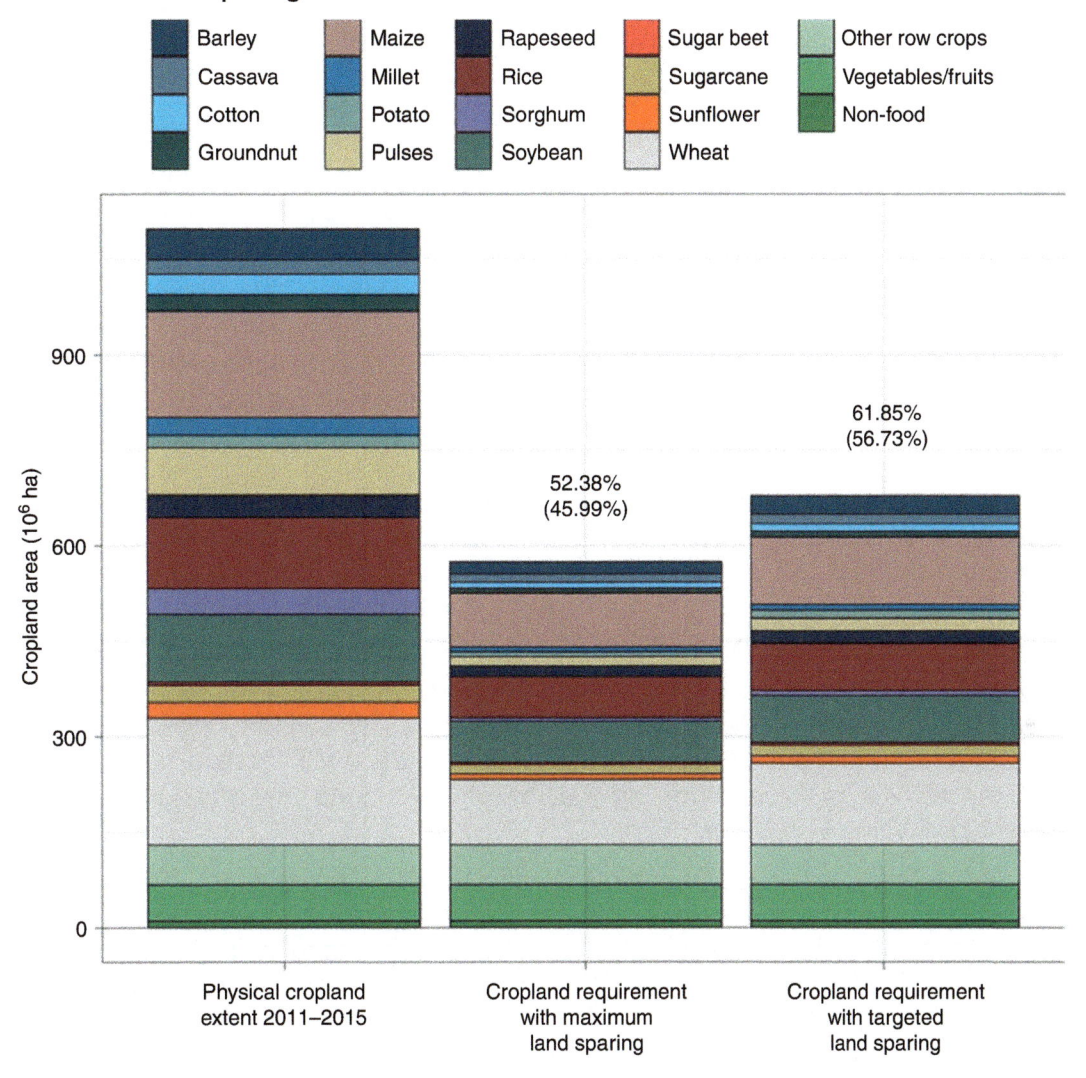

Source: Folberth et al. (2020). *Permission to reproduce from Springer Nature.* Note: Cropland area needed to grow 16 major crops (plus other row crops, fruit and vegetables, and non-food crops) with existing cropland (left bar), and under two scenarios. The middle bar shows proportion of existing cropland optimised for maximum land sparing potential, and the right bar shows optimised cropland with universal sparing of at least 20%, and complete absence of production from biodiversity hotspots.

Unintended consequences from maximising yield are possible. Yield increases can potentially undermine making space for Nature if they lead to lower prices which stimulate demand; the increased profits could encourage agricultural expansion and result in further biodiversity loss (Lambin and Meyfroidt, 2011; Phalan et al. 2016). Measures to avoid this include institutions discussed in Chapters 7, 8, 18, 19 and 20, such as land-use zoning (introducing clear boundaries for conservation and for agriculture); payments to avoid habitat conversion; strategically deploying technology, infrastructure or knowledge; and standards and certification schemes (Phalan et al. 2016). Reducing food loss and waste would also make a significant contribution to improving land-use efficiency, as discussed in Box 16.1.

Chapter 16: Demand for Provisioning Goods and Its Consequences

> **Box 16.1**
> **Food Loss and Waste**
>
> Production of food which is never consumed has significant environmental impact: greenhouse gas (GHG) emissions from the production of lost and wasted food amount to 8% of total GHG emissions, making food waste the third-largest carbon emitter in the world, behind the US and China (Hanson and Mitchell, 2017). The position of food loss and waste along supply chains varies depending on global context: loss is substantial in handling, storage and transport stages in regions where these are difficult to achieve (FAO, 2019). For aquatic food, post-harvest loss is a particular problem for small-scale fisheries. Large quantities of fish are lost because of post-harvest mishandling in transport, storage, processing and in waiting for sale, especially where access to electricity for cold storage is a difficulty (Béné et al. 2015). Post-harvest losses were estimated at 10% of global capture fishery and aquaculture production in 2005 (Béné, Macfadyen, and Allison, 2007).
>
> Three major types of footprint of food loss and waste are quantifiable: GHG emissions, pressures on land, and pressures on water. These all have an impact on biodiversity. Meeting environmental objectives through reducing food loss and waste requires an understanding of where in the supply chain food waste occurred, the footprints of the commodities which are wasted, and the costs of intervening to reduce waste at different points (FAO, 2019). For example, to reduce the land footprint, interventions should be aimed at reducing animal product waste because 60% of the land footprint of food loss and waste is for livestock production. To target water scarcity and GHG emissions, waste of cereals and pulses should be tackled, as at least 70% of food waste's water footprint and more than 60% of its GHG footprint originates from these. However, environmental footprints for commodities vary depending on region due to differences in yields and production techniques, so interventions must be context-specific (FAO, 2019).
>
> Some countries are already achieving marked reductions in food waste – the UK is one example, as documented in the Waste and Resources Action Programme's (WRAP) most recent Courtauld Commitment progress report: between 2015 and 2018 the UK saw a 7% reduction in food waste (480,000 tonnes), mainly through working in partnership with food retailers and tackling waste at the household stage (WRAP, 2020).
>
> It is unlikely that food waste could be completely eliminated, but waste which cannot be designed out of production practices and supply chains can be used: there is increasing pressure for food waste to be fed to pigs. The alternative is feed specifically produced for animals, which has its own production footprint (Salemdeeb et al. 2017). Although using food waste for animal feed was all but banned in the European Union after 2001's foot and mouth outbreak (thought to be caused by a UK farmer feeding uncooked food waste to pigs), re-legislating to allow food as feed enjoys widespread support among farmers and agricultural stakeholders (zu Ermgassen et al. 2018).

Agricultural systems can be net providers of ecosystem services, including regulating and maintenance services, beyond food production (Thorn et al. 2016). One example is agroforestry, which can provide food and other ecosystem services, such as sequestering carbon in vegetation and soil (Ango et al. 2014). Agroforestry has been found to improve multiple biodiversity indicators, and is particularly good for soil microbial diversity (reviewed in Udawatta, Rankoth and Jose, 2019). Farmers can also supplement their income with products from wood (Prasad et al. 2012). On-farm and near-farm plant and insect biodiversity in agroforestry have been found to lead to increased yields of pollinator dependent crops: a study of coffee farms in Brazil found that farms nearest forest fragments had a 14.6% increase in production related to pollination services (De Marco and Coelho, 2004). In North

Chapter 16: Demand for Provisioning Goods and Its Consequences

Eastern South Africa, it was found that mango yields were enhanced by the addition of patches of wild native flowers in agricultural fields (Carvalheiro et al. 2012).

Organic agriculture eliminates the damage done to ecosystems by harmful agricultural inputs (as it refrains from using synthetic fertilisers and pesticides, avoids overtilling, and thereby prevents the loss of organic matter),[496] but there is debate about its potential for feeding the world's population without causing further agricultural expansion (IAASTD, 2009; Foley et al. 2011; Chan et al. 2019). Its environmental benefits are widely reported (Tuomisto et al. 2012; Meier et al. 2015; Reganold and Wachter, 2016), however yields are lower, meaning larger land areas are required to produce the same volumes of food as conventional industrial agricultural systems (Connor, 2008; Seufert, Ramankutty, and Foley, 2012; Connor, 2013). Muller and colleagues assessed the role of organic agriculture in sustainable food systems and found that organic agriculture could feed the world and simultaneously provide environmental benefits associated with decreased pesticide use if food waste and consumption of animal products were reduced (Muller et al. 2017). Figure 16.10 shows environmental outcomes for a 100% organic agriculture scenario compared with 0% organic agriculture in 2050, with and without impacts of climate change on yields. Nutrient loading and pesticide use have a much lower environmental impact under organic agriculture, however, land use, deforestation, and erosion are slightly increased compared to 0% organic agriculture (Muller et al. 2017). This suggests that organic agriculture can make a contribution to sustainable food production, but is not the only answer.

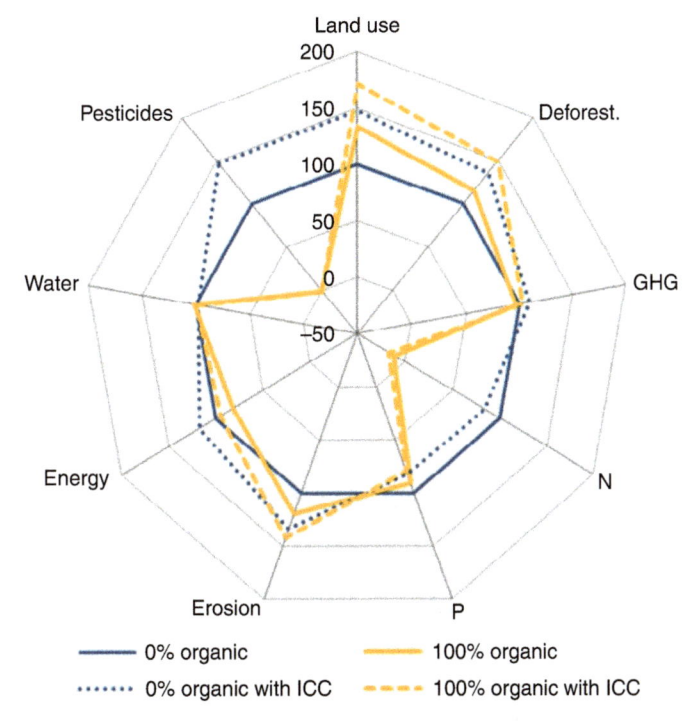

Figure 16.10 Year 2050 Environmental Impacts of a Full Conversion to Organic Agriculture

Source: Muller et al. (2017). Note: Year 2050 environmental impacts of a full conversion to organic agriculture. Environmental impacts of organic scenarios (100% organic agriculture, yellow lines) are shown relative to the reference scenario (0% organic agriculture, blue lines), with (dotted lines) and without (solid lines) impacts of climate change on yields; Calories are kept constant for all scenarios. Indicators displayed: cropland use, deforestation, GHG emissions (including. deforestation and organic soils), N-surplus and P-surplus, water use, non-renewable energy use, soil erosion, pesticide use.

---

[496] For correspondence on the subject, we are grateful to Tid Morton, who practises organic farming in Norfolk, UK.

Regenerative agriculture can also contribute to multiple ecosystem services. Practices such as zero- or low-tillage crop production have been adopted in parts of the USA where soil erosion and poor soil structure resulted from extremely intense farming. These approaches can improve regulating services of water retention and carbon sequestration, as well as soil biodiversity (Margulies, 2012).

Marine production systems can also be managed to deliver provisioning goods while simultaneously providing regulating and maintenance services. One important example is low trophic level aquaculture: producing species which do not depend on feed inputs, such as seaweed and bivalves. This can increase production of nutritious food without negative impacts on marine ecosystems, and is capable of enhancing capture fishery yield by creating artificial habitats (Costello et al. 2019; Theuerkauf et al. 2019). Bivalve and seaweed species absorb nutrients from water, which decreases coastal eutrophication and provide the important regulating service of improving seawater quality and clarity (Higgins, Stephenson, and Brown, 2011; Rabiei et al. 2014; Schroder et al. 2014; Rose, Bricker, and Ferreira, 2015). There are significant opportunities for several regions to benefit from the food production and environmental benefits which development of unfed aquaculture would present (Theuerkauf et al. 2019). Barriers to large-scale implementation of unfed aquaculture include regulatory difficulties and cultural acceptance (Costello et al. 2019).

Further examples of sustainable production systems that balance the provision of multiple ecosystems services over time are provided in Chapter 19.

16.3 Technology to Increase Efficiency in Our Use of the Biosphere

As mentioned in Chapter 4, various technologies are capable of increasing the efficiency in our use of the biosphere, in terms of minimising damage to ecosystems from extraction of provisioning goods, as well as waste and pollution from consumption and production. Here, we discuss the potential of technologies to reduce harm to the biosphere, with a focus on food production. These technologies have potential, but also face barriers to their large-scale implementation.

Innovations in food production can reduce agriculture's contribution to climate change and land-use change (through methods such as precision agriculture, vertical farming and cellular agriculture), reducing environmentally damaging inputs (using precision agriculture and integrated pest management for example), reducing bycatch in fisheries, and to improve production system resilience (through methods such as increasing diversity with molecular breeding techniques). Genetically modified crops can also contribute to several of these objectives. Effective institutions are important for technology uptake, and for minimising unintended consequences from technological improvement (Chapters 7–9).

16.3.1 Genetically Modified Crops

Changing the biological capabilities of crops offers the possibility of using marginal land for production, improving crop resistance to pathogens, obtaining higher yield on existing farmland, and enhancing nutritional quality. Although it remains controversial, genetic modification (GM) is the quickest and most targeted way to change an organism's properties currently available. It has been estimated that an additional 20 million hectares of land would have been needed to produce without GM the global harvest of maize and soybean produced with GM technology (Barrows, Sexton, and Zilberman, 2014). There are numerous examples of how classic GM and the newer, faster and more precise technique of gene-editing with CRISPR/Cas9 have conferred desirable traits on crops.

Gene-editing is carried out through the clustered, regularly interspaced, short palindromic repeat (CRISPR/*Cas9*) system. Based on bacterial immune systems, this technology uses targeted molecular guides to alter the genetic code of an organism in a heritable way (Arora and Narula, 2017). Changes leave no trace that DNA has been edited, causing global debate among regulators about whether to classify the technology as genetic modification or not. In 2018, the EU ruled that organisms altered through CRISPR/*Cas9* were genetically modified, and therefore could not be used for food. However, the potential of the technology to mitigate losses from crop disease is significant.

Two of the best-studied examples of the economic impact of GM are insect-resistant *Bacillus thuringiensis* (Bt) (most famously cotton) and herbicide tolerant (HT) crops. At the micro level, adoption of HT crops has led to decreased farm expenditure on herbicide, labour, machinery, and fuel, and increases in no-till practices which benefit soil quality (Trigo, 2011). However, as crops have been developed by private companies, fees charged for seeds can offset the cost reduction. Early studies for HT soybeans in the US found that the overall cost impact was small or negative (Fernandez-Cornejo, Klotz-Ingram, and Jans, 2002; Qaim, 2009). Bt crops produce proteins which are toxic to the larvae of some groups of insects, and can act as a substitute for chemical insecticides. Adoption of Bt crops reduces the need for insecticide reduction and increases yield. Farmers who face high pest pressure, but only use small amounts of insecticide, will see considerable yield benefits. For farms using high volumes of chemical pesticide the insecticide reduction effect will be dominant: their yield will not vary substantially, but the insecticide action of the Bt crop will reduce the need for applied pesticide (Qaim and Zilberman, 2003).

Economic impacts of the first decade of GM crop adoption have been analysed using partial and general equilibrium approaches. Using a partial equilibrium approach to examine macrolevel impacts, Price et al. (2003) found that Bt cotton generated an economic surplus gain of around US$164 million annually in the US by the late 1990s. Of this, 45% went to innovating companies, 37% to farmers and the remaining 18% to consumers. Distribution of economic benefits varies depending on the strength of a country's intellectual property regulation: in the US, the agricultural biotechnology company Monsanto gained 60% of the benefit from HT soybean adoption, with farmers receiving 20%, whereas in Argentina, 90% of the benefit went to farmers in the early 2000s (Qaim and Traxler, 2005). General equilibrium models have been used to analyse direct and indirect effects of GM crop adoption: Subramanian and Qaim (2010) built a detailed model of a typical cotton-growing village economy (based on census data from the Indian state of Maharashtra) to examine effects of Bt cotton uptake. They found that Bt technology produces higher incomes than conventional cotton (all types of household benefitted from Bt cotton compared to conventional cotton, and 60% of benefits were captured by extremely or moderately poor households), and generates employment, especially for female agricultural workers (Subramanian and Qaim, 2010).

Another example of transformational crop change through genetic modification relates to rice: changing its primary metabolism to increase yield dramatically (C_4 rice). As rice makes up 19% of global calorie intake, improvements to its growth efficiency have far-reaching consequences (Elert, 2014). Plants can carry out photosynthesis using a basic C_3 or C_4 molecule; the C_4 route is much more efficient. Yields of species currently using C_3 photosynthesis would be considerably higher if they could be re-engineered to use C_4. Rice is currently a C_3 plant, but a global research consortium has been working since 2009 to introduce genes from maize and change its photosynthetic pathway to C_4. The researchers estimate that rice yields could be improved by up to 50%, and that water use efficiency could be doubled (Rizal et al. 2012; Ermakova et al. 2019).

Disease resistance has been conferred on plants through genetic engineering and gene-editing: examples include resistance against the rice blast disease and rice bacterial blight, powdery mildew in wheat, and citrus canker (Jiang et al. 2013; Wang et al. 2014; Peng et al. 2017).

Chapter 16: Demand for Provisioning Goods and Its Consequences

The rice blast fungus (*Magnaporthe oryzae*) alone is responsible for 30% of global rice losses each year, enough to feed 60 million people (Skamnioti and Gurr, 2009); yield gains from deploying resistant varieties would be substantial. However, widespread use of individual strains could deepen problems caused by the lack of genetic diversity in crops; introducing resistance into a wide variety of cultivars would counter this.

There are also examples of gene-editing improving nutritional quality: golden rice is a significant example, and work has also manipulated tomato metabolism to improve vitamin A content, and content of the antioxidant molecules (with protective properties against serious conditions including cardiovascular disease and cancer) carotenoids and anthocyanins (Čermák et al. 2015; Li et al. 2018).

GM crops remain controversial and face high regulatory barriers. This has become a real hurdle to the further development of GM crops, and economics has an important role to play in designing efficient regulatory and innovation diffusion mechanisms (Qaim, 2009). The costs in lost benefits of not overcoming these barriers may be large, especially for low income countries (Qaim, 2009). Currently, regulatory systems constrain the introduction of new GM crop varieties. Regulation is not always internally consistent: for example, India allows production of GM cotton, but not GM rice (Zilberman, Holland, and Trilnick, 2018). Zilberman argues that strict regulation of GM crops was justified when the technology was in its infancy and surrounded by high levels of uncertainty, but that a more appropriate response given the decades of evidence now available would be regulation which weighed up benefits, costs and risks, now much better understood as uncertainty has decreased (Zilberman, Holland, and Trilnick, 2018).

16.3.2 Precision Agriculture

Precision agriculture consists of remote sensing, information systems, embedded machinery, and informed land management. The overall aim is to optimise production by accounting for the variability of agricultural systems, and to reduce input use by eliminating unnecessary application (Gebbers and Adamchuk, 2010). The benefits are reduced fertiliser and pesticide use (saving a farm money and lessening harmful environmental impact), optimised yield, and enhanced soil health (The Food and Land Use Coalition, 2019).

Remote sensing is an important aspect of precision agriculture: it shows the variability which land managers need to harness. Remote sensing applications for precision farming began in the 1980s with sensors for soil organic matter, and have since diversified into satellite, aerial, and hand-held or tractor mounted sensors (Mulla, 2013). Precision agriculture is not just about understanding variability in plants and soil: it is also applied to livestock, enabling farmers to understand the health and needs of individual animals. This information can be used to improve the specificity of pharmaceutical application, and to hone breeding programmes for maximum yield.

Precision agriculture can help maintain biodiversity and ecosystem services while producing food, by coupling understanding of biological systems and the natural environment with precise management. For example, by implementing data-driven, precise soil management, both production and soil quality can be enhanced, along with other benefits including reduced pollution and minimised flood risk (IAgrE, 2012).

Costs of investing in precision farming machinery can be prohibitively high; at present, large farms in developed countries are the main adopters of precision agriculture. For example, 84% of grain farmers surveyed in Australia had adopted some form of Global Navigation Satellite System (GNSS), compared to only 5% surveyed in Turkey (Keskin and Sekerli, 2016; Bramley and Ouzman, 2018; Lowenberg-DeBoer and Erickson, 2019). For a financially stretched farm anywhere in the world, investing in expensive new machinery is challenging or even impossible without assistance. Before a farm invests in equipment, its effectiveness needs to be demonstrated. Demonstration farms proving the long-term cost reduction benefits of these technologies have helped with this. For example, in Ghana, the Ministry of Food and Agriculture has established over 1,200 community demonstration projects showcasing new agricultural technologies (Ngumbi, 2017).

16.3.3 Vertical Farming

With the aim of increasing crop yield per unit of land area, vertical farming aims to reduce pressure on traditional agricultural land by using soil-free growth systems in tightly controlled environments. The technique has the potential to reduce not only land use, but also water use, and use of intensive inputs. Nutrients are delivered through hydroponic (delivering nutrients through formulated liquid solutions), aquaponic (using fish to deliver nutrients) or aeroponic (delivering nutrient solutions in as a fine mist) systems; none are wasted as water and air are cycled through plant growth chambers. This is the equivalent to applying fertiliser to soil. There is no need for pesticides, as plants are housed in controlled growth facilities. The environmental benefits of reducing inputs and land use through growing crops in this way could be substantial (Beacham, Vickers, and Monaghan, 2019).

However, questions remain about its energy use intensity, and the variety of crops it can be used to grow. Vertical farming is only currently used commercially for leafy, high-value salad crops: the light-emitting diodes (LEDs) providing crops with light are expensive to run and companies cannot afford to waste plant biomass (for example, growing strawberries requires providing enough energy to support the entire plant, but only the fruits are harvested). Further, data and modelling studies aiming to quantify the potential environmental benefits of using vertical farming for a number of crops on a large scale are lacking.

16.3.4 Meat Analogues

Producing meat and meat-like products without animal agriculture is a rapidly growing field, with the potential to reduce significantly land use and environmentally damaging inputs (compared with conventional meat production). Cellular agriculture and plant-based meat are the two major meat analogues.

Numerous companies are developing plant-based meat analogues. Based on comparative assessments of the lifecycles of plant-based and beef burgers, plant-based burger production generates 90% less GHG emissions, needs 46% less energy and 99% less water, and has 93% less impact on land use (Heller and Keoleian, 2018).

Cellular agriculture is the production of animal products from cell cultures using a combination of biotechnology, tissue engineering, molecular biology and synthetic biology. The most well-known application is the production of cultured meat, but cellular agriculture is also developing dairy products, eggs, leather and bone among other things. Products are valued for their environmental, ethical and in some cases health and safety advantages, over their animal-derived equivalents. Cellular agriculture circumvents many of the environmentally damaging aspects of conventional animal agriculture, such as habitat invasion, antibiotic-resistant pathogen strains and methane (CH_4) emissions contributing to global warming (Fayaz Bhat, Kumar, and Fayaz, 2015).

Initially, the cost of producing meat in a lab was extremely high, but it has reduced significantly since 2013 (when Mark Post, a Dutch pharmacologist, bit into the first lab-cultured burger). The cost of production continues to fall, but challenges in achieving the flavour, texture, and nutritional profile of conventional meat remain. Culturing a recognisable cut of meat is a complex technological process, involving engineering multiple cell types and organising their structure (Fayaz Bhat, Kumar, and Fayaz, 2015).

Environmental impact life cycle analysis (LCA) has suggested that, compared to conventionally produced European meat, cultured meat production could result in a 7–45% reduction in energy use,

78–96% less GHG emissions, 99% less land use, and 82–96% less water use (Tuomisto and de Mattos, 2011). A more recent LCA compared CO_2 equivalent emissions from cultured meat production scenarios and conventional meat production across multiple timeframes, looking up to 1,000 years into the future. The authors found that, while in many cases cultured meat gives better environmental outcomes than conventional meat production, there could be scenarios in which this is not the case. These scenarios are characterised by CO_2 emissions from energy for cultured meat production overtaking CH_4 emissions from conventional meat production, when overall meat consumption declines (Lynch and Pierrehumbert, 2019).

Deep-set personal preferences attached to meat are expected to be a barrier to widespread demand for cultured meat. These preferences are often culturally defined and socially embedded. Chapter 9 gave detail on socially embedded preferences, and how the institutions shaping them can play roles in shifts towards sustainable choices.

Chapter 17
Managing Nature-Related Financial Risk and Uncertainty

Introduction

Chapters 4 and 4* discussed how the embeddedness of the global economy in the biosphere means that degradation or depletion of natural assets results in losses and disruption to economic activity. This has macroeconomic and financial implications for businesses and financial institutions, via, for example, reduced commodity yields, disrupted supply chains, output losses due to natural disasters such as droughts, and the loss of potential new sources of products and services, such as medicines and other pharmaceutical products (McCraine et al. 2019; OECD, 2019a; Rudgley and Seega, 2020). In addition, the existence of ecosystem tipping points and regime shifts mean that changes to the productivity of ecosystems, and the quantity and quality of the services they provide, can be both abrupt and long-lasting (Chapter 3). Both are aspects which would have significant implications for a range of economic activities and livelihoods (Johnson et al. (forthcoming); World Economic Forum, 2020a) (Chapters 5 and 14).

Loss of biodiversity – which is an important characteristic of natural assets as discussed in Chapters 1 and 2 – results in greater volatility and uncertainty around the goods and services ecosystems provide. In the parlance of finance, greater biodiversity reduces the *beta* component (the risk or unpredictability) of a yield.[497] Chapter 5 discussed the general risk and uncertainty associated with biodiversity loss. This chapter looks at the associated *financial* risks and uncertainty in greater depth. To date, most of the literature on financial risks associated with changes in natural capital has focussed on climate change. But changes associated with other aspects of natural capital, such as biodiversity loss, water stress, and resource scarcity, also have significant financial implications (Dempsey, 2013).[498] More generally, a range of anthropogenic changes in Nature – such as droughts, erosion, invasive species, air pollution, and contamination of water bodies and soil – have had identifiable adverse financial effects, such as declines in real estate prices, stock prices and bank defaults (Bassen et al. 2019). The COVID-19 pandemic, which has its roots in disrupted ecosystems,[499] has had global macroeconomic and financial implications.

Financial actors typically view the biosphere through a distinct lens, such as habitat loss, water supply, waste disposal, invasive species, deforestation or climate change (Natural Capital Coalition, 2018a). Indeed, in recent years, there has been increased understanding and awareness of risks related to climate change among financial actors. But, as Chapters 2 and 3 discussed, all processes associated with the biosphere should be considered together, given their interconnectedness and inextricable interlinkages.[500]

---

[497] Chapter 2 discussed in depth the ways in which more diverse ecosystems are more stable and function more effectively, and thus the services they provide are more reliable.

[498] Natural capital depletion and its implications for the real economy are relatively well documented. But there is limited empirical academic research on the links between biodiversity loss, natural capital and financial risks (Koumbarakis et al. 2020).

[499] Studies have shown that deforestation and loss of wildlife increases the emergence and spread of infectious diseases. For more, see Chapters 4 and 13.

[500] Biodiversity loss reduces resilience to climate change and opportunities for adaptation to inevitable climate change, and can also increase greenhouse gas emissions, for example through deforestation. Conversely, climate change is itself a major driver of Nature loss, as it accelerates the extinction of species and leads to rapid changes in ecosystems, therefore exacerbating Nature-related risks. This, in turn, drastically affects the carbon storage and sequestration capacities of ecosystems, which again worsens climate change.

Chapter 17: Managing Nature-Related Financial Risk and Uncertainty

At present, however, the management of financial risks from broader changes in natural capital remains relatively limited among the vast majority of financial actors (WWF France and AXA, 2019; Nagrawala and Springer, 2020). In addition, several current climate-change financial risk models do not consider other forms of natural capital and ecosystem services depletion, and their feedback loops (Koumbarakis et al. 2020), which will considerably undervalue the monetary costs of climate change.

17.1 Nature-Related Financial Risks

Nature-related financial risks are financial risks that arise from changes in either the stock or condition (or combination of both) of natural capital and from societal responses to changes in the state or quality of natural capital. As defined here, Nature-related risks include risks related to climate change and other environmental financial risks.[501] Changes in the stock and condition of natural capital alter its ability to provide the goods and services upon which businesses depend, and therefore have implications for the operations and profitability of businesses and financial institutions.

As discussed in Chapter 4, changes in natural capital include the loss and degradation of ecosystems, biodiversity loss, species population decline, and pollution. The prominent changes associated with a warming climate result in more frequent occurrence of extreme weather events, such as major storms, flooding, and droughts. Nature-related risks stemming from societal responses to changes in natural capital also have implications for both the operations and profitability of businesses and financial institutions. Societal responses include regulation and pricing of externalities, technological changes, evolving social norms and consumer preferences, and the threat of legal liabilities and litigation (McCraine et al. 2019).

Nature-related financial risks vary in type and severity depending on several factors, including where they stem from (i.e. changes in natural capital or societal responses) and the time period over which the risk materialises. One approach to categorising these risks is to frame them by building on the terminology used for risks related to climate change (Task Force on Climate-related Financial Disclosures, 2017). In this typology, Nature-related financial risks can arise from three channels or 'risk factors': physical, transition and litigation (Figure 17.1). These risk factors are classified separately, but are likely to interact with each other.[502]

Physical risks are strongly related to human dependence on Nature, and refer to the financial impact of changes in natural capital (Box 17.1). Loss and degradation of ecosystems can lead to disruption and even collapse of ecosystem services, such as pollination, disease control, climate regulation, flood and storm protection, and water regulation. Reductions in the quantity and quality of ecosystem services can damage fixed assets and infrastructure, and disrupt supply chains and business operations by affecting resource dependency, scarcity and quality. This causes direct economic and financial losses for businesses and financial institutions. Physical risks can be acute, short-term event-based risks (such as damage from catastrophic crop loss, damage from extreme weather events and flooding, or disruption from infectious diseases) or chronic, long-term changes from changes in environmental conditions (such as reduced suitability of land for crop cultivation).

---

[501] There is a wide related literature (Caldecott et al. 2013; WWF, 2019; Bank of International Settlements, 2020a; The Sustainable Finance Platform, 2020; WEF, 2020; Koumbarakis et al. 2020).

[502] For example, strong and immediate actions to mitigate biodiversity loss would increase transition risks and limit physical risks. In contrast, delayed and weak action to mitigate biodiversity loss would not necessarily eliminate transition risks, but would lead to higher and potentially catastrophic physical risks (Bank of International Settlements, 2020a).

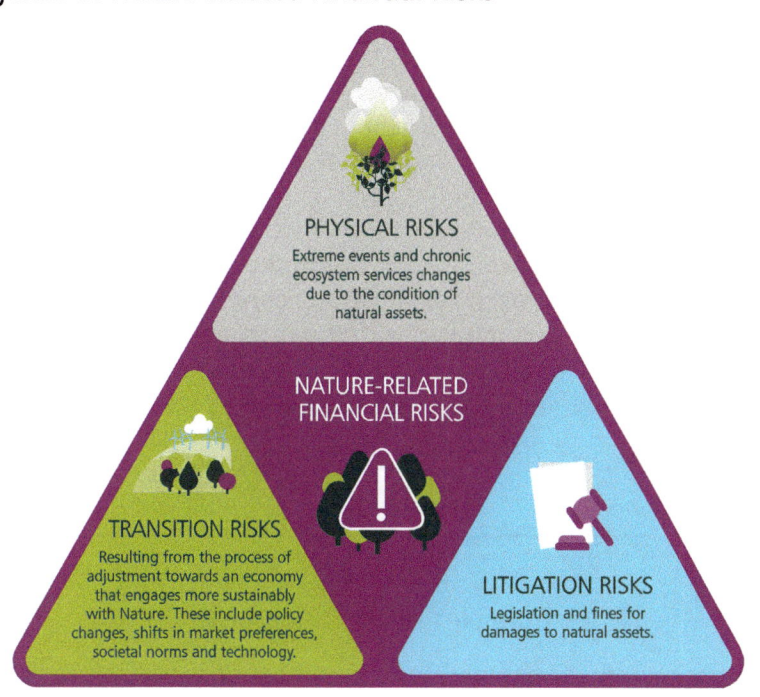

Figure 17.1 Categories of Nature-Related Financial Risks

PHYSICAL RISKS
Extreme events and chronic ecosystem services changes due to the condition of natural assets.

NATURE-RELATED FINANCIAL RISKS

TRANSITION RISKS
Resulting from the process of adjustment towards an economy that engages more sustainably with Nature. These include policy changes, shifts in market preferences, societal norms and technology.

LITIGATION RISKS
Legislation and fines for damages to natural assets.

Box 17.1
Examples of Physical Risks

Fourteen of India's 20 largest thermal power utility companies experienced disruptions due to water shortages at least once between 2013 and 2016, losing more than US$1.4 billion in total potential revenue. A key reason for the shortages was that a proportion of the water used was not being returned to the original source after being withdrawn (Ecologist, 2014).

By providing protection from coastal flooding and storm surge, mangroves reduce losses to economic activity from damage and disruption (Spalding et al. 2014). If all the mangroves in existence today were lost, it is estimated that around 18 million more people would be affected by flooding every year (a 39% increase) and that annual damages to property would increase by 16% (US$82 billion) (World Economic Forum, 2020a).

Wetlands provide many ecosystem services, including water filtration and flood control. During Hurricane Sandy in 2012, wetlands are estimated to have reduced the costs of flood damage by more than US$625 million (Narayan et al. 2017). Studies have shown that the total economic impact of Hurricane Katrina (estimated at US$150 billion) would have been significantly reduced if coastal wetlands in the region had been preserved (UNEP Finance Initiative, 2008). It is estimated that protecting coastal wetlands could save the insurance industry US$52 billion a year through reduced losses from storm and flood damage (Barbier et al. 2018).

Degradation of forests can threaten the availability and long-term security of valuable commodities on which the €200 billion global cosmetics market depends (World Economic Forum, 2020a). For example, the supply of shea butter used in various cosmetics products is reliant on the shea tree, which currently is threatened by deforestation, parasites and pollinator loss.

Chapter 17: Managing Nature-Related Financial Risk and Uncertainty

Transition risks result, either directly or indirectly, from the process of adjustment towards a more sustainable economy (Box 17.2). Losses result from societal change, and can be triggered, for example, by a relatively abrupt adoption of regulatory policies, technological progress or changes in market sentiment and preferences. Specifically, transition risks are related to the transition to an economy that is sustainable in terms of the Impact Equation. For example, in order to change key elements in the Impact Equation on both demand and supply sides, governments and financial regulators may implement regulatory changes such as land use restrictions, quotas and thresholds, disclosure requirements, compensation costs and taxes, procurement standards, licensing and permitting procedures, or prohibitions and outright bans (Chapters 7 and 8). Technological innovations towards more sustainable technologies can also have implications for the business processes of sectors and industries that damage ecosystems. For example, sectors or businesses that do not adopt technologies or adapt business processes may be at risk of adverse financial impacts from either regulatory changes or changing market landscapes (or a combination of both).

Box 17.2
Examples of Transition Risks

In 2018, Indonesia issued a three-year moratorium on clearing primary forests and peatlands, for activities such as palm oil plantations and logging. This was made permanent in 2019. The moratorium is expected to reduce certain business activities, particularly in Sumatra (Indonesia's largest palm oil-producing region), and will have implications for financial institutions that are directly or indirectly exposed to investment in those activities (McCraine et al. 2019).

In 2012, the Canadian gold mining company, Infinito Gold, was not given permission by the Costa Rican government to develop a mine as a result of its potentially significant impacts on agriculture, forests and endangered species. This led to a decrease in share value of 50% and a reference in the annual report to material uncertainties regarding the company's ability to continue as a going concern (Bonner et al. 2012).

In 2014, a Coca-Cola bottling plant in northern India was ordered to close after locals blamed it for local water shortages. A decade earlier, another of the company's bottling plants has been closed in southern India for similar reasons (Ecologist, 2014).

In 2010, Greenpeace launched a campaign against Nestlé's KitKat brand to raise awareness about its use of palm oil from Indonesian rainforests. Nestlé's stock subsequently decreased in value by 4% (McCraine et al. 2019).

In 2008, the Norwegian Pension Fund withdrew its £500 million stake in the mining business Rio Tinto and excluded the business from its funds over concerns it was causing "severe environmental damage" through a joint mining operation in Indonesia (Stewart, 2008).

The European pharmaceutical company Bayer lost almost 40% of its market capitalisation in less than one year, causing shareholders billions in monetary losses, after acquiring an agrochemical company accused of adversely affecting honeybee populations (Bender, 2019).

Litigation risks are related to firms' impacts on natural capital and the breaching of legal frameworks. They arise if parties who have suffered losses from either physical or transition risks seek to recover these losses from those they view as responsible. A litigation risk may arise, for example, if a firm's operations lead to unlawful biodiversity loss and another party seeks compensation from the firm they hold responsible. The risk of legal suits founded in biodiversity loss may increase as disclosure and external reporting on companies' biodiversity and environmental impact assessments increase. For example, investors in bonds issued by the Pacific Gas and Electric Company filed securities action against the utility company for misrepresenting efforts to address wildfire risks in California in 2017 and 2018 (Koumbarakis et al. 2020).

Physical, transition and litigation risks affect real economic activities, which in turn affect financial institutions. These impacts can occur directly, for example lower profitability or the devaluation of assets, or indirectly through macro-financial changes. These risk factors are drivers of prudential financial risk, in particular credit risk, market risk, and operational risk.[503] Table 17.1 gives an overview of the ways in which these risk factors relate to standard forms of financial risk.

Financial institutions face *credit risk* because they have exposures to investees that may default on their obligations. Severe disruption or collapse of ecosystems can disrupt supply chains, leading to asset quality deterioration and non-performing assets. This reduces both the debt-servicing capacity and the collateral of the financial institution. This increases the credit risk on their loan books, as both the probability of default and the loss given default increase. In addition, if damages from these physical risks are not insured, then the financial burden can fall onto other market participants, further increasing credit exposures. The crystallisation of extreme acute physical risks may even lead to bank defaults (Klomp, 2014; Schüwer et al. 2019).

Financial institutions may also have credit exposures to businesses with models that are not aligned with sustainably managing natural capital, and which therefore could face a high risk of reduced corporate earnings and business disruption over time. This may leave them unable to repay loans or meet their obligations on other financial transactions, at the same time as reducing their value. This could result in businesses in some sectors or industries facing increased cost of capital or lending requirements that are conditional on either their dependency or management of natural assets (or a combination of both); poor management of natural assets could reduce the ability of businesses to access finance (Chapter 20 discusses how financial actors can monitor management of natural assets).

---

[503] There are other financial risks, such as liquidity risk and business model risk, for financial institutions (European Central Bank, 2020).

Chapter 17: Managing Nature-Related Financial Risk and Uncertainty

Table 17.1 An Illustrative Framework for Nature-Related Financial Risks

	Credit	Market	Operational
Physical	• Changes in the provision of ecosystem services present risks to fixed investments and real estate • Falls in output from sectors highly dependent on natural capital increases default rates • Revaluation of debt-servicing capacity and collateral • Increased insurance claims resulting from changes in rainfall and flood patterns	• Rating downgrades and share price losses after ecosystem disruption or tipping points	• Severe Nature-related events affect production process via supply chains • Ecosystem disruption or regime shift affects balance sheet
Transition	• Investee suffers substantial losses due to sanctions, damages or increased taxes stemming from its negative impact on natural capital • Increased environmental standards lead to changes in lending, with increased cost of capital for some industries • Potential mispricing of new insurance products covering greener technologies • Stranded assets adversely affect portfolios	• Long-term profitability changes due to market shifts as a result of actions to address biodiversity loss • Natural capital regulatory policies lead to re-pricing of assets	• Changing sentiment and behaviour towards Nature leads to reputational risks for financial institutions and their investees
Litigation	• Costs from breaching legal frameworks with activities • Damages due to false reporting of Nature-related risks		

Source: Adapted and extended from Koumbarakis et al. (2020) and European Central Bank (2020).

The transition to correct the overshoot in the Impact Equation could also lead to stranded assets. These are produced capital assets whose operation damages ecosystems, which can suffer from unanticipated or premature write-offs, downward revaluations or be converted to liabilities. For example, biodiversity loss and declining ecosystem services could have significant consequences for the value of agricultural assets which, as regulation intensifies, may lead to stranded assets (Caldecott et al. 2013). This would have an adverse impact on the portfolio value of financial institutions exposed to such assets. In the case of stranded assets in the energy sector, the exposure of the European financial sector to high-carbon assets was estimated to have been over €1 trillion (Weyzig et al. 2014).

The insurance sector is particularly vulnerable to credit risk. The occurrence of events related to ecosystem degradation and climate change could adversely affect both their investment returns and their underwriting profitability – for example, floods due to deforestation may lead to insured or uninsured losses. In a similar compound effect to a rise in weather-related insurance claims, if insurance premiums rise due to Nature-related issues and insurance companies have more to pay out, this will result in increased premiums for households and businesses over time. Nature-dependent industries (such as agriculture, pharmaceuticals and tourism) in particular would be affected adversely. The insurance sector is also exposed through increases in their protection gap – that is the gap between economic and insured losses – as unidentified losses rise due to higher claims related to severe events that are caused, or exacerbated, by the depletion of natural assets.

Market risk is the risk of losses arising from adverse changes in market prices. It is relevant to several aspects of financial institutions' activities. The increasing frequency of severe events related to changes in ecosystems, such as floods and droughts, can affect underlying macroeconomic conditions through sustained damage to infrastructure and weaken fundamental factors such as growth, employment, and inflation. Depletion of natural assets is a particularly significant macroeconomic risk to income, growth and stability in economies that rely on natural resources.

For countries endowed with a high supply of natural assets, there are a range of financial implications associated with the depletion of natural resources. For example, if a country or region experiences significant degradation of natural assets, financial flows may divert away from this area as investors reallocate current and planned investments or divest from existing investments, or as firms reorient operations to new nodes of production with reliable ecosystem service flows (Caldecott and McDaniels, 2014). For those countries most susceptible to Nature-related risks, this can have implications for the market price of sovereign debt, as the value of sovereign bonds relies in part on the management of natural assets.[504] Recent research has highlighted the risks associated with sovereign bonds and metrics have been developed to help increase understanding of these risks for countries (Box 17.3). However, at present, the risks associated with natural assets for sovereign bonds appear to be largely ignored or mispriced in sovereign bond markets (Pinzón et al. 2020).

Box 17.3
Climate and Nature Sovereign Index

Developed jointly by WWF and Ninety One, the Climate and Nature Sovereign Index (CNSI) assesses the exposure of countries to risks related to climate change and biodiversity loss that could impede their macroeconomic performance (Patterson et al. 2020).

The CNSI combines conventional economic and financial factors that currently inform risk-modelling at the country level with the risks related to natural capital that a country is exposed to. To capture a wide range of economic and financial channels of exposure to physical and transition risks arising from climate change, biodiversity loss, and other changes in natural capital, the CNSI uses 85 indicators grouped into the following broad pillars:

(1) Biodiversity and natural capital

(2) Physical risks (chronic and acute)

(3) Transition risks

(4) Financial and socio-economic resilience

The framework used to construct the CNSI assigns a single risk score to each country, on a scale of 0–1, with 0 being highest risk and 1 being lowest (Figure 17.2). It also provides sub-indices to enhance understanding of the complex dynamics from the types of Nature-related risk a country faces. The CNSI is a dynamic risk index: risk scores change as the underlying data are updated on an ongoing basis.

---

[504] On food chain disruption, for example, if government subsidies are required to reduce the impacts of imported inflation on consumers, this could put pressure on public finances, leading to higher borrowing, an inflated ratio of public debt to GDP, increased probability of a sovereign credit downgrade, and second-order financial risks to municipal and sovereign bond holders.

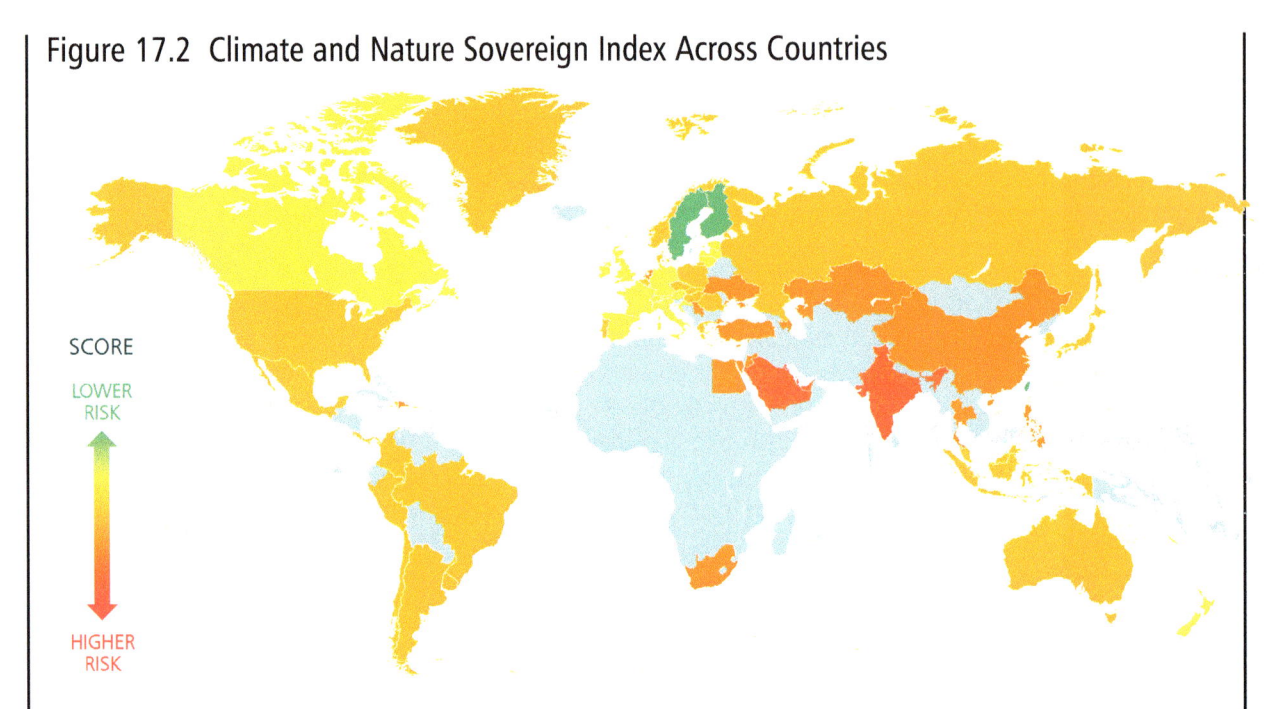

Chapter 17: Managing Nature-Related Financial Risk and Uncertainty

Figure 17.2 Climate and Nature Sovereign Index Across Countries

Source: Based on Patterson et al. (2020), with Review calculations

The CNSI is aimed for use by actors in the sovereign-debt market, including investors and governments who would benefit from a single, coherent framework through which to assess long-term risk related to climate change and natural capital at a country level. The CNSI builds on existing indices – such as the Yale Environmental Performance Index (Wendling et al. 2020) – by enhancing the set of indicators used to increase applicability to portfolio risk assessment, and by incorporating real-time data and forward-looking projections (Patterson et al. 2020). The CNSI and other related indicators have the potential both to support investors in integrating Nature-related factors into their overall risk management and to encourage countries to establish institutions and policy mechanisms that support greater investment in natural assets and address the Impact Inequality.

The transition to address overshoot in human demands on the biosphere in the Impact Inequality will also be associated with market risks, as the transition affects a range of financial asset classes, such as energy and commodity prices, corporate bonds, equities and derivative contracts. While the risk of a sudden and significant system-wide adjustment may not be immediate, the risk from an abrupt transition can increase if, over the coming years, portfolios are not aligned with expected pathways. Shifts in consumer behaviour and preferences could be reflected in changes in the demand for commodities, products and services due to concerns about our natural assets. This may result in financial losses arising from fluctuations in the market values of positions held by financial institutions. For example, Exxon experienced a drop in stock price after the oil spill from its Deepwater Horizon oil rig in 2010 (Humphrey et al. 2016; Heflin and Wallace, 2017).

Operational risk is the risk of loss resulting from external events or from inadequate or failed internal processes, staff capacity, or systems. Loss could occur due to severe Nature-related events which may impact business continuity, including branch networks, offices, infrastructure, processes, and employment. The pricing of inputs such as energy, water and insurance could increase, affecting profitability. On the transition side, reputational risk can arise from shifting sentiment among customers, or from increasing attention and scrutiny from other stakeholders on the financial system's response to biodiversity loss.

In recent years, the Union for Ethical Biotrade's (UEBT) *Biodiversity Barometer* – an annual global survey of consumers on biodiversity awareness – has highlighted increased awareness and understanding of biodiversity among consumers over the past decade, particularly among young people, and that people increasingly expect businesses to respect biodiversity but do not trust them to do so (Union for Ethical Biotrade 2019, 2020). For example, in the 2020 survey, 82% of respondents suggested that businesses have a moral obligation to ensure a positive impact on people and biodiversity, but only 41% of those surveyed felt confident that businesses are paying serious attention to the ethical sourcing of biodiversity (Union for Ethical Biotrade, 2020) (see Chapter 6 for more on trust and social capital).

17.2 Uncertainty and Short-Termism

As outlined above, Nature-related financial risks can arise through multiple pathways, depend on multiple nonlinear interacting dynamics (i.e. natural, technological, societal, regulatory and cultural systems) (Wiedmann et al. 2020). Nature-related risks are characterised by deep uncertainty, in the sense that they are subject to uncertainties related to tipping points and regime shifts, complex transmission channels, and potentially far-reaching impacts on all agents in the economy (Chapters 3 and 5). Chapter 5 noted that while there is significant uncertainty over the exact timing of some Nature-related risks, they are nonetheless either likely or quite certain to occur (such as pandemics and climate change), as confirmed by scientific evidence. These characteristics have led these types of risks to be labelled as 'green swans' in the parlance of modern finance (Box 17.4).

Box 17.4
White, Black and Green Swans

'Black swan' events can take many shapes, from terrorist attacks to disruptive technologies. These events typically fit fat-tailed probability distributions, i.e. they exhibit greater kurtosis than a normal distribution.[505] Unlike other types of risk events which are relatively certain and predictable, such as car accidents and health events ('white swans'), 'black swans' cannot be predicted by relying on backward-looking probabilistic approaches that assume normal distributions (Bank of International Settlements, 2020a).

Some in the finance community have adopted this framework of thinking about risks associated with the biosphere, terming them 'green swans' (or environmental black swans) (Bank of International Settlements, 2020a). 'Green swans' present many features of typical 'black swans'; in that they are unexpected when they occur by most agents (who regard the past as a good proxy of the future); they feature non-linear propagation; impacts are significant in magnitude and intensity; and they entail large negative externalities at a global level.[506]

However, despite several common features, 'black swans' and 'green swans' differ in several key aspects. A key difference is their likelihood of occurrence. 'Green swans' are either likely or quite certain to occur (e.g. increased droughts, water stress, flooding, and heat waves), but their timing and form of occurrence are uncertain. By contrast, 'black swans' do not manifest themselves with high likelihood or quasi-certainty.

---

[505] The term is based on an ancient saying that presumed black swans did not exist; it was reinterpreted to teach a different lesson after black swans were discovered in the wild.

[506] This feature is crucial, as it means that, for both types of swan, neither the private sector nor government can ever adequately 'price' the associated risks in terms of financial or macroeconomic costs. Luiz Awazu Pereira da Silva (2020) gives an example of the macroeconomic/financial risk of global pandemics that has previously been at best minimised or ignored.

Chapter 17: Managing Nature-Related Financial Risk and Uncertainty

> 'Black swans' are severe and unexpected events that can only be rationalised and explained after their occurrence. While for 'green swans', the likelihood of occurrence means the case for preventative action, despite prevailing uncertainty regarding the timing and nature of impacts of these events, is strong (it should be noted that Chapter 5 highlighted how the existence of option value is a key factor in the case for preventative action, irrespective of the risk tolerance).
>
> Other differences include who provides the main explanation for the events and their reversibility. Explanation for 'black swans' tend to come from economists and financial analysts, while for 'green swans' understanding comes from ecologists and earth scientists. The impacts of 'green swans' are, in most cases, irreversible (Chapter 3), whereas for 'black swan' events – such as typical financial crises – have effects that are persistent, but have the potential to be reversed over time.

Uncertainty over both timing and impact of Nature-related risks has profound implications for action to mitigate such risks (Chapter 5). For financial actors, specifically, there are likely to be large time lags before natural capital changes become apparent and irreversible, and so the most damaging effects will be felt beyond the typical time horizons in which financial actors operate in. The physical impacts of changes to natural capital will be felt over a long-term horizon, with potentially significant economic costs and impacts over many generations (Chapter 5), but the time horizon in which financial actors plan and act is much shorter.

Estimates suggest that equity fund managers turn over their portfolios on average every 1.7 years, with around four-fifths of them doing so within three years (Bernhardt et al. 2017). Most portfolio managers' incentives are yearly (Thomä et al. 2015), and financial analysis is generally limited to 3 to 5 years (Dupré and Chenet, 2012; Naqvi et al. 2017). There has been a decline in the length of time over which investments are supposed to generate an acceptable level of financial return, from 8 years to 8 months, over the course of the last two decades (Kim and Asuncion, 2019). The time horizons of central banks and financial regulators to maintain both monetary and financial stability are of a broadly similar length.[507]

An emphasis on short-term financial performance can result in agents foregoing financial investment opportunities with a positive long-term net present value and that reduce risks over a longer time horizon, including investments in more sustainable activities. This 'short-termism' – which is ultimately a function of institutional failure (Chapter 7) – is a well-established impediment to addressing longer-term issues by financial markets (Zadek and Robins, 2018; Kim and Asuncion, 2019; Chenet, 2019; Suttor-Sorel, 2019; Kedward et al. 2020). A significant implication of 'short-termism' is that even if there is acceptance that Nature-related risks will manifest themselves at some point in the future, acceptance alone is not necessarily enough to trigger a reaction from financial actors if the occurrence of risks does not coincide with their own typical time horizon. Put simply, if the time horizon attached to a material risk does not align with the agent's typical time horizon, then a risk will be ignored or at least not acted on (Chenet, 2019).[508] This is commonly referred to today as the 'tragedy of the horizon' (Carney, 2015).

One example of the manifestation of 'short-termism' can be seen in financial flows to industries that are more 'unsustainable', such as non-renewable energy. Despite unsustainability in such industries (and thus profitability) over the longer-run, while they remain profitable for over a shorter time horizon, investors will still have an incentive to invest. Indeed they still attract still sizeable financial investments

---

[507] The horizon for monetary policy extends to 2–3 years. For financial stability, it is longer, but typically only to the outer boundaries of the credit cycle – about a decade (Carney, 2015).

[508] In theory, financial markets can allocate financial capital efficiently if the market signals they receive are material and relevant for agents and are integrated into their decision-making processes.

(Christophers, 2019). But if such financial investments were clearly considered by governments to be undesirable from a social welfare perspective, then actual or expected regulations could help to shift pricing signals in the near-term. As long as risks emerging from our unsustainable use of the biosphere, and opportunities to correct this overshoot, are not reflected in market prices, there will be few incentives for these risks to be incorporated in financial decisions. This means there is a role for precautionary policy intervention by governments and financial regulators, to compensate for the inability of markets to react in the face of potentially catastrophic losses related to tipping points (Bahaj and Foulis, 2016; Cullen, 2018; Ryan-Collins, 2019; Kedward et al. 2020).

17.3 Assessing Nature-Related Financial Risks

Existing research suggests that the risks stemming from Nature have the potential to be material in a range of sectors, geographies, and financial asset classes (UNEP, UNEP Finance Initiative, and Global Canopy, 2020; World Economic Forum 2020a, 2020c). For example, it has been estimated suggest that around US$44 trillion of global economic value generation – more than half of nominal global GDP in 2019 – is moderately or highly dependent on natural assets and their ecosystem services (World Economic Forum 2020a). As another example, recent estimates suggest that more than 25% of financial investments by development financial institutions (around US$3 trillion out of around US$11 trillion) are directed to projects that are highly dependent on vulnerable ecosystems (Finance for Biodiversity, 2020).

The uncertainty associated with, and complexity of, Nature-related financial risks means that increased awareness and understanding of natural capital by financial actors is fundamental to managing and mitigating these risks. This includes greater understanding of natural capital dependencies and impacts, and utilising a range of methods to identify and measure both physical and transition Nature-related financial risks. The potential for several Nature-related risks, not only climate change, to have systemic implications has been recognised by the Central Banks and Supervisors Network for Greening the Financial System (NGFS). In 2019, the NGFS – a global group of central banks and financial supervisors currently consisting of over 80 members – indicated that environment-related risks are posed by the exposure of financial firms and the financial sector to activities that may either cause or be affected (or combination of both) by, degradation of natural assets (including air pollution, water pollution and scarcity of fresh water, land contamination, biodiversity loss and deforestation) (Network for Greening the Financial System, 2019). Mispricing of Nature-related risks at the portfolio level could lead to an institutional level of risk mispricing and inaccurate levels of capital cushions to buffer against financially material risks.[509] The potentially compounding effect could hinder regulators' prudential ability to ensure the safety and soundness of financial systems.

For financial institutions, the failure to incorporate Nature-related risks into risk assessments is of strategic significance for two reasons (Cambridge Centre for Sustainable Finance, 2016). First, all credit and investments are deployed on the basis of expected 'risk-adjusted' returns. If Nature-related risk is underestimated, credit will be over-allocated to higher risk activities. Nature-related risk analysis is needed to support more efficient allocation of investments for long-term stability, and to prevent excessive investment in unsustainable, riskier activities and under-investment in sustainable, lower-risk activities. Second, managing risk is central to the functioning and stability of financial institutions. Inadequate understanding of Nature-related sources of risk could allow threats to financial institutions to accumulate.

---

[509] Maintaining financial stability relies on financial institutions holding sufficient capital and having adequate risk controls in place. This, in turn, informs the risk coefficients of risk-weighted assets at a transactional and portfolio level, and of countercyclical capital buffers at an institutional level.

At the micro level, the extent to which Nature-related risks can become financially material for a financial institution depends on its level of exposure and vulnerability. Exposure is a function of several factors, such as a firm's industry and stage of the supply chain, the geographic spread of its supply chain and customer base, and its degree of reliance on natural assets for ecosystem services. For financial institutions, exposure stems from a range of channels such as bank loans, equities and other major asset classes, including sovereign and corporate bonds, commodities, and real estate. Vulnerability is the extent to which a business or financial institution can adapt to the threat at hand; it is determined by factors such as expendable capital, risk management practices along the value chain, risk awareness, the degree of operational and managerial resilience, product diversification, and market power.

17.3.1 Tools and Methods for Risk Assessments

Several approaches have been developed to support financial actors in understanding the potential materiality of Nature-related financial risks, based on their exposure in terms of dependencies. These tools and methods can be used either to assess the activities of financial institutions as they relate to Nature-related risks, or to assess Nature-related capital risks of their investees directly. Tools and methods (outlined in Table 17.2) support Nature-related financial risk assessments of financial institutions and businesses by either identifying the relative risk to sectors or industries from disruption to natural assets and ecosystem services based on their dependencies; or choosing a potential driver of ecosystem change and assessing its potential impact on a sector or asset class, based on the likelihood of disruption of relevant ecosystem services.

Table 17.2 Tools to Support Nature-Related Risk Assessment[510]

Tool	Overview
Exploring Natural Capital Opportunities, Risks and Exposure (ENCORE)	ENCORE provides financial institutions with systematic information for assessing their exposure to Nature-related risks, exploring businesses' dependencies on natural capital, and the effects of changes in the condition of natural assets businesses rely on.
	ENCORE looks at the effects of drivers of environmental change on natural assets and the ecosystem services these assets supply. Potential disruptions in the provision of ecosystem services are linked to sector production processes per sector by an ecosystem service materiality assessment. This evaluates the extent of the resulting loss of functionality in production processes and the resulting financial losses.
	An ongoing initiative is enhancing how ENCORE assesses impacts of sectors on biodiversity.
The Integrated Biodiversity Assessment Tool (IBAT)	IBAT is a web-based map and reporting tool that provides rapid access to three of the world's biggest biodiversity-related datasets – the World Database on Protected Areas, the IUCN Red List of Threatened Species, and the World Database of Key Biodiversity Areas. It enables users to create bespoke reports based on these datasets, defined by geographical area.

---

[510] It should be recognised that both the quantity and quality of these tools will evolve over time.

Table 17.2 (cont.)

Tool	Overview
SCRIPT	SCRIPT benchmarks companies on the strength of their soft commodity production, and assesses portfolio exposure to deforestation, biodiversity loss and other soft commodity sector risks. It aims to help financial institutions understand and mitigate the risks associated with financing companies in soft commodity supply chains.
	SCRIPT contains two tools, one allowing financial institutions to benchmark their policies on deforestation against those of their peers, and one allowing institutions to assess how they may be exposed to deforestation risk in their investment portfolios.
Natural Capital Protocol	The Natural Capital Protocol is a standardised framework for conducting natural capital assessments, by which businesses and financial institutions can identify, measure and value their direct and indirect impacts and dependencies on natural capital.
	The Protocol allows for measurement, valuation and integration of natural capital dependencies and impacts into existing business processes, such as risk mitigation, sourcing, supply chain management and product design.
	A supplement specifically for financial institutions was developed to support assessments of natural capital impacts and dependencies of their investments and portfolios.

Nature-related financial risk assessments focus on the state of natural assets and the severity of the drivers of environmental changes that affect these assets, both of which determine the risk of disruption or changes to the continued provision of ecosystem services that economic sectors are materially dependent upon. These risk assessments allow financial organisations to identify the dependence on natural assets of real economy activities they provide financial services, and thus provide insight into potential physical risks (Natural Capital Coalition, 2018a). For example, analysis by the Natural Capital Finance Alliance has shown that 13 of the 18 sectors that made up the FTSE 100 in 2018 (a total of US$1.6 trillion in net market capitalisation) were associated with production processes that have high (or very high) material dependence on natural capital – equivalent to almost three quarters of the market capitalisation of the FTSE 100 (UNEP Finance Initiative, 2018). Examples of such high economic dependencies highlighted included the harvesting of cereals and their reliance on pollination, and metal processing and its reliance on ground water provision. Risks assessments have been conducted by certain banks in Colombia, South Africa and Peru and covered assessments including the implications of droughts for the agriculture sector, water stress for the manufacturing sector, and wind speed increases for the tourism sector.

Risk assessments that only consider current dependencies or use historical trends (or combination of both) can only lead to mispricing of Nature-related financial risks, as these risks have only relatively recently started to materialise (Bank of International Settlements, 2020a). Physical and litigation risks will depend on future biodiversity loss and global warming, and transition risks depend on the scale and ambition of government policies. Therefore, other methods, which involve forward-looking, scenario-based analysis, are needed. One method is stress testing: a scenario analysis exercise in which the financial implications of one or more future scenarios are assessed, including a plausible but relatively extreme adverse scenario. Stress testing does not explicitly consider the likelihood of particular events or impacts occurring, but rather looks at the magnitude of the impacts, should the event occur.

Stress testing can help inform financial actors at both the micro level (i.e. individual financial institutions) and the macro level (i.e. central banks and regulators). At the micro level, financial institutions can use this approach to test the resilience of businesses in their investment portfolios to potential materialisations of physical and transition risks, their impact on key performance indicators and the adaptive capacities of these firms. At the macro level, this approach can be used by central banks and regulators to test the resilience of the financial system to Nature-related risks and tail-events. The risks of environmental tipping points and regime shifts, resulting from natural capital depletion, could be financially systemic risks because of complexity, interdependence and interconnectedness within the financial system (Bank of International Settlements, 2020a). For example, a rapid and ambitious transition to a shift to actions that achieve Impact Equality could involve the introduction of regulation to prevent the extraction of a range of provisioning goods (including fossil fuels). Financial assets linked to extractive activities would become stranded (discussed previously) – as a *fire sale* might result as these financial assets suddenly lose value – potentially triggering losses throughout the system.[511]

Since the global financial crisis of 2007 to 2008, central banks and financial regulators have used stress testing to determine the resilience of the financial sector in relation to financial shocks. More recently, central banks and regulators have started to use stress testing in relation to climate change, and it appears to be facilitating systematic integration of risks related to climate change in the financial sector (Bank of International Settlements, 2020a).[512] The adoption of this approach for climate change has led to calls for stress tests to be used for the assessment of other Nature-related risks, such as air pollution, carbon emissions, natural hazards, and water stress. In practice, this would mean assessing scenarios that could cover both transition risks (such as stranded assets) and physical risks. These include chronic stresses to natural assets – such as increases in the spread of pests and diseases – or particular acute stress – for example droughts and water shortages – or combination of both types of stress. Some preliminary stress tests have been focused on physical risks; for example, the UK natural capital stress test (White et al. 2017) and drought stress tests (Carter and Moss, 2017).

17.3.2 The Role of Impact Assessments to Inform Risk Assessments

An assessment of a financial institution's impacts, in addition to its dependencies, on natural capital is important in understanding exposures to both transition and physical risks. Through identifying both the positive and negative impacts their activities have on elements of natural capital, financial institutions can identify where they need to change the type and provision of their financial services and channel financial flows in a way that support natural assets, and their sustainable use (Chapter 20). This in turn can help mitigate their exposure to potential Nature-related risks (UN Environment Programme, UNEP Finance Initiative, and Global Canopy, 2020). An assessment can also support understanding of the impacts of a financial institution's portfolios. This can help identify changes needed to align activities with sustainable use of natural assets and reduce the institution's ecological impacts.

Some financial institutions have already started to increase their awareness of their impacts on biodiversity specifically. However, this has mainly been in the context of institutions seeking to make a positive impact, rather than as part of their risk management.[513] Footprinting methods can play a central role in impact assessments.[514] For a subset of natural capital issues, such as greenhouse gas emissions and water

---

[511] It should be noted that some responses by financial actors to specific Nature-related risks also have the potential to amplify potential systemic risk. For example, while development of financial products in response to climate-related risks, such as weather derivatives, may help individual institutions hedge against specific risks, they can also amplify systemic risk (Bank of International Settlements, 2020a).

[512] For example, some insurance companies are reassessing the cost of insuring physical risk; some rating agencies are increasingly re-evaluating credit risks in the light of climate change; and some asset managers are becoming more selective in their choice of financial assets, in relation to their impact on the biosphere, in their financial portfolio (Bank of International Settlements, 2020a).

[513] For example, Dutch financial institutions do not yet conduct any systematic risk analyses, but biodiversity, deforestation and forest conversion increasingly feature in their sustainability policies. A few institutions have already set specific goals, for example by restricting their lending to businesses that do not cause deforestation (Schellekens and van Toor, 2019).

[514] Financial institutions take two main approaches to assessing their impact on natural capital. The first is based on disclosure by businesses who have assessed their impacts on Nature. The second involves financial institutions themselves assessing their impacts directly, through their lending decisions and portfolios.

consumption, footprinting metrics are already relatively familiar to businesses (WWF France and AXA, 2019).

Unlike for carbon or water footprinting, there is no globally accepted metric for biodiversity footprinting at present, but various methods have emerged, in part spurred on by some proactive financial institutions interested in knowing the biodiversity performance of their portfolios (Lammerant et al. 2019).[515] The Biodiversity Footprint for Financial Institutions (BFFI) and the Global Biodiversity Score (GBS) were developed by ASN Bank, the Netherlands, and CDC Biodiversité, France, respectively, to determine their impacts on biodiversity (Boxes 17.5 and 17.6).[516] Both methods link a species indicator (for example mean species abundance (MSA) defined as the fraction of naturally present biodiversity that still remains) to the area where the impact is felt (spatial element) and the assessment period (time factor). They use an unaffected area of Nature as a reference point, and express impact in terms of an increase or decrease in the number of species (Berger et al. 2018).

Box 17.5
CDC Biodiversité and the Global Biodiversity Score

CDC Biodiversité is a private subsidiary of the Caisse des Dépôts Group (CDC). The CDC Group is a French public financial institution that has a focus on long-term investment supporting territorial, ecological, energy, digital, demographic and social transitions. Since 2008, the company has conducted biodiversity initiatives and biodiversity offsets, participating in some 40 offset projects, and contributing to the conservation and restoration of almost 2,000 hectares of land (Natural Capital Coalition, 2018b).

CDC Biodiversité's ambition with the Global Biodiversity Score (GBS) was to create a quantitative, global, spatial, transparent, cross-sectoral, scientific biodiversity footprint tool. The metric describes biodiversity changes with reference to the original state of the ecosystem, defined as the abundance of observed species relative to their abundance in the undisturbed ecosystem. Thus, the MSA of an area is between 0 (no originally occurring species are still present in the ecosystem) and 1 (all originally occurring species are still present in the ecosystem). The unit used in GBS is the km^2MSA, the surface area equivalent of the MSA; a footprint of $1km^2$MSA corresponds to the complete destruction of $1km^2$ of intact natural area. This unit can thus be compared to an 'artificialisation' measurement, with the benefit of being understandable by employees in businesses. CDC Biodiversité worked on projects with Michelin, Solvay and BNP Paribas Asset Management during the GBS development phases. The project with BNP Paribas aimed to compute the biodiversity footprint of one of their portfolios of listed equities. BNP Paribas had a portfolio of 10 food and agro-business companies with a total turnover of €467.6 billion. Through using partial elements of the GBS tool, the project concluded that the overall negative impact of BNP's portfolio on biodiversity was limited (Mission Économie De La Biodiversité, 2019).

Financial institutions can use footprinting methods to screen their activities directly, by assessing the impacts generated by an activity – product manufacturing, business, value chain, equity portfolio – on an indicator of biodiversity (for example MSA, the fraction of naturally present biodiversity that still

---

[515] A biodiversity footprint is usually based on monitoring actual changes in biodiversity through time (i.e. assessment of actual impact), or on assessing the potential or expected impact from the contribution of real economy activities to drivers of biodiversity loss or gain (i.e. assessment of potential impact).

[516] Climate change is an important pressure on biodiversity and is taken into account in both the GBS and BFFI methodology. The first step in the pressure-impact relation, i.e. the relation between green-house gas emissions and the rise of temperatures, is the same in both methodologies.

remains) linked to surface area (for example km^2). This approach has several benefits for a financial institution. It can be used to monitor the footprint of products and services produced by businesses within its portfolio, assess sustainable investment decisions and policies, and for external reporting, for example to inform stakeholders about the overall footprint of the portfolio. Footprinting methods can also help financial institutions produce and set quantitative targets for biodiversity, to assess whether they are contributing to 'No Net Loss' or 'Net Gain' (WWF France and AXA 2019; UN Environment Programme, UNEP Finance Initiative, and Global Canopy, 2020). There are several hurdles to translating biodiversity targets into portfolio allocation targets. For example, there is no agreed quantifiable goal for biodiversity globally, and no commonly agreed definition to measure biodiversity or specific biodiversity loss scenarios. That said, in principle, biodiversity footprinting in this way can support a shift to greater sustainability in terms of the Impact Equation.

Box 17.6

ASN Bank's Biodiversity Footprint

ASN Bank, the Netherlands, has calculated its biodiversity footprint using the Biodiversity Footprint for Financial Institutions (BFFI) methodology since 2016. The BFFI methodology expresses impact, in terms of an increase or decrease in the number of species, as the 'potentially disappeared fraction of species' (PDF). The footprint results show how biodiversity impact hotspots relate to the various investments in the bank's portfolio, and where, in the corresponding value chains, the impact is highest. This provides the bank with an overview of the material elements that have to be considered when managing its negative and positive impact. ASN Bank has calculated its biodiversity impact per investment category (government bonds, mortgages, equity, etc.) using the BFFI. In 2017, ASN Bank calculated that it was responsible for a 64,849 hectare loss of biodiversity due to its investments (Berger et al. 2018). The bank aims to reduce ecological damage stemming from its loans and investments and to boost biodiversity, for example by investing in wildlife conservation, sustainable energy and the circular economy.

Management and mitigation of Nature-related financial risks can lead to increased resilience of balance sheets for financial institutions. By channelling credit and investments toward projects that enhance, rather than degrade, natural capital and ecosystem services, financial institutions are exposed to investments with relatively lower risk and greater certainty around yields. For example, a bank may lend money to an agricultural business. The business may take actions to reduce soil erosion on its land, leading to increased yield, thus enhancing the ability of the business to service its debt. This, in turn, improves the credit quality of the bank's portfolio, and thus lowers risk. There is evidence that businesses with strong sustainability performance deliver improved long-term returns (Waygood, 2014). High sustainability businesses – defined as those with a substantial number of environmental and social policies adopted for a significant number of years – significantly outperform their peers over the long term, in both stock market valuation and accounting terms (Fulton, Kahn and Sharples, 2012; Eccles, Ioannou and Serafeim, 2014).

There have been signs that policymakers and financial regulators will increasingly demand that financial institutions systematically assess both Nature-related financial risks and their own impacts on Nature. For example, following recommendations from the EU High-Level Expert Group on Sustainable Finance, the European Commission released an Action Plan on sustainable finance in 2018 (European Commission, 2018). An important element of this plan is enhancing non-financial information disclosure; the legislative text on disclosure not only focuses on information about climate change, but also aims to address other adverse impacts on natural capital. Separately, France has already put in place measures to address imported deforestation, and extended Article 173 of their Energy Transition Law to include biodiversity and ecosystem requirements (Green Finance Platform, 2019).

Several initiatives are working towards greater Nature-related disclosure and reporting. For example, the EU Commission and the EU@Biodiversity Initiative have been working on schemes that businesses can use to disclose their impacts on biodiversity. In 2020, the first steps to the creation of a Task Force on Nature-related Financial Disclosures (TNFD) were announced (Task Force on Nature-related Financial Disclosures, 2020). The aim of the TNFD is to build awareness and capacity on Nature-related dependencies, impacts, and financial risks, among financial institutions. Its work should complement the ongoing work by the Task Force on Climate-related Financial Disclosures (TCFD) (Box 17.7).

Box 17.7

Task Force on Climate-Related Financial Disclosures

In 2015, G20 finance ministers and Central Bank governors asked the Financial Stability Board (FSB) to review how the financial sector could take account of issues related to climate change. The Board established the Task Force on Climate-related Financial Disclosures (TCFD) to develop recommendations for consistent disclosures that will help financial market participants understand their climate-related risks. The Task Force was industry-led, and included 32 international members, including providers of capital, insurers, large non-financial companies, accounting and consulting firms, and credit rating agencies.

The TCFD's recommendations are structured around four thematic areas representing core elements of how organisations operate: governance, strategy, risk management, and metrics and targets (Task Force on Climate-related Financial Disclosures, 2017). To underpin its recommendations and help guide current and future developments in climate-related financial reporting, the Task Force developed a set of principles for effective disclosure.

Since the publication of the TCFD's recommendations, support has increased rapidly. The TCFD's 1,030 supporting organisations span the public and private sectors, and represent over 80 industries in 50 countries, including seven national governments. Current supporters of the TCFD control balance sheets totaling US$120 trillion and include the world's top banks, asset managers, pension funds, insurers, credit rating agencies, accounting firms and shareholder advisory services (Task Force on Climate-related Financial Disclosures, 2020).

Central banks and financial regulators are starting to assess and manage Nature-related risks. The Bank of England, De Nederlandsche Bank, the European Central Bank and the Banco de Mexico have all published reports on financial risks due to climate change and other Nature-related risks (Bank of England, 2018; Schellekens and Toor, 2019; Banco de México and UNEP, 2020; European Central Bank, 2020). In 2020, the NGFS outlined how risks due to climate change and other Nature-related risks could be integrated into prudential supervision (Network for Greening the Financial System, 2020).

However, to date, assessments by central banks and supervisors of Nature-related risks has largely focused on climate change (Bank of International Settlements, 2020a). Given the range and potential interplay of these risks, financial regulators and central banks need to explore both the microprudential and macroprudential consequences of broader Nature-related financial risks. This could involve further exploration of stress tests, based on aligned and common practices, to analyse the impact of physical, transition, litigation risks and systemic risks stemming from biodiversity loss and ecosystem degradation on the financial sector and on individual financial institutions. The first central bank to have actively assessed the potential financial implications of biodiversity loss is De Nederlandsche Bank (Box 17.8), providing both precedence and inspiration for the central banking community.

Chapter 17: Managing Nature-Related Financial Risk and Uncertainty

> ### Box 17.8
> ### Biodiversity Loss and Risks for the Dutch Financial Sector
>
> In 2020, De Nederlandsche Bank together with the Netherlands Environmental Assessment Agency (PBL) assessed how and to what extent Dutch financial institutions were exposed to risks from biodiversity loss (van Toor et al. 2020).[517] The joint DNB-PBL study explored the dependence and impact of Dutch financial institutions on ecosystem services and biodiversity and thus exposure to potential associated financial risks.
>
> On the dependency side – using the ENCORE tool (Table 17.2) – they found that at least 36% (€510 billion) of the €1.4 trillion in financial investments held by Dutch financial institutions, were highly or very highly dependent on one or more ecosystem services (Toor et al. 2020). The highest dependence was on ecosystems that provide groundwater and surface water, as for every euro invested by Dutch financial institutions, around one-quarter was dependent on these ecosystems. One ecosystem service examined in greater depth was pollination, where the study found that the Dutch financial sector is exposed to around €28 billion to products that depend on pollination (van Toor et al. 2020).
>
> On the impact side, the GLOBIO model (developed by PBL) was used to assess the biodiversity footprint of Dutch financial institutions, as a consequence of changing land use and greenhouse gas emissions. The study found that the biodiversity footprint of Dutch financial institutions was comparable with the loss of over 58,000 km² of pristine Nature – an area more than 1.7 times the land surface of the Netherlands. Around half of this was the result of changing land use, and the other half was due to greenhouse gas emissions.
>
> Given this footprint, the study explored the extent of potential exposure to transition risks through two potential scenarios: (i) the possible expansion of Protected Areas, and (ii) changes that mean business would have to limit their total nitrogen emissions. The first scenario highlighted that the Dutch financial sector has €15 billion in exposure to companies that are active in already Protected Areas. The second scenario showed that three large Dutch banks have granted loans totaling €81 billion to nitrogen-emitting sectors, and that these sectors account for around 39% of total lending in the Netherlands. Both scenarios highlighted the importance of Dutch financial institutions to gain a holistic picture of the sensitivity of their portfolio to transition risks (van Toor et al. 2020).

---

[517] The study only identified a limited number of Nature-related risks, due to the data limitations, and therefore the results are lower bounds of the total exposure of Dutch financial institutions to such risks.

Chapter 18
Conservation of Nature

Introduction

Directly investing in our stock of natural assets can help to reduce the Impact Inequality. Conservation of functioning ecosystems and restoration of degraded ecosystems are two important approaches to maintain, improve and increase our stock of natural assets and their associated biodiversity. This chapter and Chapter 19 focus on approaches to conservation and restoration of our stock of ecosystems, the living organisms that live in them and the genes they contain.

Conservation seeks to prevent the degradation of ecosystems and the associated species and genes within them, whereas restoration seeks to recover ecosystems from degraded states (Figure 18.1). Ecosystems are dynamic systems that are not fixed in time and space; they interact in multiple ways. Chapter 2 highlighted evidence of how biodiversity enables ecosystems to be resilient, adapt and evolve, and how it supports their productivity in providing ecosystem goods and services. Chapter 2 also described how ecosystems are the key components of the biosphere in the economics of biodiversity and showed that biodiversity is an enabling characteristic. As noted in Chapter 2, both restoration and conservation of ecosystems and their biodiversity count as investment because ecosystems regenerate.

Figure 18.1 Conservation and Restoration

18.1 Ecosystem Assets

Three attributes of ecosystems influence the flow of ecosystem services and their contribution to inclusive wealth: their quantity, quality and distribution (Natural Capital Committee, 2014). The quantity or extent of natural assets within an asset portfolio strongly influences the flow of services to people. Similarly, the quality of assets makes a significant difference to the benefits provided to people, and the ability of the assets to regenerate and persist. Biodiversity is a useful indicator of ecosystem

Chapter 18: Conservation of Nature

quality. The location and distribution of the stock also matter both in relation to other natural assets and the people who benefit from the flows of services. For some ecosystem services, including many of the regulating and maintenance services and cultural services, the spatial configuration of the assets determines whether benefits are derived or not. For example, the restoration of a riverine flood plain could reduce peak flows of water and potential flood risk faced by conurbations down-stream. The spatial location of the portfolio of natural assets also matters when considering issues of equity and environmental justice (Chapter 14).

As noted in Annexes 2 and 3 of Chapter 2, a number of approaches and initiatives are contributing to our understanding of the quantity and quality of our stock of natural assets (for example Ichii et al. 2019; Nature Map, 2020). Figure 18.2 presents terrestrial global habitats in 2015, based on IUCN habitat classes, and the percentages of land allocated to different habitat types globally (Jung et al. 2020). Only 13% of the ocean (K. Jones et al. 2018) and 23% of the land are sufficiently free of human influence that they can still be considered wilderness (Watson et al. 2016). Over 40% of the planet is used for agriculture and urban settlements, and there are ongoing declines in areas of other ecosystems, such as temperate grasslands, subtropical forests, coral reefs and mangroves (Ichii et al. 2019). Biodiversity, as an enabling characteristic of ecosystems, indicates ecosystem quality and resilience but measuring and quantifying biodiversity as an indicator is difficult. There is no definitive list of species that exist on the planet, but measures of species abundance, extinction risk and composition are all currently declining, and estimates suggest that 515 vertebrate species are on the brink of extinction (Ichii et al. 2019; Ceballos, Ehrlich, and Raven, 2020) (Chapter 2, Annex 3; Chapter 4).

Given the significant risk and uncertainty in relation to unpredictable regime shifts in ecosystems and unknowable demands on the biosphere from future generations, it makes sense to conserve ecosystems to maintain their option values and support well-being of present and future generations (Chapter 3 and 5). In light of the non-linearity of ecosystems, as we have seen in Chapter 7, quantity restrictions are a more effective policy approach than pricing mechanisms, to prevent possible regime shifts in ecosystems. This chapter explores how conservation can make a meaningful contribution to preventing further biosphere degradation and maintain our natural assets.

Chapter 18: Conservation of Nature

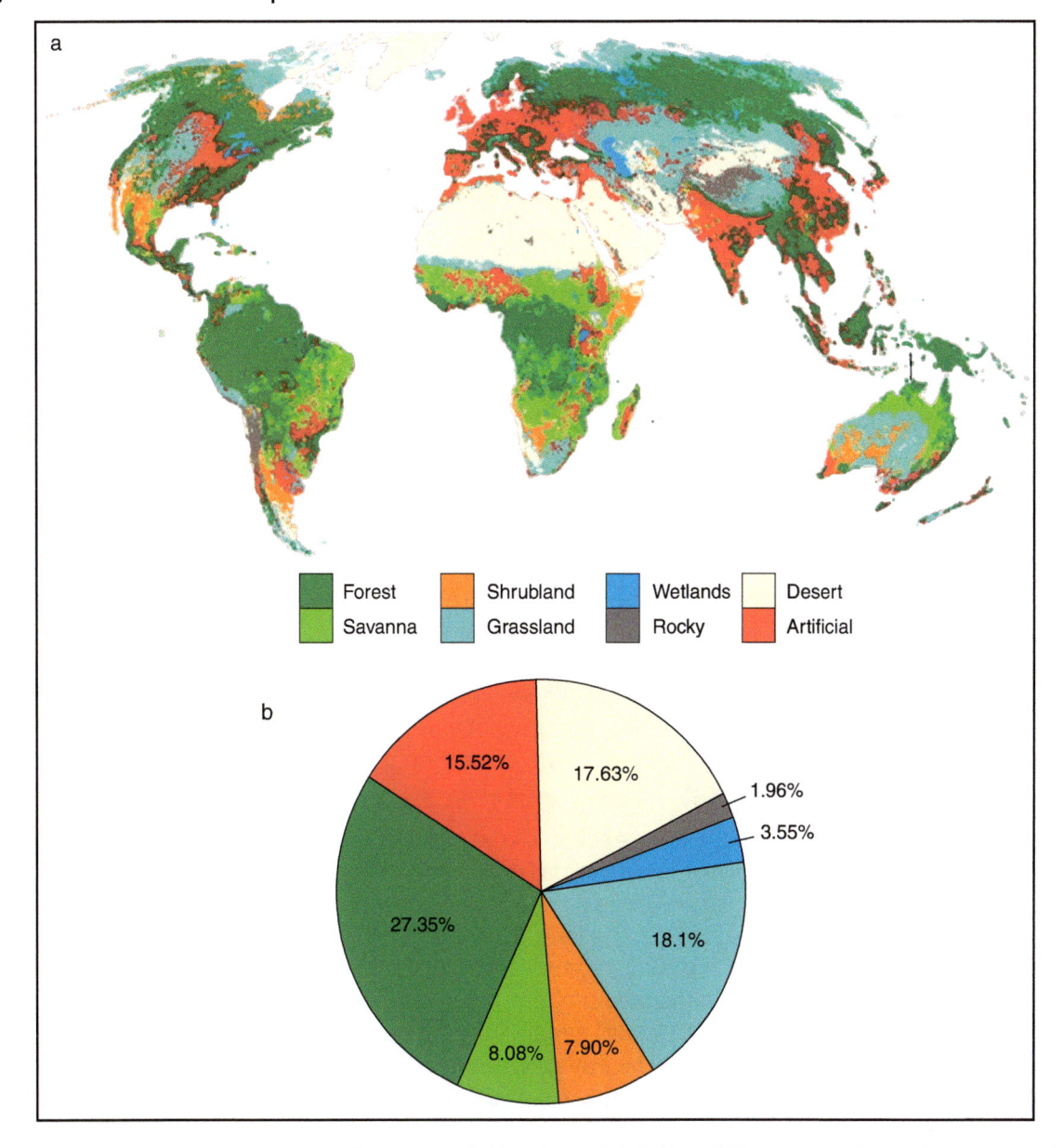

Figure 18.2 A Global Map of Terrestrial Habitats in 2015

Source: Jung et al. (2020). Note: Distribution of Level 1 IUCN habitat classes globally (a), and (b) percentage of global land area occupied by each Level 1 IUCN habitat class.

18.2 How Much Ecosystem Stock Do We Need?

The Impact Inequality describes the imbalance between our demands on the biosphere and its ability to regenerate (Chapter 4). Increasing the quantity and quality of our ecosystem stocks tackles the supply-side to help balance the equation but how much more do we need? As noted in Chapter 4, there is no definitive answer to this question, but a number of studies have explored the level of protection needed to meet specific human objectives. In the marine environment, O'Leary et al. (2016) calculated that 37% of the world's ocean needs protection to achieve a suite of biodiversity and fisheries objectives (O'Leary et al. 2016). A similar analysis on land suggests that 30% of the Earth's surface requires

Chapter 18: Conservation of Nature

protection by 2030 if we wish to maintain biodiversity and secure ecosystem services. The same study suggested that an additional 20% needs to be designated as climate stabilisation areas to help stay below the Intergovernmental Panel on Climate Change's 1.5°C target for global warming (Dinerstein et al. 2019). Analysis of the economic feasibility of expanding the Protected Area network suggests that global benefits are substantial, and a number of positive financial and social outcomes could flow from this approach (Box 18.1).

Box 18.1
Costs and Benefits of Expanding Protected Areas

There are costs and benefits of expanding the Protected Area network. A study of the economic feasibility of expanding the area of no-take Marine Protected Areas showed that the global benefits exceed the costs by a factor of 1.4–2.7, depending on the extent of expansion and the location (Brander et al. 2020).

Analysis of the global impacts of terrestrial Protected Areas and Marine Protected Areas on conservation, agriculture, forestry, and fisheries sectors showed that benefits of protecting 30% of land and ocean would exceed costs, and would provide better financial outcomes and higher non-monetary benefits than currently (Waldron et al. 2020). A number of financial and economic benefits have been found from expanding Protected Areas: eco-tourism income, the provision of health clinics, education, improved health outcomes, and other forms of support to local communities, and the avoidance of spending on natural disaster prevention and recovery (e.g. flood defences, storm damage mitigation, for example Barbier, 2016; Reguero et al. 2018). In all scenarios studied by Waldron et al. overall gross economic output was found to be US$64–454 billion per year higher by 2050 than in the counterfactual of non-expansion of Protected Areas (Waldron et al. 2020).

The largest gains were found in the Nature sector of the economy, which was predicted to expand by 4–6% per year, with substantial revenue increases expected from eco-tourism (Figure 18.3). Modest increases in agriculture and forestry revenue were predicted even though production may decline, as this was assumed to be outweighed by smaller reductions in prices for food and higher prices for timber. Fisheries were projected to decline regardless of the size of Marine Protected Areas, driven by overfishing and climate change. In the medium term, the study found that if no-catch zones are part of Marine Protected Area policy, there could be spill-over effects into catch zones which would result in net gains for the fisheries industry. The authors found that the ocean economy is likely to grow overall, as decreases in wild-capture fisheries are offset by eco-tourism in Marine Protected Areas (Waldron et al. 2020).

Waldron et al. found that social benefits of expanding Protected Areas are likely to be large. A partial economic analysis, focusing only on forests and mangroves, showed that there would be overall benefits of US$170–534 billion per year by 2050,[518] from avoided flooding, climate change, soil loss and coastal storm damage. These benefits are weighed against the direct investment required to expand the Protected Areas on land and sea, which is estimated to cost US$103–178 billion per year by 2030 (Waldron et al. 2020). These costs will be offset by the financial and economic benefits within the Nature and conservation sector alone.

---

[518] If world Gross Domestic Product (GDP) is US$160 trillion in 2050, roughly doubling from 2020, this represents c.0.2%-0.3% of global GDP per year in 2050.

Overall, Waldron et al. (2020) provide empirical evidence for large social benefits from expanding and enhancing Protected Areas, and lower indirect costs on related industries than often expected. With no major financial losses predicted and the potential for small net financial gains globally, their analysis suggests that concerns over losses of jobs and livelihoods from Protected Areas are misplaced.

Figure 18.3 Revenue growth Projected in 2030 to 2050 in the Scenario 30% of Land and Ocean is Protected

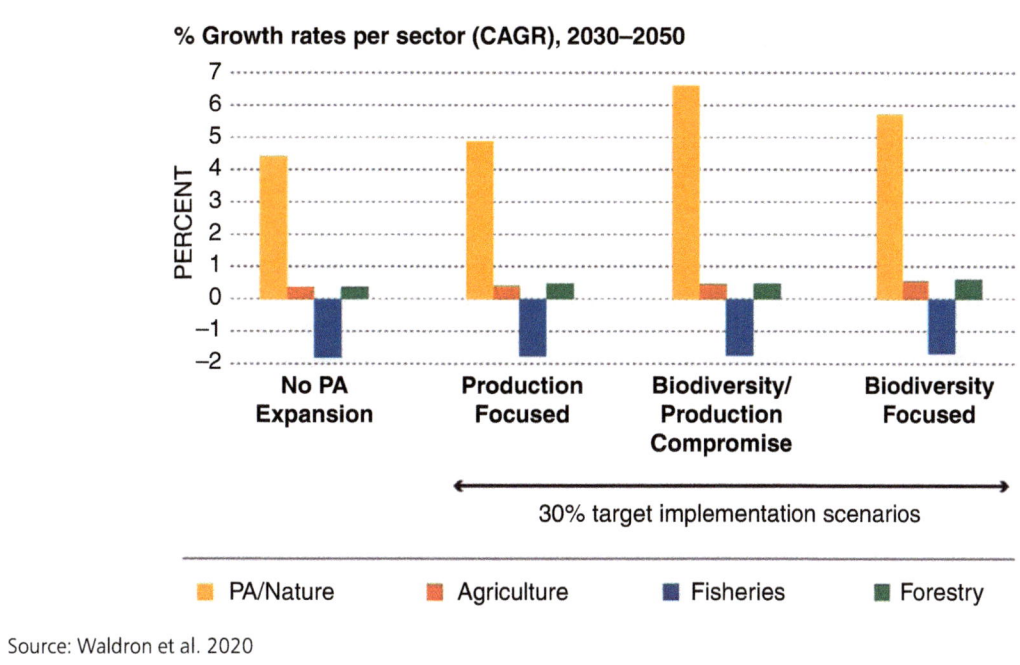

Source: Waldron et al. 2020

Another approach to identifying the amount of natural assets we require is to consider constraints within which humanity must operate to avoid ecosystems shifting into new stable states that are less beneficial for human societies (Chapters 3 and 4). Researchers have identified 'critical' natural capital, which cannot be replaced by produced capital (Ekins, 2003), which includes ecosystem functions and processes that underpin the provision of ecosystem services (Fitter, 2013; Ekins, Milligan and Usubiago-Liano, 2019). Alongside setting targets for ecosystem assets in terms of quantity and quality, setting standards and quantity restrictions for safe operating spaces can also help us to avoid tipping points of ecosystems to new stable states (as discussed in Chapters 3 and 7).

18.3 What Kind of Stock Do We Need?

The world has a spectrum of ecosystems with different ecological outcomes that people engage with – and benefit from – in different ways: intact ecosystems with high biodiversity; marine and terrestrial ecosystems that are used for a range of purposes but still support biodiversity and ecosystem services; and monoculture ecosystems that have been heavily modified for human use with a single purpose in mind, such as growing crops (a provisioning good). Monocultures are invariably poor in terms of biodiversity and ecosystem function, and consequently are not a target for conservation. The purpose of conservation and restoration is to maintain or improve ecosystems with some desired state in mind that is defined by biotic and geographical parameters, and by societal choice and purpose. The spectrum of possible states that have been created by human modification of intact ecosystems for various purposes is presented in Figure 18.4. Opportunities exist along this spectrum to consider and enable synergies between ecosystem services and biodiversity outcomes.

Chapter 18: Conservation of Nature

Figure 18.4 Conservation and Restoration Spectrum, Relating Modification and Purpose

Despite our growing population and consumption, there is space to conserve and restore ecosystem assets both on land and at sea. For example, more efficient farming with higher yields could free up space for ecosystem conservation and restoration (as discussed in Chapter 16) (Strassburg et al. 2014; Balmford, Green, and Phalan, 2015; Phalan et al. 2016; Pollock, Thuiller, and Jetz, 2017). Multifunctional areas can provide benefits for people and support ecosystems and biodiversity. They can contribute significantly to the mitigation and adaptation of climate change (Section 19.3, Chapter 19). For example, saltmarshes provide coastal protection, sequester carbon, provide food and act as nurseries for fish. The global portfolio of ecosystems is currently a mix of intact biodiverse places and multifunctional areas managed to support biodiversity and provide multiple benefits for people. Ongoing conservation and restoration of Nature will need to continue to provide a mix of assets across the spectrum to provide a portfolio that supports the biosphere and our present and future well-being. Conservation and restoration of natural assets also support jobs and livelihoods. The International Labour Organization (ILO) (2018) estimates that 1.2 billion jobs rely on effective management and sustainability of ecosystems. Ecosystem services such as water purification, pollination, soil renewal, pest control, and flood protection all underpin sectors such as farming, fishing and forestry which are vulnerable to ecosystem degradation. Environmental sustainability is critical for these jobs but is also expected to result in more and better jobs in the future (for example see Section 19.3 on Nature-based Solutions, Chapter 19) (ILO, 2018).

18.4 How Can We Improve and Increase Our Stocks?

Two broad conservation approaches exist: place-based and species-led conservation. Within each of these are a myriad of methods. Protected Areas have been the cornerstone of place-based conservation efforts, but other place-based approaches, such as community-led protection and practices of indigenous people and local communities, have also proved highly effective for protection of ecosystems (as discussed in Chapter 8). Species-led conservation involves approaches such as conservation to meet individual species' requirements, conservation of charismatic, keystone or flagship species, regulation of wildlife trade, and conservation of genetic diversity through seed banks.

18.4.1 Place-Based Conservation

Protected Areas are widely used for the conservation of habitats and species and have been shown to be successful on land and at sea. Without Protected Areas, global loss of biodiversity would be even greater (Laurance et al. 2012; Gill et al. 2017; Geldmann et al. 2018; UNEP-WCMC, IUCN, and NGS, 2018). A recent global review of terrestrial Protected Areas showed that species richness and abundance are higher inside Protected Areas than outside them (Gray et al. 2016). Marine Protected Areas have been increased from 3.2 million km^2 in 2000, to 26.9 million km^2 in 2020; globally, they now make up 7.4% of the ocean area (Duarte et al. 2020a, 2020b). On land, around 15 % of the Earth's surface is within one of 238,563 individual Protected Areas. Some countries and regions contain very large Nature reserves (e.g. Africa, South America, Australia, Greenland and Russia), whereas others have higher numbers of smaller Protected Areas (e.g. Europe) (UNEP-WCMC, IUCN, and NGS, 2018). Protected Areas can act as a form of quantity restrictions as alternative approach to market mechanisms to prevent degradation of our natural assets (Chapter 7).

Alongside biodiversity conservation, Protected Areas have been found to secure the provision of a range of local ecosystem services, help reduce poverty, provide food and water security, and reduce the risk of natural disasters (Larsen, Turner, and Brooks, 2012; Hanauer and Canavire-Bacarreza, 2015; UNEP-WCMC and IUCN, 2016; Jepson et al. 2017). A study in Nepal found that Protected Areas reduce overall poverty and extreme poverty, and do not increase inequality; tourism was found to be a key driver of poverty alleviation, but Protected Areas also reduce extreme poverty in less-visited areas (den Braber, Evans, and Oldekop, 2018). A review of the well-being of people living near over 600 Protected Areas in 34 countries found that Protected Areas associated with tourism provide households with higher wealth (by 17%) and lower poverty (by 16%) than similar households far from Protected Areas. Children under the age of five years living near multiple-use Protected Areas are taller (by 10%) and are less likely to be stunted (by 13%) than similar children far from Protected Areas (Naidoo et al. 2019). Protected Areas are estimated to deliver 20% of the total continental rainwater runoff globally, providing freshwater downstream to nearly two-thirds of the world population (Harrison et al. 2016). Worryingly, the COVID-19 pandemic is having a negative impact on the management capacity, budgets and effectiveness of Protected Areas as well as impacts on the livelihoods of communities living in and around these sites due in part to loss of tourism income (Hockings et al. 2020).

The success of Protected Areas for conservation depends on factors such as local support, good management and sufficient resourcing and state support (Watson et al. 2014; Di Minin and Toivonen, 2015; Oldekop et al. 2016; Gill et al. 2017; Coad et al. 2019). There is considerable room for improvement, as only 20% of Protected Areas are being managed well (UNEP-WCMC, IUCN, and NGS, 2018). There are concerns over the quality and effectiveness of Marine Protected Areas, particularly in relation to fishing practices (Sala et al. 2018). While 71% of the reserves have improved fish populations, the level of protection is often considered weak (Costello and Ballantine, 2015). On land, only half of 60 Protected Areas in tropical forests were considered to be effective or 'performed passably'; the rest experienced biodiversity decline both taxonomically and functionally. The strongest predictors of declines were habitat disruption, hunting and forest-product exploitation (Laurance et al. 2012). The wider context within which each Protected Area sits is critical. A study in Spain highlighted the importance of understanding the flow of ecosystem services both into and out of Protected Areas. Most of the negative drivers of change, such as the transformation of land associated with agriculture and tourism, were outside the boundaries of the Protected Areas, as were most areas that benefited from the ecosystem services, highlighting the need for broader planning strategies including the Protected Area and surrounding areas (Palomo et al. 2013). There are concerns that human pressures on Protected Areas are increasing, particularly on those close to urban areas, due to increasing urbanisation. It is estimated that the growth of urban areas could threaten 29 million hectares of natural habitat; 40% of strictly Protected Areas will be within 50 km of a city between 2000 and 2030 (McDonald et al. 2018).

Chapter 18: Conservation of Nature

There is strong evidence of Protected Area downgrading, downsizing and degazettement[519] (known by the acronym 'PADDD') (Golden Kroner et al. 2019; Qin et al. 2019). Between 1892 and 2018 there were 3,749 PADDD events in 73 countries, in which 519,857km^2 were removed from protection, and regulations were reduced in an additional 1,659,972km^2 (Golden Kroner et al. 2019). An alarming 78% of these events have happened since 2000, and a large proportion were associated with resource extraction and development (Golden Kroner et al. 2019). The designation of a Protected Area is clearly not sufficient to conserve Nature within it; rather, ongoing resources, both human and financial,[520] and strong institutional and government support are required to avoid so-called 'paper parks'.

The fixed boundaries around Protected Areas can be an issue when conserving dynamic ecosystems. In some cases, features of interest for which the site was originally established (e.g. eroding coastlines or meandering rivers) may move outside site boundaries. Climate change is transforming the climate spaces available within Protected Areas. Shifting climatically suitable areas for ecosystems and their associated species may undermine the effectiveness of the areas for the protection of ecosystems and their biodiversity (Batllori et al. 2017). Biases in the geographical locations of Protected Areas towards colder and wetter ecosystems reduce opportunities for species to move to appropriate climates and maintain adequate protection (Elsen et al. 2020). Building considerations of climate in the Protected Area network is essential to safeguard biodiversity over time and adapt to a changing climate (Elsen et al. 2020). Complementing place-based approaches with species-led conservation can help to overcome these issues.

Approximately 40% of all terrestrial Protected Areas and ecologically intact land (for example, boreal and tropical primary forests, savannahs and marshes) overlap with land managed by indigenous people and local communities (Garnett et al. 2018). As noted in Chapter 8, indigenous people and local communities often support biodiversity conservation through their knowledge, practices and institutions (Berkes, 2017). Biodiversity indicators show declines of 30% less and 30% more slowly in indigenous lands than in lands not managed by indigenous people (Ichii et al. 2019). Values of connectivity, reciprocity and trust in relationships with all species have provided the basis of effective institutions for environmental management (for more, Chapter 6 looked in detail at trust and social capital, Chapter 8 looked at community institutions for managing Common Pool Resources and Chapter 12 examined non-anthropocentric perspectives on human connections with ecosystems) (Whyte, Brewer, and Johnson, 2016).

Aotearoa (the Māori name for New Zealand) has granted legal personhood to national park Protected Areas. Te Urewera National Park in Tuhoe Tribal Lands is now recognised as a 'legal entity' with 'all the rights, powers, duties and liabilities of a legal person'. Governance of the park has shifted to the Te Urewera Board, which uses Tuhoe values, laws and concepts of management in their decisions and practices, restoring their roles as kaitiaki (guardians) (Muller et al. 2019).

Chapter 6 notes that for conservation and restoration to be successful, engagement of communitarian institutions (either communities or civil society or both) is essential. Analysis of Protected Areas found that sites which are co-managed, empower local people, reduce economic inequalities, and maintain local livelihoods successfully deliver both conservation and socio-economic outcomes (Oldekop et al. 2016). A global meta-analysis of 171 studies involving 165 Protected Areas showed that co-management of Protected Areas by local communities and conservation agencies was associated

---

[519] Downgrading is a decrease in the legal restrictions on the number, magnitude, or extent of human activities within a Protected Area; downsizing is a decrease in the size of a Protected Area as a result of removal of an area of land or sea area through a legal boundary change; and degazettement is a loss of legal protection for an entire Protected Area (Mascia and Pailler, 2011).

[520] Significant increases in financing are essential to enable significant investment in natural assets. Chapter 20 provides details on finance for both conservation and restoration.

with delivery of greater local benefits than management by community or state alone (Oldekop et al. 2016). Co-management between indigenous people and government institutions has brought together traditional knowledge and values with western science in Gwaii Haanas National Park Reserve in Canada and Cabo Pulmo National Park in Mexico (Box 18.2).

Box 18.2
Cases of Management by Indigenous Peoples and Local Communities

In 1993, the Gwaii Haanas National Park Reserve and Haida Heritage Site were established on the island of Haida Gwaii in Canada, in response to Haida nation concerns over the damage and destruction to ancestral sites. The park consists of pacific temperate rainforest that stretches from the wildlife rich sea up the slopes of the San Christoval Mountains. Humans are thought to have inhabited the area for 12,500 years, and the area is rich in cultural sites. The park is co-managed by the Council of the Haida nation and Parks Canada, with the goal to maintain and restore the "rich cultural and ecological heritage of the Gwaii Haanas" for the benefit of present and future generations. Both traditional knowledge and western science are used in decision-making and planning. In 2010, the Gwaii Haanas National Marine Conservation Area Reserve and Haida Heritage Site was established adjacent to the terrestrial Protected Area. Management is based on *yahguudang* (respect for all living things) and aims to balance protection of the area with Haida food, cultural, economic and ceremonial needs. The Haida nation established the Haida Gwaii Watchmen, who monitor and steward the Protected Area and are formally integrated into its management (Stephenson et al. 2014).

Cabo Pulmo, in Baja California, Mexico, was once a thriving fishing village. However, overfishing emptied the coral reef and destroyed livelihoods of the villagers. The community decided to petition the Mexican government to set up a Marine Protected Area, and in 1995, Cabo Pulmo National Park was established. Commercial fishing was banned, and the local people oversaw the management and enforcement of the no-take marine reserve. Between 1999 and 2009, total fish biomass increased by over 450%, from 0.75 t ha to 4.24 t ha, and the biomass of top predators and carnivores increased by 11 and 4 times, respectively. The recovery of top predators and carnivores is particularly significant, as this suggests that Cabo Pulmo National Park is approaching an ecosystem assemblage that is seen in systems with little or no fishing pressure. The combination of a governance model combining a nationally designated Protected Area with local community management, effective enforcement, social cohesion, and local benefits was critical to success and recovery (Aburto-Oropeza et al. 2011). The local people benefit from the spill-over of fish from the reserve to their fisheries; eco-tourism has boomed, providing livelihoods for local people and incentives to continue to maintain and invest in Cabo Pulmo National Park.

In some cases, successful communitarian management in conservation can deliver a range of different outcomes on the conservation spectrum. For example, in Germany, the Spreewald Biosphere Reserve strictly protects some areas but enables sustainable resource use of other areas for both the local community and visitors from further afield. The Reserve is an inland delta of streams around the River Spree, composed of meadows, forests, fens and 296 km of navigable channels. It is an important site for a range of habitats and species, including fire-bellied toads, white storks, otters, sundews, sedge meadows and alder forests. Within the 475km^2 are 10 core areas which are strictly protected and can only be entered by people with permission to undertake research. The core areas are embedded in a buffer zone which seeks to maintain habitats and species but allows tourist access on land and water. Finally, there is a wider transition area, where traditional farming practices take place and where restoration of formerly intensively managed land is progressing. The transition area can be used for recreation, agriculture, forestry, fishing and hunting. The aim of the Biosphere Reserve is for sustainable economic, social, cultural and ecological

Chapter 18: Conservation of Nature

development. Regional products are marketed under the Spreewald umbrella brand which certifies companies in the Spreewald economic area, using ingredients predominantly from the area and with strict environmental standards (State Office for the Environment (LfU), 2020).

There is no doubt that Protected Areas already play an important role in maintaining our current stocks of biodiverse Nature. Further improvements can be made by expanding their coverage, ensuring that they are well connected and integrated into the surrounding land and sea, are governed equitably, including indigenous people and local communities, are sufficiently resourced, and are managed effectively (Watson et al. 2014; UNEP-WCMC, IUCN, and NGS, 2018).[521]

Other effective area-based conservation measures (OECMs)[522] offer a complementary approach to Protected Area conservation (Jonas et al. 2017; Dudley et al. 2018). Conservation can be achieved by many different actors, management systems, and achieved through institutions in which conservation is not the primary focus (for example Box 18.3). In the ocean, opportunities for – and risks to – conservation have been identified with providers of offshore wind turbines, coastal and marine tourism, and oil and gas decommissioning (Friends of Ocean Action, 2020). For example, offshore wind farms may create opportunities for conservation (or arguably, restoration) by creating artificial reefs. Multifunctional lands and seas that provide benefits for people and biodiversity form an important part of the portfolio of natural assets promoting the co-existence of people and wildlife (Tyrrell, Toit, and Macdonald, 2020).

Box 18.3
The Coastal Cloud Forest of Loma Alta, Ecuador

The coastal cloud forests of Western Ecuador are incredibly diverse. In Loma Alta, there are over 300 species of bird, of which 79 are endemic (Becker et al. 2005). The Loma Alta community was given legal tenure to the land in 1937 and governed it through democratic decision making. Day-to-day decisions were made on the community's behalf by a small cabildo (council), and conditions were right for sustainable management – clear property rights and good local governance. Yet despite this, the community gave some individuals rights to log for timber and clear the forest for charcoal production. Areas were cleared to grow plants to make Panama hats, and the community let ranchers from a neighbouring province convert forest to pasture for cattle.

Local people had not realised that the forest was providing an essential ecosystem service for the community – water provision. The vegetation of the cloud forest was collecting the moisture from the fog that descended each day, keeping soils wet and allowing the streams to flow, even during the dry season. The pasture did not perform this function. Analysis of water collection data suggested that the conversion to pasture cost the villagers 38 million litres of water every dry season, with an estimated annual value of US$130,000 (Becker, 1999). Presentations of this to the community alarmed them. The villagers took a vote on establishing a forest reserve and banning clearance, logging and hunting. The cabildo had to consider and address the concerns of the villagers who made a living from the forest through hunting, timber and charcoal, and did so by

---

[521] Proposals for 30% of the Earth's surface to be in Protected Areas have social, political and practical implications. Concerns have been raised from the perspectives of equity, human rights, governance and livelihoods (see, for example, Büscher et al. 2017).

[522] An OECM is a "geographically defined area other than a Protected Area, which is governed and managed in ways that achieve positive and sustained long-term outcomes for the in situ conservation of biodiversity with associated ecosystem functions and services and where applicable, cultural, spiritual, socio–economic, and other locally relevant values" (IUCN-WCPA Task Force on OECMs, 2019).

offering them jobs associated with the reserve, such as forest guards. In 1996, after three months and four meetings, the community voted unanimously in favour of the forest reserve (Becker, 1999; Becker and Ghimire 2003; Becker et al. 2005; Balmford, 2012). It has since been expanded twice so that it now covers 40% of their land. Alongside the work and commitment of the local community, there has been important backing from NGOs. For example, Earthwatch and People Allied for Nature supported the science, and People Allied for Nature installed a water pump, supported legal help to end disputes with the ranchers, funded assistant teachers, and paid for children to attend high school. In this case, the community recognised the connection between the forest as an asset providing regulating services that were important for their own livelihoods, and had the institutions that enabled them to do something about it.

18.4.2 Species-Led Conservation

Place-based approaches to conservation are not always sufficient for addressing species conservation problems (Geldmann et al. 2013, 2018; Gray et al. 2016; Gill et al. 2017). As of January 2018, 35% of Key Biodiversity Areas[523] were not covered by any form of Protected Area legislation. The geographic ranges of only 22% of terrestrial amphibian species, 56% of bird species and 46% of mammal species were sufficiently represented in the global Protected Area system (Butchart et al. 2015; UNEP-WCMC, IUCN, and NGS, 2018). Only 4–9% of terrestrial mammal, amphibian and bird species' ranges overlap with Protected Areas that are adequately resourced[524] (Coad et al. 2019).

The vertebrate species data from the IUCN Red List show that 7% of species (64/928) have improved in conservation status by changing category due to successful conservation action over the last approximately 20 years prior to 2010 (Hoffmann et al. 2010). More recent estimates suggest that conservation action prevented 21–32 birds and 7–16 mammal extinctions between 1993 and 2020 (Bolam et al. 2020). The number of marine species assessed by the IUCN Red List as threatened with global extinction has declined from 18% in 2000 to 11% in 2019, due to conservation actions (Figure 18.5, based on Duarte et al. 2020a).

International agreements have been established to reduce or control the hunting of marine species, such as the Convention on the International Trade of Endangered Species – see Box 18.4 and the global Moratorium on Commercial Whaling (International Whaling Commission, 1982). Other interventions to protect marine species include: water quality improvements to reduce nutrients and sewage inputs; management of fisheries through catch and effort restrictions, closed areas, regulation of capacity and fishing gear, catch shares, and co-management arrangements; and the creation of Marine Protected Areas. For marine mammals, 47% of 124 well-assessed populations showed significant increases over the past decades. For example, since the whaling ban in 1986, humpback whales have increased to 25,000 individuals in the South Atlantic between South America and Antarctica: a significant comeback. Exploitation had pushed the numbers from an estimated 27,000 in 1830 to just 440 in the mid-1950s. Now the species has bounced back to 93% of its pre-exploitation population size (Zerbini et al. 2019)[525]. Northern elephant seals were almost driven to extinction by commercial sealers, but have recovered to more than 200,000 today in the USA and Mexico (Lowry et al. 2014).

---

[523] Key Biodiversity Areas are areas of global importance for biodiversity (approximately 15,000 sites have been identified) (UNEP-WCMC, IUCN, and NGS, 2018).

[524] As defined by the Protected Area Management Effectiveness data agreed upon by Convention on Biological Diversity signatory nations.

[525] Chapter 13, Annex 2, contains a study that predicted that even from the purely commercial point of view, the ban would be profitable.

Chapter 18: Conservation of Nature

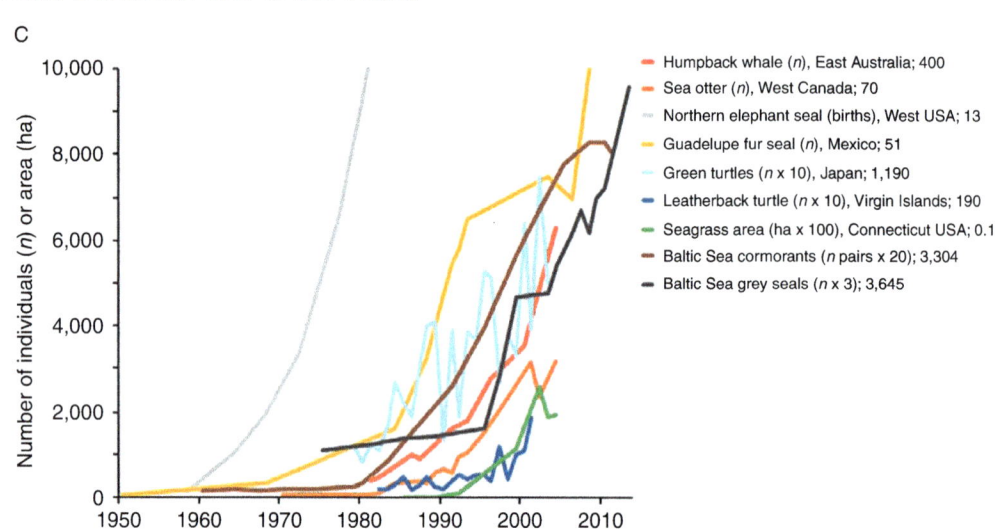

Figure 18.5 Recovery Trends of Marine Populations: Sample Trajectories of Recovering Species From Different Parts of the World

Legend:
- Humpback whale (*n*), East Australia; 400
- Sea otter (*n*), West Canada; 70
- Northern elephant seal (births), West USA; 13
- Guadelupe fur seal (*n*), Mexico; 51
- Green turtles (*n* x 10), Japan; 1,190
- Leatherback turtle (*n* x 10), Virgin Islands; 190
- Seagrass area (ha x 100), Connecticut USA; 0.1
- Baltic Sea cormorants (*n* pairs x 20); 3,304
- Baltic Sea grey seals (*n* x 3); 3,645

Source: Duarte et al. (2020a). *Permission to reproduce from Springer Nature.* Note: Units were adjusted to a common scale as indicated in the legend (n x), numbers at the end of the legends indicate the initial count at the beginning of time series.

In species-led conservation, bespoke actions can help flagship species while also driving wider conservation efforts. For example, poaching, habitat loss and human-wildlife conflict caused tiger numbers to drop from approximately 100,000 individuals in 1910 to 3,200 in 2010, occupying only 7% of their former range (IUCN 2014). In the Manas Tiger Conservation Programme of Aaranyak, Guwahati, India, conservation efforts included the development of alternative livelihoods for the people who were dependent on the Manas National Park (IUCN, 2016b). The programme supported farm-based livelihoods for homesteads to grow crops such as a potatoes, mushrooms and pulses; develop nurseries for areca nuts or lemons; support livestock keeping; and develop rural enterprises such as weaving and tailoring. These efforts led to considerable improvements in average annual income of local communities, increased tiger numbers, and reduced human pressure on the park by providing alternative sources of fuel through community nurseries and plantation of fuelwood species as local source of fuelwood supply outside the park (Firoz Ahmed pers. comm).

As noted in early chapters of the Review, over-exploitation is the removal or harvesting from the wild of living resources, such as plants, fungi or animals, at rates that exceed their capacity to regenerate, often to the point of diminishing returns; these stocks would ordinarily be renewable. Over-exploitation of animals, plants and other wildlife has continued to drive biodiversity loss. It is the greatest cause of loss in the seas, principally through fisheries, and the second greatest on land (Díaz et al. 2019).

Over-exploitation has been directly responsible for the extinction of species, and for reducing once abundant populations to levels at which they have effectively become economically extinct as they are so diminished in numbers that they are no longer worth harvesting. At their reduced levels, they are often no longer able to fill their role in the ecosystem, and they become vulnerable to other chance factors that might cause their global or local extinction. Harvesting of wildlife does not have to be for trade; much might be for local use or for personal consumption. Harvesting by indigenous people or local communities may have been practised for millennia, and it is often, but not always, sustainable. Domestic or international trade can shift the balance from sustainable to unsustainable exploitation (wider issues around trade are discussed in Chapter 15). Pressures from trade typically result from a combination of other indirect drivers (Díaz et al. 2019), including the expansion of infrastructure, such as roads or flights, and/or improvements in technology; these may make access to, and transport of, produce from formerly remote areas simpler, reduce costs, or enable species to be exploited in ways that were not previously possible. Linked to this is the increasing scope for the long-range, high-volume, rapid transportation of goods to all parts of the world. With the world being increasingly better connected and *per capita* consumption increasing, market demand can be satisfied or stimulated for wildlife products that were previously beyond the reach of many consumers.

Measures to regulate trade in wildlife have been introduced nationally and globally, the latter notably through the Convention on International Trade in Endangered Species of Wild Flora and Fauna (CITES), since 1975 (see Box 18.4). As its title shows, CITES regulates only international and not domestic trade, but measures required by it often have domestic implications.

Box 18.4
International Trade in Endangered Species

With 183 Parties, the Convention on International Trade in Endangered Species of Wild Flora and Fauna (CITES) is one of the most widely adopted of multilateral environmental agreements (MEAs); unlike other MEAs, CITES includes mechanisms to achieve compliance. Trade is regulated through permits applied to species listed in one of its three Appendices; currently trade is regulated in some 30,000 plant and 5,800 animal species.

However, far more species are not listed than are listed, and the volume and value of trade in commercial fisheries and timber probably far exceeds those in CITES-listed species. CITES has focused on traded tropical timber species and commercially harvested sharks in recent additions to the Appendices, which suggests increasing support by Parties to regulate such trade where CITES can add weight to other mechanisms.

Trade regulation, especially through CITES, has played a significant role in conserving stocks of wild species, and the market has also provided incentives, when aligned with appropriate governance and community involvement, to re-build and restore stocks, thus enabling increased economic benefits to be derived from trade. However, Parties have not achieved universal success in their implementation of CITES, and the loss of stocks has still occurred despite it. Illegal trade and shifts to captive production may each interact differently with markets and so with the incentives they provide for conserving species in the wild.

In parallel with legal regulation of trade, there is a significant volume of illegal wildlife trade. Legal and illegal markets may be distinct and operate independently of one another, or they may be linked and merge (UNODC, 2016). Significant claims are made as to the value of illicit wildlife trade; some estimates suggest that it is worth US$7–23 billion annually (Nellemann et al. 2016). This figure is based on the best sources and criminal intelligence from INTERPOL, but the lack of statistics means there is still much uncertainty. In addition, forestry crimes including corporate crimes and illegal logging account for an estimated US$51–152 billion; and illegal fisheries an estimated US$11–24 billion (Nellemann et al. 2016). However, the UN Office on Drugs and Crime (UNODC) (2016) notes: "it is nearly impossible to give an accurate and consistent estimate of the criminal revenues generated by wildlife trafficking". Regardless of its total value, illegal wildlife trade (and related issues such as illegal, unregulated and unreported fishing) place wildlife stocks at risk by avoiding regulation designed to achieve sustainability. They also deprive governments of revenue streams derived from the use of wildlife (such as taxes, tariffs and payments for permits), impact on the well-being of people who depend on wild resources and provide income to criminals who continue to exploit wildlife. Businesses that rely on wild plants and animals have sought to encourage sustainable wildlife management through certification, and traceability along supply paths to ensure goods are legal. Box 18.5 describes how regulation of wildlife trade can support conservation of species and provide an income to the communities that live alongside.

Chapter 18: Conservation of Nature

> **Box 18.5**
> **Trade in Vicuña Fibre in South America's Andes Region**
>
> The vicuña, a small member of the camelid family, is one of the most valuable and highly prized sources of animal fibre on the international market. Luxury garments made from vicuña fibre are sold in exclusive fashion houses around the world; a scarf can sell for several thousand pounds. Once hunted to near extinction, the vicuña now thrives in the high-elevation puna grasslands of the Andes. The decision to grant usufructuary rights to communities to shear live vicuña and sell vicuña fibre increased their economic incentive to manage the species sustainably and protect it (Kasterine and Lichtenstein, 2018). As a result, vicuña populations have recovered,[526] and between 2007 and 2016, trade increased by 78% (by volume), and the export value in 2016 was approximately US$3.2 million per annum (Cooney, 2019). Vicuña have become an asset to some of the most isolated and poorest Andean rural communities, rather than being seen as a competitor for pasture with domestic livestock, thus reducing illegal killing and motivating communities to carry out anti-poaching and protection measures. Economic returns from vicuña fibre trade, regulated by CITES, have motivated more communities to start management, extending protection across a large area that central governments could not police effectively. Broader benefits to habitats from decreased grazing have also resulted. However, while this is generally seen as a conservation success story, the equitable distribution of benefits remains a challenge, and communities only receive a small share of the final product value. Efforts are being made to find ways to add value to the fibre that benefits communities.

There is an established link between wildlife trade and human disease through human contact with wild species exploited for trade and domestic use (Pavlin et al. 2009). Both illegal and legal wildlife harvest can involve transmission risk (Carreira et al. 2020). COVID-19 is the latest zoonotic disease that is linked to wildlife exploitation and trade (Wu et al. 2020). Curtailing the emergence of zoonotic diseases is complex and requires a holistic approach. Measures to prevent the unsustainable exploitation of wildlife are part of the solution. Other factors include better wildlife trade regulation in high risk species for zoonotic disease emergence. Improvements in the enforcement and regulation of wildlife conservation, animal welfare and public health regulations are also part of the solution (ZSL, 2020).

Genetic diversity is one of the important attributes of biodiversity which imparts resilience to ecosystems (Chapter 2 and Annex 2.2). With rapid anthropogenic change altering selection pressures on species, maintaining genetic diversity to enable adaption is particularly important. Maintaining genetic diversity keeps future options open to respond to unpredictable changes and reduces ecosystem vulnerability. As discussed in Chapter 5, keeping our options open in this way secures option value of ecosystems. For example, future crop security in agriculture and industry is reliant on maintaining plant genetic diversity (Jump, Marchant, and Peñuelas, 2009). Another example of keeping our options open is the development of seed banks. Seed banks store the living genetic diversity of plants, in the form of seeds, to enable future use. Various types of seed bank exist, to support different sectors and interventions, e.g. agriculture, forestry, restoration and conservation. They provide a secure and relatively low-cost method of conserving a large amount of genetic material in a relatively small space.

The Millennium Seed Bank, coordinated by the Royal Botanic Gardens, Kew, London, is a conservation seed bank that aims to protect wild plant diversity. The most diverse wild-species seed bank in the world, it is associated with a global network of scientists, botanists, technicians, educators and collectors who form the Millennium Seed Bank Partnership, covering 100 countries and territories. The partnership

---

[526] The total South American vicuña population is between 473,297 and 527,691, i.e. ca. 500,494 animals, with 46% in Perú, 29% in Bolivia, 21% in Argentina, 3% in Chile, and 2% in Ecuador. (IUCN Red List, 2019).

works to build and improve seed banks around the world, in order to conserve species in their countries of origin; duplicate seeds are sent to the Millennium Seed Bank for safety. To date, 16% of global plant species diversity is conserved by the Millennium Seed Bank Partnership. These species now have a reduced threat of extinction and are held as a significant resource and insurance against global biodiversity loss; they also provide a repository of genetic material for research, breeding and habitat restoration.

The Adapting Agriculture to Climate Change Project was a global initiative covering 24 countries that focused on the seeds of wild relatives of 29 of the world's most important crop species (Castañeda-Álvarez et al. 2016). Participants in this project have conserved 242 taxa of crop wild relatives. These collections will be used to identify traits of value in crop breeding, such as tolerance of heat, drought, salinity and waterlogging, resistance to pests and diseases, resistance to root rot, and yield.

18.5 Conservation Planning and Evaluation

We demand far more from our conservation approaches now than the conservation of biodiversity alone. In an increasingly crowded world, they need to be as effective as possible at delivering biodiversity gains, support ecosystem functions and processes as well as ecosystem services, and may also be asked to provide jobs and livelihoods (see Chapter 16 for more on this). Conservation planning and modelling can help to understand this complex set of requirements, select optimal places for conservation, and allocate resources appropriately. For example, the Grassland Programme in South Africa used systematic conservation planning to identify priority areas for ecosystem services and biodiversity conservation, and to combine them into one plan to focus conservation effort (Egoh et al. 2011). Evaluation of conservation and restoration interventions and policies is not yet common practice but helps to design effective programmes and use scarce resources efficiently (Baylis et al. 2016). Evaluation can help provide evidence to improve the current intervention and future policy making; and help to understand what is working, for whom, where and when (HM Treasury, 2020). Impacts of Protected Areas in Brazil, Costa Rica, Indonesia and Thailand were studied in relation to their contribution to carbon storage in forests (Ferraro et al. 2015). Protected Areas stored an additional 1,000 Mt of CO_2 and delivered ecosystem services worth at least US$5 billion compared to unprotected areas. Impacts on carbon are associated with poverty exacerbation in some settings but with poverty reduction in others, and these were identified spatially. Combining evaluation with ecosystem service modelling, valuation and conservation planning can help achieve multiple demands as effectively as possible, improve their efficacy and maximise our inclusive wealth.

18.6 Multilateral Environmental Agreements

As noted in Chapter 8, one of the particular difficulties with the conservation and restoration of Nature is how to deal with the externalities that arise from open-access resources and the global commons. Issues such as climate change and biodiversity loss in global public goods such as the high seas cannot be addressed by a single nation. Degradation of natural assets unilaterally can have implications well beyond a nation's borders. Effective global conservation and restoration action therefore requires international efforts, transboundary measures and international co-operation. Under these circumstances, Multilateral Environmental Agreements (MEAs) are a tool through which shared common purpose has been expressed. For example, the United Nations Convention on the Law of the Sea (UNCLOS) seeks to address conservation and sustainable use of marine biodiversity beyond national jurisdictions (Box 18.6). Some MEAs take the form of environmental protocols, which are a type of treaty bound in international law.

Chapter 18: Conservation of Nature

Criticisms of MEAs include the difficulty of enforcing them, their benefits for non-participating countries, that scientific uncertainty can be used to block action, and the lack of monitoring (Kanie, 2018; Mitchell et al. 2020). Public opinion and interest in the effectiveness of MEAs and appropriate resourcing have been brought into the mainstream by climate change activists.

Public pressure to ensure agreements are adhered to has been holding governments to account for unfulfilled pledges. The UN Framework Convention on Climate Change, the Convention on Biological Diversity and the Convention on International Trade in Endangered Species are MEAs that are particularly relevant to biodiversity loss and ecosystem degradation, among many others.[527] MEAs can provide helpful international frameworks with aims and goals that other regional, national and local governance bodies can support and deliver. They can be an important part of a polycentric governance approach, where purposes align to deliver shared goals.

Box 18.6
Conservation of Biodiversity in the High Seas

Marine areas beyond national jurisdiction (ABNJ), or high seas, are international waters outside the 200 nautical mile limits of national jurisdiction. They make up 61% of the oceans. Despite their isolation, human activities have reached even into the deepest areas of the seas, as evident from plastic bags found in the Mariana Trench (Morelle, 2019), and human impacts on the marine environment are increasing (Halpern et al. 2019). Currently ABNJs are collectively managed by countries under the United Nations Convention on the Law of the Sea (UNCLOS) (see also Chapter 8). Concerns of ongoing declines in the ocean has resulted in the development of an international legally binding instrument for the conservation and sustainable use of biodiversity beyond national jurisdiction: the United Nations Convention on the Law of the Sea (UNCLOS) (UNGA, 2015; Wright et al. 2015). Negotiations have been focused around four elements: marine genetic resources, including benefits sharing; area-based management tools, including Marine Protected Areas; environmental impact assessments; and capacity building and the transfer of marine technology (UNGA, 2015). UNCLOS intends to improve co-operation and co-ordination for the conservation and sustainable use of marine biodiversity. Challenges remain around mandate, responsibilities and enforcement (O'Leary et al. 2020).

---

[527] The International Environmental Agreements Database lists over 1,300 multilateral environmental agreements, over 2,200 bilateral environmental agreements and 250 other environmental agreements (Mitchell, 2020).

Chapter 19
Restoration of Nature

Introduction

Restoration seeks to assist Nature recover from a degraded, damaged – even dilapidated – state. Potential benefits from restoring our ecosystems have been discussed earlier in the Review (see in particular Chapter 2). Restored ecosystems are more effective at providing regulating services, which have been lost in many places due to overharvesting of provisioning goods (Chapters 2 and 16). Restoration can make a significant contribution to the mitigation of, and adaptation to, climate change. Successful restoration requires a wider understanding of the relationship between people, Nature and livelihoods, and effective institutions. It relies on sound science, clear objectives, community engagement, monitoring and evaluation (Wortley, Hero, and Howes, 2013; Perring et al. 2015; Gann et al. 2019).

While the concept of restoration is quite straight-forward to understand, in practice it raises many questions: What is the goal? Who are we restoring for? How can restoration be achieved? In some places, ecosystems may have been degrading for thousands of years (Chapter 0). If we wish to undertake restoration, are we aiming for a past reference point in history, a new target based on current use, or potential future needs, which may require creating new ecosystems?

Are we seeking biodiversity outcomes, ecosystem services outcomes, both or something else? The answers to these sorts of questions are context-specific and need careful consideration to plan and restore to the best of our ability. Often conservation and restoration go hand in hand with both approaches combined to maintain and improve our natural assets.

Much of global biodiversity and many of our ecosystems lie outside Protected Areas, and the integration of biodiversity into multifunctional lands and seas is a vital but challenging activity. The ongoing extraction of provisioning goods beyond the regenerative capacity of the biosphere has led to ecosystem degradation and traded off regulating services with provisioning goods (Chapters 4 and 16). Restoration can redress the decline in the quantity and quality of our stock of natural assets, improve ecosystem resilience and increase the potential supply of regulating services. This chapter looks at some useful approaches to restoration (with some relevance to conservation also): rewilding, Nature-based Solutions (NbS), sustainable production, removing invasive non-native species and improved spatial planning.

19.1 The Role of Ecosystem Restoration to Improve and Increase Our Stocks

Evidence on the efficacy of restoration is mixed: some projects have been successful, others less so. Different ecosystems, species and associated ecosystem services take different amounts of time to recover. As noted in Chapters 3 and 5, ecosystems that have undergone regime shifts to new stable states may never return to their former state due to hysteresis. A meta-analysis of 89 restoration assessments worldwide showed that restoration had increased biodiversity and ecosystem services by 44% and 25% respectively (Benayas et al. 2009), although not to a level seen in intact ecosystems. Significant recovery debt[528] in recovering and restored ecosystems has been documented in a worldwide study of over 3,000 plots that found deficits in species abundance, species diversity, carbon

---

[528] Recovery debt is the interim reduction of biodiversity and biogeochemical functions occurring during ecosystem recovery. This debt affects the quantity and quality of ecosystem services provided by the recovering systems.

Chapter 19: Restoration of Nature

cycling and nitrogen cycling (Moreno-Mateos et al. 2017). In addition, a meta-analysis of 400 studies worldwide found that ecosystems that have suffered from large-scale disturbances, such as oil spills, agriculture and logging, rarely recover completely (Jones, K. et al. 2018). While restoration is important, these findings suggest that avoiding degradation should be a higher priority, as already suggested in Chapter 5. A recent study by Leclère et al. (2020) demonstrated that both ecological restoration and providing food for the growing population is possible but ambitious (Leclère et al. 2020), but a further analysis suggests that we need to act urgently if we are to succeed (Box 19.1).

Box 19.1
The Urgent Case for Conservation and Restoration

A recent analysis of an ensemble of land-use and biodiversity models found that it is possible to reverse the downward global terrestrial biodiversity trends caused by land-use change while providing food for the growing human population but it requires "unprecedented ambition" (Leclère et al. 2020). Seven scenarios were explored in the models to assess how future biodiversity would be affected by trends in habitat destruction and degradation. Their analysis suggests that actions to retain existing biodiversity, restore degraded ecosystems and implement landscape level conservation planning could shift biodiversity trends to be positive from the middle of the 21st century. Tackling the drivers of loss through reducing food waste, more plant-based human diets, and further sustainable intensification and trade that supports equitable global distribution of produce can help to enable affordable food provision and reduce the negative environmental impacts of the food system (Leclère et al. 2020).

Further analysis of three linked models (one land-use and two biodiversity models) found that immediate additional action, over and above current climate change and biodiversity policies, is required to maintain biodiversity intactness even at today's depleted levels (Vivid Economics and Natural History Museum, forthcoming). This analysis focused on the outcome of restoration through reforestation to generate positive outcomes for terrestrial biodiversity. Without additional action, more endemic species will go extinct in the coming 30 years than appear to have died out in the entire period 850–1850 CE. If this action is delayed by just 10 years, the costs required for incentives for reforestation double and the scale of change required is so large, that it is probably infeasible to achieve.

By acting now, the cumulative social cost[529] of stabilising biodiversity intactness by 2050 is estimated to be US$7 trillion dollars (equivalent to around 8% of global GDP in 2019). Delaying action by 10 years would more than double the social cost, at approximately US$15 trillion (equivalent to around 17% of global GDP in 2019). The difference in costs between acting now and later is equivalent to 9% of global GDP (in 2019) – almost 40% of the GDP of the US in 2019. The total global cost of food and materials production from 2021 to 2050 is lower under immediate action and is 8% higher if action is delayed (Vivid Economics and Natural History Museum, forthcoming).

Increasing the evidence base for future restoration project design is important to maximise success. Currently, many restoration projects have inadequate objectives and poor implementation, monitoring and evaluation (Cooke et al. 2018; Cooke, Bennett, and Jones, 2019). For example, reported restoration gains in mangrove habitats in south-east Asia were often in single-species plantations, sometimes in seagrass beds (which themselves are threatened), and had mixed success with establishment (Lee et al. 2019). Having explicit objectives for both ecosystem services and biodiversity leads to more effective restoration than those simply hoping that restoration will deliver ecosystem services (Bullock et al. 2011). Clear objectives also

---

[529] This social cost represents the incentives required to achieve sufficient land-use change to reach the necessary levels of biodiversity intactness.

enable overt consideration of synergies and trade-offs, and have been found to inform better project design. As our understanding of the importance of ecosystem functions and processes has increased (Chapter 2), there has been an increase in their explicit inclusion in restoration interventions (Wortley, Hero, and Howes, 2013; Perring et al. 2015; Kollmann et al. 2016). As noted in Chapter 2, restoration does not always require active interventions, and can be successfully achieved through natural regeneration if we give Nature time and space to recover.

Restoration is as much about people as it is about Nature. As we have seen in Chapter 8 with regard to governance of common pool resources, understanding the context and the relationship people have with Nature is important for successful restoration outcomes and improvements to human well-being. In addition, community and civil engagement is essential for success (Chapter 6). For example, the re-introduction of the Hima grazing system in northern Jordan was strongly based on community participation to ensure restoration and poverty improvements (Box 19.2). While it has been recognised that the inclusion of social considerations and economics is essential for success in restoration, there is still limited evidence of realised social and economic outcomes or impacts from restoration to inform future restoration design (Aronson et al. 2010; Wortley, Hero, and Howes, 2013; Perring et al. 2015).

Box 19.2

Restoration of Drylands and Rangelands in the Zarqa River Basin, Jordan

The Zarqa River Basin in northern Jordan is home to approximately half the Jordanian population. It has been subject to severe environmental degradation, resulting in high levels of poverty and unemployment. Traditional land management had shifted to high-intensity agriculture due to contested land tenure, departure of tribal land ownership to private ownership and the introduction of government subsidies for dry-season cropping. The change resulted in overexploitation of natural resources, such as water, in a region already subject to desertification and unpredictable rainfall. The Hima grazing system that had been used by the Bedouin people involved setting aside heavily grazed areas seasonally, to allow recovery. The Bedouin migrated through vast areas and across borders, to enable ecosystem regeneration. Sites were governed by villages or clans, through consensus. Now Bedouins live in Jordan permanently, but the revival of the Hima system was proposed as an opportunity for greater community cohesion and sustainable land management to support livelihoods and reduce poverty (Haddad, 2014).

The IUCN, in collaboration with the Ministry of Agriculture and the Arab Women Organization (AWO) in Jordan, implemented a four-year project to restore and manage dryland and rangeland ecosystems in the Zarqa River Basin as a Nature-based Solution to address poverty reduction and biodiversity protection (IUCN, 2012). Through participatory approaches, bringing together national, regional and community groups, objectives for location-specific, long-term land management visions and strategies were developed to combat desertification. A small-scale Hima system was implemented, alongside the transfer of management rights to communities in the project area, to enable sustainable grazing and associated enterprises, such as collection of indigenous medicinal herbs by local women. As a result, the sites have improved environmentally and socially: tribal conflicts over natural resources have reduced, grazing is better managed, and biodiversity has revived (IUCN, 2012). The Hima system has now been incorporated as the basic approach in the governance of Jordan's rangelands by the Ministry of Agriculture (Cohen-Shacham et al. 2016).

Restoration of the marine environment has focused both on the reduction of anthropogenic pressures; and direct habitat management particularly concerning seagrass beds, coral reefs, saltmarshes, oyster reefs and mangrove forests (Box 19.3) (Lester et al. 2020). Pressures have been reduced through

Chapter 19: Restoration of Nature

sustainable management of fisheries, reductions in hunting, improvements in water quality, mitigation of climate change, and designation of Marine Protected Areas. Evidence from previous recovery rates suggests that substantial recovery of abundance, structure and function of marine life could occur by 2050 if anthropogenic pressures, including climate change, are addressed (Duarte et al. 2020a).

Box 19.3
Coral Reef Restoration

Coral reefs have been declining for centuries due to over-fishing and pollution. Now ocean warming and acidification, brought on by climate change, are compounding the impacts and exerting huge pressure on these ecosystems (Hughes et al. 2017; Duarte et al. 2020a). There has been some success with coral reef restoration (for example Pulau Badi Island in Indonesia; Williams et al. 2019), but many attempts have been expensive, slow and difficult to scale up, and only tens of hectares have been restored so far (Bayraktarov et al. 2015; Duarte et al. 2020a). Coral reefs are not only important for their biodiversity, but provide essential ecosystem services, such as food, coastal protection, recreation for coastal communities and are a big draw for tourism. Solutions for reef restoration need to address both the science and socio-ecological systems associated with the reefs. Researchers are identifying corals that are tolerant of high temperatures and acidity and considering innovative techniques, such as managed selection, genetic manipulation, managed relocations, and algal symbiont manipulation, but many of these techniques are in their infancy and are unproven with respect to reef restoration (National Academies of Sciences Engineering and Medicine, 2019). Alongside the science, as noted in Chapters 6–9, community and civil society engagement, as well as polycentric groups of institutions are required to enable and coordinate action both locally and globally, to address the range of anthropocentric drivers and allow the persistence of coral reefs into the future (Hughes et al. 2017).

Agriculture and food production drive most land-use change (Chapter 16). While a case has been made for making space for biodiverse ecosystems through 'land sparing' (Phalan et al. 2011, 2016; Balmford, Green, and Phalan, 2015), regions with a long history of agriculture focus more on 'land sharing' as a pragmatic approach to maintaining biodiversity in multifunctional ecosystems (Section 19.4). Restoration of agricultural systems to enable them to support biodiversity and multiple ecosystem services is an essential part of the global portfolio of natural assets and part of the conservation spectrum (Chapter 18).

The United Nations (UN) General Assembly adopted a resolution in 2019 that proclaimed 2021–2030 as the Decade of Ecosystem Restoration (UNGA, 2019). This aims to focus attention on the substantial scaling up of restoration efforts required to recover the biosphere. In particular, the initiative seeks to: share knowledge about what works; restore ecosystems through extensive and proactive restoration; create links between business and restoration for sustainable production; and increase the actors involved. This ambitious agenda relies on the critical components identified in this Review around balancing the supply of, and demands we make on, the biosphere and the choices we have as citizens to create a sustainable future.

19.2 Rewilding

Rewilding is a broad approach to restoration that seeks to recover natural processes without predetermined targets other than the restoration of functioning ecosystems. There is no single definition of rewilding. Rewilding can be very large-scale, but can also be applied on individual farms. Concepts around rewilding are broad; different interpretations have developed with mixed success

(Lorimer et al. 2015; Nogués-Bravo et al. 2016). Some approaches to rewilding use baselines that are much further back in time than most conservation projects, such as the Pleistocene and Holocene.

One well-known example of rewilding is in the Netherlands, at the Oostvaardersplassen (Vera, 2009). The site was created as a polder in 1968 for industrial development in the 1970s, but was abandoned when the Dutch economy fell into recession; it is now a Nature Reserve and Ramsar Wetland. The Oostvaardersplassen consist of 6,000 ha of open water, marshland, wet and dry open grasslands and flowering communities with trees and shrubs. It is grazed by large mammals including Heck cattle, red deer and Konik ponies to keep the site open and the sward short for a range of wetland birds. The ideas driving the restoration were to work at a large scale, use large grazing herbivores to encourage openness (mimicking Holocene ecosystems), and let the site self-determine the outcomes. Over 250 species of bird have been recorded and hundreds of them breed there, including Europe's largest raptor, the white-tailed eagle, which had disappeared from the Netherlands. The approach in the Oostvaardersplassen is not without controversy. The herbivores are not fed, so they are limited by food availability and die at times of scarcity – some people object to this on welfare grounds. There are disagreements about the degree of openness in Holocene ecosystems, and hence the role of grazing. And allowing the site to develop naturally has meant that some species have become less common as time has gone on. Nevertheless, it has shown what can be achieved by allowing ecosystems to regenerate with minimal intervention (Balmford, 2012; Vera, 2009). Natural regeneration has increasingly been shown to be a successful and cost-effective approach to restoration (Box 19.4).

Other examples of rewilding include: the reintroduction of wolves into Yellowstone National Park, USA, to help manage increasing populations of elk; introduction of tortoises to Mauritius and the Galapagos Islands to restore historic vegetation; and the development of the Pleistocene Park in Siberia, which seeks to mimic the mammoth steppe ecosystem through reintroductions of large herbivores (Lorimer et al. 2015). In Britain, the Knepp Castle Estate of 1,400 ha converted an unprofitable farm to wildlands focused on natural processes with free-roaming grazing animals. The diversification of spin off enterprises has made the site profitable and the site has had significant biodiversity gains (Tree, 2017).

Rewilding has the potential to create novel ecosystems. For example, the introduction of African elephants to the Australian outback has been suggested to control invasive species that are found in the elephants' natural geographic range (Bowman, 2012), although the introduction of non-native species has its risks (see section on invasive non-native species). If we accept that some ecosystems cannot be restored to their previous state, rewilding approaches may enable us to create sustainable functioning ecosystems which do not mimic a particular reference state, but nevertheless support biodiversity and mitigate against the impacts of climate change.

Box 19.4
Natural Regeneration or Active Restoration

A meta-analysis of 133 studies of active versus natural regeneration in tropical forests showed that natural regeneration surpassed active regeneration in terms of plant, bird and invertebrate species richness and abundance, and for measures of vegetation structure in this habitat (Crouzeilles et al. 2017).

A spatial analysis of cost-effectiveness of dryland forest restoration for multiple ecosystem services found that natural regeneration provided positive net social benefits, worth between US$1 million and US$42 million, and a benefit-cost ratio of above four. Active restoration was less cost-effective due to the additional labour, materials and opportunity costs from lower agricultural production. In all areas, carbon sequestration was estimated to provide the largest economic benefit after restoration (Birch et al. 2010).

Chapter 19: Restoration of Nature

19.3 Nature-Based Solutions

Actions to protect, sustainably manage and restore natural or modified ecosystems while simultaneously providing benefits for human well-being and biodiversity, have been referred to as 'Nature-based Solutions' (NbS) (Cohen-Shacham et al. 2016). NbS can contribute to a wide range of ecosystem services, through the restoration of biodiversity and ecosystems, if implemented with these goals in mind (Seddon et al. 2020) (Figure 19.1). In contrast, technological solutions often provide only one benefit. For example, coastal defence can be achieved by restoring or creating biodiverse intertidal mudflats and sandbanks, or by the technological solution of building sea walls (Box 19.6; Fitter, 2013; Cohen-Shacham et al. 2016; Keesstra et al. 2018). There is potential for conflict between different objectives, and risks remain around simplistic single-objective approaches to NbS, which could result in perverse outcomes and trade-offs, and are less resilient in the long term. For example, unsophisticated tree-planting approaches to address climate change can result in afforestation of monocultures of non-native tree species which, over the long term, may not be resilient, may compromise carbon storage, and may deliver little or nothing for biodiversity. To ensure that NbS deliver biodiversity gains as well as ecosystem services requires careful planning (Box 19.5).

Box 19.5

Restoration of Forests to Store Carbon – Good or Bad for Biodiversity?

The IPCC suggests that increasing the total area of the world's forests, woodlands and woody savannahs could store roughly a quarter of atmospheric carbon necessary to limit global warming to 1.5°C (IPCC 2018). To do so would mean adding an additional 24 million ha of forest every year until 2030 (Lewis et al. 2019). Many countries are responding with restoration plans, but 45% of all commitments involve planting vast monocultures of trees. Reforestation of Eucalyptus and Acacia trees in plantations only offers a temporary solution to carbon storage, as once the trees are harvested, the carbon is released again by the decomposition of plantation waste and products (predominantly paper and woodchip boards).

Lewis et al. (2019) calculated carbon uptake under four restoration scenarios that were pledged by 43 countries under the Bonn challenge, which seeks to restore 350 million ha of forest by 2030. They found that natural forests were six times better than agroforestry and 40 times better than plantations at storing carbon (Lewis et al. 2019). Furthermore, these have greater associated biodiversity and ecosystem services. The pledged mix of natural forest restoration, plantation and agroforestry would sequester only a third of the carbon sequestered by a natural forest restoration scenario. The authors recommended four ways to increase the potential for carbon sequestration by forests: increase the proportion of land restored to forests; prioritise natural regeneration in the Tropics; target degraded forests and partly wooded areas for regeneration; and protect natural forests once they are restored (Lewis et al. 2019).

NbS build on extensive research on ecosystem service management and on the principles of the ecosystem approach advocated by the Convention of Biological Diversity.[530] Measures include catchment restoration and regeneration of wetlands to reduce flood risk (Dixon et al. 2016); restoration of savannahs to support soil carbon, protect water resources (Honda and Durigan, 2016) and reduce the risk of serious fires (Buisson et al. 2019); and restoration of mangroves for coastal protection and fisheries (Lee et al. 2019). For other examples of NbS, see Box 19.6.

---

[530] The ecosystem approach is a strategy for the integrated management of land, water and living resources that promotes conservation and sustainable use in an equitable way, enabled by a suite of principles (Convention on Biological Diversity, 1995, 1998a, b).

Chapter 19: Restoration of Nature

NbS are often closely associated with places and the local communities who manage and use the sites, and benefit from some of the resulting ecosystem services. In common with the conservation of our natural assets, community and civil society engagement and good governance is essential to bring together the breadth of experiences and knowledge across institutions (Chapters 6 and 8; Hughes et al. 2017; Seddon et al. 2020).

Figure 19.1 The Benefits of Nature-based Solutions

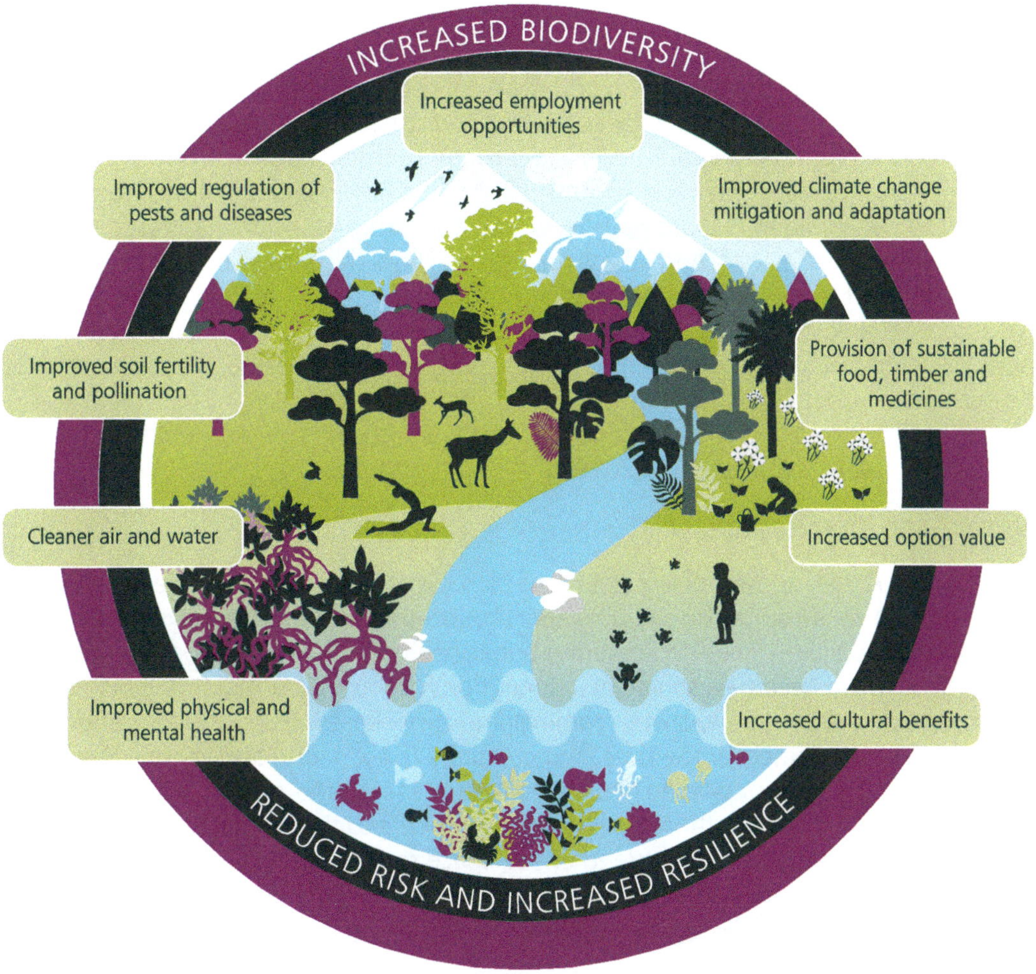

Box 19.6
Examples of Nature-Based Solutions

MarineVegetated Habitats for Coastal Protection

Conserving and restoring vegetated coastal habitats can help mitigate climate change and reduce the risks from sea-level rise. Highly productive seagrasses, salt marshes and mangroves enhance sediment accretion, build carbon stores and sequester carbon. It is estimated that marine vegetated habitats (seagrasses, salt-marshes, macroalgae and mangroves) occupy 0.2% of the ocean surface, but contribute 50% of carbon burial in marine sediments (Duarte et al. 2013). The physical structure of the submerged canopies reduces water flow and turbulence, and semi-submerged vegetation affects flow speed, reduces wave action and encourages sediment deposition, all of which contribute to coastal protection (Duarte et al. 2013). For example,

The Economics of Biodiversity: The Dasgupta Review

managed realignment on the Humber estuary, the Blackwater estuary and at Medmerry on the east coast of England created extensive intertidal mudflats, saltmarsh and freshwater habitats, to reduce the risk of coastal flooding and storm surges (Luisetti et al. 2011; Burgess, Kilkie, and Callaway, 2016).

Wallasea Island Restoration in the UK

The UK coast is changing due to accelerating sea-level rise and coastal erosion, resulting in loss of intertidal habitats and increasing pressure on existing flood defences. The Essex coast in south-east England has lost 91% of its intertidal salt marsh over the past 400 years due to past land claims for agriculture and increasing coastal erosion and sea-level rise. Wallasea Island was enclosed in sea walls and used for grazing marsh until it was drained and converted to arable land in the 1930s. The site is in an important estuary for biodiversity, close to the Thames Gateway – one of Europe's largest economic regeneration areas. Between 2009 and 2016, the Royal Society for the Protection of Birds (RSPB) undertook a managed realignment on the site to restore intertidal habitat, creating more space for sea water in the estuary and access for visitors. More than 3 million metric tonnes of earth were brought by boat from the tunnels of a large rail infrastructure project in London to help create a 115 ha intertidal area of saltmarsh, islands and mudflats. The reserve covers more than 740 ha, two-thirds of which have now been transformed from arable farmland to saltmarsh, mudflats, lagoons and grazing marsh. Visitor numbers for 12 months to end March 2017 were 21,000, representing a 40% increase on the previous year. Wallasea Island is now a wildlife-rich habitat and a popular site for people to visit (RSPB, 2020).

Cheonggyecheon River Restoration

Seoul, Republic of Korea, undertook an ambitious plan to restore the Cheonggyecheon River, which runs from east to west through the city. The area around the river was densely populated, and the river became highly polluted, thus endangering public health. In the late 1950s, it was covered with a four-lane road, which became a source of air pollution for the local population who were twice as likely as people living elsewhere to suffer from respiratory diseases. The river restoration project removed the highway to uncover the river, and created a 5.4 km long section of green and blue infrastructure with terraces for wildlife and pathways for access and recreation; around 20 million people visit each year. The restoration provides many benefits to the inhabitants of Seoul, including flood protection, reduced urban heat-island effect, space for recreation and recuperation, tourism, reduced particulate pollution, and areas for wildlife and biodiversity. Species richness along the restored river increased by 639% between 2003 and 2008 (Asian Development Bank, 2016; Ryu and Kwon, 2016).

Economic evaluation of NbS has rarely been conducted (the Ganges case study in Chapter 13 is an exception). But restored rivers have been shown to provide large benefits. Eight restored rivers in Europe, for instance, were found to provide net social economic benefits over unrestored rivers, of €1,400 ± 600 ha^{-1} yr^{-1} associated with an increase in cultural and regulating services, while provisioning goods remained the same (Vermaat et al. 2016). Analysis of the flood alleviation potential of ecosystem-based adaptation of two river catchments in Fiji found that NbS were more cost-effective than hard infrastructure options. Planting riparian buffer vegetation was the most cost-effective option, yielding benefit-cost ratios of between 2.8 and 21.6. Upland afforestation provided the greatest benefits overall, yielding net present values between 12.7 and 101.8 million Fijian dollars (approximately US$6.1–48.9 million). Of the hard infrastructure options, river dredging provided the greatest benefits, but costs were high relative to the benefits, and the benefits were only accrued in part of the catchment downstream (Daigneault, Brown, and Gawith, 2016).

An analysis of the benefits of ecosystem restoration considered the capital investment and maintenance costs (94 studies) compared to the monetary value of ecosystem services provided (225 studies). The study found that benefit cost ratios ranged from 0.5:1 (coral reefs and coastal systems, worst-case scenario) to as high as 35:1 (grasslands, best-case scenario) and reported that most of the studies analysed provided net benefits (De Groot et al. 2013).

Mangroves provide large social benefits in terms of coastal defence compared to artificial offshore breakwater structures. A synthesis of the costs and benefits of Nature-based coastal defence for 52 projects showed that salt marshes and mangroves can be two to five times cheaper than submerged breakwaters designed to reduce wave heights by up to half a metre. NbS provide other benefits, including reduced damage from storm events, reduced erosion, and reduced costs of coastal defence (Narayan et al. 2016).

There is evidence that NbS can provide large social benefits by reducing coastal risks due to climate change and development. In the Gulf of Mexico, USA, where such risks are increasing, the cost-effectiveness of adaptation measures were compared, including oyster reef or wetland restoration, grey infrastructure, and policy measures such as home elevation (Reguero et al. 2018). Flooding costs were predicted to be US$134–176.6 billion in 2030, with annual costs expected to double by 2050 due to increasing risks. The NbS compared favourably with engineered solutions; average benefit-cost ratios for NbS were above 3.5. Cost-effective coastal adaptation measures could prevent US$57–101 billion in losses; NbS could avert more than US$50 billion of these costs (Reguero et al. 2018).

Changes in net wealth due to the storm protection benefits provided by coastal and estuarine restoration have been calculated (Barbier, 2016). The creation of approximately 100km2 of coastal marsh wetlands in 2012 to 2031 was proposed by the 2012 Master Plan for the Louisiana Coast, USA (Coastal Protection and Restoration Authority of Louisiana, 2012). Including the risk of restoration failure, this was calculated to result in a net wealth increase in real terms of around US$33 per person (2005 $) over 20 years, due to the protection benefits of the wetlands (Barbier, 2016). This net benefit in wealth was assessed using one ecosystem service; the inclusion of additional services would likely lead to greater positive impacts on wealth.

Restoration of Nature and NbS protects many existing jobs which are threatened by the increasing frequency and severity of natural disasters and impacts of climate change on productivity and working conditions (ILO, 2018). NbS can create a significant number of new jobs globally to support economic recovery and address biodiversity loss and climate change together (Box 19.7).

Box 19.7

Job Opportunities From Nature Conservation and Restoration

Compared to other sectors in the economy, investing in Nature may have higher employment returns. Investments in 'Nature-based solutions' (NbS) create jobs that typically have low training and education requirements, are fast to establish and require relatively little produced capital for each worker. This means on average, for every US$1 million invested in NbS, close to 40 jobs are created, which is equivalent to around 10 times the job creation rate of investments in fossil fuels (Levy, Brandon, and Studart, 2020). ILO (2012; 2020) corroborate these estimates. They have found that green infrastructure programmes (e.g. reforestation, watershed management and agricultural land improvement) are likely to create jobs. These studies all note the large economic multipliers of these types of investments in Nature, with evidence that 70% spent in local economies is invested or spent locally.

The World Economic Forum (2020b) conducted a literature review of business opportunities from investing in Nature. The authors attempted to quantify job creation by estimating annual savings or

Chapter 19: Restoration of Nature

revenue generated from Nature investments, based on regional labour productivity rates. They estimate that ecosystem restoration and avoided land and ocean use expansion could deliver 11 million more jobs by 2030 through opportunities such as ecotourism, sustainable forestry management and NbS for mitigating climate change. They also estimate that using Nature as 'infrastructure' – which includes incorporating existing or restored ecosystems, such as floodplains, wetlands and forests, into built environments to mitigate against extreme events, for flood control or water filtration – could create an additional 4 million jobs by 2030.

More immediately, and in the context of the post-COVID recovery, recent publications propose that natural capital investments (e.g. afforestation, expanding parkland, enhancing rural ecosystems) should have high priority (UNEP, 2020b), as they can be implemented quickly and meet social distancing norms (Hepburn et al. 2020). The publications report that many countries have 'shovel-ready' projects through current institutional structures, including through National Determined Contributions (NDCs), which are programmes set aside to achieve goals set out in international climate agreements.

19.4 Sustainable Production Landscapes and Seas

Agricultural land is the largest terrestrial biome covering around 30% of the ice-free planet and has driven environmental declines (Chapters 4 and 16; Ellis and Ramankutty, 2008; Ramankutty et al. 2008; IPBES, 2018;). In the marine environment at least a third of fish stocks are over-fished (FAO, 2018a). While monoculture systems can increase food production, this is at the expense of biodiversity, ecosystem functioning and other ecosystem services – particularly regulating services (Rasmussen et al. 2018). Restoration can shift monocultures and degraded lands and seas onto the conservation spectrum providing multiple ecosystem services (Chapter 18), balancing provisioning goods with regulating services and underpinning natural assets and biodiversity.

In Europe, concerns have been raised that increasing intensification of farming has reduced the capacity of rural areas to provide multiple ecosystem services to local communities whereas low-input, small-scale and traditional farming practices were associated with providing multiple benefits (Plieninger, Ramond, and Oteros-Rozas, 2016; Fischer, Meacham, and Queiroz, 2017). One policy response to this market failure has been to establish agri-environment schemes that offer incentives to farmers to adopt practices that support biodiversity and ecosystem services. Agri-environment schemes can be considered to be payments for ecosystems services (PES) in which governments are buying public goods and services, but governments rarely make payments based on ecosystem service provision. For example, European Union schemes are based on income foregone. There is a rich literature on the design, efficiency and effectiveness of these schemes and whether or not they lead to long-term sustainability of asset management (see Chapters 7 and 20 for a further discussion concerning PES, and also for examples, Kleijn and Sutherland, 2003; Kleijn et al. 2006; Ansell et al. 2016). Evidence for European agri-environment scheme effectiveness is mixed, with general increases in farmland biodiversity but with significant dependencies on aspects of scheme design, targeting, provision of advice and scale of funding (Batáry et al. 2015).

Box 19.8
Environmental Land Management in England

Approximately 70% of land in England is used for agriculture, and therefore a land-sharing approach is important to allow space for both agriculture and biodiversity (Defra, 2020b). Bateman et al. (2013) assessed the consequences of alternative land-use futures in the United Kingdom to 2060 (Bateman et al. 2013). They found that highly significant value increases can be obtained from targeted planning for a mix of potential ecosystem services and their values, not only market values of agricultural, whilst also conserving wild species diversity. A targeted biodiversity constraint only marginally reduced the ecosystem service and agricultural production gains. The findings suggest that decisions considering the market prices of agricultural production alone deprive society of many other ecosystem services, biodiversity and risks leaving the United Kingdom worse, rather than better, off. In contrast, an approach that considers ecosystem services yields net benefits in almost all areas, with the largest gains in areas of high population.

A new policy for land-use decision-making is under development in England and seeks to recognise and reward the provision and value of ecosystem services from sustainable land management. The Environmental Land Management (ELM) scheme is an ambitious new agri- environment scheme, akin to a national Payment for Ecosystem Services (PES) mechanism (Defra, 2020a, 2020b). It aims to pay farmers public money for delivering environmental public goods and services, rather than other approaches that have used income foregone or land area as a basis for payment. Through ELM, farmers, foresters and other land managers will be paid for deliver ecosystem goods and services. The public goods which ELM will pay for include primarily: clean air, clean and plentiful water, thriving plants and wildlife, protection from and mitigation of environmental hazards, beauty, heritage and engagement with the environment, mitigation of and adaption to climate change. The planned reforms under ELM support the delivery of the objectives set out in the UK government's 25 Year Environment Plan, and commitment to net-zero emissions by 2050, while also contributing to a productive, competitive and resilient farming sector (Defra, 2018).

It is estimated that 3.2 billion people live in areas that are degraded through soil erosion, salinisation, compaction and contamination, organic matter decline, forest fires and overgrazing (Stoett et al. 2019). Restoring these degraded areas to make them more productive in a sustainable way would have multiple positive benefits. For example, the South African Grasslands Programme sought to balance biodiversity conservation and development in an area used for agricultural production. This was done through careful planning, knowledge exchange and support for communities including a biodiversity stewardship scheme and development of community forests (SANBI, 2021).

There are long-standing forms of sustainable land management such as shifting cultivation and crop rotations (to maintain and increase soil fertility and reduce pests). Sustainable intensification seeks to use contemporary methods to increase crop yields. For example, maintaining soil fertility, improving water use efficiency and reducing chemical inputs can be achieved through zero tillage or intercropping with two or more crops. Other approaches include plant breeding for temperature and pest tolerance, creating bio-controls for crop pests and pathogens, and reducing fossil fuel use in agriculture (see Chapter 16 for further detailed examples).

Other marine and land-use systems such as unfed aquaculture and agroforestry can support biodiversity and ecosystem services as well as the provision of food (Chapter 16). In the marine environment, a number of restoration approaches seek to maintain sustainable ocean systems for the continued support of the economy and people. Marine and coastal spatial planning that explicitly considers ecosystem service provision and trade-offs is one example (Box 19.10); protection and restoration of mangroves and other habitats for coastal protection is another.

19.5 Invasive Non-Native Species

Non-native species or alien species are organisms that are accidentally or intentionally introduced by humans to areas outside their natural geographic range. Introductions are the result of increasing globalisation and associated travel, tourism and trade. Some non-native species become invasive and

Chapter 19: Restoration of Nature

have significant impacts on native species, ecosystems and their functions, potentially causing regime shifts. Marine environments are particularly susceptible to invasive species, through aquaculture and shipping (Molnar et al. 2008). For example, the lionfish spread rapidly from the Indian and Pacific Oceans through to the Atlantic Ocean and the Mediterranean Sea, causing declines in native reef fish populations and danger to humans from their venomous spines (Green et al. 2012; Andradi-Brown, 2019). In freshwater systems, the water hyacinth (*Eichhornia crassipes*), a South American native, is now found in more than 50 countries on five continents. It is an extremely fast-growing floating plant that prevents sunlight and oxygen from reaching the water column and submerged plants. Invasive non-native species (INNS) are major drivers of biodiversity loss at a global level (Díaz et al. 2019), and a number of international agreements seek to manage their spread.[531]

Estimates of costs associated with INNS are substantial, and affect a range of sectors, including agriculture, forestry and fisheries. In the USA, losses were estimated to be almost US$120 billion per year (Pimentel, Zuniga, and Morrison, 2005). In the European Union, annual costs of over €12 billion were estimated due to the impact of INNS (Kettunen et al. 2008). The negative effects of INNS are exacerbated by climate change, pollution, habitat loss and human-induced disturbance. Increasing dominance of a few invasive species increases global homogenisation of biodiversity, reducing local diversity and distinctiveness. INNS can directly affect human health: infectious diseases can be caused by INNS that are imported by travellers or vectored by non-native species. INNS can exacerbate poverty, when they affect livelihoods associated with agriculture, forestry and fisheries. Crops such as maize, rice, cassava, sorghum and millet, which people in many regions rely on, are all affected by INNS as pests or pathogens (Perrings, 2005). In East Africa, the economic impacts of five major INNS on maize crops for small-holders was calculated as combined annual losses of US$0.9–1.1 billion, with future annual losses (over the next 5–10 years) of US$1.0–1.2 billion in six countries (Pratt, Constantine, and Murphy, 2017).

Some INNS have been successfully eradicated, with substantial benefits to native species, particularly on islands (Jones et al. 2016). For example, rat eradication was undertaken on Great Bird Island, one of Antigua's offshore islands, which is home to the world's rarest snake, the critically endangered Antiguan racer (*Alsophis antiguae*). The eradication resulted in the population of this snake increasing 20 times on four islands (Lawrence and Daltry 2015). Holmes et al. (2019) identified 169 islands where eradication was feasible, could be initiated by 2020 or 2030, and would improve the survival of 9.4% of the most highly threatened terrestrial island vertebrates (Holmes et al. 2019). However, the numbers of INNS establishing themselves is increasing, suggesting that efforts to mitigate their spread are not keeping up with increasing globalisation (Seebens et al. 2017).

Box 19.9
Working for Water in South Africa

Since the arrival of European settlers to the Cape of South Africa in 1652, people have been introducing crop species, including trees, from around the world. Some of these plants have spread alarmingly across vast areas of South Africa including the fynbos, a unique low shrub habitat with over 6,000 endemic plants. Concerns that the INNS trees were using valuable water resources were first mooted in the 1920s by farmers, who blamed them for the rivers drying up (Balmford, 2012).

---

[531] These include the Convention on Biological Diversity Article 8(h) that states that "Each contracting Party shall, as far as possible and as appropriate, prevent the introduction of, control or eradicate those alien species which threaten ecosystems, habitats or species" as agreed by the Conference of the Parties (COP) (decision IV/1), in 1998 (Convention on Biological Diversity, 1998b). And the United Nations Convention on the Law of the Sea (UNCLOS) Article 196 requires Member States to take all measures necessary to prevent, reduce and control the intentional or accidental introduction of species (non-native or new) to a particular part of the marine environment, which may cause significant and harmful changes (United Nations, 1992).

> An increasing body of research since that time has shown that tree plantations do use up significant water resources to the detriment of fynbos and open grasslands. By 1994, data suggested that, if unchecked, INNS would reduce water supplies to the city of Cape Town by 30% (Le Maitre et al. 1996). Further work showed that more water could be delivered per unit cost by integrating INNS management into water supply infrastructure (van Wilgen, Cowling, and Burgers, 1996).
>
> Understanding the benefits of INNS management for water resources and biodiversity convinced the South African government to create the Working for Water programme. The programme employs unskilled, previously unemployed people from impoverished rural communities to cut and control 18 invasive tree plant species. By 2015, Working for Water had 300 clearing projects spread across all nine provinces in South Africa, employing over 25,000 people and had cleared 2.5 million ha (van Wilgen and Wannenburgh, 2016). Each of the projects is run locally. People are employed for up to 24 months, and are not just trained to handle chainsaws and chemicals, but also to use spreadsheets and assist with family planning; the programme provides people with transferable skills, raises awareness of HIV/AIDS, and provides health and community centres where possible (Balmford, 2012). There are two main criticisms of the programme: firstly, there is inadequate funding for monitoring and evaluation, so that the impact of the programme on water resources is not known; and secondly, resources have not been focused on areas most at risk from INNS, because these areas do not necessarily overlap with priority areas for employment (van Wilgen and Wannenburgh, 2016). These issues are difficult, but not unsurmountable, and Working for Water has demonstrated how ecosystem restoration, unemployment and poverty can be addressed together.

19.6 Bringing Natural Capital into Spatial Planning

Humans have influenced and changed many ecosystems around the world. Some severely affected ecosystems have been re-categorised as anthromes, reflecting the extent of human activity within them (Ellis et al. 2010). One way in which we can manage this influence and activity, including for conservation and restoration, is through careful land-use and marine spatial planning to balance economic, social and environmental trade-offs. Spatial planning aims to develop a long-term vision and framework for citizens that considers multiple scales, balancing competing demands and directing resource-allocation decisions (Albrechts, 2004). Spatial plans are usually legally binding instruments led by the public sector, but involving communities and stakeholders such as businesses.

Building consideration of Nature and ecosystem services into the planning framework alongside other concerns such as public health, water management, housing, economic growth, and climate change can lead to tensions, but it can also help resolve those tensions by facing them explicitly (Hislop, Scott, and Corbett, 2019). Spatial plans can help define human activity which is permissible (or not) in geographical areas and potentially clarifying property rights where they may be weak or contested. Spatial planning informed by natural capital offers huge opportunities to conserve and restore Nature, and to ensure that existing conservation and restoration is as effective as possible in an increasingly crowded world.

Spatially explicit modelling and conservation planning can inform option appraisal contributing to spatial planning, as well as sustainability assessment and policy analysis covered in Chapter 13, by exploring the consequences of different scenarios. Modelling approaches have been developed that incorporate biophysical data with environmental economic analyses (Bateman et al. 2013; Rieb et al. 2017; Nayak and Smith, 2019). The Integrated Valuation of Ecosystem Services and Tradeoffs (InVEST) model maps and values ecosystem services and goods from Nature (Kareiva et al. 2011; Sharp et al. 2018) enabling decision-makers to assess quantified trade-offs associated with alternative management choices and to identify areas where investment in natural capital can enhance ecosystem

Chapter 19: Restoration of Nature

service benefits for people. The Restoration Opportunities Optimization Tool (ROOT) was designed to help people make decisions around forest restoration, and was used, for example, to identify reforestation areas in Malawi based on ecosystem services and their benefits to people (Beatty et al. 2018). Other models have looked at the consequences of different policy actions or drivers of change. Bateman et al. (2016) used an integrated model of the economic consequences of climate change in the UK, assessing the impacts from climate change on farm gross margins, land use, water quality and recreation, at the scale of both individual farms and the catchment (Bateman et al. 2016).

Embedding the outcomes of these modelling approaches into spatial planning is not common practice, although it has been done (Box 19.9). For example, the Chinese government has used models to develop national ecological function zoning to conserve and restore natural capital (Annex 12.1) (Ouyang et al. 2019). A recent review of the application of spatial planning principles to marine ecosystem restoration found that they are rarely incorporated despite having the potential to improve restoration success and ecosystem service provision (Lester et al. 2020). The increasing need for large scale and costly restoration interventions is expected to drive further development of tools, data and integrated spatial planning approaches to increase natural capital stocks.

Box 19.10
Building Natural Capital into Spatial Planning

Sustainable Infrastructure Framework Applied to Andros Island, Bahamas

Awareness of the role that ecosystems play in shielding coastal communities from adverse climatic events has increased due to major catastrophic hurricanes in recent decades. However, the implementation of NbS as a protective mechanism against these disasters is still quite limited (Arkema et al. 2017). Andros Island, in the Bahamas, is a notable exception where the development of infrastructure services has incorporated natural capital and sustainability. This was based on the creation of a national cross-sectoral planning framework and the development of a Sustainable Development Master Plan specifically tailored for Andros, driven by a stakeholder-led process and informed by sound scientific advice. Lessons from this case are useful for Small Island Developing States aiming for similar processes in which natural assets are particularly important for sustainable development in the face of climate change.

Coastal Zone Management in Belize

Belize's eastern coast has marine and coastal ecosystems that are important for a diversity of endangered species, such as sea manatees. The coast is also home to 35% of the human population of Belize, and many people rely on fishing for their livelihoods. Each year, around 800,000 tourists come to snorkel and dive in the rich waters, and the coral reefs, mangroves and seagrass beds also provide the coast with protection from flooding and erosion. Various economic development proposals, for activities such as such as aquaculture, tourism and marine transportation, were threatening the ecosystems that underpin the economy. The government of Belize decided to undertake their first Integrated Coastal Zone Management (ICZM) plan to achieve conservation, restoration and development goals. The Belize Coastal Zone Management Authority and Institute worked with the Natural Capital Project in 2010 to use InVEST ecosystem service models to consider options and scenarios for zoning ocean and coastal activities. This process involved significant participatory approaches and iterative discussions of various model runs to reach a consensus for a preferred zoning scheme for the ICZM plan that helped facilitate finance for investments in the restoration and conservation of the marine and coastal ecosystems. The preferred plan led to greater returns from coastal protection and tourism than other scenarios that were focused on either only conservation or coastal development. The plan also reduced impacts on habitats and increased revenues from lobster fishing compared to the current management. Consideration of the impacts of alternatives on a wide range of services provided by marine and coastal ecosystems enabled the creation of an integrated plan; this would not have been possible via the separate evaluation of sectors (Verutes et al. 2017; Arkema, 2019).

Development control within the planning process can help to maintain or increase stocks of natural capital through a requirement for increased space for Nature to compensate for planning losses (i.e. 'net gain'). These 'biodiversity offsets' compensate for residual impacts on biodiversity of development projects, as a practical way to reconcile development with biodiversity conservation. Projects, programmes or developments are required to follow a mitigation hierarchy in which impacts should be avoided if possible, mitigated for if they cannot be avoided (through minimisation or on-site rehabilitation); lastly, any residual impacts should be compensated for. Impacts on biodiversity are measured in standardised ways identified by the scheme, and these measures are used to decide on the size, content and position of offsets.

Offsets typically seek 'no net loss' of biodiversity, but in some countries, net gain in biodiversity is required (for example, in the UK). Between 74 and 100 countries have laws or policies on biodiversity offsets (ten Kate, von Hase, and Maguire, 2018). A recent review by Bull and Strange (2018) identified 12,983 offset projects that covered over 153,679km2 in 37 countries, mostly emerging economies, particularly in South America. Most of the offsets (99.7%) were small, and were stimulated by regulatory requirements (Bull and Strange, 2018). The use of biodiversity offsets has been criticised, because offsets may not be effective, and because their use may provide a 'licence' to degrade the environment (Bull et al. 2013; Maron et al. 2015). The IUCN published a policy on biodiversity offsets to address some of these issues (IUCN, 2016a). If we are to improve and increase our stock of Nature via restoration, biodiversity offsets need to provide more than no net loss: net gain in biodiversity will be required to improve both Nature's quantity and quality.

Box 19.11
Biodiversity Offsets for a High-Speed Railway Line, Eiffage, France

The Bretagne-Pays de la Loire high-speed railway line is a new transport link 182 km long between the French cities of Le Mans and Rennes. Construction of the line took four years (2012 to 2016), but it took two decades of planning prior to construction to agree the plan, route, and mitigation hierarchy. The project used iterative environmental and technical studies to inform planning at an early stage. Impacts were avoided where possible, for example through route planning to avoid the most sensitive habitats. Unavoidable impacts were mitigated by undertaking measures to preserve the connectivity and ecological function of habitats, in particular the wetlands, ponds and watercourses. Targeted ecological engineering measures were used to create riverbanks for certain insects and artificial shelters for reptiles. The residual impacts were compensated for through a biodiversity offset scheme (Table 19.1).

Compensation measures were implemented on 242 sites representing 920 ha in total, in a 2 km band along the railway line. To ensure the sustainability of the measures, a long-term management and monitoring plan has been scheduled until 2036 (Bourge and Aubrat, 2018).

Chapter 19: Restoration of Nature

Table 19.1 Offset Areas Implemented for Each Biodiversity Habitat During Construction of the Bretagne-Pays de la Loire High-Speed Railway Line, France

Item	Effective offset implemented compared to the offset required
Bird habitats	+ 3%
Bat habitats	+ 25%
Insect habitats	=
Semi-aquatic mammal habitats	=
Fish habitats	=
Flora (habitats)	+ 200%
Wetlands	+3%
Ponds	+29%
Watercourses	+67%
Riparian linears	=
Spawning grounds	=

Source: Bourge and Aubrat (2018)

Habitat banking is an extension of biodiversity offsetting. In habitat banking, an organisation, landowner or private company restores, creates, enhances or conserves habitat to sell as credits to developers or permittees as compensation for their impacts elsewhere. Often the requirement to buy credits by the developer and the number of credits required is regulated through a process similar to the mitigation hierarchy applied in offsetting. Permittees are released of their obligation to produce compensatory mitigation by purchasing credits. The USA's wetland mitigation scheme in the 1970s was one of the first that used this concept to compensate for wetland losses (McKenney and Kiesecker, 2010). Habitat banking allows developers to avoid case-by-case responses, and can allow them to make meaningful contributions to bigger conservation projects. This is particularly pertinent for small developments, each of which may degrade the environment only in a small way, but collectively they may have a large cumulative impact. Buying credits exempts developers from providing offsets, thus reducing the burden on developers by shifting responsibility to providers (Carroll et al. 2008). The use of biodiversity credits can reduce time lags when restoring habitats or species, and may enable optimisation of habitat connectivity, by concentrating mitigation in large areas (Fox and Nino-Murci, 2005). A clear enforceable regulatory approach has been found to be needed to ensure the success of habitat banking, and certain issues, similar to those applying to biodiversity offsets, need to be addressed: monitoring, enforcement, equivalence, longevity and equity of outcomes (Santos et al. 2015).

Chapter 20
Finance for Sustainable Engagement with Nature

Introduction

Finance is an enabling asset that facilitates investments in capital assets, including natural assets, and plays a role in influencing both sides of the Impact Equation (Chapters 1 and 13 discussed the distinction between capital goods and enabling assets in more depth, respectively). In other words, finance plays a role in determining both the stock of natural capital and the extent of human demands on the biosphere.

On the supply side of the Impact Equation (Chapter 4), finance enables investment in conservation and restoration of ecosystems and their biodiversity (Chapters 18 and 19). This investment enhances our stock of natural capital and their regenerative rate (i.e. S and G(S)). Current global estimates of financial investments in natural capital suggest that they are small, both in absolute terms, ranging from around US$78 to US$143 billion per year (equivalent to around 0.1% of global nominal GDP), and relative to estimates of what is required to prevent further declines in the stock of natural assets (Deutz et al. 2020; OECD, 2020a; Seidl et al. 2020).

On the demand side of the Impact Equation, finance plays a role in influencing human demands through the channelling of financial flows to different real economy activities. Moreover, finance influences the efficiency of our use of the biosphere, by channelling flows to enable research and development. Chapter 8 highlighted how government subsidies are amplifying our demands on the biosphere and estimates of financial flows for activities that are harmful to natural assets significantly outstrip finance for investments in those assets. For example, recent estimates suggest that governments globally spend around US$500 billion per year on support that is potentially harmful to biodiversity (OECD, 2020a).[532] More broadly, combined public and private financial flows estimated to be harmful to natural assets significantly dwarf those flows devoted to enhancing these assets (Figure 20.1) (Deutz et al. 2020; OECD, 2020a; Seidl et al. 2020).

A sustained shift from Impact Inequality to Equality will, therefore, require financial flows both to increase our stock of natural assets, and their sustainable use, and reduce or mitigate the negative impact on the biosphere of real economy activities. Global finance is made up of both public and private sources, and there are several mechanisms through which each can support and create incentives for the actions needed. This chapter discusses these mechanisms, providing examples and discussing existing limitations.

While financial actors have a key role to play in shifting from Impact Inequality to Equality – through greater channelling of financial flows towards natural assets and their sustainable use – it should be stressed that their role is ultimately bound by broader government and regulatory policies to correct for institutional failures. Chapters 7 to 9 discussed the failure of governments to internalise externalities fully, through fiscal measures, standards, regulations and market mechanisms. This failure means that financial markets cannot incorporate these costs into pricing, and therefore into credit allocation and lending decisions. Until this fundamental failure is addressed, pricing and allocation of financial flows alone will not be sufficient to enable a sustainable engagement with Nature.

---

[532] Annex 8.1 discussed how the total cost of subsidies is much larger when taking into account the externalities.

Chapter 20: Finance for Sustainable Engagement with Nature

Figure 20.1 Balance of Nature Positive and Negative Financial Flows

20.1 Public Finance

Governments are key financial investors in domestic and foreign natural assets, and their sustainable use, given the societal benefits of such investments (Chapter 7). The open-access nature of many ecosystems and the fact that accounting prices are not reflected in market prices for many ecosystem services were covered in detail in Chapters 1, 7 and 10. Both of these have implications for the degree and type of private investment which takes place (Section 20.2). Governments can channel financial flows through a range of mechanisms, change incentive structures around financial investments, develop policy and undertake financial de-risking, which can increase the amount of private finance invested in natural assets (UNDP, 2020).

Most current sources of finance devoted to supporting our stock of natural assets are public funds. Funds to address particular environmental issues – such as climate change – have been mobilised significantly over recent years (Buchner et al. 2019),[533] but only a small percentage (around 3%) of these funds are allocated to biodiversity and ecosystems specifically, despite the potential cost-effectiveness of these funds relative to other interventions (Griscom et al. 2017).[534]

Given the multifaceted nature of public finance in relation to the biosphere, some countries have adopted approaches that attempt to pull different mechanisms together. In one example, the UNDP's Global Biodiversity Finance Initiative (BIOFIN) supports governments to identify the amount of finance currently being devoted to conservation and restoration, and where expenditures are directly harming natural assets. This approach aims to make resources available to enhance natural assets, and to identify

---

[533] It is estimated that annual flows to climate finance were US$579 billion in 2017 and 2018, representing a 25% per cent increase from 2015 and 2016.

[534] Chapter 19 discussed how, relative to other interventions, Nature-based solutions have the potential to be cost-effective and provide multiple benefits beyond climate adaptation and disaster risk reduction.

and reform current expenditures to ensure they are working effectively to enhance natural assets (UNDP, 2018). Currently, 36 countries are using this approach to assess the current state of play in their public finances.

20.1.1 Domestically Focused Mechanisms

Public finance can influence the condition of natural assets and the extent of the sustainability of their use by providing incentives for more sustainable production or consumption patterns, or generating revenue that can be used to support conservation and restoration initiatives. Across countries, the majority of public finance for enhancing domestic natural assets and their sustainable use is allocated via domestic budgets and tax policies (Deutz et al. 2020), so that concessional public financing, grants and donations are essential contributors to the financing of investments in ecosystems and their biodiversity (UNDP, 2020).

Fiscal instruments, such as taxes, fees and charges, can help reflect the social value of natural assets in market prices. Chapter 7 highlighted how these fiscal instruments can be used to place costs on natural resources in order to reflect the negative externalities generated through their use. The revenue generated from fiscal instruments can be channelled back into the conservation, restoration and sustainable use of natural assets (OECD, 2019a). Globally, the number of biodiversity-relevant taxes[535] – defined as taxes aimed at activities that have a proven, specific negative impact on Nature – have steadily increased over recent decades; there have been 229 biodiversity-relevant taxes, of which 206 are in force today, in 59 countries[536] (see Box 20.1 for an example). But the total amount of resource collected from biodiversity-relevant taxes is small. Recent estimates suggest they generate only around US$7.5 billion a year in revenue, equivalent to around 1% of the total revenue from environmentally relevant taxes[537] (OECD, 2020c).

Box 20.1

Denmark's Pesticide Tax

In 1996, a Pesticide Tax was introduced in Denmark, with the aim of achieving a 50% reduction in pesticide use. Prior to the tax, there were growing concerns about the negative health effects of polluted water, and about residue in crops, water courses, lakes, soil and rainwater. The tax was phased in with the aim of moving towards an optimal tax rate: a general ad valorem rate in 1996 and a reformed differentiated rate with higher levies on more toxic agents in 2013 (Lago et al. 2015; Pedersen, 2016). Ex-ante the tax was estimated to be able to reduce pesticide use by 18–20%, assuming a price elasticity of demand between –0.5 to –1.0. The overall aim was to reduce the Treatment Frequency Index (TFI) to below 2, which is a measure of the average number of pesticide applications per cultivated area per calendar year, based on total cultivated area and pesticide use.

Denmark's pesticide tax has had a mixed impact in terms of reducing the frequency of use of pesticides. The TFI fell to below the target of 2 in 1996, but has since fluctuated, and even rose to above 3 between 2010 and 2012. Pedersen (2016) argues that it is difficult to identify the tax effect among external effects on pesticide use, such as other agricultural policies and the price of grain. It is

---

[535] For example, these include taxes on pesticides, fertilisers, forest products and timber harvests.

[536] Some governments have used ecosystem valuation to determine environmental externality costs to set taxes, determine compensation payments for damage to natural resources, and inform cost-benefit analyses for policies and projects (OECD, 2019a).

[537] Defined as taxes that are aimed at activities that have a proven, specific negative impact on the environment.

> likely that the Ministry of Tax underestimated the necessity of pesticides for agriculture, and therefore the price inelasticity of demand.
>
> The Pesticide Tax has highlighted several administrative and practical challenges with applying these sorts of market-based instruments. There was evidence of hoarding behaviour prior to the tax being introduced. However, the Ministry for Taxation considers it a cost-effective instrument for reducing pesticide pollution. Pedersen (2016) estimates that the tax raised nearly 600 million DKK (or €80 million) in 2016, and suggests the reformed, more optimally designed tax rate introduced in 2013 is more likely to have a behavioural effect. If higher prices still do not achieve the policy objectives, stronger incentives may need to be considered, including higher prices or quotas.

Subsidies can be used to enhance natural assets and support their sustainable use. Currently, an estimated 146 environmentally motivated subsidies relevant to biodiversity are in place in 24 countries (OECD, 2020b).[538] Biodiversity-relevant subsidies – defined as subsidies that are aimed at reducing (directly or indirectly) activities that have a proven, specific adverse impact on biodiversity and ecosystems – include subsidies for forest management and reforestation, organic or environmentally supporting agriculture, pesticide-free cultivation, and land conservation. But, in many countries, most of the public finance directed towards subsidies goes to activities that are harmful to ecosystems and biodiversity, through the extraction of provisioning goods and pollution (Chapter 8, Annex 8.1). Harmful subsidies include financial support for sectors such as fossil fuels, agriculture, fisheries and mining.

Estimates of the true environmental cost of some subsidies are much larger in magnitude than their monetary cost. Coady et al. (2019) estimate that, when accounting for the negative externalities arising from fossil fuel subsidies, the aggregate cost of these subsidises is around US$5.2 trillion annually. As noted in Chapter 8, government support in the form of subsidies can exacerbate price distortions and give free goods a *negative* price, in the form of producer, consumer and market price support. In some countries, work is being done to identify the most harmful subsidies and promote their reform (Box 20.2). Beyond subsidies, other public finance, such as procurement spending and large-scale infrastructure projects, also enable activities that are harmful to the biosphere (OECD, 2020a).

> ### Box 20.2
> ### Agricultural Subsidies Reform in Switzerland
>
> Over the past two decades, Switzerland has undertaken a series of reforms to the system of agricultural subsidies and introduced direct payments for ecosystem services (PES) (OECD, 2017b). The primary aim of the reforms was to align the direct payment system better with meeting policy goals, including for biodiversity.
>
> The reforms entailed both removing direct payments to livestock farmers and increasing payments to farmers able to change their practices to meet biodiversity targets. Reform was embodied in the new Agricultural Policy, which sets out clear goals and targets, including transition payments to reduce the negative impact on farmers' incomes. The policy underwent broad consultation involving a wide range of stakeholders, such as the Farmers' Union, economic institutions and environmental NGOs. An impact assessment was conducted to examine the expected environmental and biodiversity implications of the policy, as well as impacts on production and income.

---

[538] In 2019, these subsidies globally totalled around US$0.89 billion per year (OECD, 2019a).

Chapter 20: Finance for Sustainable Engagement with Nature

> The impact assessment showed that the 'policy' scenario (with introduction of PES) was better than the 'business-as-usual' scenario (with existing agricultural subsidies) (OECD, 2017b). As a result of the reforms, the agricultural sector as a whole received a slight increase in budgetary payments over the 2014 to 2017 period (OECD, 2017b). Participation in voluntary programmes funded by the biodiversity direct payments has exceeded expectations. Incomes and productivity in the sector are expected to be higher as a result of the reforms.

In addition to conventional fiscal instruments, there are other mechanisms through which public finance can influence the condition of natural assets and the extent of the sustainability of their use. One instrument is payments for ecosystem services (PES) (see Chapters 7 and 19), which are made in over 550 active programmes around the world, amounting to an estimated US$36–42 billion in annual transactions (Salzman et al. 2018). The most common examples are payments for carbon storage, biodiversity conservation and watershed services (Box 20.3 discusses a study of PES in Colombia). While much faith has been placed in PES as a source of private finance, in reality PES have largely been found to be another form of public subsidy; more than 90% of PES are estimated to be funded through public sources (Gómez-Baggethun and Muradian, 2015).

Biodiversity offsets are another mechanism through which fund actions are designed to compensate for significant, residual biodiversity loss from development projects, by achieving biodiversity outcomes elsewhere (Chapter 19). Biodiversity offsets are attracting increasing interest: over 100 countries have laws or policies in place which require or enable the use of biodiversity offsets (OECD, 2016a).

Box 20.3
PES in Colombia

In the past 25 years, Colombia has lost 5.2 million hectares of forest, of which 3 million has been deforested in municipalities affected by armed conflict. Post-conflict Colombia is now pursuing conservation strategies to enhance livelihoods whilst conserving natural assets and the supply of ecosystem services (Alpízar et al. 2020). In this context, the Colombian government has proposed PES schemes as a way of promoting sustainable economic alternatives to populations affected by the conflict. The proposal for a PES program specifically aims to preserve 500,000 hectares of highly biodiverse forested ecosystems and to mitigate deforestation. The scheme estimates the value of the payments for specific ecosystem services based on the opportunity cost of agriculture and cattle ranching, and then pays 75% of that opportunity cost of foregone land-use.

Banerjee et al. (2020) examine the potential impacts of implementing PES and other conservation strategies in Colombia over the next two decades (2020 to 2040) on ecosystem services, GDP and Inclusive Wealth (Chapter 13). The PES scheme has the effect of marginally reducing GDP by constraining future agricultural land availability, and so on the basis of simple net present value (NPV) analysis, the PES investment would be rejected. However, once the value of changes in natural capital and ecosystem services are included (Table 20.1), the analysis suggests that the PES scheme leads to increases in Inclusive Wealth in the order of around US$14 billion. The NPV analysis including changes in natural capital and future ecosystem service flows suggests that the PES scheme would provide around US$4.4 billion in net present value terms.[539]

---

[539] This compares favourably to other options considered, including (i) implementation of silvopastoral systems (SPS) to restore degraded pasture lands and enhance livestock productivity on 125,000 hectares of degraded areas; and (ii) expansion of Colombia's Habitat Banking system, a type of biodiversity offsetting mechanism, by 500,000 hectares for preservation and ecosystem restoration, particularly when combined with the SPS scheme.

Chapter 20: Finance for Sustainable Engagement with Nature

Table 20.1 Estimated Impact on Ecosystem Services from PES Scheme

	% increase
Soil erosion mitigation	3.3
Carbon storage	6.3
Nutrient (nitrogen) storage	7.3
Nutrient (phosphorus) storage	4.9
Annual water yield	6.4
Biodiversity intactness	6.4

Source: Banerjee et al. (2020)

Governments can raise finance to invest in natural assets and encourage more sustainable use via sovereign green and blue bonds. All funds invested in these interest-bearing bonds are used for funding projects that are considered to be 'environment-friendly' and 'climate-responsible'. Green bonds have, so far, primarily funded projects which address climate-change mitigation and adaptation, but can also address natural resource depletion, loss of biodiversity, and water and air pollution (Climate Bonds Initiative, 2017).

While there is no comprehensive reporting on the percentage of green bonds that benefit biodiversity directly, recent estimates suggest that, globally, biodiversity conservation receives 4% of bond proceeds, and sustainable land use benefits from only 2% of bond proceeds (Convention on Biological Diversity, 2017). In 2019, 12 governments had issued 23 green bonds, totalling US$47.5 billion (Climate Bonds Initiative, 2019). In 2018, Seychelles launched the world's first sovereign blue bond, demonstrating the potential for countries to harness capital markets to finance the sustainable use of marine resources, and marine restoration and conservation (Box 20.4).

Box 20.4
Seychelles Blue Bond

Seychelles is a small archipelagic island country with an economy that is highly dependent on the ocean. After tourism, the fisheries sector is the most important industry, employing about 17% of the population. The World Bank has supported Seychelles' efforts to build a diversified blue economy (broadly defined as encompassing all real economic activities related to oceans, seas and coasts), in part through a model of financial markets using an approach similar to green bond financing, which the World Bank pioneered with the first green bond issuance in 2008.

In 2018, the World Bank helped the Government of Seychelles issue the world's first blue bond. The World Bank Treasury put together an innovative financing package that mobilised US$15 million of private-sector investment to support the ocean economy, and helped the Seychelles' government save over US$8 million in interest charges over the next 10 years (World Bank, 2018). Proceeds from the bond will support projects including the expansion of marine protected areas and improved governance of priority fisheries (World Bank, 2018). Grants and loans for the individual projects will be channelled through the Blue Grants Fund and Blue Investment Fund, managed respectively by the Seychelles' Conservation and Climate Adaptation Trust (SeyCCAT) and the Development Bank of Seychelles (DBS).

20.1.2 Globally Focused Mechanisms

Governments can also enhance foreign natural assets through a range of mechanisms. Present estimates suggest that, globally, governments spend 5–12% of their funds for biodiversity on international projects to protect foreign natural assets (OECD, 2020a). The bulk of current international public expenditure devoted to biodiversity is provided as Official Development Assistance (ODA): aid, disbursed by countries either directly (bilaterally) or through multilateral institutions (multilaterally), designed to support and promote the economic development and welfare of developing countries.[540] A subset of overall ODA is devoted to biodiversity: financial flows to developing countries with explicit goals relating to conservation and sustainable management of biodiversity. This type of ODA can be used by governments to enhance natural assets situated in other countries and to support their sustainable use.

Sources of bilateral biodiversity-related ODA are concentrated among a few donors, with the USA, Germany, France and Japan estimated to account for over half of committed bilateral biodiversity-related ODA between 2012 and 2016. The largest share of ODA was received by African countries, followed by Asia, the Americas, Europe and Oceania (Deutz et al. 2020). The majority (around 75%) of bilateral biodiversity-related ODA has been concentrated in a small number of sectors – including forestry, water supply and sanitation, agriculture and fishing – and aid has been channelled into projects specifically focused on conservation and sustainable management of natural resources (Deutz et al. 2020). In a global study of biodiversity-related ODA, Miller et al. (2013) conclude that such financial flows have been relatively well-targeted historically, as the allocation of biodiversity aid is positively associated with the number of threatened species in recipient countries, after controlling for other factors.[541] However, they also point out that flows to date have been insufficient to meet conservation needs in developing countries. There is also evidence that biodiversity-related ODA has been an important mechanism to catalyse additional financial resources for biodiversity in developing countries (Drutschinin and Ockenden, 2015).

Multilateral development banks (MDBs), bilateral development banks and development finance institutions (DFIs), along with other public development finance institutions, play a central role in securing multilateral ODA. These institutions catalyse finance, both public and private, to increase the resources available for conservation and restoration projects, and for sustainable economic activities.[542] They do this in part through blended finance, as both facilitators of blended transactions and providers of development finance that is used to mobilise private capital (Section 20.2.2). Concessional capital for blended finance transactions focused on conservation has been most commonly provided by development agencies and multi-donor funds. Estimates suggest that biodiversity-related multilateral flows are currently around US$565 million per year (OECD, 2020a).[543]

MDBs and other global funds can take more risk and accept lower financial returns than private investors, thus reducing the overall cost of finance, particularly for training and capacity-building activities that are often required in the early stages of project development. Investments occur directly and through partnerships with other multilateral financial institutions. For example, the World Bank is the lead agency of the Global Wildlife Programme, a grant programme of the Global Environment

---

[540] ODA includes concessional finance, grants and the provision of technical assistance. The UN target, since 1970, has been for developed countries to provide 0.7% of Gross National Income (GNI) as ODA to developing countries. This target was reconfirmed in 2015; however, only five countries met the 0.7% target in 2019 (Deutz et al. 2020).

[541] Such as country size, national population, and wealth.

[542] Unlike private financial institutions, and in line with their mandates and wider development agenda, MDBs do not seek to maximise profits for their shareholders (Engen and Prizzon, 2018).

[543] MDBs do not report data on biodiversity finance in a comprehensive or consistent way, meaning that estimates of the current level of investment are incomplete. Given the lack of available data, this is likely to be an underestimate (OECD, 2020a).

Chapter 20: Finance for Sustainable Engagement with Nature

Facility (GEF). The GEF has invested more than US$3.5 billion to conserve and restore the biosphere and use it in a more sustainable way, and this finance has leveraged over US$10 billion in additional funds, supporting 1,300 projects in more than 155 countries (World Bank and GEF, 2018). Regional development banks have also supported investments in natural assets in similar ways. For example, the Inter-American Development Bank's (IDB) Natural Capital Lab has, since 2016, leveraged an additional US$55 million for projects, to be deployed alongside new IDB resources, much of it to develop new financing models with the private sector.[544] Such projects included improving land-use, agriculture and marine ecosystems.

While MDBs play some role in enhancing natural assets, it is recognised that there is scope for them to increase their contribution, both in terms of channelling financial flows and increasing understanding about the impact different economic activities have on natural assets (World Bank, 2020b). On channelling financial flows, MDBs are a central part of the international climate-change finance architecture (EBRD, 2020), and thus are in a position to ensure climate-change funds are used to support natural assets more broadly, such as through Nature-based solutions (Chapter 19). On increasing understanding, while there have been some efforts among MDBs to take account of their impact on natural assets, primarily through safeguarding investments to ensure that they are environmentally and socially sustainable (ADB, 2004; EIB, 2006), considerations within MDBs of broader natural capital issues are not systematic across analysis, policy advice and financing (Humphrey, 2016).

In addition to ODA, countries can use other mechanisms to enhance natural assets abroad; for example, debt-for-Nature swaps can be used to help support the conditions of foreign natural assets (OECD, 2019a). These swaps are agreements that reduce a developing country's debt stock or debt servicing costs in exchange for a commitment from the debtor-government to protect Nature (UNDP, 2017). Swaps are voluntary transactions, whereby the donor cancels the debt owned by a developing country's government. The savings from the reduced debt service are then invested in conservation and restoration projects. Currently, debt-for-Nature swaps carried out by all high-income nations have resulted in US$1 billion of debt being cancelled, and have generated about US$500 million for conservation (Sommer, Restivo, and Shandra, 2019).

20.2 Private Finance

It is widely acknowledged that using and reforming public financing mechanisms alone will not be enough to ensure that current deteriorating trends in the biosphere are slowed and reversed (UNDP, 2018; OECD, 2019a; World Bank, 2020b), and the global pool of private finance is much larger than public finance. For example, total credit to the non-financial private sector globally amounted to around US$120 trillion in 2019, compared to around US$60 trillion for total credit to the public sector (Bank of International Settlements, 2020b).

As in public finance, existing private financial flows that are adversely affecting the biosphere outstrip those that are enhancing natural assets, and there is a need to identify and reduce financial flows that directly harm and deplete natural assets. For example, recent research suggests that private finance devoted to biodiversity ranges from US$6.6 billion to US$13.6 billion per year (Deutz et al. 2020; OECD, 2020a). But that private financial flows directed to activities that are harmful to natural assets are much greater. For example, it has been estimated that in 2019 the world's largest financial institutions provided more than US$2.6 trillion worth of loans and underwriting services to sectors which have been identified as primary drivers of biodiversity loss and ecosystem disruption, including food, forestry, mining, and fossil fuels sectors among others (Chapter 16) (Portfolio Earth, 2020). Separately, it has

---

[544] These estimates were provided by the IDB, for which we are grateful for.

been estimated that around US$44 billion was channelled into businesses that were directly or indirectly involved in deforestation in the Amazon, Congo Basin and Papua New Guinea in 2019 (Global Witness, 2019). Despite some divestment in fossil fuel businesses over recent years (Nauman, 2019), there remains significant investment in the sector (Influence Map, 2019). As market prices have continued to stray from accounting prices for ecosystem services (Chapters 7 to 9), this has created incentives for private financial flows to be directed towards continued degradation of the biosphere.

The case for private investment in conservation and restoration, and in activities that adopt a more sustainable engagement with the biosphere, is not only one of wider societal benefits. As discussed in Chapter 7, there are limits to current pricing mechanisms; however, even so, there are markets for a range of ecosystem services, and there is potential to generate profitable opportunities (Cooper and Trémolet, 2019). As discussed in Chapter 17, the condition (and societal responses to such condition) of natural assets can influence the resilience of balance sheets for financial institutions, given the potential materiality of a range of natural-related financial risks. For asset managers in particular, there is an argument that environmental issues should be part of their fiduciary duty (Box 20.5).[545]

Indeed, appetite among some private financial investors for channelling finance towards restoration and conservation projects – although still relatively small – has grown in recent years. For example, the Global Impact Investing Network (GIIN) 2018 impact investor survey – a global survey of over 250 impact investors – indicated that US$3.2 billion (3% of impact assets under management) was associated with conservation (Bass, Dithrich, and Mudaliar, 2018). Investments in conservation were made by 16% of survey respondents, and a third of respondents intended to increase their investment in conservation (Bass, Dithrich, and Mudaliar, 2018).

Box 20.5
Fiduciary Duty and the Biosphere

Fiduciary duties are imposed upon a person or an organisation who exercises some discretionary power in the interests of another person, in circumstances that give rise to a relationship of trust and confidence (UNEP, and PRI, 2019). They are of particular importance in asymmetrical relationships: situations where there are imbalances in information and expertise, and where the beneficiaries have limited ability to monitor or oversee the actions of the entity acting in their interests. In terms of investment, fiduciary duties exist to ensure that those who manage other people's money act in the interests of the beneficiaries, rather than serving their own interests.

As currently defined in most jurisdictions, fiduciary duties do not typically require an asset manager to account for the impact of their investment activity on natural capital (or other Environmental, Social and Governance (ESG) factors), beyond its financial performance. As of 2019, no jurisdictions[546] have rules exhaustively prescribing how investors should go about integrating ESG opportunities and risks in their investment practices and processes, or set timeframes over which they define their investment goals. Over the past decade, there has been relatively little change in laws globally relating to fiduciary duty. But multilateral global agreements (such as the 2015 Paris Climate Agreement) and growing evidence that ESG issues have financial implications, particularly in relation to climate change (Task Force on Climate-related Financial Disclosures, 2020) have led to

---

[545] More work is needed to explore ways in which natural capital and biodiversity can be explicitly included in fiduciary duties, alongside other environmental issues, such as climate change (Chenet, 2019).

[546] Jurisdiction is the practical authority granted to a legal body to administer justice within a defined field of responsibility.

Chapter 20: Finance for Sustainable Engagement with Nature

> increased discussion among investors of the extent to which environmental issues are part of their fiduciary duties.
>
> In 2019, the Principles for Responsible Investment (PRI) – an investor initiative in partnership with UNEP Finance Initiative and UN Global Compact – assessed the extent to which there should be incorporation of ESG standards into regulatory conceptions of fiduciary duty (PRI, 2019). They concluded that not considering long-term investment value drivers, which include environmental, social and governance issues, in investment practices was a failure of fiduciary duty on the basis that (i) ESG incorporation was becoming an investment norm, (ii) ESG issues are financially material, and (iii) policy and regulatory frameworks were changing to require ESG incorporation (UNEP and PRI, 2019).

20.2.1 Mechanisms and Investment Approaches

As noted at the beginning of this chapter, private finance can be directed into supporting investments in natural assets through conservation and restoration, and to real economic activities that minimise harm to the biosphere and are thus sustainable. While the transition of many real economic activities to more sustainable practices will also involve elements of conservation and restoration processes, this distinction between potential financial investment in more sustainable real economy activities and conservation and restoration projects is important, as they represent different types of private financial investment and face different challenges in terms of acquiring greater private financial support.

Currently, private investments in enhancing natural assets and in their sustainable use are mostly restricted to projects on existing activities transitioning to more sustainable practices. Examples include sustainable agriculture, low-carbon energy production and processes, green infrastructure, and eco-tourism (Suttor-Sorel and Hercelin, 2020). These activities differ from pure restoration and conservation activities, such as protecting a natural habitat through the creation of protected areas or rewilding zones, or re-naturalising degraded soils; given market pricing and profitability prospects, there is no case for private financial investment in restoration and conservation activities.

At present, private financial investments in natural capital are typically viewed as a sub-set of financial investments in broader investment categories, such as 'sustainable' finance and 'green' finance (Figure 20.2). 'Sustainable' investing is an umbrella term that covers approaches to investment where broader non-financial factors, including environmental ones, in combination with financial considerations, guide the selection and management of investments (Suttor-Sorel and Hercelin, 2020).[547] 'Green' finance is a subset of sustainable finance, and is more narrowly defined as concentrating on Nature (Cooper and Trémolet, 2019).[548] There are a range of mechanisms and instruments subsumed under 'green finance', which include green bonds, sustainability-linked loans, private equity funds in supporting biodiversity, environmental impact bonds and other insurance products (The Nature Conservancy, 2019; Deutz et al. 2020; OECD, 2020b). Carbon markets (or emissions trading schemes) are another potential mechanism for supporting conservation and restoration projects (Unger and Emmer, 2018), and thus natural assets.

---

[547] The roots of sustainable investing derive from religious convictions of entrepreneurs and industrialists who wished to see their concerns reflected in the businesses they invested in. In its original form, 'responsible investing' was carried out by the Quakers who divested from businesses associated with slavery, and by colleges who divested from certain companies to protest against the South African Apartheid regime (Molthan, 2003).

[548] While there are no clear definitions of 'sustainable' and 'green' investing, there is some harmonisation of the different investment practices and strategies that these approaches entail.

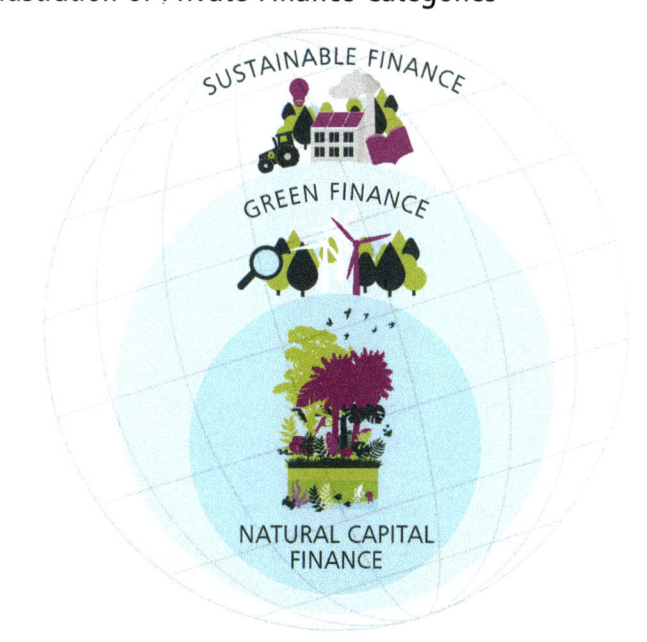

Figure 20.2 Simple Illustration of Private Finance Categories

Despite the rise in 'sustainable' and 'green' financial investments over recent years, financial investments in natural capital represent a small percentage of overall markets.[549] There is uncertainty over the exact magnitude of private finance for conservation and restoration of natural capital, but it is likely to represent a small percentage of the overall green finance market at around US$10 billion.[550] For example, only a minority of green bond investments go directly towards enhancing the condition of natural assets, for example to finance projects that improve water quality, protect coastlines or support vulnerable forests (Cooper and Trémolet, 2019). In addition, asset managers have yet to develop a sophisticated approach to biodiversity loss. In an assessment of 75 of the world's largest asset managers, ShareAction found that none had a comprehensive investment policy on biodiversity, while less than 5% integrated biodiversity into policies for high-risk sectors (Nagrawala and Springer, 2020).

One of the most prominent sustainable financing strategies is ESG investing. ESG investing differs from other forms of sustainable investing (such as Socially Responsible Investing) in that ESG factors are chosen for their financial, rather than ethical or social, considerations. ESG issues are those that are deemed to have financial relevance but are not part of traditional financial analysis, such as the water management practices, supply chain management or response to climate change of a business. Private financial institutions, including banks, asset managers and institutional investors, have increasingly used ESG approaches. For example, the number of private investment funds with an ESG mandate was 1,868 in 2018, more than double the number a decade earlier (IMF, 2019). The lack of consistent definitions makes it difficult to estimate the global asset size related to ESG, with estimates ranging from US$3 trillion to US$31 trillion (IMF, 2019).

---

[549] Sustainable financial investment assets totalled around US$30 trillion at the start of 2018, after a >30% increase in two years, and included around 35% of all professionally managed assets (Cooper and Trémolet, 2019).

[550] Existing estimates for the magnitude of private investment in natural capital vary widely, due to the lack of adequate metrics and robust methods to track these investments. In 2016, Credit Suisse and McKinsey estimated that conservation finance was attracting about US$52 billion per year, of which around US$40 billion (the vast majority) came from public and philanthropic sources (Huwlyer, Käppeli, and Tobin, 2016). According to Ecosystem Marketplace, private investments in conservation totalled just US$8.2 billion over the period 2004 to 2015 (Forest Trends, 2016).

Chapter 20: Finance for Sustainable Engagement with Nature

Impact investing, an approach through which private finance is directly channelled towards enhancing both the condition and use of natural assets, actively seeks to attain a tangible social or environmental impact together with a financial return (OECD, 2020a). Total impact investments committed since 2004 increased by 62% in just two years, from US$5.1 billion in 2013 to US$8.2 billion in 2015 (Hamrick, 2016). Sustainable food and fibre value chains accounted for the bulk of impact investments, while less finance was devoted to habitat conservation and improving water quality or quantity (Hamrick, 2016).

Despite the recent rise in private investments aimed at improving the condition and use of natural assets, three barriers – characteristics of projects and of financial capital markets – limit the extent to which private actors alone finance conservation and restoration projects at scale (Suttor-Sorel and Hercelin, 2020; UNDP, 2020; World Bank, 2020b).[551]

First, a profitable financial return is not always possible. Directly enhancing the quality and quantity of natural assets, for example through the creation of protected areas, rewilding or enhancing degraded soils, are not activities that produce reliable revenue streams, unless market prices are in line with accounting prices (Chapters 1, 7 and 10). Even conservation and restoration projects that are deemed profitable with market prices often take time to generate revenue streams, and that timeframe can reduce the attractiveness of the projects by impacting the liquidity of these financial assets (Suttor-Sorel and Hercelin, 2020).

Second, typical conservation and restoration projects are often too small to attract financial investment. Though most projects related to climate change mitigation do match the size requirements of large asset owners, the mean average value of broader nature-related projects in Europe has been around €7.4 million (Neate, Hime, and Paschos, 2014). A similar pattern is seen globally: only a few projects are estimated to be globally scalable beyond US$5 million (Huwlyer, Käppeli, and Tobin, 2016). Therefore, private financial investment has tended to be tailor-made for specific local conditions, and thus in small-scale interventions. The small size of projects has implications for the riskiness of each financial investment and for the time taken to set up each project (Cooper and Trémolet, 2019; World Bank, 2020b).

Third, a lack of standardised data and transparency on financial investments has contributed to a lack of proof-of-concept and track record, so that potential investors lack information about the project returns and impact. More generally, the lack of information or information asymmetry regarding the outcome of investments has been identified as a barrier to the deployment of private finance for sustainability (G20 Sustainable Finance Study Group, 2018). For example, in a 2019 global survey of 62 asset-owners and managers (who jointly manage more than US$3 trillion in assets), the majority of respondents said that they relied on in-house research for information on natural capital investment opportunities (Cooper and Trémolet, 2019).

These barriers to investment significantly reduce the current ability of private finance to provide significant funding to invest in natural capital. However, several mechanisms and initiatives have been and are being developed to try to overcome the barriers. One example is blended finance, which uses public finance to mobilise sources of private funding.[552] In blended finance, governments generally provide both grants and guarantees to cover or lower the risks related to loans and equities. While guarantees are usually provided to cover potential first losses, grants are used to support initial finance, provide venture funding, undertake result-payments or provide technical assistance. Through their structure, blended finance mechanisms can signal to investors the financial returns of a project, de-risk it and develop proof-of-concept for innovative projects (Suttor-Sorel and Hercelin, 2020). Blended

---

[551] There is survey evidence that private investors with an interest in natural capital have been deterred from investing in this area due to the perceived difficulty of finding projects of a suitable scale, a lack of data to measure the impact of their investments and, in some cases, difficulties in working alongside governments or other public sector investors.

[552] Defined as the use of catalytic capital from public or philanthropic sources to increase private sector investment in sustainable development.

finance can thereby attract investors requiring higher rates of financial return and reduce the risks of investment (Box 20.6). Blended finance has channelled finance to a range of projects that have worked to enhance natural assets and their use, including by land restoration, reforestation, improved ocean waste management, sustainable agriculture and sustainable forestry (Convergence, 2019).

Box 20.6
EcoEnterprises Fund

The EcoEnterprises Fund is a pioneer in channelling financial flows to support natural assets. This venture fund works with the private sector to invest in small businesses where long-term natural resource management is essential for their financial success. The Interamerican Development Bank (IDB) Group is a key investor in the Fund, which is focused on Latin America and has attracted private investment from over 40 companies, mobilising US$140 million for sustainable agriculture, agroforestry and ecotourism. The Fund has helped to protect or sustainably manage between 5.5 and 6.5 million hectares of land (IDB (forthcoming)).

Investment in natural assets has typically been considered high-risk in terms of financial returns. To overcome this issue, the Fund has adopted five core strategies:

(1) Blended finance: working with small businesses leads to challenges of aggregation, technical assistance and capacity building; blended finance reduces the risk to private investors

(2) Diversified portfolio and experienced portfolio management: aggregating projects with similar cash flow and risk profiles reduces the risk of single transactions to investors; this requires strong management and structuring, and a good understanding of the projects involved

(3) Venture financing: the Fund applies the tools and principles of venture financing; and brings together investors, advisors and technical experts through a collaborative model

(4) Monitoring: the Fund has developed diligence, evaluation, monitoring and measurement impact systems, which are critical for attracting private investment

(5) Third-party certifications: the Fund recognises that certifications are important for the independent monitoring of performance.

Other mechanisms have been developed, in the absence of a correction of market pricing, to reduce uncertain profitability and risk premiums of financial investments in natural assets (UNDP, 2020). One example is pooled funds, which aggregate several projects into one fund, enabling project and cash flow aggregation into one financial vehicle so as to diversify risk. A mature market fund would typically invest in projects in the most mature sustainable markets, such as sustainable agriculture, sustainable forestry and ecotourism, and several mature market funds have successfully invested in conservation (Box 20.6).

Some financial market innovations also support investments directly in natural assets and in their sustainable use. For example, in 2020, HSBC Global Asset Management and Pollination Group Holdings announced their intention to form a natural capital 'fund manager' (HSBC Global Asset Management and Pollination Group, 2020). The funds – which aim to raise US$6 billion from institutional investors, including pension funds, sovereign wealth funds and insurers – will invest in a range of projects that will conserve and restore a range of natural assets, and support their sustainable use. These include sustainable forestry, regenerative agriculture, Nature-based biofuels and other projects that reduce carbon dioxide and generate carbon credits. There are also separate ongoing efforts to establish

Chapter 20: Finance for Sustainable Engagement with Nature

a financial market exchange to buy and sell equity into natural capital – via business ownership – to support the conservation and restoration of natural assets that the businesses manage as part of their economic activities (Box 20.7).[553]

In order to combat the issue of lack of information and monitoring, several initiatives have sought to increase awareness about potential financial investments in natural asset projects and monitor their outcomes. The lack of understanding about local natural assets is starting to be overcome through a range of initiatives (Cooper and Trémolet, 2019). These include satellite imaging, the collection and processing of 'big data', and tools to determine natural capital dependencies (Chapter 17 discussed the extent and range of tools available to assess the state of natural assets and their degradation). UNDP has developed platforms and tools to enhance understanding of potential private investment mechanisms, such as the BIOFIN catalogue, the Financing Solutions for Sustainable Development online platform, and the Conservation Finance Alliance Guide (UNDP, 2020).[554]

Box 20.7
Intrinsic Value Exchange

Intrinsic Value Exchange (IVE), an exchange platform, is establishing a new asset class based on natural assets for financial markets, with support from the IDB Group. The IVE aims to generate private finance at the volume needed to address the degradation of natural resources (IDB (forthcoming)). The core IVE instrument is a 'natural equity'. Ecosystem service values are measured using standard valuation methods (Chapter 12), and their value informs the natural asset's value, much like any equity.

IVE aims to elevate natural assets onto a level playing field with other financial asset classes. Taking public 'natural asset companies' on the exchange means that anyone can buy and sell shares of such businesses, thus generating financial capital and revenue to support the conservation and restoration of the natural assets the businesses manage (for example, a national park).

In principle, the IVE could be used to protect and expand existing natural areas, restore degraded landscapes, and finance changes to existing production practices. For example, Costa Rica has a legal framework that requires financial sustainability for its national parks. IVE is working with public authorities to form a natural asset company from one or more of its national parks, and to take that company public in order to create resources to maintain and improve the park.

20.2.2 Investment Approaches and the Impact Equation

In principle, 'sustainable' financial investment strategies can support a shift towards sustainability as defined in terms of the Impact Equation. That can be achieved by reducing financial flows to activities that have identifiable adverse impacts on the biosphere (reducing y in the Impact Equation), identifying investments in technologies that could enhance our use of the biosphere (increasing α), and channelling financial flows in a way that enhances natural assets directly (increasing S). However, common financial investment approaches, as outlined above, do not do this in their current form, for three reasons.

---

[553] In 2016, Credit Suisse set out a blueprint to increase investments in conservation through building new financial products (Huwlyer, Käppeli, and Tobin, 2016).

[554] In addition, the Coalition for Private Investment in Conservation (CPIC) aims to facilitate the scaling of conservation investment by creating blueprints for investable priority conservation projects to showcase potential investments, connect those running the projects with technical support, and connect those delivering conservation projects with investors to implement investable deals (CPIC, 2017).

First, evidence suggests that those conducting ESG investing are primarily concerned with the impact of the environment on financial returns, as opposed to the investment's impact on the environment. In a global survey of over 400 mainstream investors in 2017, around two-thirds of respondents who used ESG information in investment decisions said that they did so because they considered it financially material to investment performance, while only around a third said that they did so for ethical reasons (Amel-Zadeh and Serafeim, 2018). A focus only on ESG factors that may have a *material* financial impact does not systematically ensure that there is consideration of environmental impacts in all investment decisions.[555] Indeed, there is little evidence that the financial capital allocation decisions of sustainable investors have led to significant positive environmental impacts to date (Kölbel et al. 2018).

Second, negative or exclusionary screening practices that exclude certain sectors, businesses or practices from a fund or portfolio, based on norms or specific ESG criteria have not included consideration of Nature. The screening approach is sometimes presented as a way of mainstreaming divestment, but it has largely been focused on exclusions either to conform with international conventions (for example, controversial weapons) or with relatively consensual issues (for example, tobacco or gambling). Screening so far has generally not included criteria relating to harmful impacts on natural assets. However, the case has been made for screening investments for their impacts on natural assets in line with international conventions. In 2015, WWF set out the case that investors should exclude potential financial investments that may damage natural World Heritage Sites. They found that a third of 229 natural World Heritage Sites were subject to extractive activity in some form, either with active operations already within their boundaries or through concessions that might bring such operations in the foreseeable future. Intrusion into these sites was especially high in Africa, where around 60% of World Heritage Sites were subject to some form of extractive concession or activity (WWF, 2015).

Third, current limitations in terms of lack of harmonisation and standardisation create barriers to a universal take-up of ESG investing. There are currently no harmonised or standardised metrics for ESG factors (WWF France, and AXA, 2019), which has led to a range of competing firms providing ratings and data on various indicators.[556] Legislation in some jurisdictions is increasingly demanding more transparency about ESG methodologies, and so some businesses specifically market themselves on the ability of their metrics to plug gaps in pre-existing metrics (Suttor-Sorel and Hercelin, 2020). This has led, in recent years, to concern by some financial actors about 'greenwashing', where funds make an unsubstantiated or misleading claim about the environmental benefits of a product, service, technology or company practice (SCM Direct, 2019).

In principle, a taxonomy approach has the potential to bring consistency to the way private and public financial actors assess environmental impacts (OECD, 2020c). Over time, a harmonised and defined classification system should influence the cost of funding of those activities, as well as their ability to attract public subsidies and private funding. In the EU, there are efforts ongoing to address concerns over harmonisation by legislating for a common taxonomy to describe the environmental impacts of the underlying activities targeted by financial investments (Box 20.8). The benefits of a taxonomy approach are also being explored by others (OECD, 2020c).[557] For example, in 2015, the People's Bank of China issued the first iteration of its Green Bond Endorsed Project Catalogue, commonly referred to as 'the Chinese taxonomy'. While in 2017, the Ministry of the Environment of Japan launched Japan's green bond guidelines.

---

[555] Indeed ESG investing – along with other sustainable investment approaches – has been criticised for being too incremental, and for its failure to take a systematic approach (Thurm, Baue, and van der Lugt, 2018).

[556] Firms include Sustainalytics, MSCI, Refinitiv, Trucost, and ISS ESG (Suttor-Sorel and Hercelin, 2020).

[557] Other countries expressing interest on sustainable finance taxonomies include Canada, Kazakhstan and Indonesia.

> **Box 20.8**
> **An EU Taxonomy for Sustainable Activities**
>
> The EU Taxonomy is a tool to help investors, companies, issuers and project promoters navigate the transition to a low-carbon, resilient and resource-efficient economy (EU Technical Expert Group on Sustainable Finance, 2020). It helps to plan and report the transition to an economy that is consistent with the EU's environmental objectives. To be defined as sustainable, an activity has to make a substantive contribution to one of the six environmental objectives of the Taxonomy: (1) climate-change mitigation; (2) climate-change adaptation; (3) sustainable protection of water and marine resources; (4) transition to a circular economy; (5) pollution prevention and control; and (6) protection and restoration of biodiversity and ecosystems. The objectives span the demand and supply sides of the Impact Equation by working towards both reducing ecological footprints and enhancing our stock of natural assets.
>
> The EU Taxonomy sets performance thresholds (referred to as 'technical screening criteria') for economic activities which make a substantive contribution to one of the six environmental objectives; do no significant harm to the other five; and, where relevant, meet minimum safeguards (e.g. OECD Guidelines on Multinational Enterprises and the UN Guiding Principles on Business and Human Rights). The Taxonomy categorises activities in a binary way – they are either Taxonomy-compliant or not. The Taxonomy will serve as the basis for other sustainable finance initiatives, including EU labels and standards for financial products and ESG benchmarks. In addition, the Taxonomy is likely to be used to build environmental tax incentives, secure green public procurements, and to ensure the sustainability-proofing of public investments.

20.2.3 Engagement, Monitoring and Influence

As well as through investments, private finance actors can contribute to broader adoption of sustainable processes through stakeholder or corporate engagement.[558] There is some evidence that engagement by investors with businesses in relation to sustainability issues can influence business activities and processes (Kölbel et al. 2018). Through the production of information and monitoring of financial investments, financial institutions can also influence the businesses in which they invest and encourage sustainable activities (Schoenmaker and Schramade, 2018).

For example, following fires in the Amazon rainforest in 2019, 246 investors representing approximately US$17.5 trillion in assets signed a statement on deforestation and forest fires (Pinzón et al. 2020). The statement asks investee companies to increase their efforts to eliminate deforestation from their supply chains, including by disclosing and implementing zero-deforestation policies, assessing and minimising deforestation risks in their operations, establishing transparent monitoring systems, and reporting on the management of their deforestation risk. Some investors, such as Nordea, have indicated that they will extend their focus on deforestation to their sovereign bond holdings by quarantining Brazilian government bond purchases and revising existing holdings.

However, current overall engagement by investors with businesses in relation to the sustainable use of natural assets is low. A 2020 survey of 75 of the world's largest asset managers found that less than half of those surveyed engaged with firms they invested in on corporate strategy on biodiversity. Even fewer asked for better disclosure of the impacts of supply chains on ecosystems (Nagrawala and Springer, 2020). Only

---

[558] Engagement can include supporting greater take-up of corporate accounting standards to include natural capital considerations (Suttor-Sorel, 2019).

Chapter 20: Finance for Sustainable Engagement with Nature

7% of analysed voting policies included a commitment to vote in favour of increased transparency around the wider environmental impacts of business operations (Nagrawala and Springer, 2020). Only 11% of respondents stated in their policies that they expect the businesses they invest in to mitigate the negative impacts of their operations on Nature (Nagrawala and Springer, 2020).

Recent advances in technology and *'big data'* have the potential to enhance the ability of financial actors to monitor the impacts of businesses on natural assets, and thus to account for and mitigate the risks (see Chapter 6 for more on the importance of transparency and verification of impacts). Technological advances present opportunities to address the issues faced in existing asset-level datasets – these include time lags, measurement inconsistency and data gaps (Caldecott, 2019). For example, progress in distributed cloud computing has increased the processing capacity available to implement machine learning and artificial intelligence techniques, while geospatial data – including Earth observation, remote sensing and telecommunications – are now more widely available (Box 20.9).

Box 20.9
Spatial Finance

Effective institutions are a precondition for sustainability (Chapter 6), and one way to support building such institutions is by ensuring sufficient transparency, verifying who is breaking rules, and enforcing those rules. Spatial finance – the integration of geospatial data and analysis into financial theory and practice – can support transparency around natural assets in the financial system. Spatial finance can include overlaying business operations with protected areas and protected species' ranges, and monitoring land-use impacts of business operations and supply chains (Christiaen, 2020). Spatial finance analysis can reduce information asymmetries that exist both between businesses and their investors, and between financial institutions and their regulators. Spatial finance can also be used by financial actors to measure, monitor and manage impacts that investees, and financial portfolios, are having on local and global natural assets.

This approach has already been used to estimate a global figure for the potential threat of extractive activities – such as oil and gas extraction, mining, commercial logging, fishing and intensive agriculture – to natural World Heritage Sites (WWF, and Swiss Re Institute, 2020). Spatial finance also supports the assessment and management of different Nature-related financial risks (as discussed in Chapter 17) in sovereign debt investing (WWF, and Investec, 2019). Recent initiatives aim to mainstream geospatial capabilities from space technology and data science into financial decision-making (Spatial Finance Initiative, 2020).

Part III – The Road Ahead

Chapter 21
Options for Change

Introduction

At their core, the problems we face today are no different from those our ancestors faced: how to find a balance between what we take from the biosphere and what we leave behind for our descendants. But while our distant ancestors were incapable of affecting the Earth System as a whole, we are not only able to do that, we are doing it. Humanity now faces a choice: we can continue down a path where our demands on Nature far exceed Nature's capacity to supply them on a sustainable basis; or we can take a different path, one where our engagements with Nature are not only sustainable but also enhance our collective well-being and the well-being of our descendants.

To pursue a sustainable future will require a transformative change in our mode of thinking and acting. The appearance of COVID-19 has obliged us to make transformative changes to our daily lives, but because in many ways we were unprepared for the pandemic, there was little time for deliberation over ways to meet it. Biodiversity loss in contrast has sent us warnings (Chapters 0–4). The extent to which we have collectively degraded the biosphere has created extreme risks and uncertainties, endangered our economies and livelihoods, and given rise to existential risks for humanity.

That we have not meaningfully responded to the warnings so far is a sign of institutional failure writ large – it is not merely a case of market failure (Chapter 6).[559] It is also a sign of the failure of contemporary conceptions of economic possibilities to acknowledge that we are embedded in Nature, we are not external to it (Chapter 4*). That we have not responded is, above all, a sign that our thoughts are rigid. The near-universal conception we hold today of *economic* progress is wildly misleading. In correcting that, the Review has developed a unified framework for the economics of biodiversity (Chapters 0–13) and extended it by exploring its implications for practice (Chapters 14–20). The grammar we have developed enables economics to serve our values, not direct them.

That the causes of biodiversity loss are diverse means that ushering transformative change requires action not only by governments, but also by businesses, intergovernmental organisations and communities. And because all too often the desecration of Nature cannot be traced to those who are responsible, the ability of human institutions to implement sustainable development is limited. We need more than mere institutions of laws and social norms to curb our excesses. We will need to learn to practise self-restraint.

Because biodiversity varies geographically, success will look different from one region, ecosystem or country to the next (Chapter 14). Environmental problems also require action that is sensitive to local socio-ecological conditions. In some places, the prime recommendation of a concerned citizen – we have called her collectively the *social evaluator* – would be to let their communities manage their local ecosystems without external interference, while in other places, well-meaning central governance has weakened the place of local knowledge in people's lives. Then there are places where, with every good intention, the state has embraced policies that may have worked elsewhere but are unsuitable there. And, of course, societies differ also in their conception of what enables lives to flourish. That is why we

---

[559] The failure writ large includes especially our failure to create appropriate institutions. We are genetically and culturally a small group animal that in an evolutionary eyeblink has gone from tens of hundreds to groups of millions to billions. We have developed technologies in a few generations that enable us to alter the biosphere dramatically, potentially exterminate ourselves, but we have not developed the supra-national institutions that are increasingly necessary. We are most grateful to Paul Ehrlich for framing this problem for us.

Chapter 21: Options for Change

do not attempt to produce a blueprint of policies appropriate in different locations. What follows instead seeks to guide the reader through *options* humanity has for achieving the necessary change.

The options for change are geared towards three broad, interconnected transitions, requiring humanity to (i) ensure that our demands on Nature do not exceed its supply, and that we increase Nature's supply relative to its current level; (ii) change our measures of economic success to help to guide us on a more sustainable path; and (iii) transform our institutions and systems – in particular our finance and education systems – to enable these changes and sustain them for future generations (Figure 21.1).

Figure 21.1 Summary of Options for Change

The success stories from around the world that have been recounted in the Review not only show us what is possible, they also demonstrate that the same ingenuity – some would say the innate capability – that has led us to make demands on the biosphere so large, so damaging and so quickly relative to the history of the biosphere, can be redeployed to bring about transformative change, perhaps even in just as short a time. Although time is not on our side, it is not too late for us, both individually and collectively, to make the conscious decision to change paths. Our descendants deserve nothing less.

21.1 Address the Imbalance Between Our Demand and Nature's Supply, and Increase Nature's Supply

In the wake of the Second World War, the Marshall Plan[560] was launched to rebuild Western Europe. While most historians agree that the recovery experienced in Europe cannot be attributed to the Marshall Plan alone, there is little doubt that it hastened the recovery: industrial production in recipient

---

[560] The brainchild of US Secretary of State George C. Marshall, whom it was named after.

European countries leapt by 55% in just four years (1947 to 1951). By the effective end of the Marshall Plan in 1951, national per capita incomes in Britain, France and West Germany were more than 10% above pre-World War II levels; and the resumption of growth was sustained over the decades that followed (De Long and Eichengreen, 1991; Eichengreen, 2010). If we are to enhance the supply of natural capital and reduce our demands on the biosphere, large-scale changes will be required, underpinned by levels of ambition, coordination and political will at least as great as those of the Marshall Plan.[561]

21.1.1 Nature's Supply: Conservation and Restoration of Ecosystems

In Part I, we identified reasons it is less costly to conserve Nature than it is to restore it, other things equal. It was noted that markets alone are inadequate for protecting ecosystems from overuse. Uncertainty in our knowledge of ecosystem tipping points, the irreversibility of ecosystem processes, and imperfections in verifying one another's activities, when taken together, mean that quantity restrictions (e.g. on extraction or pollution) may be a better instrument than taxation. In the context of conservation, it follows that quantity restrictions, informed by science and supported by legislation, will help to correct the externalities pervasive in our engagements with Nature.

Protected Areas have an essential role in conserving and restoring our natural capital, but it has been estimated that only 20% of Protected Areas are being managed well.[562] Improvements can be made by ensuring that Protected Areas (i) are extended and integrated into the surrounding land and sea; (ii) involve indigenous people and local communities; and (iii) receive sufficient resources for their effective management (Chapter 18). The *Review* points to successes from around the world to demonstrate what is possible and what works. Examples include the co-management of the Gwaii Haanas National Park Reserve and Haida Heritage Site in Canada, the community-led management of Cabo Pulmo National Park in Mexico, and increasing the global designation of Marine Protected Areas, from 3.2 million km2 in 2000 to 26.9 million km2 today (Duarte et al. 2018; Duarte et al. 2020a, 2020b).

More investment in Protected Areas is needed. It has been estimated that to protect 30% of the world's land and ocean and manage these areas effectively by 2030 would require an average investment of US$140 billion annually, equivalent to only 0.16% of global GDP and less than a third of the global government subsidies currently supporting activities that destroy Nature (Waldron et al. 2020). The benefits of such levels of protection, even when confined to financial benefits, are estimated to exceed the costs significantly (Waldron et al. 2020). But there are wider benefits too, including lowering the risks of societal catastrophes in relation to human health, not least the risks of the emergence and spread of infectious diseases. Dobson et al. (2020) have estimated that the associated costs over a 10-year period of efforts to monitor and prevent disease spillover (which is driven by wildlife trade and by loss and fragmentation of tropical forests) would represent just 2% of the estimated costs of COVID-19.

There is the fear, though, that biodiversity conservation afforded by marine and terrestrial Protected Areas would be neutralised by disruptions caused by climate change. And that should remind us that Nature's regulating and maintenance services are complementary to one another. Neither climate change nor biodiversity loss can be tempered on its own and efforts should be complementary.

While avoiding degradation of Nature should be the priority, restoration through habitat management, rewilding, allowing natural regeneration and creating sustainably productive lands and seas also play an essential role in improving the health of the biosphere (Chapter 19). Restoration can also help us to

---

[561] Such a comparison has been made by others, including in Al Gore's *Earth in Balance* (1992), and more recently in a speech by HRH The Prince of Wales to mark the start of Climate Week NYC, 2020.

[562] See UNEP-WCMC, IUCN and NGS (2018).

Chapter 21: Options for Change

address the imbalance between our growing demands for the biosphere's provisioning goods on the one hand, and its depleted supply of regulating, maintenance and cultural services on the other.

Much of global biodiversity and many of our ecosystems lie outside Protected Areas. Approaches to increase biodiversity within lands and seas that also provide benefits to people are vital. Modern agriculture has driven much environmental decline (Chapter 16). Even though monoculture systems have raised food production, they have diminished biodiversity. Restoration can shift monocultures and degraded lands and seas to provide multiple ecosystem services, balancing provisioning goods with regulating services. In addition to long-standing forms of sustainable land management, such as shifting cultivation and crop rotations which increase soil fertility and reduce pests, offering incentives to farmers to adopt practices that support biodiversity and ecosystem services – for example through agri-environment schemes and payments for ecosystems services (PES) – should be further developed. As noted in Chapter 7, the effectiveness of PES schemes has proved to be mixed in low income countries, but even there such schemes hold great potential. And wherever they are adopted, their success depends on their design and scale of funding.

Considerably better land-use planning and marine spatial planning, in the form of legally binding instruments, can help to provide a long-term framework for balancing the competing demands we make of our ecosystems. By requiring that more space be given over to Nature, the planning process can also help to maintain, even increase, stocks of natural capital. Lessons can be drawn from successes, from sustainable infrastructure development in the Bahamas, to coastal zone management in Belize (Chapter 19).

Ecological solutions (often referred to as Nature-based Solutions) have the potential to provide multiple benefits. Restoring ecosystems not only addresses biodiversity loss and climate change, it can also deliver wider economic benefits (Chapter 19). Nature-based Solutions have frequently been found to be more cost-effective than engineered solutions and have far fewer unexpected consequences. They also create employment. As part of fiscal stimulus packages and public investments, investment in natural capital has high social value and the potential for quick returns. Recent research suggests that ecological investments such as afforestation, parkland expansion and restoration of rural ecosystems should have high priority as part of COVID-19 recovery stimuli (UNEP, 2020b). Hepburn et al. (2020) have pointed to three reasons for investing in such activities. First, training requirements are minimal for many 'green' projects, which means they can be implemented quickly. Second, the work meets social distancing norms. Third, many countries have blueprints of projects in existing mandates (e.g. in programmes designed to meet international agreements on climate change).

A deeper case can be made for why we should expect a positive link between employment and 'green investment'. If natural capital were valued at accounting prices, we would expect green investment to increase substantially, possibly compensating for declines in produced or human capital accumulation that would be required in order to finance the investment. Moving toward Nature-based economic development will lead to greater returns to human capital. That in turn would lead to a greater demand for investment in human capital and for employment.

21.1.2 Our Demand: Shifting to a Sustainable Ecological Footprint

If we are to avoid exceeding the limits to what Nature can provide on a sustainable basis, consumption, production and supply chains have to be fundamentally restructured. By quantifying the resources required for meeting basic human needs for over 150 countries, a recent study found that while needs such as nutrition, sanitation and the elimination of extreme poverty could be met for all people without transgressing planetary boundaries, the universal achievement of more qualitative goals such as attaining the contemporary lifestyles in high income countries would require resources several times the sustainable level (O'Neill et al. 2018).

The estimates are very much in line with the crude estimates we reported in Chapter 4 of our ecological footprint.

21.1.2.1 Changing Consumption and Production Patterns

Estimates of our current and predicted future use of provisioning goods including food, fibre, biofuels, timber, water, and fishery and aquaculture output tell a clear story of escalating demands, and corresponding declines in regulating and maintenance services (Chapter 16).

Several approaches are available to correct this. In addition to changing the balance of crops intended for human food and animal feed, closing gaps in agricultural yield could go some way towards addressing the demands without expanding agricultural land further (Foley et al. 2011). Establishing clear boundaries for conservation and agriculture (known as 'land-use zoning'); making payments to avoid habitat conversion; reducing food waste; strategically deploying technology, infrastructure or knowledge; and introducing standards and certification schemes are examples of schemes that can help (Phalan et al. 2016). As well as strategies for avoiding habitat conversion, sustainable production systems can effectively deliver multiple ecosystem services: regenerative agriculture, organic agriculture, agroforestry and low-trophic level aquaculture are examples of production systems capable of enhancing regulating services (such as pollination and air quality regulation) alongside provision of food.

As Chapters 4 and 4* discussed, our ecological footprint is not only made up of the material we take from the biosphere, but also of the transformed material we deposit into it as waste. Enforcing standards for re-use, recycling and sharing has an important role to play in reducing such waste, and is likely to have a positive economic impact, including the creation of jobs.

Technological innovations can contribute enormously to reducing our footprint. Genetically modified crops can increase yields even while reducing the contribution food production makes to climate change and biodiversity loss. Other technologies such as vertical farming, and meat analogues can increase yields, while reducing the contribution food production makes to climate change and biodiversity loss. In addition, there are ways to reduce environmentally damaging inputs, through such methods as precision agriculture and integrated pest management. Technological innovations can also help to reduce bycatch in fisheries.

While these technologies have potential, they will only emerge if institutional changes provide incentives to develop and establish them on a large scale. Historically, the private sector has relied on the state to provide the investments in research and development (R&D) that raise the productivity of its own investment in R&D. That there are synergies to be exploited between the two has been much discussed among historians of science. The state has an enormous role here for helping to finance and coordinate the investment that will prove to be necessary to help shift to a sustainable future. The *Review* makes clear in no uncertain terms, however, that we cannot rely on technology and human ingenuity alone, we need also to change our production and consumption patterns. The human economy is bounded, so it would be entirely counterproductive to seek economic growth that damages Nature in order to provide the necessary finance for investment in R&D.

If 'business' were to continue as usual, consumption in high income countries – and emerging upper-middle and lower-middle income countries – is projected to remain the key factor in driving the world's ecological footprint ever larger. An important feature of our ecological footprint is our diet (Chapter 16). Diets rich in animal products have much higher footprints than those based on plant products. If pastureland and land used for livestock feed are combined, animal agriculture uses nearly 80% of global agricultural land. Moreover, greenhouse gas emissions from plant-based food are 10 to 50 times less than those from animal products (Poore and Nemecek, 2018). Regions differ substantially in their relative footprints. Estimates suggest that it would be possible to feed the world's present population with as little as 50% of current agricultural land if diets shifted away from animal products.

Chapter 21: Options for Change

Estimates also suggest that it would not be possible to supply the world with environmentally damaging diets even if the Earth's entire land surface was converted to agriculture (Alexander et al. 2016).

21.1.2.2 Supply Chains and Trade

The expansion in global trade over recent decades has run parallel with the increase in global GDP (Chapters 4 and 14). That has increased our ecological footprint greatly; it has also given rise to greater transfers of wealth from primary producers to importing countries (Chapter 13). It is therefore useful to study the effect of trade on the biosphere by tracking entire supply chains to inform consumption and production decisions.[563] A shift to sustainable patterns of consumption and production will require us to embed environmental considerations along entire supply chains. Transparency and the sharing of information across supply chains is needed, and such information should be verifiable and support enforcement of standards (Chapter 8). Novel technologies can help. Improvements in geospatial data and implementation of 'blockchains' along entire supply chains can help to raise transparency, for they display the impact of commodity production on local ecosystems and individual species. Geospatial data combined with machine learning have been used to estimate a global figure for the potential threat posed by extractive activities to natural World Heritage Sites (WWF and Swiss Re Institute, 2020). They have also been used to monitor Nature-related risks in sovereign debt investments (WWF and Investec, 2019). Certification schemes can make a difference, but existing schemes differ in their effectiveness. WWF, for example, compared various voluntary certification schemes in 2013 and concluded that their socio-ecological performance varied substantially between schemes.[564]

Changes in broader trade practices and policies can support a shift to sustainability (Chapter 15). While there are limits to what market-correcting measures, such as taxes, are able to achieve for reducing our ecological footprint, they can make a difference if they are applied widely and designed well. Border Adjustment Taxes are an important example, but face both technical and political problems. Sustainability provisions in Regional Trade Agreements also have potential, and there are encouraging signs that the number of Regional Trade Agreements featuring environmental provisions has increased over time. Similarly, financial commitments to environmental objectives as part of aid expenditure directed at supporting trade have also increased; today they account for approximately a third of total aid (UNEP/WTO, 2018).

21.1.2.3 Pricing

The accounting price of an asset or service is the sum of its market price and the tax that ought to be imposed on it.[565] The gap between accounting prices and market prices is therefore a measure of inefficiency in the allocation of goods and services: the gap reflects waste in our use of resources. But unlike food waste, the gap is not visible. The *Review* discusses various ways available for estimating accounting prices. Open access resources such as ground water and ocean fisheries are free, so their accounting prices are the taxes that ought to be imposed on their use. Chapter 7 discussed this in more detail.

When estimates of accounting prices are available, taxes can be a useful instrument for reducing environmentally damaging activities (Chapter 7). At present, no OECD or G20 country collects more than 1% of its GDP in environmental taxes, beyond those related to energy or motor vehicles (OECD, 2019d). There is scope, therefore, to raise further revenue through environmental taxation. Such taxes need to be designed carefully, to avoid potential shifts to other activities that damage Nature (leakage),

---

[563] Green et al. (2019), for example, have traced the habitat loss caused by consumption of soy to the product's final destination.

[564] Perhaps not surprisingly, in countries with weak governance there has been rapid growth of ineffective certification schemes (WWF, 2013).

[565] In a case where the ideal policy would be to subsidise the good, the accounting price would be the market price less the subsidy.

to avoid the risk of crossing tipping points (if behaviour is insufficiently influenced by the tax) and to ensure the environmental harm being taxed is measured to a reasonable degree of accuracy.

There is also an urgent need to tackle perverse subsidies, which in total are equivalent to some 5–7% of global GDP (Annex 8.1), implying their accounting value must be greater still. All prevailing subsidies have a historical rationale – distributional justice, national food sufficiency, political pressure from powerful lobbies, and so on – which is why they prove difficult to dislodge. But the resources that would become available to governments if they were removed could be used to finance programmes that benefit not only populations at large, but in particular the most vulnerable in society. Correcting inefficient economic distortions to resolve institutional failures can only serve the common good.

21.1.2.4 Future Population

Expanding human numbers have had significant implications on our global footprint, and the global population is only expected to continue to rise (Chapter 4). In Part I, we explained how fertility choices are influenced by the choices of others. We explored ways in which a society can become embedded in a self-sustaining mode of behaviour characterised by high fertility and stagnant living standards, even when there are potentially self-sustaining modes of behaviour that are characterised by low fertility and rising living standards (Chapter 9).

As well as improving women's access to finance, information and education, greater access to community-based modern family planning and reproductive health programmes is a means for women to have greater control over their lives, shift behaviours and improve the chance of having healthy babies.

If the benefits of modern family planning and reproductive health programmes are high, the costs are low. By one estimate (Guttmacher Institute, 2020), expanding and improving services to meet women's needs for modern contraception in developing countries would cost under US$2 a year per person. It has even been suggested that a dollar spent on family planning and reproductive health is more beneficial than a dollar spent on agricultural research, rotavirus vaccination, preschool education, trade facilitation or even mosquito nets (Kohler and Behrman, 2014). But there has been significant underinvestment in family planning to date. OECD estimates suggest that less than 1% of Official Development Assistance (ODA) from the EU to Africa is directed towards family planning. As a route to accelerating the demographic transitions, investment in community-based family planning and reproductive health programmes should now be regarded as essential.

21.2 Changing Our Measures of Economic Progress

Standard economic measures such as GDP can mislead badly. If the societal goal is to protect and promote well-being across the generations (i.e. 'social well-being'), governments should measure inclusive wealth (societal means to those ends). Inclusive wealth is the sum of the accounting values of produced capital, human capital and natural capital. The measure corresponds directly to well-being across the generations (Chapter 13): if a change enhances social well-being, it raises inclusive wealth; if the change diminishes social well-being, it reduces inclusive wealth. Social well-being and inclusive wealth are not the same object, but they move in tandem. There lies the value of inclusive wealth in economic accounts.

Natural capital accounting serves as a necessary step towards the creation of inclusive wealth accounts. It enables us to understand and appreciate the place of Nature's services in our economies, including the services that are otherwise overlooked; it enables us to track the movement of natural capital over time

Chapter 21: Options for Change

(a prerequisite for sustainability assessment); and it offers us a way to estimate the impact of policies on natural capital (a prerequisite for policy analysis).

Frameworks for natural capital accounting and its counterpart, economic evaluation, are being developed, in many cases through the UN's System of Environmental and Economic Accounts. Countries are beginning to incorporate natural capital and the flow of ecosystem services into national economic metrics of success. China's Gross Ecosystem Product, and New Zealand's Living Standards Framework are examples (Annex 12.1 and Chapter 13). These are early days for natural capital accounts. Increased investment in physical accounts and their valuation would improve them. Moreover, international cooperation in the construction of national accounts and the sharing of data would improve decision-making around the world. Harmonisation of national accounts should be coupled with technical assistance. Incorporating natural capital accounts in macroeconomic surveillance undertaken by international financial institutions – for example, the International Monetary Fund's Article IV surveillance activities (IMF, 2020) – would also send a strong signal, inspiring governments' reform agendas to reflect the scale and urgency of the problems societies face.

Accounting for Nature in economic measures is a key to interpreting productivity.[566] As Chapter 13 discussed, contemporary models of economic growth and development tend only to consider produced and human capital as primary factors of production, and natural capital enters weakly, if at all. The absence of natural capital means that typical Total Factor Productivity estimates are biased and should be treated with scepticism. Improving and using measures of productivity that account for the use of, and impact on, Nature are therefore crucial for understanding the productivity of capital goods. There are several initiatives, such as the OECD's *greening productivity measurement* workstream (OECD, 2017a), that have made a start.

21.3 Transforming Our Institutions and Systems

Our global collective failure to achieve sustainability has its roots in our institutions. Many of the institutions we have built have proved to be wholly unfit to curb our excesses; worse, they have helped to widen the gap between what we are led to believe is possible and Nature's bounded capacity to respond to our demands. Effective institutions are the foundations on which to rebuild our engagement with Nature and manage our assets. Changes in three other systems will prove to be especially important: the protection of public goods, the financial system and education.

In Chapter 6, we presented a general finding: neither top-down nor bottom-up institutional structures work well. What the inhabitant of an ecosystem knows and can observe differs from what an agent from the national government knows and can observe. Moreover, institutions that work well are neither entirely rigid nor entirely flexible; they are both 'polycentric' and 'layered', meaning that knowledge and perspectives at all levels from different organisations, communities and individuals are pooled and spread.

It is a commonplace assertion that we live in a highly interconnected world today. That has been applauded over the years because the fruits of labour at one place have been transmitted to other places in short order. Ideas and practices in one location are transmitted rapidly to other locations through movements of people and goods. But adverse disturbances are also transmitted rapidly. That has become cruelly evident in the rapid spread of COVID-19.

---

[566] The technical term for productivity of aggregate capital is Total Factor Productivity (Chapter 13).

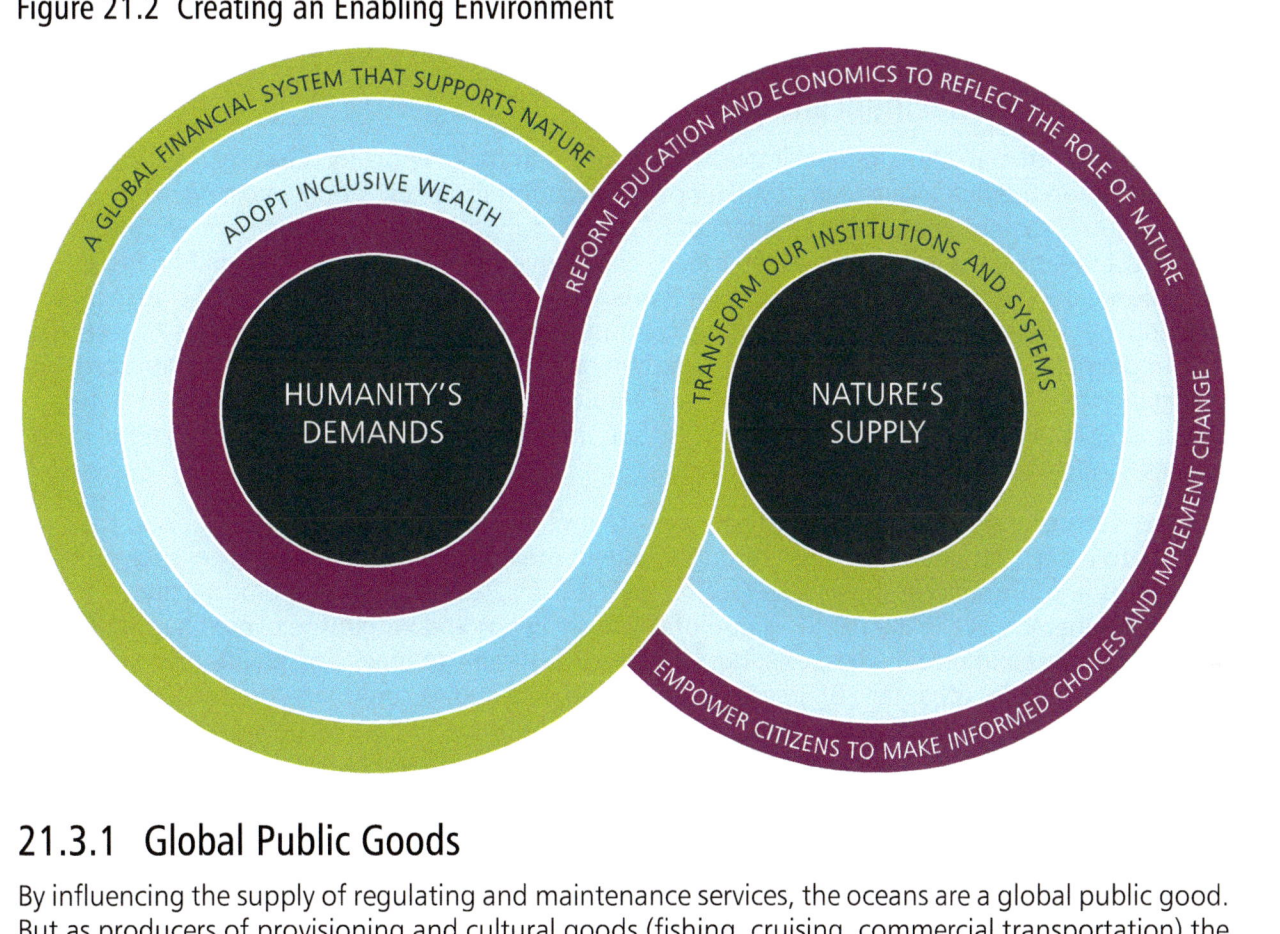

Figure 21.2 Creating an Enabling Environment

A global financial system that supports nature
Adopt inclusive wealth
HUMANITY'S DEMANDS
Reform education and economics to reflect the role of nature
Transform our institutions and systems
NATURE'S SUPPLY
Empower citizens to make informed choices and implement change

21.3.1 Global Public Goods

By influencing the supply of regulating and maintenance services, the oceans are a global public good. But as producers of provisioning and cultural goods (fishing, cruising, commercial transportation) the oceans beyond Exclusive Economic Zones are open access resources.[567] As they are a global public good (in the 1970s, they were called a Common Heritage of Mankind), the accounting rents from fishing, cruising and transporting goods could be shared among nations. A lively discussion took place in the 1970s of the amount of global rents that could be collected in the form of a tax on ocean resources, the idea being that the tax could be used as development aid.[568] The proposed Global Oceans Treaty, currently under negotiation among UN members, presents an opportunity to address biodiversity monitoring and conservation in areas beyond national jurisdiction.

Some ecosystems have the features of a global public good but are not a global common because they are located within national boundaries. Tropical rainforests are a global example. Management of river basins that cover many countries are a regional example. A fair approach would be to pay those countries, both for the protection of the ecosystems on whose services we all rely. The 15th Conference of Parties of the Convention on Biological Diversity will agree on global goals post-2020. Nature is also an important theme for the 26th Conference of Parties of the United Nations Framework Convention on Climate Change. These conferences provide opportunities to set a new direction for the coming decade. But boldness is elusive. We all could benefit if we disavowed our individual rights over the use of the global commons and permitted

---

[567] By creating Exclusive Economic Zones (EEZs) of up to 200 miles, the UN Convention of the Law of the Sea sought to eliminate the incentives nations have to deplete ocean resources. The fish catch in international waters beyond EEZs is regulated by Regional Fisheries Management Organisations and their member countries. However, gaps in governance and technological and geographical advantages mean that adverse incentives remain.

[568] Manganese nodules on the sea bed were of particular interest.

Chapter 21: Options for Change

supra-national institutions to enforce mutually beneficial patterns of use. But the idea remains alien to international thinking.

21.3.2 The Global Financial System

Finance plays a crucial role in influencing both sides of the Impact Inequality (Chapter 4). The problem is to encourage the financial system to take note of accounting prices, *if only implicitly*. The qualification is required because market prices of natural capital are far from their accounting prices. So indirect means are required. One such means is for governments and international financial institutions such as Multilateral Development Banks to invest in Nature. That could of course involve directing investment *away* from projects that contribute to unsustainable resource use. Projects that are complementary to public investment would then be attractive to private financial institutions. The financial system would also shift its lending and credit activities towards the protection of Nature if consumers signal their distaste for investments that are rapacious in the use of Nature's goods and services.

To leave Nature alone so that it is able to thrive is to invest in it. Governments have tools at their disposal to make that happen, even if through indirect means. They range from taxes, subsidies, regulations and prohibitions to Nature-specific mechanisms, such as PES, and biodiversity offsetting schemes. While the relative share of biodiversity funding within the overall budget for ODA has increased in recent years, existing flows are insufficient to meet conservation needs in developing countries (Chapter 20). Increasing those financial flows could take the form of debt forgiveness, direct grants or technical assistance. The creation of binding targets on public investments in natural capital to ensure that globally agreed objectives are met would go an important step further.

The risks associated with biodiversity loss – reductions in the productivity and resilience of ecosystems along supply chains – have significant macroeconomic and financial implications. Far more global support is needed for initiatives directed at enhancing the understanding and awareness among financial institutions of such Nature-related financial risks, learning and building on the advances on climate-related financial risks. Central banks and financial supervisors can support this by assessing the systemic extent of Nature-related financial risks. A set of global standards is required. They should be underpinned by data that are both credible and useful for decision-making. Businesses and financial institutions could then be obliged to integrate Nature-related considerations within their objectives. The idea ultimately is to have them assess and disclose their use of natural capital. The Task Force on Nature-related Financial Disclosures, established in 2020, is a step in that direction.

There is growing evidence that individual investors want investment providers to consider sustainability and Nature in their investment decisions (UNEP/PRI, 2019). Integrating the protection of biodiversity with the fiduciary duties of institutional investors and asset managers would be a way to ensure their investment policies account for natural capital. One barrier to this is myopia. Impacts of Nature's diminution can be felt over long time-horizons. Fisheries practices in the oceans will leave an imprint that will be felt by generations of future people. Destruction of tropical rainforests is to all intents and purposes irreversible. The time horizon in which financial actors plan and act is, unhappily, not more than a few years. Financial regulators and supervisors can play a key role in the necessary shift by changing their own assessment horizons and using their regulatory powers.

Many of the risks to life and property associated with ecological degradation are *positively correlated* among those who are affected. If the loss of mangrove forests makes someone more vulnerable to cyclones, it makes their neighbours more vulnerable also. Which is why there is a need for global, regional and national insurance funds. Although there are examples of regional insurance schemes for environmental disasters (e.g. the African Risk Capacity and the Caribbean Catastrophe Risk Insurance Facility), there is currently no global insurance scheme. In principle, a global risk pool – with contributions from all countries – could help protect vulnerable countries against such shocks following extreme events. Emergency relief is not uncommon even at the global level, but its volume is always uncertain. And that comes *after* a disaster strikes a community. Insurance in contrast is a security against disaster. Investing in Nature is a reliable form of insurance.

21.3.3 Empowered Citizenship

Ultimately though, it is we citizens who can bring about such changes. As citizens, we need to demand and shape the change we seek. We can do this, for example, by insisting that financiers invest our money sustainably, that firms disclose environmental conditions along their supply chains (product labelling is a partial method for doing that), and even boycotting products that do not meet standards. Reputation matters, and that can be exploited by citizens (Chapter 6). If we are not acting now, it may in part be because we have grown distant from Nature. Such detachment is in part a symptom of societal change, including growing urbanisation, the profusion of technology, and reduced access to green spaces. Detachment from Nature has meant a loss in our physical and emotional state.

Chapter 11 reported the role played by our direct experiences of biodiversity in our personal well-being. Two aspects of those experiences were distinguished: *contact* with Nature and *connectedness* with Nature. As well as direct interaction with Nature, the former could even involve interaction with the natural world via indoor plants or from virtual representations of Nature such as photographs or paintings of natural landscapes. The latter refers to a person's sense of connectedness with the natural world, it reflects the extent to which she internalises the experiences she has with Nature. If contact with the natural world is a means to furthering personal well-being, connectedness with Nature is an aspect of well-being itself.

Access to green spaces (they are local public goods) can also reduce socio-economic inequalities in health. Interventions to increase people's contact and connectedness with Nature would not only improve our health and well-being, there is a growing body of evidence to suggest that those interventions would also motivate us to make informed choices and demand change.

There are glimmers of hope here. Examples of Nature renewal in urban environments are one. That our desires and wants are to a significant extent socially embedded is also a cause for hope that economising on the use of Nature's provisioning goods – among the many throughout the world who enjoy high material standards of living – would prove to be personally less costly than they fear, provided of course the economy is shared. Contemporary conceptions of the human person cut against the grains of our sociability; we are encouraged to imagine that the thriftiness we may feel should be practised by all will fall on us unilaterally. That probably explains why most of us do not act with our neighbours to practise it.

So, there are grounds for hope. The grounds no doubt have involved small initiatives so far, but the economics of biodiversity is not the preserve of the large. The conception we would all wish to adopt is grand, but it is ultimately we citizens who will determine whether we are able live in peace with Nature.

21.3.4 Education

Our increasing detachment from Nature has accompanied the increasing detachment from Nature of economic reasoning. Many view Nature almost entirely through an anthropocentric lens, even while our physical interaction with and emotional attachment to Nature declines. It would seem that if we are

Chapter 21: Options for Change

to appreciate our place in Nature, we have to educate ourselves.[569] We would only then begin to appreciate the infinitely beautiful tapestry of Nature's processes and forms.

Every child in every country is owed the teaching of natural history, to be introduced to the awe and wonder of the natural world, and to appreciate how it contributes to our lives. Establishing the natural world within educational policy would contribute to countering the shifting baseline, whereby we progressively redefine ourselves as inhabitants of an emptying world and believe that what we see is how it is and how it will continue to be. This shifting baseline has been termed the 'extinction of experience' (Pyle, 1993).

Achieving tangible effects, however, is not straightforward. The development and design of environmental education programmes can be directed to overcome the problems. In their wide-ranging survey documenting direct impacts of environmental education, Ardoin, Bowers, and Gaillard (2020) have suggested that focusing on local issues or locally relevant dimensions of broader issues, such as collaborating with scientists, resource managers and community organisations, is of enormous help. Our emphasis in the Review on the role of communities and civil societies in the economics of biodiversity is consonant with that line of thinking.[570]

But even that would not be enough. Connecting with Nature needs to be woven throughout our lives. The connection has been found to decline from childhood to an overall low in the mid-teens, followed by a steady rise that reaches a plateau lasting the rest of one's life (Hughes et al. 2019). It is a cruel irony that we surround children with pictures and toys of animals and plants, only to focus subsequently on more conceptual knowledge, marginalising environmental education relative to the wider curriculum. Even if we had studied Nature in primary school, we may not have encountered the subject subsequently.[571] In universities in the United States, it was common practice to require first-year students to complete a course on a broad-brush history of civilisation. There is every reason universities should require new students to attend a course on basic ecology. Field studies that would accompany such a course would be a way to connect students with Nature, in particular those who may have grown up in an urban environment. Understanding even the simplest of the biosphere's processes may well be the first step toward developing a love of Nature.

There is a further reason for connecting people with Nature at an early age. The three pervasive features of Nature – mobility, silence, and invisibility – mean that the consequences of actions which desecrate Nature are often untraceable to those who are responsible. Neither the rule of law nor the dictates of social norms are sufficient to make us account for Nature in our daily practices. Institutional rules, no matter how well designed, would be insufficient for eliminating environmental externalities. We will have to rely also on *self-enforcement*, that is, be our own judge and jury. And that cannot happen unless we create an environment in which, from an early age, we are able to connect with Nature.

Correct economic reasoning is entangled with our values. Biodiversity does not only have instrumental value, it also has existence and intrinsic value, perhaps even moral worth. Each of these senses is enriched when we recognise that we are embedded in Nature. To detach Nature from economic reasoning is to imply that we consider ourselves to be external to Nature. The fault is not in economics; it lies in the way we have chosen to practise it.

---

[569] We are grateful to Mary Colwell for preparing a note for us on the place of education in the economics of biodiversity. This section is adapted from her piece.

[570] In a study of institutions of higher education in India, Bawa et al. (2020) have advocated that agricultural colleges should now include biodiversity science in their curricula.

[571] There are signs of change. The charity iAfrica is experimenting with a digital education programme on Nature conservation in schools in sub-Saharan Africa. In the UK, a GCSE in Natural History is expected to be introduced in England in 2022.

Appendix

Acronyms

This is a list of acronyms which are used frequently in the Review.

BCE	Before Common Era
BCR	Benefit Cost Ratio
BII	Biodiversity Intactness Index
CBD	Convention on Biological Diversity
CDR	Consumption Discount Rate
CE	Common Era
CICES	Common International Classification of Ecosystem Services
CITES	Convention on International Trade in Endangered Species of Wild Fauna and Flora
CO_2	Carbon dioxide
CPRs	Common Pool Resources
CVM	Contingent Valuation Method
DM	Decision-maker
EBV	Essential Biodiversity Variable
EEZ	Exclusive Economic Zone
E/MSY	Extinctions per Million Species per Year
GDP	Gross Domestic Product
GM	Genetically Modified
GHG	Greenhouse Gas
GEP	Gross Ecosystem Product
GNP	Gross National Product
HMT	Her Majesty's Treasury
INNS	Invasive Non-Native Species
IPBES	Intergovernmental Science-Policy Platform on Biodiversity and Ecosystem Services
IPCC	Intergovernmental Panel on Climate Change
LPI	Living Planet Index
MA	Millennium Ecosystem Assessment
MDB	Multilateral Development Bank
MEA	Multilateral Environmental Agreement

Acronyms

MSA	Mean Species Abundance
NbS	Nature-based Solutions
NDP	Net Domestic Product
NGO	Non-Governmental Organisation
NPP	Net Primary Productivity
NPV	Net Present Value
ODA	Official Development Assistance
PDV	Present Discounted Value
PES	Payment for Ecosystem Services
PPP	Purchasing Power Parity
R&D	Research and Development
RLI	Red List Index
RCPs	Representative Concentration Pathways
SDGs	Sustainable Development Goals
SSPs	Shared Socio-economic Pathways
TFP	Total Factor Productivity
TFR	Total Fertility Rate
UNCLOS	United Nations Convention on the Law of the Sea
UN DESA	United Nations Department of Economic and Social Affairs
UNDP	United Nations Development Programme
UNEP	United Nations Environment Programme
UNFCCC	United Nations Framework Convention on Climate Change
UNFPA	United Nations Population Fund
UNPD	United Nations Population Division
UN SEEA	UN System of Environmental-Economic Accounting
UN SNA	UN System of National Accounts
VSL	Value of a Statistical Life
WTFR	Wanted Total Fertility Rate

Glossary

Abiotic – Having to do with the chemical, geological, and physical aspects of an entity i.e. the non-living components (Levin, 2009).

Accounting price – The contribution that an additional unit of a good, service or asset makes to intergenerational well-being, other things equal. In simple terms, accounting prices reflect the true value to society of any good, service or asset. Also called 'shadow price' (*Review* definition).

Agent – A person, company, or organisation that has an influence on the economy by producing, buying, or selling (Cambridge Dictionary, 2020).

Aichi (Biodiversity) Targets – The 20 targets set by the Conference of the Parties to the Convention on Biological Diversity (CBD) at its tenth meeting, under the Strategic Plan for Biodiversity 2011–2020 (Convention on Biological Diversity, 2010).

Anthropocene – The proposed name for the current geological age in which human activity has become the dominant influence on the biosphere (Crutzen and Stoermer, 2000).

Arbitrage conditions – Rules governing portfolio selection are summarised in 'arbitrage conditions'. A portfolio is the best for the agent only if the assets in it have the same rate of return (adjusted for risk). An asset that has a lower rate of return than another will not be chosen (*Review* definition).

Assemblage (in ecology) – A taxonomically related group of species populations that occur together in space (Shantz and Sweatman, 2015).

Asset – A durable object, which produces a flow of goods and/or services over time (*Review* definition).

Biodiversity – The variety of life in all its forms, and at all levels, including genes, species, and ecosystems. The CBD defines biodiversity as 'the variability among living organisms from all sources including, inter alia, terrestrial, marine and other aquatic ecosystems and the ecological complexes of which they are part; this includes diversity within species, between species and of ecosystems' (Convention on Biological Diversity).

Biomass – The total mass of living biological material in a given ecosystem at a given time (Levin, 2009).

Biome – Any various generalised regional or global community types, such as tundra or tropical forest, that are characterised by dominant plant life forms and prevailing climate (Levin, 2009).

Biosphere – The living world; the total area of the Earth that is able to support life (Levin, 2009).

Biotic – Having to do with or involving living organisms (Levin, 2009).

Capital goods – There are three categories of capital goods: produced capital, human capital and natural capital (see below) (*Review* definition).

Climate change – The change of climate which is attributed directly or indirectly to human activity that alters the composition of the global atmosphere and which is in addition to natural climate variability observed over comparable time periods (UNFCCC, 1994).

Common pool resources – Spatially-confined ecosystems, such as woodlands and water sources, that are common property of a community, but access by outsiders is not permitted without the community's consent (*Review* definition).

Glossary

Conservation – An action taken to promote the persistence of ecosystems and biodiversity (Adapted from Levin, 2009).

Cultural services – All the non-material, and normally non-rival and non-consumptive, outputs of ecosystems (biotic and abiotic) that affect physical and mental states of people (CICES, 2018).

Depreciation – Decline in the value of an asset over time. In the case of natural capital, when humanity's demand exceeds the regeneration rate, ecosystems depreciate (*Review* definition).

Ecological footprint – The *Review* defines the global ecological footprint as humanity's demands on the biosphere per unit of time (also referred to as 'impact' and 'demand' in the Review). The ecological footprint is affected by the size and composition of our individual demands, the size of the human population, and the efficiency with which we both convert Nature's services to meet our demands and return our waste back into Nature (*Review* definition). The *Global Footprint Network* defines ecological footprint as a measure of how much area of biologically productive land and water an individual, population or activity requires to produce all the resources it consumes and to absorb the waste it generates, using prevailing technology and resource management practices (Global Footprint Network 2020).

Economic evaluation – The Review uses the term 'economic evaluation' to refer to two processes required for managing a society's portfolio of assets: 1) assessing whether economies achieve 'progress' over time (sustainability assessment) and 2) assessing whether an investment, policy or plan will contribute to 'progress' (policy analysis). The Review shows that the index for both is an inclusive measure of wealth (*Review* definition).

Economic model – A theoretical construct representing economic processes by a set of variables and a set of logical and/or quantitative relationships.

Ecosystem – A natural unit consisting of all the plants, animals, and microorganisms (biotic) factors in a given area, interacting with all of the non-living physical and chemical (abiotic) factors of this environment (Levin, 2009).

Ecosystem function – The flow of energy and materials through the biotic and abiotic components of an ecosystem. It includes many processes such as biomass production, trophic transfer through plants and animals, nutrient cycling, water dynamics and heat transfer. Ecosystem functions and processes underpin the production of ecosystem services (adapted from IPBES, 2019b).

Ecosystem services – The contributions that ecosystems make to human well-being. The Review classifies these into provisioning goods, regulating and maintenance services, and cultural services (MA, 2005a; Haines-Young and Potschin, 2018).

Efficient portfolio – This occurs when assets in a portfolio yield the same rate of return, as estimated by the manager, corrected for risk.

Eukaryote – An organism that consists of one or more cells each of which has a nucleus and other specialised organelles. It includes animals, plants and fungi (Cole, 2016).

Enabling assets – Assets that confer value to other capital goods by facilitating their use. They are not included in the Review's three-way classification of capital goods for Inclusive Wealth (human capital, natural capital and produced capital) but endow them with their social worth (*Review* definition).

Expected utility – The subjective weighted average of all possible outcomes for utility, with the weights being assigned by the likelihood, or probability, that any particular event will occur (Oxford Dictionary, 2020).

Glossary

Externality – A positive or negative consequence (benefits or costs) of an action that affects someone other than the agent undertaking that action and for which the agent is neither directly compensated nor penalised.

Factors of production – The inputs needed for the creation of a final good or service (output).

Functional diversity – The variety and number of species that fulfil different functional roles in a community or ecosystem (Levin, 2009).

Gross Domestic Product (GDP) – The market value of the flow of all final goods and services produced within a country in a given year.

Gross National Product (GNP) – GDP plus incomes earned by residents from overseas earnings, minus incomes earned within the economy by overseas residents.

Holocene – The current geological age, which began approximately 11,000 years ago, after the last glacial period.

Human capital – This refers to the productive wealth embodied in labour, skills and knowledge.

Hysteresis (ecological) – When the return trajectory from one stable ecological state to another is different from the outgoing trajectory. It may require more 'work' to recover and might even be irreversible (Beisner, 2012).

Impact Inequality – The Review's term for the overshoot in humanity's demands on the biosphere and its supply. That difference has been widening in recent decades. Because demand is decomposed into its several factors, the Impact Inequality points to policy levers that can help steer the global economy toward equality between supply and demand on a sustainable basis (*Review* definition).

Inclusive investment – The change in inclusive wealth over a period of time, which is Net Domestic Product minus aggregate consumption (Dasgupta, 2004). Hamilton and Clemens (1999) called it 'genuine saving'.

Inclusive wealth – The social value (based on accounting prices) of an economy's total stock of natural, produced and human capital assets.

Intergenerational well-being – The (possibly discounted) sum of the well-being of all who are here today and all who will ever be born.

Institutional failure – These include (i) law and policy failures (e.g. perverse subsidies), (ii) market failures (arising from externalities), (iii) organisational failures (e.g. lack of transparency and political legitimacy in decision making) and (iv) informal institutional failures (e.g. breakdown of social norms due to erosion of trust).

Institution – An established law, custom, usage, practice, organisation, or other element in the political or social life of a people (Oxford English Dictionary, 2020). More broadly, institutions are the arrangements that govern collective undertakings, including legal entities like the modern firm, communitarian associations, markets, rural networks, households and governments.

International dollars – An international dollar would buy in a specific place a comparable amount of goods and services as a US dollar in the United States (i.e. used interchangeably with dollars in Purchasing Power Parity terms). It provides a way of comparing incomes or prices that accounts for the fact that prices differ from place to place.

Keystone species – A species that has a disproportionately large impact on ecosystem structure and function relative to its own abundance (Levin 2009).

Glossary

Marginal product – The change in output as a result of one more unit of a particular input.

Market price – The price at which a good, service or asset is exchanged in a market.

Nash equilibrium – A course of actions (strategy, in game theoretic parlance) per party, such that no party would have any reason to deviate from his or her course of actions if all other parties were to pursue their courses of actions.

Natural assets – Naturally occurring living and non-living entities that together comprise ecosystems and deliver ecosystem services that benefit current and future generations (*Review* definition).

Natural capital – The stock of renewable and non-renewable natural assets (e.g. ecosystems) that yield a flow of benefits to people (i.e. ecosystem services). The term 'natural capital' is used to emphasise it is a capital asset, like produced capital (roads and buildings) and human capital (knowledge and skills).

Natural resources – Resources which are naturally occurring, including renewable resources such as forests and non-renewable resources such as minerals.

Nature – The Review uses the term 'Nature' to refer to the natural world.

Nature-related financial risks – Financial risks that arise from changes in the stock and/or condition of natural capital and from societal responses to those changes. These risks can arise from three channels or 'risk factors': physical, transition and litigation.

Net Domestic Product (NDP) – Gross Domestic Product (GDP) less the depreciation of all capital assets.

Net Primary Productivity (NPP) – The biological productivity, that is the rate of conversion in a given location of physical energy (sunlight) into biological energy (through photosynthesis) in the form of organic carbon that becomes available for other trophic levels in the ecosystem (Levin, 2009). In the Review's terminology, NPP is the regenerative rate of a stock of primary producers.

Non-linearity – In a non-linear relationship, when a process is disrupted, the relationship between one variable and another does not increase or decrease proportionately.

Numeraire – A unit of account in terms of which the relative price of other traded goods, services or assets is expressed. The price of the good chosen as numeraire is by definition 1 (i.e. unity).

Open access resources – Natural assets that are open to use by all free of charge, for example fisheries in waters beyond national jurisdiction (*Review* definition).

Own rate of return – An asset's own rate of return is its marginal yield per unit of time (*Review* definition).

Perverse subsidies – Government payments to activities that exploit the biosphere, thereby reducing the price users pay for the global commons from zero to negative figures.

Portfolio – A grouping of assets. Assets in an efficient portfolio yield the same rate of return, as estimated by the manager, corrected for risk.

Present discounted value – The discounted flow of the future net value of an action or good, service or asset. Also referred to as Net Present Value (NPV).

Planetary boundaries – Earth system processes critical for maintaining the stable state of the Holocene, such as biosphere integrity, land-use change and climate change. Although not all these processes have definable single thresholds, crossing the boundaries increases the risk of large-scale, potentially irreversible, environmental changes (Rockström et al. 2009; Steffen et al. 2015).

Glossary

Polycentric institutions – Polycentric structures (that are best placed to protect and promote biodiversity) are layered institutions: global, regional, national, and communitarian. Each layer requires an authority at the apex to achieve coordination below and with other layers laterally (*Review* definition, based on Ostrom, 2010b).

Primary producer – An organism capable of converting atmospheric carbon dioxide into organic matter (Levin, 2009).

Prokaryotes – A cellular organism that lacks a nucleus. It includes the two domains of Bacteria and Archaea (Cole, 2016).

Produced capital – Capital goods embodied in human-made goods or structures, such as roads, buildings, machines, and equipment.

Protected Area – A clearly defined geographical space, recognised, dedicated and managed, through legal or other effective means, to achieve the long-term conservation of Nature and associated ecosystem services and cultural values (IUCN definition, 2008b).

Provisioning goods – The vast range of goods we obtain from ecosystems e.g. food, freshwater, fuel, fibre, medicines, genetic resources and ornamental resources.

Public goods – Goods or services that are neither rivalrous (access to a public good by any one group of people has no effect on the quantity available to others) nor excludable (no one can be excluded from access to the good).

Rate of return – The rate of return on an asset (as opposed to the asset's own rate of return) is its yield plus the capital gains it enjoys over a unit of time (*Review* definition).

Reciprocal externalities – An externality where each party inflicts an unaccounted-for harm (or confers an unaccounted-for benefit) on all others in a defined population (*Review* definition).

Regenerative rate (of an ecosystem) – The rate at which an ecosystem forms new organic matter per unit of item. The regenerative rate of an ecosystem can be measured using the proxy of Net Primary Productivity (NPP).

Regime shift – Substantial reorganisation in ecosystem structure, functions and feedbacks that often occurs abruptly and persists over time (Crépin et al. 2012).

Regulating and maintenance services – All ways in which ecosystems control or modify biotic or abiotic parameters that define the environment of people. These are ecosystem outputs that are not consumed but affect the performance of people and their activities (CICES, 2018).

Replacement Fertility Rate – The total fertility rate (see below) at which a region's population would remain constant in the long run. As a rule of thumb, the replacement fertility rate is taken to be 2.1.

Resilience – The magnitude of disturbance that an ecosystem or society can undergo without crossing a threshold to a situation with different structure or outputs i.e. a different state. Resilience depends on factors such as ecological dynamics as well as the organisational and institutional capacity to understand, manage, and respond to these dynamics (IPBES, 2019b).

Restoration – Any intentional activities that initiate or accelerate the recovery of an ecosystem from a degraded state (IPBES, 2019b).

Risk – This is the probability that an outcome (or investment's actual gains) will differ from an expected outcome (or return).

Glossary

Social capital – Mutual trust and associated norms of reciprocity that enable people to engage with one another (Helliwell and Putnam, 2004).

Social cost-benefit analysis – A systematic approach to estimating the strengths and weaknesses of alternative options for society based on estimating the benefits and costs over time.

Social norm – A mutually accepted and reinforcing rule of behaviour.

Socially embedded preferences – Human preferences which are significantly influenced by the choices of others (e.g. peers, community, family). This can include the desire to compete or conform with others (Barrett et al. 2020).

Species – A fundamental category for the classification and description of organisms, defined in various ways but typically on the basis of reproductive capacity; i.e. the members of a species can reproduce with each other to produce fertile offspring but cannot do so with individuals outside the species (Levin, 2009).

Stability regime – A stable state, in terms of a set of unique biotic and abiotic conditions, in which an ecosystem can exist. Ecosystems can exist under multiple alternative stability regimes. These alternative states are considered stable over ecologically relevant timescales. Ecosystems may transition from one stability regime to another, in what is known as a 'regime shift' when perturbed (Beisner, 2012).

Sustainable – A situation is sustainable if it can persist indefinitely. An unsustainable state of affairs cannot persist indefinitely (*Review* condition).

Sustainable development – Development that meets the needs of the present without compromising the ability of future generations to meet their own needs (Brundtland report) i.e. by bequeathing to its successor at least as large a productive base as it had inherited from its predecessor (*Review* definition).

Sustainable Development Goals – A set of goals adopted by the United Nations in 2015 to end poverty, protect the planet, and ensure prosperity for all, as part of the 2030 Agenda for Sustainable Development.

Tipping point (ecological) – A set of conditions of an ecological system where further perturbation will cause change to a new state and prevent the system from returning to its former state (Lenton, 2012).

Total Factor Productivity (TFP) – A measure of productivity calculated by dividing economy-wide total production by the weighted average of capital inputs. Typically, TFP is the portion of growth in output not explained by growth in traditionally measured inputs of human and produced capital (but not natural capital) used in production. TFP is also referred to as multi-factor productivity (MFP) or Solow residual (Adapted from Hulten (2001).

Total Fertility Rate (TFR) – A region's TFR is the average number of births per woman over her reproductive years (taken to be 15–49).

Trophic level – The position of a given species in the chain of energy or nutrients. In a three-level chain, the top level is taken by predators, the second level by herbivores, and the bottom level by plants (Levin, 2009).

Uncertainty – Any situation in which the current state of knowledge is such that: the order or nature of things is unknown; the consequences, extent, or magnitude of circumstances, conditions, or events is unpredictable; and, credible probabilities to possible outcomes cannot be assigned. Uncertainty can result from lack of information or from disagreement about what is known or even knowable.

Glossary

Unidirectional externalities – Externalities where one agent (or a group of agents) inflicts an unaccounted-for damage or confers an unaccounted-for benefit on another (or others) (*Review* definition).

Unmet need for family planning services – Number of women in a region aged 15 to 49 who are seeking to stop or delay childbearing but are not using modern methods of contraception (Bradley et al. 2012).

Utilitarian/Utilitarianism – The doctrine that an action is right in so far as it promotes happiness or well-being, and that the greatest well-being of the greatest number should be the guiding principle of conduct (Sen and Williams, 1982).

Well-being – A measure of the extent to which a person's informed desires are realised (*Review* definition).

Zoonotic – Describing a disease that can be spread from animals to people (Levin, 2009).

References

Abel, G. J., B. Barakat, S. KC, and W. Lutz (2016), 'Meeting the Sustainable Development Goals Leads to Lower World Population Growth', *Proceedings of the National Academy of Sciences*, 113(50), 14294–14299.

Aburto-Oropeza, O., B. Erisman, G. R. Galland, I. Mascareñas-Osorio, E. Sala, and E. Ezcurra (2011), 'Large Recovery of Fish Biomass in a No-Take Marine Reserve', *PLoS ONE*, 6(8), e23601.

Acemoglu, D. (2008), *Introduction to Modern Economic Growth* (Princeton, NJ: Princeton University Press).

ADB (2004), *African Development Bank Group's Policy on the Environment*.

Adelman, I. and C.T. Morris (1965), "A Factor Analysis of the Interrelationship between Social and Political Variables and per Capita Gross National Product," *Quarterly Journal of Economics*, 79(3), 555–578.

Adler, M. D., and M. Fleurbaey (2016), *The Oxford Handbook of Well-Being and Public Policy* (New York, NY: Oxford University Press).

Agarwal, A., and S. Narain (1992), 'Towards Green Villages, A Strategy for Environmentally Sound and Participatory Rural Development', *Environment and Urbanization*, 4(1), 53–64.

Agarwal, B. (2001), 'Participatory Exclusions, Community Forest, and Gender: An Analysis for South Asia and Conceptual Framework', *World Development*, 29(10), 1623–1648.

Aghion, P., and P. Howitt (1998), *Endogenous Growth Theory* (Cambridge, MA: MIT Press).

Agliardi, E. (2011), 'Sustainability in Uncertain Economies', *Environmental and Resource Economics*, 48(1), 71–82.

Aguilar-Rodríguez, J., J. Payne, and A. Wagner (2017), 'A Thousand Empirical Adaptive Landscapes and Their Navigability', *Nature Ecology and Evolution*, 1(2), 0045.

Ai, N., and K. R. Polenske (2008), 'Socioeconomic Impact Analysis of Yellow-Dust Storms: An Approach and Case Study for Beijing', *Economic Systems Research*, 20(2), 187–203.

Akerlof, G. A. (1982), 'Labor Contracts as Partial Gift Exchange', *The Quarterly Journal of Economics*, 97(4), 543–569.

Albrechts, L. (2004), 'Strategic (Spatial) Planning Re-Examined', Environment and Planning *B: Planning and Design*, 31(5), 743–758.

Alcalá, F., and A. Ciccone (2004), 'Trade and Productivity', *The Quarterly Journal of Economics*, 119(2), 613–646.

Alder, J., and D. Pauly eds. (2008), 'A Comparative Assessment of Biodiversity, Fisheries and Aquaculture in 53 Countries' Exclusive Economic Zones', *Fisheries Centre Research Reports*, 16(7).

Aldy, J. E., S. Barrett and R. N. Stavins (2003), 'Thirteen Plus One: A Comparison of Global Climate Policy Architectures', *Climate Policy*, 3(4), 373–397.

Aldy, J. E., and W. Kip Viscusi (2008), 'Adjusting the Value of a Statistical Life for Age and Cohort Effects', *The Review of Economics and Statistics*, 90(3), 573–581.

References

Alexander, P., A. Berri, D. Moran, D. Reay, and M. D. A. Rounsevell (2020), 'The Global Environmental Paw Print of Pet Food', *Global Environmental Change*, 65, 102153.

Alexander, P., C. Brown, A. Arneth, J. Finnigan, and M. D. A. Rounsevell (2016), 'Human Appropriation of Land for Food: The Role of Diet', *Global Environmental Change*, 41, 88–98.

Alkema, L., V. Kantorova, C. Menozzi, and A. Biddlecom (2013), 'National, Regional, and Global Rates and Trends in Contraceptive Prevalence and Unmet Need for Family Planning between 1990 and 2015: A Systematic and Comprehensive Analysis', *The Lancet*, 381(9878), 1642–1652.

Allcott, H. (2011), 'Social Norms and Energy Conservation', *Journal of Public Economics*, 95 (9-10), 1082–1095.

Alpízar, F., F. Carlsson, and O. Johansson-Stenman (2005), 'How Much Do We Care About Absolute Versus Relative Income and Consumption?', *Journal of Economic Behavior and Organization*, 56(3), 405–421.

Alpízar, F., Madrigal, R., Alvarado, I., Brenes Vega, E., Camhi, A., Maldonado, J. H., Marco, J., Martínez, A., Pacay, E., and G. Watson (2020), 'Mainstreaming of Natural Capital and Biodiversity into Planning and Decision-Making : Cases from Latin America and the Caribbean', *IDB Working Paper Series*, No. IDB-WP-01193.

Alston, L. J., G. D. Libecap, and B. Mueller (1999), *Titles, Conflict, and Land Use: The Development of Property Rights and Land Reform on the Brazilian Amazon Frontier* (Ann Arbor, MI: University of Michigan Press).

Amel-Zadeh, A., and G. Serafeim (2018), 'Why and How Investors Use ESG Information: Evidence from a Global Survey', *Financial Analysts Journal*, 74(3), 87–103.

Anderson, R., and R. M. May (1991), *Infectious Diseases of Humans: Dynamics and Control* (Oxford: Oxford University Press).

Andradi-Brown, D. A. (2019), 'Invasive Lionfish (*Pterois volitans* and *P. miles*): Distribution, Impact, and Management', in Y. Loya, K. A. Puglise, and T. Bridge, eds., *Mesophotic Coral Ecosystems* (Cham: Springer).

Andres, L. A., M. Thibert, C. L. Cordoba, A. V. Danilenko, G. Joseph, and C. Borja-Vega (2019), *Doing More with Less: Smarter Subsidies for Water Supply and Sanitation.*

Angel, S., Parent, J., Civco, D. L., Blei, A., and D. Potere (2011), 'The Dimensions of Global Urban Expansion: Estimates and Projections For all Countries, 2000-2050', *Progress in Planning*, 75(2), 53–107.

Ango, T. G., L. Börjeson, F. Senbeta, and K. Hylander (2014), 'Balancing Ecosystem Services and Disservices: Smallholder Farmers' Use and Management of Forest and Trees in an Agricultural Landscape in Southwestern Ethiopia', *Ecology and Society*, 19(1), 30.

Ansar, A., B. Flyvbjerg, A. Budzier, and D. Lunn (2014), 'Should We Build More Large Dams? The Actual Costs of Hydropower Megaproject Development', *Energy Policy*, 69, 43–56.

Ansell, D., D. Freudenberger, N. Munro, and P. Gibbons (2016), 'The Cost-Effectiveness of Agri-Environment Schemes for Biodiversity Conservation: A Quantitative Review', *Agriculture, Ecosystems and Environment*, 225, 184–191.

Araújo, M. B., F. Ferri-Yáñez, F. Bozinovic, P. A. Marquet, F. Valladares, and S. L. Chown (2013), 'Heat Freezes Niche Evolution', *Ecology Letters*, 16(9), 1206–1219.

Ardoin, N. M., A. W. Bowers, and E. Gaillard (2020), 'Environmental Education Outcomes for Conservation: A Systematic Review' *Biological Conservation*, 241, e108224.

Aredo, D. (2010), 'The Iddir: An Informal Insurance Arrangement in Ethiopia', *Savings and Development*, 34(1), 53–72.

Arkema, K. (2019), 'Caribbean: Implementing Successful Development Planning and Investment Strategies', in L. Mandle, Z. Ouyang, J. Salzman, and G. C. Daily, eds., *Green Growth That Works* (Washington, DC: Island Press).

Arkema, K. K., R. Griffin, S. Maldonado, J. Silver, J. Suckale, and A. D. Guerry (2017), 'Linking Social, Ecological, and Physical Science to Advance Natural and Nature-Based Protection for Coastal Communities', *Annals of the New York Academy of Sciences*, 1399(1), 5–26.

Arnott, R., and J. E. Stiglitz (1991), 'Moral Hazard and Nonmarket Institutions: Dysfunctional Crowding Out of Peer Dysfunctional Crowding Out or Peer Monitoring?', *The American Economic Review*, 81(1), 179–190.

Aronson, J., J. N. Blignaut, S. J. Milton, D. Le Maitre, K. J. Esler, A. Limouzin, C. Fontaine, M. P. De Wit, W. Mugido, P. Prinsloo, L. Van Der Elst, and N. Lederer (2010), 'Are Socioeconomic Benefits of Restoration Adequately Quantified? A Meta-Analysis of Recent Papers (2000-2008) in Restoration Ecology and 12 Other Scientific Journals', *Restoration Ecology*, 18(2), 143–154.

Arora, L., and A. Narula (2017), 'Gene Editing and Crop Improvement Using CRISPR-Cas9 System', *Frontiers in Plant Science*, 8, 1932.

Arrow, K.J. (1951), *Social Choice and Individual Values* (New York: Wiley).

Arrow, K. J. (1965), *Aspects of the Theory of Risk Bearing* (Helsinki: Yrjö Jahnssonin Säätiö). Arrow, K. J. (1970), 'Political and Economic Evaluation of Social Effects and Externalities', in J. Margolis, ed., *The Analysis of Public Output* (National Bureau Committee for Economic Research).

Arrow, K. J. (1973), 'Rawls's Principle of Just Saving', *The Swedish Journal of Economics*, 75(4), 323–335.

Arrow, K. J. (1999), 'Discounting, Morality and Gaming', in P. R. Portney and J. P. Weyant, eds., *Discounting and Intergenerational Equity* (Washington, DC: Resources for the Future).

Arrow, K. J. (2000), 'Observations on Social Capital', in P. Dasgupta and I. Serageldin, eds., *Social Capital: A Multifaceted Perspective* (Washington, DC: World Bank).

Arrow, K. J., B. Bolin, R. Costanza, P. Dasgupta, C. Folke, C. S. Holling, B-O. Jansson, S. Levin, K.-G. Mäler, C. Perrings, and D. Pimentel (1995), 'Economic Growth, Carrying Capacity, and the Environment', *Science*, 268(5210), 520–521.

Arrow, K. J., M. L. Cropper, C. Gollier, B. Groom, G. M. Heal, R. G. Newell, W. D. Nordhaus, R. S. Pindyck, W. A. Pizer, P. R. Portney, T. Sterner, R. S. J. Tol, and M. L. Weitzman (2014), 'Should Governments Use a Declining Discount Rate in Project Analysis?', *Review of Environmental Economics and Policy*, 8(2), 145–163.

Arrow, K. J., G. Daily, P. Dasgupta, P. Ehrlich, L. Goulder, G. Heal, S. Levin, K. -G. Mäler, S. Schneider, D. Starrett, and B. Walker, (2007), 'Consumption, Investment, and Future Well-Being: Reply to Daly et al.', *Conservation Biology*, 21(5), 1363–1365.

Arrow, K. J., and P. Dasgupta (2009), 'Conspicuous Consumption, Inconspicuous Leisure', *The Economic Journal*, 119(541), F497–F516.

References

Arrow, K. J., P. Dasgupta, L. Goulder, G. Daily, P. Ehrlich, G. Heal, S. Levin, K.-G. Mäler, S. Schneider, D. Starrett, and B. Walker (2004), 'Are We Consuming Too Much?', *Journal of Economic Perspectives*, 18(3), 147–172.

Arrow, K. J., P. Dasgupta, L. H. Goulder, K. J. Mumford, and K. Oleson (2012), 'Sustainability and the Measurement of Wealth', *Environment and Development Economics*, 17(3), 317–353.

Arrow, K. J., P. Dasgupta, L. H. Goulder, K. J. Mumford, and K. Oleson (2013), 'Sustainability and the Measurement of Wealth: Further Reflections', *Environment and Development Economics*, 18(4), 504–516.

Arrow, K. J., P. Dasgupta, and K.-G. Mäler (2003a), 'Evaluating Projects and Assessing Sustainable Development in Imperfect Economies', *Environmental and Resource Economics*, 26, 647–685.

Arrow, K. J., P. Dasgupta, and K.-G. Mäler (2003b), 'The Genuine Savings Criterion and the Value of Population', *Economic Theory*, 21(2/3), 217–225.

Arrow, K. J., and A. C. Fisher (1974), 'Environmental Preservation, Uncertainty, and Irreversibility', *The Quarterly Journal of Economics*, 88(2), 312–319.

Arrow, K. J., and L. Hurwicz (1977), 'Appendix: An Optimality Criterion for Decision-Making Under Ignorance' in K. J. Arrow and L. Hurwicz, eds., *Studies in Resource Allocation Processes* (Cambridge: Cambridge University Press).

Arrow, K. J., and M. Kurz (1970), *Public Investment, the Rate of Return, and Optimal Fiscal Policy* (Baltimore, MD: Johns Hopkins Press) 1st Edition.

Arrow, K. J., and M. Kurz (2013), *Public Investment, the Rate of Return, and Optimal Fiscal Policy* (New York, NY: Earthscan).

Arrow, K. J., and M. Priebsch (2014), 'Bliss, Catastrophe, and Rational Policy', *Environmental and Resource Economics*, 58(4), 491–509.

Arrow, K. J., R. Solow, P. R. Portney, E. E. Leamer, R. Radner, and H. Schuman (1993), 'Report of The NOAA Panel on Contingent Valuation', *Federal register*, 58, 4601–4614.

Asian Development Bank (2016), *Nature-based Solutions for Building Resilience in Towns and Cities: Case studies from the Greater Mekong Subregion*.

Asian Development Bank (2018), *The Korea Emissions Trading Scheme: Challenges and Emerging Opportunities*.

Aspers, P., and F. Godart (2013), 'Sociology of Fashion: Order and Change', *Annual Review of Sociology*, 39, 171–192.

Asquith, N. M., M. T. Vargas, and S. Wunder (2008), 'Selling Two Environmental Services: In- Kind Payments for Bird Habitat and Watershed Protection in Los Negros, Bolivia', *Ecological Economics*, 65(4), 675–684.

Atkinson, A. B (1970), 'On the Measurement of Inequality', *Journal of Economic Theory*, 2(3), 244–263.

Attanasio, O. P., and G. Weber (2010), 'Consumption and Saving: Models of Intertemporal Allocation and Their Implications for Public Policy', *Journal of Economic Literature*, 48(3), 693–751.

Aunger, R., and V. Curtis (2016), 'Behaviour Centred Design: Towards an Applied Science of Behaviour Change', *Health Psychology Review*, 10(4), 425–446.

Babalola, S., L. Folda, and H. Babayaro (2008), 'The Effects of a Communication Program on Contraceptive Ideation and Use Among Young Women in Northern Nigeria', *Studies in Family Planning*, 39(3), 211–220.

Baccini, A., W. Walker, L. Carvalho, M. Farina, D. Sulla-Menashe, and R. A. Houghton (2017), 'Tropical Forests are a Net Carbon Source Based on Above Ground Measurements of Gain and Loss', *Science*, 358(6360), 230–234.

Bahaj, S., and A. Foulis (2016), 'Macroprudential Policy Under Uncertainty', *Bank of England Staff Working Paper*, 584.

Baland, J.-M., and J.-P. Platteau (1996), *Halting Degradation of Natural Resources: Is There a Role for Rural Communities?* (Oxford: Oxford University Press).

Baland, J.-M., and J.-P. Platteau (1999), 'The Ambiguous Impact of Inequality and Collective Action in Local Resource Management', *World Development*, 27(5), 773–788.

Balasubramanian, R. (2008), 'Community Tanks vs Private Wells: Coping Strategies and Sustainability Issues in South India' in R. Ghate, N.S. Jodha, and P. Mukhopadhyay, eds., *Promise, Trust and Evolution: Managing the Commons of South Asia* (Oxford: Oxford University Press).

Baldwin, R. (2016), *The Great Convergence: Information Technology and the New Globalization* (Cambridge, MA: Harvard University Press).

Baldwin, R., and P. Martin (1999), 'Two Waves of Globalisation: Superficial Similarities, Fundamental Differences', *National Bureau of Economic Research Working Paper*, No. 6904.

Balian, E. V., H. Segers, C. Lévèque, and K. Martens (2008), 'The Freshwater Animal Diversity Assessment: An Overview of the Results', *Hydrobiologia*, 595(1), 627–637.

Baliga, S., and E. Maskin (2003), 'Mechanism Design for the Environment', in K. -G. Mäler and J. R. Vincent, eds., *Handbook of Environmental Economics, Vol. I: Environmental Degradation and Institutional Responses* (Amsterdam: Elsevier).

Balmford, A. (2012), *Wild Hope: On the Front Line of Conservation Success* (Chicago: The University of Chicago Press).

Balmford, A., R. Green, and B. Phalan (2015), 'Land for Food and Land for Nature?', *Daedalus*, 144(4), 57–75.

Balvanera, P., and A. Pfaff (2019), *The IPBES Global Assessment on Biodiversity and Ecosystem Services – Chapter 2. Status and Trends; Chapter 2.1. Status and Trends – Drivers of Change (Draft)*.

Banco de México and UNEP (2020), *Climate and Environmental Risks and Opportunities in Mexico's Financial System from Diagnosis to Action*.

Bandyopadhyay, S., and P. Shyamsundar (2004), 'Fuelwood Consumption and Participation in Community Forestry in India', *World Bank Policy Research Working Paper*, No. 3331.

Banerjee, O., M. Cicowiez, Ž. Malek, P. H. Verburg, R. Vargas, and S. Goodwin (2020), 'The Value of Biodiversity in Economic Decision Making: Applying the IEEM+ESM Approach to Conservation Strategies in Colombia', *IDB Working Paper Series 01193*.

Bank of England (2018), *Transition in Thinking: The Impact of Climate Change on the UK Banking Sector*.

Bank of England (2020), *Bank of England Monetary Policy Report August 2020*.

References

Bank of International Settlements (2020a), *The Green Swan: Central Banking and Financial Stability in The Age of Climate Change*.

Bank of International Settlements (2020b), *Credit to the Non-Financial Sector*.

Baragwanath, K., and E. Bayi (2020), 'Collective Property Rights Reduce Deforestation in the Brazilian Amazon', *Proceedings of the National Academy of Sciences*, 117(34), 20495–20502.

Barbarossa, V., R. J. P. Schmitt, M. A. J. Huijbregts, C. Zarfl, H. King, and A. M. Schipper (2020), 'Impacts of Current and Future Large Dams on the Geographic Range Connectivity of Freshwater Fish Worldwide', *Proceedings of the National Academy of Sciences*, 117(7), 3648–3655.

Barbier, E. B. (1994), 'Valuing Environmental Functions: Tropical Wetlands', *Land Economics*, 70(2), 155–173.

Barbier, E. B. (2005), *Natural Resources and Economic Development* (Cambridge: Cambridge University Press).

Barbier, E. B. (2011), *Scarcity and Frontiers: How Economies Have Developed Through Natural Resource Exploitation* (Cambridge: Cambridge University Press).

Barbier, E. B. (2016), 'The Protective Value of Estuarine and Coastal Ecosystem Services in a Wealth Accounting Framework', *Environmental and Resource Economics*, 64(1), 37–58.

Barbier, E. B. (2019), *The Water Paradox: Overcoming the Global Crisis in Water Management* (New Haven, CT: Yale University Press).

Barbier, E. B., and M. Cox (2003), 'Does Economic Development Lead to Mangrove Loss? A Cross Country Analysis', *Contemporary Economic Policy*, 21(4), 418–432.

Barbier, E. B., J. C. Burgess, and T. J. Dean (2018), 'How to Pay for Saving Biodiversity', *Science*, 360(6388), 486–488.

Bardhan, P. K., and J. Dayton-Johnson (2007), 'Inequality and the Governance of Water Resources in Mexico and South India', in J. -M. Baland, P. Bardhan, and S. Bolwes, eds., *Inequality, Cooperation, and Environmental Sustainability* (Princeton, NJ: Princeton University Press).

Bardhan, P. K., and C. Udry (1999), *Development Microeconomics* (Oxford: Oxford University Press).

Barnosky, A. D. (2008), 'Megafauna Biomass Tradeoff as a Driver of Quaternary and Future Extinctions', *Proceedings of the National Academy of Sciences*, 105(1), 11543–11548.

Bar-On, Y. M., R. Phillips, and R. Milo (2018), 'The Biomass Distribution on Earth', *Proceedings of the National Academy of Sciences*, 115(25), 6506–6511.

Barrett, S. (2003), *Environment and Statecraft: The Strategy of Environmental Treaty-Making* (Oxford: Oxford University Press)

Barrett, S. (2012), 'Credible Commitments, Focal Points and Tipping: The Strategy of Climate Treaty Design', in R. Hahn and A. Ulph, eds., *Climate Change and Common Sense: Essays in Honour of Tom Schelling* (Oxford: Oxford University Press).

Barrett, S. (2015), 'Why Have Climate Negotiations Proved So Disappointing?' in P. Dasgupta, R. Ramanathan, and S. Sorondo, eds., *Sustainable Humanity, Sustainable Nature: Our Responsibility* (Vatican City: Vatican Press).

Barrett, S., and A. Dannenberg (2012), 'Climate Negotiations Under Scientific Uncertainty', *Proceedings of the National Academy of Sciences*, 109(43), 17372–17376.

Barrett, S., and A. Dannenberg (2014a), 'Negotiating to Avoid "Gradual" versus "Dangerous" Climate Change: An Experimental Test of Two Prisoners' Dilemmas', in T. L. Cherry, J. Hovi, and D. M. McEvoy, eds., *Toward a New Climate Agreement: Conflict, Resolution and Governance*(Abingdon: Routledge).

Barrett, S., and A. Dannenberg (2014b), 'Sensitivity of Collective Action to Uncertainty About Climate Tipping Points', *Nature Climate Change*, 4, 36–39.

Barrett, S., A. Dasgupta, P. Dasgupta, W. N. Adger, J. Anderies, J. van den Bergh, C. Bledsoe, J. Bongaarts, S. Carpenter, F. S. Chapin III, A.-S. Crépin, G. Daily, P. Ehrlich, C. Folke, N. Kautsky, E. F. Lambin, S. A. Levin, K.-G. Mäler, R. Naylor, K. Nyborg, S. Polasky, M. Scheffer, J. Shogren, P. S. Jørgensen, B. Walker, and J. Wilen (2020), 'Social Dimensions of Fertility Behavior and Consumption Patterns in the Anthropocene', *Proceedings of the National Academy of Sciences*, 117(12), 6300–6307.

Barrett, S., and M. Hoel (2007), 'Optimal Disease Eradication', *Environment and Development Economics*, 12(5), 627–652.

Barrett, S., and R. Stavins (2003), 'Increasing Participation and Compliance in International Climate Change Agreements', *International Environmental Agreements*, 3, 349–376.

Barro, R. J., and X. Sala-i-Martin (2003), *Economic Growth* (Cambridge, MA: MIT Press).

Barrows, G., S. Sexton, and D. Zilberman (2014), 'Agricultural Biotechnology: The Promise and Prospects of Genetically Modified Crops', *Journal of Economic Perspectives*, 28(1), 99–120.

Barry, J. P., C. Lovera, K. R. Buck, E. T. Peltzer, J. R. Taylor, P. Walz, P. J. Whaling, and P. G. Brewer (2014), 'Use of a Free Ocean CO2 Enrichment (FOCE) System to Evaluate the Effects of Ocean Acidification on the Foraging Behavior of a Deep-Sea Urchin', *Environmental Science and Technology*, 48(16), 9890–9897.

Bass, R., H. Dithrich, and A. Mudaliar (2018), *Global Impact Investing Network: 2018 Annual Impact Investor Survey*.

Bassen, A., T. Busch, K. Lopatta, E. Evans, and O. Opoku (2019), *Nature Risks Equal Financial Risks: A Systematic Literature Review*.

Basu, A. M., and S. Amin (2000), 'Conditioning Factors for Fertility Decline in Bengal: History, Language Identity, and Openness to Innovations', *Population and Development Review*, 26(4), 761–794.

Batáry, P., L. V. Dicks, D. Kleijn, and W. J. Sutherland (2015), 'The Role of Agri-Environment Schemes in Conservation and Environmental Management', *Conservation Biology*, 29(4), 1006–1016.

Bateman, I. J., M. Agarwala, A. Binner, E. Coombes, B. Day, S. Ferrini, C. Fezzi, M. Hutchins, A. Lovett, and P. Posen (2016), 'Spatially Explicit Integrated Modeling and Economic Valuation of Climate Driven Land Use Change and its Indirect Effects', *Journal of Environmental Management*, 181, 172–184.

Bateman, I. J., A. R. Harwood, G. M. Mace, R. T. Watson, D. J. Abson, B. Andrews, A. Binner, A. Crowe, B. H. Day, S. Dugdale, C. Fezzi, J. Foden, D. Hadley, R. Haines-Young, M. Hulme, A. Kontoleon, A. A. Lovett, P. Munday, U. Pascual, J. Paterson, G. Perino, A. Sen, G. Siriwardena, D. Van Soest, M. Termansen (2013), 'Bringing Ecosystem Services into Economic Decision-Making: Land Use in the United Kingdom', *Science*, 341(6141), 45–50.

Bateman, I. J., and Mace, G. M. (2020), The Natural Capital Framework for Sustainably Efficient and Equitable Decision Making', *Nature Sustainability*, 3(10), 776–783.

References

Batllori, E., M. A. Parisien, S. A. Parks, M. A. Moritz, and C. Miller (2017), 'Potential Relocation of Climatic Environments Suggests High Rates of Climate Displacement Within the North American Protection Network', *Global Change Biology*, 23(8), 3219–3230.

Bauer, P. T. (1981), *Equality, the Third World and Economic Delusion* (London: Weidenfeld and Nicolson).

Baumgärtner, S., A. M. Klein, D. Thiel, and K. Winkler (2015), 'Ramsey Discounting of Ecosystem Services', *Environmental and Resource Economics*, 61(2), 273–296.

Bavel, J. J. V., K. Baicker, P. S. Boggio, V. Capraro, A. Cichocka, M. Cikara, M. J. Crockett, A. J. Crum, K. M. Douglas, J. N. Druckman, J. Drury, O. Dube, N. Ellemers, E. J. Finkel, J. H. Fowler, M. Gelfand, S. Han, S. A. Haslam, J. Jetten, S. Kitayama, D. Mobbs, L. E. Napper, D. J. Packer, G. Pennycook, E. Peters, R. E. Petty, D. G. Rand, S. D. Reicher, S. Schnall, A. Shariff, L. J. Skitka, S. S. Smith, C. R. Sunstein, N. Tabri, J. A. Tucker, S. van der Linden, P. van Lange, K. A. Weeden, M. J. A. Wohl, J. Zaki, S. R. Zion and R. Willer (2020), 'Using Social and Behavioural Science to Support COVID-19 Pandemic Response', *Nature Human Behaviour*, 4(5), 460–471.

Bawa, K. S., N. Nawn, R. Chellam, J. Krishnaswamy, V. Mathur, S. B. Olsson, N. Pandit, P. Rajagopal, M. Sankaran, R. U. Shaankar, D. Shankar, U. Ramakrishnan, A. T. Vanak, and S. Quader (2020), 'Opinion: Envisioning a Biodiversity Science for Sustaining Human Well-Being', *Proceedings of the National Academy of Sciences*, 117(42), 25951–25955.

Baylis, K., J. Honey-Rosés, J. Börner, E. Corbera, D. Ezzine-de-Blas, P. J. Ferraro, R. Lapeyre, U. M. Persson, A. Pfaff, and S. Wunder (2016), 'Mainstreaming Impact Evaluation in Nature Conservation', *Conservation Letters*, 9(1), 58–64.

Bayraktarov, E., M. I. Saunders, S. Abdullah, M. Mills, J. Beher, H. P. Possingham, P. J. Mumby, and C. E. Lovelock (2015), 'The Cost and Feasibility of Marine Coastal Restoration', *Ecological Applications*, 26(4), 1055–1074.

Beach, T., S. Luzzadder-Beach, and N. P. Dunning (2019), 'Out of the Soil: Soil (Dark Matter Biodiversity) and Societal 'Collapses' from Mesoamerica to Mesopotamia and Beyond', in P. Dasgupta, P. H. Raven, and A. L. McIvor, eds., *Biological Extinction: New Perspectives* (Cambridge: Cambridge University Press).

Beach, T., S. Luzzadder-Beach, S. Krause, T. Guderjan, F. Valdez Jr., J. C. Fernandez-Diaz, S. Eshleman, and C. Doyle (2019), 'Ancient Maya Wetland Fields Revealed Under Tropical Forest Canopy from Laser Scanning and Multiproxy Evidence', *Proceedings of the National Academy of Sciences*, 116(43), 21469–21477.

Beacham, A. M., L. H. Vickers, and J. M. Monaghan (2019), 'Vertical Farming: A Summary of Approaches to Growing Skywards', *Journal of Horticultural Science and Biotechnology*, 94(3), 277–283.

Beatty, C. R., L. Raes, A. L. Vogl, P. L. Hawthorne, M. Moraes, J. L. Saborio, and K. M. Prado (2018), *Landscapes, at your service: Applications of the Restoration Opportunities Optimization Tool (ROOT)*.

Beck, T., and C. Nesmith (2001), 'Building on Poor People's Capacities: The Case of Common Property Resources in India and West Africa', *World Development*, 29(1), 119–133.

Becker, C D. (1999), 'Protecting a Garua Forest in Ecuador: The Role of Institutions and Ecosystem Valuation', *Ambio*, 28(2), 156–161.

Becker, C. D., A. Agreda, E. Astudillo, M. Costantino, and P. Torres (2005), 'Community-Based Monitoring of Fog Capture and Biodiversity at Loma Alta, Ecuador Enhance Social Capital and Institutional Cooperation', *Biodiversity and Conservation*, 14(11), 2695–2707.

Becker, C. D, and K. Ghimire (2003), 'Synergy Between Traditional Ecological Knowledge and Conservation Science Supports Forest Preservation in Ecuador', *Conservation Ecology*, 8(1).

Becker, G. S., T. J. Philipson, and R. R. Soares (2005), 'The Quantity and Quality of Life and The Evolution of World Inequality', *American Economic Review*, 95(1), 277–291.

Behrens, K.G. (2012), "Moral Obligations Towards Future Generations in African Thought," *Journal of Global Ethics*, 8(2-3), 179–191.

Beisner, B. E. (2012), 'Alternative Stable States' *Nature Education Knowledge* 3(10), 33.

Bellmann, C., B. Lee, and J. Hepburn (2019), *Delivering Sustainable Food and Land Use Systems: The Role of International Trade* (London: The Hoffmann Centre for Sustainable Resource Economy).

Bello, C., M. Galetti, M. A. Pizo, L. F. S. Magnago, M. F. Rocha, R. A. F. Lima, C. A. Peres, O. Ovaskainen, and P. Jordano (2015), 'Defaunation Affects Carbon Storage in Tropical Forests', *Science Advances*, 1(11), e1501105.

Benayas, J. M. R., A. C. Newton, A. Diaz, and J. M. Bullock (2009), 'Enhancement of Biodiversity and Ecosystem Services by Ecological Restoration: A Meta-Analysis', *Science*, 325(5944), 1121–1124.

Bender, R. (2019), 'Bayer Shareholders Signal Loss of Confidence in CEO', *Wall Street Journal*, 'https://www.wsj.com/articles/bayer-ceo-faces-shareholder-ire-over-monsanto-deal-11556292088'.

Bender, S. F., C. Wagg, and M. G. A. van der Heijden (2016), 'An Underground Revolution: Biodiversity and Soil Ecological Engineering for Agricultural Sustainability', *Trends in Ecology and Evolution*, 31(6), 440–452.

Béné, C., M. Barange, R. Subasinghe, P. Pinstrup-Andersen, G. Merino, G. I. Hemre, and M. Williams (2015), 'Feeding 9 Billion by 2050 – Putting Fish Back on the Menu', *Food Security*, 7, 261–274.

Béné, C., G. Macfadyen, and E. H. Allison (2007), 'Increasing the Contribution of Small-Scale Fisheries to Poverty Alleviation and Food Security', *FAO Fisheries Technical Paper*, No. 481.

Benjamin, D. J., O. Heffetz, M. S. Kimball, and N. Szembrot (2014), 'Beyond Happiness and Satisfaction: Toward Well-Being Indices Based on Stated Preference', *American Economic Review*, 104(9), 2698–2735.

Bentham, J. (1789), *An Introduction to the Principles of Morals and Legislation* (London:W. Pickering).

Berger, J., M. J. Goedkoop, W. Broer, R. Nozeman, C. D. Grosscurt, M. Bertram, and F., Cachia (2018), *Common Ground in Biodiversity Footprint Methodologies for the Financial Sector*.

Berkes, F. (2017), *Sacred Ecology* (New York, NY: Routledge) 4th Edition.

Berkes, F., R. Mahon, R. Pollnac, and R. Pomeroy (2001), *Managing Small-Scale Fisheries: Alternative Directions and Methods* (Ottowa: International Development Research Centre).

Berman, M. G., J. Jonides, and S. Kaplan (2008), 'The Cognitive Benefits of Interacting with Nature', *Psychological Science*, 19(12), 1207–1212.

Bernauer, T., T. Böhmelt, and V. Koubi. (2012). 'Environmental Changes and Violent Conflict', *Environmental Research Letters*, 7(1), 015601.

Bernhardt, A., R. Dell, J. Ambachtsheer, and R. Pollice (2017), *The Long and Winding Road: How Long-Only Equity Managers Turn Over Their Portfolios Every 1.7 Years*.

Berthe, A., and L. Elie. (2015). 'Mechanisms Explaining the Impact of Economic Inequality on Environmental Deterioration', *Ecological Economics*, 116, 191–200.

References

Béteille, A. ed. (1983), *Equality and Inequality: Theory and Practice* (Oxford: Oxford University Press).

Beugelsdijk, S., H. L. F. de Groot, and A. B. T. M. van Schaik (2004), 'Trust and Economic Growth: A Robustness Analysis', *Oxford Economic Papers*, 56(1), 118–134.

Bibby, R., P. Cleall-Harding, S. Rundle, S. Widdicombe, and J. Spicer (2007), 'Ocean Acidification Disrupts Induced Defences in the Intertidal Gastropod Littorina littorea', *Biology Letters*, 3(6), 699–701.

Biesmeijer, J. C., S. P. M. Roberts, M. Reemer, R. Ohlemüller, M. Edwards, T. Peeters, A. P. Schaffers, S. G. Potts, R. Kleukers, C. D. Thomas, J. Settele, and W. E. Kunin (2006), 'Parallel Declines in Pollinators and Insect-Pollinated Plants in Britain and the Netherlands', *Science*, 313(5785), 351–354.

Biggs, D., R. Cooney, D. Roe, H. T. Dublin, J. R. Allan, D. W. S. Challender, and D. Skinner (2017), 'Developing a Theory of Change for a Community-Based Response to Illegal Wildlife Trade', *Conservation Biology*, 31(1), 5–12.

Bilby, R. E., B. R. Fransen, P. A. Bisson, and J. K. Walter (1998), 'Response of Juvenile Coho Salmon (Oncorhynchus kisutch) and Steelhead (Oncorhynchus mykiss) to the Addition of Salmon Carcasses to Two Streams in Southwestern Washington, U.S.A.', *Canadian Journal of Fishery and Aquatic Science*, 55, 1909–1918.

Binmore, K. (2007), *Playing for Real: A Text on Game Theory* (Oxford: Oxford University Press).

Binmore, K., and P. Dasgupta (1986), 'Introduction' in K. Binmore and P. Dasgupta, eds., *Economic Organizations as Games* (Oxford: Blackwell).

Binswanger, H. P. (1991), 'Brazilian Policies That Encourage Deforestation in the Amazon', *World Development*, 19(7), 821–829.

Birch, J. C., A. C. Newton, C. A. Aquino, E. Cantarello, C. Echeverría, T. Kitzberger, I. Schiappacasse, and N. T. Garavito (2010). 'Cost-Effectiveness of Dryland Forest Restoration Evaluated by Spatial Analysis of Ecosystem Services', *Proceedings of the National Academy of Sciences*, 107(50), 21925–21930.

BIT (2019), *Behavior Change for Nature: A Behavioral Science Toolkit for Practitioners*.

Black, D. (1948), "On the Rationale of Group Decision-Making," *Journal of Political Economy*, 56(1), 23–34.

Blackburn, S. (2003), *Ethics: A Very Short Introduction* (Oxford: Oxford University Press).

Blanchflower, D. G., and A. J. Oswald (2004), 'Well-Being Over Time in Britain and the USA', *Journal of Public Economics*, 88(7–8), 1359–1386.

Bledsoe, C. (1990), 'The Politics of Children: Fosterage and the Social Management of Fertility Among the Mende of Sierra Leone', in W. P. Handwerker ed., *Births and Power: The Politics of Reproduction* (Boulder, CO: Westview Press).

Bledsoe, C. (1994), ''Children are Like Young Bamboo Trees': Potentiality and Reproduction in Sub-Saharan Africa', in K. Lindahl-Kiessling and H. Landberg, eds., *Population, Economic Development and the Environment: The Making of Our Common Future* (Oxford: Oxford University Press).

Bledsoe, C. (1996), 'Contraception and "Natural" Fertility in America', *Population and Development Review*, 22, 297–324.

Bledsoe, C. H., and G. Pison (1994), *Nuptiality in Sub-Saharan Africa: Contemporary Anthropological and Demographic Perspectives* (Oxford: Clarendon Press).

Blok, D., G. Schaepman-Strub, H. Bartholomeus, M. M. P. D. Heijmans, T. C. Maximov, and F. Berendse (2011), 'The Response of Arctic Vegetation to the Summer Climate: Relation Between Shrub Cover, NDVI, Surface Albedo and Temperature', *Environmental Research Letters*, 6(3), 035502.

Blume, L. E., W. A. Brock, S. N. Durlauf, and Y. M. Ioannides (2011), 'Identification of Social Interactions', in J. Benhabib, M. O. Jackson, and A. Bisin, eds., *Handbook of Social Economics* (Amsterdam: North Holland).

Blume, L. E., and S. N. Durlauf (2001), 'The Interactions-Based Approach to Socioeconomic Behaviour', in S. N. Durlauf and H. Peyton Young, eds., *Social Dynamics* (Cambridge, MA: MIT Press).

Bodirsky, B. L., S. Rolinski, A. Biewald, I. Weindl, A. Popp, and H. Lotze-Campen (2015), 'Global Food Demand Scenarios for the 21st Century', *PLoS ONE*, 10(11), e0139201.

Bogahawatte, C., and J. Herath (2011), 'Air Quality and Cement Production: Examining the Implications of Point Source Pollution in Sri Lanka', A. K. E. Haque, M. N. Murty, P. Shyamsundar, eds., *Environmental Valuation In South Asia* (New Delhi: Cambridge University Press).

Bolam, F., L. Mair, M. Angelico, T. Brooks, M. Burgman, C. Hermes, M. Hoffmann, R. Martin, P. McGowan, A. Rodrigues, C. Rondinini, H. Wheatley, Y. Bedolla-Guzmán, J. Calzada, M. Child, P. Cranswick, C. Dickman, B. Fessl, D. Fisher, S. T. Garnett, J. J. Groombridge, C, N. Johnson R. J. Kennerley, S. R. B. King, J. F. Lamoreux, A. C. Lees, L. Lens, S. P. Mahood, D. P. Mallon, E. Meijaard, F. Méndez-Sánchez, A. Reis Percequillo, T. J. Regan, L. M. Renjifo, M. C. Rivers, N. S. Roach, L. Roxburgh, R. J. Safford, P. Salaman, T. Squires, E. Vázquez-Domínguez, P. Visconti, J. C. Z. Woinarski, R. P. Young, S. H. M. Butchart (2020), 'How Many Bird and Mammal Extinctions has Recent Conservation Action Prevented?', *Conservation Letters*, e12762.

Bolt, J., R. Inklaar, H. de Jong and J. L. van Zanden (2018), 'Rebasing 'Maddison': New Income Comparisons and the Shape of Long-run Economic Development', *Maddison Project Working Paper 10*.

Bongaarts, J. (2011), 'Can Family Planning Programs Reduce High Desired Family Size in Sub-Saharan Africa?', *International Perspectives on Sexual and Reproductive Health*, 37(4), 209–216.

Bongaarts, J. (2016), 'Development: Slow Down Population Growth', *Nature*, 530(7591), 409–412.

Bongaarts, J., and M. Cain (1981), 'Demographic Responses to Famine', *Population Council, Center for Policy Studies Working Paper*, No. 77.

Bongaarts, J., and J. Casterline (2013), 'Fertility Transition: Is Sub-Saharan Africa Different?',*Population and Development Review*, 38(Suppl 1), 153–168.

Bongaarts, J., O. Frank, and R. Lesthaeghe (1984). 'The Proximate Determinants of Fertility in Sub-Saharan Africa', *Population and Development Review*, 10(3), 511–537.

Bonner, J., A. Grigg, S. Hime, G. Hewitt, R. Jackson, and M. Kelly (2012), *Is Natural Capital a Material Issue?*.

Boorstin, D. J. (1983), *The Discoverers* (New York, NY: Random House).

Borer, E. T., and D. S. Gruner (2009), 'Top-Down and Bottom-Up Regulation of Communities', in S. A. Levin, S. R. Carpenter, H. C.J. Godfray, A. P. Kinzig, M. Loreau, J. B. Losos, B. Walker, and D. S. Wilcove, eds., *The Princeton Guide to Ecology* (Princeton, NJ: Princeton University Press).

Boserup, E. (1965), *The Conditions of Agricultural Growth: The Economics of Agrarian Change Under Population Pressure* (London: G. Allen and Unwin).

References

Bourge, C., and M. Aubrat (2018), *The Ecological Offset on the BPL HSL, a French Linear Railway Infrastructure: Eiffage, France – A Case Study (2018)*.

Bowman, D. (2012), 'Conservation: Bring Elephants to Australia?', *Nature*, 482(7383), 30.

Bowman, W. D., S. D. Hacker, and M. L. Cain (2018), *Ecology* (Oxford: Oxford University Press).

Bradley, S. E. K., T. N. Croft, J. D. Fishel, and C. F. Westoff (2012), 'Revising Unmet Need for Family Planning', *DHS Analytical Studies*, No. 25.

Bradshaw, C. J. A., N. S. Sodhi, K. S.-H. Peh, and B. W. Brook (2007), 'Global Evidence that Deforestation Amplifies Flood Risk and Severity in the Developing World', *Global Change Biology*, 13(11), 2379–2395.

Bramley, R. G. V., and J. Ouzman (2018), 'Farmer Attitudes to the Use of Sensors and Automation in Fertilizer Decision-Making: Nitrogen Fertilization in the Australian Grains Sector', *Precision Agriculture*, 20(1), 157–175.

Brander, L. M., P. van Beukering, L. Nijsten, A. McVittie, C. Baulcomb, F. V. Eppink, and J. A. Cado van der Lelij (2020), 'The Global Costs and Benefits of Expanding Marine Protected Areas', *Marine Policy*, 116, 103953.

Bratman, G. N., C. B. Anderson, M. G. Berman, B. Cochran, S. de Vries, J. Flanders, C. Folke, H. Frumkin, J. J. Gross, T. Hartig, P. H. Kahn Jr., M. Kuo, J. J. Lawler, P. S. Levin, T. Lindahl, A. Meyer-Lindenberg, R. Mitchell, Z. Ouyang, J. Roe, L. Scarlett, J. R. Smith, M. van den Bosch, B. W. Wheeler, M. P. White, H. Zheng, G. C. Daily (2019), 'Nature and Mental Health: An Ecosystem Service Perspective', *Science Advances*, 5(7), eaax0903.

Brauman, K., L. Garibaldi, S. Polasky, and C. Zayas (2019), *The IPBES Global Assessment on Biodiversity and Ecosystem Services – Chapter 2. Status and Trends; Chapter 2.3. Status and Trends- Nature's Contributions to People (Draft)*.

Breeze, T. D., S. P. M. Roberts, and S. G. Potts (2012), *The Decline of England's Bees: Policy Review and Recommendations* (London: Friends of the Earth).

Breitburg, D., L. A. Levin, A. Oschlies, M. Grégoire, F. P. Chavez, D. J. Conley, V. Garçon, D. Gilbert, D. Gutiérrez, K. Isensee, G. S. Jacinto, K. E. Limburg, I. Montes, S. W. A. Naqvi, G. C. Pitcher, N. N. Rabalais, M. R. Roman, K. A. Rose, B. A. Seibel, M. Telszewski, M. Yasuhara, J. Zhang (2018), 'Declining Oxygen in the Global Ocean and Coastal Waters', *Science*, 359(6371), eaam7240.

Breslin, P., and M. Chapin (1984), 'Conservation Kuna-style.', *Grassroots Development*, 8(2), 26–35.

Brewer, N. T., M. G. Hall, S. M. Noar, H. Parada, A. Stein-Seroussi, L. E. Bach, S. Hanley, and K.M. Ribisl (2016), 'Effect of Pictorial Cigarette Pack Warnings on Changes in Smoking Behavior: A Randomized Clinical Trial', *JAMA Internal Medicine*, 176(7), 905–912.

Briggs, S. V., and N. Taws (2003), 'Impacts of Salinity on Biodiversity – Clear Understanding or Muddy Confusion?' *Australian Journal of Botany*, 51(6), 609–617.

Bright, G. (2017), *Natural Capital Restoration Project Report*.

Bright, G., E. Connors, and J. Grice (2019), 'Measuring natural capital: Towards Accounts for the UK and a Basis for Improved Decision-Making', *Oxford Review of Economic Policy*, 35(1), 88–108.

Broadberry, S., J. Custodis, and B. Gupta (2015), 'India and the Great Divergence: An Anglo-Indian Comparison of GDP per capita, 1600-1871', *Explorations in Economic History*, 55, 58–75.

Brock, W. A., and S. A. Durlauf (2001), 'Interactions-Based Models', in J. J. Heckman and E. E. Leamer, eds., *Handbook of Econometrics*, Vol. 5 (Amsterdam: North Holland).

Brock, W. A., and D. Starrett (2003), 'Managing Systems with Non-convex Positive Feedback', *Environmental and Resource Economics*, 26(4), 575–602.

Brock, W. A., and M. S. Taylor (2010), 'The Green Solow Model', *Journal of Economic Growth*, 15, 127–153.

Bromley, D. W. ed. (1992), *Making the Commons Work: Theory, Practice, and Policy* (San Francisco, CA: ICS Press).

Brondizio, E. S., E. Ostrom, and O. R. Young (2009), 'Connectivity and the Governance of Multilevel Social-Ecological Systems: The Role of Social Capital', *Annual Review of Environment and Resources*, 34, 253–278.

Broome, J. (1992), *Counting the Cost of Global Warming: A report to the Economic and Social Research Council on Research by John Broome and David Ulph* (Cambridge: The White Horse Press).

Broome, J. (2004), *Weighing Lives* (Oxford: Oxford University Press).

Broome, J. (2012), *Climate Matters: Ethics in a Warming World* (London: W.W. Norton).

Brown, G., and C. B. McGuire (1967), 'A Socially Optimum Pricing Policy for a Public Water Agency', *Water Resources Research*, 3(1), 33–43.

Bruckner, M., T. Häyhä, S. Giljum, V. Maus, G. Fischer, S. Tramberend, and J. Börner (2019), 'Quantifying the Global Cropland Footprint of the European Union's Non-Food Bioeconomy', *Environmental Research Letters*, 14(4), 045011.

Bruins, R. J. F., T. J. Canfield, C. Duke, L. Kapustka, A. M. Nahlik, and R. B. Schäfer (2017), 'Using Ecological Production Functions to Link Ecological Processes to Ecosystem Services', *Integrated Environmental Assessment and Management*, 13(1), 52–61.

Brundtland, G. H. (1987), *Our Common Future: Report of the World Commission on Environment and Development* (Oxford: Oxford University Press).

Bruno, J. F., and E. R. Selig (2007), 'Regional Decline of Coral Cover in the Indo-Pacific: Timing, Extent, and Subregional Comparisons', *PLoS ONE*, 2(8), e711.

Bryan, B. A., L. Gao, Y. Ye, X. Sun, J. D. Connor, N. D. Crossman, M. Stafford-Smith, J. Wu, C. He, D. Yu, Z. Liu, A. Li, Q. Huang, H. Ren, X. Deng, H. Zheng, J. Niu, G. Han, and X. Hou. (2018), 'China's Response to a National Land-System Sustainability Emergency', *Nature*, 559, 193–204.

Buchner, B., A. Clark, A. Falconer, C. Macquarie, R. Tolentino, and C. Watherbee (2019), *Global Landscape of Climate Finance 2019*.

Buisson, E., S. Le Stradic, F. A. O. Silveira, G. Durigan, G. E. Overbeck, A. Fidelis, G. W. Fernandes, W. J. Bond, J. -M. Hermann, G. Mahy, S. T. Alvarado, N. P. Zaloumis, and J. W. Veldman, (2019), 'Resilience and Restoration of Tropical and Subtropical Grasslands, Savannas, and Grassy Woodlands', *Biological Reviews*, 94(2), 590–609.

Bull, J. W., K. B. Suttle, A. Gordon, N. J. Singh, and E. J. Milner-Gulland (2013), 'Biodiversity Offsets in Theory and Practice', *Oryx*, 47(3), 369–380.

Bull, J. W., and N. Strange (2018), 'The Global Extent of Biodiversity Offset Implementation Under No Net Loss Policies', *Nature Sustainability*, 1(12), 790–798.

References

Bullock, J. M., J. Aronson, A. C. Newton, R. F. Pywell, and J. M. Rey-Benayas (2011), 'Restoration of Ecosystem Services and Biodiversity: Conflicts and Opportunities', *Trends in Ecology and Evolution*, 26(10), 541–549.

Bulte, E. H., L. Lipper, R. Stringer, and D. Zilberman (2008), 'Payments for Ecosystem Services and Poverty Reduction: Concepts, Issues, and Empirical Perspectives', *Environment and Development Economics*, 13(3), 245–254.

Bureau of Economic Analysis (2020), *Gross Domestic Product by Industry, 2nd Quarter 2020 and Annual Update*.

Burek, P., Y. Satoh, G. Fischer, T. Kahil, A. Scherzer, S. Tramberend, L. F. Nava, Y. Wada, S. Eisner, M. Flörke, N. Hanasaki, P. Magnuszewski, B. Cosgrove, and D. Wiberg (2016), 'Water Futures and Solution – Fast Track Initiative (Final Report)', *IIASA Working Paper*, WP 16-006.

Burgess, H., P. Kilkie, and T. Callaway (2016), 'Understanding the Physical Processes Occurring Within a New Coastal Managed Realignment Site, Medmerry, Sussex, *Coastal Management Conference Proceedings*, 263–272.

Burgin, C. J., J. P. Colella, P. L. Kahn, and N. S. Upham (2018), 'How Many Species of Mammals Are There?', *Journal of Mammalogy*, 99(1), 1–14.

Büscher, B., R. Fletcher, D. Brockington, C. Sandbrook, W. M. Adams, L. Campbell, C. Corson, W. Dressler, R. Duffy, N. Gray, G. Holmes, A. Kelly, E. Lunstrum, M. Ramutsindela, and K. Shanker (2017), 'Half-Earth or Whole Earth? Radical Ideas for Conservation, and Their Implications', *Oryx*, 51(3), 407–410.

Butchart, S. H. M., H. R. Akçakaya, J. Chanson, J. E. M. Baillie, B. Collen, S. Quader, W. R. Turner, R. Amin, S. N. Stuart, and C. Hilton-Taylor (2007), 'Improvements to the Red List Index', *PLoS ONE*, 2(1), e140.

Butchart, S. H. M., M. Clarke, R. J. Smith, R. E. Sykes, J. P. W. Scharlemann, M. Harfoot, G. M. Buchanan, A. Angulo, A. Balmford, B. Bertzky, T. M. Brooks, K. E. Carpenter, M. T. Comeros-Raynal, J. Cornell, G. F. Ficetola, L. D. C. Fishpool, R. A. Fuller, J. Geldmann, H. Harwell, C. Hilton-Taylor, M. Hoffmann, A. Joolia, L. Joppa, N. Kingston, I. May, A. Milam, B. Polidoro, G. Ralph, N. Richman, C. Rondinini, D. B. Segan, B. Skolnik, M. D. Spalding, S. N. Stuart, A. Symes, J. Taylor, P. Visconti, J. E. M. Watson, Louisa Wood, and N. D. Burgess (2015). 'Shortfalls and Solutions for Meeting National and Global Conservation Area Targets', *Conservation Letters*, 8(5), 329–337.

Butzer, K. W. (2012). 'Collapse, Environment, and Society', *Proceedings of the National Academy of Sciences*, 109(10), 3632–3639.

Butzer, K. W., and G. H. Endfield. (2012). 'Critical Perspectives on Historical Collapse',*Proceedings of the National Academy of Sciences*, 109(10), 3628–3631.

Cai, Y., T. M. Lenton, and T. S. Lontzek (2016), 'Risk of Multiple Interacting Tipping Points Should Encourage Rapid CO2 Emission Reduction', *Nature Climate Change*, 6(5), 520–525.

Cain, M. T. (1977), 'The Economic Activities of Children in a Village in Bangladesh', *Population and Development Review*, 3(3), 201–227.

Caldecott, B. (2019), 'Viewpoint: Spatial Finance has a Key Role', *Investment and Pensions Europe*, November.

Caldecott, B., N. Howarth, and P. McSharry (2013), *Stranded Assets in Agriculture*: Protecting Value from Environment-Related Risks.

Caldecott, B., and J. McDaniels (2014), *Financial Dynamics of the Environment: Risks, Impacts, and Barriers to Resilience*.

Caldwell, J. C. (1990), 'The Soft Underbelly of Development: Demographic Transition in Conditions of Limited Economic Change', *The World Bank Economic Review*, 4(Suppl 1), 207–254.

Caldwell, J. C., and P. Caldwell (1990), 'High Fertility in Sub-Saharan Africa', *Scientific American*, 262(5), 118–125.

Caldwell, J. C., and P. Caldwell (1992), 'What Does the Matlab Fertility Experience Really Show?', *Studies in Family Planning*, 23(5), 292–310.

Cambridge Centre for Sustainable Finance (2016), *Environmental Risk Analysis by Financial Institutions – A Review of Global Practice*.

Camerer, C., and M. Weber (1992), 'Recent Developments in Modeling Preferences: Uncertainty and Ambiguity', *Journal of Risk and Uncertainty*, 5(4), 325–370.

Campbell, B., A. Mandondo, N. Nemarundwe, B. Sithole, W. De Jong, M. Luckert, and F. Matose (2001), 'Challenges to Proponents of Common Property Recource Systems: Despairing Voices from the Social Forests of Zimbabwe', *World Development*, 29(4), 589–600.

Cantril, H. (1965), *The Pattern of Human Concern* (New Brunswick, NJ: Rutgers University Press).

Capaldi, C. A., H.-A. Passmore, E. K. Nisbet, J. M. Zelenski, and R. L. Dopko (2015), 'Flourishing in Nature: A Review of the Benefits of Connecting with Nature and its Application as a Wellbeing Intervention', *International Journal of Wellbeing*, 5(4), 1–16.

Cardinale, B. J., J. E. Duffy, A. Gonzalez, D. U. Hooper, C. Perrings, P. Venail, A. Narwani, G. M. Mace, D. Tilman, D. A. Wardle, A. P. Kinzig, G. C. Daily, M. Loreau, J. B. Grace, A. Larigauderie, D. S. Srivastava, and S. Naeem (2012), 'Biodiversity Loss and its Impact on Humanity', *Nature*, 486, 59–67.

Carney, M. (2015), *Speech: Breaking the Tragedy of the Horizon – Climate Change and Financial Stability*.

Carpenter, S. R., W. A. Brock, C. Folke, E. H. van Nes, M. Scheffer, and S. Polasky (2015), 'Allowing Variance May Enlarge the Safe Operating Space for Exploited Ecosystems', *Proceedings of the National Academy of Sciences*, 112(46), 14384–14389.

Carpenter, S. R., N. F. Caraco, D. L. Correll, R. W. Howarth, A. N. Sharpley, and V. H. Smith (1998), 'Nonpoint Pollution of Surface Waters with Phosphorus and Nitrogen', *Ecological Applications*, 8(3), 559–568.

Carpenter, S. R., and K. L. Cottingham (1997), 'Resilience and Restoration of Lakes', *Conservation Ecology*, 1(1).

Carpenter, S. R., D. Ludwig, and W. A. Brock (1999), 'Management of Eutrophication for Lakes Subject to Potentially Irreversible Change', *Ecological Applications*, 9(3), 751–771.

Carpenter, S. R., E. H. Stanley, and M. J. Vander Zanden (2011) 'State of the World's Freshwater Ecosystems: Physical, Chemical, and Biological Changes', *Annual Review of Environment and Resources*, 36, 75–99.

Carreira, J.C.A., C. Bueno, and A.V.M. da Silva (2020) 'Wild Mammal Translocations: A Public Health Concern', *Open Journal of Animal Sciences*, 10 (1), 64–133.

Carroll, N., J. Fox, R. Bayon eds. (2008), *Conservation and Biodiversity Banking: A Guide to Setting Up and Running Biodiversity Credit Trading Systems* (London: Earthscan).

References

Carson, R. (1962), *Silent Spring* (Boston, MA: Houghton Mifflin).

Carter, L., and S. Moss (2017), *Drought Stress Testing, Making Financial Institutions More Resilient to Environmental Risks*.

Carus, M., and L. Dammer (2013), 'Food or Non-Food: Which Agricultural Feedstocks Are Best for Industrial Uses?', *Industrial Biotechnology*, 9(4), 171–176.

Carvalheiro, L. G., C. L. Seymour, S. W. Nicolson, and R. Veldtman (2012), 'Creating Patches of Native Flowers Facilitates Crop Pollination in Large Agricultural Fields: Mango as a Case Study', *Journal of Applied Ecology*, 49, 1373–1383.

Cashdan, E. (1983), 'Territoriality Among Human Foragers: Ecological Models and an Application to Four Bushman Groups with Comments and Reply', *Current Anthropology*, 24(1), 47–66.

Castañeda-Álvarez, N. P., C. K. Khoury, H. A. Achicanoy, V. Bernau, H. Dempewolf, R. J. Eastwood, L. Guarino, R. H. Harker, A. Jarvis, N. Maxted, J. V. Müller, J. Ramirez-Villegas, C. C. Sosa, P. C. Struik, H. Vincent, and J. Toll (2016), 'Global Conservation Priorities for Crop Wild Relatives', *Nature Plants*, 2, 16022.

Casterline, J. B., and S. W. Sinding (2000), 'Unmet Need for Family Planning in Developing Countries and Implications for Population Policy', *Population and Development Review*, 26(4), 691–723.

Cavender-Bares, J., P. Balvanera, E. King, and S. Polasky (2015), 'Ecosystem Service Trade-offs Across Global Contexts and Scales', *Ecology and Society*, 20(1), 22.

Cavendish, W. (2000), 'Empirical Regularities in the Poverty-Environment Relationship of Rural Households: Evidence from Zimbabwe', *World Development*, 28(11), 1979–2003.

Ceballos, G., A. H. Ehrlich, and P. R. Ehrlich (2015), *The Annihilation of Nature: Human Extinction of Birds and Mammals* (Baltimore, MD: John Hopkins University Press).

Ceballos, G., P. R. Ehrlich, and P. H. Raven (2020), 'Vertebrates on the Brink as Indicators of Biological Annihilation and the Sixth Mass Extinction', *Proceedings of the National Academy of Sciences*, 117(24), 13596–13602.

Centre for Science and Environment (1990), *Human-Nature Interactions in a Central Himalayan Village: A Case Study of Village Bemru* (New Delhi: Centre for Science and Environment).

Čermák, T., N. J. Baltes, R. Čegan, Y. Zhang, and D. F. Voytas (2015), 'High-frequency, Precise Modification of the Tomato Genome', *Genome Biology*, 16, 232.

Chace, J. F., and J. J. Walsh (2006), 'Urban Effects on Native Avifauna: A Review', *Landscape and Urban Planning*, 74(1), 46–69.

Chami, R., T. Cosimano, C. Fullenkamp, and S. Oztosun (2019), 'Saving the Whale: How Much do you Value your Next Breath?', *International Workshop on Financial System Architecture and Stability*, April, 1–7.

Chan, K. M. A., J. Agard, J. Liu, A. P. de A. D., D. Armenteras, A. K. Boedhihartono, W. W. L. Cheung, S. Hashimoto, G. C. H. Pedraza, T. Hickler, J. Jetzkowitz, M. Kok, M. M. H. P. O'Farrell, T. Satterfield, A. K. Saysel, R. Seppelt, B. Strassburg, and D. Xue (2019), *IPBES GlobalAssessment on Biodiversity and Ecosystem Services – Chapter 5. Pathways towards a Sustainable Future (Draft)*.

Chancel, L. and T. Piketty (2015), *Carbon and Inequality: from Kyoto to Paris Trends in the Global Inequality of Carbon Emissions (1998-2013) and Prospects for an Equitable Adaptation Fund* (Paris: Paris School of Economics).

Chanthorn, W., F. Hartig, W. Y. Brockelman, W. Srisang, A. Nathalang, and J. Santon (2019), 'Defaunation of Large-Bodied Frugivores Reduces Carbon Storage in a Tropical Forest of Southeast Asia', *Scientific Reports*, 9(1), 1–9.

Chaplin-Kramer, R., R. P. Sharp, C. Weil, E. M. Bennett, U. Pascual, K. K. Arkema, K. A. Brauman, B. P. Bryant, A. D. Guerry, N. M. Haddad, M. Hamann, P. Hamel, J. A. Johnson, L. Mandle, H. M. Pereira, S. Polasky, M. Ruckelshaus, M. R. Shaw, J. M. Silver, A. L. Vogl, and G. C. Daily (2019), 'Global Modeling of Nature's Contributions to People Supplementary Information', *Science*, 366(6462), 255–8.

Chapman, A. D. (2009), *Numbers of Living Species in Australia and the World* (Toowoomba: Australian Biodiversity Information Services).

Chaudhary, A., and T. M. Brooks (2019), 'National Consumption and Global Trade Impacts on Biodiversity', *World Development*, 121, 178–187.

Chaudhary, A., Z. Burivalova, L. P. Koh, and S. Hellweg (2016), 'Impact of Forest Management on Species Richness: Global Meta-Analysis and Economic Trade-Offs', *Nature Scientific Reports*, 6, 23954.

Chaves, L. S. M., J. Fry, A. Malik, A. Geschke, M. A. M. Sallum, and M. Lenzen (2020), 'Global Consumption and International Trade in Deforestation-associated Commodities Could Influence Malaria Risk', *Nature Communications*, 11, 1258.

Chen, X., F. Lupi, G. He, and J. Liu (2009), 'Linking Social Norms to Efficient Conservation Investment in Payments for Ecosystem Services', *Proceedings of the National Academy of Sciences*, 106(28), 11812–11817.

Chenet, H. (2019), *Planetary Health and the Global Financial System*.

Cheung, S. N. S. (1973), 'The Fable of the Bees: An Economic Investigation', *The Journal of Law and Economics*, 16(1), 11–33.

Chichilnisky, G. (1994), 'North-South Trade and the Global Environment', *The American Economic Review*, 84(4), 851–874.

Chichilnisky, G., and G. M. Heal (1998), 'Economic Returns from the Biosphere', *Nature*, 391(6668), 629–630.

Chopra, K., and S. C. Gulati (1998), 'Environmental Degradation, Property Rights and Population Movements: Hypotheses and Evidence from Rajasthan (India)', *Environment and Development Economics*, 3(1), 35–57.

Chopra, K., G. K. Kadekodi, and M. N. Murty (1990), *Participatory Development: People and Common Property Resources* (New Delhi and London: Sage).

Chowdhury, M. S. H., M. Koike, S. Akther, and M. D. Miah (2011), 'Biomass Fuel Use, Burning Technique and Reasons for the Denial of Improved Cooking Stoves by Forest User Groups of Rema-Kalenga Wildlife Sanctuary, Bangladesh', *International Journal of Sustainable Development and World Ecology*, 18(1), 88–97.

Christensen, L. B. (2006), 'Marine Mammal Populations: Reconstructing Historical Abundances at the Global Scale', *Fisheries Centre Research Reports*, 14(9), 167.

Christiaen, C. (2020), 'Using Spatial Finance for Sustainable Development', *Refinitiv*, June.

Christiaensen, L. (2017), 'Agriculture in Africa – Telling Myths from Facts: A Synthesis', *Food Policy*, 67, 1–11.

References

Christie, M., N. Hanley, J. Warren, K. Murphy, R. Wright, and T. Hyde (2006), 'Valuing the Diversity of Biodiversity', *Ecological Economics*, 58(2), 304–317.

Christophers, B. (2019), 'Environmental Beta or How Institutional Investors Think about Climate Change and Fossil Fuel Risk', Annals of the American Association of *Geographers*, 109(3), 754–774.

CISL (2020), *Biodiversity Loss and Land Degradation: An Overview of the Financial Materiality*.

CITES (1975), 'CITES Convention', https://www.cites.org/eng/disc/text.php.

Clark, A. E., P. Frijters, and M. A. Shields (2008), 'Relative Income, Happiness, and Utility: An Explanation for the Easterlin Paradox and Other Puzzles', *Journal of Economic Literature*, 46(1), 95–144.

Clark, C. W. (1976), *Mathematical Bioeconomics: The Optimal Management of Renewable Resources* (New York, NY: John Wiley).

Clark, P., and L. Martínez (2016), 'Local Alternatives to Private Agricultural Certification in Ecuador: Broadening Access to "New Markets"?', *Journal of Rural Studies*, 45, 292–302.

Cleland, J., S. Bernstein, A. Ezeh, A. Faundes, A. Glasier, and J. Innis (2006), 'Family Planning: The Unfinished Agenda', *The Lancet*, 368(9549), 1810–1827.

Cleland, J., and C. Wilson (1987), 'Demand Theories of the Fertility Transition: An Iconoclastic View', *Population Studies*, 41(1), 5–30.

Climate Bonds Initiative (2017), *Sovereign Green Bonds Briefing*.

Climate Bonds Initiative (2019). *2019 Green Bonds Market Summary*.

Cline, W. R. (1992), *The Economics of Global Warming* (Washington, DC: Institute for International Economics).

Coad, L., J. E. M. Watson, J. Geldmann, N. D. Burgess, F. Leverington, M. Hockings, K. Knights, and M. Di Marco (2019), 'Widespread Shortfalls in Protected Area Resourcing Undermine Efforts to Conserve Biodiversity', *Frontiers in Ecology and the Environment*, 17(5), 259–264.

Coady, D., I. Parry, N.-P. Le, and B. Shang (2019), 'Global Fossil Fuel Subsidies Remain Large. An Update Based on Country-Level Estimates', *IMF Working Paper 19/89*.

Coase, R. H. (1960), 'The Problem of Social Cost', *The Journal of Law and Economics*, 3(1), 1–44.

Coastal Protection and Restoration Authority of Louisiana (2012), *Louisiana's Comprehensive Master Plan for a Sustainable Coast*.

Cochrane, S. H., and S. M. Farid (1989), 'Fertility in Sub-Saharan Africa: Analysis and Explanation', *World Bank Discussion Paper No. 43*.

Coe, D. T., E. Helpman, and A. W. Hoffmaister (1997), 'North-South R and D Spillovers', *The Economic Journal*, 107(440), 134–149.

Coe, D. T., E. Helpman, and A. W. Hoffmaister (2009), 'International R and D Spillovers and Institutions', *European Economic Review*, 53(7), 723–741.

Cohen-Shacham, E., G. Walters, C. Janzen, S. Maginnis, eds. (2016), *Nature-based Solutions to Address Global Societal Challenges*.

Cole, L. A. (2016), 'Chapter 13 Evolution of Chemical, Prokaryotic, and Eukaryotic Life', in L. A. Cole ed., *Biology of Life*, (New York, NY: Academic Press).

Colchester, M. (1994), 'Sustaining the Forests: The Community-based Approach in South and South-East Asia', *Development and Change*, 25(1), 69–100.

Colchester, M., A. P. Wee, M. C. Wong, and T. Jalong (2007), *Land is Life: Land Rights and Oil Palm Development in Sarawak*.

Coleman, J. S. (1988), 'Social Capital in the Creation of Human Capital', *American Journal of Sociology*, 94, S95–S120.

Connor, D. J. (2008), 'Organic Agriculture Cannot Feed the World', *Field Crops Research*, 106, 187–190.

Connor, D. J. (2013). 'Organically Grown Crops do not a Cropping System Make and nor can Organic Agriculture Nearly Feed the World', *Field Crops Research*, 144, 145–147.

Convention on Biological Diversity (1995), *COP2 Decision11/8*.

Convention on Biological Diversity (1998a), *SBSTTA 5 Recommendation V/10*. Convention on Biological Diversity (1998b), *COP 4 Decision IV/1*.

Convention on Biological Diversity (2010), *The Strategic Plan for Biodiversity 2011-2020 and the Aichi Biodiversity Targets, Decision X.2 UNEP/CBD/DEC/X/2*.

Convention on Biological Diversity (2017), 'Green Bonds', 'https://www.cbd.int/financial/ greenbonds .shtml'.

Convergence (2019), *Blending in Conservation Finance*.

Cooke, S. J., J. R. Bennett, and H. P. Jones (2019), 'We Have a Long Way to Go if We Want to Realize the Promise of the "Decade on Ecosystem Restoration"', *Conservation Science and Practice*, 1(12), e129.

Cooke, P., G. Köhlin, and W. F. Hyde (2008), 'Fuelwood, Forests and Community Management – Evidence from Household Studies', *Environment and Development Economics*, 13(1), 103–135.

Cooke, S. J., A. M. Rous, L. A. Donaldson, J. J. Taylor, T. Rytwinski, K. A. Prior, K. E. Smokorowski, and J. R. Bennett (2018), 'Evidence-based Restoration in the Anthropocene—from Acting with Purpose to Acting for Impact', *Restoration Ecology*, 26(2), 201–205.

Cooney, R. (2019), 'Harvest and Trade of Vicuña Fibre in Bolivia', 'https://cites.org/sites/default/files/eng/ prog/Livelihoods/case_studies/2.%20Bolivia_vicuna_long_Aug2.pdf'.

Cooper, G., and S. Trémolet (2019), *Investing in Nature: Private Finance for Nature-based Resilience*.

Cordell, J., and M. McKean (1985), 'Sea tenure in Bahia, Brazil', *Proceedings of the Conference on Common Property Resource Management*, April, 85–114.

Costanza, R., R. d'Arge, R. De Groot, S. Farber, M. Grasso, B. Hannon, K. Limburg, S. Naeem, R. V. O'Neill, J. Paruelo, R. G. Raskin, P. Sutton, and M. van den Belt (1997), 'The Value of the World's Ecosystem Services and Natural Capital', *Nature*, 387(6630), 253–260.

Costanza, R., R. de Groot, P. Sutton, S. van der Ploeg, S. J. Anderson, I. Kubiszewski, S. Farber, and R. K. Turner (2014), 'Changes in the Global Value of Ecosystem Services', *Global Environmental Change*, 26(1), 152–158.

Costello, C., D. Ovando, R. Hilborn, S. D. Gaines, O. Deschenes, and S. E. Lester (2012), 'Status and Solutions for the World's Unassessed Fisheries', *Science*, 338(6106), 517–520.

Costello, C., D. Ovando, T. Clavelle, C. Kent Strauss, R. Hilborn, M. C. Melnychuk, T. A. Branch, S. D. Gaines, C. S. Szuwalski, R. B. Cabral, D. N. Rader, and A. Leland (2016), 'Global Fishery Prospects

References

Under Contrasting Management Regimes', *Proceedings of the National Academy of Sciences*, 113(18), 5125–5129.

Costello, C., L. Cao, and S. Gelcich (2019), *The Future of Food from the Sea* (Washington, DC: World Resources Institute).

Costello, M. J., and B. Ballantine (2015), 'Biodiversity Conservation Should Focus on No-take Marine Reserves: 94% of Marine Protected Areas Allow Fishing', *Trends in Ecology and Evolution*, 30(9), 507–509.

Cowie, C. (2020), *The Repugnant Conclusion: A Philosophical Inquiry* (London: Routledge).

Coyle, D. (2014), *GDP: A Brief but Affectionate History* (Princeton, NJ: Princeton University Press).

CPIC (2017), *Statement of Intent*.

Crépin, A. S., R. Biggs, S. Polasky, M. Troell and A. De Zeeuw (2012), 'Regime Shifts and Management', *Ecological Economics*, 84, 15–22.

Crona, B. I., T. M. Daw, W. Swartz, A. V. Norström, M. Nyström, M. Thyresson, C. Folke, J. Hentati-Sundberg, H. Österblom, L. Deutsch, and M. Troell (2016), 'Masked, Diluted and Drowned Out: How Global Seafood Trade Weakens Signals from Marine Ecosystems', *Fish and Fisheries*, 17(4), 1175–1182.

Cropper, M. L. (2014), 'How Should Benefits and Costs Be Discounted in an Inter-generational Context?', in D. Southerton and A. Ulph, eds., *Sustainable Consumption: Multi-Disciplinary Perspectives in Honour of Professor Sir Partha Dasgupta* (Oxford: Oxford University Press).

Cropper, M., and C. Griffiths (1994), 'The Interaction of Population Growth and Environmental Quality', *American Economic Review*, 84(2), 250–254.

Crouzeilles, R., M. S. Ferreira, R. L. Chazdon, D. B. Lindenmayer, J. B. B. Sansevero, L. Monteiro, A. Iribarrem, A. E. Latawiec, and B. B. N. Strassburg (2017), 'Ecological Restoration Success is Higher for Natural Regeneration than for Active Restoration in Tropical Forests', *Science Advances*, 3(11), e1701345.

Crutzen, P. J. and E. F. Stoermer (2000), 'The "Anthropocene"', *Global Change Newsletter*, 41, 17–18.

Cuff, D. J., and A. S. Goudie (2009), *Oxford Companion to Global Change* (Oxford: Oxford University Press).

Cullen, J. (2018), 'After "HLEG": EU Banks, Climate Change Abatement and the Precautionary Principle', *Cambridge Yearbook of European Legal Studies*, 20, 61–87.

Curtis, P. G., C. M. Slay, N. L. Harris, A. Tyukavina, and M. C. Hansen (2018), 'Classifying Drivers of Global Forest Loss', *Science*, 361(6407), 1108–1111.

Cushing, L., R. Morello-Frosch, M. Wander, and M. Pastor (2015), 'The Haves, the Have-Nots, and the Health of Everyone: The Relationship Between Social Inequality and Environmental Quality', *Annual Review of Public Health*, 36, 193–209.

Cutter, S. L. (2006), *Hazards, Vulnerability and Environmental Justice* (London: Routledge).

Czakó, Á., and E. Sik (1988), 'Managers' Reciprocal Transactions', *Connections*, 11(3), 23–32.

Daigneault, A., P. Brown, and D. Gawith (2016), 'Dredging Versus Hedging: Comparing Hard Infrastructure to Ecosystem-Based Adaptation to Flooding', *Ecological Economics*, 122, 25–35.

Daily, G. C., and P. R. Ehrlich (1996), 'Global Change and Human Susceptibility to Disease', *Annual Review of Energy and the Environment*, 21, 125–44.

Dakos, V., and A. Hastings (2013), 'Editorial: Special Issue on Regime Shifts and Tipping Points in Ecology', *Theoretical Ecology*, 6(3), 253–254.

Dalin, C., Y. Wada, T. Kastner, and M. J. Puma (2017), 'Groundwater Depletion Embedded in International Food Trade', *Nature*, 543(7647), 700–704.

Daly, H. E., B. Czech, D. L. Trauger, W. E. Rees, M. Grover, T. Dobson, and S. C. Trombulak (2007), 'Are We Consuming Too Much – For What?', *Conservation Biology*, 21(5), 1359–1362.

Das, S. (2011), 'Examining the Storm Protection Services of Mangroves of Orissa During the 1999 Cyclone', *Economic and Political Weekly*, 46(24), 60–68.

Das, S., and J. R. Vincent (2009), 'Mangroves Protected Villages and Reduced Death Toll During Indian Super Cyclone', *Proceedings of the National Academy of Sciences*, 106(18), 7357–7360.

Dasgupta, Aisha (2021), 'Contraception, Avortement, Droits Reproductifs', in Y. Charbit, ed., *Dynamiques démographiques et développement* (Paris/London: ISTE), 107–138.

Dasgupta, Aisha and P. Dasgupta (2017), 'Socially Embedded Preferences, Environmental Externalities, and Reproductive Rights', *Population and Development Review*, 43(3), 405–441.

Dasgupta, Aisha and P. Dasgupta (2021), 'Population Overshoot', in G. Arrhenius, K. Bykvist, and T. Campbell, eds., *Oxford Handbook of Population Ethics* (Oxford: Oxford University Press).

Dasgupta, (Amiya) A. K. (1985), *Epochs of Economic Theory* (Oxford: Basil Blackwell).

Dasgupta, P. (1969), 'On the Concept of Optimum Population', *The Review of Economic Studies*, 36(3), 295–318.

Dasgupta, P. (1974), 'On Some Alternative Criteria for Justice Between Generations', *Journal of Public Economics*, 3(4), 405–423.

Dasgupta, P. (1982), *The Control of Resources* (Cambridge, MA: Harvard University Press).

Dasgupta, P. (1988), 'Trust as a Commodity,' in D Gambetta, ed., *Trust: Making and Breaking Cooperative Relations* (Oxford: Blackwell).

Dasgupta, P. (1990), 'The Environment as a Commodity', *Oxford Review of Economic Policy*, 6(1), 51–67.

Dasgupta, P. (1993), *An Inquiry into Well-Being and Destitution* (Oxford: Clarendon Press).

Dasgupta, P. (2000), 'Economic Progress and the Idea of Social Capital', in P. Dasgupta and I. Serageldin eds., *Social capital: A Multifaceted Perspective* (Washington, DC: World Bank).

Dasgupta, P. (2002), 'A Model of Fertility Transition', in B. Kriström, P. Dasgupta, and K. G. Löfgren, eds., *Economic Theory for the Environment: Essays in Honour of Karl-Göran Mäler* (Cheltenham: Edward Elgar).

Dasgupta, P. (2004), *Human Well-Being and the Natural Environment* (Oxford: Oxford University Press).

Dasgupta, P. (2007), *Economics: A Very Short Introduction* (Oxford: Oxford University Press).

Dasgupta, P. (2010), 'The Place of Nature in Economic Development', in D. Rodrik and M. R. Rosenzweig, eds., *Handbook of Development Economics*, Vol. 5 (Amsterdam: Elsevier).

Dasgupta, P. (2013), 'Getting India Wrong', *Prospect*, August, 64–66.

Dasgupta, P. (2014), 'Measuring the Wealth of Nations', *Annual Review of Resource Economics*, 6(1), 17–31.

References

Dasgupta, P. (2019), *Time and the Generations: Population Ethics for a Diminishing Planet* (New York, NY: Columbia University Press).

Dasgupta, P. (2022), "The Economics of Biodiversity: Afterword," *Environmental and Resource Economics* (Symposium on The Economics of Biodiversity: The Dasgupta Review), 83(4), pp. 1017–1039. https://doi.org/10.1007/s10640-022-00731-9.

Dasgupta, P., Aisha Dasgupta, and S. Barrett (2023), "Population, Ecological Footprint, and the Sustainable Development Goals," *Environmental and Resource Economics*, 84(2), pp. 659–675.

Dasgupta, P. and P.R. Ehrlich (2013), "Pervasive Externalities at the Population, Consumption, and Environment Nexus," *Science*, 2013, 19 (340), 324–328.

Dasgupta, P., and S. Goyal. (2019), 'Narrow Identities', *Journal of Institutional and Theoretical Economics*, 175(3), 395–419.

Dasgupta, P., and G. M. Heal (1974), 'The optimal depletion of exhaustible resources', *The Review of Economic Studies*, 41(1), 3–28.

Dasgupta, P., and G. M. Heal (1979), *Economic Theory and Exhaustible Resources* (Cambridge: Cambridge University Press).

Dasgupta, P. and S. Levin (2023), "Economic Factors Underlying Biodiversity Loss," *Philosophical Transactions of the Royal Society B*, 378(1881, 17 July), 20220197, pp. 1–15. https://doi.org/10.1098/rstb.2022.0197.

Dasgupta, P. and K. -G. Mäler, (2000), 'Net National Product, Wealth, and Social Well-Being', *Environment and Development Economics*, 5(1/2), 69–93.

Dasgupta, P. and K. -G. Mäler, (2003), 'Introduction' in Dasgupta, P., and K. -G. Mäler, eds., *The Economics of Non-Convex Ecosystems* (Dordrecht: Springer).

Dasgupta, P., S. A. Marglin, and A. K. Sen (1972), *Guidelines for Project Evaluation* (New York, NY: United Nations).

Dasgupta, P. and E. Maskin (2004), "The Fairest Vote of All," *Scientific American*, 290(3), 92–97.

Dasgupta, P. and E. Maskin (2005), 'Uncertainty and Hyperbolic Discounting', *American Economic Review*, 95(4), 1290–1299.

Dasgupta, P. and E. Maskin (2008), "On the Robustness of Majority Rule," *Journal of the European Economic* Association, 6(5), 949–973.

Dasgupta, P. and E. Maskin (2020), "Strategy-Proofness, Independence of Irrelevant Alternatives, and Majority Rule," *American Economic Review: Insights*, 2(4), 459–474.

Dasgupta, P., T. Mitra, and G. Sorger (2019), 'Harvesting the Commons', *Environmental and Resource Economics*, 72(3), 613–636.

Dasgupta, P., and P. H. Raven (2019), 'Introduction', in P. Dasgupta, P. H. Raven, and A. L. McIvor, eds., *Biological Extinction: New Perspectives* (Cambridge: Cambridge University Press).

Dasgupta, P., P. H. Raven, and A. McIvor (2019), *Biological Extinction: New Perspectives* (Cambridge: Cambridge University Press).

Dasgupta, P., and D. Ray (1986), 'Inequality as a Determinant of Malnutrition and Unemployment: Theory', *The Economic Journal*, 96(384), 1011–1034.

Dasgupta, P., and D. Ray (1987), 'Inequality as a Determinant of Malnutrition and Unemployment: Policy', *The Economic Journal*, 97(385), 177–188.

Dasgupta, P., and I. Serageldin, eds. (2000), *Social Capital: A Multifaceted Perspective* (Washington, DC: World Bank).

Dasgupta, P., D. Southerton, A. Ulph, and D. Ulph (2016), 'Consumer Behaviour with Environmental and Social Externalities: Implications for Analysis and Policy', *Environmental and Resource Economics*, 65, 191–226.

Dasgupta, P., and J. Stiglitz (1980), 'Industrial Structure and the Nature of Innovative Activity', *The Economic Journal*, 90(358), 266–293.

Dasgupta, Shamik (2020), '*Undoing the Truth Fetish: The Normative Path to Pragmatism*', Typescript, Department of Philosophy, University of California, Berkeley.

Daskalov, G. M., A. N. Grishin, S. Rodionov, and V. Mihneva (2007), 'Trophic Cascades Triggered by Overfishing Reveal Possible Mechanisms of Ecosystem Regime Shifts', *Proceedings of the National Academy of Sciences*, 104(25), 10518–10523.

Davis, A. Y., J. A. Belaire, M. A. Farfan, D. Milz, E. R. Sweeney, S. R. Loss, and E. S. Minor (2012), 'Green Infrastructure and Bird Diversity Across an Urban Socioeconomic Gradient', *Ecosphere*, 3(11), 105.

Daw, T., K. Brown, S. Rosendo, and R. Pomeroy (2011), 'Applying the Ecosystem Services Concept to Poverty Alleviation: The Need to Disaggregate Human Well-Being', *Environmental Conservation*, 38(4), 370–379.

Dawson, M. N. (2012), 'Species Richness, Habitable Volume, and Species Densities in Freshwater, the Sea, and on Land', *Frontiers of Biogeography*, 4(3), 105–116.

Day, R. H. (1983), 'The Emergence of Chaos from Classical Economic Growth', *The Quarterly Journal of Economics*, 98(2), 201–213.

De Groot, R. S., J. Blignaut, S. Van Der Ploeg, J. Aronson, T. Elmqvist, and J. Farley (2013), 'Benefits of Investing in Ecosystem Restoration', *Conservation Biology*, 27(6), 1286–1293.

De Marco, P., and F. M. Coelho (2004), 'Services Performed by the Ecosystem: Forest Remnants Influence Agricultural Cultures' Pollination and Production', *Biodiversity and Conservation*, 13, 1245–1255.

De Palma, A., A. Hoskins, R. E. Gonzalez, L. Börger, T. Newbold, K. Sanchez-Ortiz, S. Ferrier, and A. Purvis (2018), 'Annual Changes in the Biodiversity Intactness Index in Tropical and Subtropical Forest Biomes, 2001-2012', *bioRxiv*.

De Prato, G., and D. Nepelski (2014), 'Global Technological Collaboration Network: Network Analysis of International Co-Inventions', *The Journal of Technology Transfer*, 39, 358–375.

De Silva, M. M. G. T., and A. Kawasaki (2018), 'Socioeconomic Vulnerability to Disaster Risk: A Case Study of Flood and Drought Impact in a Rural Sri Lankan Community', *Ecological Economics*, 152, 131–140.

De Vos, J. M., L. N. Joppa, J. L. Gittleman, P. R. Stephens, and S. L. Pimm (2014), 'Estimating the Normal Background Rate of Species Extinction', *Conservation Biology*, 29(2), 452–462.

Deaton, A. (2008), 'Income, Health, and Well-Being Around the World: Evidence from the Gallup World Poll', *Journal of Economic Perspectives*, 22(2), 53–72.

Deaton, A. and A. A. Stone (2013), 'Two Happiness Puzzles', *American Economic Review*, 103(3), 591–597.

References

Debreu, G. (1959), *Theory of Value: An Axiomatic Analysis of Economic Equilibrium* (New Haven, CT and London: Yale University Press).

Defra (2018) *A Green Future: Our 25 Year Plan to Improve the Environment*. Defra (2020a), *Environmental Land Management: Policy Discussion Document*.

Defra (2020a), *Farming for the future: Policy and progress update*.

Defries, R. S., T. Rudel, M. Uriarte, and M. Hansen (2010), 'Deforestation Driven by Urban Population Growth and Agricultural Trade in the Twenty-First Century', *Nature Geoscience*, 3, 178–181.

Deininger, K. and D. Byerlee – with J. Lindsay, A. Norton, H. Selod, and M. Stickler (2011), *Rising Interest in Farmland: Can It Yield Sustainable and Equitable Benefits?* (Washington, DC: World Bank).

De Long, J. B. and B. Eichengreen (1991), 'The Marshall Plan: History's Most Successful Structural Adjustment Program', *National Bureau of Economic Research Working Paper*, No. 3899.

Dempsey, J. (2013), 'Biodiversity Loss as Material Risk: Tracking the Changing Meanings and Materialities of Biodiversity Conservation', *Geoforum*, 45, 41–51.

den Braber, B., K. L. Evans, and J. A. Oldekop (2018), 'Impact of Protected Areas on Poverty, Extreme Poverty, and Inequality in Nepal', *Conservation Letters*, 11(6), e12576.

Dennett, D. C. (1993), *Consciousness Explained* (London: Penguin).

Depledge, M. (2011), 'Reduce Drug Waste in the Environment', *Nature*, 478), 36.

Deutz, A., G. M. Heal, R. Niu, E. Swanson, T. Townsend, L. Zhu, A. Delmar, A. Meghji, S. A. Sethi, and J. Tobin-de la Puente (2020), *Financing Nature: Closing the Global Biodiversity Financing Gap*.

Di Minin, E. and T. Toivonen (2015), 'Global Protected Area Expansion: Creating more than Paper Parks', *BioScience*, 65(7), 637–638.

Diamond, J. (2005), *Collapse: How Societies Choose to Fail or Succeed* (London: Viking Press).

Díaz, S., J. Settele, E. S. Brondizio, H. T. Ngo, J. Agard, A. Arneth, P. Balvanera, K. A. Brauman, S. H. M. Butchart, K. M. A. Chan, L. A. Garibaldi, K. Ichii, J. Liu, S. M. Subramanian, G. F. Midgley, P. Miloslavich, Z. Molnár, D. Obura, A. Pfaff, S. Polasky, A. Purvis, J. Razzaque, B. Reyers, R. R. Chowdhury, Y. -J. Shin, I. Visseren-Hamakers, K. J. Willis, and C. N. Zayas. (2019), 'Pervasive Human-Driven Decline of Life on Earth Points to the Need for Transformative Change', *Science*, 366(6471), eaax3100.

Dickson, B., R. Blaney, L. Miles, E. Regan, A. Van Soesbergen, E. Väänänen, S. Blyth, M. Harfoot, C. Martin, C. Mcowen, and T. Newbold (2014), *Towards a Global Map of Natural Capital: KeyEcosystem Assets* (Nairobi: UNEP).

Diener, E., J. F. Helliwell, and D. Kahneman, eds (2010), *International Differences in Well-Being*(Oxford: Oxford University Press).

Diffenbaugh, N. S., and M. Burke (2019), Global Warming has Increased Global Economic Inequality. *Proceedings of the National Academy of Sciences*, 116 (20), 9808–9813.

Diffey, B. L. (2011), 'An Overview Analysis of The Time People Spend Outdoors', *British Journal of Dermatology*, 164(4), 848–854.

Dinerstein, E., C. Vynne, E. Sala, A. R. Joshi, S. Fernando, T. E. Lovejoy, J. Mayorga, D. Olson, G. P. Asner, J. E. M. Baillie, N. D. Burgess, K. Burkart, R. F. Noss, Y. P. Zhang, A. Baccini, T. Birch, N. Hahn, L. N. Joppa, and E. Wikramanayake (2019), 'A Global Deal for Nature: Guiding Principles, Milestones, and Targets', *Science Advances*, 5(4), eaaw2869.

Dixon, S. J., D. A. Sear, N. A. Odoni, T. Sykes, and S. N. Lane (2016), 'The Effects of River Restoration on Catchment Scale Flood Risk and Flood Hydrology', *Earth Surface Processes and Landforms*, 41(7), 997–1008.

Dobson, A. P., S. L. Pimm, L. Hannah, L. Kaufman, J. A. Ahumada, A. W. Ando, A. Bernstein, J. Busch, P. Daszak, J. Engelmann, M. F. Kinnaird, B. V. Li, T. Loch-Temzelides, T. Lovejoy, K. Nowak, P. R. Roehrdanz, and M. M. Vale (2020), 'Ecology and Economics for Pandemic Prevention',*Science*, 369(6502), 379–381.

Donati, P. (2011), *Relational Sociology: A New Paradigm for the Social Sciences*(London: Routledge).

Douglas, M. and B. Isherwood (1979), *World of Goods: Towards an Anthropology of Consumption* (London: Allen Lane).

Downey, S. S., W. R. Haas, and S. J. Shennan (2016), 'European Neolithic Societies Showed Early Warning Signals of Population Collapse', *Proceedings of the National Academy of Sciences*, 113(35), 9751–9756.

Drupp, M. A. (2018), 'Limits to Substitution Between Ecosystem Services and Manufactured Goods and Implications for Social Discounting', *Environmental and Resource Economics*, 69(1), 135–158.

Drupp, M. A., M. C. Freeman, B. Groom, and F. Nesje (2018), 'Discounting Disentangled',*American Economic Journal: Economic Policy*, 10(4), 109–134.

Drutschinin, A., and S. Ockenden (2015), 'Financing for Development in Support of Biodiversity and Ecosystem Services', *OECD Development Co-Operation Working Paper* No. 23.

Duarte, C. M., S. Agusti, E. Barbier, G. L. Britten, J. C. Castilla, J.-P. Gattuso, R. W. Fulweiler, T. P. Hughes, N. Knowlton, C. E. Lovelock, H. K. Lotze, M. Predragovic, E. Poloczanska, C. Roberts, and B. Worm (2020a), 'Rebuilding Marine Life', *Nature*, 580(7801), 39–51.

Duarte, C. M., S. Agusti, E. Barbier, G. L. Britten, J. C. Castilla, J.-P. Gattuso, R. W. Fulweiler, T. P. Hughes, N. Knowlton, C. E. Lovelock, H. K. Lotze, M. Predragovic, E. Poloczanska, C. Roberts, and B. Worm (2020b), 'Rebuilding Marine Life: Supplementary Information', *Nature*, 580(7801), 39–51.

Duarte, C. M., I. J. Losada, I. E. Hendriks, I. Mazarrasa, and N. Marbà (2013), 'The Role of Coastal Plant Communities for Climate Change Mitigation and Adaptation', *Nature Climate Change*, 3(11), 961–968.

Duarte, C.M., I. Poiner and J. Gunn (2018), 'Perspectives on a Global Observing System to Assess Ocean Health', *Frontiers in Marine Science*, 5, 265.

Dudley, N., H. Jonas, F. Nelson, J. Parrish, A. Pyhälä, S. Stolton, and J. E. M. Watson (2018), 'The Essential Role of Other Effective Area-Based Conservation Measures in Achieving Big Bold Conservation Targets', *Global Ecology and Conservation*, 15, e00424.

Duesenberry, J. S. (1949), *Income, Saving and the Theory of Consumer Behavior* (Cambridge, MA: Harvard University Press).

Dupré, S., and H. Chenet (2012), *Connecting the Dots Between Climate Goals, Portfolio Allocation and Financial Regulation*.

Easterlin, R. A. (1974). 'Does Economic Growth Improve the Human Lot? Some Empirical Evidence,' in P. A. David and M. Reder, eds., *Nations and Households in Economic Growth. Essays in Honor of Moses Abramovitz*. (New York, NY: Academic Press).

References

EAT-Lancet Commission (2019). *Healthy Diets from Sustainable Food Systems: Food Planet Health.* Summary Report of the EAT-Lancet Commission on Healthy Diets from Sustainable Food Systems.

Ebeling, F., and S. Lotz (2015), 'Domestic Uptake of Green Energy Promoted by Opt-out Tariffs', *Nature Climate Change*, 5, 868–871.

EBRD (2020), *Joint Report on Multilateral Development Banks' Climate Finance 2019*.

Eccles, R. G., I. Ioannou, and G. Serafeim (2014), 'The Impact of Corporate Sustainability on Organizational Processes and Performance', *Management Science*, 60(11), 2835–2857.

Ecologist (2014), 'India: Coca Cola Bottling Plant Shut Down', *The Ecologist*, https://theecologist.org/2014/jun/19/india-coca-cola-bottling-plant-shut-down.

Edgeworth, F. Y. (1881), *Mathematical Psychics: An Essay on the Application of Mathematics to the Moral Sciences* (London: C. Kegan Paul and Co.).

Edwards, D. P., J. A. Hodgson, K. C. Hamer, S. L. Mitchell, A. H. Ahmad, S. J. Cornell, and D. S. Wilcove (2010), 'Wildlife-friendly Oil Palm Plantations Fail to Protect Biodiversity Effectively', *Conservation Letters*, 3(4), 236–242.

Edwards, D. P., and S. G. Laurance (2012), 'Green Labelling, Sustainability and the Expansion of Tropical Agriculture: Critical Issues for Certification Schemes', *Biological Conservation*, 151(1), 60–64.

Edwards, S. (1998), 'Openness, Productivity and Growth: What do We Really Know?', *The Economic Journal*, 108(447), 383–398.

Egoh, B. N., B. Reyers, M. Rouget, and D. M. Richardson (2011), 'Identifying Priority Areas for Ecosystem Service Management in South African Grasslands', *Journal of Environmental Management*, 92(6), 1642–1650.

Ehrlich, P. R., and A. H. Ehrlich (1981), *Extinction: The Causes and Consequences of the Disappearance of Species* (New York, NY: Random House).

Ehrlich, P. R., and A. H. Ehrlich (2008), *The Dominant Animal: Human Evolution and the Environment* (Washington, DC: Island Press).

Ehrlich, P. R., and J. P. Holdren (1971), 'Impact of Population Growth', *Science*, 171(3977), 1212–1217.

EIB (2006), 'Environmental and Social Safeguards', *European Investment Bank*, 'https://www.eib.org/en/press/news/environmental-and-social-safeguards'.

Eichengreen, B. (2010), 'Lessons from the Marshall Plan', *World Bank*, *World Development Report 2011 Background Paper*.

Ekins, P. (2003), 'Identifying Critical Natural Capital: Conclusions about Critical Natural Capital', *Ecological Economics*, 44(2–3), 277–292.

Ekins, P., B. Milligan and A. Usubiaga-Liaño (2020), 'A Single Indicator of Strong Sustainability for Development: Theoretical Basis and Practical Implementation.' *AFD Research papers*, No.2019-112.

Elert, E. (2014), 'Rice by the Numbers: A Good Grain', *Nature*, 514(7524), S50–S51.

Ellis, E. C., K. K. Goldewijk, S. Siebert, D. Lightman, and N. Ramankutty (2010), 'Anthropogenic Transformation of the Biomes, 1700 to 2000', *Global Ecology and Biogeography*, 19(5), 589–606.

Ellis, E. C., and N. Ramankutty (2008). 'Putting People in the Map: Anthropogenic Biomes of the World', *Frontiers in Ecology and the Environment*, 6(8), 439–447.

Ellsberg, D. (1961), 'Risk, Ambiguity, and the Savage Axioms', *The Quarterly Journal of Economics*, 75(4), 643–669.

Elmqvist, T., C. Folke, M. Nyström, G. Peterson, J. Bengtsson, B. Walker, and J. Norberg (2003), 'Response Diversity, Ecosystem Change, and Resilience', *Frontiers in Ecology and the Environment*, 1(9), 488–494.

Elsen, P. R., W. B. Monahan, E. R. Dougherty, and A. M. Merenlender (2020), 'Keeping Pace with Climate Change in Global Terrestrial Protected Areas', *Science Advances*, 6(25), eaay0814.

Engel, S., S. Pagiola, and S. Wunder (2008), 'Designing Payments for Environmental Services in Theory and Practice: An Overview of the Issues', *Ecological Economics*, 65(4), 663–674.

Engen, L., and A. Prizzon (2018), *A Guide to Multilateral Development Banks*, (London: Overseas Development Institute).

Ensminger, J. (1992), *Making a Market: The Institutional Transformation of an African Society* (Cambridge: Cambridge University Press).

Erb, K.-H., T. Kastner, C. Plutzar, A. L. S. Bais, N. Carvalhais, T. Fetzel, S. Gingrich, H. Haberl, C. Lauk, M. Niedertscheider, J. Pongratz, M. Thurner, and S. Luyssaert (2018), 'Unexpectedly Large Impact of Forest Management and Grazing on Global Vegetation Biomass', *Nature*, 553(7686), 73–76.

Ermakova, M., F. R. Danila, R. T. Furbank, and S. von Caemmerer (2019), 'On the Road to C4 Rice: Advances and Perspectives', *The Plant Journal*, 101(4), 940–950.

Estevadeordal, A., B. Frantz, and A. M. Taylor (2003), 'The Rise and Fall of World Trade, 1870-1939', *The Quarterly Journal of Economics*, 118(2), 359–407.

EU Technical Expert Group on Sustainable Finance (2020), *Taxonomy: Final report of the Technical Expert Group on Sustainable Finance*.

European Central Bank (2020), *Guide on Climate-Related and Environmental Risks. Supervisory Expectations Relating to Risk Management and Disclosure*.

European Commission (2018), *Renewed Sustainable Finance Strategy and Implementation of the Action Plan on Financing Sustainable Growth*, 'https://ec.europa.eu/info/publications/sustainable-finance-renewed-strategy_en'.

European Commission (2020), *EU Emissions Trading System*, 'https://ec.europa.eu/clima/policies/ets_en#tab-0-1'.

European Investment Bank (2006), *Environmental and Social Safeguards*.

Fabic, M. S., Y. Choi, J. Bongaarts, J. E. Darroch, J. A. Ross, J. Stover, A. O. Tsui, J. Upadhyay, and E. Starbird (2015), 'Meeting Demand for Family Planning Within a Generation: The Post-2015 Agenda', *The Lancet*, 385(9981), 1928–1931.

Falconer, J., and C. R. S. Koppell (1990), 'The Major Significance of "Minor" Forest Products: The Local Use and Value of Forests in the West African Humid Forest Zone', *Community Forestry Note 6* (Rome: FAO).

FAO (n.d.), 'Women Feed the World', 'http://www.fao.org/NEWS/1998/980305-e.htm'.

FAO, and ITPS (2015), *Status of the World's Soil Resources (SWSR) – Main Report* (Rome: FAO).

FAO (2009), *The State of Food Insecurity in the World 2009: Economic Crises – Impacts and Lessons Learned* (Rome: FAO).

References

FAO (2015a), *Global Forest Resources Assessment 2015: How are the World's Forests Changing?* (Rome: FAO).

FAO (2015b), *Measuring Sustainability in Cotton Farming Systems: Towards a Guidance Framework* (Rome: FAO).

FAO (2016), *The State of Food and Agriculture 2016: Climate Change, Agriculture and Food Security* (Rome: FAO).

FAO (2017), *The Future of Food and Agriculture – Trends and Challenges* (Rome: FAO)

FAO (2018a), *State of the World's Fisheries and Aquaculture – Meeting the Sustainable Development Goals* (Rome: FAO).

FAO (2018b), *The State of Agricultural Commodity Markets 2018. Agricultural Trade, Climate Change and Food Security* (Rome: FAO).

FAO (2019), *The State of Food and Agriculture 2019. Moving Forward on Food Loss and Waste Reduction* (Rome: FAO)

FAO (2020a), *FAO Yearbook of Forest Products 2018* (Rome: FAO).

FAO (2020b), *Global Forest Resources Assessment 2020 – Key Findings* (Rome: FAO).

FAO (2020c), *The State of World Fisheries and Aquaculture 2020. Sustainability in Action* (Rome: FAO).

FAOSTAT, 'http://www.fao.org/faostat/en/#home'.

Farooq, G. M., and I. I. Ekanem (1987), 'Family Size Preferences and Fertility in Southwestern Nigeria', Oppong C. ed., *Sex Roles, Population and Development in West Africa*, (Portsmouth, NH: Heinemann).

Fayaz Bhat, Z., S. Kumar, and H. Fayaz (2015), 'In Vitro Meat Production: Challenges and Benefits Over Conventional Meat Production', *Journal of Integrative Agriculture*, 14(2), 241–248.

Feder, G., and R. Noronha (1987), 'Land Rights Systems and Agricultural Development in Sub-Saharan Africa', *The World Bank Research Observer*, 2(2), 143–169.

Feenstra, R. C., R. Inklaar, and M. P. Timmer (2015), 'The Next Generation of the Penn World Table', *American Economic Review*, 105(10) 3150-3182, 'www.ggdc.net/pwt'.

Feeny, D., F. Berkes, B. J. McCay, and J. M. Acheson (1990), 'The Tragedy of the Commons: Twenty-Two years later', *Human Ecology*, 18(1), 1–19.

Femenia, F. (2019), 'A Meta-Analysis of the Price and Income Elasticities of Food Demand', *German Journal of Agricultural Economics*, 68(2), 77–98.

Feng, S., S. Liu, Z. Huang, L. Jing, M. Zhao, X. Peng, W. Yan, Y. Wu, Y. Lv, A. R. Smith, M. A. McDonald, S. D. Patil, A. J. Sarkissian, Z. Shi, J. Xia, and U. S. Ogbodo (2020), 'Reply to Zhang et al.: Using Long-Term All-Available Landsat Data to Study Water Bodies Over Large Areas Represents a Paradigm Shift', *Proceedings of the National Academy of Sciences*, 117(12), 6310–6311.

Fenichel, E. P., J. K. Abbott, J. Bayham, W. Boone, E. M. K. Haacker, and L. Pfeiffer (2016), 'Measuring the Value of Groundwater and Other Forms of Natural Capital', *Proceedings of the National Academy of Sciences*, 113(9), 2382–2387.

Fernandez-Cornejo, J., C. Klotz-Ingram, and S. Jans (2002), 'Farm-Level Effects of Adopting Herbicide-Tolerant Soybeans in the U.S.A', *Journal of Agricultural and Applied Economics*, 34(1), 149–163.

Ferraro, P. J., M. M. Hanauer, D. A. Miteva, J. L. Nelson, S. K. Pattanayak, C. Nolte, and K. R. E. Sims (2015), 'Estimating the Impacts of Conservation on Ecosystem Services and Poverty by Integrating Modeling and Evaluation', *Proceedings of the National Academy of Sciences*, 112(24), 7420–7425.

Ferraro, P. J., K. Lawlor, K. L. Mullan, and S. K. Pattanayak (2012), 'Forest Figures: Ecosystem Services Valuation and Policy Evaluation in Developing Countries', *Review of Environmental Economics and Policy*, 6(1), 20–44.

Ferraro, P. J., and M. K. Price (2013), 'Using Nonpecuniary Strategies to Influence Behavior: Evidence from a Large-Scale Field Experiment', *Review of Economics and Statistics*, 95(1), 64–73.

Field, C. B., M. J. Behrenfeld, J. T. Randerson, and P. Falkowski (1998), 'Primary Productivity of the Biosphere: Integrating Terrestrial and Oceanic Components', *Science*, 281(5374), 237–240.

Field-Dodgson, M. S. (1987), 'The Effect of Salmon Redd Excavation on Stream Substrate and Benthic Community of Two Salmon Spawning Streams in Canterbury, New Zealand', *Hydrobiologia*, 154, 3–11.

Finance for Biodiversity (2020), *Aligning Development Finance with Nature's Needs: Protecting Nature's Development Dividend*.

Finance Watch. (2019), *Making Finance Serve Nature*.

Finley, M. I. (1982), *Economy and Society in Ancient Greece* (New York, NY: Viking Press).

Finley, M. I. (2002), *The World of Odysseus*, (London: Folio Society) 2nd Edition.

Fischer, J., M. Meacham, and C. Queiroz (2017), 'A Plea for Multifunctional Landscapes', *Frontiers in Ecology and the Environment*, 15(2), 59.

Fisher, B. (2010), 'African Exception to Drivers of Deforestation', *Nature Geoscience*, 3, 375–376.

Fisheries New Zealand, (2020), *Fishery – Hoki (Including Key Bycatch Stocks)*, 'https://fs.fish.govt.nz/Page.aspx?pk=5&fpid=16'.

Fitter, A. H. (2013), 'Are Ecosystem Services Replaceable by Technology?', *Environmental and Resource Economics*, 55(4), 513–524.

Folberth, C., N. Khabarov, J. Balkovi, R. Skalský, P. Visconti, P. Ciais, I. A. Janssens, J. Peñuelas, and M. Obersteiner (2020), 'The Global Cropland-Sparing Potential of High-Yield Farming', *Nature Sustainability*, 3, 281–289.

Foley, J. A., R. DeFries, G. P. Asner, C. Barford, G. Bonan, S. R. Carpenter, F. S. Chapin, M. T. Coe, G. C. Daily, H. K. Gibbs, J. H. Helkowski, T. Holloway, E. A. Howard, C. J. Kucharik, C. Monfreda, J. A. Patz, I. C. Prentice, N. Ramankutty, and P. K. Snyder (2005), 'Global Consequences of Land Use', *Science*, 309(5734), 570–574.

Foley, P., and B. McCay (2014), 'Certifying the Commons: Eco-Certification, Privatization, and Collective Action', *Ecology and Society*, 19(2), 28.

Foley, J. A., N. Ramankutty, K. A. Brauman, E. S. Cassidy, J. S. Gerber, M. Johnston, N. D. Mueller, C. O'Connell, D. K. Ray, P. C. West, C. Balzer, E. M. Bennett, S. R. Carpenter, J. Hill, C. Monfreda, S. Polasky, J. Rockström, J. Sheehan, S. Siebert, D. Tilman, D. Zaks (2011), 'Solutions for a Cultivated Planet', *Nature*, 478(7369), 337–342.

Folke, C., S. Carpenter, B. Walker, M. Scheffer, T. Elmqvist, L. Gunderson, and C. S. Holling (2004), 'Regime Shifts, Resilience, and Biodiversity in Ecosystem Management', *Annual Review of Ecology, Evolution, and Systematics*, 35(1), 557–581.

References

Fong, W. K., M. Sotos, M. Doust, S. Schultz, A. Marques, and C. Deng-Beck (2014). *Global Protocol for Community-Scale Greenhouse Gas Emission Inventories: An Accounting and Reporting Standard for Cities*. World Resources Institute.

Forrester, J. W. (1971), *World Dynamics* (Cambridge, MA: Wright-Allen Press) 2nd Edition.

Fortes, M. (1978), 'Parenthood, Marriage and Fertility in West Africa', *The Journal of Development Studies*, 14(4), 121–149.

Fothergill, A., and L. A. Peek (2004), 'Poverty and Disasters in the United States: A Review of Recent Sociological Findings', *Natural Hazards*, 32(1), 89–110.

Fouquet, R., and S. Broadberry (2015), 'Seven Centuries of European Economic Growth and Decline', *Journal of Economic Perspectives*, 29(4), 227–244.

Fox, J., and A. Nino-Murci (2005), 'Status of Species Conservation Banking in the United States', *Conservation Biology*, 19(4), 996–1007.

FP2020 (2019), *Women at the Center 2018–2019*.

Frank, R. H. (1997), 'The Frame of Reference as a Public Good', *The Economic Journal*, 107(445), 1832–1847.

Frankel, J. A., and D. Romer (1999), 'Does Trade Cause Growth?', *The American Economic Review*, 89(3), 379–399.

Freedom House (2019), *Freedom in the World 2019: Democracy in Retreat*.

Freeman III, A. M. (1982), *Air and Water Pollution Control: A Benefit-Cost Assessment* (New York, NY: Wiley).

Freeman III, A. M. (1993), *The Measurement of Environmental and Resource Values: Theory and Methods* (Washington, DC: Resources for the Future) 1st Edition.

Freeman III, A. M. (2003), *The Measurement of Environmental and Resource Values: Theory and Methods* (Washington, DC: Resources for the Future) 2nd Edition.

Freeman III, A. M., Herriges, J. A., and Kling, C. L. (1996). The Measurement of Environmental and Resource Values: Theory and Methods. *Land Economics*, 72(2).

Friends of Ocean Action. (2020), *The Business Case for Marine Protection and Conservation*.

Frost, P. G. H., and I. Bond (2008), 'The CAMPFIRE Programme in Zimbabwe: Payments for Wildlife Services', *Ecological Economics*, 65(4), 776–787.

Fudenberg, D., and E. Maskin (1986), 'The Folk Theorem in Repeated Games with Discounting or with Incomplete Information', *Econometrica*, 54(3), 533–554.

Fukuyama, F. (1995), *Trust: The Social Virtues and the Creation of Prosperity* (London: Hamish Hamilton).

Fuller, R. A., and K. N. Irvine (2013), 'Interactions Between People and Nature in Urban Environments', in K. J. Gaston, ed., *Urban Ecology* (Cambridge: Cambridge University Press).

Fulton, M., B. M. Kahn, and C. Sharples (2012), *Sustainable Investing: Establishing Long-Term Value and Performance*.

Furtado, C. (1964), *Development and Underdevelopment* (Berkeley, CA: University of California Press).

References

G20 Sustainable Finance Study Group (2018), *Sustainable Finance Synthesis Report*. Galbraith, J. K. (1958), *The Affluent Society* (Boston, MA: Houghton Mifflin).

Gaertner, W. (2009), *A Primer in Social Choice Theory* (Oxford: Oxford University Press).

Galor, O. (2011), *Unified Growth Theory* (Princeton, NJ: Princeton University Press).

Gambetta, D. (1993), *The Sicilian Mafia: The Business of Private Protection* (Cambridge, MA: Harvard University Press).

Gann, G., T. McDonald, B. Walder, J. Aronson, C. Nelson, J. Jonson, J. Hallett, C. Eisenberg, M. Guariguata, J. Liu, F. Hua, C. Echeverría, E. Gonzales, N. Shaw, K. Decleer, and K. Dixon (2019), 'International Principles and Standards for the Practice of Ecological Restoration. Second Edition', *Restoration Ecology*, 27(S1), 1–46.

Garibaldi, L. A., M. A. Aizen, A. M. Klein, S. A. Cunningham, and L. D. Harder (2011), 'Global Growth and Stability of Agricultural Yield Decrease with Pollinator Dependence', *Proceedings of the National Academy of Sciences*, 108(14), 5909–5914.

Garibaldi, L. A., I. Steffan-Dewenter, R. Winfree, M. A. Aizen, R. Bommarco, S. A. Cunningham, C. Kremen, L. G. Carvalheiro, L. D. Harder, O. Afik, I. Bartomeus, F. Benjamin, V. Boreux, D. Cariveau, N. P. Chacoff, J. H. Dudenhöffer, B. M. Freitas, J. Ghazoul, S. Greenleaf, J. Hipólito, A. Holzschuh, B. Howlett, R. Isaacs, S. K. Javorek, C. M. Kennedy, K. Krewenka, S. Krishnan, Y. Mandelik, M. M. Mayfield, I. Motzke, T. Munyuli, B. A. Nault, M. Otieno, J. Petersen, G. Pisanty, S. G. Potts, R. Rader, T. H. Ricketts, M. Rundlöf, C. L. Seymour, C. Schüepp, H. Szentgyörgyi, H. Taki, T. Tscharntke, C. H. Vergara, B. F. Viana, T. C. Wanger, C. Westphal, N. Williams and A. M. Klein (2013), 'Wild Pollinators Enhance Fruit Set of Crops Regardless of Honey Bee Abundance', *Science*, 339(6127), 1608–1611.

Garnett, S. T., N. D. Burgess, J. E. Fa, Á. Fernández-Llamazares, Z. Molnár, C. J. Robinson, J. E. M. Watson, K. K. Zander, B. Austin, E. S. Brondizio, N. F. Collier, T. Duncan, E. Ellis, H. Geyle, M. V. Jackson, H. Jonas, P. Malmer, B. McGowan, A. Sivongxay and I. Leiper (2018), 'A Spatial Overview of the Global Importance of Indigenous Lands for Conservation', *Nature Sustainability*, 1(7), 369–374.

Gebbers, R., and V. Adamchuk (2010), 'Precision Agriculture and Food Security', *Science*, 327(5967), 828–831.

Gebrehiwot, S. G., D. Ellison, W. Bewket, Y. Seleshi, B. Inogwabini, and K. Bishop (2019), 'The Nile Basin Waters and the West African Rainforest: Rethinking the Boundaries', *Wiley Interdisciplinary Reviews: Water*, 6(1), e1317.

Geertz, C. (1962), 'The Rotating Credit Association: A "Middle Rung" in Development', *Economic Development and Cultural Change*, 10(3), 241–263.

Geldmann, J., L. Coad, M. D. Barnes, I. D. Craigie, S. Woodley, A. Balmford, T. M. Brooks, M. Hockings, K. Knights, M. B. Mascia, L. McRae, and N. D. Burgess (2018), 'A Global Analysis of Management Capacity and Ecological Outcomes in Terrestrial Protected Areas', *Conservation Letters*, 11(3), 1–10.

Geldmann, J., M. Barnes, L. Coad, I. D. Craigie, M. Hockings, and N. D. Burgess (2013), 'Effectiveness of Terrestrial Protected Areas in Reducing Habitat Loss and Population Declines', *Biological Conservation*, 161: 230–238.

Gibb, R., D. W. Redding, K. Q. Chin, C. A. Donnelly, T. M. Blackburn, T. Newbold, and K. E. Jones (2020), 'Zoonotic Host Diversity Increases in Human-Dominated Ecosystems', *Nature*, 584(7821), 398–402.

References

Gibbard, A. (1973), "Manipulation of Voting Schemes: A General Result," *Econometrica*, 41(4), 587–603.

Gibbs, H. K., A. S. Ruesch, F. Achard, M. K. Clayton, P. Holmgren, N. Ramankutty, and J. A. Foley (2010), 'Tropical Forests Were the Primary Sources of New Agricultural Land in the 1980s and 1990s', *Proceedings of the National Academy of Sciences*, 107(38), 16732–16737.

Gilboa, I., and D. Schmeidler (1989), 'Maxmin Expected Utility with Non-Unique Prior', *Journal of Mathematical Economics*, 18(2), 141–153.

Gilchrist, D. S., and E. G. Sands (2016), 'Something to Talk About: Social Spillovers in Movie Consumption', *Journal of Political Economy*, 124(5), 1339–1382.

Gill, D. A., M. B. Mascia, G. N. Ahmadia, L. Glew, S. E. Lester, M. Barnes, I. Craigie, E. S. Darling, C. M. Free, J. Geldmann, S. Holst, O. P. Jensen, A. T. White, X. Basurto, L. Coad, R. D. Gates, G. Guannel, P. J. Mumby, H. Thomas, S. Whitmee, S. Woodley and H. E. Fox (2017), 'Capacity Shortfalls Hinder the Performance of Marine Protected Areas Globally', *Nature*, 543(7647), 665–669.

Gilman, E., K. Passfield, and K. Nakamura (2014), 'Performance of Regional Fisheries Management Organizations: Ecosystem-Based Governance of Bycatch and Discards', *Fish and Fisheries*, 15(2), 327–351.

Global Footprint Network (2019), 'National Footprint and Biocapacity Accounts', 'https://www.footprintnetwork.org/licenses/public-data-package-free/'.

Global Footprint Network (2020), 'Glossary', 'https://www.footprintnetwork.org/resources/glossary/'.

Global Resources Initiative (2020), *Global Resource Initiative: Final Recommendations Report*.

Global Witness (2019), *Money to Burn: How Iconic Banks and Investors Fund the Destruction of the World's Largest Rainforests*.

Glynn, P. W. (1993), 'Coral Reef Bleaching: Ecological Perspectives', *Coral Reefs*, 12(1), 1–17.

Godfray, H. C. J., P. Aveyard, T. Garnett, J. W. Hall, T. J. Key, J. Lorimer, R. T. Pierrehumbert, P. Scarborough, M. Springmann, and S. A. Jebb (2018), 'Meat Consumption, Health, and the Environment', *Science*, 361(6399), eaam5324.

Golden Kroner, R. E., S. Qin, C. N. Cook, R. Krithivasan, S. M. Pack, O. D. Bonilla, K. A. Cort-Kansinally, B. Coutinho, M. Feng, M. I. M. Garcia, Y. He, C. J. Kennedy, C. Lebreton, J. C. Ledezma, T. E. Lovejoy, D. A. Luther, Y. Parmanand, C. A. Ruíz-Agudelo, E. Yerena, V. M. Zambrano and M. B. Mascia (2019), 'The Uncertain Future of Protected Lands and Waters', *Science*, 364(6443), 881–886.

Golet, G. H., C. Low, S. Avery, K. Andrews, C. J. McColl, R. Laney, and M. D. Reynolds (2018), 'Using Ricelands to Provide Temporary Shorebird Habitat During Migration', *Ecological Applications*, 28(2), 409–426.

Gollier, C. (2007), 'The Consumption-based Determinants of the Term Structure of Discount Rates', *Mathematics and Financial Economics*, 1(2), 81–101.

Gollier, C. (2008), 'Discounting with Fat-Tailed Economic Growth', *Journal of Risk and Uncertainty*, 37(2-3), 171–186.

Gómez-Baggethun, E., and R. Muradian (2015), 'In Markets we Trust? Setting the Boundaries of Market-Based Instruments in Ecosystem Services Governance', *Ecological Economics*, 117, 217–224.

Gonzalez, A., N. Mouquet, and M. Loreau (2009), 'Biodiversity as Spatial Insurance: The Effects of Habitat Fragmentation and Dispersal on Ecosystem Functioning' in S. Naeem, D. E. Bunker, A. Hector,

M. Loreau, and C. Perrings, eds., *Biodiversity, Ecosystem Functioning, and Human Wellbeing: An Ecological and Economic Perspective* (Oxford: Oxford University Press).

Gonzalez, M. T., T. Hartig, G. G. Patil, E. W. Martinsen, and M. Kirkevold (2010), 'Therapeutic Horticulture in Clinical Depression: A Prospective Study of Active Components', *Journal of Advanced Nursing*, 66(9), 2002–2013.

Gonzalez, P., R. P. Neilson, J. M. Lenihan, and R. J. Drapek (2010), 'Global Patterns in the Vulnerability of Ecosystems to Vegetation Shifts Due to Climate Change', *Global Ecology and Biogeography*, 19, 755–768.

Goody, J. (1989), 'Futures of the Family in Rural Africa', *Population and Development Review*, 15, 119–144.

Gordon, H. S. (1954), 'The Economic Theory of a Common-Property Resource: The Fishery', *Journal of Political Economy*, 62(2), 124–142.

Gore, A. (1992), *Earth in the Balance: Ecology and the Human Spirit* (New York, NY: Houghton Mifflin).

Gore, T. (2015), *Extreme Carbon Inequality: Why the Paris Climate Deal Must Put the Poorest, Lowest Emitting and Most Vulnerable People First*.

Gössling, S., and A. Humpe (2020), 'The Global Scale, Distribution and Growth of Aviation: Implications for Climate Change', *Global Environmental Change*, 65, 102194.

Gottdenker, N. L., D. G. Streicker, C. L. Faust, and C. R. Carroll (2014), 'Anthropogenic Land Use Change and Infectious Diseases: A Review of the Evidence', *Ecohealth*, 11(4), 619–632.

Gould, R. K., S. C. Klain, N. M. Ardoin, T. Satterfield, U. Woodside, N. Hannahs, G. C. Daily, and K. M. Chan (2015), 'A Protocol for Eliciting Nonmaterial Values through a Cultural Ecosystem Services Frame', *Conservation Biology*, 29(2), 575–586.

Granato, J., R. Inglehart, and D. Leblang (1996), 'The Effect of Cultural Values on Economic Development: Theory, Hypotheses, and Some Empirical Tests', *American Journal of Political Science*, 40(3), 607–631.

Gray, C. L., S. L. L. Hill, T. Newbold, L. N. Hudson, L. Boïrger, S. Contu, A. J. Hoskins, S. Ferrier, A. Purvis, and J. P. W. Scharlemann (2016), 'Local Biodiversity is Higher Inside Than Outside Terrestrial Protected Areas Worldwide', *Nature Communications*, 7(1), 12306.

Green Finance Platform (2019), *France's Law on Energy and Climate Adds Coverage of Biodiversity, Ecosystems, and Renewable Energy to Investors' Non-Financial Reporting*.

Green, J. M. H., S. A. Croft, A. P. Durán, A. P. Balmford, N. D. Burgess, S. Fick, T. A. Gardner, J. Godar, C. Suavet, M. Virah-Sawmy, L. E. Young, and C. D. West (2019), 'Linking Global Drivers of Agricultural Trade to On-The-Ground Impacts on Biodiversity', *Proceedings of the National Academy of Sciences*, 116(46), 23202–23208.

Green, S., J. L. Akins, A. Maljković, and I. M. Côté (2012), 'Invasive Lionfish Drive Atlantic Coral Reef Fish Declines', *PLoS ONE*, 7(3), 332596.

Gregr, E. J., V. Christensen, L. Nichol, R. G. Martone, R. W. Markel, J. C. Watson, C. D. G. Harley, E. A. Pakhomov, J. B. Shurin, and K. M. A. Chan (2020), 'Cascading Social-Ecological Costs and Benefits Triggered by a Recovering Keystone Predator', *Science*, 368(6496), 1243–1247.

Greif, A. (1993), 'Contract Enforceability and Economic Institutions in Early Trade: The Maghribi Traders' Coalition', *The American Economic Review*, 83(3), 525–548.

References

Greif, A., P. Milgrom, and B. R. Weingast (1994), 'Coordination, Commitment, and Enforcement: The Case of the Merchant Guild', *Journal of Political Economy*, 102(4), 745–776.

Greiner, R., and O. Cacho (2001), 'On the Efficient Use of a Catchment's Land and Water Resources: Dryland Salinization in Australia', *Ecological Economics*, 38(3), 441–458.

Griffin, J. T. (1986), *Well-Being: Its Meaning, Measurement, and Moral Importance* (Oxford: Clarendon Press).

Griliches, Z. (1957), 'Hybrid Corn: An Exploration in the Economics of Technological Change', *Econometrica*, 25(4) 501–522.

Griliches, Z. (1987a), 'R&D and Productivity: Measurement Issues and Econometric Results', *Science*, 237(4810), 31–35.

Griliches, Z. (1987b), 'R&D and Productivity', in P. Dasgupta, and P. Stoneman, eds., *Economic Policy and Technological Performance* (Cambridge: Cambridge University Press).

Griscom, B. W., J. Adams, P. W. Ellis, R. A. Houghton, G. Lomax, D. A. Miteva, W. H. Schlesinger, D. Shoch, J. V Siikamäki, P. Smith, P. Woodbury, C. Zganjar, A. Blackman, J. Campari, R. T. Conant, C. Delgado, P. Elias, T. Gopalakrishna, M. R. Hamsik, M. Herrero, J. Kiesecker, E. Landis, L. Laestadius, S. M. Leavitt, S. Minnemeyer, S. Polasky, P. Potapov, F. E. Putz, J. Sanderman, M. Silvius, E. Wollenberg and J. Fargione (2017), 'Natural Climate Solutions', *Proceedings of the National Academy of Sciences*, 114(44), 11645–11650.

Groom, B. (2020), *Biodiversity Measures*. Unpublished.

Grossman, G. M., and A. B. Krueger (1991), 'Environmental Impacts of a North American Free Trade Agreement', *National Bureau of Economic Research Working Paper*, No. 3914.

Grossman, G. M., and A. B. Krueger (1995), 'Economic Growth and the Environment', *The Quarterly Journal of Economics*, 110(2), 353–377.

Grossman, S. J. (1977), 'The Existence of Futures Markets, Noisy Rational Expectations and Informational Externalities', *Review of Economic Studies*, 44(3), 431–449.

Grossman, S. J. (1989), *The Informational Role of Prices* (Cambridge, MA: MIT Press).

Grove-White, R. (1992), 'The Christian "Person" and Environmental Concern', *Studies in Christian Ethics*, 5(2), 1–17.

Gruère, G., C. Ashley, and J. J. Cadilhon. (2018). 'Reforming Water Policies in Agriculture: Lessons from Past Reforms', *OECD Food, Agriculture and Fisheries Papers*, No. 113.

The Guardian (2013), 'Connecticut's Wealthy Gold Coast: Where Life is Good, if You can Afford it', https://www.theguardian.com/world/2013/feb/15/connecticut-gold-coast-life-afford.

Gupta, U. (2008), 'Valuation of Urban Air Pollution: A Case Study of Kanpur City in India', *Environmental and Resource Economics*, 41, 315–326.

Gupta, U. (2011), 'Estimating Welfare Losses from Urban Air Pollution Using Panel Data from Household Health Diaries', in A. K. E. Haque, M. N. Murty and P. Shyamsundar, eds., *Environmental Valuation in South Asia* (New Delhi: Cambridge University Press).

Gutiérrez Rodríguez, L., N. J. Hogarth, W. Zhou, C. Xie, K. Zhang, and L. Putzel (2016), 'China's Conversion of Cropland to Forest Program: A Systematic Review of the Environmental and Socioeconomic Effects', *Environmental Evidence*, 5(21).

Guttmacher Institute (2020), *Adding It Up: Investing in Sexual and Reproductive Health 2019* (New York, NY: Guttmacher Institute).

Guyer, J. I., and S. M. E. Belinga (1995), 'Wealth in People as Wealth in Knowledge: Accumulation and Composition in Equatorial Africa', *The Journal of African History*, 36(1), 91–120.

Haberl, H., K. -H. Erb, and F. Krausmann, (2014), 'Human Appropriation of Net Primary Production: Patterns, Trends, and Planetary Boundaries', *Annual Review of Environment and Resources* 39(1), 363–391.

Haddad, F. F. (2014), 'Rangeland Resource Governance – Jordan', in P.M. Herrera, J. Davies and P. M. Baena, eds., *Governance of Rangelands: Collective Action for Sustainable Pastoralism* (New York, NY: Routledge).

Haddad, N. M., L. A. Brudvig, J. Clobert, K. F. Davies, A. Gonzalez, R. D. Holt, T. E. Lovejoy, J. O. Sexton, M. P. Austin, C. D. Collins, W. M. Cook, E. I. Damschen, R. M. Ewers, B. L. Foster, C. N. Jenkins, A. J. King, W. F. Laurance, D. J. Levey, C. R. Margules, B. A. Melbourne, A. O. Nicholls, J. L. Orrock, D.-X. Song, and J. R. Townshend (2015), 'Habitat Fragmentation and its Lasting Impact on Earth's Ecosystems', *Science Advances*, 1(2), e1500052.

Haines-Young, R. and M. B. Potschin (2018), *Common International Classification of Ecosystem Services (CICES) V5.1 and Guidance on the Application of the Revised Structure*.

Hale, L. Z., and J. Rude (2017), *Learning from New Zealand's 30 Years of Experience Managing Fisheries Under a Quota Management System*.

Hall, R.E. and C.I. Jones (1999), "Why Do Some Countries Produce So Much More Output Per Worker Than Others?," *Quarterly Journal of Economics*, 114(1), 83–116.

Hallegatte, S., M. Bangalore, L. Bonzanigo, M. Fay, T. Kane, U. Narloch, J. Rozenberg, D. Treguer, and A. Vogt-Schilb (2016), *Shock Waves: Managing the Impacts of Climate Change on Poverty* (Washington, DC: World Bank)

Halpern, B. S., M. Frazier, J. Afflerbach, J. S. Lowndes, F. Micheli, C. O'Hara, C. Scarborough, and K. A. Selkoe (2019), 'Recent Pace of Change in Human Impact on the World's Ocean', *Scientific Reports*, 9(1), 11609.

Hamann, M., K. Berry, T. Chaigneau, T. Curry, R. Heilmayr, P. J. G. Henriksson, J. Hentati- Sundberg, A. Jina, E. Lindkvist, Y. Lopez-Maldonado, E. Nieminen, M. Piaggio, J. Qiu, J. C. Rocha, C. Schill, A. Shepon, A. R. Tilman, I. van den Bijgaart, and T. Wu (2018), 'Inequality and the Biosphere', *Annual Review of Environment and Resources*, 43(1), 61–83.

Hamilton, S. E., and D. Casey (2016), 'Creation of a High Spatio-Temporal Resolution Global Database of Continuous Mangrove Forest Cover for the 21st Century (CGMFC-21)', *Global Ecology and Biogeography*, 25(6), 729–738.

Hamilton, K., and M. Clemens (1999), 'Genuine Savings Rates in Developing Countries', *World Bank Economic Review*, 13(2), 333–356.

Hammond, P. J. (1976), 'Equity, Arrow's Conditions, and Rawls' Difference Principle',*Econometrica*, 44(4), 793–804.

Hammond, P., and Malinvaud, E. (1973). *Lectures on Microeconomic Theory. The Economic Journal*, 83 (330).

Hamrick, K. (2016), *State of Private Investment in Conservation 2016: A Landscape Assessment of an Emerging Market*.

References

Hanauer, M. M., and G. Canavire-Bacarreza (2015), 'Implications of Heterogeneous Impacts of Protected Areas on Deforestation and Poverty', *Philosophical Transactions of the Royal Society B*, 370(1681), 20140272.

Hanemann, W. M. (2006), 'The Economic Conception of Water', in P. P. Rogers, M. Ramón Llamas and L. Martínez- Cortina, eds., *Water Crisis: Myth or Reality?* (London: Taylor and Francis).

Hanley, N., and C. Perrings (2019), 'The Economic Value of Biodiversity', *Annual Review of Resource Economics*, 11, 355–375.

Hanson, A. (2019), Ecological Civilization in the People's Republic of China: Values, Action, and Future Needs, *ADB East Asia Working Paper Series*, 21.

Hanson, C., and P. Mitchell (2017), *The Business Case for Reducing Food Loss and Waste: A Report on Behalf of Champions* 12.3.

Haque, A. K. E., M. N. Murty, and P. Shyamsundar (2011), *Environmental Valuation in South Asia* (Cambridge: Cambridge University Press).

Hardin, G. (1968), 'The Tragedy of the Commons', *Science*, 162(3859), 1243–1248.

Harfoot, M., S. A. M. Glaser, D. P. Tittensor, G. L. Britten, C. McLardy, K. Malsch, and N. D. Burgess (2018), 'Unveiling the Patterns and Trends in 40 years of Global Trade in CITES-listed Wildlife', *Biological Conservation*, 223, 47–57.

Harris, G., S. Thirgood, J. G. C. Hopcraft, J. P. G. M. Cromsigt, and J. Berger (2009), 'Global Decline in Aggregated Migrations of Large Terrestrial Mammals', *Endangered Species Research*, 7(1), 55–76.

Harrison, I. J., P. A. Green, T. A. Farrell, D. Juffe-Bignoli, L. Sáenz, and C. J. Vörösmarty (2016), 'Protected Areas and Freshwater Provisioning: a Global Assessment of Freshwater Provision, Threats and Management Strategies to Support Human Water Security', *Aquatic Conservation: Marine and Freshwater Ecosystems*, 26(S1), 103–120.

Harrod, R. F. (1948), *Towards a Dynamic Economics: Recent Developments of Economic Theory and their Application to Policy* (London: Macmillan).

Harsanyi, J. C. (1955), 'Cardinal Welfare, Individualistic Ethics, and Interpersonal Comparisons of Utility', *Journal of Political Economy*, 63(4), 309–321.

Hartig, T., R. Mitchell, S. de Vries, and H. Frumkin (2014), 'Nature and Health', *Annual Review of Public Health*, 35, 207–228.

Hausman, J. (2012), 'Contingent Valuation: From Dubious to Hopeless', *Journal of Economic Perspectives*, 26(4), 43–56.

Hawksworth, D., and R. Lücking (2017), 'Fungal Diversity Revisited: 2.2 to 3.8 Million Species', *Microbiology Spectrum*, 5(4), 1–17.

Heal, G. M. (1969), 'Planning without Prices', *The Review of Economic Studies*, 36(3), 347–362.

Heal, G. M. (1998), *Valuing the Future: Economic Theory and Sustainability* (New York: Columbia University Press).

Heal, G., and A. Millner (2014), 'Reflections: Uncertainty and Decision Making in Climate Change Economics', *Review of Environmental Economics and Policy*, 8(1), 120–137.

Heath, J., and H. Binswanger (1996), 'Natural Resource Degradation Effects of Poverty and Population Growth are Largely Policy-Induced: The Case of Colombia', *Environment and Development Economics*, 1(1), 65–84.

Hecht, S. B., A. B. Anderson, and P. May (1988), 'The Subsidy from Nature: Shifting Cultivation, Successional Palm Forests, and Rural Development', *Human Organization*, 47(1), 25–35.

Heflin, F., and D. Wallace (2017), 'The BP Oil Spill: Shareholder Wealth Effects and Environmental Disclosures', *Journal of Business Finance and Accounting*, 44(3–4), 337–374.

Heller, M. C., and G. A. Keoleian (2018), *Beyond Meat's Beyond Burger Life Cycle Assessment: A Detailed Comparison between a Plant-Based and an Animal-Based Protein Source*.

Helliwell, J. F. (2003), 'How's Life? Combining Individual and National Variables to Explain Subjective Well-Being', *Economic Modelling*, 20(2), 331–360.

Helliwell, J. F., and H. Huang (2010), 'How's the Job? Well-Being and Social Capital in the Workplace', *ILR Review*, 63(2), 205–227.

Helliwell, J. F., R. Layard, and J. D. Sachs (2013), *World Happiness Report 2013* (New York, NY: UN Sustainable Development Solutions Network).

Helliwell, J. F., and R. D. Putnam (2004), 'The Social Context of Well-Being', Philosophical Transactions of the Royal Society of London, B, 359(1449), 1435–1446.

Helliwell, J. F., and S. Wang (2012), 'The State of World Happiness', in J. F Helliwell, R. Layard, and J. Sachs, eds., *World Happiness Report 2012* (New York, NY: The Earth Institute).

Helliwell, J. F., and S. Wang (2013), 'World Happiness: Trends, Explanations and Distribution', in J. F. Helliwell, R. Layard, and J. Sachs, eds., *World Happiness Report 2013* (New York, NY: UN Sustainable Development Solutions Network).

Hellmann, J. J. (2002), 'The Effect of an Environmental Change on Mobile Butterfly Larvae and the Nutritional Quality of their Hosts', *Journal of Animal Ecology*, 71, 925–936

Helpman, E. (2004), *The Mystery of Economic Growth*, (Cambridge, MA: Harvard University Press).

Henry, C. (1974), 'Investment Decisions under Uncertainty: The "Irreversibility Effect"', *The American Economic Review*, 74(6), 1006–1012.

Hepburn, C., B. O'Callaghan, N. Stern, J. Stiglitz, D. Zenghelis (2020), 'Will COVID-19 Fiscal Recovery Packages Accelerate or Retard Progress on Climate Change?' *Oxford Review of Economic Policy*, 36(S1), S359–S381.

Heron, T., P. Prado, and C. West (2018), 'Global Value Chains and the Governance of "Embedded" Food Commodities: The Case of Soy', *Global Policy*, 9(52), 29–37.

Hicks, J. R. (1940), 'The Valuation of the Social Income', *Economica*, 7(26), 105–124.

Higgins, C. B., K. Stephenson, and B. L. Brown (2011), 'Nutrient Bioassimilation Capacity of Aquacultured Oysters: Quantification of an Ecosystem Service', *Journal of Environmental Quality*, 40(1), 271–277.

Hill, S. L. L., R. Gonzalez, K. Sanchez-Ortiz, E. Caton, F. Espinoza, T. Newbold, J. Tylianakis, J. P. W. Scharlemann, A. De Palma, and A. Purvis (2018), 'Worldwide Impacts of Past and Projected Future Land-Use Change on Local Species Richness and the Biodiversity Intactness Index', *bioRxiv*.

References

Hirota, M., M. Holmgren, E. H. Van Nes, and M. Scheffer (2011), 'Global Resilience of Tropical Forest and Savanna to Critical Transitions', *Science*, 334(6053), 232–235.

Hirsch, F. (1976), *Social Limits to Growth* (Cambridge, MA: Harvard University Press).

Hislop, M., A. J. Scott, and A. Corbett (2019), 'What Does Good Green Infrastructure Planning Policy Look Like? Developing and Testing a Policy Assessment Tool within Central Scotland UK', *Planning Theory and Practice*, 20(5), 633–655.

HM Government (2018), *A Green Future: Our 25 Year Plan to Improve the Environment*.

HM Treasury (2018), *The Green Book: Central Government Guidance on Appraisal and Evaluation* (London: OGL Press).

HM Treasury (2020), *Magenta Book: Central Government Guidance on Evaluation* (London: OGL Press).

Hobsbawm, E. J. (1959), *Primitive Rebels: Studies in Archaic Forms of Social Movement in the 19th and 20th Centuries* (London: W.W. Norton).

Hochard, J. P., S. Hamilton, and E. B. Barbier (2019), 'Mangroves Shelter Coastal Economic Activity from Cyclones', *Proceedings of the National Academy of Sciences*, 116(25), 12232–12237.

Hock, R., G. Rasul, C. Adler, B. Cáceres, S. Gruber, Y. Hirabayashi, M. Jackson, A. Kääb, S. Kang, S. Kutuzov, A. Milner, U. Molau, S. Morin, B. Orlove, and H. Steltzer (2019), 'High Mountain Areas', in H.-O. Pörtner, D. C. Roberts, V. Masson-Delmotte, P. Zhai, M. Tignor, E. Poloczanska, K. Mintenbeck, A. Alegría, M. Nicolai, A. Okem, J. Petzold, B. Rama, and N. M. Weyer, eds., *IPCC Special Report on the Ocean and Cryosphere in a Changing Climate*.

Hockings, M., N. Dudley, W. Elliott, M. N. Ferreira, K. Mackinnon, M. K. S. Pasha, A. Phillips, S. Stolton, S. Woodley, M. Appleton, O. Chassot, J. Fitzsimons, C. Galliers, R. G. Kroner, J. Goodrich, J. Hopkins, W. Jackson, H. Jonas, B. Long, and M. Mumba (2020), 'Editorial essay: Covid-19 and Protected and Conserved Areas', *Parks*, 26(1), 7–24.

Hoegh-Guldberg, O. (1999), 'Climate Change, Coral Bleaching and The Future of The World's Coral Reefs', *Marine and Freshwater Research*, 50(8), 839–866.

Hoegh-Guldberg, O., P. J. Mumby, A. J. Hooten, R. S. Steneck, P. Greenfield, E. Gomez, C. D. Harvell, P. F. Sale, A. J. Edwards, K. Caldeira, N. Knowlton, C. M. Eakin, R. Iglesias-Prieto, N. Muthiga, R. H. Bradbury, A. Dubi, and M. E. Hatziolos (2007), 'Coral Reefs Under Rapid Climate Change and Ocean Acidification', *Science*, 318(5857), 1737–1742.

Hoekstra, A. Y., and M. M. Mekonnen (2012), 'The Water Footprint of Humanity', *Proceedings of the National Academy of Sciences*, 109(9), 3232–3237.

Hoel, M., and T. Sterner (2007), 'Discounting and Relative Prices', *Climatic Change*, 84(3-4), 265–280.

Hof, C., A. Voskamp, M. F. Biber, K. Böhning-Gaese, E. K. Engelhardt, A. Niamir, S. G. Willis, and T. Hickler (2018), 'Bioenergy Cropland Expansion May Offset Positive Effects of Climate Change Mitigation for Global Vertebrate Diversity', *Proceedings of the National Academy of Sciences*, 155(52), 13294–13299.

Hoffmann, A. A., and C. M. Sgró (2011). 'Climate Change and Evolutionary Adaptation'. *Nature*, 470(7335), 479–485.

Hoffmann, M., C. Hilton-Taylor, A. Angulo, M. Böhm, T. M. Brooks, S. H. M. Butchart, K. E. Carpenter, J. Chanson, B. Collen, N. A. Cox, W. R. T. Darwall, N. K. Dulvy, L. R. Harrison, V. Katariya, C. M. Pollock, S. Quader, N. I. Richman, A. S. L. Rodrigues, M. F. Tognelli, J.-C. Vie, J. M. Aguiar, D. J. Allen, G. R. Allen,

G. Amori, N. B. Ananjeva, F. Andreone, P. Andrew, A. L. Aquino Ortiz, J. E. M. Baillie, R. Baldi, B. D. Bell, S. D. Biju, J. P. Bird, P.Black-Decima, J. J. Blanc, F. Bolaños, W. Bolivar-G., I. J. Burfield, J. A. Burton, D. R. Capper, F. Castro, G. Catullo, R. D. Cavanagh, A. Channing, N. L. Chao, A. M. Chenery, F. Chiozza, V. Clausnitzer, N. J. Collar, L. C. Collett, B. B. Collette, C. F. Cortez Fernandez, M. T. Craig, M. J. Crosby, N. Cumberlidge, A. Cuttelod, A. E. Derocher, A. C. Diesmos, J. S. Donaldson, J. W. Duckworth, G. Dutson, S. K. Dutta, R. H. Emslie, A. Farjon, S. Fowler, J. Freyhof, D. L. Garshelis, J. Gerlach, D. J. Gower, T. D. Grant, G. A. Hammerson, R. B. Harris, L. R. Heaney, S. B. Hedges, J.-M. Hero, B. Hughes, S. A. Hussain, J. Icochea M., R. F. Inger, N. Ishii, D. T. Iskandar, R. K. B. Jenkins, Y. Kaneko, M. Kottelat, K. M. Kovacs, S. L. Kuzmin, E. La Marca, J. F. Lamoreux, M. W. N. Lau, E. O. Lavilla, K. Leus, R. L. Lewison, G. Lichtenstein, S. R. Livingstone, V. Lukoschek, D. P. Mallon, P. J. K. McGowan, A. McIvor, P. D. Moehlman, S. Molur, A. M. Alonso, J. A. Musick, K. Nowell, R. A. Nussbaum, W. Olech, N. L. Orlov, T. J. Papenfuss, G. Parra-Olea, W. F. Perrin, B. A. Polidoro, M. Pourkazemi, P. A. Racey, J. S. Ragle, M. Ram, G. Rathbun, R. P. Reynolds, A. G. J. Rhodin, S. J. Richards, L. O. Rodríguez, S. R. Ron, C. Rondinini, A. B. Rylands, Y. S. de Mitcheson, J. C. Sanciangco, K. L. Sanders, G. Santos-Barrera, J. Schipper, C. Self-Sullivan, Y. Shi, A. Shoemaker, F. T. Short, C. Sillero-Zubiri, D. L. Silvano, K. G. Smith, A. T. Smith, J. Snoeks, A. J. Stattersfield, A. J. Symes, A. B. Taber, B. K. Talukdar, H. J. Temple, R. Timmins, J. A. Tobias, K. Tsytsulina, D. Tweddle, C. Ubeda, S. V. Valenti, P. P. van Dijk, L. M. Veiga, A. Veloso, D. C. Wege, M. Wilkinson, E. A. Williamson, F. Xie, B. E. Young, H. Resit Akçakaya, L. Bennun, T. M. Blackburn, L. Boitani, H. T. Dublin, G. A. B. da Fonseca, C. Gascon, T. E. Lacher Jr., G. M. Mace, S. A. Mainka, J. A. McNeely, R. A. Mittermeier, G. McG. Reid, J. P. Rodriguez, A. A. Rosenberg, M. J. Samways, J. Smart, B. A. Stein, S. N. Stuart, (2010), 'The Impact of Conservation on the Status of the World's Vertebrates', *Science*, 330(6010), 1503–1509.

Holland, D. S. (2010), 'Management Strategy Evaluation and Management Procedures: Tools for Rebuilding and Sustaining Fisheries', *OECD Food, Agriculture and Fisheries Working Papers*, 25, 1–66.

Hollands, G. J., G. Bignardi, M. Johnston, M. P. Kelly, D. Ogilvie, M. Petticrew, A. Prestwich, I. Shemilt, S. Sutton, and T. M. Marteau (2017), 'The TIPPME Intervention Typology for Changing Environments to Change Behaviour', *Nature Human Behaviour*, 1, 0140.

Hollands, G. J., E. Cartwright, M. Pilling, R. Pechey, M. Vasiljevic, S. A. Jebb, and T. M. Marteau (2018), 'Impact of Reducing Portion Sizes in Worksite Cafeterias: a Stepped Wedge Randomised Controlled Pilot Trial', *International Journal of Behavioral Nutrition and Physical Activity*, 15, 78.

Hollands, G. J., A. Prestwich, and T. M. Marteau (2011), 'Using Aversive Images to Enhance Healthy Food Choices and Implicit Attitudes: An Experimental Test of Evaluative Conditioning', *Health Psychology*, 30(2), 195–203.

Holling, C. S. (1973), 'Resilience and Stability of Ecological Systems', *Annual Review of Ecology and Systematics*, 4(1), 1–23.

Hollinger, D. A. (2006), 'From Identity to Solidarity', *Daedalus*, 135(4), 23–31.

Holmes, N. D., D. R. Spatz, S. Oppel, B. Tershy, D. A. Croll, B. Keitt, P. Genovesi, I. J. Burfield, D. J. Will, A. L. Bond, A. Wegmann, A. Aguirre-Muñoz, A. F. Raine, C. R. Knapp, C. -H. Hung, D. Wingate, E. Hagen, F. Méndez-Sánchez, G. Rocamora, H. -W. Yuan, J. Fric, J. Millett, J. Russell, J. Liske-Clark, E. Vidal, H. Jourdan, K. Campbell, K. Springer, K. Swinnerton, L. Gibbons-Decherong, O. Langrand, M. L. de Brooke, M. McMinn, N. Bunbury, N. Oliveira, P. Sposimo, P. Geraldes, P. McClelland, P. Hodum, P. G. Ryan, R. Borroto-Páez, R. Pierce, R. Griffiths, R. N. Fisher, R. Wanless, S. A. Pasachnik, S. Cranwell, T. Micol and S. H. M. Butchart (2019), 'Globally Important Islands where Eradicating Invasive Mammals will Benefit Highly Threatened Vertebrates', *PLoS ONE*, 14(3), e0212128.

References

Holmlund, C. M., and M. Hammer (1999), 'Ecosystem Services Generated by Fish Populations', *Ecological Economics*, 29(2), 253–268.

Homer-Dixon, T. F. (1994), 'Environmental Scarcities and Violent Conflict: Evidence from Cases', *International Security*, 19(1), 5–40.

Honda, E. A., and G. Durigan (2016), 'Woody Encroachment and its Consequences on Hydrological Processes in the Savannah', *Philosophical Transactions of the Royal Society B: Biological Sciences*, 371(1703), 20150313.

Hooke, R. L. (2000), 'On the History of Humans as Geomorphic Agents', *Geology*, 28(9), 843–846.

Hooper, D. U., F. S. Chapin, J. J. Ewel, A. Hector, P. Inchausti, S. Lavorel, J. H. Lawton, D. M. Lodge, M. Loreau, S. Naeem, B. Schmid, H. Setälä, A. J. Symstad, J. Vandermeer, and D. A. Wardle (2005), 'Effects of Biodiversity on Ecosystem Functioning: a Consensus of Current Knowledge', *Ecological Monographs*, 75(1), 3–35.

Hooper, D. U., and P. M. Vitousek (1997), 'The Effects of Plant Composition and Diversity on Ecosystem Processes', *Science*, 277(5330), 1302–1305.

Horan, R. D., J. F. Shogren, and B. M. Gramig (2008), 'Wildlife Conservation Payments to Address Habitat Fragmentation and Disease Risks', *Environment and Development Economics*, 13(3), 415–439.

Houngbo, E. N. (2019), 'Box 2.1 Sacred Spaces: A Tradition of Forest Conservation in Benin', in J. W. Wilson, and R. B. Primack, eds., *Conservation Biology in Sub-Saharan Africa* (Cambridge: Open Book Publishers).

Howard, P. H., and T. Sterner (2017), 'Few and Not So Far Between: A Meta-Analysis of Climate Damage Estimates', *Environmental and Resource Economics*, 68(1), 197–225.

Howe, J. (1986), *The Kuna Gathering: Contemporary Village Politics in Panama* (Austin, TX: University of Texas Press).

HSBC Global Asset Management and Pollination Group (2020), *HSBC Global Asset Management and Pollination Launch: Partnership to Create World's Largest Natural Capital Manager*.

Hubbell, S. P. (2015), 'Estimating the Global Number of Tropical Tree Species, and Fisher's Paradox', *Proceedings of the National Academy of Sciences*, 112(4), 7343–7344.

Hufbauer, G. C., and Z. Lu (2016), *Increased Trade: A Key to Improving Productivity*. Hughes, T. P., M. L. Barnes, D. R. Bellwood, J. E. Cinner, G. S. Cumming, J. B. C. Jackson, J. Kleypas, I. A. van de Leemput, J. M. Lough, T. H. Morrison, S. R. Palumbi, E. H. van Nes, and M. Scheffer (2017), 'Coral reefs in the Anthropocene', *Nature*, 546(7656), 82–90.

Hughes, J., M. Rogerson, J. Barton, and R. Bragg (2019), 'Age and Connection to Nature: When is Engagement Critical?', *Frontiers in Ecology and the Environment*, 17(5), 265–269.

Hulten, C. R. (2001), 'Total Factor Productivity: A Short Biography', in C. R. Hulten, E. R. Dean, and M. J. Harper, M J eds., *New Developments in Productivity Analysis*, (Chicago, IL: Chicago University Press).

Humphrey, C. (2016), *Time for a New Approach to Environmental and Social Protection at Multilateral Development Banks*.

Humphrey, P., D. A. Carter, and B. Simkins (2016), 'The Market's Reaction to Unexpected, Catastrophic Events: The Case of Oil and Gas Stock Returns and the Gulf Oil Spill', *Journal of Risk Finance*, 17(1), 2–25.

Hunter, C., and J. Shaw (2005), 'Applying the Ecological Footprint to Ecotourism Scenarios', *Environmental Conservation*, 32(4), 294–304.

Huwlyer, F., J. Käppeli, and J. Tobin (2016), *Conservation Finance from Niche to Mainstream: The Building of an Institutional Asset Class*.

Iannaccone, L. R (1998), 'Introduction to the Economics of Religion', *Journal of Economic Literature*, 36(3), 1465–1495.

Ichii, K., Z. Molnár, D. O. Obura, A. Purvis, and K. Willis (2019), *IPBES Global Assessment on Biodiversity and Ecosystem Services – Chapter 2.2 Status and Trends – Nature (draft.)*

IDH The Sustainable Trade Initiative (2020a), *A New Collaboration Platform for Supply Chain Sustainability Changemakers*, 'https://www.idhsustainabletrade.com/approach/sourceup/'.

IDH The Sustainable Trade Initiative (2020b), *Verified Sourcing Areas: An IDH Developed Concept*.

ILO (2012), *Working Towards Sustainable Development: Opportunities for Decent Work and Social Inclusion in a Green Economy*.

ILO (2018), *World Employment and Social Outlook 2018: Greening with Jobs*.

ILO (2020), *Employment-Intensive Investment Programme (EIIP) Guidance*.

IMF (2020), 'IMF Surveillance', 'https://www.imf.org/en/About/Factsheets/IMF-Surveillance'. IMF (2019), *Global Financial Stability Report (October 2019): Lower for Longer*.

The Indian Express (2016), *Delhi Pollution Level High Despite Strong Wind*, 'https://indianexpress.com/article/delhi/delhi-pollution-level-high-despite-strong-wind-4445039/'.

Influence Map (2019), 'Asset Managers and Climate Change', 'https://influencemap.org/report/FinanceMap-Launch-Report-f80b653f6a631cec947a07e44ae4a4a7'.

Inglehart, R. F (2010), 'Faith and Freedom: Traditional and Modern Ways to Happiness', in Diener E., Helliwell J. F., and Kahneman D., eds, *International Differences in Well-Being*, 351–397 (Oxford: Oxford University Press).

Inglehart, R., R. Foa, C. Peterson, and C. Welzel (2008), 'Development, Freedom, and Rising Happiness: A Global Perspective (1981–2007)', *Perspectives on Psychological Science*,3(4), 264–285.

Inglehart, R., and C. Welzel (2005), '*Modernization, Cultural Change, and Democracy: The Human Development Sequence*' (Cambridge: Cambridge University Press).

Inter-American Development Bank (2020), 'Sustainability and Safeguards', 'https://www.iadb.org/en/about-us/sustainability-and-safeguards'.

Inter-American Development Bank, (2021), *Innovative Finance for Conservation in Latin America and the Caribbean Report*.

International Assessment of Agricultural Knowledge, Science and Technology for Development (IAASTD) (2009), *Agriculture at a Crossroads: Global Report*.

International Energy Agency (2019), *Renewables Information 2019*.

International Institute for Sustainable Development (2019), *Institutional Finance Update: Mobilising Private Sector to Reach 2020 Global Goal on Climate Finance*.

International Resource Panel (IRP) (2019), *Global Resources Outlook 2019: Natural Resources for the Future We Want* (Nairobi: UNEP).

References

International Whaling Commission (1982), 'Moratorium on Commercial Whaling', 'https://iwc.int/commercial'.

Investment and Pensions Europe (2008), 'Norway Excludes Rio Tinto Over Environmental Damage', 'https://www.ipe.com/norway-excludes-rio-tinto-over-environmental-damage/29077.article'.

IPBES (2018), *The IPBES Assessment Report on Land Degradation and Restoration*, L. Montanarella, R. Scholes, and A. Brainich eds., *Secretariat of the Intergovernmental Science- Policy Platform on Biodiversity and Ecosystem Services*. (Bonn: IPBES Secretariat).

IPBES (2019a), *Summary for Policymakers of the Global Assessment Report on Biodiversity and Ecosystem Services of the Intergovernmental Science-Policy Platform on Biodiversity and Ecosystem Services*, S. Díaz, J. Settele, E. S. Brondízio, H. T. Ngo, M. Guèze, J. Agard, A. Arneth, P. Balvanera, K. A. Brauman, S. H. M. Butchart, K. M. A. Chan, L. A. Garibaldi, K. Ichii, J. Liu, S. M. Subramanian, G. F. Midgley, P. Miloslavich, Z. Molnár, D. Obura, A. Pfaff, S. Polasky, A. Purvis, J. Razzaque, B. Reyers, R. Roy Chowdhury, Y. J. Shin, I. J. Visseren-Hamakers, K. J. Willis, and C. N. Zayas, eds. (Bonn: IPBES Secretariat).

IPBES (2019b), 'Glossary', https://ipbes.net/glossary.

IPCC (2000), *Summary for Policymakers: Emissions Scenarios – A Special Report of Working Group III of the Intergovernmental Panel on Climate Change* (Geneva: IPCC).

IPCC (2014), *Climate Change 2014: Synthesis Report. Contribution of Working Groups I, II and III to the Fifth Assessment Report of the Intergovernmental Panel on Climate Change*, R. K. Pachauri, and L. Meyer, eds., with Core Writing Team (Geneva: IPCC).

IPCC (2018), 'Summary for Policymakers', in *Global Warming of 1.5°C. An IPCC Special Report on The Impacts Of Global Warming of 1.5°C Above Pre-Industrial Levels and Related Global Greenhouse Gas Emission Pathways, in the Context of Strengthening the Global Response to the Threat of Climate Change, Sustainable Development, and to Eradicate Poverty*, V. Masson-Delmotte, P. Zhai, H. -O. Pörtner, D. Roberts, J. Skea, P. R. Shukla, A. Pirani, W. Moufouma-Okia, C. Péan, R. Pidcock, S. Connors, J. B. R. Matthews, Y. Chen, X. Zhou, M. I. Gomis, E. Lonnoy, T. Maycock, M. Tignor, and T. Waterfield eds. (Geneva: World Meteorological Organization).

IPNI (2020), 'International Plant Names Index', 'http://www.ipni.org'.

IQAir (2020), *2019 World Air Quality Report*.

Irwin, E. G., S. Gopalakrishnan, and A. Randall (2016), 'Welfare, Wealth, and Sustainability', *Annual Review of Resource Economics*, 8, 77–98.

Isham, J., D. Kaufmann, and L. Pritchett (1997), 'Civil Liberties, Democracy, and the Performance of Government Projects', *World Bank Economic Review*, 11(2), 219–242.

Isham, J., D. Narayan, and L. Pritchett (1995), 'Does Participation Improve Performance? Establishing Causality with Subjective Data', *World Bank Economic Review*, 9(2), 175–200.

IUCN (2008a), *The 2008 Review of the IUCN Red List of Threatened Species* (Gland: IUCN). IUCN (2008b), 'Protected Areas', 'https://www.iucn.org/theme/protected-areas/about'.

IUCN (2012), *The Zarqa River Basin, Jordan – Reviving Hima Sites*.

IUCN (2014), 'Integrated Tiger Habitat Conservation Programme', 'https://www.iucn.org/theme/species/our-work/action-ground/integrated-tiger-habitat-conservation-programme'.

IUCN (2016a), *IUCN Policy on Biodiversity Offsets*.

IUCN (2016b), *Integrated Tiger Habitat Conservation Programme – Project Portfolio Snapshots*.

IUCN (2020), *IUCN Red List of Threatened Species, Version 2020-2* 'https://www.iucnredlist.org/'.

IUCN-WCPA Task Force on OECMs (2019), *Recognising and Reporting other Effective Area-based Conservation Measures*.

Jack, B. K., C. Kouskya, and K. R. E. Simsa (2008), 'Designing Payments for Ecosystem Services: Lessons from Previous Experience with Incentive-Based Mechanisms', *Proceedings of the National Academy of Sciences*, 105(28), 9465–9470.

Jactel, H., J. Koricheva, and B. Castagneyrol (2019), 'Responses of Forest Insect Pests to Climate Change: Not so Simple', *Current Opinion in Insect Science*, 35, 103–108.

Jalan, J., and E. Somanathan (2008), 'The Importance of Being Informed: Experimental Evidence on Demand for Environmental Quality', *Journal of Development Economics*, 87(1), 14–28.

Jamison, D. T., L. H. Summers, G. Alleyne, K. J. Arrow, S. Berkley, A. Binagwaho, F. Bustreo, D. Evans, R. G. A. Feachem, J. Frenk, G. Ghosh, S. J. Goldie, Y. Guo, S. Gupta, R. Horton, M. E. Kruk, A. Mahmoud, L. K. Mohohlo, M. Ncube, A. Pablos-Mendez, K. Srinath Reddy, H. Saxenian, A. Soucat, K. H. Ulltveit-Moe, G. Yamey (2013), 'Global Health 2035: A World Converging within a Generation', *The Lancet*, 382(9908), 1898–1955.

Janoski, T. (1998), *Citizenship and Civil Society: A Framework of Rights and Obligations in Liberal, Traditional, and Social Democratic Regimes* (Cambridge: Cambridge University Press).

Jensen, R., and E. Oster (2009), 'The Power of TV: Cable Television and Women's Status in India', *The Quarterly Journal of Economics*, 124(3), 1057–1094.

Jepson, P. R., B. Caldecott, S. F. Schmitt, S. H. C. Carvalho, R. A. Correia, N. Gamarra, C. Bragagnolo, A. C. M. Malhado, and R. J. Ladle (2017), 'Protected Area Asset Stewardship', *Biological Conservation*, 212(A), 183–190.

Jiang, W., H. Zhou, H. Bi, M. Fromm, B. Yang, and D. P. Weeks (2013), 'Demonstration of CRISPR/Cas9/sgRNA-Mediated Targeted Gene Modification in Arabidopsis, Tobacco, Sorghum and Rice', *Nucleic Acids Research*, 41(20), e188.

Jodha, N. S. (1986), 'Common Property Resources and Rural Poor in Dry Regions of India', *Economic and Political Weekly*, 21(27), 1169–1181.

Jodha, N. S. (2001), *Life on The Edge: Sustaining Agriculture and Community Resources in Fragile Environments* (Oxford: Oxford University Press).

Johnson, J. A., U. Baldos, R. Cervigni, E. Corong, J. Gerber, T. Hertel, C. Nootenboom, S. Polasky, and G. Ruta (Forthcoming), 'Making the Economic Case for Nature: Integrated Economic and Ecosystem Services Modeling', *World Bank Publications*.

Johnston, R. J., K. J. Boyle, W. Adamowicz, J. Bennett, R. Brouwer, T. A. Cameron, W.M. Hanemann, N. Hanley, M. Ryan, R. Scarpa, R. Tourangeau, and C. A. Vossler (2017), 'Contemporary Guidance for Stated Preference Studies', *Journal of the Association of Environmental and Resource Economists*, 4(2), 319–405.

Jonas, H. D., E. Lee, H. C. Jonas, C. Matallana-Tobon, K. S. Wright, F. Nelson, and E. Enns (2017), 'Will 'Other Effective Area-Based Conservation Measures' Increase Recognition and Support for ICCAs?', *Parks*, 23(2), 63–78.

References

Jones, B. A., D. Grace, R. Kock, S. Alonso, J. Rushton, M. Y. Said, D. McKeever, F. Mutua, J. Young, J. McDermott, and D. U. Pfeiffer (2013), 'Zoonosis Emergence Linked to Agricultural Intensification and Environmental Change', *Proceedings of the National Academy of Sciences*, 110(21), 8399–8404.

Jones, C. I., and P. M. Romer (2010), 'The New Kaldor Facts: Ideas, Institutions, Population, and Human Capital', *American Economic Journal: Macroeconomics*, 2(1), 224–245.

Jones, C. I., and J. C. Williams (1998), 'Measuring the Social Return to R&D', *The Quarterly Journal of Economics*, 113(4), 1119–1135.

Jones, H. P., N. D. Holmes, S. H. M. Butchart, B. R. Tershy, P. J. Kappes, I. Corkery, A. Aguirre-Muñoz, D. P. Armstrong, E. Bonnaud, A. A. Burbidge, K. Campbell, F. Courchamp, P. E. Cowan, R. J. Cuthbert, S. Ebbert, P. Genovesi, G. R. Howald, B. S. Keitt, S. W. Kress, C. M. Miskelly, S. Oppel, S. Poncet, M. J. Rauzon, G. Rocamora, J. C. Russell, A. Samaniego-Herrera, P. J. Seddon, D. R. Spatz, D. R. Towns, D. A. Croll (2016), 'Invasive Mammal Eradication on Islands Results in Substantial Conservation Gains', *Proceedings of the National Academy of Sciences*, 113(15), 4033–4038.

Jones, H. P., P. C. Jones, E. B. Barbier, R. C. Blackburn, J. M. R. Benayas, K. D. Holl, M. McCrackin, P. Meli, D. Montoya, and D. M. Mateos (2018), 'Restoration and Repair of Earth's Damaged Ecosystems', *Proceedings of the Royal Society B*, 285(1873), 20172577.

Jones, K. E., N. G. Patel, M. A. Levy, A. Storeygard, D. Balk, J.L. Gittleman, and P. Daszak (2008), 'Global Trends in Emerging Infectious Diseases', *Nature*, 451(7181), 990–993.

Jones, K. R., C. J. Klein, B. S. Halpern, O. Venter, H. Grantham, C. D. Kuempel, N. Shumway, A. M. Friedlander, H. P. Possingham, and J. E. M. Watson (2018), 'The Location and Protection Status of Earth's Diminishing Marine Wilderness', *Current Biology*, 28(15), 2506–2512.

Joosten, H. (2010), *The Global Peatland CO2 Picture: Peatland Status and Drainage Related Emissions in All Countries of the World* (Ede: Greifswald University).

Joosten, H. (2016), 'Peatlands Across the Globe', in A. Bonn and R. Stoneman, eds., *Peatland Restoration and Ecosystem Services: Science, Policy, and Practice* (Cambridge: Cambridge University Press).

Jordà, Ò., K. Knoll, D. Kuvshinov, M. Schularick, and A. M. Taylor (2019), 'The Rate of Return on Everything, 1870-2015', *The Quarterly Journal of Economics*, 134(3), 1225–1298.

Jorgenson, A. A., and J. Rice (2005), 'Structural Dynamics of International Trade and Material Consumption: A Cross-National Study of the Ecological Footprints of Less-Developed Countries', *Journal of World-Systems Research*, 11(1), 57–77.

Jorgenson, A., and E. Kick eds. (2006), *Globalization and the Environment* (Leiden and Boston, MA: Brill).

Joshi, S., and T. P. Schultz (2013), 'Family Planning and Women's and Children's Health: Long-Term Consequences of an Outreach Program in Matlab, Bangladesh', *Demography*, 50(1), 149–180.

Juma, C. (2019), 'Game Over?: Drivers of Biological Extinction in Africa', in P. Dasgupta, P. H. Raven, and A. L. McIvor, eds., *Biological Extinction: New Perspectives* (Cambridge: Cambridge University Press).

Jump, A. S., R. Marchant, and J. Peñuelas (2009), 'Environmental Change and the Option Value of Genetic Diversity', *Trends in Plant Science*, 14(1), 51–58.

Jung, M., P. R. Dahal, S. H. M. Butchart, P. F. Donald, X. De Lamo, M. Lesiv, V. Kapos, C. Rondinini, and P. Visconti (2020), 'A Global Map of Terrestrial Habitat Types', *Scientific Data*, 7, 256.

Kahneman, D., J. L. Knetsch, and R. H. Thaler (1990), 'Experimental Tests of the Endowment Effect and the Coase Theorem', *Journal of Political Economy*, 98(6),1325–1348.

Kahneman, D. and A. B. Krueger (2006), 'Developments in the Measurement of Subjective Well- Being', *Journal of Economic Perspectives*, 20(1), 3–24.

Kahneman, D., A. B. Krueger, D. A. Schkade, N. Schwarz, and A. A. Stone (2004), 'A Survey Method for Characterizing Daily Life Experience: The Day Reconstruction Method', *Science*, 306(5702), 1776–1780.

Kahneman, D., and J. Riis (2005), 'Living, and Thinking About It: Two Perspectives on Life' in F. A. Huppert, N. Baylis and B. Keverne, eds., *The Science of Well-Being* (Oxford: Oxford University Press).

Kahneman, D., P. Slovic, and A. Tversky (1982), 'Introduction', in D. Kahneman, P. Slovic, and A. Tversky, eds., *Judgment Under Uncertainty: Heuristics and Biases* (Cambridge: Cambridge University Press).

Kahneman, D., and A. Tversky (1979), 'Prospect Theory: An Analysis of Decision Under Risk', *Econometrica*, 47(2), 263–292.

Kalnay, E., and M. Cai (2003), 'Erratum: Corrigendum: Impact of Urbanization and Land-Use',*Nature*, 425(6953), 102–102.

Kanie, N. (2018), 'Governance with Multilateral Environmental Agreements: A Healthy or Ill-Equipped Fragmentation?' in K. Conca and G Dabelko eds., *Green Planet Blues: Critical Perspectives on Global Environmental Politics* (New York, NY: Routledge).

Kant, R. (2012), 'Textile Dyeing Industry an Environmental Hazard', *Natural Science*, 4(1), 22–26.

Kaplan, S., and R. Kaplan (2009), 'Creating a Larger Role for Environmental Psychology: The Reasonable Person Model as an Integrative Framework', *Journal of Environmental Psychology*, 29(3), 329–339.

Kapteyn, A., S. van de Geer, H. Van de Stadt, and T. Wansbeek (1997), 'Interdependent Preferences: An Econometric Analysis', *Journal of Applied Econometrics*, 12(6), 665–686.

Karacaoglu, G. (2020), 'I LOVE YOU – Investing for Intergenerational Wellbeing', *Real-World Economics Review*, 92, 207–227.

Kareiva, P., H. Tallis, T. H. Ricketts, G. C. Daily, and S. Polasky (2011), *Natural Capital: The Theory and Practice of Mapping Ecosystem Services* (Oxford: Oxford University Press).

Karousakis, K. and S. Yamaguchi (2020), 'Trade Policy and The Post-2020 Global Biodiversity Framework', *Trade for Development*, 'https://trade4devnews.enhancedif.org/en/news/trade-policy-and-post-2020-global-biodiversity-framework'.

Kasterine, A. and G. Lichtenstein (2018), *Trade in Vicuña: The Implications for Conservation and Rural Livelihoods*.

Kastner, T., K. -H. Erb and H. Haberl, (2014), 'Rapid Growth in Agricultural Trade: Effects on Global Area Efficiency and the Role of Management', *Environmental Research Letters*, 9(3), 034015.

Kaya, Y., and K. Yokobori (1997), *Environment, Energy and Economy: Strategies for Sustainability*, (New York: United Nations University Press).

Keats, S., and S. Wiggins (2014), *Future Diets: Implications for Agriculture and Food Prices* (London: Overseas Development Institute).

References

Kedward, K., J. Ryan-Collins, and H. Chenet (2020), *Managing Nature-Related Financial Risks: A Precautionary Policy Approach for Central Banks and Financial Supervisors*.

Keesstra, S., J. Nunes, A. Novara, D. Finger, D. Avelar, Z. Kalantari, and A. Cerdà (2018), 'The Superior Effect Of Nature Based Solutions in Land Management for Enhancing Ecosystem Services', *Science of the Total Environment*, 610–611, 997–1009.

Kéfi, S., M. Rietkerk, C. L. Alados, Y. Pueyo, V. P. Papanastasis, A. Elaich, and P. C. de Ruiter (2007), 'Spatial Vegetation Patterns and Imminent Desertification in Mediterranean Arid Ecosystems', *Nature*, 449(7159), 213–217.

Kellert, S. R., and E. O. Wilson eds. (1993), *The Biophilia Hypothesis* (Washington, DC: Island Press).

Kerr, S., R. G. Newell, and J. N. Sanchirico (2004), 'Evaluating the New Zealand Individual Transferable Quota Market for Fisheries Management', in *Tradeable Permits Policy Evaluation, Design and Reform* (Paris: OECD).

Keskin, M., and Y. E. Sekerli (2016), 'Awareness and Adoption of Precision Agriculture in the Cukurova Region of Turkey', *Agronomy Research*, 14(4), 1307–1320.

Kettunen, M., E. Bodin, E. Davey, S. Gionfra, and C. Charveriat (2020), *An EU Green Deal for Trade Policy and the Environment: Aligning Trade with Climate and Sustainable Development Objectives*.

Kettunen, M., P. Genovesi, S. Gollasch, S. Pagad, U. Starfinger, P. ten Brink, and C. Shine (2008), *Technical Support to EU Strategy on Invasive Species (IAS) – Assessment of the Impacts of IAS in Europe and the EU (Final Module Report for the European Commission)*.

Keynes, J. M. (1931), *Essays in Persuasion* (London: Macmillan).

Khan, H. (2011), 'Valuation of Recreational Amenities from Environmental Resources: The Case of Two National Parks in Northern Pakistan', in A. K. E. Haque, M. N. Murty and P. Shyamsundar, eds., *Environmental Valuation in South Asia* (Cambridge: Cambridge University Press).

Khoury, C. K., A. D. Bjorkman, H. Dempewolf, J. Ramirez-Villegas, L. Guarino, A. Jarvis, L. H. Rieseberg, and P. C. Struik (2014), 'Increasing Homogeneity in Global Food Supplies and the Implications for Food Security', *Proceedings of the National Academy of Sciences*, 111(11), 4001–4006.

Kidd, K. A., P. J. Blanchfield, K. H. Mills, V. P. Palace, R. E. Evans, J. M. Lazorchak, and R. W. Flick (2007), 'Collapse of a Fish Population After Exposure to a Synthetic Estrogen', *Proceedings of the National Academy of Sciences*, 104(21), 8897–8901.

Kim, R., and J. Asuncion (2019), *Investing in Our Future: Is Time the Most Potent Prism for Climate Action?*

Kintisch, E. (2016), 'The Lost Norse', *Science*, 354(6313), 696–701.

Kinzig, A. P., S. W. Pacala, and D. Tilman (2002), *The Functional Consequences of Biodiversity: Empirical Progress and Theoretical Extensions* (Princeton, NJ: Princeton University Press).

Kiontke, K., and D. H. A. Fitch (2013), 'Nematodes', *Current Biology*, 23(19), R862–R864.

Klasing, M. J., and P. Milionis (2014), 'Quantifying the Evolution of World Trade, 1870-1949', *Journal of International Economics*, 92(1), 185–197.

Kleijn, D., and W. J. Sutherland (2003), 'How Effective are European Agri-Environment Schemes in Conserving and Promoting Biodiversity?', *Journal of Applied Ecology*, 40(6), 947–969.

Kleijn, D., R. A. Baquero, Y. Clough, M. Díaz, J. De Esteban, F. Fernández, D. Gabriel, F. Herzog, A. Holzschuh, R. Jöhl, E. Knop, A. Kruess, E. J. P. Marshall, I. Steffan-Dewenter, T. Tscharntke, J. Verhulst, T. M. West, and J. L. Yela (2006), 'Mixed Biodiversity Benefits of Agri-Environment Schemes in Five European Countries', *Ecology Letters*, 9(3), 243–254.

Klein, A.-M., B. E. Vaissière, J. H. Cane, I. Steffan-Dewenter, S. A. Cunningham, C. Kremen, and T. Tscharntke (2007), 'Importance of Pollinators in Changing Landscapes for World Crops', *Proceedings of the Royal Society B: Biological Sciences*, 274, 303–313.

Klenow, P. J. and A. Rodriguez-Clare (2005), 'Externalities and Growth', in P. Aghion and S. N. Durlauf, eds., *Handbook of Economic Growth* (Dordrecht: Elsevier).

Kling, C. L., D. J. Phaneuf, and J. Zhao (2012), 'From Exxon to BP: Has Some Number Become Better than No Number?', *Journal of Economic Perspectives*, 26(4), 3–26.

Klomp, J. (2014), 'Financial Fragility and Natural Disasters: An Empirical Analysis', *Journal of Financial Stability*, 13(C), 180–192.

Knack, S., and P. Keefer (1997), 'Does Social Capital Have an Economic Payoff? A Cross-Country Investigation', *The Quarterly Journal of Economics*, 112(4), 1251–1288.

Kneese, A.V. (1970), *Economics and the Environment: A Materials Balance Approach* (Washington, DC: Resources for the Future).

Kneese, A. V., R. U. Ayres, and R. C. d'Arge (1970), *Economics and the Environment: A Materials Balance Approach* (Washington, DC: Resources for the Future).

Kohler, H.-P., J. R. Behrman, and S. C. Watkins (2001), 'The Density of Social Networks and Fertility Decisions: Evidence from South Nyanza District, Kenya', *Demography*, 38, 43–58.

Kohler, H.-P., J. R. Behrman (2014), 'Benefits and Costs of the Population and Demography Targets for the Post-2015 Development Agenda' (Copenhagen Consensus Center).

Kohn, R. E., and P. D. Capen (2002), 'Optimal Volume of Environmentally Damaging Trade', *Scottish Journal of Political Economy*, 49(1), 22–38.

Kölbel, J., F. Heeb, F. Paetzold, and T. Busch (2018), 'Can Sustainable Investing Save the World? Reviewing the Mechanisms of Investor Impact', *SSRN Electronic Journal*, 328544.

Kolbert, E. (2013a), 'The Lost World: Parts I and II', *The New Yorker*.

Kolbert, E. (2013b), 'The Lost World: The Mastodon's Molars (Annals of Extinction), Part I', *The New Yorker*.

Kolbert, E. (2014), *The Sixth Extinction: An Unnatural History* (New York: Henry Holt and Company).

Kollmann, J., S. T. Meyer, R. Bateman, T. Conradi, M. M. Gossner, M. de Souza Mendonça Jr, G. W. Fernandes, J. -M. Hermann, C. Koch, S. C. Müller, Y. Oki, G. E. Overbeck, G. B. Paterno, M. F. Rosenfield, T. S. P. Toma, and W. W. Weisser (2016), 'Integrating Ecosystem Functions into Restoration Ecology—Recent Advances and Future Directions', *Restoration Ecology*, 24(6), 722–730.

Kolm, S. -C. (1969), 'The Optimal Production of Social Justice', in J. Margolis and H. Guitton, eds., *Public Economics: An Analysis of Public Production and Consumption and Their Relations to the Private Sectors* (London: Macmillan).

Koopmans, T. C. (1957), 'Allocation of Resources and the Price System' in T. C. Koopmans, *Three Essays on the State of Economic Science* (New York, NY: McGraw Hill Book Company).

References

Koopmans, T. C. (1960), 'Stationary Ordinal Utility and Impatience', *Econometrica*, 28(2), 287–309.

Koopmans, T. C. (1965), 'On the Concept of Optimal Economic Growth', in *Study Week on the Econometric Approach to Development Planning, October 7-13, 1963, Part I (Pontificiae Academiae Scientiarum Scripta Varia 28)* (Vatican City: Pontifical Academy of Sciences).

Koopmans, T. C. (1967), 'Objectives, Constraints, and Outcomes in Optimal Growth Models', *Econometrica*, 35(1), 1–15.

Koopmans, T. C. (1972), 'Representation of Preference Orderings over Time', in C. B. McGuire and R. Radner, eds., *Decision and Organization* (Amsterdam: North Holland Publishing Co.).

Koopmans, T. C. (1977), 'Concepts of Optimality and Their Uses', *The American Economic Review*, 67(3), 261–274.

Kotsadam, A., E. H. Olsen, C. H. Knutsen, and T. Wig (2015), 'Mining and Local Corruption in Africa', *Memorandum*, No. 9/2015 (Oslo: University of Oslo).

Koumbarakis, A., S. Hirschi, K. Meier, S. Tsankova, A. Favier, G. Duyck, I. Mugglin, and M. Tormen (2020), *Nature is too Big to Fail: Biodiversity: The Next Frontier in Financial Risk Management*.

Kousky, C., and S. E. Light (2019), 'Insuring Nature', *Duke Law Journal*, 69(2), 323–376.

Krausmann, F., K. -H. Erb, S. Gingrich, C. Lauk, H. Haberl, (2008), 'Global Patterns of Socioeconomic Biomass Flows in the Year 2000: A Comprehensive Assessment of Supply, Consumption and Constraints', *Ecological Economics*, 65(3), 471–487.

Krausmann, F., K. -H. Erb, S. Gingrich, H. Haberl, A. Bondeau, V. Gaube, C. Lauk, C. Plutzar, and T. D. Searchinger (2013), 'Global Human Appropriation of Net Primary Production Doubled in the 20th Century', *Proceedings of the National Academy of Sciences*, 110(25), 10324–10329.

Krausmann, F., H. Haberl, K. -H. Erb, M. Wiesinger, V. Gaube, and S. Gingrich, (2009), 'What Determines Geographical Patterns of the Global Human Appropriation of Net Primary Production?', *Journal of Land Use Science*, 4(1–2), 15–33.

Krenn, S., L. Cobb, S. Babalola, M. Odeku, and B. Kusemiju (2014), 'Using Behavior Change Communication to Lead a Comprehensive Family Planning Program: The Nigerian Urban Reproductive Health Initiative', *Global Health: Science and Practice*, 2(4), 427–443.

Kriegler, E., J. W. Hall, H. Held, R. Dawson, and H. J. Schellnhuber (2009), 'Imprecise Probability Assessment of Tipping Points in the Climate System', *Proceedings of the National Academy of Sciences*, 106(13), 5041–5046.

Krugman, P. R. (1991). *Geography and Trade* (Cambridge, MA: MIT Press).

Kumar, P., ed. (2010), *The Economics of Ecosystems and Biodiversity: Ecological and Economic Foundations* (London: Earthscan).

Kummu, M., and O. Varis (2011), 'The World by Latitudes: A Global Analysis of Human Population, Development Level and Environment Across the North-South Axis Over the Past Half Century', *Applied Geography*, 31(2), 495–507.

Kurz, M. (1968), 'Optimal Economic Growth and Wealth Effects', *International Economic Review*, 9(3), 348–357.

Kuznets, S., L. Epstein and E. Jenks (1941), *National Income and Its Composition, 1919-1938* (New York, NY: National Bureau of Economic Research).

Lago, M., J. Mysiak, M. Lago, C. M. Gómez, G. Delacámara, and A. Maziotis (2015), *Use of Economic Instruments in Water Policy* (Heidelberg: Springer).

Lam, D. (2011), 'How the World Survived the Population Bomb: Lessons From 50 Years of Extraordinary Demographic History', *Demography*, 48,1231–1262.

Lambin, E. F., and P. Meyfroidt (2011), 'Global Land Use Change, Economic Globalization, and the Looming Land Scarcity', *Proceedings of the National Academy of Sciences*, 108(9), 3465–3472.

Lambsdorff, J. G. (2003), 'How Corruption Affects Productivity', *Kyklos*, 56(4), 457–474.

Lammerant, J., A. Grigg, J. Dimitrijevic, K. Leach, S. Brooks, A. Burns, J. Berger, J. Houdet, M. van Oorschot, and M. Goedkoop (2019), *Assessment of Biodiversity Measurement Approaches for Businesses and Financial Institutions*.

Landes, D. S. (1998), *The Wealth and Poverty of Nations: Why Some Are So Rich and Some So Poor* (London: W.W. Norton).

Lange, O. (1936), 'On the Economic Theory of Socialism: Part One', *The Review of Economic Studies*, 4(1), 53–71.

Lange, O. (1937). 'On the Economic Theory of Socialism: Part Two', *The Review of Economic Studies*, 4(2), 123–142.

Larsen, B. B., E. C. Miller, M. K. Rhodes, and J. J. Wiens (2017), 'Inordinate Fondness Multiplied and Redistributed: The Number of Species on Earth and the New Pie of Life', *The Quarterly Review of Biology*, 92(3), 229–265.

Larsen, F. W., W. R. Turner, and T. M. Brooks (2012), 'Conserving Critical Sites for Biodiversity Provides Disproportionate Benefits to People', *PLoS ONE*, 7(5), e36971.

Lartigue, J., and J. Cebrian (2009), 'Ecosystem Productivity and Carbon Flows: Patterns across Ecosystems', in S. A. Levin, ed., *The Princeton Guide to Ecology* (Princeton, NJ: Princeton University Press).

Laurance, W. F., T. E. Lovejoy, H. L. Vasconcelos, E. M. Bruna, R. K. Didham, P. C. Stouffer, C. Gascon, R. O. Bierregaard, S. G. Laurance, and E. Sampaio (2002) 'Ecosystem Decay of Amazonian Forest Fragments: A 22-Year Investigation', *Conservation Biology*, 16(3), 605–618.

Laurance, W. F., S. Sloan, L. Weng, and J. A. Sayer (2015), 'Estimating the Environmental Costs of Africa's Massive "Development Corridors"', *Current Biology*, 25(24), 3202–3208.

Laurance, W. F., D. C. Useche, J. Rendeiro, M. Kalka, C. J. A. Bradshaw, S. P. Sloan, S. G. Laurance, M. Campbell, K. Abernethy, P. Alvarez, V. Arroyo-Rodriguez, P. Ashton, J. Benítez- Malvido, A. Blom, K. S. Bobo, C. H. Cannon, M. Cao, R. Carroll, C. Chapman, R. Coates, M. Cords, F. Danielsen, B. De Dijn, E. Dinerstein, M. A. Donnelly, D. Edwards, F. Edwards, N. Farwig, P. Fashing, P-M. Forget, M. Foster, G. Gale, D. Harris, R. Harrison, J. Hart, S. Karpanty, W. J. Kress, J. Krishnaswamy, W. Logsdon, J. Lovett, W. Magnusson, F. Maisels, A. R. Marshall, D. McClearn, D. Mudappa, M. R. Nielsen, R. Pearson, N. Pitman, J. van der Ploeg, A. Plumptre, J. Poulsen, M. Quesada, H. Rainey, D. Robinson, C. Roetgers, F. Rovero, F. Scatena, C. Schulze, D. Sheil, T. Struhsaker, J. Terborgh, D. Thomas, R. Timm, J. N. Urbina-Cardona, K. Vasudevan, S. Joseph Wright, J. C. Arias-G., L. Arroyo, M. Ashton, P. Auzel, D. Babaasa, F. Babweteera, P. Baker, O. Banki, M. Bass, I. Bila-Isia, S. Blake, W. Brockelman, N. Brokaw, C. A. Brühl, S. Bunyavejchewin, J-T. Chao, J. Chave, R. Chellam, C. J. Clark, J. Clavijo, R. Congdon, R. Corlett, H. S. Dattaraja, C. Dave, G. Davies, B. de Mello Beisiegel, R. de Nazaré Paes da Silva, A. Di Fiore, A. Diesmos, R. Dirzo, D. Doran-Sheehy, M. Eaton, L. Emmons, A. Estrada, C. Ewango, L. Fedigan, F. Feer, B. Fruth, J. G. Willis, U. Goodale, S. Goodman, J. C. Guix, P. Guthiga, W. Haber, K. Hamer,

References

I. Herbinger, J. Hill, Z. Huang, I. Fang Sun, K. Ickes, A. Itoh, N. Ivanauskas, B. Jackes, J. Janovec, D. Janzen, M. Jiangming, C. Jin, T. Jones, H. Justiniano, E. Kalko, A. Kasangaki, T. Killeen, H-b. King, E. Klop, C. Knott, I. Koné, E. Kudavidanage, J. L. da Silva Ribeiro, J. Lattke, R. Laval, R. Lawton, M. Leal, M. Leighton, M. Lentino, C. Leonel, J. Lindsell, L. Ling-Ling, K. E. Linsenmair, E. Losos, A. Lugo, J. Lwanga, A. L. Mack, M. Martins, W. S. McGraw, R. McNab, L. Montag, J. M. Thompson, J. Nabe-Nielsen, M. Nakagawa, S. Nepal, M. Norconk, V. Novotny, S. O'Donnell, M. Opiang, P. Ouboter, K. Parker, N. Parthasarathy, K. Pisciotta, D. Prawiradilaga, C. Pringle, S. Rajathurai, U. Reichard, G. Reinartz, K. Renton, G. Reynolds, V. Reynolds, E. Riley, M-O. Rödel, J. Rothman, P. Round, S. Sakai, T. Sanaiotti, T. Savini, G. Schaab, J. Seidensticker, A. Siaka, M. R. Silman, T. B. Smith, S. Soares de Almeida, N. Sodhi, C. Stanford, K. Stewart, E. Stokes, K. E. Stoner, R. Sukumar, M. Surbeck, M. Tobler, T. Tscharntke, A. Turkalo, G. Umapathy, M. van Weerd, J. Vega Rivera, M. Venkataraman, L. Venn, C. Verea, C. Volkmer de Castilho, M. Waltert, B. Wang, D. Watts, W. Weber, P. West, D. Whitacre, K. Whitney, D. Wilkie, S. Williams, D. D. Wright, P. Wright, L. Xiankai, P. Yonzon and F. Zamzani (2012), 'Averting Biodiversity Collapse in Tropical Forest Protected Areas', *Nature*, 489(7415), 290–294.

Lavorel, S., J. Storkey, R. D. Bardgett, F. De Bello, M. P. Berg, X. Le Roux, M. Moretti, C. Mulder, R. J. Pakeman, S. Díaz, and R. Harrington (2013), 'A Novel Framework for Linking Functional Diversity of Plants with Other Trophic Levels for the Quantification of Ecosystem Services', *Journal of Vegetation Science*, 24(5), 942–948.

Lawrence, D., and K. Vandecar (2015), 'Effects of Tropical Deforestation on Climate and Agriculture', *Nature Climate Change*, 5, 27–36.

Lawrence, S. N., and J. C. Daltry (2015), 'Antigua Announces 15th Island Cleared of Invasive Alien Mammals.', *Oryx*, 49(3), 389.

Layard, R. (2011), *Happiness: Lessons from a New Science* (London: Penguin) 2nd Edition.

Layard R., A. Clark, and C. Senik (2012), 'The Causes of Happiness and Misery', in J. F. Helliwell, R. Layard, J. Sachs eds., *World Happiness Report 2012* (New York, NY: The Earth Institute).

Layard, R., G. Mayraz, and S. Nickell (2010), 'Does Relative Income Matter? Are the Critics Right?' in E. Diener, J. F. Helliwell, and D. Kahneman, eds., *International Differences in Well- Being* (Oxford: Oxford University Press).

Le Maitre, D. C., B. W. Van Wilgen, R. A. Chapman, and D. H. McKelly (1996), 'Invasive Plants and Water Resources in the Western Cape Province, South Africa: Modelling the Consequences of a Lack of Management', *Journal of Applied Ecology*, 33(1), 161–172.

Leclère, D., M. Obersteiner, M. Barrett, S. H. M. Butchart, A. Chaudhary, A. De Palma, F. A. J. DeClerck, M. D. Marco, J. C. Doelman, M. Dürauer, R. Freeman, M. Harfoot, T. Hasegawa, S. Hellweg, J. P. Hilbers, S. L. L. Hill, F. Humpenöder, N. Jennings, T. Krisztin, G. M. Mace, H. Ohashi, A. Popp, A. Purvis, A. M. Schipper, A. Tabeau, H. Valin, H. van Meijl, W. -J. van Zeist, P. Visconti, R. Alkemade, R. Almond, G. Bunting, N. D. Burgess, S. E. Cornell, F. D. Fulvio, S. Ferrier, S. Fritz, S. Fujimori, M. Grooten, T. Harwood, Petr Havlík, M. Herrero, A. J. Hoskins, M. Jung, T. Kram, H. Lotze-Campen, T. Matsui, C. Meyer, D. Nel, T. Newbold, G. Schmidt-Traub, E. Stehfest, B. B. N. Strassburg, D. P. van Vuuren, C. Ware, J. E. M. Watson, W. Wu, Lucy Young (2020), 'Bending the Curve of Terrestrial Biodiversity Needs an Integrated Strategy', *Nature*, 585(7826), 551–556.

Lee, S. Y., S. Hamilton, E. B. Barbier, J. Primavera, and R. R. Lewis III (2019), 'Better Restoration Policies Are Needed to Conserve Mangrove Ecosystems', *Nature Ecology and Evolution*, 3(6), 870–872.

Leisher, C., G. Temsah, F. Booker, M. Day, L. Samberg, D. Prosnitz, B. Agarwal, E. Matthews, D. Roe, D. Russell, T. Sunderland, and D. Wilkie (2016), 'Does the Gender Composition of Forest And Fishery

Management Groups Affect Resource Governance and Conservation Outcomes? A Systematic Map', *Environmental Evidence*, 5, 6.

Leite, C., and J. Weidmann (1999), 'Does Mother Nature Corrupt? Natural Resources, Corruption and Economic Growth', *IMF Working Paper No. WP/99/85*.

Lele, S., I. Patil, S. Badiger, A. Menon, and R. Kumar (2011), 'Forests, Hydrological Services, and Agricultural Income: A Case Study from the Western Ghats of India' in A. K. E. Haque, M. N. Murty, and P. Shyamsundar, eds., *Environmental Valuation in South Asia* (Cambridge: Cambridge University Press).

Lemay, M. A., P. Henry, C. T. Lamb, K. M. Robson, and M. A. Russello. (2013). 'Novel Genomic Resources for a Climate Change Sensitive Mammal: Characterization of the American Pika Transcriptome', *BMC Genomics*, 14(1), 311.

Lenton, T. M., H. Held, E. Kriegler, J. W. Hall, W. Lucht, S. Rahmstorf, and H. J. Schellnhuber (2008), 'Tipping Elements in the Earth's Climate System', *Proceedings of the National Academy of Sciences*, 105(6), 1786–1793.

Lenton, T. M. (2013) 'Environmental Tipping Points,' *Annual Review of Environment and Resources*, 38, 1–29.

Lenzen, M., D. Moran, K. Kanemoto, B. Foran, L. Lobefaro, and A. Geschke (2012), 'International Trade Drives Biodiversity Threats in Developing Nations', *Nature*, 486(7401), 109–112.

Lerner, A. P (1934), 'Economic Theory and Socialist Economy', *Review of Economic Studies*, 2(1), 51–61.

Leshan, J., I. Porras, A. Lopez, and P. Kazis (2017), *Sloping Lands Conversion Programme, People's Republic of China*.

Leshan, J., I. Porras, P. Kazis, and A. Lopez (2018), 'China's Eco Compensation Programme', in I. Porras and N. Asquith, eds., *Ecosystems, Poverty Alleviation and Conditional Transfers* (London: International Institute for Environment and Development).

Lester, S. E., A. K. Dubel, G. Hernán, J. McHenry, and A. Rassweiler (2020), 'Spatial Planning Principles for Marine Ecosystem Restoration', *Frontiers in Marine Science*, 7, 328.

Levhari, D., and T. N. Srinivasan (1969), 'Optimal Savings under Uncertainty', *The Review of Economic Studies*, 36(2),153–163.

Levin, S. A. (1992), 'The Problem of Pattern and Scale in Ecology', *Ecology*, 73(6),1943–1967.

Levin, S. A. (1999), *Fragile Dominion: Complexity and the Commons* (Reading, MA: Perseus Books).

Levin, S. A. (2000), 'Multiple Scales and the Maintenance of Biodiversity', *Ecosystems*, 3(6), 498–506.

Levin, S. A. ed. (2009) *The Princeton Guide to Ecology* (Princeton, NJ: Princeton University Press).

Levin, S.A. (2020), *Emergent and Vanishing Biodiversity, and Evolutionary Suicide*, 'https://royalsociety.org/topics-policy/projects/biodiversity/emergent-and-vanishing-biodiversity-and-evolutionary-suicide/'.

Levin, S A., and J. Lubchenco (2008), 'Resilience, Robustness, and Marine Ecosystem-Based Management', *BioScience*, 58(1), 27–32.

Levin, S. A., and R. T. Paine (1974), 'Disturbance, Patch Formation, and Community Structure', *Proceedings of the National Academy of Sciences*, 71(7), 2744–2747.

References

Levy, B. J., C. Brandon, and R. Studart (2020), *Designing the COVID-19 Recovery for a Safer and More Resilient World*.

Lewis, S. L., C. E. Wheeler, E. T. A. Mitchard, and A. Koch (2019), 'Restoring Natural Forests is The Best Way to Remove Atmospheric Carbon', *Nature*, 568, 25–28.

Lewis, W. A. (1988), 'The Roots of Development Theory', in H. Chenery and T. N. Srinivasan, eds., *Handbook of Development Economics, Vol. 1* (Amsterdam: North Holland).

Li, X., Y. Wang, S. Chen, H. Tian, D. Fu, B. Zhu, Y. Luo, and H. Zhu (2018), 'Lycopene is Enriched in Tomato Fruit by CRISPR/Cas9-Mediated Multiplex Genome Editing', *Frontiers in Plant Science*, 9, 559.

Lin, D., L. Hanscom, A. Murthy, A. Galli, M. Evans, E. Neill, M. Mancini, J. Martindill, F.-Z. Medouar, S. Huang, and M. Wackernagel (2018), 'Ecological Footprint Accounting for Countries: Updates and Results of the National Footprint Accounts, 2012–2018', *Resources*, 7(3), 58.

Lin, D., L. Wambersie, M. Wackernagel, and P. Hanscom (2020), 'Calculating Earth Overshoot Day 2020: Estimates Point to August 22nd', 'https://www.overshootday.org/content/uploads/2020/06/Earth-Overshoot-Day-2020-Calculation-Research-Report.pdf'.

Lindahl, E. (1958), 'Just Taxation—A Positive Solution' in R.A. Musgrave and A. T. Peacock, eds, *Classics in the Theory of Public Finance* (London: Palgrave Macmillan)

Little, I. M. D., and J. A. Mirrlees (1969), *Manual of Industrial Project Analysis in Developing Countries, Vol. I: Methodology and Case Studies* (Paris: Development Centre of the Organisation for Economic Co-operation and Development).

Little, I. M. D., and J. A. Mirrlees (1974), *Project Appraisal and Planning for Developing Countries* (London: Heinemann Educational Books).

Liu, L. (2018), *The Global Collaboration Against Transnational Corruption: Motives, Hurdles and Solutions* (Singapore: Palgrave Macmillan).

Liu, Z., and J. Lan (2015), 'The Sloping Land Conversion Program in China: Effect on the Livelihood Diversification of Rural Households', *World Development*, 70(C), 147–161.

Lock, K., and S. Leslie (2007), *New Zealand's Quota Management System: A History of the First 20 Years*.

Lomborg, B. (2001), *The Skeptical Environmentalist: Measuring the Real State of the World* (Cambridge: Cambridge University Press).

Lomborg, B. (2013), *How Much Have Global Problems Cost the World?: A Scorecard from 1900 to 2050*. (Cambridge: Cambridge University Press).

López, R. (1998), 'The Tragedy of the Commons in Côte d'Ivoire Agriculture: Empirical Evidence and Implications for Evaluating Trade Policies', *World Bank Economic Review*, 12(1),105–131.

López-Feldman, A., and J. E. Wilen (2008), 'Poverty and Spatial Dimensions of Non-Timber Forest Extraction', *Environment and Development Economics*, 13(5), 621–642.

Loreau, M., S. Naeem, P. Inchausti, J. Bengtsson, J. P. Grime, A. Hector, D. U. Hooper, M. A. Huston, D. Raffaelli, B. Schmid, D. Tilman, and D. A. Wardle (2001), 'Biodiversity and Ecosystem Functioning: Current Knowledge and Future Challenges', *Science*, 294(5543), 804–808.

Lorimer, J., C. Sandom, P. Jepson, C. Doughty, M. Barua, and K. J. Kirby (2015), 'Rewilding: Science, Practice, and Politics', *Annual Review of Environment and Resources*, 40(1), 39–62.

Lovejoy, T. E., and C. Nobre (2018), 'Amazon Tipping Point', *Science Advances*, 4(2), eaat2340.

Lovejoy, T. E., and C. Nobre (2019), 'Amazon Tipping Point: Last Chance for Action', *Science Advances*, 5(12), eaba2949.

Lovejoy, T. E., and L. Hannah (2019), *Biodiversity and Climate Change: Transforming the Biosphere* (New Haven, CT: Yale University Press).

Lovelock, J. E. (1995), *The Ages of Gaia: A Biography of Our Living Earth* (New York, NY: W. W. Norton).

Lowenberg-DeBoer, J., and B. Erickson (2019), 'Setting the Record Straight on Precision Agriculture Adoption', *Agronomy Journal*, 111(4),1552–169.

Lowry, M. S., R. Condit, B. Hatfield, S. G. Allen, R. Berger, P. A. Morris, B. J. Le Boeuf, and J. Reiter (2014), 'Abundance, Distribution, and Population Growth of the Northern Elephant Seal (Mirounga angustirostris) in the United States from 1991 to 2010', *Aquatic Mammals*, 40(1),20–31.

Lu, Y., Y. Zhang, X. Cao, C. Wang, Y. Wang, M. Zhang, R. C. Ferrier, A. Jenkins, J. Yuan, M. J. Bailey, D. Chen, H. Tian, H. Li, E. U. von Weizsäcker, and Z. Zhang (2019), 'Forty Years of Reform and Opening Up: China's Progress Toward a Sustainable Path', *Science Advances*, 5(8), eaau9413.

Luce, R. D., and H. Raiffa. (1957), *Games and Decisions: Introduction and Critical Survey* (New York, NY: Wiley).

Ludwig, D., S. Carpenter, and W. Brock (2003), 'Optimal Phosphorus Loading for a Potentially Eutrophic Lake', *Ecological Applications*, 13(4),1135–1152.

Lugato, E., P. Smith, P. Borrelli, P. Panagos, C. Ballabio, A. Orgiazzi, O. Fernández-Ugalde, L. Montanarella, and A. Jones (2018), 'Soil Erosion is Unlikely to Drive a Future Carbon Sink in Europe', *Science Advances*, 4(11), eaau3523.

Lugauer, S., R. Jensen, and C. Sadler (2014), 'An Estimate of the Age Distribution's Effect on Carbon Dioxide Emissions', *Economic Inquiry*, 52(2), 914–929.

Lughadha, E. N., R. Govaerts, I. Belyaeva, N. Black, H. Lindon, R. Allkin, R. E. Magill, and N. Nicolson (2016), 'Counting Counts: Revised Estimates of Numbers of Accepted Species of Flowering Plants, Seed Plants, Vascular Plants and Land Plants with a Review of Other Recent Estimates', *Phytotaxa*, 272(1), 82–88.

Luisetti, T., R. K. Turner, I. J. Bateman, S. Morse-Jones, C. Adams, and L. Fonseca (2011), 'Coastal and Marine Ecosystem Services Valuation for Policy and Management: Managed Realignment Case Studies in England', *Ocean and Coastal Management*, 54(3), 212–224.

Lutz, W., W. P. Butz, and S. KC eds. (2014), *World Population and Human Capital in the Twenty- First Century* (Oxford: Oxford University Press).

Lynch, J., and R. Pierrehumbert (2019), 'Climate Impacts of Cultured Meat and Beef Cattle', *Frontiers in Sustainable Food Systems*, 3, 5.

MA – Millennium Ecosystem Assessment – eds., R. Hassan, R. Scholes, and N. Ash (2005a), *Ecosystems and Human Well-Being, I: Current State and Trends* (Washington, DC: Island Press).

MA – Millennium Ecosystem Assessment – eds., S.R. Carpenter, P.L. Pingali, E.M. Bennet, and M. B. Zurek (2005b), *Ecosystems and Human Well-Being, II: Scenarios* (Washington, DC: Island Press).

MA – Millennium Ecosystem Assessment – eds., K. Chopra, R. Leemans, P. Kumar, and H. Simmons (2005c), *Ecosystems and Human Well-Being, III: Policy Responses* (Washington, DC: Island Press).

References

MA – Millennium Ecosystem Assessment – eds., D. Capistrano, C. Samper K., M.J. Lee, and C. Randsepp-Hearne (2005d), *Ecosystems and Human Well-Being, IV: Multiscale Assessments* (Washington, DC: Island Press).

Maas, J., R. A. Verheij, S. de Vries, P. Spreeuwenberg, F. G. Schellevis, and P. P. Groenewegen (2009), 'Morbidity is Related to a Green Living Environment', *Journal of Epidemiology and Community Health*, 63(12), 967–973.

MacArthur, R. H., and E. O. Wilson (1967), *The Theory of Island Biogeography* (Princeton, NJ: Princeton University Press).

Mace, G. M., M. Barrett, N. D. Burgess, S. E. Cornell, R. Freeman, M. Grooten, and A. Purvis (2018), 'Aiming Higher to Bend the Curve of Biodiversity Loss', *Nature Sustainability*, 1, 448–451.

Mace, G. M., B. Reyers, R. Alkemade, R. Biggs, F. S. Chapin, S. E. Cornell, S. Díaz, S. Jennings, P. Leadley, P. J. Mumby, A. Purvis, R. J. Scholes, A. W. R. Seddon, M. Solan, W. Steffen, and G. Woodward (2014), 'Approaches to Defining a Planetary Boundary for Biodiversity', *Global Environmental Change*, 28(1), 289–297.

Maddison, A. (2001), *The World Economy: A Millennial Perspective* (Paris: Development Centre of the Organisation for Economic Co-operation and Development).

Maddison, A. (2010). *Maddison Project Database* 2010.

Maddison, A. (2018). *Maddison Project Database* 2018.

Mailath, G. J., and L. Samuelson. (2006), *Repeated Games and Reputations: Long-Run Relationships* (Oxford: Oxford University Press).

Mäler, K. - G. (1974), *Environmental Economics: A Theoretical Inquiry* (Baltimore, MD: Johns Hopkins University Press).

Malhi, Y. (2012), 'The Productivity, Metabolism and Carbon Cycle of Tropical Forest Vegetation' *Journal of Ecology*, 100(1), 65–75

Malhi, Y., T. A. Gardner, G. R. Goldsmith, M. R. Silman, and P. Zelazowski (2014), 'Tropical Forests in the Anthropocene', *Annual Review of Environment and Resources*, 39(1), 125–159.

Malinowski, B. (1926), *Crime and Custom in Savage Society* (London: Kegan Paul).

Malinvaud, E. (1967), 'Decentralized Procedures for Planning', in E. Malinvaud and M. Bacharach, eds., *Activity Analysis in the Theory of Growth and Planning* (Amsterdam: Springer).

Malinvaud, E. (1972), *Lectures on Microeconomic Theory* (Amsterdam: North Holland).

Managi, S., and P. Kumar (2018), *Inclusive Wealth Report 2018: Measuring Progress Towards Sustainability* (New York, NY: Routledge).

Marglin, S. A. (1963a), 'The Social Rate of Discount and The Optimal Rate of Investment', *The Quarterly Journal of Economics*, 77(1), 95–111.

Marglin, S. A. (1963b), 'The Opportunity Costs of Public Investment', *The Quarterly Journal of Economics*, 77(2), 274–289.

Margulies, J. (2012), 'No-Till Agriculture in the USA'. In E. Lichtfouse, ed., *Organic Fertilisation, Soil Quality and Human Health, Vol. 9* (Dordrecht: Springer).

Markandya, A., and M. N. Murty (2000), *Cleaning-up the Ganges: A Cost-Benefit Analysis of the Ganga Action Plan* (New Delhi: Oxford University Press).

Markandya, A., and M. N. Murty (2004), 'Cost–Benefit Analysis of Cleaning the Ganges: Some Emerging Environment and Development Issues', *Environment and Development Economics*, 9(1), 61–81.

Maron, M., A. Gordon, B. G. Mackey, H. P. Possingham, and J. E. M. Watson (2015), 'Conservation: Stop Misuse of Biodiversity Offsets', *Nature*, 523(7561), 401–403.

Marques, A., I. S. Martins, T. Kastner, C. Plutzar, M. C. Theurl, N. Eisenmenger, M. A. J. Huijbregts, R. Wood, K. Stadler, M. Bruckner, J. Canelas, J. P. Hilbers, A. Tukker, K. -H. Erb and, H. M. Pereira (2019), 'Increasing Impacts of Land Use on Biodiversity and Carbon Sequestration Driven by Population and Economic Growth', *Nature Ecology and Evolution*, 3(4), 628–637.

Marrocoli, S., T. T. Gatiso, D. Morgan, M. Reinhardt Nielsen, and H. Kühl (2018), 'Environmental Uncertainty and Self-monitoring in the Commons: A Common-pool Resource Experiment Framed Around Bushmeat Hunting in the Republic of Congo', *Ecological Economics*, 149, 274–284.

Marvier, M., P. Kareiva, and M. G. Neubert (2004), 'Habitat Destruction, Fragmentation, and Disturbance Promote Invasion by Habitat Generalists in a Multispecies Metapopulation', *Risk Analysis*, 24(4), 869–878.

Mascia, M. B., and S. Pailler (2011), 'Protected Area Downgrading, Downsizing, and Degazettement (PADDD) and its Conservation Implications', *Conservation Letters*, 4(1), 9–20.

Maskin, E. (2021), "Arrow's IIA Condition, May's Axioms, and the Borda Count," Discussion Paper, Department of Economics, Harvard University.

Maurer, J., and A. Meier (2008), 'Smooth it Like the "Joneses"? Estimating Peer-Group Effects in Intertemporal Consumption Choice', *The Economic Journal*, 118(527), 454–476.

Mauro, P (1995), 'Corruption and Growth', *The Quarterly Journal of Economics*, 110(3), 681–712.

Maxwell, S. L., R. A. Fuller, T. M. Brooks, and J. E. M. Watson (2016), 'Biodiversity: The Ravages of Guns, Nets and Bulldozers', *Nature*, 536(7615), 143–145.

May, K.O. (1952), "Set of Necessary and Sufficient Conditions for Simple Majority Decisions," *Econometrica*, 20(4), 680–684.

Mayer, F. S., C. M. Frantz, E. Bruehlman-Senecal, and K. Dolliver (2009), 'Why Is Nature Beneficial? The Role of Connectedness to Nature', *Environment and Behavior*, 41(5), 607–643.

McCraine, S., C. Anderson, C. Weber, and M. R. Shaw (2019), *The Nature of Risk: A Framework for Understanding Nature-related Risk to Business*.

McDonald, R. I., M. Colbert, M. Hamann, R. Simkin, and B. Walsh (2018), *Nature in the Urban Century: A Global Assessment of Where and How to Conserve Nature for Biodiversity and Human Wellbeing*.

McDonnell, M. J., and A. K. Hahs (2009), 'Comparative ecology of cities and towns: Past, present and future', in M. J. McDonnell, and A. K. Hahs, eds., *Ecology of Cities and Towns: A Comparative Approach* (Cambridge, Cambridge University Press).

McKean, M. A. (1992), 'Success on the Commons: A Comparative Examination of Institutions for Common Property Resource Management', *Journal of Theoretical Politics*, 4(3), 247–281.

McKenney, B. A., and J. M. Kiesecker (2010), 'Policy Development for Biodiversity Offsets: A Review of Offset Frameworks', *Environmental Management*, 45(1), 165–176.

McMahan, E. A., and D. Estes (2015), 'The Effect of Contact with Natural Environments on Positive and Negative Affect: A Meta-Analysis', *The Journal of Positive Psychology*, 10(6), 507–519.

References

Meade, J. E. (1952), 'External Economies and Diseconomies in a Competitive Situation', *The Economic Journal*, 62(245), 54–67.

Meade, J. E. (1978), *The Structure and Reform of Direct Taxation: The Meade Report* (London: George Allen & Unwin).

Meadows, D. H., D. L. Meadows, J. Randers, and W. W. Behrens III (1972), *The Limits to Growth: A report for the Club of Rome's project on the Predicament of Mankind* (New York, NY: Universe Books).

Meier, M. S., F. Stoessel, N. Jungbluth, R. Juraske, C. Schader, and M. Stolze (2015), 'Environmental Impacts of Organic and Conventional Agricultural Products – Are the Differences Captured by Life Cycle Assessment?', *Journal of Environmental Management*, 149, 193–208.

Mekonnen, M. M., and A. Y. Hoekstra (2010), 'The Green, Blue and Grey Water Footprint of Crops and Derived Crop Products', *Value of Water Research Report Series*, No. 47.

Mendelsohn, R., A. Dinar, and L. Williams (2006), 'The Distributional Impact of Climate Change on Rich and Poor Countries', *Environment and Development Economics*, 11(2), 159–178.

Meyfroidt, P., E. F. Lambin, K. -H. Erb, and T. W. Hertel, (2013), 'Globalization of Land Use: Distant Drivers of Land Change and Geographic Displacement of Land Use', *Current Opinion in Environmental Sustainability*, 5(5), 438–444.

Micklethwait, J., and A. Wooldridge (2003), *A Future Perfect: The Challenge and Promise of Globalization* (New York, NY: Random House).

Milgrom, P. R., D. C. North, and B. R. Weingast (1990). 'The Role of Institutions in the Revival of Trade: The Law Merchant, Private Judges, and the Champagne Fairs', *Economics and Politics*, 2(1), 1–23.

Miller, D. C., A. Agrawal, and J. T. Roberts (2013), 'Biodiversity, Governance, and the Allocation of International Aid for Conservation', *Conservation Letters*, 6(1), 12–20.

Miller, G., and K. S. Babiarz (2016), 'Family Planning Program Effects: Evidence from Microdata', *Population and Development Review*, 42(1), 7–26.

Ministry of Fisheries (2009), *Fisheries 2030. New Zealanders Maximising Benefits from the Use of Fisheries within Environmental Limits*.

Mirrlees, J. A. (1967), 'Optimum Growth when Technology is Changing', *The Review of Economic Studies*, 34(1), 95–124.

Mirrlees, J. A. (1969). 'The Evaluation of National Income in an Imperfect Economy', *The Pakistan Development Review*, 9(1), 1–13.

Mirrlees, J. A., S. Adam, T. Besley, R. Blundell, S. Bond, R. Chote, M. Gammie, P. Johnson, G. Myles and J. M. Poterba (2011), *Tax by Design: The Mirrlees Review* (Oxford: Oxford University Press).

Misak, C. J. (2016), *Cambridge Pragmatism: from Peirce and James to Ramsey and Wittgenstein* (Oxford: Oxford University Press).

Mission Économie de la Biodiversité (2019), *Global Biodiversity Score: A Tool to Establish and Measure Corporate and Financial Commitments for Biodiversity*.

Mitchell, G., and D. Dorling (2003), 'An Environmental Justice Analysis of British Air Quality', *Environment and Planning A*, 35(5), 909–929.

Mitchell, G., and P. Norman (2012), 'Longitudinal Environmental Justice Analysis: Co-Evolution of Environmental Quality and Deprivation in England, 1960-2007', *Geoforum*, 43(1), 44–57.

Mitchell, G., P. Norman, and K. Mullin (2015), 'Who Benefits from Environmental Policy? An Environmental Justice Analysis of Air Quality Change in Britain, 2001-2011', *Environmental Research Letters*, 10(10), 105009.

Mitchell, R. B. (2020), 'International Environmental Agreements (IEA) Database Project', '*https://iea.uoregon.edu/*'.

Mitchell, R. B., L. B. Andonova, M. Axelrod, J. Balsiger, T. Bernauer, J. F. Green, J. Hollway, R. E. Kim, and J. F. Morin (2020), 'What We Know (and Could Know) About International Environmental Agreements', *Global Environmental Politics*, 20(1), 103–121.

Mitchell, R. J., E. A. Richardson, N. K. Shortt, and J. R. Pearce (2015), 'Neighbourhood Environments and Socioeconomic Inequalities in Mental Well-Being.', *American Journal of Preventive Medicine*, 49(1), 80–84.

Mittermeier, R. A., W. Turner, F. W. Larsen, T. M. Brooks, and C. Gascon (2011), 'Global Biodiversity Conservation: The Critical Role of Hotspots', In F. Zachos, J. and Habel eds., *Biodiversity Hotspots* (Berlin and Heidelberg: Springer).

Mo, P. H. (2001), 'Corruption and Economic Growth', *Journal of Comparative Economics*, 29(1), 66–79.

Molnar, J. L., R. L. Gamboa, C. Revenga, and M. D. Spalding (2008), 'Assessing the Global Threat of Invasive Species to Marine Biodiversity', *Frontiers in Ecology and the Environment*, 6(9), 485–492.

Molthan, P. (2003), 'Introduction', in D. Broadhurst, J. Watson, and J. Marshall, eds., *Ethical and Socially Responsible Investment – A Reference Guide for Researchers* (Munich: K.G. Saur).

Montgomery, M. R., and J. B. Casterline (1998), '*Social Networks and the Diffusion of Fertility Control*', Population Council, Policy Research Division Working Paper, No. 119.

Mora, C., A. G. Frazier, R. J. Longman, R. S. Dacks, M. M. Walton, E. J. Tong, J. J. Sanchez, L. R. Kaiser, Y. O. Stender, J. M. Anderson, C. M. Ambrosino, I. Fernandez-Silva, L. M. Giuseffi, and T. W. Giambelluca (2013), 'The Projected Timing of Climate Departure from Recent Variability', *Nature*, 502(7470), 183–187.

Mora, C., D. P. Tittensor, S. Adl, A. G. B. Simpson, and B. Worm (2011), 'How Many Species Are There on Earth and in the Ocean?', *PLoS Biology*, 9(8), e1001127.

Moran, D., and K. Kanemoto (2017), 'Identifying Species Threat Hotspots from Global Supply Chains', *Nature Ecology and Evolution*, 1, 0023.

Moran, D. D., M. Lenzen, K. Kanemoto, and A. Geschke (2013), 'Does Ecologically Unequal Exchange Occur?', *Ecological Economics*, 89, 177–186.

Morelle, R. (2019), 'Mariana Trench: Deepest-Ever Sub Dive Finds Plastic Bag', *BBC News*.

Moreno-Mateos, D., E. B. Barbier, P. C. Jones, H. P. Jones, J. Aronson, J. A. López-López, M. L. McCrackin, P. Meli, D. Montoya, and J. M. R. Benayas (2017), 'Anthropogenic Ecosystem Disturbance and the Recovery Debt', *Nature Communications*, 8(1), 14163.

Mori Junior, R., D. M. Franks, and S. H. Ali (2016), 'Sustainability Certification Schemes: Evaluating Their Effectiveness and Adaptability', *Corporate Governance*,16(3), 579–592.

Morikawa, M. K., and S. R. Palumbi (2019), 'Using Naturally Occurring Climate Resilient Corals to Construct Bleaching-Resistant Nurseries', Proceedings of the National Academy of *Sciences*, 116(21), 10586–10591.

References

Mueller, G. M., and J. P. Schmit (2007), 'Fungal Biodiversity: What Do we Know? What Can we Predict?', *Biodiversity Conservation*, 16(1), 1–5.

Mukhopadhyay, P. (2008), 'Heterogeneity, Commons, and Privatization: Agrarian Institutional Change in Goa', in R. Ghate, N.S. Jodha, and P. Mukhopadhyay, eds., (2008), *Promise, Trust and Evolution: Managing the Commons of South Asia* (Oxford: Oxford University Press).

Mukhopadhyay, P., S. Ghosh, V. Da Costa, and S. Pednekar (2020), 'Recreational Value of Coastal and Marine Ecosystems in India: A Macro Approach', *Tourism in Marine Environments*, 15(1) 11–27.

Mulla, D. J. (2013), 'Twenty-Five Years of Remote Sensing in Precision Agriculture: Key Advances and Remaining Knowledge Gaps', *Biosystems Engineering*, 114(4), 358–371.

Müller, A., C. Schader, N. El-Hage Scialabba, J. Brüggemann, A. Isensee, K. -H. Erb, P. Smith, P. Klocke, F. Leiber, M. Stolze, and U. Niggli, (2017), 'Strategies for Feeding the World More Sustainably with Organic Agriculture', *Nature Communications*, 8(1), 1290.

Müller, A., and P. Sukhdev (2018), *Measuring What Matters in Agriculture and Food Systems: A Synthesis of the Results and Recommendations of TEEB for Agriculture and Food's Scientific and Economic Foundations Report*.

Müller, S., S. Hemming, and D. Rigney (2019), 'Indigenous Sovereignties: Relational Ontologies and Environmental Management', *Geographical Research*, 57(4), 399–410.

Mullin, K., G. Mitchell, N. R. Nawaz, and R. D. Waters (2018), 'Natural Capital and the Poor in England: Towards an Environmental Justice Analysis of Ecosystem Services in a High Income Country', *Landscape and Urban Planning*, 176(January), 10–21.

Munshi, K., and J. Myaux (2006), 'Social Norms and the Fertility Transition', *Journal of Development Economics*, 80(1), 1–38.

Murina, M., and A. Nicita (2017), 'Trading with Conditions: The Effect of Sanitary and Phytosanitary Measures on the Agricultural Exports from Low-income Countries', *World Economy*, 40(1), 168–181.

Nagel, T. (1979), "Death," in *Mortal Questions* (Cambridge: Cambridge University Press). Originally published in *Nous*, 1970, 4(1), 73–80.

Nagel, T. (2012), *Mind and Cosmos: Why the Materialist Neo-Darwinian Conception of Nature is Almost Certainly False* (Oxford: Oxford University Press).

Nagrawala, F., and K. Springer (2020), *Point of No Returns: A Ranking of 75 of the World's Largest Asset Managers' Approaches to Responsible Investment*.

Naidoo, R., D. Gerkey, D. Hole, A. Pfaff, A. M. Ellis, C. D. Golden, D. Herrera, K. Johnson, M. Mulligan, T. H. Ricketts, and B. Fisher (2019), 'Evaluating the Impacts of Protected Areas on Human Well-Being Across the Developing World', *Science Advances*, 5(4), eaav3006.

Naqvi, M., B. Burke, S. Hector, T. Jamison, and S. Dupré (2017), *All Swans Are Black in the Dark: How the Short-Term Focus of Financial Analysis Does Not Shed Light on Long Term Risks*.

Narayan, S., M. W. Beck, B. G. Reguero, I. J. Losada, B. Van Wesenbeeck, N. Pontee, J. N. Sanchirico, J. C. Ingram, G. M. Lange, and K. A. Burks-Copes (2016), 'The Effectiveness, Costs and Coastal Protection Benefits of Natural and Nature-Based Defences', *PLoS ONE*, 11(5), e0154735.

Narayan, S., M. W. Beck, P. Wilson, C. J. Thomas, A. Guerrero, C. C. Shepard, B. G. Reguero, G. Franco, J. C. Ingram, and D. Trespalacios (2017), 'The Value of Coastal Wetlands for Flood Damage Reduction in the Northeastern USA', *Scientific Reports*, 7, 9463.

Nasar, S. (1998), *A Beautiful Mind* (London: Faber & Faber).

National Academies of Sciences Engineering and Medicine (2019), *A Research Review of Interventions to Increase the Persistence and Resilience of Coral Reef* (Washington, DC: The National Academies Press).

National Research Council (2007), *Status of Pollinators in North America* (Washington, DC: The National Academies Press).

Natural Capital Coalition (2018a), *Connecting Finance and Natural Capital: A Supplement to the Natural Capital Protocol*.

Natural Capital Coalition (2018b), *Connecting Finance and Natural Capital: Case Study for CDC Biodiversité*.

Natural Capital Committee (2014), *Towards a Framework for Defining and Measuring Change in Natural Capital*.

Natural Capital Committee (2018), *The Green Book Guidance: Embedding Natural Capital into Public Policy Appraisal*.

Natural Capital Committee (2019), *State of Natural Capital Annual Report 2019*.

Natural Capital Committee (2020), *The Green Book Guidance: Embedding Natural Capital into Public Policy Appraisal – November 2020 update*.

Natural Capital Finance Alliance (2020), *Beyond 'Business as Usual': Biodiversity Targets and Finance*.

Natural England (2010), *England's Peatlands: Carbon Storage and Greenhouse Gases*.

Natural England (2019), *Monitor of Engagement with the Natural Environment: The National Survey on People and the Natural Environment*.

Nature Map (2020), 'Nature Map Earth', 'https://naturemap.earth/'.

Nauman, B. (2019), *Sharp Rise in Number of Investors Dumping Fossil Fuel Stocks*.

Nayak, D. R., and P. Smith (2019), *Review and Comparison of Models used for Land Allocation and Nature Valuation*.

Neate, V., S. Hime, and V. Paschos (2014), *Study to Support an Ex Ante Assessment for A Natural Capital Financing Facility*.

Nellemann, C., R. Henriksen, A. Kreilhuber, D. Stewart, M. Kotsovou, P. Raxter, E. Mrema, and S. Barrat, eds., (2016), *The Rise of Environmental Crime: A Growing Threat to Natural Resources, Peace, Development and Security* (Nairobi: UNEP).

Nelson, E., H. Sander, P. Hawthorne, M. Conte, D. Ennaanay, S. Wolny, S. Manson, and S. Polasky (2010), 'Projecting Global Land-Use Change and Its Effect On Ecosystem Service Provision And Biodiversity With Simple Models', *PLoS ONE*, 5(12), e14327.

Nelson, G. C., H. Valin, R. D. Sands, P. Havlík, H. Ahammad, D. Deryng, J. Elliott, S. Fujimori, T. Hasegawa, E. Heyhoe, P. Kyle, M. Von Lampe, H. Lotze-Campen, D. Mason D'croz, H. Van Meijl, D. Van Der Mensbrugghe, C. Müller, A. Popp, R. Robertson, S. Robinson, E. Schmid, C. Schmitz, A. Tabeau, and D. Willenbockel (2014), 'Climate Change Effects On Agriculture: Economic Responses To Biophysical Shocks', *Proceedings of the National Academy of Sciences*, 111(9), 3274–3279.

References

Nepal, P., J. Korhonen, J. P. Prestemon, and F. W. Cubbage (2019), 'Projecting Global Planted Forest Area Developments and the Associated Impacts on Global Forest Product Markets', *Journal of Environmental Management*, 240, 421–430.

Netting, R. McC. (1968), *Hill Farmers of Nigeria: Cultural Ecology of the Kofyar of the Jos Plateau* (Seattle: University of Washington Press).

Netting, R. McC. (1981), *Balancing on an Alp: Ecological Change and Continuity in a Swiss Mountain Community* (Cambridge: Cambridge University Press).

Network for Greening the Financial System (2019), *A Call for Action, Climate Change as a Source of Financial Risk*.

Network for Greening the Financial System (2020), *Guide for Supervisors: Integrating Climate- Related and Environmental Risks into Prudential Supervision*.

Newbold, T. (2018), 'Future Effects of Climate and Land-Use Change on Terrestrial Vertebrate Community Diversity Under Different Scenarios', *Proceedings of the Royal Society B*, 285(1881), 20180792.

Newbold, T., L. N. Hudson, S. L. L. Hill, S. Contu, I. Lysenko, R. A. Senior, L. Börger, D. J. Bennett, A. Choimes, B. Collen, J. Day, A. De Palma, S. Díaz, S. Echeverria-Londoño, M. J. Edgar, A. Feldman, M. Garon, M. L. K. Harrison, T. Alhusseini, D. J. Ingram, Y. Itescu, J. Kattge, V. Kemp, L. Kirkpatrick, M. Kleyer, D. L. Correia, C. D. Martin, S. Meiri, M. Novosolov, Y. Pan, H. R. Phillips, D. W. Purves, A. Robinson, J. Simpson, S. L. Tuck, E. Weiher, H. J. White, R. M. Ewers, G. M. Mace, J. P. Scharlemann, and A. Purvis (2015), 'Global Effects of Land Use on Local Terrestrial Biodiversity', *Nature*, 520(7545), 45–50.

Newman, D. J., and G. M. Cragg (2016), 'Natural Products as Sources of New Drugs from 1981 to 2014', *Journal of Natural Products*, 79(3), 629–661.

Ng, S., and A. Y. Lee (2015), *Handbook of Culture and Consumer Behavior* (Oxford: Oxford University Press).

Ng, Y.-K. (1975), 'Bentham or Bergson? Finite Sensibility, Utility Functions and Social Welfare Functions', *The Review of Economic Studies*, 42(4), 545–569.

Ngumbi, E. N. (2017), *Demonstration Farms Can Help Revolutionise African Agriculture*.

Niinimäki, K., G. Peters, H. Dahlbo, P. Perry, T. Rissanen, and A. Gwilt (2020), 'The Environmental Price of Fast Fashion', *Nature Reviews Earth and Environment*, 1(4),189–200.

Nilsson, C., C. A. Reidy, M. Dynesius, and C. Revenga (2005), 'Fragmentation and Flow Regulation of the World's Large River Systems', *Science*, 308(5720), 405–408.

Nobles, J., E. Frankenberg, and D. Thomas (2015), 'The Effects of Mortality on Fertility: Population Dynamics After a Natural Disaster', *Demography*, 52(1),15–38.

Nogués-Bravo, D., D. Simberloff, C. Rahbek, and N. J. Sanders (2016), 'Rewilding is the New Pandora's Box in Conservation', *Current Biology*, 26(3), 87–91.

Norberg, J. (2016), *Progress: Ten Reasons to Look Forward to The Future* (London: Oneworld Publications).

Nordhaus, W. D. (1973), 'World Dynamics: Measurement Without Data', *The Economic Journal*, 83(332),1156–1183.

Nordhaus, W. D. (1994), *Managing the Global Commons: The Economics of Climate Change* (Cambridge, MA: MIT Press).

Nordhaus, W. D. (2007), 'A Review of the Stern Review on the Economics of Climate Change', *Journal of Economic Literature*, 45(3), 686–702.

Nordhaus, W.D. and J. Boyer (2000), *Warming the World: Economic Models of Global Warming* (Cambridge, MA: MIT Press).

North, D. C. (1990), *Institutions, Institutional Change and Economic Performance*. (Cambridge: Cambridge University Press).

North, D. C., and R. P. Thomas eds. (1973), *The Rise of The Western World: A New Economic History* (Cambridge: Cambridge University Press).

Nussbaum, M. C., and A. Sen eds. (1993), *The Quality of Life* (Oxford: Clarendon Press).

NYDF Assessment Partners. (2019), *Protecting and Restoring Forests: A Story of Large Commitments yet Limited Progress*.

O'Brien, P. K. (2010), 'Ten Years of Debate on the Origins of the Great Divergence', *Reviews in History*, 1008, 1–15.

O'Leary, B. C., G. Hoppit, A. Townley, H. L. Allen, C. J. McIntyre, and C. M. Roberts (2020), 'Options for Managing Human Threats to High Seas Biodiversity', *Ocean and Coastal Management*, 187, 105110.

O'Leary, B. C., M. Winther-Janson, J. M. Bainbridge, J. Aitken, J. P. Hawkins, and C. M. Roberts (2016), 'Effective Coverage Targets for Ocean Protection', *Conservation Letters*, 9(6), 398–404.

O'Neill, B. C., M. Dalton, R. Fuchs, L. Jiang, S. Pachauri, and K. Zigova (2010), 'Global Demographic Trends and Future Carbon Emissions', *Proceedings of the National Academy of Sciences*, 107(41), 17521–17526.

O'Neill, D. W., A. L. Fanning, W. F. Lamb, J. K. Steinberger (2018), 'A Good Life for All Within Planetary Boundaries', *Nature Sustainability*, 1, 88–95.

OECD (2013a), *How's Life? 2013: Measuring Well-Being*.

OECD (2013b), *Policy Instruments to Support Green Growth in Agriculture*.

OECD (2016a), *Biodiversity Offsets: Effective Design and Implementation*.

OECD (2016b), *The Ocean Economy in 2030*.

OECD (2017a), *Green Growth Indicators*.

OECD (2017b), *Reforming Agricultural Subsidies to Support Biodiversity in Switzerland*.

OECD (2019a), *Biodiversity: Finance and the Economic and Business Case for Action*.

OECD (2019b), *Agricultural Policy Monitoring and Evaluation 2019*.

OECD (2019c), *Global Material Resources Outlook to 2060: Economic Drivers and Environmental Consequences*.

OECD (2019d), *Policy Instruments for the Environment*, 'https://pinedatabase.oecd.org/'.

OECD (2020a), *A Comprehensive Overview of Global Biodiversity Finance*.

OECD (2020b), *Developing Sustainable Finance Definitions and Taxonomies*.

OECD (2020c), *Tracking Economic Instruments and Finance for Biodiversity 2020*.

References

Ogilvie, S. (1997), *State Corporatism and Proto-Industry: the Württemberg Black Forest, 1580-1797* (Cambridge: Cambridge University Press).

Oldekop, J. A., G. Holmes, W. E. Harris, and K. L. Evans (2016), 'A Global Assessment of the Social and Conservation Outcomes of Protected Areas', *Conservation Biology*, 30(1), 133–141.

Olmstead, S. M. (2014), 'Climate Change Adaptation and Water Resource Management: A Review of the Literature', *Energy Economics*, 46, 500–509.

ONS (2019), *UK Natural Capital: Peatlands*.

Ord, T. (2020), *The Precipice: Existential Risk and the Future of Humanity* (London: Bloomsbury).

Orgiazzi, A., C. Ballabio, P. Panagos, A. Jones, and O. Fernández-Ugalde (2018), 'LUCAS Soil, the Largest Expandable Soil Dataset for Europe: A Review', *European Journal of Soil Science*, 69(1), 140–153.

Orgiazzi, A., R. D. Bardgett, E. Barrios, V. Behan-Pelletier, M. J. I. Briones, J. L. Chotte, G. B. de Deyn, P. Eggleton, N. Fierer, T. Fraser, K. Hedlund, S. Jeffery, N. C. Johnson, A. Jones, E. Kandeler, N. Kaneko, P. Lavelle, P. Lemanceau, L. Miko, L. Montanarella, F. M. S. Moreira, K. S. Ramirez, S. Scheu, B. K. Singh, J. Six, W. H. van der Putten, and D. H. Wall (2016), *Global Soil Biodiversity Atlas* (Luxembourg: European Union).

Ortiz, J.-C., N. H. Wolff, K. R. N. Anthony, M. Devlin, S. Lewis, and P. J. Mumby (2018), 'Impaired Recovery of the Great Barrier Reef Under Cumulative Stress', *Science Advances*, 4(7), eaar6127.

Ostrom, E. (1990), *Governing the Commons: The Evolution of Institutions for Collective Action* (Cambridge: Cambridge University Press).

Ostrom, E. (1992), *Crafting Institutions for Self-Governing Irrigation Systems*. (San Francisco, CA: ICS Press).

Ostrom, E. (1995), 'Incentives, rules of the game, and development', *Proceedings of the Annual World Bank Conference on Development Economics, 1995* (Supplement to the *World Bank Economic Review* and the *World Bank Research Observer*), 207–248.

Ostrom, E. (2010a), 'Beyond Markets and States: Polycentric Governance of Complex Economic Systems', *American Economic Review*, 100(3), 641–672.

Ostrom, E. (2010b), 'Polycentric Systems for Coping with Collective Action and Global Environmental Change', *Global Environmental Change*, 20(4), 550–557.

Ostrom, E., and E. Schlager (1996), 'The Formation of Property Rights', in S. Hannah, C. Folke, and K. -G. Mäler, eds., *Rights to Nature: Ecological, Economic, Cultural, and Political Principles of Institutions for the Environment* (Washington, DC: Island Press).

Our World in Data (2018), *Trade and Globalization*, 'https://ourworldindata.org/trade-and-globalization'.

Ouyang, Z., C. Song, C. Wong, G. C. Daily, J. Liu, J. E. Salzman, L. Kong, H. Zheng, and C. Li (2019), 'China: Designing Policies to Enhance Ecosystem Services', in L. Mandle, Z. Ouyang, J. E. Salzman, and G. C. Daily, eds., *Green Growth That Works* (Washington, DC: Island Press).

Ouyang, Z., C. Song, H. Zheng, S. Polasky, Y. Xiao, I. J. Bateman, J. Liu, M. Ruckelshaus, F. Shi, Y. Xiao, W. Xu, Z. Zou, and G. C. Daily (2020), 'Using Gross Ecosystem Product (GEP) to Value Nature in Decision Making', *Proceedings of the National Academy of Sciences*, 117(25) 14593–14601.

Ouyang, Z., W. Xu, and Y. Xiao (2017), *Ecosystem Pattern, Quality, Services and their Changes in China* (Beijing: Science Press).

Ouyang, Z., H. Zheng, Y. Xiao, S. Polasky, J. Liu, W. Xu, Q. Wang, L. Zhang, Y. Xiao, E. Rao, L. Jiang, F. Lu, X. Wang, G. Yang, S. Gong, B. Wu, Y. Zeng, W. Yang, and G. C. Daily (2016), 'Improvements in Ecosystem Services from Investments in Natural Capital', *Science*, 352(6292), 1455–1459.

Pagiola, S. (2008), 'Payments for Environmental Services in Costa Rica', *Ecological Economics*, 65(4), 712–724.

Pagiola, S., A. R. Rios, and A. Arcenas (2008), 'Can the Poor Participate in Payments for Environmental Services? Lessons from the Silvopastoral Project in Nicaragua', *Environment and Development Economics*, 13(3), 299–325.

Palomo, I., B. Martín-López, M. Potschin, R. Haines-Young, and C. Montes (2013), 'National Parks, Buffer Zones and Surrounding Lands: Mapping Ecosystem Service Flows', *Ecosystem Services*, 4,104–116.

Parfit, D. (1984), *Reasons and Persons* (Oxford: Clarendon Press).

Parfit, D. (2012), 'We are not Human Beings', *Philosophy*, 87(1), 5–28.

Parfit, D. (2016), "Can We Avoid the Repugnant Conclusion?" *Theoria*, 82(2), 110–127.

Parikh, K. S., J. Parikh, and T. L. R. Ram (1999), 'Air and Water Quality Management: New Initiatives Needed', in K. S. Parikh, ed., *India Development Report 1999-2000* (New Delhi: Oxford University Press).

Parmesan, C (2006), 'Ecological and Evolutionary Responses to Recent Climate Change', *Annual Review of Ecology*, 37, 637–669.

Parmesan, C., N. Ryrholm, C. Stefanescu, J. K. Hill, C. D. Thomas, H. Descimon, B. Huntley, L. Kaila, J. Kullberg, T. Tammaru, W. J. Tennent, J. A. Thomas, and M. Warren (1999), 'Poleward Shifts in Geographical Ranges of Butterfly Species Associated with Regional Warming', *Nature*, 399, 579–583.

Pattanayak, S. K. (2004), 'Valuing Watershed Services: Concepts and Empirics from Southeast Asia', *Agriculture, Ecosystems and Environment*, 104(1), 171–184.

Pattanayak, S. K., and R. A. Kramer (2001), 'Worth of Watersheds: A Producer Surplus Approach for Valuing Drought Mitigation in Eastern Indonesia', *Environment and Development Economics*, 6(1), 123–146.

Pattanayak, S. K., and E. O. Sills (2001), 'Do Tropical Forests Provide Natural Insurance? The Microeconomics of Non-Timber Forest Product Collection in the Brazilian Amazon', *Land Economics*, 77(4), 595–612.

Pattanayak, S. K., S. Wunder, and P. J. Ferraro (2010), 'Show Me the Money: Do Payments Supply Environmental Services in Developing Countries?', *Review of Environmental Economics and Policy*, 4(2), 254–274.

Patterson, D., S. Schmitt, S. Singh, P. Eerdmans, M. Hugman, and A. Roux (2020), *Climate and Nature Sovereign Index*.

Pavlin, B.I., L.M. Schloegel, and P. Daszak (2009), 'Risk of Importing Zoonotic Diseases Through Wildlife Trade, United States', *Emerging infectious diseases*, 15(11), 1721.

Payn, T., J. -M. Carnus, P. Freer-Smith, M. Kimberley, W. Kollert, S. Liu, C. Orazio, L. Rodriguez, L. N. Silva, and M. J. Wingfield, (2015), 'Changes in Planted Forests and Future Global Implications', *Forest Ecology and Management*, 352, 57–67.

References

Pearson, A. L., A. Rzotkiewicz, J. L. Pechal, C. J. Schmidt, H. R. Jordan, A. Zwickle, and M. E. Benbow (2019), 'Initial Evidence of the Relationships between the Human Postmortem Microbiome and Neighborhood Blight and Greening Efforts', *Annals of the American Association of Geographers*, 109(3), 958–978.

Pechey, R., E. Cartwright, M. Pilling, G. J. Hollands, M. Vasiljevic, S. A. Jebb, and T. M. Marteau (2019), 'Impact of Increasing the Proportion of Healthier Foods Available on Energy Purchased in Worksite Cafeterias: A Stepped Wedge Randomized Controlled Pilot Trial', *Appetite*, 133, 286–296.

Pecl, G. T., M. B. Araújo, J. D. Bell, J. Blanchard, T. C. Bonebrake, I. C. Chen, T. D. Clark, R. K. Colwell, F. Danielsen, B. Evengård, L. Falconi, S. Ferrier, S. Frusher, R. A. Garcia, R. B. Griffis, A. J. Hobday, C. Janion-Scheepers, M. A. Jarzyna, S. Jennings, J. Lenoir, H. I. Linnetved, V. Y. Martin, P. C. McCormack, J. McDonald, N. J. Mitchell, T. Mustonen, J. M. Pandolfi, N. Pettorelli, E. Popova, S. A. Robinson, B. R. Scheffers, J. D. Shaw, C. J. B. Sorte, J. M. Strugnell, J. M. Sunday, M. Tuanmu, A. Vergés, C. Villanueva, T. Wernberg, E. Wapstra, and S. E. Williams (2017). 'Biodiversity Redistribution Under Climate Change: Impacts on Ecosystems and Human Well-Being', *Science*, 355(6332), eaai9214.

M. Pedersen, A. B. (2016), *Pesticide Tax in Denmark*.

Pendrill, F., U. M. Persson, J. Godar, T. Kastner, D. Moran, S. Schmidt, and R. Wood (2019), 'Agricultural and Forestry Trade Drives Large Share of Tropical Deforestation Emissions', *Global Environmental Change*, 56, 1–10.

Peng, A., S. Chen, T. Lei, L. Xu, Y. He, L. Wu, L. Yao, and X. Zou (2017), 'Engineering Canker-Resistant Plants Through CRISPR/Cas9-targeted Editing of the Susceptibility Gene CsLOB1 Promoter in Citrus', *Plant Biotechnology Journal*, 15(12), 1509–1519.

Pepper, I. L., C. P. Gerba, D. T. Newby, and C. W. Rice (2009), 'Soil: A Public Health Threat or Savior?', Critical Reviews in Environmental Science and *Technology*, 39(5), 416–432.

Pereira da Silva, L. A. (2020), *Green Swan 2 – Climate Change and Covid-19: Reflections on Efficiency Versus Resilience*.

Pereira, H. M., S. Ferrier, M. Walters, G. N. Geller, R. H. G. Jongman, R. J. Scholes, M. W. Bruford, N. Brummitt, S. H. M. Butchart, A. C. Cardoso, N. C. Coops, E. Dulloo, D. P. Faith, J. Freyhof, R. D. Gregory, C. Heip, R. Höft, G. Hurtt, W. Jetz, D. S. Karp, M. A. McGeoch, D. Obura, Y. Onoda, N. Pettorelli, B. Reyers, R. Sayre, J. P. W. Scharlemann, S. N. Stuart, E. Turak, M. Walpole, and M. Wegmann (2013), 'Essential Biodiversity Variables', *Science*, 339(6117), 277–278.

Pereira, H. M., L. M. Navarro, and I. S. Martins (2012), 'Global Biodiversity Change: The Bad, the Good, and the Unknown', *Annual Review of Environment and Resources*, 37, 25–50.

Perring, M. P., R. J. Standish, J. N. Price, M. D. Craig, T. E. Erickson, K. X. Ruthrof, A. S. Whiteley, L. E. Valentine, and R. J. Hobbs (2015), 'Advances in Restoration Ecology: Rising to the Challenges of the Coming Decades', *Ecosphere*, 6(8), 1–25.

Perrings, C. (2005), *The Socioeconomic Links Between Invasive Alien Species and Poverty. Report to the Global Invasive Species Program*.

Perrings, C. (2014), *Our Uncommon Heritage: Biodiversity Change, Ecosystem Services, and Human Wellbeing* (Cambridge: Cambridge University Press).

Perrings, C., S. Burgiel, M. Lonsdale, H. Mooney, and M. Williamson (2010), 'International Cooperation in the Solution to Trade-related Invasive Species Risks', Annals of the New York *Academy of Sciences*, 1195, 198–212.

Peters, G. M., G. Sandin, and B. Spak (2019), 'Environmental Prospects for Mixed Textile Recycling in Sweden', *ACS Sustainable Chemistry and Engineering*, 7(13), 11682–11690.

Peterson, M. (2019), 'The St. Petersburg Paradox', 'https://plato.stanford.edu/archives/fall2019/entries/paradox-stpetersburg/'.

Petraglia, M. D., H. S. Groucutt, M. Guagnin, P. S. Breeze, and N. Boivin (2020), 'Human Responses to Climate and Ecosystem Change in Ancient Arabia', *Proceedings of the National Academy of Sciences*, 117(15), 8263–8270.

Phalan, B., R. E. Green, L. V. Dicks, G. Dotta, C. Feniuk, A. Lamb, B. B. N. Strassburg, D. R. Williams, E. K. H. J. zu Ermgassen, and A. Balmford (2016), 'How Can Higher-yield Farming Help to Spare Nature?', *Science*, 351(6272), 450–451.

Phalan, B., M. Onial, A. Balmford, and R. E. Green (2011), 'Reconciling Food Production and Biodiversity Conservation: Land Sharing and Land Sparing Compared', *Science*, 333(6047), 1289–1291.

Pigou, A. C. (1920), *The Economics of Welfare* (London: Macmillan).

Pimentel, D. (2006), 'Soil Erosion: A Food and Environmental Threat', *Environment, Development and Sustainability*, 8(1), 119–137.

Pimentel, D., and M. Burgess (2013), 'Soil Erosion Threatens Food Production', *Agriculture*, 3(3), 443–463.

Pimentel, D., and N. Kounang (1998), 'Ecology of Soil Erosion in Ecosystems', *Ecosystems*, 1(5), 416–426.

Pimentel, D., R. Zuniga, and D. Morrison (2005), 'Update on the Environmental and Economic Costs Associated with Alien-invasive Species in the United States', *Ecological Economics*, 52(3), 273–288.

Pimm, S. L., C. N. Jenkins, and B. V. Li (2018), 'How to Protect Half of Earth to Ensure it Protects Sufficient Biodiversity', *Science Advances*, 4(8).

Pimm, S. L., C. N. Jenkins, R. Abell, T. M. Brooks, J. L. Gittleman, L. N. Joppa, P. H. Raven, C. M. Roberts, and J. O. Sexton (2014), 'The Biodiversity of Species and their Rates of Extinction, Distribution, and Protection', *Science*, 344(6187), 987–997.

Pimm, S. L, and P. H. Raven (2019), 'The State of the World's Biodiversity', In Dasgupta P., Raven P. H., and McIvor A. , eds., *Biological Extinction: New Perspectives* (Cambridge: Cambridge University Press).

Pimm, S. L., and P. H. Raven (2000), 'Extinction by Numbers', *Nature*, 403(6772), 843–845.

Pindyck, R. S. (2011), 'Fat Tails, Thin Tails, and Climate Change Policy', *Review of Environmental Economics and Policy*, 5(2), 258–274.

Pinker, S. (2018), *Enlightenment Now: The Case for Reason, Science, Humanism and Progress* (London: Penguin).

Pinzón, A., N. Robins, M. McLuckie, and G. Thoumi (2020), *The Sovereign Transition to Sustainability: Understanding The Dependence Of Sovereign Debt On Nature*.

Plank, B., N. Eisenmenger, A. Schaffartzik, and D. Wiedenhofer (2018), 'International Trade Drives Global Resource Use: A Structural Decomposition Analysis of Raw Material Consumption from 1990-2010', *Environmental Science and Technology*, 52(7), 4190–4198.

References

Plieninger, T., C. M. Raymond, and E. Oteros-Rozas (2016), 'Cultivated Lands', In M. Potschin, R. Haines-Young, R. Fish, and R.K. Turner eds., *Routledge Handbook of Ecosystem Services* (New York, NY: Routledge).

Polasky, S., C. Costello, and C. McAusland (2004), 'On Trade, Land-use, and Biodiversity', *Journal of Environmental Economics and Management*, 48(2), 911–925.

Pollock, L. J., W. Thuiller, and W. Jetz (2017), 'Large Conservation Gains Possible for Global Biodiversity Facets', *Nature*, 546(7656), 141–144.

Poloczanska, E. S., C. J. Brown, W. J. Sydeman, W. Kiessling, D. S. Schoeman, P. J. Moore, K. Brander, J. F. Bruno, L. B. Buckley, M. T. Burrows, C. M. Duarte, B. S. Halpern, J. Holding, C. V. Kappel, M. I. O'Connor, J. M. Pandolfi, C. Parmesan, F. Schwing, S. A. Thompson, and A. J. Richardson (2013). 'Global Imprint of Climate Change on Marine Life', *Nature Climate Change*, 3(10), 919–925.

Pomeranz, K. (2000), *The Great Divergence: China, Europe and the Making of the Modern World Economy* (Princeton, NJ: Princeton University Press).

Poore, J., and T. Nemecek (2018), 'Reducing Food's Environmental Impacts Through Producers and Consumers', *Science*, 360(6392), 987–992.

Popkin, B. M., L. S. Adair, and S. W. Ng (2012), 'Now and Then: The Global Nutrition Transition: The Pandemic of Obesity in Developing Countries', *Nutrition Reviews*, 70(1), 3–21.

Portfolio Earth (2020), *Bankrolling Extinction: The Banking Sector's Role in the Global Biodiversity Crisis*.

Pörtner, H. - O., D. M. Karl, P. W. Boyd, W. W. L. Cheung, S. E. Lluch-Cota, Y. NojiriD. N. Schmidt, and P. O. Zavialov, (2014), 'Climate Change 2014: Impacts, Adaptation, and Vulnerability. 6 Ocean Systems in IPCC', (2014) in *Climate Change 2014: Synthesis Report. Contribution of Working Groups I, II and III to the Fifth Assessment Report of the Intergovernmental Panel on Climate Change* (Geneva: IPCC).

Potschin, M., and R. Haines-Young (2016), 'Defining and Measuring Ecosystem Services', in M. Potschin, R. Haines-Young, R. Fish, and R. K. Turner, eds., *Routledge Handbook of Ecosystem Services* (London: Routledge).

Potts, S. G., S. P. M. Roberts, R. Dean, G. Marris, M. A. Brown, R. Jones, P. Neumann, and J. Settele (2010), 'Declines of Managed Honey Bees and Beekeepers in Europe', *Journal of Apicultural Research*, 49(1), 15–22.

Potts, S. G., V. Imperatriz-Fonseca, H. T. Ngo, M. A. Aizen, J. C. Biesmeijer, T. D. Breeze, L. V Dicks, L. A. Garibaldi, R. Hill, J. Settele, and A. J. Vanbergen (2016), 'Safeguarding Pollinators and their Values to Human Well-Being', *Nature*, 540(7632), 220–229.

Power, M. E., D. Tilman, J. A. Estes, B. A. Menge, W. J. Bond, L. S. Mills, G. Daily, J. C. Castilla, J. Lubchenco, and R. T. Paine (1996), 'Challenges in the Quest for Keystones: Identifying Keystone Species is Difficult – but Essential to Understanding How Loss of Species Will Affect Ecosystems', *BioScience*, 46(8), 609–620.

Prasad, J. V. N. S., K. Srinivas, C. S. Rao, C. Ramesh, K. Venkatravamma, and B. Venkateswarlu (2012), 'Biomass Productivity and Carbon Stocks of Farm Forestry and Agroforestry Systems of Leucaena and Eucalyptus in Andhra Pradesh, India', *Current Science*, 103(5), 536–540.

Pratt, C. F., K. L. Constantine, and S. T. Murphy (2017), 'Economic Impacts of Invasive Alien Species on African Smallholder Livelihoods', *Global Food Security*, 14, 31–37.

Pratt, J. (1964), 'Risk Aversion in the Small and in the Large', *Econometrica*, 32(1), 122–136.

Prest, A. R., and R. Turvey (1965), 'Cost-Benefit Analysis: A Survey', *The Economic Journal*, 75(300), 683–735.

Price, G., W. W. Lin, J. B. Falck-Zepeda, and J. Fernandez-Cornejo (2003), 'Size and Distribution of Market Benefits from Adopting Biotech Crops', *USDA-ERS Technical Bulletin No. 1906*.

PRI (2019), *The Final Chapter of Fiduciary Duty in the 21st Century*.

Pritchett, L. H. (1994), 'Desired Fertility and the Impact of Population Policies', *Population and Development Review*, 20(1), 1–55.

Property Rights Alliance. (2019), *International Property Rights Index 2019 2: Property Rights and Social Complexity*.

Puettmann, K. J., S. M. Wilson, S. C. Baker, P. J. Donoso, L. Drössler, G. Amente, B. D. Harvey, T. Knoke, Y. Lu, S. Nocentini, F. E. Putz, T. Yoshida, and J. Bauhus (2015), 'Silvicultural Alternatives to Conventional Even-aged Forest Management – What Limits Global Adoption?', *Forest Ecosystems*, 2, 8.

Purvis, A., T. Newbold, A. De Palma, S. Contu, S. L. L. Hill, K. Sanchez-Ortiz, H. R. P. Phillips, L. N. Hudson, I. Lysenko, L. Börger, and J. P. W. Scharlemann (2018), 'Modelling and Projecting the Response of Local Terrestrial Biodiversity Worldwide to Land Use and Related Pressures: The PREDICTS Project', *Advances in Ecological Research*, 58, 201–241.

Putnam, H. (2004), *The Collapse of the Fact/Value Dichotomy and Other Essays* (Cambridge, MA: Harvard University Press).

Putnam, R. D. – with R. Leonardi, and R. Y. Nanetti – (1993), *Making Democracy Work: Civic Traditions in Modern Italy* (Princeton, NJ: Princeton University Press).

Pyle, R. M. (1993), *The Thunder Tree: Lessons from an Urban Wildland* (Boston, MA: Houghton Mifflin)

Qaim, M (2009), 'The Economics of Genetically Modified Crops', *Annual Review of Resource Economics*, 1, 665–694.

Qaim, M., and G. Traxler (2005), 'Roundup Ready Soybeans in Argentina: Farm Level and Aggregate Welfare Effects', *Agricultural Economics*, 32(1), 73–86.

Qaim, M., and D. Zilberman (2003), 'Yield Effects of Genetically Modified Crops in Developing Countries', *Science*, 299(5608), 900–902.

Qin, S., R. E. Golden Kroner, C. Cook, A. T. Tesfaw, R. Braybrook, C. M. Rodriguez, C. Poelking, and M. B. Mascia (2019), 'Protected Area Downgrading, Downsizing, And Degazettement as a Threat to Iconic Protected Areas', *Conservation Biology*, 33(6), 1275–1285.

Rabiei, R., S. M. Phang, H. Y. Yeong, E. P. Lim, D. Ajdari, G. Zarshenas, and J. Sohrabipour (2014), 'Bioremediation Efficiency and Biochemical Composition of Ulva reticulata forsskal (Chlorophyta) Cultivated in Shrimp (Penaeus monodon) Hatchery Effluent', *Iranian Journal of Fisheries Sciences*, 13(3), 621–639.

Radner, R. (1963), *Notes on the Theory of Economic Planning* (Berkeley, CA :University of California).

Raiffa, H. (1968), *Decision Analysis: Introductory Lectures on Choices under Uncertainty* (Reading, MA: Addison-Wesley).

Ranger, N., Hallegatte, S., Bhattacharya, S., Bachu, M., Priya, S., Dhore, K., Rafique, F., Mathur, P., Naville, N., Henriet, F., Herweijer, C., Pohit, S., and J. Corfee-Morlot (2011), 'An assessment of the potential impact of climate change on flood risk in Mumbai', *Climatic Change*, 104(1), 139–67.

References

Ramankutty, N., A. T. Evan, C. Monfreda, and J. A. Foley (2008), 'Farming the Planet: 1. Geographic Distribution of Global Agricultural Lands in the Year 2000', *Global Biogeochemical Cycles*, 22(1), 1–19.

Ramankutty, N., and J. A. Foley (1999), 'Estimating Historical Changes in Global Land Cover: Croplands from 1700 to 1992', *Global Biogeochemical Cycles*, 13(4), 997–1027.

Ramankutty, N., Z. Mehrabi, K. Waha, L. Jarvis, C. Kremen, M. Herrero, and L. H. Rieseberg (2018), 'Trends in Global Agricultural Land Use: Implications for Environmental Health and Food Security', *Annual Review of Plant Biology*, 69, 1–27.

Ramsey, F. P. (1928), 'A Mathematical Theory of Saving', *The Economic Journal*, 38(152), 543–559.

Rasmussen, L. V., B. Coolsaet, A. Martin, O. Mertz, U. Pascual, E. Corbera, N. Dawson, J. A. Fisher, P. Franks, and C. M. Ryan (2018), 'Social-ecological Outcomes of Agricultural Intensification', *Nature Sustainability*, 1(6), 275–282.

Raven, P. H. (2020), 'Biological Extinction and Climate Change', in W. K. Al-Delaimy, V. Ramanathan, and M. Sánchez Sorondo, eds., *Health of People, Health of Planet and Our Responsibility* (Cham: Springer).

Raven, P. H., R. E. Gereau, P. B. Phillipson, C. Chatelain, C. N. Jenkins, and C. U. Ulloa (2020), 'The Distribution of Biodiversity Richness in the Tropics', *Science Advances*, 6, 5–10.

Rawls, J. (1972), *A Theory of Justice* (Oxford: Clarendon Press).

Raymond, C. M., S. Gottwald, J. Kuoppa, and M. Kyttä (2016), 'Integrating Multiple Elements of Environmental Justice into Urban Blue Space Planning Using Public Participation Geographic Information Systems', *Landscape and Urban Planning*, 153, 198–208.

Rees, M. J. (2003), *Our Final Century: A Scientist's Warning; How Terror, Error and Environmental Disaster Threaten Humankind's Future in this Century – on Earth and Beyond* (London: Heinemann).

Reganold, J. P., and J. M. Wachter (2016), 'Organic Agriculture in the Twenty-first Century', *Nature Plants*, 2(15221), 1–8.

Reguero, B. G., M. W. Beck, D. N. Bresch, J. Calil, and I. Meliane (2018), 'Comparing the Cost Effectiveness of Nature-Based and Coastal Adaptation: A Case Study from the Gulf Coast of the United States', *PLoS ONE*, 13(4), e0192132.

Reisch, L. A., and C. R. Sunstein (2016), 'Do Europeans Like Nudges?', *Judgment and Decision Making*, 11(4), 310–325.

Renard, D., and D. Tilman (2019), 'National Food Production Stabilized by Crop Diversity', *Nature*, 571, 257–260.

Reny, P. (2001), "Arrow's Theorem and the Gibbard-Satterthwaite Theorem: A Unified Approach," *Economics Letters*, 70(1), 99–105.

Repetto, R., W. Magrath, M. Wells, C. Beer, and F. Rossini (1989), *Wasting Assets: Natural Resources in the National Income Accounts* (Washington DC: World Resources Institute).

Research and Markets (2020), *Fast Fashion Global Market Report (2020-2030) – Covid 19 Growth and Change – Summary*.

Research Northwest, and Morrison Hershfield (2017), *Yukon 'State of Play': Analysis of Climate Change Impacts and Adaptation*.

Reynolds, M. D., B. L. Sullivan, E. Hallstein, S. Matsumoto, S. Kelling, M. Merrifield, D. Fink, A. Johnston, W. M. Hochachka, N. E. Bruns, M. E. Reiter, S. Veloz, C. Hickey, N. Elliott, L. Martin, J. W. Fitzpatrick, P. Spraycar, G. H. Golet, C. McColl, C. Low and S. A. Morrison (2017), 'Dynamic Conservation for Migratory Species', *Science Advances*, 3(8).

Ricardo, D. (1817), *On the Principles of Political Economy and Taxation* (London: John Murray).

Ricketts, T. H., G. C. Daily, P. R. Ehrlich, and C. D. Michener (2004), 'Economic Value of Tropical Forest to Coffee Production', *Proceedings of the National Academy of Sciences*, 101(34), 12579–12582.

Ricketts, T. H., J. Regetz, I. Steffan-Dewenter, S. A. Cunningham, C. Kremen, A. Bogdanski, B. Gemmill-Herren, S. S. Greenleaf, A. M. Klein, M. M. Mayfield, L. A. Morandin, A. Ochieng, and B. F. Viana (2008), 'Landscape Effects on Crop Pollination Services: Are there General Patterns?', *Ecology Letters*, 11(5), 499–515.

Ridley, M. (2010), *The Rational Optimist: How Prosperity Evolves* (London: Fourth Estate).

Rieb, J. T., R. Chaplin-Kramer, G. C. Daily, P. R. Armsworth, K. Böhning-Gaese, A. Bonn, G. S. Cumming, F. Eigenbrod, V. Grimm, B. M. Jackson, A. Marques, S. K. Pattanayak, H. M. Pereira, G. D. Peterson, T. H. Ricketts, B. E. Robinson, M. Schröter, L. A. Schulte, R. Seppelt, M. G. Turner, and E. M. Bennett (2017), 'When, Where, and How Nature Matters for Ecosystem Services: Challenges for the Next Generation of Ecosystem Service Models', *BioScience*, 67(9), 820–833.

Rights and Resources Initiative (2015), *Who Owns the World's Land? A Global Baseline of Formally Recognized Indigenous and Community Land Rights*. (Washington, DC: Rights and Resources Initiative).

Riker, W.H. (1986), *The Art of Political Manipulation* (New Haven, CT: Yale University Press).

Riley, J. C. (2005), 'Estimates of Regional and Global Life Expectancy, 1800-2001', *Population and Development Review*, 31(3), 537–543.

Riley, S. P. (1998), 'The Political Economy of Anti-Corruption Strategies in Africa', *The European Journal of Development Research*, 10(1), 129–159.

Ripple, W. J., C. Wolf, T. M. Newsome, M. G. Betts, G. Ceballos, F. Courchamp, M. W. Hayward, B. Van Valkenburgh, A. D. Wallach, and B. Worm (2019), 'Are We Eating the World's Megafauna to Extinction?', *Conservation Letters*, 12(3), e12627.

Rizal, G., S. Karki, V. Thakur, J. Chatterjee, R. A. Coe, S. Wanchana, and W. P. Quick (2012), 'Towards a C4 Rice', *Asian Journal of Cell Biology*, 7(2), 13–31.

Roberts, C. M., C. J. McClean, J. E. N. Veron, J. P. Hawkins, G. R. Allen, D. E. McAllister, C. G. Mittermeier, F. W. Schueler, M. Spalding, F. Wells, C. Vynne, and T. B. Werner (2002), 'Marine Biodiversity Hotspots and Conservation Priorities for Tropical Reefs', *Science*, 295(5558), 1280–1284.

Rockström, J., W. Steffen, K. Noone, Å. Persson, F. S. Chapin III, E. F. Lambin, T. M. Lenton, M. Scheffer, C. Folke, H. J. Schellnhuber, B. Nykvist, C. A. de Wit, T. Hughes, S. van der Leeuw, H. Rodhe, S. Sörlin, P. K. Snyder, R. Costanza, U. Svedin, M. Falkenmark, L. Karlberg, R. W. Corell, V. J. Fabry, J. Hansen, B. Walker, D. Liverman, K. Richardson, P. Crutzen, and J. A. Foley (2009), 'A Safe Operating Space for Humanity', *Nature*, 461(7263), 472–475.

Rogers, A. D., O. Aburto-Oropeza (2020), *Critical Habitats and Biodiversity: Inventory, Thresholds and Governance* (Washington, DC: World Resources Institute).

Rohitha, W. R. (2011), 'Evaluating Gains from De-Eutrophication of the Dutch Canal in Sri Lanka', in A. K.E. Haque, M.N. Murty, and P. Shyamsundar, eds., *Environmental Valuation in South Asia* (New Delhi: Cambridge University Press).

References

Romer, P. M. (1986), 'Increasing Returns and Long-Run Growth', *Journal of Political Economy*, 94(5), 1002–1037.

Rosa, E. A., and T. Dietz (2012). 'Human Drivers of National Greenhouse-Gas Emissions', *Nature Climate Change*, 2(8), 581–586.

Rose, J. M., S. B. Bricker, and J. G. Ferreira (2015), 'Comparative Analysis of Modeled Nitrogen Removal by Shellfish Farms', *Marine Pollution Bulletin*, 91(1), 185–190.

Rosenzweig, C., J. Elliott, D. Deryng, A. C. Ruane, C. Müller, A. Arneth, K. J. Boote, C. Folberth, M. Glotter, N. Khabarov, K. Neumann, F. Piontek, T. A. M. Pugh, E. Schmid, E. Stehfest, H. Yang, and J. W. Jones (2014), 'Assessing Agricultural Risks of Climate Change in the 21st Century in a Global Gridded Crop Model Intercomparison', *Proceedings of the National Academy of Sciences*, 111(9), 3268–3273.

Rosenzweig, C., A. Iglesias, X. B. Yang, P. R. Epstein, and E. Chivian (2001), 'Climate Change and Extreme Weather Events – Implications for Food Production, Plant Diseases, and Pests', *Global Change and Human Health*, 2(2), 90–104.

Rosenzweig, M. L. (1995), *Species Diversity in Space and Time* (Cambridge: Cambridge University Press).

Rothschild, M., and J. E. Stiglitz (1970), 'Increasing Risk: I. A Definition', *Journal of Economic Theory*, 2(3), 225–243.

Rounsevell, M. D. A., M. Harfoot, P. A. Harrison, T. Newbold, R. D. Gregory, and G. M. Mace (2020), 'A Biodiversity Target Based on Species Extinctions', *Science*, 368(6496), 1193–1195.

Rousseau, Y., R. A. Watson, J. L. Blanchard, and E. A. Fulton (2019), 'Evolution of Global Marine Fishing Fleets and The Response of Fished Resources', *Proceedings of the National Academy of Sciences*, 116(25), 12238–12243.

Royal Society (2009), *Reaping the Benefits: Science and the Sustainable Intensification of Global Agriculture* (London: Royal Society).

Royal Society for the Protection of Birds (2020), 'Wallasea Island Wild Coast Project', 'https://www.rspb.org.uk/our-work/our-positions-and-casework/casework/cases/wallasea-island/#qpqs3lHghgOr2z8Y.99'.

Rudgley, G., and N. Seega (2020), *Biodiversity Loss and Land Degradation: An Overview of the Financial Materiality*.

Rudra, A. (1982), *Indian Agricultural Economics: Myths and Realities* (New Delhi: Allied Publishers).

Rutt, C. L., V. Jirinec, M. Cohn-Haft, W. F. Laurance, and P. C. Stouffer (2019), 'Avian Ecological Succession in the Amazon: A Long-Term Case Study Following Experimental Deforestation', *Ecology and Evolution*, 9(24), 13850–13861.

Ryan-Collins, J. (2019), 'Beyond Voluntary Disclosure: Why a "Market – Shaping" Approach to Financial Regulation is Needed to Meet the Challenge of Climate Change', *SUERF (Société Universitaire Européenne de Recherches Financières) Policy Note*,61.

Ryder Jr., H. E., and G. M. Heal (1973), 'Optimal Growth with Intertemporally Dependent Preferences', *The Review of Economic Studies*, 40(1), 1–31.

Ryu, C., and Y. Kwon (2016), 'How Do Mega Projects Alter the City to Be More Sustainable? Spatial Changes Following the Seoul Cheonggyecheon Restoration Project in South Korea', *Sustainability*, 8(11), 1178.

Sachs, J. D., and A. M. Warner (1995), 'Natural Resource Abundance and Economic Growth', *National Bureau of Economic Research Working Paper Series*, No. 5398.

Sachs, J. D., and A. M. Warner (2001), 'The Curse of Natural Resources', *European Economic Review*, 45, 827–838.

Sahlins, M. (1972), *Stone Age Economics* (Chicago: Aldine-Atherton).

Sala, E., J. Lubchenco, K. Grorud-Colvert, C. Novelli, C. Roberts, and U. R. Sumaila (2018), 'Assessing Real Progress Towards Effective Ocean Protection', *Marine Policy*, 91, 11–13.

Salemdeeb, R., E. K. H. J. zu Ermgassen, M. H. Kim, A. Balmford, and A. Al-Tabbaa (2017), 'Environmental and Health Impacts of Using Food Waste as Animal Feed: A Comparative Analysis of Food Waste Management Options', *Journal of Cleaner Production*, 140(2), 871–880.

Salzman, J., G. Bennett, N. Carroll, A. Goldstein, and M. Jenkins (2018), 'The Global Status and Trends of Payments for Ecosystem Services', *Nature Sustainability*, 1(3), 136–144.

Samraj, P., V. N. Sharda, S. Chinnamani, V. Lakshmanan, and B. Haldorai (1988), 'Hydrological Behaviour of the Nilgiri Sub-Watersheds as Affected by Bluegum Plantations, Part I. The Annual Water Balance', *Journal of Hydrology*, 103(3-4), 335–345.

Samuelson, P. A. (1954), 'The Pure Theory of Public Expenditure', *The Review of Economics and Statistics*, 36(4), 387–389.

Samuelson, P. A. (1961), 'The Evaluation of "Social Income": Capital Formation and Wealth', in D. C. Hague ed., *The Theory of Capital*, (London: Macmillan).

SANBI. (2021), *Grasslands Programme* 'https://www.sanbi.org/biodiversity/science-into-policy-action/mainstreaming-biodiversity/grasslands-programme/'.

Sanderman, J., T. Hengl, and G. J. Fiske (2017), 'Soil Carbon Debt of 12,000 Years of Human Land Use', *Proceedings of the National Academy of Sciences*, 114(36), 9575–9580.

Santos, R., C. Schröter-Schlaack, P. Antunes, I. Ring, and P. Clemente (2015), 'Reviewing the Role of Habitat Banking and Tradable Development Rights in the Conservation Policy Mix', *Environmental Conservation*, 42(4), 294–305.

Satterthwaite, M.A. (1975), "Strategy-Proofness and Arrow's Conditions: Existence and Correspondence Theorems for Voting Procedures and Social Welfare Functions," *Journal of Economic Theory*, 10(2), 187–217.

Savage, L. J. (1954), *The Foundations of Statistics* (New York, NY: John Wiley and Sons).

Scanlon, T. M., K. K. Caylor, S. A. Levin, and I. Rodriguez-Iturbe (2007), 'Positive Feedbacks Promote Power-Law Clustering of Kalahari Vegetation', *Nature*, 449(7159), 209–212.

Scarborough, P., A. Matthews, H. Eyles, A. Kaur, C. Hodgkins, M. M. Raats, and M. Rayner (2015), 'Reds are More Important Than Greens: How UK Supermarket Shoppers Use the Different Information on a Traffic Light Nutrition Label in a Choice Experiment', *International Journal of Behavioral Nutrition and Physical Activity*, 12, 151.

Schaafsma, M., and B. Fisher (2016), 'What are the Links Between Poverty and Ecosystem Services?', In M. Potschin, R. Haines-Young, R. Fish, and R. K. Turner, eds., *Routledge Handbook of Ecosystem Services*, (New York, NY: Routledge).

Schaafsma, M., S. Morse-Jones, P. Posen, R. D. Swetnam, A. Balmford, I. J. Bateman, N. D. Burgess, S. A. O. Chamshama, B. Fisher, T. Freeman, V. Geofrey, R. E. Green, A. S. Hepelwa,

References

A. Hernández-Sirvent, S. Hess, G. C. Kajembe, G. Kayharara, M. Kilonzo, K. Kulindwa, J.F. Lund, S. S. Madoffe, L. Mbwambo, H. Meilby, Y. M. Ngaga, I. Theilade, T. Treue, P. van Beukering, V. G. Vyamana, and R. K. Turner (2014), 'The Importance of Local Forest Benefits: Economic Valuation of Non-Timber Forest Products in the Eastern Arc Mountains in Tanzania', *Global Environmental Change*, 24(1), 295–305.

Schama, S. (1987), *The Embarrassment of Riches: An Interpretation of Dutch Culture in the Golden Age* (London: Collins).

Schama, S. (1995), *Landscape and Memory* (London: Harper Collins).

Scheffer, M. (2009), *Critical Transitions in Nature and Society* (Princeton, NJ: Princeton University Press).

Scheffer, M. (2016), 'Anticipating Societal Collapse; Hints from the Stone Age', *Proceedings of the National Academy of Sciences*, 113(39), 10733–10735.

Scheffer, M., and S. R. Carpenter (2003), 'Catastrophic Regime Shifts in Ecosystems: Linking Theory to Observation', *Trends in Ecology and Evolution*, 18(12), 648–656.

Scheffer, M., S. R. Carpenter, V. Dakos, and E. H. van Nes (2015), 'Generic Indicators of Ecological Resilience: Inferring the Chance of a Critical Transition', *Annual Review of Ecology, Evolution, and Systematics*, 46(1), 145–167.

Scheffer, M., S. R. Carpenter, T. M. Lenton, J. Bascompte, W. Brock, V. Dakos, J. Van De Koppel, I. A. van de Leemput, S. A. Levin, E. H. van Nes, M. Pascual, and J. Vandermeer (2012), 'Anticipating Critical Transitions', *Science*, 338(6105), 344–348.

Scheffler, S. (2013), *Death & the Afterlife* (Oxford: Oxford University Press).

Schell, J. (1982), *The Fate of the Earth* (New York: Avon).

Schellekens, G., and J. van Toor (2019), *Values at Risk? Sustainability Risks and Goals in The Dutch Financial Sector*.

Schipper, A., J. Hilbers, J. Meijer, L. Antão, A. Benítez-López, M. De Jonge, L. Leemans, E. Scheper, R. Alkemade, J. Doelman, S. Mylius, E. Stehfest, D. Van Vuuren, W. Van Zeist, and M. Huijbregts (2019), 'Projecting Terrestrial Biodiversity Intactness with GLOBIO 4', *Global Change Biology*, 26(2), 760–771.

Schipper, A., M. Bakkenes, J. Meijer, R. Alkemade, and M. Huijbregts (2016), *The GLOBIO Model. A Technical Description of Version 3.5*.

Schoenmaker, D., and G. Zachmann (2015), *Can a Global Climate Risk Pool Help the Most Vulnerable Countries?*

Schoenmaker, D., and W. Schramade (2018), *Principles of Sustainable Finance* (Oxford: Oxford University Press).

Scholes, R. J., and R. Biggs (2005), 'A Biodiversity Intactness Index', *Nature*, 434, 45–49.

Schroder, T., J. Stank, G. Schernewski, and P. Krost (2014), 'The Impact of a Mussel Farm on Water Transparency in the Kiel Fjord', *Ocean and Coastal Management*, 101(A), 42–52.

Schultz, T. W. (1961), 'Investment in Human Capital', *The American Economic Review*, 51(1), 1–17.

Schulze, K., Ž. Malek, and P. H. Verburg (2019), 'Towards Better Mapping of Forest Management Patterns: A Global Allocation Approach', *Forest Ecology and Management*, 432, 776–785.

Schüwer, U., C. Lambert, and F. Noth (2019), 'How do Banks React to Catastrophic Events? Evidence from Hurricane Katrina', *Review of Finance*, 23(1), 75–116.

Schwindt, A. R., D. L. Winkelman, K. Keteles, M. Murphy, and A. M. Vajda (2014), 'An Environmental Oestrogen Disrupts Fish Population Dynamics through Direct and Transgenerational Effects on Survival and Fecundity', *Journal of Applied Ecology*, 51(3), 582–591.

SCM Direct. (2019), *Greenwashing: Misclassification and Mis-selling of Ethical Investments*.

Seabright, P. (1997), 'Is Co-operation Habit-Forming?', in P. Dasgupta and K. -G. Mäler, eds., *The Environment and Emerging Development Issues, Vol. 2*, (Oxford: Oxford University Press).

Seddon, N., A. Chausson, P. Berry, C. A. J. Girardin, A. Smith, and B. Turner (2020), 'Understanding the Value and Limits of Nature-Based Solutions to Climate Change and Other Global Challenges', *Philosophical Transactions of the Royal Society B*, 375(1794), 20190120.

Seebens, H., T. M. Blackburn, E. E. Dyer, P. Genovesi, P. E. Hulme, J. M. Jeschke, S. Pagad, P. Pyšek, M. Winter, M. Arianoutsou, S. Bacher, B. Blasius, G. Brundu, C. Capinha, L. Celesti- Grapow, W. Dawson, S. Dullinger, N. Fuentes, H. Jäger, J. Kartesz, M. Kenis, H. Kreft, I. Kühn, B. Lenzner, A. Liebhold, A. Mosena, D. Moser, M. Nishino, D. Pearman, J. Pergl, W. Rabitsch, J. Rojas-Sandoval, A. Roques, S. Rorke, S. Rossinelli, H. E. Roy, R. Scalera, S. Schindler, K. Štajerová, B. Tokarska-Guzik, M. van Kleunen, K. Walker, P. Weigelt, T. Yamanaka, and F. Essl (2017), 'No Saturation in the Accumulation of Alien Species Worldwide', *Nature Communications*, 8, 14435.

Seidl, A., K. Mulungu, M. Arlaud, O. van den Heuvel, and M. Riva, (2020), 'Finance for Nature: A Global Estimate of National Biodiversity Investments', *Ecosystem Services*, 46, 101216.

Selig, E. R., D. G. Hole, E. H. Allison, K. K. Arkema, M. C. McKinnon, J. Chu, A. de Sherbinin, B. Fisher, L. Glew, M. B. Holland, J. C. Ingram, N. S. Rao, R. B. Russell, T. Srebotnjak, L. C. L. Teh, S. Troëng, W. R. Turner, and A. Zvoleff (2019), 'Mapping Global Human Dependence on Marine Ecosystems', *Conservation Letters*, 12(2), e12617.

Sen, A. (1966), "A Possibility Theorem on Majority Decisions," *Econometrica*, 34(2), 491–499.

Sen, A. (1970), *Collective Choice and Social Welfare* (San Francisco, CA: Holden-Day).

Sen, A. (1976), 'Real National Income', *The Review of Economic Studies*, 43(1), 19–39.

Sen, A. (1982), 'Approaches to the Choice of Discount Rates for Social Cost Benefit Analysis', in R. Lind, ed., *Discounting for Time and Risk in Energy Policy* (Washington, DC: Resources for the Future).

Sen, A., and B. Williams (1982), 'Introduction: Utilitarianism and Beyond', in A. Sen and B. Williams, eds., *Utilitarianism and Beyond* (Cambridge: Cambridge University Press).

Seto, K. C., R. Sánchez-Rodríguez, and M. Fragkias (2010), 'The New Geography of Contemporary Urbanization and the Environment', *Annual Review of Environment and Resources*, 35, 167–194.

Seufert, V., N. Ramankutty, and J. A. Foley (2012), 'Comparing the Yields of Organic and Conventional Agriculture', *Nature*, 485(7397), 229–232.

Shackelford, S. J. (2009), 'The Tragedy of the Common Heritage of Mankind', *Stanford Environmental Law Journal*, 27, 101–157.

Shanmugam, K. R., and S. Madheswaran (2011), 'The Value of Statistical Life', *Environmental Valuation in South Asia* (Cambridge: Cambridge University Press).

Shantz, A. A., and J. Sweatman (2015). 'Is a Community Still A Community? Reviewing Definitions of Key Terms in Community Ecology', *Ecology and Evolution*, 5(21), 4757–4765.

ShareAction. (2020), *Point of No Returns*.

References

Sharp, R., R. Chaplin-Kramer, S. Wood, A. Guerry, H. Tallis, T. Ricketts, E. Nelson, D. Ennaanay, S. Wolny, N. Olwero, K. Vigerstol, D. Pennington, G. Mendoza, J. Aukema, J. Foster, J. Forrest, D. Cameron, K. Arkema, E. Lonsdorf, C. Kennedy, G. Verutes, , C.-K. Kim, G. Guannel, M. Papenfus, J. Toft, M. Marsik, J. Bernhardt, R. Griffin, K. Glowinski, N. Chaumont, A. Perelman, M. Lacayo, L. Mandle, P. Hamel, A. L. Vogl, L. Rogers, W. Bierbower, D. Denu, and J. Douglass (2018), *InVEST User's Guide*.

Shiau, J.-T., S. Feng, and S. Nadarajah (2007), 'Assessment of Hydrological Droughts for the Yellow River, China, Using Copulas', *Hydrological Processes*, 21(16), 2157–2163.

Shove, G. F. (1942), 'The Place of Marshall's Principles in the Development of Economic Theory', *The Economic Journal*, 52(208), 294–329.

Shyamsundar, P. (2008), 'Decentralization, Devolution, and Collective Action—A Review of International Experience 1', in R. Ghate, N. Jodha, P. Mukhopadhyay, eds., *Promise, Trust and Evolution: Managing the Commons of South Asia* (Oxford: Oxford University Press).

Sidgwick, H. (1907), *The Methods of Ethics*. (London: Macmillan) 7th Edition.

Singer, M. C., and C. S. McBride (2012), 'Geographic Mosaics of Species' Association: A Definition and an Example Driven by Plant-Insect Phenological Synchrony', *Ecology*, 93(12), 2658–2673.

Singer, P. (1975), *Animal Liberation* (New York, NY: Random House).

Skamnioti, P., and S. J. Gurr (2009), 'Against the Grain: Safeguarding Rice from Rice Blast Disease', *Trends in Biotechnology*, 27(3), 141–150.

Smith, W. K., E. Nelson, J. A. Johnson, S. Polasky, J. C. Milder, J. S. Gerber, P. C. West, S. Siebert, K. A. Brauman, K. M. Carlson, M. Arbuthnot, J. P. Rozza, and D. N. Pennington (2019), 'Voluntary Sustainability Standards Could Significantly Reduce Detrimental Impacts of Global Agriculture', *Proceedings of the National Academy of Sciences*, 116(6), 2130–2137.

Sodhi, N. S., B. W. Brook, and C. J. A. Bradshaw (2009), 'Causes and Consequences of Species Extinctions', in S. A. Levin, ed., *The Princeton Guide to Ecology*, (Princeton, NJ: Princeton University Press).

Solórzano, R., R. de Camino, R. Woodward, J. Tosi, V. Watson, A. Vasquez, C. Villalobos, J. Jimenez, R. Repetto, and W. Cruz (1991), *Accounts Overdue: Natural Resource Depreciation in Costa Rica* (Washington, DC: World Resources Institute).

Solow, R. M. (1956), 'A Contribution to the Theory of Economic Growth', *The Quarterly Journal of Economics*, 70(1), 65–94.

Solow, R. M. (1957), 'Technical Change and the Aggregate Production Function', *The Review of Economics and Statistics*, 39(3), 312–320.

Solow, R. M. (1963), *Capital Theory and the Rate of Return* (Amsterdam: North Holland).

Solow, R. M. (1974a), 'Intergenerational Equity and Exhaustible Resources', *The Review of Economic Studies*, 41, 29–45.

Solow, R. M. (1974b), 'The Economics of Resources or the Resources of Economics', *The American Economic Review*, 64(2), 1–14.

Solow, R. M. (2000), 'Notes on Social Capital and Economic Performance' in P. Dasgupta, and I. Serageldin, eds., *Social Capita: A Multifaceted Perspective*, (Washington, DC: The World Bank).

Somanathan, E. (1991), 'Deforestation, Property Rights and Incentives in Central Himalaya', *Economic and Political Weekly*, 26(4), 37–46.

Somanathan, E., R. Prabhakar, and B. S. Mehta (2005), 'Does Decentralization Work? Forest Conservation in the Himalayas', *Indian Statistical Institute Discussion Paper 05-04*.

Somanathan, E., R. Prabhakar, and B. S. Mehta (2009). 'Decentralization for cost-effective conservation', *Proceedings of the National Academy of Sciences*, 106(11), 4143–4147.

Sommer, J. M., M. Restivo, and J. M. Shandra (2019), 'The United States, Bilateral Debt-for- Nature Swaps, and Forest Loss: A Cross-National Analysis', *The Journal of Development Studies*, 56(4), 748–764.

Song, X., C. Peng, G. Zhou, H. Jiang, and W. Wang (2014), 'Chinese Grain for Green Program Led to Highly Increased Soil Organic Carbon Levels: A Meta-Analysis', *Scientific Reports*, 4, 4460.

Sork, V. L., S. T. Fitz-Gibbon, D. Puiu, M. Crepeau, P. F. Gugger, R. Sherman, K. Stevens, C. H. Langley, M. Pellegrini, and S. L. Salzberg (2016). 'First Draft Assembly and Annotation of The Genome of a California Endemic Oak Quercus lobata née (Fagaceae)', *G3: Genes, Genomes, Genetics*, 6(11), 3485–3495.

Soury, A. (2007), *Sacred Forests: A Sustainable Conservation Strategy? The Case of Sacred Forests in The Ouémé Valley, Benin*.

Spalding, M. D., A. L. McIvor, M. W. Beck, E. W. Koch, I. Möller, D. J. Reed, P. Rubinoff, T. Spencer, T. J. Tolhurst, T. V. Wamsley, B. K. van Wesenbeeck, E. Wolanski, and C. D. Woodroffe (2014), 'Coastal Ecosystems: A Critical Element of Risk Reduction', *Conservation Letters*, 7(3), 293–301.

Sparkman, G., and G. M. Walton (2017), 'Dynamic Norms Promote Sustainable Behavior, Even if it is Counternormative', *Psychological Science*, 28(11), 1663–1674.

Spatial Finance Initiative (2020), '*Spatial Finance*', 'https://spatialfinanceinitiative.com/'.

Spence, A. M. (1974), 'Blue Whales and Applied Control Theory' in H. Gottinger, ed., *Systems Approaches and Environmental Problems* (Gottingen: Vandenhoeck and Ruprecht).

Spence, A. M., and D. Starrett (1975), 'Most Rapid Approach Paths in Accumulation Problems', *International Economic Review*, 16(2), 388–403.

Standards and Trade Development Facility (2013), *International Trade and Invasive Alien Species*.

Starrett, D. A. (1972), 'Fundamental Nonconvexities in the Theory of Externalities', *Journal of Economic Theory*, 4(2), 180–199.

State Office for the Environment (LfU) (2020), 'Spreewald Biosphere Reserve', 'https://www.spreewald-biosphaerenreservat.de/en/'.

Statista (2020), 'Consumption of Vegetable Oils Worldwide from 2013/14 to 2019/2020, by Oil Type', 'www.statista.com.stanford.idm.oclc.org/statistics/263937/vegetable-oils-global-consumption/'.

Staver, A. C., S. Archibald, and S. A. Levin (2011), 'Tree Cover in Sub-Saharan Africa: Rainfall and Fire Constrain Forest and Savanna as Alternative Stable States', *Ecology*, 92(5), 1063–1072.

Stavins, R., and S. Barrett (2002), 'Increasing Participation and Compliance in International Climate Change Agreements', *International Environmental Agreements*, 3(4), 349–376.

Steffen, W., W. Broadgate, L. Deutsch, O. Gaffney, and C. Ludwig (2015b), 'The Trajectory of The Anthropocene: The Great Acceleration', *Anthropocene Review*, 2(1), 81–98.

Steffen, W., J. Grinevald, P. Crutzen, and J. McNeill (2011), '*The Anthropocene: Conceptual and Historical Perspectives*', *Philosophical Transactions of the Royal Society A*, 369(1938), 842–867.

References

Steffen, W., K. Richardson, J. Rockström, S. E. Cornell, I. Fetzer, E. M. Bennett, R. Biggs, S. R. Carpenter, W. de Vries, C. A. de Wit, C. Folke, D. Gerten, J. Heinke, G. M. Mace, L. M. Persson, V. Ramanathan, B. Reyers, and S. Sörlin (2015), 'Planetary Boundaries: Guiding Human Development on a Changing Planet', *Science*, 347(6223), 1259855.

Steffen, W., J. Rockström, K. Richardson, T. M. Lenton, C. Folke, D. Liverman, C. P. Summerhayes, A. D. Barnosky, S. E. Cornell, M. Crucifix, J. F. Donges, I. Fetzer, S. J. Lade, M. Scheffer, R. Winkelmann, and H. J. Schellnhuber (2018), 'Trajectories of the Earth System in the Anthropocene', *Proceedings of the National Academy of Sciences*, 115(33), 8252–8259.

Steffen, W., R. A. Sanderson, P. D. Tyson, J. Jager, P. A. Matson, B. Moore III, F. Oldfield, K. Richardson, H. J. Schulnhuber, B. L. Turner, and R. J. Wasson (2005), *Global Change and the Earth System: A Planet Under Pressure* (Berlin Heidelburg: Springer).

Steinberger, J. K., F. Krausmann, and N. Eisenmenger (2010), 'Global Patterns of Materials Use: A Socioeconomic and Geophysical Analysis', *Ecological Economics*, 69(5), 1148–1158.

Stephenson, J., F. Berkes, N. J. Turner, and J. Dick (2014), 'Biocultural Conservation of Marine Ecosystems: Examples from New Zealand and Canada', *Indian Journal of Traditional Knowledge*, 13(2), 257–265.

Stern, N. (2006), *The Economics of Climate Change: The Stern Review* (Cambridge: Cambridge University Press).

Stern, N. (2015), *Why are we Waiting? The Logic, Urgency and Promise of Tackling Climate Change* (Cambridge, MA: MIT Press).

Stern, P. C. (2014), 'Individual and Household Interactions with Energy Systems: Toward Integrated Understanding', *Energy Research and Social Science*, 1, 41–48.

Stevenson, B., and J. Wolfers (2013), 'Subjective Well-Being and Income: Is There Any Evidence of Satiation?', *American Economic Review*, 103(3), 598–604.

Stewart, N. (2008), 'Norway Excludes Rio Tinto over Environmental Damage', *IPE*, 'https://www.ipe.com/norway-excludes-rio-tinto-over-environmental-damage/29077.article'.

Stiglitz, J. E. (1974), 'Growth with Exhaustible Natural Resources: Efficient and Optimal Growth Paths', *The Review of Economic Studies*, 41, 123–137.

Stiglitz, J. E. (1989), 'Markets, Market Failures and Development', *The American Economic Review*, 79(2), 197–203.

Stiglitz, J. E., A. Sen, and J.-P. Fitoussi (2009), *Report by the Commission on the Measurement of Economic Performance and Social Progress*.

Stigsdotter, U. K., S. S. Corazon, U. Sidenius, P. K. Nyed, H. B. Larsen, and L. O. Fjorback (2018), 'Efficacy of Nature-Based Therapy for Individuals with Stress-Related Illnesses: Randomised Controlled Trial', *British Journal of Psychiatry*, 213(1), 404–411.

Stoett, P., J. Davies, D. Armenteras, J. Hills, L. McRae, C. Zastavniouk, R. Bailey, C. Butler, I. Dankelman, K. Garcia, L. Godfrey, A. Kirilenko, P. Lemke, D. Liggett, G. Mudd, J. Seager, C.Y. Wright and C. Zickgraf. (2019), 'Chapter 6: Biodiversity', in P Ekins, J. Gupta, and P Boileau eds., *Global Environment Outlook – GEO-6: Healthy Planet, Healthy People* (Cambridge: Cambridge University Press).

Stoll, J. S., E. Fuller, and B. I. Crona (2017), 'Uneven Adaptive Capacity Among Fishers in a Sea of Change', *PLoS ONE*, 12(6), 1–13.

Stone, G. D., Netting, R. McC., and M. P. Stone (1990) 'Seasonality, Labor Scheduling, and Agricultural Intensification in the Nigerian Savanna', *American Anthropologist* 92(1), 7–23.

Stork, N. E. (2018), 'How Many Species of Insects and Other Terrestrial Arthropods Are There on Earth?', *Annual Review of Entomology*, 63(1), 31–45.

Strassburg, B. B. N., A. E. Latawiec, L. G. Barioni, C. A. Nobre, V. P. da Silva, J. F. Valentim, M. Vianna, and E. D. Assad (2014), 'When Enough Should be Enough: Improving the Use of Current Agricultural Lands Could Meet Production Demands and Spare Natural Habitats in Brazil', *Global Environmental Change*, 28(1), 84–97.

Strogatz, S. H. (1994), *Non-linear Dynamics and Chaos* (Reading, MA: Perseus Books).

Strogatz, S. H. (2004), *Sync: The Emerging Science of Spontaneous Order* (London: Penguin).

Strogatz, S. H. (2015), *Non-Linear Dynamics and Chaos: With Applications to Physics, Biology, Chemistry, and Engineering* (Cambridge, MA: Westview Press) 2nd Edition.

Stuart-Smith, S. (2020), *The Well-Gardened Mind: The Restorative Power of Nature*, (New York, NY: Schribner).

Subramanian, A., and M. Qaim (2010), 'The Impact of Bt Cotton on Poor Households in Rural India', *The Journal of Development Studies*, 46(2), 295–311.

Suhrke, A., and S. Hazarika (1993), *Pressure Points: Environmental Degradation, Migration and Conflict* (Cambridge, MA: American Academy of Arts and Sciences).

Sumaila, U. R., N. Ebrahim, A. Schuhbauer, D. Skerritt, Y. Li, H. S. Kim, T. G. Mallory, V. W. L. Lam, and D. Pauly (2019), 'Updated Estimates and Analysis of Global Fisheries Subsidies', *Marine Policy*, 109, 103695.

Sun, F., and R. T. Carson (2020), 'Coastal Wetlands Reduce Property Damage During Tropical Cyclones', *Proceedings of the National Academy of Sciences*, 117(11), 5719–5725.

Sundström, A. (2016), 'Understanding Illegality and Corruption in Forest Management: A Literature Review'. *QoG Working Paper Series* 2016, No. 1. (Gothenburg: University of Gothenburg).

Sunstein, C. R., and E. Ullmann-Margalit (2001), 'Solidarity Goods', *The Journal of Political Philosophy*, 9(2), 129–149.

Sushinsky, J. R., J. R. Rhodes, H. P. Possingham, T. K. Gill, and R. A. Fuller (2013), 'How Should we Grow Cities to Minimize Their Biodiversity Impacts?', *Global Change Biology*, 19(2), 401–410.

Suttor-Sorel, L. (2019), *Making Finance Serve Nature*.

Suttor-Sorel, L., and N. Hercelin (2020), *Nature's Return: Embedding Environmental Goals at the Heart of Economic and Financial Decision-making*.

Swann, A. L., I. Y. Fung, S. Levis, G. B. Bonan, and S. C. Doney (2010), 'Changes in Arctic Vegetation Amplify High-Latitude Warming Through the Greenhouse Effect', *Proceedings of the National Academy of Sciences*, 107(4), 1295–1300.

Taillie, L. S., M. Reyes, M. A. Colchero, B. Popkin, and C. Corvalán (2020), 'An Evaluation of Chile's Law of Food Labeling and Advertising on Sugar-Sweetened Beverage Purchases from 2015 to 2017: A Before-and-After Study', *PLoS Medicine*, 17(2), e1003015.

Tallis, H., S. Pagiola, W. Zhang, S. Shaikh, E. Nelson, C. Stanton, and P. Shyamsundar (2011), 'Poverty and the Distribution of Ecosystem Services' in P. Kareiva, H. Tallis, T. H. Ricketts, G. C. Daily and

References

S. Polasky, eds., *Natural Capital: Theory and Practice of Mapping Ecosystem Services* (Oxford: Oxford University Press).

Tan-Soo, J.-S., N. Adnan, I. Ahmad, S. K. Pattanayak, and J. R. Vincent (2016), 'Econometric Evidence on Forest Ecosystem Services: Deforestation and Flooding in Malaysia', *Environmental and Resource Economics*, 63, 25–44.

Task Force on Climate-related Financial Disclosures (2017), *Recommendations of the Task Force on Climate-related Financial Disclosures*.

Task Force on Climate-related Financial Disclosures (2020), 'Task Force on Climate-related Financial Disclosures: About', 'https://www.fsb-tcfd.org/about/'.

Task Force on Nature-related Financial Disclosures (2020), *Bringing Together a Task Force on Nature-related Financial Disclosures* 'https://tnfd.info/'.

Tchakatumba, P. K., E. Gandiwa, E. Mwakiwa, B. Clegg, and S. Nyasha (2019), 'Does the CAMPFIRE Programme Ensure Economic Benefits from Wildlife to Households in Zimbabwe?', *Ecosystems and People*, 15(1), 119–135.

Teixidó-Figueras, J., J. K. Steinberger, F. Krausmann, H. Haberl, T. Wiedmann, G. P. Peters, J. A. Duro, and T. Kastner (2016), 'International Inequality of Environmental Pressures: Decomposition and Comparative Analysis', *Ecological Indicators*, 62, 163–173.

Temin, P. (2013), *The Roman Market Economy* (Princeton, NJ: Princeton University Press).

ten Kate, K., A. von Hase, and P. Maguire (2018), 'Principles of the Business and Biodiversity Offsets Programme', in W. Wende, G. -M. Tucker, F. Quétier, M. Rayment, M. Darbi, eds., *Biodiversity Offsets* (Cham: Springer).

Thaler, R. H., and C. R. Sunstein (2008), *Nudge: Improving Decisions About Health, Wealth, and Happiness* (New Haven, CT: Yale University Press).

The Food and Land Use Coalition (2019), *Growing Better: Ten Critical Transitions to Transform Food and Land Use*.

The Institution of Agricultural Engineers (IAgrE) (2012), *Agricultural Engineering: A Key Discipline for Agriculture to Deliver Global Food Security*.

The Nature Conservancy (2018), 'Migration Moneyball' 'https://www.nature.org/en-us/about-us/where-we-work/united-states/california/stories-in-california/migration-moneyball/'.

The Nature Conservancy (2019), 'Insuring Nature to Ensure a Resilient Future', 'https://www.nature.org/en-us/what-we-do/our-insights/perspectives/insuring-nature-to-ensure-a-resilient-future/'. The Patents Act (1977).

The Sustainable Finance Platform (2020), *Biodiversity Opportunities and Risks for the Financial Sector*.

Theuerkauf, S. J., J. A. Morris Jr., T. J. Waters, L. C. Wickliffe, H. K. Alleway, and R. C. Jones (2019), 'A Global Spatial Analysis Reveals Where Marine Aquaculture Can Benefit Nature and People', *PLoS ONE*, 14(10), e0222282.

Thomä, J., C. Weber, S. Dupré, and M. Naqvi (2015), *The Long-Term Risk Signal Valley of Death: Exploring the Tragedy of the Horizon*.

Thomas, R., and N. Dimsdale (2017), 'A Millennium of Macroeconomic Data for the UK', *Bank of England OBRA Dataset*.

Thomson, J. A. K., ed. (1976), *The Ethics of Aristotle: The Nicomachean Ethics* (Harmondsworth, NY: Penguin).

Thomson, J. T., D. Feeny, and R. J. Oakerson (1986), 'Institutional Dynamics: The Evolution and Dissolution of Common-Property Resource Management' in National Research Council, *Proceedings of the Conference on Common Property Resource Management*. (Washington, DC: National Academy Press).

Thorn, J. P. R., R. Friedman, D. Benz, K. J. Willis, and G. Petrokofsky (2016), 'What Evidence Exists for the Effectiveness of On-Farm Conservation Land Management Strategies for Preserving Ecosystem Services in Developing Countries? A Systematic Map', *Environmental Evidence*, 5, 13.

Thurm, R., B. Baue, and C. van der Lugt (2018), *Blueprint 5: The Transformation Journey A Step- By-Step Approach to Organizational Thriveability and System Value Creation*.

Tietenberg, T. (2003), 'The Tradable-Permits Approach to Protecting the Commons: Lessons for Climate Change', *Oxford Review of Economic Policy*, 19(3), 400–419.

Tilburt, J.C, and T. J. Kaptchuk (2008), 'Herbal medicine research and global health: an ethical analysis', *Bulletin of the World Health Organization*, 86(8), 594–599.

Tilman, D. (1997), 'Biodiversity and Ecosystem Functioning', in G. C. Daily, ed., *Nature's Services: Societal Dependence on Natural Ecosystems* (Washington, DC: Island Press).

Tilman, D., K. G. Cassman, P. A. Matson, R. Naylor, and S. Polasky (2002), 'Agricultural Sustainability and Intensive Production Practices', *Nature*, 418, 671–677.

Tilman, D., and M. Clark (2014), 'Global Diets Link Environmental Sustainability and Human Health', *Nature*, 515(7528), 518–522.

Tilman, D., and J. A. Downing (1994), 'Biodiversity and Stability in Grasslands', *Nature*, 367, 363–365.

Tilman, D., J. Fargione, B. Wolff, C. D'Antonio, A. Dobson, R. Howarth, D. Schindler, W. H. Schlesinger, D. Simberloff, and D. Swackhamer (2001), 'Forecasting Agriculturally Driven Global Environmental Change', *Science*, 292(5515), 281–284.

Tilman, D., F. Isbell, and J. M. Cowles (2014), 'Biodiversity and Ecosystem Functioning', *Annual Review of Ecology, Evolution, and Systematics*, 45, 471–493.

Tilman, D., J. Knops, D. Wedin, P. Reich, M. Ritchie, and E. Siemann (1997), 'The Influence of Functional Diversity and Composition on Ecosystem Processes', *Science*, 277(5330), 1300–1302.

Tilman, D., R. M. May, C. L. Lehman, and M. A. Nowak (1994), 'Habitat Destruction and the Extinction Debt', *Nature*, 371(6492), 65–66.

Tilman, D., P. B. Reich, and J. M. H. Knops (2006), 'Biodiversity and Ecosystem Stability in a Decade-Long Grassland Experiment', *Nature*, 441, 629–632.

Tol, R. S. J. (2008), 'The Social Cost of Carbon: Trends, Outliers and Catastrophes', *Economics: The Open-Access, Open-Assessment E-Journal*, 2, 2008-25.

Tomlinson, J. (2018), 'Nigerian Briefing: How Engineers Can Help Secure a Sustainable Economy', *Proceedings of Institution of Civil Engineers–Energy*, 171(3), 121–128.

Traeger, C. P. (2011), 'Sustainability, Limited Substitutability, and Non-Constant Social Discount Rates', *Journal of Environmental Economics and Management*, 62(2), 215–228.

TRAFFIC (2017), *Briefing Paper: World Rhino Day 2017*.

References

Transparency International (2019), 'Corruption Perceptions Index', 'https://www.transparency.org/en/cpi/2019/index/'.

Transparency International (n.d.), 'Petty Corruption – Corruptionary A-Z', 'https://www.transparency.org/en/corruptionary/petty-corruption'.

TRASE and Forest 500 (2019), *Eliminating Deforestation from Supply Chains by 2020: A Review of the Amsterdam Declaration Countries*.

Tree, I. (2017), 'The Knepp Wildland Project', *Biodiversity*, 18(4), 206–209.

Tree, I. (2018), *Wilding: The Return of Nature to an English Farm* (London: Picador – Pan Macmillan).

Trentmann, F., ed. (2012), *The Oxford Handbook of the History of Consumption* (Oxford: Oxford University Press).

Trentmann, F. (2016), *Empire of Things: How We Became a World of Consumers, from the Fifteenth Century to the Twenty-First* (London: Allen Lane).

Trigo, E. (2011), *Fifteen Years of Genetically Modified Crops in Argentine Agriculture*.

Trumbore, S., P. Brando, and H. Hartmann (2015), 'Forest Health and Global Change', *Science*, 349(6250), 814–818.

Tscharntke, T., J. C. Milder, G. Schroth, Y. Clough, F. Declerck, A. Waldron, R. Rice, and J. Ghazoul (2015), 'Conserving Biodiversity Through Certification of Tropical Agroforestry Crops at Local and Landscape Scales', *Conservation Letters*, 8(1), 14–23.

Tuomisto, H. L., and M. J. T. de Mattos (2011), 'Environmental Impacts of Cultured Meat Production', *Environmental Science & Technology*, 45(14), 6117–6123.

Tuomisto, H. L., I. D. Hodge, P. Riordan, and D. W. Macdonald (2012), 'Does Organic Farming Reduce Environmental Impacts? A Meta-analysis of European Research', *Journal of Environmental Management*, 112, 309–320.

Turner, B. L., and A. M. Shajaat Ali (1996) Induced intensification: Agricultural change in Bangladesh with implications for Malthus and *Boserup. Proceedings of the National Academy of Sciences*, 93, 14984–14991.

Turner, M. G., W. J. Calder, G. S. Cumming, T. P. Hughes, A. Jentsch, S. L. LaDeau, T. M. Lenton, B. N. Shuman, M. R. Turetsky, Z. Ratajczak, J. W. Williams, A. P. Williams, and S. R. Carpenter (2020), 'Climate Change, Ecosystems and Abrupt Change: Science Priorities', *Philosophical Transactions of the Royal Society B*, 375(1794), 20190105.

Turner, R. K. (2011), 'Box 2.1 – Designing Coastal Protection Based on the Valuation of Natural Coastal Ecosystems' in P. Kareiva, H. Tallis, T. H. Ricketts, G. C. Daily, and S. Polasky, eds., *Natural Capital: Theory and Practice of Mapping Ecosystem Services* (Oxford: Oxford University Press).

Turner, R. K., and M. Schaafsma (2015), *Coastal Zones Ecosystem Services: From Science to Values and Decision Making* (Cham: Springer).

Tversky, A., and D. Kahneman (1981), 'The Framing of Decisions and the Psychology of Choice', *Science*, 211(4481), 453–458.

Tversky, A., and D. Kahneman (1986), 'Rational Choice and the Framing of Decisions', *The Journal of Business*, 59(4, 2), S251–S278.

Tversky, A., and D. Kahneman (1988), 'Rational Choice and the Framing of Decisions', in D. E. Bell, H. Raiffa and A. Tversky, eds., *Decision Making: Descriptive, Normative, and Prescriptive Interactions* (Cambridge: Cambridge University Press).

Tversky, A., and D. Kahneman (1992), 'Advances in Prospect Theory: Cumulative Representation of Uncertainty', *Journal of Risk and Uncertainty*, 5, 297–323.

Twohig-Bennett, C., and A. Jones (2018), 'The Health Benefits of the Great Outdoors: A Systematic Review and Meta-Analysis of Greenspace Exposure and Health Outcomes', *Environmental Research*, 166, 628–637.

Tyrrell, P., J. T. Toit, and D. W. Macdonald (2020), 'Conservation beyond Protected Areas: Using Vertebrate Species Ranges and Biodiversity Importance Scores to Inform Policy for an East African Country in Transition', *Conservation Science and Practice*, 2(1), e136.

U.S. Fish and Wildlife Service (1973), *Endangered Species Act of 1973*.

Udawatta, R. P., L. M. Rankoth, and S. Jose (2019), 'Agroforestry and Biodiversity', *Sustainability*, 11(10), 2879.

Udry, C. (1990), 'Credit Markets in Northern Nigeria: Credit as Insurance in a Rural Economy', *The World Bank Economic Review*, 4(3), 251–269.

UK Natural Capital Committee (2019), *State of Natural Capital Annual Report 2019: Sixth Report to the Economic Affairs Committee of the Cabinet*.

Umamaheswari, L., K. O. Hattab, P. Nasurudeen, and P. Selvaraj (2011), 'Should Shrimp Farmers Pay Paddy Farmers?: The Challenges of Examining Salinization Externalities in South India', in A. K. E. Haque, M. N. Murthy, and P. Shyamsunder, eds., *Environmental Valuation in South Asia* (New Delhi: Cambridge University Press).

UNAIDS (2020), *Global HIV and AIDS statistics – 2020 Fact Sheet*, 'https://www.unaids.org/en/resources/fact-sheet'.

UNCTAD (2019), *Trade and Development Report 2019: Financing a Global Green New Deal*.

UNDP (1990), *Human Development Report 1990. Concept and Measurement of Human Development*.

UNDP (2017), *Financing Solutions for Sustainable Development: Debt for Nature Swaps*.

UNDP (2018), *The BIOFIN Workbook 2018: Finance for Nature*.

UNDP (2020), *Moving Mountains: Unlocking Private Capital for Biodiversity and Ecosystems*.

UNEP (2016), *Green is Gold: The Strategy and Actions of China's Ecological Civilization*.

UNEP (2020a), *Aid for Trade: A Vehicle to Green Trade and Build Climate Resilience*.

UNEP (2020b), *Building Back Better: The Role of Green Fiscal Policies (Policy Brief)*.

UNEP Finance Initiative (2008), *Biodiversity and Ecosystem Services: Bloom or Bust?*

UNEP Finance Initiative (2018), *Groundbreaking New Tool Enables Financial Institutions to See Their Exposure to Natural Capital Risk*.

UNEP, and PRI (2019), *Fiduciary Duty in the 21st Century*.

UNEP, UNEP Finance Initiative and Global Canopy (2020), *Beyond 'Business as Usual': Biodiversity Targets and Finance*.

References

UNEP/WTO (2018), *Making Trade Work for the Environment, Prosperity and Resilience*.

UNEP-WCMC, and IUCN (2016), *Protected Planet Report 2016. How Protected Areas Contribute to Achieving Global Targets for Biodiversity*.

UNEP-WCMC, IUCN, and NGS (2018), *Protected Planet Report 2018.*

UNFPA (1995), 'Program of Action of the 1994 International Conference on Population and Development (Chapters I-VIII)', *Population and Development Review*, 21(1), 187–213.

UNFPA (2019), *State of World Population Report 2019: Unfinished Business: The Pursuit of Rights and Choices for All*.

UNFSS (2018), *Voluntary Sustainability Standards, Trade and Sustainable Development: 3rd Flagship Report of the United Nations Forum on Sustainability Standards*.

UNGA (1970), *Declaration of Principles Governing the Sea-Bed and the Ocean Floor, and the Subsoil Thereof, beyond the Limits of Nations Jurisdiction*.

UNGA (2015), *Resolution 69/292. Development of an International Legally Binding Instrument under the United Nations Convention on the Law of the Sea on the Conservation and Sustainable Use of Marine Biological Diversity of Areas beyond National Jurisdiction*.

UNGA (2019), *Resolution 73/284 United Nations Decade on Ecosystem Restoration (2021-2030)*.

UNGA, UNFCCC (1994) *Resolution Adopted by the General Assembly*, *20 January 1994, A/RES/48/189*.

Unger, M. von, and I. Emmer (2018), *Carbon Market Incentives to Conserve, Restore and Enhance Soil Carbon*.

Union of Concerned Scientists (2020), *Reviving the Dead Zone: Solutions to Benefit Both Gulf Coast Fishers and Midwest Farmers*.

Union for Ethical Biotrade (2019), *UEBT Biodiversity Barometer 2019*.

Union for Ethical Biotrade (2020), *UEBT Biodiversity Barometer 2020*.

United Nations (1992), *United Nations Convention on the Law of the Sea*.

United Nations (2015), *The Sustainable Development Goals*.

United Nations Climate Change (2018), *UN Helps Fashion Industry Shift to Low Carbon*.

United Nations, European Commission, FAO, IMF, OECD, and World Bank (2013), *System of Environmental-Economic Accounting 2012: Central Framework* (New York, NY: United Nations).

United Nations, European Commission, FAO, OECD, and World Bank (2014), *System of Environmental-Economic Accounting 2012: Experimental Ecosystem Accounting* (New York, NY: United Nations).

United Nations Statistical Division (2017), *National Account Statistics: Analysis of Main Aggregates* (New York, NY: United Nations).

UNODC (2016) *World Drug Report 2016*.

UNPD (2018), *The World's Cities in 2018 – Data Booklet* (New York, NY: United Nations).

UNPD (2019a), *World Population Prospects 2019 – Data booklet* (New York, NY: United Nations).

UNPD (2019b), *World Population Prospects 2019 Highlights* (New York, NY: United Nations).

UNPD (2019c), *World Urbanization Prospects: The 2018 Revision* (New York, NY: United Nations).

References

UNPD (2020), *Estimates and Projections of Family Planning Indicators 2020* (New York, NY: United Nations).

UNU-IHDP and UNEP (2012), Inclusive *Wealth Report 2012: Measuring Progress Toward Sustainability* (Cambridge: Cambridge University Press).

UNU-IHDP and UNEP (2014), *Inclusive Wealth Report 2014. Measuring Progress Towards Sustainability* (Cambridge: Cambridge University Press).

UNWTO (2019), *International Tourism Highlights*.

Van Dam, Y. K., and J. De Jonge (2015), 'The Positive Side of Negative Labelling', *Journal of Consumer Policy*, 38, 19–38.

van der Linden, S. (2015), 'Green Prison Programmes, Recidivism and Mental Health: A Primer', *Criminal Behaviour and Mental Health*, 25(5), 338–342.

van Klink, R., D. E. Bowler, K. B. Gongalsky, A. B. Swengel, A. Gentile, and J. M. Chase (2020), 'Meta-Analysis Reveals Declines in Terrestrial but Increases in Freshwater Insect Abundances', *Science*, 368(6489), 417–420.

van Toor, J., D. Piljic, G. Schellekens, M. van Oorschot, and M. Kok (2020), *Indebted to Nature: Exploring Biodiversity Risks for the Dutch Financial Sector.*van

van Wilgen, B. W., R. M. Cowling and C. J. Burgers (1996), 'Valuation of Services Ecosystem: A Case Study from South African Fynbos Ecosystems', *BioScience*, 46(3), 184–189.

van Wilgen, B. W., and A. Wannenburgh (2016), 'Co-facilitating Invasive Species Control, Water Conservation and Poverty Relief: Achievements and Challenges in South Africa's Working for Water programme', *Current Opinion in Environmental Sustainability*, 19, 7–17.

Veblen, T. (1899), *The Theory of the Leisure Class: An Economic Study of Institutions* (New York, NY: Macmillan) 1925 Edition.

Veenhoven, R (2010), 'How Universal is Happiness?', in E. Diener, J. F. Helliwell, and D. Kahneman, eds., *International Differences in Well-Being* (Oxford: Oxford University Press).

Vera, F. W. M (2009), 'Large-scale Nature Development – the Oostvaardersplassen', *British Wildlife*, June, 28–36.

Vermaat, J. E., A. J. Wagtendonk, R. Brouwer, O. Sheremet, E. Ansink, T. Brockhoff, M. Plug, S. Hellsten, J. Aroviita, L. Tylec, M. Giełczewski, L. Kohut, K. Brabec, J. Haverkamp, M. Poppe, K. Böck, M. Coerssen, J. Segersten, and D. Hering (2016), 'Assessing the Societal Benefits of River Restoration Using the Ecosystem Services Approach', *Hydrobiologia*, 769(1), 121–135.

Vermeulen, F (2012), 'Foundations of Revealed Preference: Introduction', *The Economic Journal*, 122(560), 287–294.

Verones, F., D. Moran, K. Stadler, K. Kanemoto, and R. Wood (2017), 'Resource Footprints and their Ecosystem Consequences', *Scientific Reports*, 7, 40743.

Verutes, G. M., K. K. Arkema, C. Clarke-Samuels, S. A. Wood, A. Rosenthal, S. Rosado, M. Canto, N. Bood, and M. Ruckelshaus (2017), 'Integrated Planning that Safeguards Ecosystems and Balances Multiple Objectives in Coastal Belize', *International Journal of Biodiversity Science, Ecosystem Services and Management*, 13(3), 1–17.

References

Vilela, T., A. M. Harb, A. Bruner, V. L. Da Silva Arruda, V. Ribeiro, A. A. C. Alencar, A. J. E. Grandez, A. Rojas, A. Laina, and R. Botero (2020), 'A Better Amazon Road Network for People and the Environment', *Proceedings of the National Academy of Sciences*, 117(13), 7095–7102.

Vincent, J. R. (2011), 'Valuing the Environment as a Production Input', in A. K. E. Haque, M. N. Murty, and P. Shyamsundar, eds., *Environmental Valuation in South Asia* (Cambridge: Cambridge University Press).

Vincent, J. R., I. Ahmad, N. Adnan, W. B. Burwell, S. K. Pattanayak, J.-S. Tan-Soo, and K. Thomas (2016), 'Valuing Water Purification by Forests: An Analysis of Malaysian Panel Data', *Environmental and Resource Economics*, 64, 59–80.

Vincent, J. R., R. M. Ali, C. Y. Tan, J. Yahaya, K. A. Rahim, L. T. Ghee, A. S. Meyer, M. S.H. Othman, and G. Sivalingam (1997), *Environment and Development in a Resource-rich Economy : Malaysia Under the New Economic Policy* (Harvard Institute for International Development, Harvard University).

Vincent, J. R., R. T. Carson, J. R. DeShazo, K. A. Schwabe, I. Ahmad, S. K. Chong, Y. T. Chang, and M. D. Potts (2014), 'Tropical Countries May Be Willing to Pay More to Protect Their Forests', *Proceedings of the National Academy of Sciences*, 111(28), 10113–10118.

Vincent, J. R., O. Nabangchang, and C. Shi (2020), 'Is the Distribution of Ecosystem Service Benefits Pro-Poor? Evidence from Water Purification by Forests in Thailand', *Water Economics and Policy*, 6(3).

Vincent, J. R., Ali, R. M., and Chang, Y. T. (1997). *Environment and Development in a Resource- Rich Economy: Malaysia Under the New Economic Policy. Harvard Studies in International Development.* (Cambridge, MA: Harvard University Press).

Viscusi, W. Kip (2018), *Pricing Lives: Guideposts for a Safer Society* (Princeton: Princeton University Press).

Viscusi, W. Kip, and J. E. Aldy (2003), 'The Value of a Statistical Life: A Critical Review of Market Estimates Throughout the World', *Journal of Risk and Uncertainty*, 27(1), 5–76.

Vitousek, P. M., H. A. Mooney, J. Lubchenco, and J. M. Melillo (1997), 'Human Domination of Earth's Ecosystems', *Science*, 277(5325), 494–499.

Vivid Economics, and Natural History Museum (2021), *The Urgency of Biodiversity Action*.

Vollset, S. E., E. Goren, C.-W. Yuan, J. Cao, A. E. Smith, T. Hsiao, C. Bisignano, G. S. Azhar, E. Castro, J. Chalek, A. J. Dolgert, T. Frank, K. Fukutaki, S. I. Hay, R. Lozano, A. H. Mokdad, V. Nandakumar, M. Pierce, M. Pletcher, T. Robalik, K. M. Steuben, H. Y. Wunrow, B. S. Zlavog, C. J. L. Murray (2020), 'Fertility, Mortality, Migration, and Population Scenarios for 195 Countries and Territories from 2017 to 2100: A Forecasting Analysis for the Global Burden of Disease Study', *The Lancet*, 396(10258), 1285–1306.

Voosen, P. (2016), 'Anthropocene Pinned to Postwar Period', *Science*, 353(6302), 852–853.

Wackernagel, M., and B. Beyers (2019), *Ecological Footprint: Managing Our Biocapacity Budget* (Gabriola Island: New Society).

Wackernagel, M., D. Lin, M. Evans, L. Hanscom, and P. Raven (2019), 'Defying the Footprint Oracle: Implications of Country Resource Trends', *Sustainability*, 11(7), 2164.

Wade, R. (1988), *Village Republics: Economic Conditions for Collective Action in South India* (Cambridge: Cambridge University Press).

Wagg, C., S. F. Bender, F. Widmer, and M. G. A. Van Der Heijden (2014), 'Soil Biodiversity and Soil Community Composition Determine Ecosystem Multifunctionality', *Proceedings of the National Academy of Sciences*, 111(14), 5266–5270.

Waldron, A., V. M. Adams, J. R. Allan, A. Arnell, G. P. Asner, S. Atkinson, A. Baccini, J. E. M. Baillie, A. Balmford, J. Austin Beau, L. Brander, E. Brondizio, A. Bruner, N. D. Burgess, K. Burkart, S. Butchart, R. Button, R. Carrasco, W. Cheung, V. Christensen, A. Clements, M. Coll, M. di Marco, M. Deguignet, E. Dinerstein, E. Ellis, F. Eppink, J. Ervin, A. Escobedo, J. Fa, A. Fernandes- Llamazares, S. Fernando, S. Fujimori, B. Fulton, S. Garnett, J. Gerber, D. Gill, T. Gopalakrishna, N. Hahn, B. Halpern, T. Hasegawa, P. Havlik, V. Heikinheimo, R. Heneghan, E. Henry, F. Humpenoder, H. Jonas, K. Jones, L. Joppa, A. R. Joshi, M Jung, N. Kingston, C. Klein, T. Krisztin, V. Lam, D. Leclere, P. Lindsey, H. Locke, T. E. Lovejoy, P. Madgwick, Y. Malhi, P. Malmer, M. Maron, J. Mayorga, H. van Meijl, D. Miller, Z. Molnar, N. Mueller, N. Mukherjee, R. Naidoo, K. Nakamura, P. Nepal, R. Noss, B. O'Leary, D. Olson, J. Palcios Abrantes, M. Paxton, A. Popp, H. Possingham, J. Prestemon, A. Reside, C. Robinson, J. Robinson, E. Sala, K. Scherrer, M. Spalding, A. Spenceley, J. Steenbeck, E. Stehfest, B. Strassborg, R. Sumaila, K. Swinnerton, J. Sze, D. Tittensor, T. Toivonen, A. Toledo, P. N. Torres, W. -J. Van Zeist, J. Vause, O. Venter, T. Vilela, P. Visconti, C. Vynne, R. Watson, J. Watson, E. Wikramanayake, B. Williams, B. Wintle, S. Woodley, W. Wu, K. Zander, Y. Zhang, and Y. P. Zhang (2020), *Protecting 30% of the Planet for Nature: Costs, Benefits and Economic Implications*.

Walker, B., A. Kinzig, and J. Langridge (1999), 'Plant Attribute Diversity, Resilience, and Ecosystem Function: The Nature and Significance of Dominant and Minor Species', *Ecosystems*, 2, 95–113.

Walker, G (2012), *Environmental Justice: Concepts, Evidence and Politics* (London: Routledge).

Walter, D. E., and H. C. Proctor eds., (2013), *Mites: Ecology, Evolution, and Behaviour*(Dordrecht: Springer) 2nd Edition.

Wang, X., A. Biewald, J. P. Dietrich, C. Schmitz, H. Lotze-Campen, F. Humpenöder, B. Leon Bodirsky, and A. Popp (2016), 'Taking Account of Governance: Implications for Land-Use Dynamics, Food Prices, and Trade Patterns', *Ecological Economics*, 122, 12–24.

Wang, X., Z. Dong, J. Zhang, and L. Liu (2004), 'Modern Dust Storms in China: An Overview', *Journal of Arid Environments*, 58(4), 559–574.

Wang, Y., X. Cheng, Q. Shan, Y. Zhang, J. Liu, C. Gao, and J.-L. Qiu (2014), 'Simultaneous Editing of Three Homoeoalleles in Hexaploid Bread Wheat Confers Heritable Resistance to Powdery Mildew', *Nature Biotechnology*, 32, 947–951.

Warde, A. (1997), 'Consumption, Taste and Social Change', in *Consumption, Food, and Taste: Culinary Antinomies and Commodity Culture* (London: SAGE Publishing).

Warde, A., and L. Martens (2000), *Eating Out: Social Differentiation, Consumption and Pleasure* (Cambridge: Cambridge University Press).

Warren, L., B. Filgueira, and I. Mason (2009), *Wild Law: Is There Any Evidence of Earth Jurisprudence in Existing Law and Practice?* (London: UK Environmental Law Association and the Gaia Foundation).

Waters, C. N., J. Zalasiewicz, C. Summerhayes, A. D. Barnosky, C. Poirier, A. Galuszka, A. Cearreta, M. Edgeworth, E. C. Ellis, M. Ellis, C. Jeandel, R. Leinfelder, J.R. McNeill, D. deB. Richter, W. Steffen, J. Syritski, D. Vidas, M. Wagreich, M. Williams, A. Zhisheng, J. Grineveld, E. Odada, N. Oreskes, and A. P. Wolfe (2016), "The Anthropocene is Functionally and Stratigraphically Distinct from the Holocene", *Science*, 351(6269), 1–10.

References

Watkins, S. C. (1990), 'From Local to National Communities: The Transformation of Demographic Regions in Western Europe 1870–1960', *Population and Development Review*, 16(2), 241–272.

Watson, J. E. M., N. Dudley, D. B. Segan, and M. Hockings (2014), 'The Performance and Potential of Protected Areas', *Nature*, 515(7525), 67–73.

Watson, J. E. M., D. F. Shanahan, M. Di Marco, J. Allan, W. F. Laurance, E. W. Sanderson, B. Mackey, and O. Venter (2016), 'Catastrophic Declines in Wilderness Areas Undermine Global Environment Targets', *Current Biology*, 26(21), 2929–2934.

Waycott, M., C. M. Duarte, T. J. B. Carruthers, R. J. Orth, W. C. Dennison, S. Olyarnik, A. Calladine, J. W. Fourqurean, K. L. Heck, A. R. Hughes, G. A. Kendrick, W. J. Kenworthy, F. T. Short, and S. L. Williams (2009), 'Accelerating Loss of Seagrasses Across the Globe Threatens Coastal Ecosystems', *Proceedings of the National Academy of Sciences*, 106(30), 12377–12381.

Waygood, S. (2014), *A Roadmap for Sustainable Capital Markets: How can the UN Sustainable Development Goals Harness the Global Capital Markets?*.

WCVP (2020), *World Checklist of Vascular Plants, Version 2.0*, 'http://wcvp.science.kew.org/'.

Weisbrod, B. A. (1964), 'Collective-Consumption Services of Individual-Consumption Goods', *The Quarterly Journal of Economics*, 78(3), 471–477.

Weitzman, M. (1970), 'Iterative Multilevel Planning with Production Targets', *Econometrica*, 38(1), 50–65.

Weitzman, M. L. (1974), 'Prices vs. Quantities', *Review of Economic Studies*, 41(4), 477–491.

Weitzman, M. L. (2009), 'On Modeling and Interpreting the Economics of Catastrophic Climate Change', *Review of Economics and Statistics*, 91(1), 1–19.

Weitzman, M. L. (2014), 'Fat Tails and the Social Cost of Carbon', *American Economic Review*, 104(5), 544–546.

Wende, W., G-M. Tucker, F. Quétier, M. Rayment, and M. Darbi, eds., *Biodiversity Offsets: European Perspectives on No Net Loss of Biodiversity and Ecosystem Services* (Cham: Springer).

Wendling, Z. A., J. W. Emerson, A. de Sherbinin, and D. C. Etsy (2020), *2020 Environmental Performance Index*.

Westphal, M. I., M. Browne, K. MacKinnon, and I. Noble (2008). 'The Link Between International Trade and the Global Distribution of Invasive Alien Species', *Biological Invasions*, 10, 391–398.

Weyzig, F., B. Kuepper, J. W. van Gelder, and R. van Tilburg (2014), *The Price of Doing Too Little Too Late*.

Wheeler, B. W., R. Lovell, S. L. Higgins, M. P. White, I. Alcock, N. J. Osborne, K. Husk, C. E. Sabel, and M. H. Depledge. (2015). 'Beyond Greenspace: An Ecological Study of Population General Health and Indicators of Natural Environment Type and Quality', *International Journal of Health Geographics*, 14, 17.

Wheeler, T., and J. von Braun (2013), 'Climate Change Impacts on Global Food Security', *Science*, 341(6145), 508–513.

White, C., C. Thoung, P. Rowcroft, M. Heaver, R. Lewney, and S. Smith (2017), *Developing and Piloting a UK Natural Capital Stress Test: Final Report*.

White, M. P., I. Alcock, B. W. Wheeler, and M. H. Depledge (2013), 'Would You Be Happier Living in a Greener Urban Area? A Fixed-Effects Analysis of Panel Data', *Psychological Science*, 24(6). 920–928.

Whiteley, P. F. (2000), 'Economic Growth and Social Capital', *Political Studies*, 48(3), 443–466.

WHO (2018), *Ambient (Outdoor) Air Pollution*, 'https://www.who.int/en/news-room/fact-sheets/detail/ambient-(outdoor)-air-quality-and-health'.

WHO and UNICEF (2017), *Progress on Drinking Water, Sanitation and Hygiene: 2017 Update and SDG Baselines* (Geneva: WHO and UNICEF).

Whyte, K. P., J. P. Brewer, and J. T. Johnson (2016), 'Weaving Indigenous Science, Protocols and Sustainability Science', *Sustainability Science*, 11(1), 25–32.

Wiedmann, T. O., H. Schandl, M. Lenzen, D. Moran, S. Suh, J. West, K. Kanemoto, and J. M. Alier (2013), 'The Material Footprint of Nations', *Proceedings of the National Academy of Sciences*, 112(20), 6271–6276.

Wiedmann, T., J. Minx, J. Barrett, and M. Wackernagel (2006), 'Allocating Ecological Footprints to Final Consumption Categories with Input-Output Analysis', *Ecological Economics*, 56(1), 28–48.

Wiedmann, T., M. Lenzen, L. T. Keyßer, and J. K. Steinberger (2020), 'Scientists' Warning on Affluence', *Nature Communications*, 11, 3107.

Wiggins, D. (1987), 'Claims of Needs' in *Needs, Values, Truth: Essays in the Philosophy of Value* (Oxford: Blackwell).

Wildavsky, A. (1987), 'Choosing Preferences by Constructing Institutions: A Cultural Theory of Preference Formation', *The American Political Science Review*, 81(1), 3–22.

Wilkinson, B. H., and B. J. McElroy (2007), 'The Impact of Humans on Continental Erosion and Sedimentation', *Geological Society of America Bulletin*, 119(1/2), 140–156.

Willer, H., G. Sampson, V. Voora, D. Dang, and J. Lernoud (2019), *The State of Sustainable Markets 2019: Statistics and Emerging Trends*.

Williams, B.A.O. (1985), *Ethics and the Limits of Philosophy* (London: Fontana Press/Collins).

Williams, B.A.O. (1993), *Shame and Necessity* (Berkeley, CA: University of California Press).

Williams, D. R., B. Phalan, C. Feniuk, R. E. Green, and A. Balmford (2018), 'Carbon Storage and Land-Use Strategies in Agricultural Landscapes Across Three Continents', *Current Biology*, 28(15), 2500–2505.e4.

Williams, J., F. Stokes, H. Dixon, and K. Hurren (2017), *The Economic Contribution of Commercial Fishing to the New Zealand Economy*.

Williams, M., J. Zalasiewicz, P. K. Haff, C. Schwägerl, A. D. Barnosky, and E. C. Ellis (2015), 'The Anthropocene Biosphere', *Anthropocene Review*, 2(3), 196–219.

Williams, S. L., C. Sur, N. Janetski, J. A. Hollarsmith, S. Rapi, L. Barron, S. J. Heatwole, A. M. Yusuf, S. Yusuf, J. Jompa, and F. Mars (2019), 'Large-Scale Coral Reef Rehabilitation After Blast Fishing in Indonesia', *Restoration Ecology*, 27(2), 447–456.

Willig, M. R., and S. J. Presley (2018), 'Latitudinal Gradients of Biodiversity: Theory and Empirical Patterns', in D. A. Dellasala, and M. I. Goldstein, eds., *Encyclopedia of the Anthropocene* (Oxford: Elsevier).

References

Willis, K. J. (2017), *State of the World's Plants 2017 Report*.

Wilson, E. O. (1984), *Biophilia* (Cambridge, MA: Harvard University Press).

Wilson, E. O. (1992), *The Diversity of Life* (Cambridge, MA: Belknap Press of Harvard University Press).

Wilson, E. O. (2002), *The Future of Life* (New York, NY: Alfred A. Knopf).

Wilson, E. O. (2016), *Half-Earth: Our Planet's Fight for Life* (New York, NY: Liveright Publishing Corporation).

Wilson, M. A., and S. R. Carpenter (1999), 'Economic Valuation of Freshwater Ecosystem Services in the United States: 1971-1997', *Ecological Applications*, 9(3), 772–783.

Wilting, H. C., A. M. Schipper, M. Bakkenes, J. R. Meijer, and M. A. J. Huijbregts (2017), 'Quantifying Biodiversity Losses Due to Human Consumption: A Global-Scale Footprint Analysis', *Environmental Science and Technology*, 51(6). 3298–3306.

WIPO (2015), *Intellectual Property and Traditional Medical Knowledge*.

Wohl, E. (2011), *A World of Rivers: Environmental Change on Ten of the World's Great Rivers* (London and Chicago, IL: University of Chicago Press).

World Bank (1992), *World Development Report 1992: Development and the Environment* (New York, NY: Oxford University Press).

World Bank (1997), *Helping Countries Combat Corruption: The Role of The World Bank*.

World Bank (2018), 'Seychelles Launches World's First Sovereign Blue Bond', 'https://www.worldbank.org/en/news/press-release/2018/10/29/seychelles-launches-worlds-first-sovereign-blue-bond'.

World Bank (2019), *World Development Indicators*.

World Bank (2020a), *World Development Indicators*.

World Bank (2020b), *Mobilizing Private Finance for Nature*.

World Bank (2020c), 'Poverty', 'https://www.worldbank.org/en/topic/poverty/overview'.

World Bank (2020d), *Purchasing Power Parities and the Size of World Economies: Results from the 2017 International Comparison Program*.

World Bank, and GEF (2018), *Global Wildlife Program*.

World Economic Forum (2016), *The Global Risks Report 2016*.

World Economic Forum (2020a), *Nature Risk Rising: Why the Crisis Engulfing Nature Matters for Business and the Economy*.

World Economic Forum (2020b), *New Nature Economy Report II: The Future of Nature and Business*.

World Economic Forum (2020c), *The Global Risks Report 2020*.

World Food Programme (2017), *108 Million People in The World Face Severe Food Insecurity– Situation Worsening*.

World Resources Institute (2005), *World Resources 2005: The Wealth of the Poor – Managing Ecosystems to Fight Poverty*.

WTTC (2015), *Travel & Tourism Economic Impact 2015*.

Worm, B., E. B. Barbier, N. Beaumont, J. E. Duffy, C. Folke, B. S. Halpern, J. B. C. Jackson, H. K. Lotze, F. Micheli, S. R. Palumbi, E. Sala, K. A. Selkoe, J. J. Stachowicz, and R. Watson (2006), 'Impacts of Biodiversity Loss on Ocean Ecosystem Services', *Science*, 314(5800), 787–790.

Wortley, L., J. M. Hero, and M. Howes (2013), 'Evaluating Ecological Restoration Success: A Review of the Literature', *Restoration Ecology*, 21(5), 537–543.

WRAP (2020), *Courtauld Commitment 2025 Milestone Progress Report*.

Wright, G., J. Rochette, E. Druel, and K. Gjerde (2015), 'The Long and Winding Road Continues: Towards a New Agreement on High Seas Governance', *IDDRI Study*, No. 01/16.

Wrigley, E. A. (2004), *Poverty, Progress, and Population* (Cambridge: Cambridge University Press).

WTO (2011), 'The Impact of Trade Opening on Climate Change', 'https://www.wto.org/english/tra top_e/envir_e/climate_impact_e.htm'.

WTO (2020), *Contribution to the 2020 High Level Political Forum*.

Wu, F., S. Zhao, B. Yu, Y.M. Chen, W. Wang, Z.G. Song, Y. Hu, Z.W. Tao, J.H. Tian, Y.Y. Pei, M.L. Yuan Y. L. Zhang, F. H. Dai, Y. Liu, Q. M. Wang, J. J. Zheng, L. Xu, E. C. Holmes and Y.-Z. Zhang, (2020), 'A New Coronavirus Associated with Human Respiratory Disease in China. *Nature*', 579(7798), 265–269.

Wu, Y., X. Zhang, and L. Shen (2011), 'The Impact of Urbanization Policy on Land Use Change: A Scenario Analysis', *Cities*, 28(2), 147–159.

WWF (1997) *Marketing Wildlife Leases*.

WWF (2013), *Searching for Sustainability*: Comparative Analysis of Certification Schemes for Biomass Used for the Production of Biofuels.

WWF (2015), *Safeguarding Outstanding Natural Value*: The Role of Institutional Investors in Protecting Natural World Heritage Sites from Extractive Activity.

WWF (2017), 'Soil Erosion and Degradation', 'www.worldwildlife.org/threats/soil-erosion-and-degradation'.

WWF (2018), M. Grooten and R. E. A. Almond, eds., *Living Planet Report 2018*: Aiming Higher.

WWF (2019), *The Nature of Risk*: A Framework for Understanding Nature-Related Risk to Business.

WWF (2020a), in R. E. A. Almond, M. Grooten, and T. Petersen, eds., *Living Planet Report 2020: Bending the Curve of Biodiversity Loss*.

WWF (2020b), 'Environmental Threats', 'https://www.worldwildlife.org/threats'.

WWF France, and AXA (2019), *Into the Wild: Integrating Nature into Investment Strategies*.

WWF, and Investec (2019), *Sustainability and Satellites: New Frontiers in Sovereign Debt Investing*.

WWF and RSPB (2017), *Risky Business. Understanding the UK's Overseas Footprint for Deforestation-Risk Commodities*.

WWF, and RSPB (2020), *Riskier Business: The UK's Overseas Land Footprint*.

WWF, and Swiss Re Institute (2020), *Conserving our Common Heritage: The Role of Spatial Finance in Natural World Heritage Protection*.

WWF, and ZSL (2020), *The Living Planet Index Database*, 'https://livingplanetindex.org/data_portal'.

References

Xepapadeas, A. (2005), 'Economic Growth and the Environment', in K. -G. Mäler and J. R. Vincent, eds., Handbook of Environmental Economics: Economywide and International Environmental Issues, Vol. III (Amsterdam: Elsevier North-Holland).

Xiao, Y., Z. Wang, Z. Li, and Z. Tang (2017), 'An Assessment of Urban Park Access in Shanghai – Implications for the Social Equity in Urban China', *Landscape and Urban Planning*, 157(1), 383–393.

Yaari, M. E. (1965), 'Uncertain Lifetime, Life Insurance, and the Theory of the Consumer', *The Review of Economic Studies*, 32(2), 137–150.

Yamaguchi, R. (2018), '*Wealth and Population Growth Under Dynamic Average Utilitarianism*', *Environment and Development Economics*, 23(1), 1–18.

Yamaguchi, R., M. Islam, and S. Managi (2019), 'Inclusive Wealth in the Twenty-first Century: A Summary and Further Discussion of Inclusive Wealth Report', *Letters in Spatial and Resources Sciences*, 12, 101–111.

Yamaguchi, R., M. Sato, and K. Ueta (2016), 'Measuring Regional Wealth and Assessing Sustainable Development: An Application to a Disaster-Torn Region in Japan', *Social Indicators Research*, 129(1), 365–389.

Yamamura, E. (2015), 'The Impact of Natural Disasters on Income Inequality: Analysis using Panel Data during the Period', *International Economic Journal*, 29(3), 359–374.

Yang, W., X. Yu, B. Zhang and Z. Huang (2019), 'Mapping the Landscape of International Technology Diffusion (1994-2017): Network Analysis of Transnational Patents', *The Journal of Technology Transfer*.

Ye, Q., and M. H. Glantz (2005), 'The 1998 Yangtze Foods: The Use of Short-term Forecasts in the Context of Seasonal to Interannual Water Resource Management', *Mitigation and Adaptation Strategies for Global Change*, 10, 159–182.

Young, M. D. (2014), 'Designing Water Abstraction Regimes for an Ever-Changing and Ever- Varying Future', *Agricultural Water Management*, 145, 32–38.

York University Ecological Footprint Initiative and Global Footprint Network (2020). 'National Footprint and Biocapacity Accounts, 2021 Edition', 'https://data.footprintnetwork.org'.

Zadek, S., and N. Robins (2018), *Making Waves: Aligning the Financial System with Sustainable Development*.

Zagheni, E. (2011), 'The Leverage of Demographic Dynamics on Carbon Dioxide Emissions: Does Age Structure Matter?', *Demography*, 48, 371–399.

Zak, P. J., and S. Knack (2001), 'Trust and Growth', *The Economic Journal*, 111(470), 295–321.

Zelazowski, P., Y. Malhi, C. Huntingford, S. Sitch, and J. B. Fisher. (2011). 'Changes in the Potential Distribution of Humid Tropical Forests on a Warmer Planet', *Philosophical Transactions of the Royal Society A*, 369(1934), 137–160.

Zeng, X., S. S. P. Shen, X. Zeng, and R. E. Dickinson (2004), 'Multiple Equilibrium States and the Abrupt Transitions in a Dynamical System of Soil Water Interacting with Vegetation', *Geophysical Research Letters*, 31(5), L05501.

Zerbini, A. N., G. Adams, J. Best, P. J. Clapham, J. A. Jackson, and A. E. Punt (2019), 'Assessing the Recovery of an Antarctic Predator from Historical Exploitation', *Royal Society Open Science*, 6 (10), 190368.

Zhang, G., W. Chen, G. Zheng, H. Xie, and C. K. Shum (2020), 'Are China's Water Bodies (Lakes) Underestimated?' *Proceedings of the National Academy of Sciences*, 117(12), 6308–6309.

Zilberman, D., T. G. Holland, and I. Trilnick (2018), 'Agricultural GMOs-What We know and Where Scientists Disagree', *Sustainability*, 10(5), 1514.

Zilberman, D., L. Lipper, and N. McCarthy (2008), 'When Could Payments for Environmental Services Benefit the Poor?', *Environment and Development Economics*, 13(3), 255–278.

Ziter, C. (2016), 'The Biodiversity-Ecosystem Service Relationship in Urban Areas: A Quantitative Review', *Oikos*, 125(6), 761–768.

ZSL (2020), 'ZSL's Position Statement on COVID-19: Wildlife Exploitation and Trade, Zoonotic Disease, and Human Health', 'https://www.zsl.org/zsls-position-statement-on-covid-19-wildlife- exploitation-and-trade-zoonotic-disease-and-human-0'.

zu Ermgassen, E. K. H. J., M. Kelly, E. Bladon, R. Salemdeeb, and A. Balmford (2018), 'Support Amongst UK Pig Farmers and Agricultural Stakeholders for the Use of Food Losses in Animal Feed', *PLoS ONE*, 13(4), e0196288.

Acknowledgements

Review team

Professor Dasgupta is grateful to his team – drawn from the public sector and based at HM Treasury – for their support on the Review: Sandy Sheard, Mark Anderson, Heather Britton, Abbas Chaudri, Dana Cybuch, Rebecca Gray, Haroon Mohamoud, Robert Marks, Emily McKenzie, Diana Mortimer, Rebecca Nohl, Felix Nugee, Ant Parham, Victoria Robb, Sehr Syed, Thomas Viegas, Ruth Waters, and Lucy Watkinson. Other colleagues involved in the early stages were David Atkinson, Tilmann Eckhardt, Tom Hegarty, Sarah Nelson, Ellen Reaich, Matt Rimmer, and Anam Shahab.

The Review was made possible thanks to the generosity and wisdom of the many people who made time to speak, write and listen to Professor Dasgupta and his team. The following individuals and organisations deserve a special mention for their contributions to the Review.

Advisory Panel

Professor Dasgupta is particularly grateful to the Advisory Panel for their expertise and wise counsel over many months. Their breadth and depth of experience was invaluable.

Inger Andersen	Professor Dame Georgina Mace[*]
Kemi Badenoch MP (Chair)[†]	Professor Dame Henrietta Moore
Juan Pablo Bonilla	Dr Cosmas Ochieng
Sir Ian Cheshire	Dame Fiona Reynolds
Dominic Christian	Charles Roxburgh[†]
Sir Roger Gifford	Professor Lord Nicholas Stern
Professor Cameron Hepburn	Kristian Teleki
David Hill[†]	Professor Sir Bob Watson
Professor Justin Lin	Kate Wylie

Past members:

Simon Clarke MP[†]   Sonia Phippard[†]

[*]Georgina Mace passed away while the Review was near completion. She was an exceptional conservation ecologist, generous to a fault, professionally impeccable and a warm personality. As member of the Advisory Panel her advice was of enormous help and has left a mark. She will be missed greatly.

[†] Acting in an ex-Officio capacity.

Acknowledgements

Reviewers

Professor Dasgupta and the Review team are very grateful to a number of people including several academic colleagues who provided valuable feedback on specific chapters, including Professor Charles Perrings for his advice and support across all of the Review's chapters.

Alexandre Antonelli
Andrew Balmford
Ed Barbier
John Bongaarts
Richard Bradbury
Steve Carpenter
William Clark
Alan Dangour
Rowan Douglas
Paul Ehrlich
Carl Folke
Ben Groom
Rashid Hassan
Carolyn Kousky
Pushpam Kumar
Richard Layard

Rebecca Lovell
Rachel Lowe
Gordon Mitchell
Rose O'Neill
Charles Perrings
Andy Purvis
Nick Robins
Alex Rogers
Jim Salzman
Nina Seega
Marco Springmann
Julia Stegemann
Will Steffen
Jeffrey Vincent
Ben Wheeler

Other contributors

We are also very grateful to James Vause, Nancy Jennings and Wanyi Li for their inputs to the Review chapters, alongside team members.

In addition, we offer thanks to the many individuals across the public sector, academia, environmental groups and the finance and business community in the UK and internationally who contributed to the Review with advice, feedback, support and information, including:

Matthew Agarwala, Firoz Ahmed, Alexandre Antonelli, Onil Banerjee, Uris Baldos, Andrew Balmford, Ian Bateman, Simon Beard, Nicola Beaumont, Tim Benton, John Bongaarts, Carter Brandon, Tom Breeze, Elinor Breman, Timothy Campbell, Raffaello Cervigni, Jagjit Chadha, Rebecca Chaplin Kramer, Ian Christie, Mike Christie, John Cleland, Mary Colwell, Ben Combes, Diane Coyle, Gretchen Daily, Aisha Dasgupta, Carol Dasgupta, Shamik Dasgupta, Ruth Davies, Raymond Dhirani, Paul Ehrlich, Karen Ellis, Will Evison, Vin Fleming, Brendan Freeman, Liz Gallagher, Dustin Garrick, Patrick Gerland, Charles Godfray, Joe Grice, Haripriya Gundimeda, William Hall, Andy Haines, Paula Harrison, Geoffrey Heal, Dieter Helm, Thomas Hertel, Rob Holland, Lauren Holt, Ian Hurst, Tim Jackson, Phillip James, Alice Jay, Dustin Johnson, Gail Kaiser, Amit Kara, Katia Karousakis, Tracey King, Pushpam Kumar, Matt Larsen-Daw, Caroline van Leenders, Tim Lenton, Simon Levin, David Lin, Alison Littlewood, Tim Littlewood, Thomas Lovejoy, Beatriz Luraschi, Shunsuke Managi, Ehsan Masood, Anthony McDonnell, Paul Morling, Sireesha Nemana, Ilan Noy, Peter Odhengo, Gus O'Donnell, Philip Osano, Linda Partridge, Unai Pascual, Linwood Pendleton, Charles Perrings, Edward Perry, Stuart Pimm, Steve Polasky, Simon Potts, Samantha Power, Peter Raven, Martin Rees, Callum Roberts, Toby Roxburgh, Mary Ruckelshaus, Caterina Ruggeri Laderchi, Giovanni Ruta, Alister Scott, Minouche Shafik, Rebecca Shaw, Robin Smale, Chris Stark, Fiona Stewart, Alfie Stirling, Isabel Studer, Pavan Sukhdev, Tim Sunderland, Chris Taylor, Dr Alex Teytelboym, Joris van Toor, Liz Varga, Mathis Wackernagel, Anthony Waldron, Gregory Watson, Chris

Weber, Dominic Whitmee, Carol Williams, Mark Wright, Graham Wynne, Ana Yang and Dimitri Zenghelis.

We are also grateful to the numerous UK government and wider public sector officials who shared their expertise and advice, including representatives from the Cabinet Office, the Department for Business Enterprise Industry and Skills (BEIS), Department for Environment Food & Rural Affairs (Defra), Department for Trade (DIT), Environment Agency (EA), Foreign, Commonwealth & Development Office (FCDO), Centre for Environment, Fisheries and Aquaculture Science (CEFAS), Forestry Commission, Government Office for Science, HM Treasury, Joint Nature Conservation Committee (JNCC), Natural England, and the Office for National Statistics (ONS).

We also extend our thanks to the many other individuals and organisations from around the world, from China, Colombia, Costa Rica, Italy, Kenya, New Zealand and elsewhere, who shared their expertise and advice and hosted us in person and through events, including the African Development Bank, the Bank of England, the City of London Corporation, the Green Finance Institute, the Inter-American Development Bank, the Natural History Museum, the OECD, the Reserve Bank of New Zealand, the Stockholm Environment Institute, the UN Environment Programme (UNEP), the World Bank, the World Economic Forum, and WWF.

Call for Evidence respondents

We are grateful to the following organisations for providing responses to our Call for Evidence over August-November 2019.

African Parks; Atkins Global; Cambridge Conversation Initiative (CCI); Centre for Ecology and Hydrology; Climate Disclosure Standards Board; Committee on Climate Change; Conservation of Arctic Flora and Fauna; Cornell University; Debating Nature's Value Network; Earlham Institute; Economics for the Environment Consultancy; Economics for the Environment Consultancy (EFTEC); Food and Drink Federation; Food and Land Use Coalition; Game & Wildlife Conservation Trust; Global Environment Facility; Global Garden Limited; Global Green Growth Institute; Grantham Research Institute; Green Alliance; Green Economy Coalition; Green Growth Knowledge Platform; Historic England; Inter-American Development Bank (IDB); International Institute for Environment and Development (IIED); International Organisation for Standardisation (ISO) sub-committee; IUCN Sustainable Use and Livelihoods Specialist Group (SULi); Issue Advocacy; Margaret Pyke Trust; Mineral Products Association; Mopane Capital; National Institute for Environmental Studies (NIES) ; Natural England; Natural Resources Wales; Netherlands Environmental Assessment Agency (PBL); New Generation Plantations (NGP); Organisation for Economic Co-operation and Development (OECD); Ornamental Aquatic Trade Association Ltd (OATA); Ornamental Fish Trade Association; Ove Arup and Associates Limited; Plymouth Marine Lab; Port of London Authority; Rothamsted Research; Royal Botanic Garden Kew; Royal Society; Royal Society of Biology; Scottish Natural Heritage; Scottish Wildlife Trust; St Helena Government and the St Helena National Trust; Stockholm Resilience Centre (SCR); Sustainable Soils Alliance; TEEB (The Economics of Ecosystem and Biodiversity) Germany; The Centre for Social and Economic Research, University of East Anglia (CSERGE); The Ex'tax Project; The International Union for Conservation of Nature (UCN), Science & Economic Knowledge Unit; The Nature Conservancy; UK Sustainable Investment and Finance Association (UKSIF); UN Environment Programme (UNEP); University of Cambridge; Wildfowl & Wetlands Trust (WWT); World Business Council For Sustainable Development (WBCSD); World Resource Institute (WRI); WWF Network and Zoological Society London – Conservation Programme (ZSL).

We also received a total of 28 responses submitted in personal capacity and thank all individuals who responded to the Call for Evidence.

Acknowledgements

Interim Report respondents

We are grateful to the following organisations for providing feedback on the Review's Interim Report, published 30 April 2020.

African Parks; African Research and Impact Network (ARIN); Australian Rainforest Conservation Society, Inc.; Born Free Foundation; Economics for Nature (E4N); Environment Agency; Environment and Climate Change Department, Government of Canada; The European Association for the Conservation of Geological Heritage (ProGEO); European Bank for Reconstruction and Development (EBRD); Green Economy Coalition (GEC); Global Footprint Network (GFN); International Institute for Environment and Development (IIED); International Union for Conservation of Nature (IUCN); Centre for Climate Change and Planetary Health, London School of Hygiene & Tropical Medicine (LSHTM); Margaret Pyke Trust; National Association for Areas of Outstanding Natural Beauty (NAAONB); Organisation for Economic Co-operation and Development (OECD); Office for National Statistics (ONS); Ornamental Aquatic Trade Association Ltd (OATA); Smith School of Enterprise and the Environment, University of Oxford (SSEE); Population Matters; Agence Française de Développement (AFD); UN Committee of Experts on Environmental Economic Accounting (UNCEEA); UNEP-WCMC and WWF.

We also received a total of 49 responses submitted in personal capacity and thank all individuals who provided feedback on the Interim report.

Review photographs

We are grateful for the permission to use the following photographs in the Review from the Natural History Museum and individual photographers in the Young Wildlife Photographer of the Year competition:

Cruz Erdmann, Will Jenkins, Carlos Pérez Naval

Author Index

Abbott, J. K. 27, 338
Abdullah, S. 460
Abel, G. J. 229
Abell, R. 88–89
Aburto-Oropeza, O. 411, 449
Acemoglu, D. 20, 47
Achard, F. 404
Acheson, J. M. 200
Achicanoy, H. A. 455
Adair, L. S. 375
Adamchuk, V., 420
Adamowicz, W. 308
Adams, G. 451
Adams, J. 474
Adams, V. M. 444–445, 495
ADB, 480
Adl, S. 63, 68, 149
Adler, C. 381
Adler, M. D. 294
Adnam, N. 49
Adnnam, N. 50
Agard, J. 382–383, 417
Agarwal, A. 200
Agarwal, B. 203
Agarwala, M. 469–470
Aghion, P. 20, 47
Agliardi, E. 337
Agrawal, A. 479
Agreda, A. 450, 451
Aguilar-Rodríguez, J. 352
Agusti, S. 381, 447, 451, 452, 460, 463–464, 495
Ahmad, I. 49, 50
Ahmadia, G. M. 447, 451
Ai, N. 318
Aizen, M. A. 413
Akçakayam, H. R. 60
Akerlof, G. A. 165
Akins, J. L. 468
Akther, S. 18
Alados, C. L. 76
Albrechts, L. 469
Alcalá, F. 386, 387
Alcock, I. 303
Alder, J. 189
Aldy, J. E. 103, 249
Alexander, P. 375, 404, 497–498
Ali, S. H. 400
Alkema,L. 237
Alkemade, R. 92
Allan, J. R. 444–445, 495
Allcott, H. 228
Alleyne, G. 9
Allison, E. H. 412, 416
Alpizar, F. 296, 477

Alston, L. J. 206
Alvarado, I. 477
Ambachtsheer, J. 432
Amel-Zadeh, A. 487
Amin, S. 233
Anderson, A. B. 202
Anderson, C. 423, 424, 426
Anderson, C. B. 303
Anderson, R. 113
Andonova, L. B. 456
Andradi-Brown, D. A. 468
Andres, L. A. 199, 210
Angel, S. 377
Angelico, M. 451
Ango, T. G. 416–417
Angulo, A. 451
Ansar, A. 69
Ansell, D. 466
Anthony, K. R. N. 63
Antunes, P. 472
Aquino, C. A. 461
Araújo, M. B. 119, 381
Arcenas, A. 195
Archibald, S. 74
Ardoin, N. M. 314, 504
Aredo, D. 172–173
Arkema, K. 470, 471
Arkema, K. K. 471
Arlaud, M. 473
Armenteras, D. 467
Arneth, A. 375, 497–498
Arnott, R. 204
Aronson, J. 458, 459
Arora, L. 419
Arrow, K. J. 25, 26, 27, 28, 83, 97, 128, 136, 147, 154, 156, 222, 240, 250, 252, 267, 278, 279, 280, 282, 283, 285, 309, 329, 334, 335, 337, 340, 342, 344, 354, 355–356, 357, 363
Ashley, C. 210
Asian Development Bank, 188, 464
Asner, G. P. 404
Aspers, P. 224
Asquith, N. M. 195
Astudillo, E. 450, 451
Asuncion, J. 432
Atkinson, A. B. 104, 257
Attanasio, O. P. 221
Aubrat, M. 471–472
Aunger, R. 227
Avery, S. 196
Aveyard, P. 404
AXA, 424, 437, 438, 487
Axelrod, M. 456
Ayres, R. U. 48, 127

Author Index

Babalola, S. 233
Babayaro, H. 233
Babiarz, K. S. 237
Baccini, A. 49, 94
Badiger, S. 47
Bahaj, S. 433
Baicker, K. 227
Bainbridge, J. M. 443
Baker, S. C. 408
Bakkenes, M. 61, 65, 391–392
Baland J.-M. 163, 200, 201, 204
Balasubramanian, R. 205
Baldos, U. 383, 384, 423
Baldwin, R. 386
Balian, E. V. 68
Baliga, S. 193
Balkovi, J. 414, 415
Ballabio, C. 45
Ballantine, B. 447
Balmford, A. 310, 414, 446, 451, 460, 461, 468–469
Balmford, A. P. 498
Baltes, N. J. 420
Balvanera, P. 387, 398, 404, 408, 413
Banco de México, 439
Bandyopadhyay, S. 203
Banerjee, O. 477–478
Bangalore, M. 377
Bank of England, 22, 439
Bank of International Settlements, 424, 431, 435, 436, 439, 480
Baragwanath, K. 204, 218
Barakat, B, 229
Barange, M. 416
Barbarossa, V. 68, 409
Barbier, E. 381, 447, 451, 452, 460, 463–464, 495
Barbier, E. B. 47, 67, 202, 310, 317, 425, 444, 458, 462, 465
Bardgett, R. D. 46, 60
Bardhan, P. K. 172, 203
Barioni, L. G. 446
Barnes, M. D. 447, 451
Barnosky, A. D. 62
Bar-On, Y. M. 41, 62
Barrett, J. 375, 387
Barrett, M. 60, 396, 457–458
Barrett, S. 98, 101, 102, 103, 113, 120, 222, 224, 319, 374, 514
Barrios, E. 46
Barro, R. J. 20, 47
Barrows, G. 418
Barry, J. P. 117
Bartholomeus, H. 381
Barton, J. 463, 504
Baruman, K. A. 404, 405, 414, 417, 497
Bass, R. 481
Bassen, A. 423
Basu, A. M. 233
Batáry, P. 466
Bateman, I. J. 330, 466, 469–470

Bateman, R. 459
Batllori, E. 448
Baue, B. 487
Bauer, P. T. 183
Baumgärtner, S. 263
Bavel, J. J. V. 227
Bawa, K. S. 504
Bayham, J. 27, 338
Bayi, E. 204, 218
Baylis, K. 455
Bayon, R. 472
Bayraktarov, E. 460
Beach, T. 11, 44, 45, 96, 317
Beacham, A. M. 421
Beatty, C. R. 469–470
Beaumont, N. 67
Beck, M. W. 425, 444, 465
Beck, T. 201–202
Becker, C. D. 356, 450, 451
Behrans III, W. W. 124
Behrenfeld, M. J. 41
Behrens, K. G. 275
Behrman, J. R. 232, 235, 499
Beisner, B. E. 511, 514
Belaire, J. A. 383
Belinga, S. M. E. 232
Bell, J. D, 381
Bellmann, C. 388, 394, 396
Bello, C. 47
Belyaeva, I. 63–64
Benayas, J. M. R. 457–458
Bender, R. 426
Bender, S. F. 45, 96
Béné, C. 416
Benjamin, D. J. 300
Bennett, G. 195, 477
Bennett, J. R. 458
Bentham, J. 292
Benz, D. 416–417
Berger, J. 437, 438
Berkes, F. 200, 201–202, 448, 449
Berman, M. 303
Berman, M. G. 302
Bernauer, T. 74
Bernhardt, A. 432
Bernstein, S. 236, 237
Berri, A. 404
Berry, K. 374, 382
Berry, P. 462, 463
Berthe, A. 374
Best, J. 451
Béteille, A. 203
Beugelsdijk, S. 159
Bewket, W, 55
Beyers, B. 106, 107, 373, 375, 387
Bi, H. 419–420
Bibby R. 117
Biber, M. F. 407
Biesmeijer, J. C. 413

Biewald, A. 378, 405, 406
Biggs, D. 177
Biggs, R. 61, 513
Bignardi, G. 228
Bilby, R. E. 413–414
Binmore, K. 42, 161
Binner, A. 469–470
Binswanger, H. P. 206
Birch, J. C. 461
Bisson, P. A. 413–414
BIT, 227
Bjorkman, A. D. 375
Black, D. 284
Blackburn, S. 182
Blackburn, T. M. 64, 468
Bladon, E. 416
Blanchard, J. L. 411
Blanchfield, P. J. 312
Blanchflower, D. G. 240
Blaney, R. 378–379
Bledsoe, C. 232
Bledsoe, C. H. 232
Blignaut, J. N. 459, 465
Blok, D. 381
Blume, L. E. 228
Bodin, E. 395
Bodirsky, B. L. 405, 406
Bogahawatte, C. 311
Boggio, P. S. 227
Böhmelt, T. 74
Bolam, F. 451
Bolin, B. 28
Bolt, J. 5, 6, 86, 87
Bond, I. 194–195
Bongaarts, J. 127, 231, 232, 237, 238, 239
Bonner, J. 426
Bonzanigo, L. 377
Booker, F. 382–383
Boorstin, D. J. 308
Borelli, P. 45
Borer, E. T. 57, 58, 59
Börjeson, L. 416–417
Börner, J. 455
Boserup, E. 10
Bourge, C. 471–472
Bowers, A. W. 504
Bowller, D. E. 64
Bowman, D. 461
Bowman, W. D. 36, 42, 51, 68
Boyd, P. W. 116
Boyer, J. 130, 355–356
Boyle, K. J. 308
Bozinovic, F. 119
Bradley, S. E. K. 237, 515
Bradshaw, C. J. A. 67, 88–89, 143
Bragg, R. 463, 504
Bramley, R. G. V. 378, 420
Brander, L. M. 444
Brando, P. 408

Brandon, C. 353, 465
Bratman, G. N. 303
Brauman, K. 380
Breeze, T. D. 30
Breitburg, D. 92, 93, 381
Bresch, D. N. 444, 465
Breslin and Chapin, 1984, 206
Brewer, J. P. 448
Brewer, N. T. 228
Bricker, S. B. 418
Briggs, S. V. 413
Bright, G. 27, 313, 348, 349
Broadberry, S. 7
Broadgate, W. 85
Brock, A. 228
Brock, W. 52
Brock, W. A. 83, 129, 140, 228, 262–263
Brockelman, W. Y. 48
Brockington, D. 450
Broer, W. 437, 438
Bromley, D. W. 203
Brondizio, E. S. 394
Brook, B. W. 67, 88–89, 143
Brooks, T. M. 391, 408, 447
Broome, J. 251
Brouwer, R. 464
Brown, A. 375, 497–498
Brown, B. L. 418
Brown, C. J. 116
Brown, G. 310
Brown, K. 383
Brown, P. 464
Browne, M. 388
Bruckner, M. 406
Brudvig, L. A. 68
Bruehlman-Senecal, E. 303
Bruins, R. J. F. 376
Brundtland, G. H. 336
Bruner, A. 69
Bruno, J. F. 116
Bryan, B. A. 317, 318
Buchner, B. 474
Buck, K. R. 117
Budzier, A. 69
Bueno, C. 454
Buisson, E. 462
Bull J. W. 471
Bullock, J. M. 457–458
Bulte, E. H. 195
Bureau of Economic Analysis, 164
Burek, P. 409, 410
Burgers, C. J. 468–469
Burgess, H. 463–464
Burgess, J. C. 425
Burgess, M. 45
Burgess, N. D. 60, 378, 448
Burgiel, S. 398
Burgin, C. J. 63
Burivalova, Z. 390, 408

Author Index

Burke, B. 432
Burke, M. 382
Busch, T. 423, 487, 488
Büscher, B. 450
Butchart, S. H. M. 60, 442, 443, 451, 468
Butz, W. P. 235
Butzer, K. W. 11
Byerlee, D. 355

Cacho, O. 413
Cadilhon, J.-J. 210
Cai, M. 377
Cai, Y. 145
Cain, M. L. 36, 42, 51, 68
Cain, M. T. 18, 127
Caldecott, B. 424, 428, 429, 447, 489
Calder, W. J. 83
Caldwell, J. C. 231, 239
Caldwell, P. 231, 239
Callaway, T. 463–464
Cambridge Centre for Sustainable Finance, 433
Camerer, C. 156
Campbell, B. 205
Canavire-Bacarreza, G, 447
Cane, J. H. 52
Canfield, T. J. 376
Cantril, H. 293
Cao, L. 411, 412, 418
Cao, X. 317
Capaldi, C. A. 301, 302
Capen, P. D. 394
Caraco, N. F. 70
Cardinale, B. J. 52, 56, 57
Carlsson, F. 296
Carney, M. 432
Carnus, J. -M. 408
Caroll, N. 195, 477
Carpenter, S. 52, 71, 78
Carpenter, S. R. 70, 71, 73, 75, 83, 140, 408
Carreira, J. C. A. 454
Carroll, C. R. 112
Carroll, N. 472
Carruthers, T. J. B. 63
Carson, R. 144
Carson, R. T. 47, 50, 309
Carter, D. A. 430
Carter, L. 436
Cartwright, E. 228
Carus, M. 406
Carvalheiro, L. G. 416–417
Carvalho, L. 49, 94
Casey, D. 63
Cashdan, E. 207
Cassman, K. G. 54, 413
Castagneyrol, B. 116
Castañeda-Álvarez, N. P. 455
Casterline, J. 231, 232, 237
Casterline, J. B. 233
Cavender-Bares, J. 413

Cavendish, W. 201, 203
Caylor, K. K. 76
Ceballos, G. 88–89, 442
Cebrian, J. 39
Čegan, R. 420
Centre for Science and Environment, 18
Čermák, T. 420
Cervigni, R. 383, 384, 423
Chace, J. F. 377
Chaikneau, T. 374, 382
Chami, R. 360
Chan, K. M. A. 382–383, 417
Chancel, L. 374
Chanson, J. 60
Chanthorn, W. 48
Chapin, F. S. 52
Chapin, M. 206
Chaplin-Kramer, R. 318, 319, 379, 382, 383, 469–470
Chapman, A. D. 63
Chapman, R. A. 468–469
Chaudhary, A. 390, 391, 408
Chausson, A. 462, 463
Chaves, L. S. M. 393
Chazdon, R. L. 461
Chellam, R. 504
Chen, S. 419–420
Chen, W. 22
Chen, X. 227
Chenet, H. 432, 433, 481
Cheng, X. 419–420
Cheung, S. N. S. 190
Chichilnisky, G. 182, 328
Chin, K. Q. 113
Chinnamani, S. 413
Choi, Y. 238
Chopra, K. 171, 200
Chowdhury, M. S. H. 18
Christensen, V. 52
Christiaen, C. 489
Christiaensen, L. 10
Christie, M. 306
Christophers, B. 433
Ciccone, A. 386, 387
Cicowiez, M. 477–478
Civco, D. L, 377
Clark, A, 474
Clark, A. 296
Clark, A. E. 294, 296
Clark, M. 375
Clark, P. 400
Clarke, M. 451
Clarke-Samuels, C. 471
Clavelle, T. 411
Cleall-Harding, P. 117
Cleland, J. 233, 236, 237
Clemens, M, 511
Climate Bonds Initiative, 478
Cline, W. R. 123, 251, 254, 257, 258
Clobert, J. 68

Coad, L. 447, 451, 476
Coady, D. 199, 209
Coase, R. H. 192
Coastal Protection and Restoration Authority of Louisiana, 465
Cobb, L. 233
Cochrane, S. H. 231
Coe, D. T. 387
Coelho, F. M. 416–417
Cohen-Shacham, E. 459, 462
Cohn-Haft, M. 68
Colbert, M. 377–378, 447–448
Colchero, M. A. 229
Colchester, M. 200, 208
Cole, L. A. 510, 513
Colella, J. P. 63
Coleman, J. S. 157
Condit, R. 451
Connor, D. J. 417
Connors, E. 27, 313, 348, 349
Constantine, K. L. 468
Convention on Biological Diversity, 462, 478, 509
Convergence, 2019, 484–485
Cook, C. 447–448
Cook, C. N. 447–448
Cooke, S. J. 202, 458
Coolsaet, B. 466
Cooney, R. 177, 454
Cooper, G. 481, 482–483, 484, 486
Corazon, S. S. 304
Corbett, A. 469
Cordell J. 202
Cordoba, C. L. 210
Correll, D. L. 70
Cosimano, T. 360
Costanza, R. 28, 29
Costello, C. 394, 411, 412, 418, 447
Côté, I. M. 468
Cottingham, K. L. 70
Cowie, C. 272
Cowles, J. M. 51, 53–54
Cowling, R. M. 468–469
Cox, M. 47
Coyle, D. 361
CPIC, 486
Cragg, G. M. 312
Crépin, A. S. 513
Croft, S. A. 397
Croft, T. N. 237, 515
Crona, B. I. 382–383, 391, 394
Cropper, M. 28
Cropper, M. L. 278, 279, 280
Crouzeilles, R. 461
Crutzen, P. 85, 87
Crutzen, P. J. 29, 509
Cubbage, F. W. 408
Cuff, D. J. 36
Cullen, J. 433
Cumming, G. S. 83

Curtis, P. G. 391
Curtis, V. 227
Cushing, L. 374
Custodis, J. 7
Cutter, S. L. 382
Czakó, Á. 169
Czech, B. 334

Da Costa, V. 310
da Silva, A. V. M. 454
Dahal, P. R. 442, 443
Dahlbo, H. 100, 407
Daigneault, A. 464
Daily, G. 335
Daily, G. C. 52, 53, 112, 379, 382, 383, 469–470
Dakos, V. 71, 75
Dalin, C. 409
Dalton, M. 230, 376
Daltry J. C. 468
Daly, H. E. 334
Dammer, L. 406
Danila, F. R. 419–420
Dannenberg, A. 319
d'Arge, R. 29
d'Arge, R. C. 48, 127
Das, S. 46, 47
Dasgupta, Aisha, 98, 100, 101, 102, 104, 120, 222, 224, 233, 234, 235, 237, 319, 374, 514
Dasgupta, Amiya K. 125
Dasgupta, P. xxiii, 19, 25, 26, 27, 31, 35, 42, 48, 76, 77, 88–89, 97, 98, 100, 101, 102, 104, 120, 126, 128, 129, 131, 157, 159, 176, 182, 183, 197, 198, 201, 202, 204, 213, 222, 224, 228, 229, 232, 233, 234, 235, 236, 237, 240, 244, 250, 260, 262, 263, 264, 267, 269, 272, 273, 280, 282, 286, 291, 292, 317, 319, 329, 330, 334, 335, 337, 340, 342, 344, 354, 355–356, 357, 367, 374, 382, 511, 514
Dasgupta, S. 254
Daskalov, G. M. 413–414
Daszak, P. 454
Davey, E. 395
Davies, J. 467
Davis, A. Y. 383
Daw, T. 383
Daw, T. M. 391, 394
Dawson, M. N. 68
Day, R. H. 308
Dayton-Johnson, J. 203
de Camino, R. 361
de Groot, H. L. F. 159
De Groot, R, 29
De Groot, R. S. 465
De Jonge, J. 229
de Kong, H. 5, 6, 86, 87
De Long, J. B. 495
De Marco, P. 416–417
de Mattos, M. J. T. 422
De Palma, A. 61, 65
De Prato, G. 387

Author Index

de Sherbinin, A. 430
De Silva, M. M. G. T. 382–383
De Vos, J. M. 88–89
de Vries, S. 301, 303
Dean, R. 413
Dean, T. J. 425
Deaton, A. 294
Debreu, G. 108
Defra, 466–467
DeFries, R. 404
Defries, R. S. 391
Deininger, K. 355
Dell, R. 432
Dempewolf, H. 375
Dempsey, J. 423
den Braber, B. 447
Dennett, D. C. 315
Depledge, M. 312
Depledge, M. H. 303
Deryng, D. 116, 406
DeShazo, J. R. 50, 309
Deutsch, L. 85
Deutz, A. 473, 475, 479, 480, 482
Di Marco, M. 63, 442
Di Minin, E. 447
Diamond, J. 11
Diaz, A. 457–458
Dick, J. 449
Dickinson, R. E. 74
Dicks, L. V. 415, 446, 460, 466, 497
Dickson, B. 378–379
Diener, E. 294, 298
Dietrich, J. P. 378
Dietz, T. 376
Diffenbaugh, N. S. 382
Diffey, B. L. 301
Dimitrijevic, J. 437
Dimsdale, N. 22
Dinar, A. 382
Dinerstein, E, 444
Dithrich, H. 481
Dixon, H. 189
Dixon, S. J. 462
Dobson, A. P. 113, 495
Dolliver, K. 303
Donaldson, L. A. 458
Donati, P. 222
Dong, Z. 318
Dorling, D. 382–383
Dougherty, E. R. 448
Douglas, M. 224
Doust, M. 376
Downey, S. S. 11
Downing, J. A. 51
Drapek, R. J. 116, 304
Druel, E. 456
Drupp, M. A. 258, 263
Drutschinin, A. 479
Duarte, C. M. 63, 381, 447, 451, 452, 460, 463–464, 495

Dubel, A. K. 459, 470
Dudley, N. 447, 450
Duesenberry, J. S. 224
Duffy, J. E. 52, 56, 57
Duke, C. 376
Dunning, N. P. 11
Dupré, S. 432
Durán, A. P. 397
Durigan, G. 462
Durlauf, S. A. 228
Durlauf, S. N. 228
Dyer, E. E. 64, 468
Dynesius,M. 409

Easterlin, R. A. 296
EAT-Lancet Commission, 405
Ebeling, F. 228
Ebrahim, N. 199, 210, 411
EBRD, 480
Eccles, R. G. 438
Ecologist, 425, 426
Edgeworth, F. Y. 246, 255, 292
Edwards, D. P. 207–208, 400
Edwards, S. 387
Egoh, B. N. 455
Ehrlich, A. H. 29, 72, 88–89, 144
Ehrlich, P. R. 13, 29, 52, 53, 72, 88–89, 98, 99, 112, 144, 319, 442
EIB, 480
Eichengreen, B. 495
Eisenmenger, N. 388
Ekanem, I. I. 235
Ekins, P. 445
Elert, E. 419
El-Hage Scialabba, N. 417
Elie, L. 374
Elliott, J. 116, 406
Elliott, W, 447
Ellis, E. C. 63, 466, 469
Ellis, P. W. 474
Ellison, D, 55
Ellsberg, D. 155
Elmqvist, T. 53, 54
Elsen, P. R. 448
Emerson, J. W. 430
Emmer, I. 482
Engel, S. 193
Engen, L. 479
Ensminger, J. 207, 208
Epstein, L. 28
Erb, K. -H. 62, 94, 377, 388, 391
Erickson, B. 378, 420
Erisman, B. 449
Ermakova, M. 419–420
Esters, J. A. 52
Estes, D. 302
Estevadeordal, A. 386
Etsy, D. C. 430
EU Technical Expert Group on Sustainable Finance, 488

European Central Bank, 427, 428, 439
European Commission, 188, 438
Evan, A. T. 466
Evans, K. L. 447, 448–449
Evans, M. 104
Ewel, J. J. 52
Eyles, H. 229
Ezeh, A. 236, 237

Fa, J. E. 378, 448
Fabic, M. S. 238
Falck-Zepeda, J. B. 419
Falconer, A. 474
Falconer, J. 200, 202
Falkowski, P. 41
Fanning, A. L. 496
FAO, 45, 109, 314, 375, 378, 381, 388, 398, 405, 406, 408, 410, 411, 416, 466
Farfan, M. A. 383
Fargione, J. 68
Farid, S. M. 231
Farooq, G. M. 235
Farrell, T. A. 447
Faust, C. L. 112
Fayaz, H. 421
Fayaz Bhat, Z. 421
Feder, G. 206, 207
Feenstra, R. C. 386
Feeny, D. 200, 204
Femenia, F. 175
Feng, S. 22, 317
Fenichel, E. P. 27, 338
Fenuik, C. 414
Fernandez-Cornejo, J. 419
Ferraro, P. J. 44, 193, 196, 228, 357, 455
Ferreira, J. G. 418
Ferreira, M. S. 461
Ferrier, S. 59
Ferri-Yáñez, F. 119
Field, C.B. 41
Field-Dodgson, M. S. 413–414
Filgueira, B. 315
Finance for Biodiversity, 433
Finley, M. I. 158, 165, 306
Fischer, G. 409, 410
Fischer, J. 466
Fishel, J. D. 237, 515
Fisher, A. C. 147
Fisher, B. 382, 391
Fisheries New Zealand, 189
Fiske, G. J. 96
Fitch, D. H. A. 90
Fitoussi, J.-P. 294
Fitter, A. H. 445, 462
Fitz-Gibbon, S. T. 117
Fletcher, R. 450
Fleurbaey, M. 294
Flyvbjerg, B. 69
Foa, R. 296, 299

Folberth, C. 414, 415
Folda, L. 233
Foley, J. A. 404, 405, 414, 417, 466, 497
Foley, P. 400
Folke, C. 53, 54, 71, 78, 140
Fong, W. K. 376
Food and Land Use Coalition, 378, 420
Forrester, J. W. 124
Fortes, M. 232
Fothergill, A. 382
Foulis, A. 433
Fouquet, R. 7
Fox, J. 472
FP2020, 183
Fragkias, M. 377
Frank, O. 232
Frank, R. H. 226
Frankel, J. A. 386
Frankenberg, E. 235
Franks, D. M. 400
Fransen, B. R. 413–414
Frantz, B. 386
Frantz, C. M. 303
Frazier, A. G. 119
Frazier, M. 381, 456
Freedom House, 291
Freeman, M. C. 258
Freeman III, A, M. 22, 139, 305, 310
Freer-Smith, P. 408
Freudenberger, D. 466
Friedman, R. 416–417
Friends of Ocean Action, 450
Frijters, P. 294, 296
Frost, P. G. H. 194–195
Frumkin, H. 303
Fry, J. 393
Fuchs, R. 230, 376
Fudenberg, D. 166
Fukuyama, F. 157
Fullenkamp, C. 360
Fuller, E. 382–383
Fuller, R. A. 377, 408
Fulton, E. A. 411
Fulton, M. 438
Fung, I. Y. 381
Furbank, R. T. 419–420
Furtado, C. 182

G20 Sustainable Finance Study Group, 484
Gaertner, W. 283
Gaillard, E. 504
Galetti, M. 47
Galland, G. R. 449
Galor, O. 20, 47
Gambetta, D. 177
Gamboa, R. L. 468
Gandiwa, E. 195
Gann, G. 457
Gao, L. 317, 318

Author Index

Gardner, T. A. 119
Garibaldi, L. A. 380, 413
Garnett, S.T. 378, 448
Garnett, T. 404
Gatiso, T. T. 227
Gawith, D. 464
Gebbers, R. 420
Gebrehiwot, S. G. 55
Geertz, C. 172–173
GEF, 480
Gelcich, S. 411, 412, 418
Geldmann, J. 447, 451, 476
Genovesi, P. 468
Gerba, C. P. 45
Gerkey, D. 447
Geschke, A. 391–392
Ghimire, K. 451
Ghosh, S. 310
Gibb, R. 113
Gibbard, A. 285
Gibbs, H. K. 404
Gilboa, I. 156
Gilchrist, D. S. 222
Giljum, S. 406
Gill, D. A. 447, 451
Gilman, E. 411
Gingrich, S. 94, 377
Giollasch, S. 468
Gittleman, J. L. 88–89
Gjerde, K. 456
Glantz, M. H. 318
Glaser, S. A. M. 388
Global Canopy, 433, 436, 438
Global Footprint Network, 104, 510
Global Resources Initiative, 395
Global Witness, 480–481
Glynn, P. W. 117
Godar, J. 17, 391, 404
Godart, F. 224
Godfray, H. C. J. 404
Goedkoop, M. J. 437, 438
Golden Kroner, R. E. 447–448
Goldewijk, K. K. 63, 469
Goldsmith, G. R. 119
Golet, G. H. 196
Gollier, C. 278, 279, 280
Gómez, C. M. 475
Gómez-Baggethun, E. 477
Gonzalez, A. 52, 56, 57, 377–378
Gonzalez, P. 116, 304
Gonzalez, R. 65
Gonzalez, R. E. 65
Goody, J. 231
Gopalakrishnan, S. 337
Gordon, A. 471
Gordon, H. S. 197
Gore, A. 495
Goren, E. 229
Gössling, S. 374

Gottdenker, N. L. 112
Gottwald, S. 383
Goudie, A. S. 36
Gould, R. K. 314
Goulder, L. 27, 337
Goulder, L. H. 25, 97, 334, 335, 337, 340, 342, 354, 355–356, 357
Govaerts, R. 63–64
Goyal, S. 176
Grace, D, 113
Gramig, B. M. 195
Granato, J. 157
Gray, C. L. 447, 451
Green, J. M. H. 397, 498
Green, P. A. 447
Green, R. 414, 446, 460
Green, R. E. 414, 415, 446, 460, 497
Green, S. 468
Green Finance Platform, 438
Gregr, E. J. 52
Greif, A. 165
Greiner, R. 413
Grice, J. 27, 313, 348, 349
Griffin, J. T. 244
Griffin, R. 470, 471
Griffiths, C. 28
Grigg, A. 426, 437
Griliches, Z. 233, 313
Grinevald, J. 85, 87
Griscom, B. W. 474
Grishin, A. N. 413–414
Groom, B. 35, 258
Grorud-Colvert, K. 447
Grossman, G. M. 28, 385
Grossman, S. J. 190
Groucutt, H. S. 10
Grove-White, R. 314
Gruère, G. 210
Gruner, D. S. 57, 58, 59
Guagnin, M. 10
Gulati, S. C. 171
Gunn, J. 495
Gupta, B. 7
Gupta, U. 311
Gurr, S. J. 419–420
Gutiérrez Rodríguez, L. 318
Guttmacher Institute, 236, 237, 499
Guyer, J. I. 232

Haas, W. R. 11
Haberl, H. 94, 377, 388
Hacker, S. D. 36, 42, 51, 68
Haddad, F. F. 459
Haddad, N. M. 68
Haff, P. K. 87, 94
Hahs, A. K. 377
Haines-Young, R. 44, 510
Hale, L. Z. 189
Hall, J. W. 145

Author Index

Hall, M. G. 228
Hallegatte, S. 377
Hallstein, E. 196
Halpern, B. S. 63, 381, 442, 456, 457–458
Hamann, M. 374, 377–378, 382, 447–448
Hamer, K. C. 207–208
Hamilton, K. 511
Hamilton, S. 47, 458, 462
Hamilton, S. E. 63
Hammer, M. 413–414
Hammond, P. J. 293
Hamrick, K. 484
Hanauer, M. M. 447, 455
Hanemann, W. M. 409
Hanley, N. 35, 306
Hannah, L. 48, 74, 95, 111, 113, 115, 116, 119, 327, 495
Hanscom, L. 373
Hanscom, P. 107, 119
Hansen, M. 391
Hanson, A. 318
Hanson, C. 416
Haque, A. K. E. 22, 139, 308
Harb, A. M. 69
Hardin, G. 24, 197
Harfoot, M. 90, 388
Harris, G. 69
Harris, N. L. 391
Harris, W. E. 447, 448–449
Harrison, I. J. 447
Harrison, P. A. 90
Harrod, R. F. 251
Harsanyi, J. C. 250, 257, 281
Hartig, F. 48
Hartig, T. 303
Hartmann, H. 408
Harwood, A. R. 330, 396
Hastings, A. 71
Hatfield, B. 451
Hattab, K. O. 310, 311
Hausman, J. 308
Hawksworth, D. 90
Hawthorne, P. 378
Häyhä, T. 406
Hazarika, S. 74
He. G. 227
Heal, G. 133, 155, 156
Heal, G. M. 126, 129, 131, 213, 244, 263, 268, 328, 352, 363, 473, 475, 479, 480, 482
Heath, J. 206
Hecht, S.B. 202
Hector, S. 432
Heeb, F. 487, 488
Heffetz, O, 300
Heflin, F. 430
Held, H. 83, 145
Heller, M. C. 421
Helliwell, J. F. 294, 295, 298, 299, 300, 514
Hellmann, J. J. 116
Helpman, E. 20, 47, 387

Hemming, S. 448
Hengl, T. 96
Henriksen, R. 177, 453
Henry, C. 147
Henry, P. 117
Hepburn, C. 466, 496
Hepburn, J. 388, 394, 396
Herath, J. 311
Hercelin, N. 482, 484–485, 487
Hernán, G. 459, 470
Hero, J. M. 457, 459
Heron, T. 397
Herriges, J. A. 139
Hertel, T. W. 391
Hicks, J. R. 331
Higgins, C. B. 418
Higgins, S. L. 303
Hilbers, J. 117, 118, 119
Hilborn, R, 411
Hill, S. L. L. 65, 403, 404, 408, 447
Hilton-Taylor, C. 451
Hime, S. 426, 484
Hirota, M. 74
Hirsch, F. 226
Hirschi, S. 423, 424, 427, 428
Hislop, M. 469
HM Treasury, 279, 329, 455
Hobsbawm, E. J. 177
Hochard, J. P. 47
Hock, R. 381
Hockings, M. 447, 450
Hodge, I. D. 417
Hodgson, J. A. 207–208
Hoegh-Guldberg, O. 116, 117
Hoekstra, A. Y. 409
Hoel, M. 113, 266, 365
Hof, C. 407
Hoffmaister, A. W. 387
Hoffmann, A. A. 117
Hoffmann, M. 451
Hogarth, N. J. 318
Holdren, J. P. 13, 319
Hole, D. 447
Hole, D. G. 412
Holland, D. S. 189
Holland, T. G. 420
Hollands, G. J. 228
Holling, C. S. 71
Hollinger, D. A. 176
Holmes, G. 447, 448–449
Holmes, N. D. 468
Holmgren, M. 74
Holmlund, C. M. 413–414
Homer-Dixon, T. F. 74
Honda, E. A. 462
Honey-Rosés, J. 455
Hooke, R. L. 96
Hooper, D. U. 51
Hooten, A. J. 116

Author Index

Hopcraft, J. G. C. 69
Hoppit, G. 456
Horan, R. D. 195
Hoskins, A. 65
Howard. P. H. 154
Howarth, N. 424, 428
Howe, J. 202
Howes, M. 457, 459
Howitt, P. 20, 47
HSBC Global Asset Management and Pollination Group, 2020), 485
Huang, H. 300
Huang, Z. 22, 387
Hubbell, S. P. 34
Hudson, L. N. 403, 404, 408
Hufbauer, G. C. 387
Hughes, J. 463, 504
Huijbregts, M. A. J. 68, 409
Humpe, A. 374
Humphrey, C. 480
Humphrey, P. 430
Hunter, C. 310
Huntingford, C. 119
Hurren, K. 189
Hurwicz, L. 156, 250
Huwlyer, F. 483, 484, 486
Hyde, W. F. 202

IAASTD (International Assessment of Agricultural Knowledge, Science and Technology for Development), 417
Iannaccone, L. R. 222
Ichii, K. 62–63, 64, 65, 381, 442, 448
IDH The Sustainable Trade Initiiative, 401
ILO, 446, 465
IMF, 483, 500
Imperatriz-Fonseca, V. 413
Inchausti, P. 67
The Indian Express, 311
Influence Map, 480–481
Inglehart, R. F. 157, 294, 296, 299
Inklaar, R. 5, 6, 86, 87, 386
Inter-American Development Bank, 480
International Assessment of Agricultural Knowledge, Science and Technology for Development (IAASTD), 417
International Energy Agency, 408
International Whaling Commission, 451
Investec, 489
Ioannou, I. 438
IPBES, 94, 95, 96, 200, 381, 466
IPCC, 87, 104, 116, 117, 119, 144, 381, 405, 412, 462
IPNI, 63–64
IQAir, 192
IRP, 389–390, 392
Irvine, K. N. 377
Irwin, E. G. 337
Isbell, F. 51, 53–54
Isham, J. 204
Isherwood, B. 224

Islam, M. 341
ITPS, 45
IUCN, 60, 64, 89, 357, 452, 454, 459, 471
IUCN-WCPA Task Force on OECMs, 2019, 450

Jack, B. K. 193
Jactel, H. 116
Jalan, J. 141
Jalong, T. 208
Jamison, D. T. 9
Janetski, N. 460
Janoski, T. 162
Jans, S. 419
Janzen, C. 459, 462
Jenkins, C. N. 34, 88–89
Jenks, E. 28
Jensen, R. 235, 376
Jepson, P. 461
Jepson, P. R. 447
Jetz, W. 446
Jiang, W. 419–420
Jirinic, V. 68
Jodha, 201
Jodha, N. S. 19
Johansson-Stenman, O. 296
Johnson, J. A. 383, 384, 400, 423
Johnson, J. T. 448
Johnston, M. 228
Johnston, R. J. 308
Jonas, H. 450
Jonas, H. C. 450
Jonas, H. D. 450
Jones, A. 380
Jones, B. A. 113
Jones, C. I. 313, 329
Jones, H. P. 458, 468
Jones, K. E. 112
Jones, K. R. 63, 442, 457–458
Jonides, J. 302
Jooston, H. 109
Joppa, L. N. 88–89
Jordà, Ò., 41, 375
Jorgenson, A. A. 391–392
Jose, S. 416–417
Joshi, S. 239
Juma, C. 10, 229
Jump, A. S. 454
Jung, M. 442, 443
Jungbluth, N. 417

Kadekodi, G. K. 200
Kahn, B. M. 438
Kahn, P. L. 63
Kahneman, D. 156, 221, 228, 292, 293, 294, 298, 309
Kalnay, E. 377
Kanemoto, K. 388, 391–392, 396
Kanie, N. 456
Kant, R. 100
Kantarova, V. 237

Kaplan, R. 302
Kaplan, S. 302
Käppeli, J. 483, 484, 486
Kaptchuk, T. J. 312
Kapteyn, A. 228
Karacaoglu, G. 301
Kareiva, P. 27, 64, 97, 318, 469–470
Karki, S. 419–420
Karl, D. M. 116
Karousakis, K. 395, 401
Kasterine, A. 454
Kastner, T. 62, 388, 392, 393, 409
Kaufmann, D. 204
Kawasaki, A. 382–383
Kaya, Y. 100
Kazis, P. 318
KC, S. 235
Keats, S. 375
Kedward, K. 432, 433
Keefer, P. 159, 171
Keesstra, S. 462
Kéfi, S. 76
Kellert, S. R. 302
Kelly, M. 416
Keoleian, G. A. 421
Kerr, S. 189
Keskin, M. 378, 420
Keteles, K. 312
Kettunen, M. 395, 468
Keynes, J. M. 9
Keyßer, L. T. 391–392, 431
Khabarov, N. 414, 415
Khan, H. 310
Khoury, C. K. 455
Kick, E. 391–392
Kidd, K. A. 312
Kiesecker, J. M. 472
Kilkie, P. 463–464
Kim, M. H. 416
Kim, R. 432
Kimball, M. S, 300
King, E. 413
Kintisch, E. 12
Kinzig, A. 53
Kinzig, A. P. 51
Kiontke, K. 90
Klain, S. C. 314
Klasing, M. J. 386
Kleijn, D. 466
Klein, A. M. 52, 263, 413
Klein, C. J. 63, 442, 457–458
Klenow, P. W. 355
Kling, C. L. 139, 308
Klomp, J. 427
Klotz-Ingram, C. 419
Knack, S. 159, 171
Kneese, A. V. 48, 127
Knetsch, J. L. 221
Knoll, K. 41

Knops, J. 51
Knops, J. M. H. 51, 53–54
Knutsen, C. H. 178
Koch, A. 462
Kock, R, 113
Koh, L. P. 390, 408
Kohler, H.-P. 232, 235, 499
Köhlin, G. 202
Kohn, R. E. 394
Koike, M. 18
Kölbel, J. 487, 488
Kolbert, E. 87, 88–89
Kollmann, J. 459
Kolm, S. -C. 104
Koopmans, T. C. 252, 253, 256, 263
Koppell, C. R. S. 200, 202
Korhonen, J. 408
Koricheva, J. 116
Kotsadam, A. 178
Koubi, V. 74
Koumbarakis, A. 423, 424, 427, 428
Kounang, N. 96
Kousky, C. 142, 143
Kouskya, C. 193
Kramer, R. A. 49
Krause, S. 317
Krausmann, F. 94, 374, 376, 377
Kreilhuber, A. 177, 453
Krenn, S. 233
Kriegler, E. 83, 145
Krost, P. 418
Krueger, A. B. 28, 293, 294, 385
Krugman, P. R. 226
Kuepper, B. 428
Kumar, P. 26, 27, 97, 98, 106, 313, 334, 337, 340, 341, 342, 380, 382
Kumar, S. 421
Kummu, M. 376
Kuoppa, J. 383
Kurz, M. 26, 83, 344, 363, 368
Kuvshinov, D. 41
Kuznets, S. 28
Kwon, Y. 464
Kyttä, M. 383

Lago, M. 475
Lam, D. 234
Lamb, C. T. 117
Lamb, W. F. 496
Lambin, E. F. 391, 415
Lambsdorff, J. G. 159
Lammerant, J. 437
Lan, J. 318
Landes, D. S. 8
Lange, O. 352
Langridge, J. 53
Larsen, B. B. 34, 63
Larsen, F. W. 378, 447
Lartigue, J. 39

Author Index

Latawiec, A. E. 446
Laurance, S. G. 400
Laurance, W. F. 68, 69, 447
Lavorel, S. 60
Lawlor, K. 44, 357
Lawrence, D. 55
Lawrence, S. N. 468
Layard, R. 292, 294, 296
Le, N.-P. 199, 209
Le Maitre, D. C. 468–469
Le Stradic, S. 462
Leblang, D. 157
Leclère, D. 396, 457–458
Lecy, M. A. 112
Lee, A. Y. 224
Lee, B. 388, 394, 396
Lee, E. 450
Lee, S. Y. 458, 462
Lehman, C. L. 52
Lei, T. 419–420
Leisher, C. 382–383
Leite, C. 160
Lele, S. 47
Lemay, M. A. 117
Lenihan, J. M. 116, 304
Lenton, T. M. 61, 65, 71, 73, 75, 83, 91, 92, 128, 145, 512, 514
Lenzen, M. 388, 391–392, 396, 431
Lerner, A. P. 352
Leshan, J. 318
Leslie, S. 189
Lester, S. E. 459, 470
Lesthaeghe, R. 232
Lévèque, C. 68
Levhari, D. 153, 278
Levin, L. A. 92, 93, 381
Levin, S. A. 36, 55, 56, 58, 74, 75, 76, 144, 509, 510, 511, 512, 513, 514, 515
Levis, S. 381
Levy, B. J. 353, 465
Lewis, S. L. 462
Lewis, W. A. 361
Li, B. V. 34
Li, X. 420
Libecap, G. D. 206
Lichtenstein, G. 454
Light, S. E. 142, 143
Lin, D. 104, 107, 119, 373
Lin, W. W. 419
Lindahl, E. 191
Lipper, L. 195
Little, I. M. D. 26, 264, 344
Liu, J. 227, 382–383, 417
Liu, L. 178, 318
Liu, S. 22
Liu, Z. 318
Lock, K. 189
Lomborg, B. 9, 144, 310
Longman, R. J. 119

Lonsdale, M. 398
Lontzek, T. S. 145
Lopatta, K. 423
Lopez, A. 318
López, R. 204
López-Feldman, A. 200
Loreau, M. 67, 377–378
Lorimer, J. 461
Lotz, S. 228
Lovejoy, T. E. 48, 55, 68, 74, 95, 111, 115, 116, 119, 327
Lovell, R. 303
Lovelock, J. E. 42, 316
Lovera, C. 117
Low, C. 196
Lowenberg-DeBoer, J. 378, 420
Lowndes, J. S. 381, 456
Lowry, M. S. 451
Lu, Y. 317
Lu, Z. 387
Lubchenco, J. 56, 94, 447
Luce, R. D. 168
Lücking, R. 90
Ludwig, D. 52, 83
Lugato, E. 45
Lugauer, S. 376
Lughadha, E. N. 63–64
Luisetti, T. 463–464
Lupi, F. 227
Lutz, S. 229
Lutz, W. 235
Luzzadder-Beach, S. 11, 44, 45, 96, 317
Lynch, J. 422

MA – Millennium Ecosystem Assessment, 28, 43, 94, 200, 510
Maas, J. 301
MacArthur, R. H. 89
Macdonald, D. W. 417, 450
Mace, G. M. 60, 92, 330
Macfadyen, G. 416
Mackey, B. G. 471
MacKinnon, K. 388
Maddison, A. 5, 6, 7, 8, 9, 85, 86, 87
Madheswaran, S. 249
Madrigal, R. 477
Maginnis, S. 459, 462
Magrath, W. 361
Maguire, P. 471
Mahon, R. 201–202
Mailath, G. J. 166
Mair, L. 451
Maldonado, S. 470, 471
Malek, Ž. 408
Mäler, K.-G. 25, 48, 76, 127, 329, 334, 337, 477–478
Malhi, Y. 119
Malik, A. 393
Malinowski, B. 164
Malinvaud, E. 137, 179, 189, 352
Maljković, A. 468

624   The Economics of Biodiversity: The Dasgupta Review

Managi, S. 26, 27, 97, 98, 106, 313, 334, 337, 340, 341, 342, 380
Mandondo, A. 205
Marchant, R. 454
Marglin, S. A. 26, 262, 264, 330, 344
Margulies, J. 418
Markandya, A. 51, 344, 345, 346
Maron, M. 471
Marques, A. 392, 393
Marrocoli, S. 227
Marteau, T. M. 228
Martens, L. 227
Martin, A. 466
Martin, P. 386
Martínez, L. 400
Martín-López, B. 447–448
Martins, I. S. 377–378, 392, 393
Marvier, M. 64
Mascia, M. B. 447, 448, 451
Maskin, E. 166, 193, 229, 280, 282, 283, 286, 287
Mason, I. 315
Matson, P. A. 54, 413
Matthews, A. 229
Maurer, J. 228
Mauro. P. 159
Maxwell, S. L. 408
May, K. O. 285, 286
May, P. 202
May, R. M. 52, 113
Mayer, F. S. 303
Mayraz, G. 292
McAusland, C. 394
McBride, C. S. 116
McCarthy, N. 195
McCay, B. 400
McCay, B. J. 200
McClean, C. J. 378
McCraine, S. 423, 424, 426
McDaniels, J. 429
McDonald, R. I. 377–378, 447–448
McDonald, T. 457
McDonnell, M. J. 377
McElroy, B. J. 96
McGuire, C. B. 310
McIvor, A. 88–89
McIvor, A. L. 425
McKean, M. 202
McKean, M. A. 203
McKelly, D. H. 468–469
McKenney, B. A. 472
McLuckie, M. 429, 488
McMahan, E. A. 302
McNeill, J. 85, 87
McSharry, P. 424, 428
Meacham, M. 466
Meade, J. E. 190
Meadows, D. H. 124
Meadows, D. L. 124
Mehrabi, Z. 403, 404, 409

Mehta, B. S. 204
Meier, A. 228
Meier, K. 423, 424, 427, 428
Meier, M. S. 417
Meijer, J. 61, 65, 117, 118, 119
Mekonnen, M. M. 409
Melillo, J. M. 94
Mendelsohn, R. 382
Menozzi, C. 237
Merenlender, A. M. 448
Meyer, S. T. 459
Meyfroidt, P. 391, 415
Miah, M. D. 18
Michener, C. D. 52
Micklethwait, J. 310
Mihneva, V. 413–414
Milder, J. C. 400
Miles, L. 378–379
Milgrom, P. 165
Milgrom, P. R. 165
Milionis, P. 386
Millennium Ecosystem Assessment, 28, 43, 94, 200, 510
Miller, D. C. 479
Miller, E. C. 34, 63
Miller, G. 237
Milligan, B. 445
Millner, A. 133, 155, 156
Mills, K. H. 312
Milo, R. 41, 62
Milton, J. 459
Ministry of Fisheries, 189
Minx, J. 375, 387
Mirrlees, J. A. 26, 129, 251, 264, 331, 344
Misak, C. J. 254
Mission Économie De La Biodiversité, 437
Mitchard, E. T. A. 462
Mitchell, G. 303, 382–383
Mitchell, P. 416
Mitchell, R. 303
Mitchell, R. B. 456
Miteva, D. A. 455
Mitra, T. 19, 198, 260, 317
Mittermeier, R. A. 378
Mo, P. H. 159
Molnar, J. L. 468
Molnár, Z. 62–63, 64, 65, 381, 442, 448
Molthan, P. 482–483
Monaghan, J. M. 421
Monahan, W. B. 448
Monfeda, C. 466
Montgomery, M. R. 233
Mooney, H. A. 94
Mora, C. 63, 68, 119, 149
Moran, D. 388, 391, 392, 396, 404
Moran, D. D. 391–392
Morelle, R. 456
Morello-Frosch, R. 374, 391
Morgan, D. 227
Mori Junior, R. 400

Author Index

Morikawa, M. K. 101
Morris Jr, J. A. 418
Morrison, D. 468
Morse-Jones, S. 382
Moss, S. 436
Mouquet, N. 377–378
Mudaliar, A. 481
Mueller, B. 206
Mueller, G. M. 90
Mukhopadhyay, P. 205, 310
Mulla, D. J. 420
Mullan, K. L. 44, 357
Muller, A. 417
Muller, S. 448
Mullin, K. 303, 382–383
Mulungu, K. 473
Mumby, P. J. 116
Munro, N. 466
Munshi, K. 235
Muradian, R. 477
Murina, M. 398
Murphy, S. T. 468
Murthy, A. 373
Murty, M. N. 22, 51, 139, 200, 308, 344, 345, 346
Mwakiwa, E. 195
Myaux, J. 235
Mysiak, J. 475

Nabangchang, O. 50
Nadarajah, S. 317
Naeem, S. 67
Nagel, T. 270, 271, 317
Nagrawala, F. 424, 483, 488
Naidoo, R. 447
Nakamura, K. 411
Naqvi, M. 432
Narain, S. 200
Narayan, D. 204
Narayan, S. 425, 465
Narula, A. 419
Nasar, S. 161
Nasurudeen, P. 310, 311
National Academies of Sciences Engineering and Medicine, 460
National Research Council, 413
Natural Capital Coalition, 423, 437
Natural Capital Committee, 27, 330, 441
Natural Capital Finance Alliance, 435
Natural England, 109, 380
The Nature Conservancy, 142, 143, 196, 482
Nature Map, 442
Nauman, B. 480–481
Navarro, L. M. 377–378
Nawaz, N. R. 383
Nawn, N. 504
Nayak, D. R. 469–470
Neate, V. 484
Neilson, R. E. 116, 304
Nellemann, C. 177, 453

Nelson, E. 400
Nelson, E. H. 378
Nelson, F. 450
Nelson, G. C. 406
Nemarundwe, N. 205
Nemecek, T. 375, 376, 388, 406, 497–498
Nepal, P. 408
Nepelski, D. 387
Nesje, F. 258
Nesmith, C. 201–202
Netting, R. McC. 165, 201, 202, 207
Network for Greening the Financial System, 433, 439
Neubert, M. G. 64
Newbold, T. 61, 115, 403, 404, 408, 451
Newby, D. T. 45
Newell, R. G. 189
Newman, D. J. 312
Newsome, T. M. 64
Newton, A. C. 457–458, 461
Ng, S. 224
Ng, S. W. 375
Ng, Y.-K. 292
Ngo, H. T. 413
Ngumbi, E. N. 421
Nichol, L. 52
Nicita, A. 398
Nickell, S. 292
Nicolson, S. W. 416–417
Niinimäki, K. 100, 407
Nijstein, L. 444
Nilsson, C. 409
Nino-Murci, A. 472
Nisbet, E. K. 301, 302
Niu, R. 473, 475, 479, 480, 482
Noar, S. M. 228
Noble, I. 388
Nobles, J. 235
Nobre, C. 55
Nogués-Bravo, D. 461
Noone, K. 91, 92, 121, 125, 128, 342, 512
Norberg, J. 310
Nordhaus, W. D. 9, 123, 124, 145, 157, 248, 254, 257, 258, 260, 350
Norman, P. 303, 382–383
Noronha, R. 206, 207
North, D. C. 163, 165, 207
Novara, A. 462
Nowak, M. A. 52
Nunes, J. 462
NYDF Assessment Partners, 397
Nyström, M. 53, 54

Oakerson, R. J. 204
Obersteiner, M. 396, 457–458
O'Brien, P. K. 308
Obura, D. O. 62–63, 64, 65, 381, 442, 448
O'Callaghan, B. 466, 496
Ockenden, S. 479
Odoni, N. A. 462

OECD, 92, 111, 185, 199, 209, 210, 211, 294, 300, 423, 473, 475, 476–477, 479, 480, 482, 484, 487, 498, 500
Ogilvie, G. 165
Oldekop, J. A. 447, 448–449
O'Leary, B. C. 443, 456
Olmstead, S. M. 409–410
Olsen, E. H. 178
O'Neill, B. C. 230, 376
O'Neill, D. W. 496
Onial, M. 414, 460
ONS, 109
Oppel, O. 468
Ord, T. 245
Orgiazzi, A. 45, 46
Ortiz, J.-C. 63
Oschlies, A. 92, 93, 381
Oster, E. 235
Ostrom, E. 108, 169, 200, 205, 208, 394
Oswald, A.J. 240
Oteros-Rozas, E. 466
Our World in Data, 386
Ouyang, Z. 318, 319, 320, 321–322, 323, 325, 470
Ouzman, J. 378, 420
Ovando, D. 411
Oztosun, S. 360

Pacala, S. W. 51
Paetzold, F. 487, 488
Pagiola, S. 19, 193, 195, 229
Pailler, S. 448
Paine, R. T. 56
Palomo, I. 447–448
Palumbi, S. R. 101
Panagos, P. 45
Parent, J. 377
Parfit, D. 251, 271, 272, 273
Parisien, M. A. 448
Parks, S. A. 448
Parmesan, C. 116
Parry, I. 199, 209
Paschos, V. 484
Passfield, K. 411
Passmore, H.-A. 301, 302
Pastor, M. 374
Patel, N. G. 112
Patil, I. 47
Pattanayak, S. K. 19, 44, 49, 193, 196, 202, 357
Patterson, D. 429, 430
Pauly, D. 189
Pavlin, B. I. 454
Payn, T. 408
Payne, J. 352
Pearson, A. L. 302
Pechal, J. L. 302
Pechey, R. 228
Pecl, G. T. 381
Pedersen, M. 475
Pednekar, S. 310
Peek, L. A. 382

Peh, K, S.-H. 143
Pendrill, F. 17, 391, 404
Peng, C. 318
Peng, A, 419–420
Peñuelas, J. 454
Pepper, I. L. 45
Pereira, H. M. 59, 377–378
Perring, M. P. 457, 459
Perrings, C. 35, 195, 306, 398, 468
Persson, U. M. 17, 391, 404
Peters, G. 100, 407
Peters, G. M. 100
Peterson, C. 296, 299
Peterson, M. 154
Petraglia, M. D. 10
Pfaff, A. 387, 398, 404, 408, 413
Phalan, B. 414, 415, 446, 460, 497
Phaneuf, D. J. 308
Phang, S. M. 418
Philipson, T. J. 356
Phillips, R. 41, 62
Pierrehumbert, R. 422
Pigou, A. C. 184
Piketty, T. 374
Piljic, D. 440
Pilling, M. 228
Pimentel, D. 45, 96, 468
Pimm, S. L. 34, 88–89, 113, 495
Pindyck, R. S. 133, 153
Pinker, S. 9, 310
Pinzón, A. 429, 488
Pison, G. 232
Pizo, M. A. 47
Plank, B. 388
Platteau, J.-P. 163, 200, 201, 204
Plieninger, T. 466
Plutzer, C. 62
Poiner, I. 495
Polasky, S. 380, 394, 413, 513
Polenske, K. R. 318
Pollice, R. 432
Pollnac, R. 201–202
Pollock, L. J. 446
Poloczanska, E. S. 116
Pomeranz, K. 8
Pomeroy, R. 201–202, 383
Poore, J. 375, 376, 388, 406, 497–498
Popkin, B. M. 375
Porras, I. 318
Portfolio Earth, 480–481
Pörtner, H. -O. 116
Portnoy, P. R. 309
Posen, P. 382
Possingham, H. P. 377–378
Potschin, M. 44, 447–448
Potschin, M. B. 510
Potts, S. G. 30, 413
Power, M. E. 52
Prabhakar, E. R. 204

Author Index

Prado, P. 397
Prasad, J. V. N. S. 416–417
Pratt, C. F. 468
Pratt, J. 136
Presley, S. J. 60
Prest, A. R. 344
Prestwich, A. 228
PRI, 482
Price, G. 419
Price, J. N. 457, 459
Price, M. K. 228
Priebsch, M. 136, 154
Pritchett, L. 204
Pritchett, L. H. 234
Prizzon, A. 479
Proctor, H. C. 90
Property Rights Alliance, 378
Puettmann, K. J. 408
Puiu, D. 117
Puma, M. J. 409
Purvis, A. 61
Putnam, H. 157, 289
Putnam, R. D. 294, 514
Pyle, R. M. 504

Qaim, M. 419, 420
Qin, S. 447–448
Queiroz, C. 466

Rabiei, R. 418
Radner, R. 352
Raes, L. 469–470
Rahbek, C. 461
Raiffa, H. 133, 168
Ramankutty, N. 403, 404, 405, 409, 414, 417, 466, 497
Ramsey, F. P. 251, 256, 261, 273
Randall, A. 337
Randers, J. 124
Randerson, J. T. 41
Rankoth, L. M. 416–417
Rao, C. S. 416–417
Rasmussen, L. V. 466
Rasul, G. 381
Raven, P. H. 34, 35, 88–89, 379, 442
Rawls, J. 114, 182, 190, 246, 249, 251, 253, 275, 290
Ray, D. 77
Raymond, C. M. 383, 466
Rebdeiro, J. 447
Redding, D. W. 113
Reemer, M. 413
Rees, M. J. 245
Reganold, J. P. 417
Regetz, J. 52
Reguero, B. G. 444, 465
Reich, P. B. 51, 53–54
Reidy, C. A. 409
Reisch, L. A. 228
Renard, D. 57
Reny, P. 282, 285

Repetto, R. 361
Research and Markets, 100
Research Northwest and Morrison Hershfield, 116
Restivo, M. 480
Revenga, C. 409, 468
Reyers, B. 92, 455
Reyes, M. 229
Reynolds, M. D. 196
Rhodes, J. R. 377–378
Rhodes, M. K. 34, 63
Ricardo, D. 386
Rice, J. 391–392
Richardson, D. M. 455
Richardson, K. 61, 65, 73, 91, 92, 128, 512
Ricketts, T. H. 27, 52, 97, 318, 469–470
Ridley, 310
Rieb, J. T. 379, 382, 383, 469–470
Rietkerk, M. 76
Rights and Resources Initiative, 378
Rigney, D. 448
Riis, J. 292
Riker, W. H. 286
Riley, J. C. 7, 308
Riley, S. P. 177
Riordan, P. 417
Rios, A. R. 195
Ripple, W. J. 64
Rizal, G. 419–420
Roberts, C. M. 378
Roberts, J. T. 479
Roberts, S. P. M. 30, 413
Robins, N. 429, 432, 488
Rochette, J. 456
Rockström, J. 73, 91, 92, 121, 125, 128, 342, 512
Rodionov, S. 413–414
Rodríguez-Clare, A. 355
Rodriguez-Iturbe, I. 76
Roe, D. 177
Rogers, A. D. 411
Rogerson, M. 463, 504
Rohitha, W. R. 310, 311
Rolinski, A. 405, 406
Romer, D. 386
Romer, P. M. 129, 329
Rosa, E. A. 376
Rose, J. M. 418
Rosendo, S. 383
Rosenzweig, C. 116, 406
Rosenzweig, M. L. 67, 89
Rothschild, M. 134
Rouget, M. 455
Rounsevell, M, D. A. 90
Rous, A. M. 458
Rousseau, Y. 411
Rowcroft, P. 436
Royal Society, 405
Royal Society for the Protection of Birds (RSPB), 387, 464
Rude, J. 189
Rudel, T. 391

Rudgley, G. 423
Rudra, A. 165
Ruesch, A. S. 404
Rundle, S. 117
Rutt, C. L. 68
Ryan-Collins, J. 432, 433
Ryder Jr., H. E. 268
Ryrholm, N. 116
Ryu, C. 464
Rzotkiewicz, A. 302

Sachs, J. D. 160, 294
Sadler, C. 376
Sahlins, M. 164
Sala, E. 444, 447
Sala-i-Martin, X. 20, 47
Salemdeeb, R. 416
Salzman, J. 195, 477
Sampson, G. 399
Samraj, P. 413
Samuelson, L. 166
Samuelson, P. A. 191, 331
SANBI, 467
Sanchez-Ortiz, K. 52, 65
Sánchez-Rodríguez, R. 377
Sanchirico, J. N. 189
Sander, H. 378
Sanderman, J. 96
Sanders, N. J. 461
Sanderson, R. A. 67
Sandin, G. 100
Sandom, C. 461
Sands, E. G. 222
Sands, R. D. 406
Santos, R. 472
Sato, M. 337
Satoh, Y. 409, 410
Satterthwaite, M. A. 285
Saunders, M. I. 460
Savage, L. J. 133, 135, 152, 155
Sayer, J. A. 69
Scanlon, T. M. 76
Scarborough, P. 229
Schaafsma, M. 94, 382
Schader, C. 417
Schaepman-Strub, G. 381
Schaffaertzik, A. 388
Schama, S. 162, 314
Scheffer, M. 11, 71, 72, 73, 74, 75
Scheffler, S. 275
Schell, J. 274
Schellekens, G. 436, 439, 440
Schernewski, G. 418
Schipper, A. 61, 65, 117, 118, 119
Schipper, A. M. 391–392
Schkade, D. A. 294
Schlager, E. 208
Schloegel, L. M. 454
Schmeidler, D. 156

Schmidt, J. P. 90
Schmitt, J. P. 68, 409
Schmitt, S. 429, 430
Schmitt, S. F. 447
Schoenmaker, D. 140, 141, 488
Scholes, R. J. 61
Schramade, W. 140, 488
Schroder, T. 418
Schröter-Schlaack, C. 472
Schroth, G. 400
Schuhbauer, A. 199, 210, 411
Schultz, T. P. 239
Schultz, T. W. 329
Schulze, K. 408
Schwindt, A. R. 312
SCM Direct, 487
Scott, A. J. 469
Seabright, P. 171
Sear, D. A. 462
Seddon, N. 462, 463
Seebens, H. 64, 468
Seega, N. 423
Segan, D. B. 447, 450
Segers, H. 68
Seidl, A. 473
Sekerli, Y. E. 378, 420
Selig, E. R. 116, 412
Selvaraj, P. 310, 311
Sen, A. 183, 248, 284, 293, 294, 331, 515
Sen, A. K. 26, 262, 264, 330, 344
Senbeta, F. 416–417
Senik, C. 296
Serafeim, G. 438, 487
Serageldin, I. xxiii
Seto, K. C. 377
Seufert, V. 417
Sexton, S. 418
Seymour, C. L. 416–417
Sgró, C. M. 117
Shackelford, S. J. 199
Shajaat Ali, A. M. 10
Shan, Q. 419–420
Shanahan, D. F. 63, 442
Shandra, J. M. 480
Shanmugam, K. R. 249
Shantz, A. A. 509
Sharda, V. N. 413
Sharp, R. 318, 319, 469–470
Sharples, C. 438
Shaw, J. 310
Shaw, M. R. 423, 424, 426
Shen, L. 377
Shen, S. S. P. 74
Shennan, S. J. 11
Shi, C. 50
Shiau, J.-T. 317
Shields, M. A. 294, 296
Shogren, J. F. 195
Shove, G. F. 306

Author Index

Shyamsundar, P. 22, 139, 203, 205, 308
Sidenious, U. 304
Sidgwick, H. 244, 248, 251, 269, 271, 294
Siebert, S. 63, 469
Sik, E. 169
Sills, E. O. 19, 202
Silveiro, F. A. O. 462
Simberloff, D. 461
Simkins, B. 430
Simsa, K. R. E. 193
Sinding, S. W. 237
Singer, M, C. 116
Singer, P. 31
Singh, S. 429, 430
Skamnioti, P. 419–420
Slay, C. M. 391
Sloan, S, 69
Slovic, P. 156
Smith, P. 45, 469–470
Smith, R. J. 451
Smith, W. K. 400
Soares, R. R. 356
Sodhi, N. S. 67, 88–89, 143
Solórzano, R. 361
Solow, R. M. 22, 129, 250, 251, 309, 329, 339
Somanathan, E. 141, 204
Sommer, J. M. 480
Song, C. 318, 319, 320, 321–322, 323
Song, X. 318
Sorger, G. 19, 198, 260, 317
Sork, V. L. 117
Sotos, M. 376
Soury, A. 314
Southerton, D. 228
Spak, B. 100
Spalding, M. D. 425, 468
Sparkman, G. 227
Spatial Finance Initiative, 489
Spatz, D. R. 468
Spence, A. M. 352, 356–358, 359, 360
Springer, K. 424, 483, 488
Srinivas, K. 416–417
Srinivasan, T. N. 153, 278
Stadler, K. 391
Standards and Trade Development Facility, 2013, 398
Standish, R. J. 457, 459
Stank, J. 418
Stanley, E. H. 408
Starrett, D. 83, 262–263, 356–358, 360
Starrett, D. A. 192
State Office for the Environment (LfU), 450
Statista, 207
Stavins, R. 74
Stavins, R. N. 103, 319
Stefanescu, C. 116
Steffan, W. 61, 65, 67, 73, 85, 87, 88, 91, 92, 121, 125, 128, 342, 512
Steffan-Dewenter, I. 52, 413
Steinberger, J. K. 374, 376, 391–392, 431

Stephenson, J. 449
Stephenson, K. 418
Stern, N. 9, 123, 130, 157, 179, 227, 248, 251, 254, 257, 258, 307, 350, 466, 496
Sterner, T. 154, 266, 365
Stevenson, B. 294
Stewart, N. 426
Stiglitz, J. E. 31, 108, 129, 134, 204, 294
Stigsdotter, U. K. 304
Stoermer, E. F. 29, 509
Stoessel, F. 417
Stoett, P. 467
Stokes, F. 189
Stoll, J. S. 382–383
Stone, A. 294
Stone, G. D. 165
Stone, M. P. 165
Stork, N. E. 63–64
Storkey, J. 60
Strange, N. 471
Strassburg, B. B. N. 446
Streiker, D. G. 112
Stringer, R. 195
Strogatz, S. H. 33
Stuart-Smith, S. 303, 304
Studart, R. 353, 465
Subasinghe, R. 416, 419
Subramanian, A. 419
Suhrke, A. 74
Sullivan, B. L. 196
Sumaila, U. R. 199, 210, 411
Summers, L. H. 9
Sun, F. 47
Sundström, A. 178
Sunstein, C. R. 176, 228
Sur, C. 460
Sushinsky, J. R. 377–378
The Sustainable Finance Platform, 424
Sutherland, W. J. 466
Suttle, K. B. 471
Suttor-Sorel, L. 432, 482, 484–485, 487, 488
Swann, A. L. 381
Swartz, W. 391, 394
Sweatman, J. 509
Swiss Re Institute, 489
Sydeman, W. J. 116

Taillie, L. S. 229
Tallis, H. 19, 27, 97, 318, 469–470
Tan-Soo, J.-S. 49
Task Force on Climate-related Financial Disclosures, 424, 481
Task Force on Nature-related Financial Disclosures, 439
Taws, N. 413
Taylor, A. M. 386
Taylor, M. S. 129
Tchakatumba, P. K. 195
Teixidó-Figueras, J. 374, 376
Temin, P. 7
Temsah, G. 382–383

Author Index

ten Kate, K. 471
Thakur, V. 419–420
Thaler, R. H. 221, 228
Theuerkauf, S. J. 418
Thibert, M. 199
Thiel, D. 263
Thirgood, S. 69
Thomä, J. 432
Thomas, D. 235
Thomas, R. 22
Thomas, R. P. 207
Thomson, J. T. 204
Thorn, J. P. R. 416–417
Thoung, C. 436
Thouni, G. 429, 488
Thuiller, W. 446
Thurm, R. 487
Tietenberg, T. 188
Tilburt, J. C. 312
Tilman, D. 51, 52, 53–54, 57, 68, 375, 413
Timmer, M. P. 386
Tittensor, D. P. 63, 68, 149, 388
Tobin, J. 483, 484, 486
Toit, J. T. 450
Toivonen, T. 447
Tol, R. S. J. 48
Tomlinson, J. 337
Townley, A. 456
Traeger, C. P. 263
TRAFFIC, 227
Transparency International, 177, 291
TRASE and Forest 500, 397
Trauger, D. L. 334
Traxler, G. 419
Tree, I. 194, 461
Trémolet, S. 481, 482–483, 484, 486
Trentmann, F. 222, 224, 226
Trigo, E. 419
Trilnick, I. 420
Trumbore, S. 408
Tscharntke, T. 400
Tuomisto, H. L. 417, 422
Turner, B. L. 10
Turner, N. J. 449
Turner, R. K. 94, 463–464
Turner, W. 378
Turner, W. R. 447
Turner M. G. 83
Turvey, R. 344
Tversky, A. 156, 221, 228, 309
Twohig-Bennett, C. 380
Tyrrell, P. 450
Tyson, P. D. 67

Udawatta, R. P. 416–417
Udry, C. 165, 172
Ueta, K. 337
Ullmann-Margalit, E. 176
Ulph, A. 228

Ulph, D. 228
Umamaheswari, L. 310, 311
UN, 319, 336, 468
UN Climate Change, 100
UN Statistical Division, 28
UNAIDS, 113
UNCTAD, 399
UNDP, 337, 474, 475, 480, 484, 485, 486
UNEP, 25, 27, 97, 313, 322, 334, 337, 401, 433, 438, 439, 466, 481, 482, 496
UNEP Finance Initiative, 425, 433, 435, 436, 438
UNEP- WCMC, IUCN, NGS, 447
UNEP/PRI, 2019, 502
UNEP-WCMC, IUCN, 447
UNEP-WCMC, IUCN, NGS, 450, 451, 495
UNEP/WTO, 395, 396, 398, 399–400, 498
UNFPA, 183, 237
UNFSS, 398
UNGA, 199, 456, 460
Unger, M. von, 482
UNICEF, 408
Union for Ethical Biotrade, 431
Union of Concerned Scientists, 50
United States Fish and Wildlife Service, 188
UNODC, 453
UNPD, 7–8, 14, 122, 183, 225, 229, 230, 231, 236, 376, 377
UNU-IHDP, 25, 27, 97, 313, 334, 337
Upham, N. S. 63
Uriarte, M. 391
Usesche, D. C. 447
Usubiago-Liano, A. 445

Vaissière, B. E. 52
Valin, H. 406
van Beukering, P. 444
van Dam, Y. K. 229
van de Geer, S. 228
van de Stadt, H. 228
van Der Heijden, M. G. A. 45, 96
van der Linden, S. 304
van der Lugt, C. 487
van Der Ploeg, S. 465
van Gelder, J. W. 428
van Klink, R. 64
van Nes, E. H. 74, 75
van Schaik, A. B. T. M. 159
van Tilburg, R. 428
van Toor, J. 436, 439, 440
van Wilgen, B. W. 468–469
van Zanden, J. L. 5, 6, 86, 87
Vandecar, K. 55
Vander Zanden, M. J. 408
Vargas, M. T. 195
Varis, O. 376
Vasconcelos, H. L. 68
Veblen, T. 222
Veenhoven, R. 293, 295, 298, 299, 300
Veldtman, R. 416–417

Author Index

Vera, F. W. M. 461
Verburg, P. H. 408
Verheij, R. A. 301
Vermaat, J. E. 464
Vermeulen, F. 221
Veron, J. E. N. 378
Verones, F. 391
Verutes, G. M. 471
Vickers, L. H. 421
Vilela, T. 69
Vincent, J. R. 46, 50, 309, 310
Viscusi, W. Kip, 249
Vitousek, P. M. 51
Vogl, A. L. 469–470
Vollset, S. E. 229
von Braun, J. 406
von Caemmerer, S. 419–420
von Hase, A. 471
Voora, V. 399
Voosen, P. 9, 87
Voskamp, A. 407
Vynne, C. 444

Wachter, J. M. 417
Wackernagel, M. 104, 106, 107, 119, 373, 375, 387
Wada, Y. 409
Wade, R. 163, 202
Wagg, C. 45, 96
Wagner, A. 352
Wagtendonk, A. J. 464
Waha, K. 403, 404, 409
Walder, B. 457
Waldron, A, 444–445, 495
Walker, B. 53, 71, 78
Walker, G. 382–383
Walker, W. 49, 94
Wallace, D. 430
Walsh, J. J. 377
Walter, D. E. 90
Walters, G. 459, 462
Walters, M. 59
Walton, G. M. 227
Wambersie, L. 107, 119
Wander, M. 374
Wang, S. 294, 295, 299
Wang, X. 318, 378, 419–420
Wang, Y. 420
Wannenburgh, A. 469
Wansbeek, T. 228
Warde, A. 227
Warner, A. M. 160
Warren, J. 306
Warren, L. 315
Waters, C. N. 87, 88
Waters, R. D. 383
Waters, T. J. 418
Watkins, S. C. 232, 233, 235
Watson, J. E. M. 63, 408, 442, 447, 450, 451, 476
Watson, R. A. 411

Waycott, M. 63
Waygood, S. 438
WCVP, 63–64
Weber, C. 423, 424, 426, 432
Weber, G. 221
Weber, M. 156
Wedin, D. 51
Wee, A. P. 208
Weidmann, J. 160
Weingast, B. R. 165
Weisbrod, B. A. 147
Weitzman, M. L. 133, 153, 188, 354
Wells, M. 361
Welzel, C. 294, 296, 299
Wendling, Z. A. 430
Weng, L. 69
West, C. 397
Westoff, C. F. 237, 515
Westphal, M. I. 388
Weyzig, F. 428
Wheeler, B. W. 303
Wheeler, C. E. 462
Wheeler, T. 406
White, C. 436
White, M. P. 303
Whiteley, P. F. 159
WHO, 311, 408
Whyte, K. P. 448
Widmer, F. 45, 96
Wiedenhofer, D. 388
Wiedmann, T. 375, 387, 391–392, 431
Wiens, J. J. 34, 63
Wig, T. 178
Wiggins, D. 237
Wiggins, S. 375
Wildavsky, A. 157
Wilen, J. E. 200
Wilkinson, B. H. 96
Willer, H. 399
Williams, B. 248, 515
Williams, B. A. O. 270, 280
Williams, D. R. 414
Williams, J. 189
Williams, J. C. 313
Williams, L. 382
Williams, M. 87, 94
Williams, S. L. 460
Willig, M. R. 60
Willis, K. J. 312
Wilson, C. 233
Wilson, E. O. 49, 88–89, 121, 144, 302
Wilson, M. A. 70
Wilson, P. 425
Wilson, S. M. 408
Wilting, H. C. 391–392
Winfree, M. A. 413
Winkelman, D. L. 312
Winther-Jansen, M, 443
WIPO, 313

Author Index

Wohl, E. 345
Wolf, C. 64
Wolfers, J. 294
Wolff, B. 68
Wolff, N. H. 63
Wong, C. 318, 322, 323
Wong, M. C. 208
Wood, S. 318, 319, 469–470
Woodward, R. 361
Wooldridge, A. 310
World Bank, 18, 27, 104, 178, 224, 225, 229, 236, 290, 291, 310, 374, 376, 382, 478, 480, 484
World Economic Forum, 408, 423, 424, 425, 433, 465
World Food Programme, 375
World Resources Institute, 201–202
Worm, B. 67
Wortley, L. 457, 459
WRAP, 179, 416
Wright, G. 456
Wrigley, E. A. 7, 19
WTO, 385, 397
WTTC, 310
Wu, F. 454
Wu, Y. 377
Wunder, S. 193, 195, 196
WWF, 63–64, 96, 195, 375, 381, 387, 409, 412, 424, 487, 489, 498
WWF France, 424, 437, 438, 487

Xepapadeas, A. 129
Xiao, Y. 319, 322, 383
Xu, W. 322

Yaari, M. E. 245
Yamaguchi, R. 334, 337, 341
Yamaguchi, S. 395, 401
Yamamura, E. 382
Yang, W. 387
Ye, Q. 318

Ye, Y. 317, 318
Yeong, H. Y. 418
Yokobori, K. 100
York University Ecological Footprint Initiative and Global Footprint Network, 373, 374
Young, M. D. 409–410
Young, O. R. 394
Yu, X. 387
Yu. B. 454
Yuan, C.-W. 229

Zachmann, G. 140, 141
Zadek, S. 432
Zagheni, E. 376
Zak, P. J. 159, 171
Zalasiewicz, J, 87, 88, 94
Zayas, C. 380
Zelazowski, P. 119
Zeng, Xiaodong, 74
Zeng, Xubin, 74
Zerbini, A. N. 451
Zhang, B. 387
Zhang, G. 22
Zhang, J. 318
Zhang, W. 19, 229
Zhang, X. 377
Zhang, Y. 317
Zhao, J. 308
Zhao, S. 454
Zheng, G. 22
Zheng, H. 318, 319, 320, 321–322
Zhou, G. 318
Zhou, H. 419–420
Zhou, W. 318
Zilberman, D. 195, 418, 419, 420
Ziter, C. 377–378
ZSL, 64, 454
zu Ermgassen, E. K. H. J. 416
Zuniga, R. 468

Subject Index

Locators in *italic* refer to figures; **bold** to tables; <u>underline</u> to glossary entries

ABNJ (areas beyond national jurisdiction), 456
abundance of species
    effects of climate change, *117*
    measures of, 60, *64*
accounting prices, 20, *21*, 24
    arbitrage conditions, 365–367
    assets, 327–328
    biodiversity, 306–308; *see also* valuing biodiversity
    biomass, 40
    capital, 278
    consumption, 261, 263–264
    definitions, 331, <u>509</u>
    dynamic economies, 264–265
    evaluation of environmental policies, 265–266
    inclusive wealth, 363–364
    optimisation problem, 364–365
    productivity as, 310
    transformative change, 498–499
acronyms, list of, 507–508
Action Plan on Sustainable Finance (EC), 438
Adapting Agriculture to Climate Change Project, 455
adverse selection, insurance markets, 137–138
advertising, to bring about behavioural change, 229
affect balance, measures of well-being, 293–294
affection, as basis of trust, 161–162
Africa
    bushmeat, 227
    contraceptive use, 235
    elephants, rewilding project, 461
    grasslands, 467
    invasive species, 468–469
    Kofyar farmers, 165
    land tenure systems, 207
    indigenous resource programmes, 194–195
African Risk Capacity (ARC), 140
aggregate demand, global economy, 98–100
aggregate supply, human impacts on the biosphere, 100–101
agriculture; *see also* food production
    agricultural subsidies in Switzerland, 476–477
    cellular agriculture, 421
    common pool resources, 201
    environmental land management in England, 466–467
    genetic modification, 418–420
    impact on the biosphere, 388, 390–391
    meat analogues, 421–422
    organic, *417*, 417
    plantations, 54–55, 207–208
    precision, 420–421
    regenerative, 418
    sustainable management, 466–467
    vertical farming, 421

agri-environment schemes, 466
agroforestry, 416–417; *see also* forests and deforestation
Aichi Biodiversity Targets, 199, <u>509</u>
Aid for Trade initiative, 401
air pollution *see* pollution
airspace, as common heritage of mankind, 198
alien species *see* invasive non-native species
Amazon rainforest
    as common heritage of mankind, 198
    forest fires, 488
    fragmented ecosystems, 68, 78
    future scenarios, 119
    ownership rights, 206
ambiguity; *see also* risk and uncertainty
    definition, 133
    dealing with, 155–156
amenity value, nature, 131, 305, 310
American pika (*Ochotona princeps*), 117
Ancient Greek mythology, 270–272
Andros Island, Bahamas, 470
Anthropocene, 9, 29, 87–89, *88*, <u>509</u>
antibiotic resistance, 53
Antiguan racer snake (*Alsophis antiguae*), 468
aquaculture *see* fisheries and fish farms
Arab Women Organization (AWO), Jordan, 459
arbitrage conditions
    accounting prices, 365–367
    definition, <u>509</u>
    nature as an asset, 23–24
    produced and natural capital, 367–368
ARC (African Risk Capacity), 140
Arctic commons, 198
areas beyond national jurisdiction (ABNJ), 456
Aristotle, on well-being, 289–290
Arrow's impossibility theorem, 282–284
ASN Bank, Netherlands, 438
aspirations, history-dependent, 224
assets, 327; *see also* economic evaluation; enabling assets; natural assets; valuing biodiversity
    amenity value, 305
    arbitrage conditions, 23–24
    biodiversity, 31–32
    capital goods, 20, *21*, 327–328, 329
    classification/categories of, 329
    conservation, 441–442
    conservation vs. pollution, 48–49
    definition, <u>509</u>
    earth systems and economic growth, 27–29
    free goods, 19
    global management, 243
    inclusive wealth, 330–335
    institutions, 30–31

Subject Index

assets (cont.)
    land, 328
    natural resource stocks, 22
    nature as an asset, 17
    portfolio management, 17–20, 22–23
    public assest management, 24–25
    rates of return, 22–23
    rural poor, 18–19
    total vs. marginal values, 29–30
    types of evaluation/comparisons, 26–27
    valuing, 327–328
    and well-being, 24–25, 27, 301–303, 327, 331–334
atmosphere, as open access resource, 24
attention-restoration theory, 302
auctions, habitats for migrating birds, 196

babassu products, common pool resources, 202
Bahamas, spatial planning, 470
Bangladesh, contraceptive uptake, 235
BECCS (Bioenergy Carbon Capture and Storage), 407
beekeeper example, mutualism, 190–191
bees, (*Apis mellifera*), pollination, 413
behavioural changes, induced, 227–229
    reproductive behaviour, 232–235, *234*
behavioural norms, 107
belief systems, 171, 173
belonging, social capital, 173–174, 175–176
BES *see* Biodiversity and Ecosystem Services
bifurcation points, 78–79, 174; *see also* regime shifts
big data, sustainability financing, 489
biodiversity, 3, 14, 33–36, 40, 56–57
    biomes, 36
    and climate change, 115–119, *407*
    definitions, 33, 34, 509
    and disturbance, 56
    as enabling asset, 25
    financial risk/uncertainty, 423, 436–437
    fragmented ecosystems, 68
    future scenarios, *407*
    geographical distribution, 378–381
    impact-based risk assessments, 436–438, 440
    indicators, 59–60, 94–95, 441–442
    and land use, *404*
    measures, 35, 62
    offsets *see below*
    portfolio management, 17–20
    and production/consumption, *393*
    and productivity, 56–57
    resilience of ecosystems, 51–54
    restoration of ecosystems, 457, 458, 462
    soils, 44–45, *46*, 52, 96
    and state corruption, 159–160
    and trade, 391–393
    transformative change, 496
    value to humans, 31–32; *see also* valuing biodiversity
    and well-being, 301–303
Biodiversity and Ecosystem Services (BES), 383, *384*
Biodiversity Finance Initiative (BIOFIN), 474
Biodiversity Footprint for Financial Institutions (BFFI), 438
Biodiversity Intactness Index (BII), 61, 64, 92
biodiversity offsets
    high speed rail project in France, 471, **472**
    restoration of ecosystems, 471–472, **472**
    sustainability financing, 477
Bioenergy Carbon Capture and Storage (BECCS), 407
biofuels
    current usage, 406–407
    future scenarios, 407
biomass, 36–37
    accounting prices, 40
    carrying capacity of ecosystems, 58, *59*
    definition, 509
    measures, 61–62
    rates of return, 40
biomes, 36, 509
biophilia, 302
biophysical processes, planetary, *91*, 92
biosphere, 33; *see also* disruptions; human impacts
    as common heritage of mankind, 198–199
    definitions, 13, 509
    ecological health, 9–10
    effects of trade, 388–394
    regenerative rates, 13–14, 42–43, 85
biotic integrity (community composition), 60, 61
bird migration, habitats for, 196
black swan (*Cygnus atratus*), 431–432
blended finance, sustainability financing, 484–485
blue bonds
    Seychelles example, 478
    sustainability financing, 478
blue whale (*Balaenoptera musculus*), 357–360
body mass index (BMI), health breakdowns, 76–77
border adjustment taxes, 394–395, 498
bounded global economy, 110–112, 123–124
    contemporary models, 129
    ecosystem goods and services, 130–131
    new model, 125–128
    substitution of goods and services, 124–125
Brazil, common pool resources, 202
Bretagne-Pays de la Loire, railway line, 471, **472**
bribery, 178; *see also* corruption
Brundtland Commission, 336–337
bushmeat, Northern Republic of Congo, 227
business-as-usual, when to stop, 144–147

Caisse des Depôts Group (CDC) Global Biodiversity Score, 437
California, habitats for migrating birds, 196
CAMPFIRE (Communal Areas Management Programme for Indigenous Resources), Zimbabwe, 194–195
Cantril ladder, measures of well-being, 294, *295*
capacity to do physical work, 76–77
capital accounting prices, 278; *see also* accounting prices
capital assets *see* assets; natural capital
capital goods, 20, 327–328
    accounting prices, 20, *21*
    classification/categories of, 329
    definition, 509

and inclusive wealth, 334–335
　India example, 356
　produced, human and natural capital, *21*
capital punishment, 251
carbon dioxide, atmospheric, 87
carbon flow, ecosystems, *39*
carbon sink, atmosphere as, 24
carbon storage
　ecosystem services, 47–48
　forests, 455, 462
cardinal functions, measures of well-being, 246–247
Caribbean Catastrophe Risk Insurance Facility, 140
Caribbean Coastal Zone Management Trust, 142–143
carrying capacity, ecosystems, 58, *59*
catastrophe, 155–156; *see also* ecosystem collapse; natural disasters
categories of consumption goods, 225–226
CBD (Convention on Biological Diversity), 395
CDRs *see* Consumption Discount Rates
cellular agriculture, 421
change *see* transformative change
Cheonggyecheon River restoration, 464
childrearing costs, 232
children as wealth perspective, 232
China
　ecosystem services, 317–318, 320–322, **323**
　Gross Ecosystem Product, 317, *319*
　National Ecosystem Assessment, 322–323
Chinese taxonomy, People's Bank of China, 487
Christianity, 315
CICES (Common International Classification of Ecosystem Services), 43–44, 50
CITES (Convention on International Trade of Endangered Species), 395–396, 451, 453
citizen investors, 243, 248, 330
citizenship, empowered, 503
civic virtues, basis of trust, 162
civil liberties, measures of well-being, **291**
civil society, 158, 175
classical utilitarianism *see* utilitarianism
Climate and Nature Sovereign Index (CNSI), 429–430
climate change, 9–10
　and biodiversity, 115–119, *407*
　biofuels/fibre for textiles, *407*
　changes already happening, 115–117
　Consumption Discount Rates, 257–258
　definition, 509
　effects of trade, 388
　food production, *406*, 406
　future scenarios, *117*, *118*, 119
　geographical distribution of impacts, 381
　protected areas, 448
　restoration of ecosystems, 457
　signals, 116
climax species, successional, 56
Coase theorem, externalities, 192–193
coastal zone management
　restoration of ecosystems, 463–464
　spatial planning in Belize example, 470–471

Coastal Zone Management Trust, Mexico, 142–143
cocoa, cultivated ecosystems, 54–55
cognitive component of happiness, 294, *295*
Colombia, payments for ecosystem services, 477, **478**
comb jelly (*Mnemiopsis leidyi*), 413
Common International Classification of Ecosystem Services (CICES), 43–44, 50
common pool resources (CPRs), 199–200
　cooperation, 199, 205, 218–219
　definition, 509
　fragility, 203–205
　geographical distribution of impacts, 382
　rural poor, 200–203
　state malfeasance, 204, 218
common pool resources, management, 213
　average and marginal productivity curves, *214*
　cooperation and non-cooperation scenarios, *215*
　modifications of the model, 217–219
　mutual enforcement, 199, 216–217
　privatisation, 217, 218
　quotas and taxes, 215–216
　timeless world example, 213
　unmanaged example, 214–215
commons, global, 31; *see also* tragedy of the commons
Communal Areas Management Programme for Indigenous Resources (CAMPFIRE), Zimbabwe, 194–195
communal property, 197
communitarian institutions, 158, 174
　iddirs, 172
　rotating savings and credit associations, 173
community composition/structure, 57–58, *59*
competition
　and GDP, 361–362
　among rival services, 28
competitive social preferences, 222–224, *223*
　consumption, 240–241
　house size, 226
　reproductive behaviour, 232
complementarities, ecosystems, 110–112, 125
composition effect, trade, 386–387
conformist social preferences, 222–224, *223*
　consumption, **241**, 241
　reproductive behaviour, 233–235, *234*
Confucianism, 315
connectedness with nature, 301, 302–303, 503, 504; *see also* contact with nature
conservation of nature
　coastal cloud forests of Western Ecuador, 450–451
　definition, 510
　ecosystem assets, 441–442
　ecosystem stocks, 443–446
　ecosystems, 48–49
　global map of terrestrial habitats, *443*
　indigenous peoples, 448, 449
　job opportunities, 465–466
　making the case for, 458
　marine conservation, 456
　Multilateral Environmental Agreements, 455–456
　place-based conservation, 447–451

Subject Index

conservation of nature (cont.)
    planning/evaluation, 455
    vs. pollution, 48–49
    protected areas, 444–445
    regime shifts/tipping points, 75
    and restoration, *441*, 441
    species-led, 451–455, *452*
    spectrum of conservation/restoration efforts, *446*
    and state corruption, 159–160
    transformative change, 495–496
Conservation status, IUCN Red List, 60
conspicuous consumption, 222, 227; *see also* competitive social preferences
consumer surplus, ecotourism example, 310
consumerism, 224, 226
consumption
    accounting price of consumption, 261, 263–264
    categories of consumption goods, 225–226
    determinants of well-being, 254–256
    discount factors, 306
    effects of trade, *393*
    and externalities, 182
    practices, 225–227
    social preferences, 222–225, *223*, 239–240, **241**
    transformative change, 497–498
    and well-being, 261–263, 289
Consumption Discount Rates (CDRs)
    intergenerational well-being, 256–258
    uncertainty conditions, 278, **279**, *280*
consumption streams, 331, 351
consumption targets, economic evaluation, 353–354
contact with nature, 302–303, 503; *see also* connectedness with nature
contingent-valuation method (CVM)
    problems with, 309
    stated preferences for public goods, 308
contraceptive use; *see also* family planning programmes
    Bangladesh, 235
    Kenya, 235
Convention on Biological Diversity (CBD), 395
Convention on International Trade of Endangered Species (CITES), 395–396, 451, 453
Convention on the Law of the Sea *see* UNCLOS
cooperative strategies, 173–174, 199, 205, 218–219
coping strategies, resource shortages, 10–12
coral reefs, 116–117, 460
corporate engagement, sustainability financing, 488–489
corruption, state, 159–160, 177–178
    and well-being, **291**, 299–300
Corruption Perception Index, 159
cost-benefit analysis, 26
    intergenerational well-being, 262, 266–267
    protected areas, 444–445
    restoration of ecosystems, 464–465
cost function, industrial pollution example, 187–188
cost of travel, ecotourism, 310
country comparisons *see* international comparisons
coupled ecosystems model, 151–152
COVID-19 pandemic, 227, 447, 493

CPRs *see* common pool resources
credit risks, nature-related, 427, **428**
CRISPR/Cas9 gene editing system, 419
crop plants, genetic modification, 418–420; *see also* agriculture; food production
Crops Shortfall Insurance (CSI) UK, 139
crying wolf, stopping business-as-usual, 144
cultivated ecosystems, 54–55; *see also* agriculture
cultural ecosystem services, 43, 54
    definition, 510
    valuation of, 321
customs, social, 164

damming of rivers, fragmented ecosystems, 68–69
Day Reconstruction Method, well-being measures, 294
dead zones, freshwater habitats, 78
death, 270–272, 273–276; *see also* human extinction
debt-for-nature swaps, sustainability financing, 480
Decade of Ecosystem Restoration (2021–2030), UN, 460
decision-makers, for transformative change, 226
decision rules, when to stop business-as-usual, 145–147
declaration of principles, use of the seabed, 199
deforestation *see* forests and deforestation
degradation of nature, and distancing from nature, 17
degraded ecosystems, cost of restoration, 70; *see also* disruptions to the biosphere
demand and supply; *see also* impact equation
    addressing imbalances, 494–495
    global variations, 373, *374*
    human impacts on the biosphere, 98–101, 374–376
democratic voting *see* voting rules
demographic transition
    family planning programmes, 235–237
    future scenarios, 229–230, *231*
    reproductive behaviour, 229, 231–232
demonstration effect, 224
Denmark, pesticide tax, 475–476
depreciation of natural capital, 97, 510
desertification, Zarqa River basin example, 459
diet, human; *see also* food production
    future scenarios, 405–406
    impacts on the biosphere, 375–376
    nutritional status, *77*
    transformative change, 497–498
disability adjusted life years (DALYs), 249
disasters, insurance markets, 139; *see also* catastrophe; ecosystem collapse
discount factors
    consumption, 306
    environmental projects, economic evaluation, 350
    intergenerational *see below*
discounting future generations, 250–252, 254–256
    directives, 258–260
    zero discounting, 252–253
diseases
    crop plants, 419–420
    and fragmented ecosystems, 68
    and land-use changes, 112–113
    spread of viruses, 112–113

and wildlife trade, 454
dispersal of species, fragmented ecosystems, 68
disruptions to the biosphere
　clues to approaching tipping points, 75–76
　conservation ecology, 75
　damming rivers, 68–69
　eutrophication of freshwater habitats, 70–71
　fencing of grasslands, 69
　fisheries, 72–73
　fragmented ecosystems, 67–68
　human body as ecosystem, 76–77
　hysteresis, 70, 73–74
　non-linearity, 67, 70, 73
　phosphorus recycling in lake systems example, 77–82
　regime shifts/tipping points, 74–75, 77–78
　stability regimes, *71*, *72*, 73
distancing from nature, and ecosystem degradation, 17; *see also* connectedness
disturbance
　and biodiversity, 56
　fragmented ecosystems, 67–68
downgrading, downsizing and degazettement (PADDD), protected areas, 448
drinking water *see* water supply
drives, behavioural, 221, 233
drugs, medical, effects on biodiversity, 312–313
durability, assets, 19–20

earth systems, economic growth, 27–29; *see also* biosphere
Easter Island, societal collapse, 11
Easterlin paradox, measures of well-being, 296–297
EBVs (Essential Biodiversity Variables), 59–60
eco-compensation programmes, Quinghai, China, **322**
EcoEnterprises Fund, sustainability financing, 485
ecological footprint, 12–15, 98; *see also* human impacts
　definition, 510
　effects of trade, 391–392
　food waste, 416
　and GDP, 29
　geographical distribution, 375
　global variations in demand and supply, 373, *374*
　impact inequality/equality, 224
　and income, *104*
　sustainable, 496–499
ecological risks; *see also* ecosystem collapse
　coupled ecosystems model, 151–152
　single ecosystems model, 151
　translation of ecological into economic risks, 150
ecological solutions *see* nature-based solutions
econometric theory, behavioural change, 227–228
economic evaluation (policy analysis and sustainability assessment), 26, 329, 335
　composition of inclusive wealth, 338
　definition, 510
　discount rates for environmental projects, 350
　economic progress concept, 105, 337–339, 360–361
　future scenarios, 368–369
　GDP and inclusive wealth, 342–343
　global changes in inclusive wealth, *341*
　inclusive wealth, 330–336, 337, 340–341, 343–344, 346–348
　intergenerational well-being, 331–334
　internal rate of return, 354
　optimum development, 351–352, *352*
　present value criterion, 346–348
　production/consumption targets, 353–354
　project complimentarities, 353
　restoration of Exmoor mires, 348–*349*, **349**
　River Ganges restoration project, 344, **345–346**
　simplifications to the model, 334
　six questions of, 335
　slow growth forests example, 350–351
　sustainable development/SDGs, 335, *336*, 337
　total factor productivity growth, 339–341
　trade, externalities and wealth transfers, 339
　types of evaluation/comparisons, 26–27
　UK Green Book, 329–330
　wealth/well-being equivalence theorem, 327, 331–334
economic growth, 7–10
　contemporary models of macroeconomic growth, 129
　and earth systems, 27–29
　idea of infinite, 103
　limits to, 9–10
　paradox of, 103
　and sustainable development, 354–355
economic history, 4–7, **5**
'Economic Possibilities for Our Grandchildren' essay (Keynes), 9
economic progress
　and GDP, 159
　as growth in wealth, 105, 337–339, 360–361
　transformative change, 493, 499–500
　and trust, 159–160
economic risk, translation of ecological risk into, 150
ecosystem/s; *see also* restoration of ecosystems
　architecture, top-down/bottom-up influences, 52
　biodiversity, 33–36, 40, 56–57
　carbon flow, *39*
　carrying capacity, 58, *59*
　complementarities, 110–112, 125
　conservation vs. pollution, 48–49
　definitions, 35, 510
　density, spatially spread ecosystems, 207
　functional diversity, 51, 53
　human body as ecosystem, 76–77
　micro/macro scales, *34*, 40
　modularity, 50, 55–56
　net primary productivity, 42
　plantations, 54–55
　population dynamics of fisheries, 37–38
　primary producers, 36–37, 38–40
　productivity and resilience, 50, 56–57
　regenerative rates, 42–43
　spatial configurations, 379–380
　structure, measures, 59
ecosystem assets *see* assets
ecosystem collapse; *see also* ecological risks
　effects of trade, 393–394

Subject Index

ecosystem collapse (cont.)
  geographical distribution, 383, *384*
  historical perspectives, 10–12
  insurance, 142
  risk reduction, 143–144
ecosystem depletion by region, Living Planet Index, *381*, 381; *see also* geographical distribution
ecosystem function, measures, 59, <u>510</u>
ecosystem services
  and biodiversity, 305
  cascade, *44*
  China example, 317–318, *319*, **323**
  classification/categories of, 43–44
  and conservation, 441–442
  definition, <u>510</u>
  eco-compensation programmes, **322**
  effects of trade, 385, 393
  geographical distribution, 379–380, 383
  invisibility and silence, 46–47
  mathematical models, 130–131
  non-anthropocentric perspectives, 51
  productivity and resilience, 50, 54
  Quinghai example, 320–322, **321**
  restoration of ecosystems, 457
  seed dispersal by large animals, 47–48
  subsidies, **209**, 211
  valuation of, 320–323
  watershed services, 49–50
ecosystem stocks
  amount needed for conservation, 443–445
  improving/increasing, 446
  quantity/quality measures, 62–65
  restoration of ecosystems, 457–460
  types, 445–446
  value of, 22
ecotourism, 305
  biodiversity, 310
  cost of travel, 310
  Quinghai, China, **321**, 321
Ecuador, cloud forests, 450–451
edges (fragmented boundaries), 68
education for transformative change, 503–504
EEZs (exclusive economic zones), 197, 198
efficiency factors, 179
  impact inequality, 15, 343
egoism, 221–222
elephants, rewilding project, 461
ELM (environmental land management), England, 466–467
embeddedness in nature, humans, 29, 32, 103
emotions, measures of well-being, 293–294
employment
  and GDP, 362
  transformative change, 496
empowered citizenship, transformative change, 503
enabling assets, 25, 328–329, 331
  biodiversity as, 25, 57, 305, 306
  definition, <u>510</u>
  finance as, 473; *see also* sustainability financing
enabling environments, transformative change, *501*

England *see* United Kingdom
environmental, social and governance (ESG) factors, 481–482, 483, 486–487
environmental collapse *see* ecosystem collapse
environmental impacts, life cycle analysis, 421–422; *see also* ecological footprint; human impacts
environmental justice, distribution of impacts, 382–383
environmental Kuznets curve, 28
environmental land management (ELM), England, 466–467
environmental policies *see* policies/regulations
equality *see* impact inequality; inequality
Essential Biodiversity Variables (EBVs), 59–60
ethics *see* moral philosophy/ethics
EU@Biodiversity Initiative, 439
EU Commission, biodiversity footprint, 439
EU Taxonomy, sustainability financing, 488
EU@Biodiversity Initiative, 439
Euphydryas editha butterfly, climate change, 116
European Social Survey (ESS), well-being, 292, 296
eutrophication, freshwater habitats, 70–71, 78–79
evolutionary perspectives, hominids/humans, 3–4
exclusive economic zones (EEZs), 197, 198
existence value, biodiversity, 305, 313–314
existential risks, human extinction, 273–276
Exmoor mires restoration project, 348–*349*, **349**
expectation, 133
expected utility theory, 133, 134–135, 155–156, <u>510</u>
exploitation of ecosystems, 178, 380–381, 452–453; *see also* human impacts
external enforcement, and trust, 161, 163–164
externalities, 30–31, 218–219; *see also* reciprocal externalities; unidirectional externalities
  definition, 179, <u>511</u>
  fisheries, 42
  insurance markets, 138, 140
  reproductive, 183–184, 224
  social, 180, 221–222; *see also* social embeddedness
  sustainable development theorem, 339
extinctions, 14, 458
  and deforestation, 89–90
  of humans, 250–252, 273–276, 305
  human impacts on, 88–89, 92
extreme events, approaching tipping points, 76

family planning programmes, 234, 235–237
  Bangladesh, 235
  Kenya, 235
  Matlab experiment, 239
  unmet needs in reproductive health, 237–239
farming, vertical, 421; *see also* agriculture
fashion industry, impacts on the biosphere, 100
fat tails, probability distributions with, 133, 152–154
fencing of grasslands, fragmented ecosystems, 69
fertility rates, measures of well-being, 290, **291**
fibre for textiles
  current usage, 406–407
  future scenarios, 407
  species-led conservation, 454
fiduciary duties, sustainability financing, 481–482

financial risk and uncertainty, 423–424, 432; *see also* risk-assessment
  categories of risk, 424, *425*
  Climate and Nature Sovereign Index, 429–430
  credit risks, 427–428
  litigation risks, 427
  market risks, 429–430
  nature-related risks, 424–431, **428**
  operational risks, 430
  physical risks, 424–425
  transition risks, 426
  uncertainty over timing and impact, 431–433
  white, black and green swans, 431–432
financial systems, transformative change, 502–503
financing sustainability *see* sustainability financing
fire sales, 436
fires, fragmented ecosystems, 68
first-best worlds, 260–261, 276–277, 307
fisheries and fish farms
  accounting prices, 310
  biodiversity, 310–311
  common pool resources, 201–202
  current usage, 410, *411*
  future scenarios, 411–412, *412*
  impact on sea turtles, 395
  low trophic level aquaculture, 418
  open access resources, 24
  overcrowding, 42
  population dynamics, 37–38
  quotas, 189
  regenerative rates, 42–43
  stability regimes, *72*, 73
  tradable permit scheme, 189
  trade-offs of provisioning/regulating services, 413–414
food production; *see also* agriculture; diet
  and biodiversity, *404*
  current harvest, 403–404
  effects of climate change, *406*, 406
  future scenarios, 405–406
  global land cover trends (1700–2007), *403*
  land sparing, 414, *415*
  land use, 404
  meat analogues, 421–422
  total calorific demand, *405*
food waste
  provisioning goods, 416
  sustainable trade, 396
footprint, ecological *see* ecological footprints
forests and deforestation; *see also* Amazon rainforest; mangrove forests
  agroforestry, 416–417
  carbon storage, 455, 462
  conservation, 450–451
  current usage, 408
  ecosystem services, 49–50
  effects of trade, 391
  future scenarios, 408
  regime shifts/tipping points, 74–75
  restoration of ecosystems, 462
  risk reduction, 143–144
  slow growth forests example, 350–351
  species extinctions, 89–90
fossil fuels, accounting prices, 307–308
fragmented ecosystems, 67–68
  palm oil plantations, 207–208
France, biodiversity offsets, 471, **472**
free goods, 19, 20
freedom of choice, determinants of well-being, 299–300
freshwater habitats; *see also* Ganges River; phosphorous recycling; wetlands
  Cheonggyecheon River example, 464
  damming rivers, 68–69
  eutrophication, 70–71, 78–79
  phosphorus recycling in lake systems, 77–82
  restoration of ecosystems, 459
fruit growing example, mutualism, 190–191
fully comparable measures of well-being, 247
fully correlated risks, 138–141; *see also* risk and uncertainty
functional diversity, ecosystems, 51, 53–54, 511
future generations, externalities, 179–180; *see also* intergenerational well-being
future scenarios; *see also* transformative change
  biodiversity, *407*
  biofuels/fibre for textiles, 407
  climate change, *117*, *118*, 119
  economic evaluation, 368–369
  fisheries, 411–412, *412*
  food production, 405–406
  inclusive wealth, 368
  intergenerational well-being, 254–256
  population size, 229–230, *231*
  timber, 408
  uncertainty conditions, 278–280
  water supply, *410*

Ganges, River, 51, 194, 344, **345–346**
gardens/gardening, and well-being, 303–304
GDP *see* gross domestic product
gene-editing, 419
General Social Survey, US, measures of well-being, 292
genetic composition, 60
genetic diversity
  conservation, 454
  Millennium Seed Bank, 454–455
genetic modification (GM), 418–420
geographical distribution of well-being, 373
  dietary impacts, 375–376
  ecological footprint rations, *374*
  ecosystem collapse, 383, *384*
  ecosystem depletion by region, *381*, 381
  global variations in demand and supply, 373
  human demands on biosphere, 374–378
  human population dynamics, 376–377
  inclusive wealth, *380*
  inequality/poverty, 382–383
  international comparisons, 294–296, 298, **299**
  Living Planet Index, *381*, 381
  natural assets and biodiversity, 378–381, *379*

Subject Index

geographical distribution of well-being (cont.)
    technological change, 378
    urbanisation, 377–378
geographical perspectives
    demand and supply, 373
    GDP impacts, *384*
    transformative change, 493–494
GEP *see* gross ecosystem product
Germany, Spreewald Biosphere Reserve, 449–450
gift-exchange, 164–165
Global Asset Management and Pollination Group, HSBC, 485
Global Biodiversity Score (GBS), Caisse des Depôts Group, 437
Global Biomass Census, 62
global commons, 31; *see also* tragedy of the commons
global competition, and GDP, 361–362
global economy, bounded *see* bounded global economy
Global Environment Facility (GEF), 479–480
global financial system, transformative change, 502–503
Global Footprint Network (GFN), 106, 107, 510; *see also* ecological footprint
global health, and population size, **7**
Global Impact Investing Network (GIIN) impact investor survey, 481
global income per person, 9, 85
global initiatives, sustainability financing, 479–480
global land cover trends (1700–2007), *403*
global natural capital accounts, 96
Global Oceans Treaty, 501
global population *see* population size
global public goods, transformative change, 501–502
Global Resources Outlook (UNEP), 388–*389*, *390*
global risk pool system, 140–141
Global Risk Report, World Economic Forum, 408
global variations, demand and supply, 373; *see also* geographical distribution
glossary of terms, 509–515
Goa, India, common pool resources, 205
governmental roles, common pool resources, 202; *see also* policies/regulations
Grassland Programme, South Africa, 455
grasslands
    common pool resources, 216–217, 218
    fragmented ecosystems, 69
    functional diversity, 53–54
    regime shifts/tipping points, 74–75
    restoration of ecosystems, 459
Great Divergence, economic, 8, 223–224
green bonds
    Seychelles example, 478
    sustainability financing, 478
Green Bond Endorsed Project Catalogue, People's Bank of China, 487
Green Book, UK, 329–330
green finance, sustainability financing, 482
green investment, transformative change, 496
green spaces, and well-being, 303–304
green swans, 431–432

grim norms, 167–169, **168**, 216
gross domestic product (GDP), *86*, *87*
    definition, 511
    ecological footprint, 29
    and economic progress, 159
    and employment, 362
    geographical distribution of impacts, *384*
    and global competition, 361–362
    impact inequality, 15
    inclusive wealth, 342–343
    India example, 356, **357**
    measures of well-being, **291**
    significance of the measure, 360–361
    standard of living measure, 4–7, **5**, *6*
gross ecosystem product (GEP)
    China, 317, *319*, 320–322, **324**–**325**
    National Ecosystem Assessment, 322–323
growth *see* economic growth

habit, 224
habitat banking, 472
habitats for migrating birds, auctions, 196
happiness, measures of well-being, 292–293, *295*
hazard rate, intergenerational well-being, 245
health
    and biodiversity, 305, 311–312, *313*
    determinants of well-being, 299–300
    human body as ecosystem, 76–77
    India example, 356
    need for nature, 301–302
hedonic price, housing, 309
herbicide tolerant (HT) crops, 419
herd sizes, regulation, 216–217, *218*
high speed rail project, France, 471, **472**
hill-climbing method, 262, 352
historical perspectives
    consumerism, 226
    economic, 4–7, **5**
    hominids/humans, 3–4
    life on earth, 3
    reproductive behaviour, 233
    social preferences, 223–224
    societal collapse and adaptation, 10–12
    history dependence *see* hysteresis
hockey stick distribution, 87
hominids, evolutionary history, 3–4
honeybee (*Apis mellifera*), pollination, 413
house size, competitive social preferences, 226
HSBC Global Asset Management and Pollination Group, 485
human body as ecosystem, 76–77
human capital, 20, *21*, 329, 511
    capital goods, 328
    India example, 356
Human Development Index (HDI) –, 26
human extinction
    and biodiversity loss, 305
    intergenerational well-being, 250–252, 273–276
human health *see* health

Subject Index

human impacts on the biosphere, 85, 113–115; *see also* ecological footprint; population size
    biodiversity, 94–95, 115–119, *117, 118*
    deforestation, 89–90
    demand and supply, 98–101, 374–376
    dietary impacts, 375–376
    economic growth, 103
    ecosystem complementarities, 110–112, 125
    fashion industry, 100
    and GDP, *86, 87*
    geographical distribution, 374–378
    global population, 85, *86*
    global wealth per capita, *98*
    impact equation, 105
    impact inequality, 101, *102*
    and income, *104*
    inequality, 103–104
    institutions and technology, 107–108
    key markers of Anthropocene, 87–89, *88*
    land-use changes, 112–113
    natural capital accounts, 96
    oceanic deoxygenation, 92–*93*
    planetary boundaries, *91*, 92
    and quality of life, *95*
    restoration of peatlands, 109
    soils, 96
    species extinctions, 88–89, 92
    Sustainable Development Goals, 106–107
    trade, 388–394
human rights; *see also* property rights
    and externalities, 181, 182–184
    and pollution, 181, 183
hysteresis
    aspirations, 224
    definition, 511
    disruptions to the biosphere, 70, 73–74
    intergenerational well-being, 268
    phosphorus in lakes, *81, 82*
    restoration of ecosystems, 457, 458
    regime shifts/tipping points, 75

iddirs, insurance against natural disasters, 172
ideation, behavioural changes, 233
identity, social capital, 173–174, 175–176
illegal trafficking, wildlife trade, 453
immune function, need for nature, 302
impact-based risk assessments
    biodiversity loss, 436–438, 440
    nature-related risks, 436–439
    Task Force on Nature-related Financial Disclosures, 439
impact equation, 105, 179, 385, 473, 486–487; *see also* demand and supply
impact inequality/equality, 15, 85, 101, *102*, 353; *see also* human impacts
    converting inequality to equality, 15, 114
    definition, 511
    ecological footprint, 224
    effects of trade, 386, 387
    efficiency parameter, 15, 343

externalities, 179, 183
    population size, 119–122
    and rights, 183
    social preferences, 223
    sustainability financing, 473
    transformative change, 502–503
    and trust, 160
impact investor survey, Global Impact Investing Network, 481
inclusive wealth, 27, 97, 330–331
    accounting prices, 363–364
    changes in, 340–341
    definition, 511
    composition of, 338
    economic evaluation, 330–336, 337, 340–341, 343–344, 346–348
    economic progress as growth in, 105, 337–339, 360–361
    future scenarios, 368
    and GDP, 342–343
    geographical distribution, *380*
    India example, 355–357
    and intergenerational well-being, 331–334
    present value criterion, 346–348
    substitutability of capital goods, 334–335
    and sustainability, 335–336
    sustainable development, 337
Inclusive Wealth Criterion, 343
income; *see also* gross domestic product
    determinants of well-being, 299–300
    and ecological footprint, *104*
    global, 9, 85
    measures of well-being, 290–292, **291**
independent risks, 138–140; *see also* risk and uncertainty
India
    common pool resources, 205
    inclusive wealth, 355–357
indicators of biodiversity, 59–60
indigenous peoples
    CAMPFIRE Programme, 194–195
    global competition, 361–362
    property rights, 378
    protected areas, 446, 448, 449
Indonesia, tsunami, 235
industrial pollution
    environmental regulations, 187–188
    Pigouvian taxes, 185–186
industrial revolution, 8
inequality; *see also* impact inequality
    geographical distribution, 382–383
    human impacts on the biosphere, 103–104
    and sustainable development, 385
infinite economic growth idea, 103
information, value of, 141–142
information campaigns, behavioural changes, 228
information gaps, 187–188, 190
informed desires, 244
INNS *see* invasive non-native species
insect-resistant cotton, genetic modification, 419
institutionism, intergenerational well-being, 253–254

Subject Index

institutions, 30–31
    communitarian, 158
    definitions, 157, 511
    effective and ineffective, *172*
    institutional failure, 114
    polycentric institutional structures, 108
    social, 157–159, *158*; *see also* social capital; trust
    and technology, 107–108
    transformative change, 500
instrumental value of biodiversity, 31
insurance
    environmental collapse insurance, 142
    iddirs, 172
    ideal, 137–138
    transformative change, 502–503
Integrated Coastal Zone Management (ICZM), Belize, 470
integrity
    basis of trust, 162, 163, 176
    ecosystems, 35
Inter-American Development Bank's (IDB), Natural Capital Lab, 480
intergenerational externalities, 179–180
intergenerational well-being, 243–244, 262
    accounting prices, 261, 263–264, 278
    consumption, 254–256
    Consumption Discount Rates, 256–258, 278, **279**, *280*
    death, 270–272
    definition, 511
    discounting directives, 258–260
    discounting future generations, 250–252, 254–256
    evaluation of environmental policies, 265–266
    existential risks, 273–276
    first-best worlds, 260–261, 276–277
    future scenarios, 254–256
    human extinction, 250–252, 273–276
    hysteresis, 268
    and inclusive wealth, 331–334
    institutionism, 253–254
    investment, 266–267
    measures of well-being, 246–247
    optimum population, 272–273
    optimum saving principle, 261–263, 276–278
    population ethics, 269–272
    pragmatism, 254, 256
    return on investment, 259–261
    second-best worlds, 244, 261, 264, 277–278
    socially embedded preferences, 267–268
    transformative change, 499
    utilitarianism, 244–248
    value of a statistical life, 248–249
    veil of ignorance, 249–250
    zero discounting, 252–253
    zero well-being, 269–270
Intergovernmental Panel for Climate Change (IPCC), 144, *405*, 406
Intergovernmental Science-Policy Platform on Biodiversity and Ecosystem Services (IPBES), 62, **63**, 65, 94–95
internal rates of return, economic evaluation, 354

international comparisons; *see also* geographical perspectives
    economic, 26
    measures of well-being, 294–296, 298, **299**
International Institute for Applied Systems Analysis the Water Futures and Solutions Initiative, 410
international trade *see* trade
Intrinsic Value Exchange (IVE), 486
intrinsic value of nature, 275, 305–306
    biodiversity, 31, 313–314
    moral worth, 314–317
invasive non-native species (INNS), 52–53
    effects of trade, 388, 398
    fragmented ecosystems, 68
    restoration of ecosystems, 467–468
    rewilding projects, 461
    water supply in South Africa example, 468–469
InVEST ecosystem service models, 470
investment
    impact equation, 486–487
    intergenerational well-being, 266–267
    sustainability financing, 482–486
    transformative change, 496
    yield on, 22–23
invisibility
    ecosystem services, 46–47
    properties of nature, 13, 30
IPBES *see* Intergovernmental Science-Policy Platform on Biodiversity and Ecosystem Services
IPCC *see* Intergovernmental Panel for Climate Change
irreversibility of ecosystem change, *81*, *82*
irrigation systems, Nepal, 205
island biogeography, 67–68, 89
IUCN (International Union for Conservation of Nature)
    conservation status, 60
    species-led conservation, 451

job opportunities, ecosystem restoration, 465–466
justice, environmental, 382–383

Kenya, contraceptive use, 235
Keynes, John Maynard, 9
keystone species, 52, 511
Kofyar farmers, Nigeria, 165

labelling, to bring about behavioural change, 229
lakes *see* freshwater habitats; phosphorus recycling
laissez faire economics, 190, 192–193
land assets, 328
land cover trends, global, *403*
land sparing, 414, *415*
land tenure systems, sub-Saharan Africa, 207
land use
    and biodiversity, *404*
    global land cover trends, *403*
    provisioning goods, 403–404
    and spread of viruses, 112–113
life cycle analysis (LCA), cultured meat, 421–422
life expectancy, 85, **291**

Subject Index

life on earth, history of, 3
life satisfaction, 293; see also well-being
lionfish (Pterois spp.), 468
litigation risks, nature-related, 427
Living Planet Index (LPI), 61, 64, 381, 381
Living Standard Framework (LSF), New Zealand, 300–301
local commons, 31; see also tragedy of the commons
logging see forests and deforestation; timber industry
loss aversion, 221
loss reduction, ecosystem collapse, 143–144; see also risk and uncertainty
lumpy investments, 353

Maddison, Angus, 4–7
Maghribi traders, 11th century, 165
Mahabharata epic poem, 226
maintenance services see regulating/maintenance services
majority rule, democratic voting rules, 285–286
Malthus, Thomas, 7–10, 15; see also population size
mangrove forests
    ecosystem services, 46
    financial risk/uncertainty associated with loss, 425
marine ecosystems; see also coastal zone management; fisheries and fish farms; oceans
    conservation, 456
    coral reefs, 460
    open access resources, 24
    restoration, 459–460, 463–464
Marine Protected Areas, 444–445, 447, 451, 495
market forces, 114
    accounting prices, 307
    externalities, 184–185, 189–190
    revealed preferences for amenities, 309–310
    well-being across the generations, 243
market risks, nature-related, **428**, 429–430
Marshall Plan, WWII, 494–495
mathematical models
    bounded global economy, 125–128
    contemporary models of macroeconomic growth, 129
    ecosystem goods and services, 130–131
Matlab experiment, family planning programmes, 239
Mayan people, societal collapse and adaptation, 12
MEAs (Multilateral Environmental Agreements), 455–456
Mean Species Abundance (MSA), effects of climate change, 117
measures of well-being, 289–290
    Cantril ladder, 295
    comparative studies, 294–296, 298, **299**
    Day Reconstruction Method, 294
    Easterlin Paradox, 296–297
    intergenerational well-being, 246–247
    objective, 289, 290–292
    relative income, 296–297
    social statistics by income level, **291**
    subjective, 289, 297–298
    surveys/questionnaires, 292–294
meat analogues, 421–422
medical drugs, effects on biodiversity, 312–313
medicines, traditional, and biodiversity, 312–313

memory of natural systems see hysteresis
Mexican Coastal Zone Management Trust, 142–143
migration
    coping strategies in face of resource stress, 10–11
    disruptions to the biosphere, 73–74
Millennium Ecosystem Assessment (MA), 94
Millennium Seed Bank, 454–455
mobility of nature, 12, 13, 30, 412–413
    common pool resources, 201
    property rights, 207
    predictability, 207
modularity
    ecosystems, 50, 55–56
    societal, 173
momentum of systems see hysteresis
moral hazard, insurance markets, 137
moral worth of nature, 314–317; see also intrinsic value
moral philosophy/ethics, 280; see also human rights; population ethics
    democratic voting rules, 281–287
    individual and social well-being, 280–281
    mosaics, ecosystem, 50, 55–56
motivations, for behavioural change, 221, 233
multilateral development banks, 479–480
Multilateral Environmental Agreements (MEAs), 455–456
mutual affection, basis of trust, 161–162
mutual enforcement
    basis of trust, 161, 163, 164–169
    common pool resources, 199, 216–217
mutualism, externalities, 190–191

Nash equilibrium, 161, <u>512</u>
National Ecosystem Assessment, China, 322–323
natural assets
    barriers to protecting, 484
    definition, <u>512</u>
    geographical distribution, 378–381, 379
natural capital, 17, 20, 21, 329
    Andros Island, Bahamas example, 470
    arbitrage conditions, 367–368
    definitions, <u>510</u>, <u>512</u>
    depreciation of, 97, <u>510</u>
    earth systems and economic growth, 27–29
    effects of trade, 385–386
    financial risk/uncertainty associated with loss of, 424
    GDP and inclusive wealth, 342–343
    geographical distribution, 380
    global accounting, 96
    human impacts on the biosphere, 96
    India example, 355, 356–357
    nature of, 30–31
    planning for, 469–472
    property rights, **209**
    restoration of Exmoor mires example, 349
    transformative change, 499–500
Natural Capital Lab, Inter-American Development Bank, 480
natural disasters, insurance markets, 139; see also catastrophe; ecosystem collapse

Subject Index

natural resources
   definition, 512
   societal collapse and adaptation, 10–12
   trade, 388–389
   value of stocks, 22
nature; see also biosphere; properties of nature
   as an asset see assets
   conservation of see conservation
   definition, 512
   human enterprise seen as external to, 29, 103
   restoration of see restoration of ecosystems
nature-based solutions (NbS)
   restoration of ecosystems, 462–465, *463*
   transformative change, 496
nature-related risks; see also risk and uncertainty
   categories of risk, 424, *425*
   definition, 512
   financial, 424–431, **428**
nature-related uncertainty, 431–433
nature reserves see protected areas
Nepal, irrigation systems, 205
net domestic product (NDP), 512
   and sustainable development, 338
net inclusive investment, 97
net present value (NPV), 343–344, 346–348, 512
net primary productivity (NPP), 39, 42
   definition, 512
   measures, 61–62
   rates of return, 40
   spatially spread ecosystems, 207
Netherlands, rewilding, 461
Network for Greening the Financial System (NGFS), 433
New Zealand
   fisheries, 189
   living standard framework, 300–301
Nicomachean Ethics (Aristotle), 290
Nigeria, Kofyar farmers, 165
Nile perch (*Lates niloticus*), non-native species, 52–53
nitrogen, ecosystem services, 50
non-government organisations (NGOs), common pool resources, 202
non-linearity, biospheric processes
   biosphere disruptions, 67, 70, 73
   common pool resources regulation, 218
   and conservation, 442
   definition, 512
   externalities, 192
   optimum development, *352*, 352
   optimum saving principle, 262
   project complimentarities, 353
   provisioning goods, 413
   social preferences, 223
non-native species see invasive non-native species
norms see social norms
Norse settlers, societal collapse and adaptation, 11–12
Northern Republic of Congo, bushmeat, 227
NPP see net primary productivity
NPV (Net Present Value), 343–344, 346–348, 512
nudging, bringing about behavioural change, 228–229

numeraires, accounting prices, 265–266, 512
nutritional quality of crop plants, gene editing, 420
nutritional status, *77*; see also diet

oak trees (*Quercus lobata*), climate change, 117
oceans; see also marine ecosystems
   ecosystem depletion by region, 381
   exclusive economic zones, 197, 198
   oxygen depletion, *92–93*
Odyssey (Homer), 226
Official Development Assistance (ODA)
   sustainability financing, 479–480
   trading, 401
Oostvaardersplassen, Netherlands, rewilding, 461
open access resources, 24, 197–198, 204, 512
operational nature-related risks, **428**, 430
opportunistic species, 56
optimisation problem, accounting prices, 364–365
optimum development, economic evaluation, 351–352, *352*
optimum saving principle, well-being, 261–263, 276–278
option values, 147, *148*, 149–150
ordering, expected utility theory, 135
organic agriculture, *417*, 417
organic economies, 19
other effective area-based conservation measures (OECMs), 450
otters, keystone species, 52
overcrowding, in fish farms, 42
over-exploitation of ecosystems, 396, 452–453
own rate of return (yield on investment), 22–23
ownership rights, 206; see also property rights
   oxygen depletion, oceanic, *92–93*

PADDD (downgrading, downsizing and degazettement of protected areas), 448
palm oil plantations, externalities, 207–208
paradox of economic growth, 103
paradoxes, St. Petersburg, 133, 136, 154–155
patchworks, ecosystems as, 50, 55–56
patents, 31
path dependence, ecosystem dynamics, 75; see also hysteresis; regime shifts
payment for ecosystem services (PES)
   behavioural changes, 227
   Colombia example, 477, **478**
   common pool resources regulation, 218
   and externalities, 193–196
   sustainability financing, 477
   sustainable management, 466
peatlands, restoration of, 109
People's Bank of China, 487
personal integrity, basis of trust, 162, 163, 176
personal relationships, determinants of well-being, 299–300
personhood, moral worth of nature, 315–317
perverse subsidies, 499, 512
PES see payment for ecosystem services
pesticide tax, Denmark, 475–476
pet food, land use, 404

pharmaceutical drugs, effects on biodiversity, 312–313
phosphorus recycling in lakes, 70–71; *see also* eutrophication
    attempts to reverse, *82*, 82
    context-specificity, 83
    irreversibility, *81*, 82
    model of, 78–79
    regime shifts/tipping points, 79–80
    two ways of regime shifts occuring, 77–78
physical nature-related risks, 424–425
Pigouvian taxes
    externalities, 184–186
    quantity restrictions, 187–188
pika (*Ochotona princeps*), 117
place-based conservation, 447–451
planetary boundaries
    definition, 512
    human impacts, *91*, 92
    safety zones, 342
planning
    nature conservation, 455
    restoration of ecosystems, 469–472
plantations
    as ecosystems, 54–55
    externalities, 207–208
policies/regulations, environmental, 187–188
    accounting prices, 265–266
    common pool resources, 202
    externalities, 179
    species-led conservation, 453
    sustainable trade, 394–395, 396–398
policy analysis *see* economic evaluation
political instability, common pool resources, 203–204
political liberties, measures of well-being, **291**
pollinators, 30, 413
pollution, 305
    ecosystems, 48–49
    environmental regulations, 187–188
    health, 311–312
    and human rights, 181, 183
    Pigouvian taxes, 185–186
polycentric institutional structures, 108, 513
pooled funds, 485
population dynamics
    fisheries, 37–38
    human populations, 376–377
population ethics, 269–272
    death, 270–272
    zero well-being, 269–270
population size, human, 119–122, 179; *see also* reproductive behaviour
    common pool resources, 204
    demographic transition, 229–230, 231–232
    drivers, 221
    externalities, 183–184
    future scenarios, 229–230, *231*
    global data, *8*, 85, *86*
    and global health, **7**
    impact inequality, 15
    impacts on biosphere, 85, *86*
    India example, 355
    Measures of trends, 60, 61
    optimum population, 272–273
    societal collapse and adaptation, 10–12
    and standard of living, 7–10, 225
    sustainability, 14
    sustainable, 119–122, 271
    transformative change, 499
portfolio management, 354
    biodiversity, 17–20
    definition, 512
    nature-related uncertainty, 432
    uncertainty conditions, 134
    yield on investment, 22–23
portion sizes, 228
potlatch, never-ending, 252
poverty
    common pool resources, 200–203
    daily lives, 18–19
    economic history, 6–7
    geographical distribution, 382–383
    protected areas, 447
power laws, clues to approaching tipping points, 76
pragmatism, intergenerational well-being, 254, 256
precision agriculture, 420–421
predictability, spatially spread ecosystems, 207
preference ordering, expected utility theory, 135
present value criterion, inclusive wealth, 346–348; *see also* Net Present Value
pricing structures
    and externalities, 182, 191–192
    transformative change, 498–499
primary producers, 33, 36–37, 38–40; *see also* net primary productivity
    accounting prices, 40
    carbon flow, *39*
    definition, 513
    rates of return, 40
Principles for Responsible Investment (PRI), 482
Prisoner's Dilemma game, 168
private finance
    sustainability financing, 480–481, *483*
    tension with public aspirations, 19, 20
privatisation, common pool resources, 217, 218
probability distributions with fat tails, 133, 152–154
produced capital, 20, *21*, 329
    arbitrage conditions, 367–368
    definition, 513
    India example, 356
production
    effects of trade, *393*
    transformative change, 497–498
production systems
    marine, 418
    sustainable, 414–418, 497
    wetlands, 310
production targets, economic evaluation, 353–354

Subject Index

productivity; *see also* net primary productivity; primary producers
    as accounting price, 310
    and biodiversity, 56–57, 305
    fragmented ecosystems, 67–68
    and resilience of ecosystems, 50, 51–54, 56–57
    spatially spread ecosystems, 207
profits, and sustainability financing, 484
progress *see* economic progress
project complimentarities, lumpy investments, 353
project evaluation, 26
properties of nature, *13*, 13, 30
property rights
    biosphere as common heritage of mankind, 198–199
    externalities, 30, 206–207
    geographical distribution, 378
    management, 208
    to natural capital, **209**
    and wealth distributions, 181
pro-social disposition, basis of trust, 161, 162
protected areas
    cost-benefit analysis, 444–445
    definition, 513
    downgrading, downsizing and degazettement, 448
    indigenous peoples, 446, 448, 449
    place-based conservation, 447–451
    transformative change, 495–496
provisioning goods, 43, 54, 403; *see also* agriculture; food production; timber industry
    agroforestry, 416–417
    balancing delivery with regulating services, 414–418
    and biodiversity, *404*, *407*
    current use, 403–404, 406–407, 408, *411*
    definition, 513
    effects of climate change, *406*, 406, *407*
    food waste, 416
    future scenarios, 405–406, *407*, 408, *410*, 411–412
    global land cover trends, *403*
    land sparing, 414, *415*
    organic agriculture, *417*, 417
    technology to increase efficiency, 418–422
    trade-offs with future provision, 413
    trade-offs with regulating services, 412–418
    valuation of, 320
public assest management, 24–25; *see also* assets
public finance
    sustainability financing, 474–478, 479–480
    tension with private financing, 19, 20
public goods, 30–31
    definition, 513
    externalities, 191–192
    transformative change, 501–502
public policies, externalities, 179; *see also* policies/regulations

quality adjusted life years (QALYs), 249
quantity restrictions, externalities, 187–188
quantity/quality of stock measures, 62–65
questionnaires, measures of well-being, 292–294

Quinghai, China, 320–322, **324–325**
Quota Management System (QMS), New Zealand fisheries, 189
quotas, common pool resources, 215–216

racer snake (*Alsophis antiguae*), 468
rainfall, 55–56, 74–75
rank-order rule, democratic voting rules, 286–287
rates of return, 22–23, 40, 513
rats, invasive species, 468
reciprocal externalities, *180*, 197; *see also* common pool resources
    biosphere as common heritage of mankind, 198–199
    definition, 513
    management of property rights, 208
    open access resources, 197–198, 204
    ownership rights, 206
    palm oil plantations, 207–208
    property rights to land, 206–207
    property rights to natural capital, **209**
    subsidies, 199, **209**, 211
recycling, 497
red noise, 140
reference streams, 255–256, 260, 331, 351
regenerative agriculture, 418
regenerative rates
    biosphere, 13–14, 42–43, 85
    definition, 513
    fisheries, 72
regime shifts, 71, 73
    and biosphere disruptions, 74–75, 77–78
    clues to approaching tipping points, 75–76
    conservation ecology, 75
    context-specificity, 83
    definition, 513
    hysteresis, *81*, *82*
    large-scale landscapes, 74–75
    phosphorus recycling in lake systems, 79–80
    societal, 174
    two ways of occuring, 77–78
Regional Trade Agreements, 399, *400*
regulating/maintenance services, 43, 54
    definition, 513
    invisibility and silence, 46–47
    trade-offs, 412–418, 495
    valuation of, 320–321
regulations *see* policies/regulations
relationships, determinants of well-being, 299–300
relative income, measures of well-being, 296–297
remote sensing, precision agriculture, 420–421
rent-seeking behaviour, 165, 177
replacement fertility rate, 230, 513
reproductive behaviour; *see also* population size
    demographic transition, 229, 231–232
    externalities, 183–184, 224
    family planning programmes, 234, 235–239
    Matlab experiment, 239
    social embeddedness, 224, 232–235, *234*
    unmet needs in reproductive health, 237–239

Subject Index

reputation, as basis of trust, 164, 169–171
residual (total factor productivity), 339–341, <u>514</u>
resilience of ecosystems
   and biodiversity, 51–54
   definition, <u>513</u>
   fragmented ecosystems, 68
   and productivity, 50, 51–54
   robustness, 56
   stability regimes, 71
resources *see* natural resources
response diversity, ecosystems, 53
restoration of ecosystems, 457
   Andros Island, Bahamas example, 470
   And biodiversity, 457, 458, 462
   biodiversity offsets, 471–472, **472**
   carbon storage, 462
   Cheonggyecheon River example, 464
   for coastal protection, 463–464
   coastal zone management in Belize example, 470–471
   and conservation, *441*, 441
   coral reefs, 460
   cost-benefit analysis, 464–465
   definition, <u>513</u>
   environmental land management, 466–467
   forests, 462
   habitat banking, 472
   invasive species, 467–468
   job opportunities, 465–466
   making the case for, 458
   natural vs. active restoration, 461
   nature-based solutions, 462–465, *463*
   peatland habitats, 109
   planning for, 469–472
   recovery trends in marine populations, *452*
   regime shifts/tipping points, 75
   rewilding, 460–461
   spectrum of conservation/restoration efforts, *446*
   stock improvements, 457–460
   sustainable management, 466–467
   transformative change, 495–496
   Wallasea Island example, 464
   water supply in South Africa example, 468–469
   Zarqa River basin example, 459
Restoration Opportunities Optimization Tool (ROOT), 470
return on investment, intergenerational well-being, 259–261
revealed preferences, 309–310
Revised Wind Erosion Equation (RWEQ) model, sandstorm prevention, 321
rewilding, 460–461
rhino horn use, encouraging behavioural changes, 227
rice, genetic modification, 419
rice blast fungus (*Magnaporthe oryzae*), 419–420
rights *see* human rights
risk and uncertainty, 133; *see also* ecological risks; financial risk and uncertainty
   comparison of independent/fully correlated risks, 138–140
   and conservation, 442

   coupled ecosystems model, 151–152
   dealing with ambiguity, 155–156
   definitions, 133, <u>513</u>
   environmental collapse insurance, 142
   expected utility theory, 133, 134–135, 155–156
   fully correlated risks, 138, 139–141
   global risk pool system to mitigate climate risk, 140–141
   independent vs. correlated risks comparison, 136
   insurance markets, 137–138
   keeping options open, 147–149
   Mexican Coastal Zone Management Trust, 142–143
   option values, 147, *148*, 149–150
   portfolio management, 134
   probability distributions with fat tails, 133, 152–154
   risk aversion, 135–136, *139*
   risk reduction, 143–144
   short-term vs. long-term predictability, 140
   single ecosystems model, 151
   St. Petersburg Paradoxes, 133, 136, 154–155
   translation of ecological into economic risks, 150
   unbounded utility functions, 133, 136, 152–154
   value of information, 141–142
   when to stop business-as-usual, 144–147
risk-assessment, for financial risk, 433–439, **434–435**; *see also* impact-based risk assessments
risk aversion
   measures, 135–136
   preferences, *139*
rivers *see* freshwater habitats
robustness of ecosystems, 56; see also resilience
rotating savings and credit associations (ROSCAs), 173
rotation of access, common pool resources, 201
Royal Society for the Protection of Birds (RSPB), 464
rule-abiders, 166–167
rural poor *see* poverty

sacred places, 31–32, 51, 251, 305–306, 313–314
safety zones, planetary boundaries, 342
St. Petersburg paradoxes, 133, 136, 154–155
sanctions, violation of social norms, 167
sandstorm prevention, 320–321
sanitary and phytosanitary standards (SPS), trade, 397–398
scale effects, trade, 385–386
scales, ecosystems, *34*, 40
sceptics, environmental, 144
SDGs *see* Sustainable Development Goals
sea otters, keystone species, 52
sea turtles, impact of shrimp fisheries, 395
second-best tax, 187
second-best worlds, intergenerational well-being, 244, 261, 264, 277–278
seed dispersal, by large animals, 47–48
selection pressures, regenerative rates of nature, 42–43
separatrix, 77
Seychelles blue bonds, sustainability financing, 478
shadow prices *see* accounting prices
short-term vs. long-term predictability, uncertainty, 140
short-termism, nature-related uncertainty, 432–433
shrimp farms, 310–311, 395; *see also* fisheries and fish farms

Subject Index

silence
  ecosystem services, 46–47
  properties of nature, 13, 30
single ecosystems model, 151
slowing down of ecosystem processes, approaching tipping points, 75
social capital, 157, 166
  as basis of societal coherence, 171, 175
  definition, 514
  and identity, 173–174, 175–176
  negatives/dark side of, 176–177
  rotating savings and credit associations, 173
social contract, basis of trust, 160, 163, 164
social cost-benefit analysis, 26, 262, 266–267, 343, 514
social costs
  carbon, 327
  shrimp farm example, 310–311
social customs, 164
social embeddedness, behaviour, 221–225, *223*; see also social preferences
social evaluators, 243, 250–252, 265–266, 330, 493; see also accounting prices
social externalities, 180, 221–222
social norms, 164, 165–167, 514
  grim norm, 167–169, **168**
social preferences (socially embedded preferences)
  behaviour, 221–225, *223*
  behavioural changes, 227–228
  competitive/conformist, 222–224, *223*, 240–241, **241**
  consumerism, 226
  consumption, 239–240
  definition, 514
  intergenerational well-being, 267–268
  moral philosophy, 281
  reproductive behaviour, 224, 232–235, *234*
  social relationships, determinants of well-being, 299–300
social status, through reproductive success, 232
social well-being, 248, 280–281, 459
social worth, 330
socially embedded preferences *see* social preferences
societal collapse, historical perspectives, 10–12
societal trust *see* trust
socio-ecological futures, 331, 351–352, 364–365
soils, 44–45
  biodiversity, 44–45, *46*, 52, 96
  degradation, 11–12
solidarity goods, 176
South African Grasslands Programme, 467
soy, sustainable trade, 397
spatial finance, 489
spatial planning
  Andros Island, Bahamas example, 470
  restoration of ecosystems, 469–472
spatially spread ecosystems, predictability, 207; see also mobility of nature
Special Report on Emissions Scenarios (SRES) storylines, IPCC, *405*, 406
species extinction *see* extinctions
species populations, 60; see also population size

species ranges, effects of climate change, 116
species traits, 60
species-led conservation, 451–455, *452*
spider monkey (Ateles geoffroyi), 396
spiritual aspects of nature, 314; see also intrinsic value of nature; sacred places
Spreewald Biosphere Reserve, Germany, 449–450
SPS (sanitary and phytosanitary) trading standards, 397–398
stability regimes; see also regime shifts
  bifurcations, 78
  definition, 514
  disruptions to the biosphere, *71*, 72–73
  fisheries, *72*, 73
stakeholder engagement, sustainability financing, 488–489
standard of living, 4–7, 9, 179
  drivers, 221
  GDP/capita, **5**, *6*
  and population size, 7–10, 225
  and well-being, 289
standards, global, transformative change, 502
standards, trading, 397–399, 400
state malfeasance, 204, 218
state of nature, statistical decision theory, 134
stocks, ecosystem *see* ecosystem stocks
stress-reduction theory, need for nature, 302
stress testing, risk-assessment, 435–436
subsidies
  agricultural subsidies in Switzerland, 476–477
  ecosystem services, **209**, 211
  open access resources, 199
  sustainability financing, 476
  and taxation, 184–185
  transformative change, 499
successional sequences, 56
sulphur oxide emissions, 28
supply, aggregate, 100–101; see also demand and supply
supply chains
  effects of trade, 387
  sustainable trade, 396–398
  transformative change, 498
surveys, measures of well-being, 292–294
sustainability, population size, 14
sustainability assessment, six questions of, 335; see also economic evaluation
sustainability financing, 473
  agricultural subsidies in Switzerland, 476–477
  balance of nature positive/negative finance, *474*
  Chinese taxonomy, 487
  EcoEnterprises Fund, 485
  ESG factors, 481–482, 483, 486–487
  EU Taxonomy, 488
  fiduciary duties, 481–482
  global initiatives, 479–480
  impact equation, 473, 486–487
  Intrinsic Value Exchange, 486
  investment approaches, 482–487
  local initiatives/individual countries, 475–478
  payments for ecosystem services, Colombia, 477, **478**
  pesticide tax in Denmark, 475–476

private finance, 480–481, *483*, 488–489
public finance, 474–478, 479–480
Seychelles blue bond example, 478
spatial finance, 489
stakeholder/corporate engagement, 488–489
sustainable
   definition, <u>514</u>
   ecological footprint, 496–499
   exploitation of ecosystems, 452–453; *see also below*
   human populations, 119–122
sustainable development, 15, 335, *336*
   definition, <u>514</u>
   and economic growth, 354–355
   as growth in inclusive wealth, 337
   and net domestic product, 338
   as positive investment, 338
   theorem, 337, 338, 339
Sustainable Development Goals (SDGs), 27, 85, 106–107, 335, *336*, <u>514</u>
sustainable management, ecosystems, 466–467
sustainable trade, 394–401
swans, white, black and green, 431–432
Switzerland, agricultural subsidies, 476–477

Task Force on Nature-related Financial Disclosures (TNFD), 439
taxation
   common pool resources, 215–216
   externalities, 184–185
   Pigouvian taxes, 184–186, 187–188
   sustainable trade, 394–395
technology
   effects of trade, 387–388
   externalities, 181–182
   geographical distribution, 378
   provisioning goods, 418–422
   restoration of ecosystems, 462
   role of institutions, 107–108
   sustainability financing, 489
   transformative change, 497
tenure, security of *see* property rights
terrestrial food production *see* food production
terrestrial habitats, global map, *443*
textiles *see* fibre for textiles
third estate, 158
tigers, species-led conservation, 452
timber industry; *see also* forests and deforestation
   current usage, 408
   externalities, 181–182, 206
   future scenarios, 408
timeless world example, common pool resources, 213
tipping points, 14, <u>514</u>; *see also* regime shifts
To Kill a Mockingbird (Lee), 281
total factor productivity (TFP) growth, 339, <u>514</u>
total vs. marginal values, nature as an asset, 29–30
tourism *see* ecotourism
tradable permit scheme, New Zealand fisheries, 189
trade, 385; *see also* CITES
   biodiversity, 391–393
   composition effect, 386–387
   ecological footprint, 391–392
   ecosystem collapse, 393–394
   externalities, 181–182
   impact equation, 385
   impact on the biosphere, 388–394
   invasive species, 388, 398
   policies/regulations, 395, 396–398, 399, *400*
   and production/consumption, *393*
   resource use, 388–*389*, *390*
   scale effect, 385–386
   standards, 397–399
   sustainable, 394–401
   Sustainable Development Theorem, 339
   technique and technology transfer, 387–388
   transformative change, 498
   US shrimp/turtle dispute, 395
   verified sourcing areas, 400–401
   World Trade Openness Index, *386*
trading standards, 397–399, 400
traditional medicines, and biodiversity, 312–313
tragedy of the commons, 24, 197, 199–200
transcendence of self, spiritual aspects of nature, 314
transformative change, options to bring about, 493–494, *494*; *see also* future scenarios
   conservation and restoration of ecosystems, 495–496
   consumption and production patterns, 497–498
   demand and supply imbalances, 494–495
   economic progress, 493, 499–500
   education for, 503–504
   empowered citizenship, 503
   enabling environments, *501*
   global financial system, 502–503
   institutions and systems, 500
   population size, 499
   pricing structures, 498–499
   public goods, 501–502
   supply chains/trade, 498
   sustainable ecological footprint, 496–499
transition nature-related risks, 426
transparency; *see also* corruption
   sustainability financing, 484
   sustainable trade, 397
   transformative change, 498
   Transparency International (NGO), 159
trees *see* forests and deforestation
trust, 160–161; *see also* social capital
   basis of, 161, 163
   breakdowns of, 176–177
   civic virtues, 162
   and economic progress, 159–160
   external enforcement, 161, 163–164
   grim norm, 167–169, **168**
   iddirs, 172
   mutual affection, 161–162
   mutual enforcement, 163, 164–169
   personal integrity, 162, 163, 176
   regime shifts/tipping points, 174
   reputation, 164, 169–171
   social contract, 160, 163, 164

Subject Index

trust (cont.)
   social norms, 165–167
tsunami, Indonesia, 235
turtles, impact of shrimp fisheries, 395

unbounded utility functions, 133, 136, 152–154
uncertainty; *see also* risk and uncertainty
   Consumption Discount Rates, 278, **279**, *280*
      definition, 133, 514
      portfolio management, 134
UNCLOS (United Nations Convention on the Law of the Sea), 198, 358, 455, 456
UNDP (United Nations Development Programme), 474, 486
UNEP (United Nations Environment Programme), 388–*389*, *390*
unemployment *see* employment
unidirectional externalities, 179–181, *180*
   Coase theorem, 192–193
   definition, 515
   environmental regulations, 187–188
   habitats for migrating birds, 196
   harmful, 192
   information gaps, 187–188, 190
   market forces, 184–185, 189–190
   mutualism, 190–191
   New Zealand fisheries, 189
   payment for ecosystem services, 193–196
   Pigouvian taxes, 184–186
   pricing structures, 182, 191–192
   property rights and wealth distributions, 181
   quantity restrictions, 187–188
   reproductive, 183–184
   and rights, 181, 182–184
   taxing and subsidising, 184–185
   trade, wealth transfers and technological change, 181–182
United Kingdom
   ecological footprint, 391–392
   environmental land management, 466–467
   Green Book, 329–330
   Waste and Resources Action Programme, 416
United Nations *see* UNCLOS; UNDP; UNEP
unmet needs, reproductive health, 237–239, 515
urbanisation
   distance between people and the natural world, 17
   geographical distribution of well-being, 377–378
United States, shrimp/turtle dispute, 395
use value, biodiversity, 305; *see also* ecosystem services
utilitarianism, 244–248
   death, 270–272
   definition, 515
   discounting future generations, 250–252
   measures of well-being, 246–247
   veil of ignorance, 249–250
   wealth/well-being equivalence theorem, 332
utility function, 135; *see also* expected utility theory

value of a statistical life (VSL), 248–249
valuing biodiversity, 305–306

accounting prices, 306–308
   China example, 317–318, *319*
   contingent-valuation method, 308, 309
   ecotourism, 310
   existence value, 305, 313–314
   Gross Ecosystem Product, 318–319
   health, 311–312, 313
   intrinsic value, 31, 313–314
   moral worth, 314–317
   productivity as accounting price, 310
   Quinghai example, 320–322
   revealed preferences for amenities, 309–310
   shrimp farm example, 310–311
   stated preferences for public goods, 308
   traditional medicines, 312–313
vegetation patches, approaching tipping points, 76
veil of ignorance, intergenerational well-being, 249–250
verified sourcing areas, trading, 400–401
vertical farming, 421
vicuña fibre, species-led conservation, 454
Vietnam, rhino horn use, 227
village tanks (artificial ponds), 205
virus spread, and land-use changes, 112–113
voluntary standards, trading, 400
voting rules, democracy, 281–287
   Arrow's theorem, 282–285
   majority rule, 285–286
   rank-order rule, 286–287
VSL (value of a statistical life), 248–249

Wallasea Island, restoration of ecosystems, 464
waste, 179; *see also* externalities
Waste and Resources Action Programme (WRAP), UK, 416
water cycle, 55–56, 74–75
Water Futures and Solutions Initiative (International Institute for Applied Systems Analysis), 410
water hyacinth (*Eichhornia crassipes*), 468
water pollution, 311–312
water supply, 320; *see also* freshwater habitats
   current usage, 408–410
   future scenarios, *410*
   irrigation systems, Nepal, 205
   South Africa example, invasive species, 468–469
waterfowl, functional diversity, 51
watershed ecosystem services, 49–50
wealth comparisons, 27
wealth distributions, and property rights, 181
wealth transfers
   externalities, 181–182
   Sustainable Development Theorem, 339
wealth/well-being equivalence theorem, 24–25, 27, 327; *see also* inclusive wealth
   capital assets, 327, 331–334
   Net Present Value, 346–348
   public assest management, 24–25
   simplifications to the model, 334
well-being, 26; *see also* geographical distribution; intergenerational; *and* measures of well-being; wealth/well-being equivalence theorem

and consumption, 261–263, 289
definitions, 244, 515
determinants of well-being, 254, 299–300
gardens/green spaces, 303–304
individual and social well-being, 280–281
need for nature, 301–303
New Zealand living standard framework, 300–301
social well-being, 248, 280–281, 459
well-functioning markets, 184, 189–190
wetlands
   biodiversity offsets, **472**
   compensation programme, **322**
   financial risk/uncertainty associated with loss of, 425
   habitats for migrating birds, 196
   historical perspectives, 12
   net present values, 344
   option values, 149–150
   production systems, 310
   reducing risks/losses, 143–144
   restoration, 109, 147–149, 348–*349*, **349**

soils, 96
whales, 357–360, 451
white swans, nature-related uncertainty, 431–432
widespread nature *see* mobility of nature
wildlife trade, illegal trafficking, 453
willingness to pay
   accounting prices, 308
   stated preferences for public goods, 308
woodlands *see* forests and deforestation
World Economic Forum, Global Risk Report, 408
World Trade Openness Index, *386*

yield on investment, rates of return, 22–23

Zarqa River basin example, Jordan, 459
Zealandia, 194
zero discounting, 252–253
zero well-being, population ethics, 269–270
Zimbabwe, CAMPFIRE Programme, 194–195